Handbook of Space Security

Kai-Uwe Schrogl
Editor-in-Chief

Maarten Adriaensen • Christina Giannopapa • Peter L. Hays • Jana Robinson
Section Editors

Ntorina Antoni
Managing Editor

Handbook of Space Security

Policies, Applications and Programs

Second Edition

Volume 1

With 248 Figures and 47 Tables

Editor-in-Chief
Kai-Uwe Schrogl
European Space Agency (ESA)
Paris, France

Section Editors
Maarten Adriaensen
European Space Agency (ESA)
Paris, France

Peter L. Hays
Space Policy Institute
George Washington University
Washington, DC, USA

Christina Giannopapa
European Space Agency (ESA)
Paris, France

Jana Robinson
Space Security Program
Prague Security Studies Institute (PSSI)
Prague, Czech Republic

Managing Editor
Ntorina Antoni
Eindhoven University of Technology
Eindhoven, The Netherlands

ISBN 978-3-030-23209-2   ISBN 978-3-030-23210-8 (eBook)
ISBN 978-3-030-23211-5 (print and electronic bundle)
https://doi.org/10.1007/978-3-030-23210-8

1st edition: © Springer Science+Business Media New York 2015
2nd edition: © Springer Nature Switzerland AG 2020
All rights are reserved by the Publisher, whether the whole or part of the material is concerned, specifically the rights of translation, reprinting, reuse of illustrations, recitation, broadcasting, reproduction on microfilms or in any other physical way, and transmission or information storage and retrieval, electronic adaptation, computer software, or by similar or dissimilar methodology now known or hereafter developed.
The use of general descriptive names, registered names, trademarks, service marks, etc. in this publication does not imply, even in the absence of a specific statement, that such names are exempt from the relevant protective laws and regulations and therefore free for general use.
The publisher, the authors, and the editors are safe to assume that the advice and information in this book are believed to be true and accurate at the date of publication. Neither the publisher nor the authors or the editors give a warranty, expressed or implied, with respect to the material contained herein or for any errors or omissions that may have been made. The publisher remains neutral with regard to jurisdictional claims in published maps and institutional affiliations.

This Springer imprint is published by the registered company Springer Nature Switzerland AG.
The registered company address is: Gewerbestrasse 11, 6330 Cham, Switzerland

Introduction

Never before has security in space been more challenged.

Never before has space been more elaborately used for military and security purposes on Earth.

And never before was it more necessary to understand and to receive orientation in the policy area of space security.

This indeed is the purpose of this second edition of the *Handbook of Space Security*, which is addressed to all persons and institutions dealing with space security on a governmental, academic, societal, international, and diplomatic level. From now, the global future will depend on the secure use of outer space for all policy areas with particular stress on climate change and environmental monitoring, telecommunications, navigation, and cyber infrastructures as well as resource management. If space utilization as critical infrastructure is disrupted, our modern societies will break down. Space security is a key factor for survival. This is why space is also contested and space assets are vulnerable.

Consequently, we see a growing "securitization" of outer space. While some decades ago, there were rather clear distinctions between military and civilian uses of space with a smaller area in between called "dual-use." This zone is ever expanding under the label of "security" diminishing exclusively military and civilian use to small fringes. Just think of environmental security, cyber security, food security, water security, etc., and it becomes clear that we have to speak of a new paradigm also for the use of space. This can also be seen as the most prominent and accelerating development since the publication of the first edition of the Handbook in 2015.

Based on this, the definitional approach we use in this Handbook in space security can be drafted as:

> "Space security" is the aggregate of all technical, regulatory and political means that aims to achieve unhindered use of outer space from any interference as well as aims to use space for achieving security on Earth.

This adds to an already existing abundance of definitions for space security. Since it is more than unlikely that a consensus would be reached, our definition stays in the practice of attempting to narrow general applicability and tailor-made approach. For us, it is a frame for the focus and the structure of the Handbook.

The structure of the Handbook comprises four parts. These four parts are edited by accomplished experts in the field. They are Peter L. Hays for Part I International Space Security Setting, Jana Robinson for Part II Space Security Policies and Strategies of States, Maarten Adriaensen for Part III Space Applications and Supporting Services for Security and Defense, and Christina Giannopapa for Part IV Space Security Programs Worldwide and Space Economy Worldwide. They have been assembling exclusive groups of contributors, which do not only reflect a broad geographic distribution (120 authors from 30 countries), but also offer an exciting diversity of practitioners and academics' approaches to the topics. An Editorial Advisory Board of 22 distinguished experts from governments and academia from some 20 countries and every continent (i.e., Africa, Asia, Europe, North America, and South America) assisted the editors in identifying and evaluating contributions to ensure a high-quality, coherent final product.

The Handbook applies political, legal, economic, and technology-oriented analysis to these topics and aims at providing a holistic understanding in each of the sections and for the theme of space security as a whole. It is a work of tertiary literature containing digested knowledge in an easily accessible format. It provides a sophisticated, cutting-edge resource on the space security–related policy portfolio and associated assets to assist fellow members of the global space community, academic audiences, and other interested parties in keeping abreast of the current and future directions of this vital dimension of international space policy. By analyzing the underlying developments in space environment and the linkages to space security from a wide range of disciplines and theoretical perspectives, this Handbook establishes itself as a leading work for reference purposes, as well as a basis for further discussion.

Furthermore, it is enriched by the account of numerous space operational applications that routinely deliver indispensable fast and reliable services for security and defense needs from and for space. The transformation of the space domain through new technological advances and types of innovation such as mega-constellations, machine learning, and artificial intelligence has resulted in major challenges. Finally, it examines how addressing these needs has led to space programs and the development of specific security space assets. In short, it examines the reciprocal relations among space policy objectives on one hand and operational capabilities on the other, allowing readers to understand the theoretical and practical interactions and limitations between them. It also features numerous recommendations concerning how to best improve the space security environment, given the often-competing objectives of the world's major space-faring nations.

The Handbook is available in both printed and electronic forms. Its success is documented by more than 100,000 chapter downloads for its first edition. The Handbook is intended to assist and promote both academic research and professional activities in this rapidly evolving field encompassing security from and for space. It aspires to remain a go-to reference manual for space policy practitioners and decision makers, scholars, students, researchers, and experts as well as the media. The Springer Publishing House has been an exemplary partner in this major undertaking. We would like to gratefully acknowledge the cooperation of Maury Solomon and Hannah Kaufmann (New York), Lydia Muller (Heidelberg), Juby George, and Sonal Nagpal (New Delhi). Ntorina Antoni, the Managing Editor of the Handbook, acted as the invaluable main liaison between our editorial team, contributors, and publisher throughout the project. It is the hope of all of us who joined together to prepare this second edition of the *Handbook of Space Security* that it will inspire those seeking to be active in shaping space security, on which we all depend.

September 2020

Kai-Uwe Schrogl
Editor-in-Chief

Advisory Board

Setsuko Aoki	Professor of Law, Keio University Law School, Tokyo, Japan
	Chair, United Nations Committee on the Peaceful Uses of Outer Space (UNCOPUOS) Legal Subcommittee, Vienna, Austria
	Vice President, International Institute of Space Law (IISL), Paris, France
Natália Archinard	Deputy Head of the Education, Science, Transport and Space Section, Directorate of Political Affairs, Federal Department of Foreign Affairs, Berne, Switzerland
	Chair, UNCOPUOS Scientific and Technical Subcommittee, Vienna, Austria
Frank Asbeck	Senior Fellow at the Prague Security Studies Institute (PSSI), Prague, Czech Republic
	Former Principle Adviser for Space and Security Policy, European External Action Service (EEAS), Brussels, Belgium
	Former Director of the European Union Satellite Centre (EU SatCen), Torrejon, Spain
Tare Brisibe	Senior Legal and Regulatory Counsel, Asia-Pacific, SES, Singapore
	Former Chair, UNCOPUOS Legal Subcommittee, Vienna, Austria
	Former Deputy Director (Legal) National Space Research and Development Agency of Nigeria, Lagos, Nigeria
Gerard Brachet	Space Policy Consultant, Paris, France
	Former Chair, UNCOPUOS, Vienna, Austria
	Former Director General, French Space Agency – CNES, Paris, France

Alice Bunn	International Director, UK Space Agency, Swindon, United Kingdom
	Co-Chair, Global Future Council on Space Technology, World Economic Forum, Cologny, Switzerland
Michael Davis	Former Chair, Space Industry Association of Australia (SIAA), Golden Grove Village, Australia
Simonetta Di Pippo	Director, United Nations Office for Outer Space Affairs (UNOOSA), Vienna, Austria
Monserrat Filho	Head, International Affairs Office of the Ministry of Science and Technology, Brasilia, Brazil
Driss El Hadani	Director, Royal Centre for Remote Sensing (CRTS), Rabat, Morocco
Armel Kerrest	Emeritus Professor of Public Law at the Universities of Western Brittany, Brest and Paris XI, France
	Vice Chairman, European Centre for Space Law of the European Space Agency (ECSL/ESA), Paris, France
Pascal Legai	Senior Adviser – Earth Observation to the European Space Agency (ESA), Frascati, Italy
	Former Director, European Union Satellite Centre (EU SatCen), Torrejon, Spain
Sergio Marchisio	Professor of International Law and Space Law, Sapienza University of Rome, Italy
	Chair, European Centre for Space Law (ECSL/ESA), Paris, France
Marian-Jean Marinescu	Member, European Parliament, Coordinator for Transport Policies, EPP Group Romania, Brussels, Belgium
	Chair, Sky and Space Intergroup in the European Parliament, Brussels, Belgium
Martha Mejía-Kaiser	Independent Researcher, Mexico/Germany
	Member of the Board of Directors, International Institute of Space Law (IISL), Paris, France
K.R. Sridhara Murthi	Professor of Aerospace Engineering and Management, Jain University, Bangalore, India
	Vice President, International Institute of Space Law (IISL), Paris, France
Rosa Ma. Ramírez de Arellano y Haro	General Director of International Affairs and Space Security at the Mexican Space Agency, Mexico City, Mexico
	Former Chair, UNCOPUOS and UNISPACE+50, Vienna, Austria
Michael Simpson	Former Executive Director, Secure World Foundation, Colorado Springs, USA
	Former President, International Space University (ISU), Strasbourg, France

Li Shouping	General-Director, Space Law Center of China National Space Administration (CNSA), Haidian District, Beijing, China
	Dean, Law School of Beijing Institute of Technology, Beijing, China
Paul Weissenberg	Senior Advisor to the Executive Director of the European GNSS Agency (GSA), Prague, Czech Republic
	Senior Advisor, The European External Action Service (EEAS), Brussels, Belgium
	Senior Advisor, The European Commission DG GROW, Brussels, Belgium
	Former Deputy Director General, European Commission for the Galileo and Copernicus space programs and Security and Defence Industries, Brussels, Belgium
Klaus-Peter Willsch	Member of the German Parliament (Deutscher Bundestag), Berlin, Germany
	Chair of the aerospace group in the German Parliament, Berlin, Germany
	Former Chair, European Interparliamentary Space Committee (EISC)
Viktor Veshchnuvov	Executive Director, Intersputnik International Organization of Space Communications, Moscow, Russian Federation

Contents

Volume 1

Part I International Space Security Setting 1

1. **International Space Security Setting: An Introduction** 3
   Peter L. Hays

2. **Definition and Status of Space Security** 9
   Ntorina Antoni

3. **Challenges to International Space Governance** 35
   Ahmad Khan and Sufian Ullah

4. **Spacepower Theory and Organizational Structures** 49
   Peter L. Hays

5. **The Laws of War in Outer Space** 73
   Steven Freeland and Elise Gruttner

6. **Arms Control and Space Security** 95
   Stacey Henderson

7. **Role of Space in Deterrence** 111
   John J. Klein and Nickolas J. Boensch

8. **Resilience of Space Systems: Principles and Practice** 127
   Regina Peldszus

9. **Space Security Cooperation: Changing Dynamics** 145
   Jessica L. West

10. **Strategic Competition for Space Partnerships and Markets** 163
    Jana Robinson, Tereza B. Kupková, and Patrik Martínek

11. **Space Export Control Law and Regulations** 185
    Ulrike M. Bohlmann and Gina Petrovici

12. **Space Systems and Space Sovereignty as a Security Issue** 211
    Annamaria Nassisi and Isabella Patatti

13	**Critical Space Infrastructures** .... Alexandru Georgescu	227
14	**Space and Cyber Threats** .... Stefano Zatti	245
15	**Space Safety** .... Joe Pelton, Tommaso Sgobba, and Maite Trujillo	265
16	**Evolution of Space Traffic and Space Traffic Management** ...... William Ailor	299
17	**Space Sustainability** .... Peter Martinez	319
18	**Security Issues with Respect to Celestial Bodies** .... George D. Kyriakopoulos	341

Part II Space Security Policies and Strategies of States **357**

19	Space Security Policies and Strategies of States: An Introduction ... Jana Robinson	359
20	**War, Policy, and Spacepower: US Space Security Priorities** ..... Everett C. Dolman	367
21	**Russia's Space Security Policy** .... Nicole J. Jackson	385
22	**Development of a Space Security Culture: Case of Western European Union** .... Alexandros Kolovos	401
23	**Strategic Overview of European Space and Security Governance** .... Ntorina Antoni, Maarten Adriaensen, and Christina Giannopapa	421
24	**European Space Security Policy: A Cooperation Challenge for Europe** .... Jean-Jacques Tortora and Sebastien Moranta	449
25	**Space and Security Policy in Selected European Countries** ...... Ntorina Antoni, Maarten Adriaensen, and Christina Giannopapa	467
26	**Poland and Space Security** .... Małgorzata Polkowska	485
27	**Space Security in the Asia-Pacific** .... Rajeswari Pillai Rajagopalan	499
28	**Chinese Space and Security Policy: An Overview** .... Zhuoyan Lu	515

29	**Chinese Concepts of Space Security: Under the New Circumstances** .... Dean Cheng	527
30	**Historical Evolution of Japanese Space Security Policy** .... Kazuto Suzuki	555
31	**India in Space: A Strategic Overview** .... Ajey Lele	571
32	**Israel's Approach Towards Space Security and Sustainability** ... Deganit Paikowsky, Tal Azoulay, and Isaac Ben Israel	589
33	**Policies and Programs of Iran's Space Activities** .... Hamid Kazemi and Mahshid TalebianKiakalayeh	601
34	**UAE Approach to Space and Security** .... Naser Al Rashedi, Fatima Al Shamsi, and Hamda Al Hosani	621
35	**Space Security in Brazil** .... Olavo de O. Bittencourt Neto and Daniel Freire e Almeida	653
36	**Space and Security Activities in Azerbaijan** .... Tarlan Mammadzada	667

Volume 2

Part III Space Applications and Supporting Services for Security and Defense **699**

37	**Introduction to Space Applications and Supporting Services for Security and Defense** .... Maarten Adriaensen	701
38	**Earth Observation for Security and Defense** .... Ferdinando Dolce, Davide Di Domizio, Denis Bruckert, Alvaro Rodríguez, and Andrea Patrono	705
39	**Satellite EO for Disasters, Risk, and Security: An Evolving Landscape** .... Helene de Boissezon and Andrew Eddy	733
40	**Space-Enabled Systems for Food Security in Africa** .... Olufunke Adebola and Simon Adebola	759
41	**Satellite Communication for Security and Defense** .... Holger Lueschow and Roberto Pelaez	779
42	**Position, Navigation, and Timing for Security** .... Jean-Christophe Martin	797

| 43 | PNT for Defense | 821 |

Marco Detratti and Ferdinando Dolce

| 44 | Space Traffic Management Through Environment Capacity | 845 |

Stijn Lemmens and Francesca Letizia

| 45 | Various Threats of Space Systems | 865 |

Xavier Pasco

| 46 | European Space Surveillance and Tracking Support Framework | 883 |

Regina Peldszus and Pascal Faucher

| 47 | China's Capabilities and Priorities in Space-Based Safety and Security Applications | 905 |

Lin Shen, Hongbo Li, Shengjun Zhang, and Xin Wang

| 48 | Cybersecurity Space Operation Center: Countering Cyber Threats in the Space Domain | 921 |

Salvador Llopis Sanchez, Robert Mazzolin, Ioannis Kechaoglou, Douglas Wiemer, Wim Mees, and Jean Muylaert

| 49 | AI and Space Safety: Collision Risk Assessment | 941 |

Luis Sanchez, Massimiliano Vasile, and Edmondo Minisci

| 50 | Space Object Behavior Quantification and Assessment for Space Security | 961 |

Moriba Jah

| 51 | Space Security and Frequency Management | 985 |

Yvon Henri and Attila Matas

| 52 | Space Weather: The Impact on Security and Defense | 1005 |

Jan Janssens, David Berghmans, Petra Vanlommel, and Jesse Andries

| 53 | Space Security in the Context of Cosmic Hazards and Planetary Defense | 1025 |

Joe Pelton

| 54 | Active Debris Removal for Mega-constellation Reliability | 1043 |

Nikita Veliev, Anton Ivanov, and Shamil Biktimirov

| 55 | Space Security and Sustainable Space Operations: A Commercial Satellite Operator Perspective | 1063 |

Jean François Bureau

| 56 | Space Debris Mitigation Systems: Policy Perspectives | 1079 |

Annamaria Nassisi, Gaia Guiso, Maria Messina, and Cristina Valente

| 57 | Security Exceptions to the Free Dissemination of Remote Sensing Data: Interactions Between the International, National, and Regional Levels | 1091 |

Philip De Man

Part IV Space Security Programs Worldwide and Space Economy Worldwide 1123

| 58 | Space Security Programs and Space Economy: An Introduction | 1125 |

Christina Giannopapa

| 59 | Satellite Programs in the USA | 1133 |

Pat Norris

| 60 | Russian Space Launch Program | 1171 |

Elina Morozova

| 61 | Institutional Space Security Programs in Europe | 1191 |

Ntorina Antoni, Maarten Adriaensen, and Christina Giannopapa

| 62 | Space and Security Programs in the Largest European Countries | 1225 |

Ntorina Antoni, Christina Giannopapa, and Maarten Adriaensen

| 63 | Space and Security Programs in Medium-Sized European Countries | 1265 |

Ntorina Antoni, Christina Giannopapa, and Maarten Adriaensen

| 64 | Space and Security Programs in Smaller European Countries | 1289 |

Ntorina Antoni, Christina Giannopapa, and Maarten Adriaensen

| 65 | Future of French Space Security Programs | 1329 |

Jean-Daniel Testé

| 66 | Italy in Space: Strategic Overview and Security Aspects | 1343 |

Francesco Pagnotta and Marco Reali

| 67 | British Spacepower: Context, Policies, and Capabilities | 1365 |

Bleddyn Bowen

| 68 | Chinese Satellite Program | 1381 |

Xiaoxi Guo

| 69 | Chinese Space Launch Program | 1401 |

Lehao Long, Lin Shen, Dan Li, Hongbo Li, Dong Zeng, and Shengjun Zhang

| 70 | Indian Space Program: Evolution, Dimensions, and Initiatives | 1421 |

Hanamantray Baluragi and Byrana Nagappa Suresh

| 71 | Australia's Space Security Program | 1441 |

Michael Davis and Chris Schacht

72	**Pakistan's Space Activities**	1455
	Ahmad Khan, Tanzeela Khalil, and Irteza Imam	
73	**Space Sector Economy and Space Programs World Wide**	1471
	Per Høyland, Estelle Godard, Marta De Oliviera, and Christina Giannopapa	
74	**The New Space Economy: Consequences for Space Security in Europe**	1499
	Jean-Pierre Darnis	
75	**Political Economy of Outer Space Security**	1511
	Vasilis Zervos	
76	**Views on Space Security in the United Nations**	1535
	Massimo Pellegrino	
77	**The Role of COSPAR for Space Security and Planetary Protection**	1559
	Leslie I. Tennen	
Index	....	1581

About the Editor-in-Chief

Prof. Dr. Kai-Uwe Schrogl is currently seconded from the European Space Agency (ESA) to the German Federal Ministry for Economic Affairs and Energy in Berlin to support the preparation of the German Presidency of the Council of the European Union in the second half of 2020. Until 2019, Dr. Schrogl was the Chief Strategy Officer of ESA (Headquartered in Paris, France). From 2007 to 2011 he was the Director of the European Space Policy Institute (ESPI) in Vienna, Austria, the leading European think tank for space policy. Prior to this, he was the Head of the Corporate Development and External Relations Department in the German Aerospace Center (DLR) in Cologne, Germany. Previously, he also worked with the German Ministry for Post and Telecommunications and the German Space Agency (DARA) in Bonn, Germany. Dr. Schrogl has been a delegate to numerous international forums and has served from 2014 to 2016 as Chairman of the Legal Subcommittee of the United Nations Committee on the Peaceful Uses of Outer Space, the highest body for space law making, comprising 73 Member States. He also was chairman of various European and global committees (ESA International Relations Committee and two plenary working groups of the UNCOPUOS Legal Subcommittee, the one on the launching State and the other on the registration practice, both leading to UN General Assembly Resolutions).

He presented, respectively testified, at hearings of the European Parliament and the U.S. House of Representatives. Dr. Schrogl is President of the International Institute of Space Law, the professional association of space law experts from 48 countries, Member of the International Academy of Astronautics (recently chairing its Commission on policy, economics, and regulations) and the Russian Academy of Cosmonautics, as well as Corresponding Member of the French Air and Space Academy. He holds a doctorate degree in political science and lectures international relations as an Honorary Professor at Tübingen University, Germany. Dr. Schrogl has written or co-edited 17 books and more than 140 articles, reports, and papers in the fields of space policy and law as well as telecommunications policy. He launched and edited until 2011 the *Yearbook on Space Policy* and the book series Studies in Space Policy, both published by ESPI at Springer-WienNewYork. He sits on editorial boards of various international journals in the field of space policy and law (*Space Policy, Zeitschrift für Luft- und Weltraumrecht*, Studies in Space Law/Nijhoff; previously also *Acta Astronautica*).

About the Section Editors

Maarten Adriaensen currently works for the Procurement Department of the European Space Agency (ESA) in Paris, France. He was previously seconded by the ESA to the European Defence Agency (EDA) in Brussels, Belgium, assigned as Policy Officer Space. In that function, he provided strategic analysis and synthesis in support of EDA Directorates, internal coordination of EDA space activities, and support to cooperation on space activities externally. Prior to that, Maarten Adriaensen worked in the Procurement Department and Policy Department of ESA in Paris; the University of Leuven, Belgium; and the European Space Policy Institute (ESPI) in Vienna, Austria. He holds an M.Sc. in Space Studies (cum laude), M.A. in European Studies: Transnational and Global Perspectives (magna cum laude), M.A. in Modern History (cum laude), LLB, and an Academic Teacher Degree from the University of Leuven. He attended the 2012 International Space University (ISU) Space Studies Program (SSP) in Florida, hosted by the Florida Institute of Technology and NASA Kennedy Space Center. The content of Maarten's contributions to the Handbook reflect his personal opinions and do not necessarily reflect the opinion of the European Space Agency.

Dr. Christina Giannopapa works at the European Space Agency (ESA) since 2007. Currently, she is seconded from ESA to the Greek Ministry of Digital Policy, Telecommunications and Media as Special Advisor on high-tech and space applications related issues. Until February 2019, Dr. Giannopapa has been ESA's Head of Political Affairs Office in the Strategy Department of the Director General's Services in Paris, being responsible for providing advice and support on political matters to the Director General. From 2012 to 2015, she has been working on Member States Relations Department of ESA. From 2010 to 2012, Dr. Giannopapa has been seconded from the Agency as Resident Fellow at the European Space Policy Institute (ESPI) in Vienna, where she has been supporting the European Interparliamentary Space Conference (EISC) and lead studies on innovation, Galileo, Copernicus, and Africa. In the policy areas she also worked briefly in DG Research, European Commission. From 2007 to 2010, Dr. Giannopapa has been working in the Mechanical Engineering Department of the Technical and Quality Management Directorate of ESA in the Netherlands. Prior to joining ESA, she has worked as a consultant to high-tech industries in research and technology development. Dr. Giannopapa held positions in academia in Eindhoven University of Technology, the Netherlands, and in the University of London, UK. She has received 14 academic scholarships and awards and has more than 60 publications in peer-reviewed journals and conferences. Dr. Giannopapa holds a Ph.D. in Engineering and Applied Mathematics, an M.Eng. in Manufacturing Systems Engineering and Mechatronics, and an M.B.A. in International Management from the University of London, UK. Additionally, she is Assistant Professor in the Department of Industrial Engineering and Innovation Sciences at Eindhoven University of Technology. Dr. Giannopapa is the Chairperson of the Committee for Liaison with International Organisations and Developing Nations (CLIODN) and the Secretary of the Space Security Committee of the International Astronautical Federation (IAF). She is also the Director for Professional Development of Women in Aerospace-Europe (WIA-E).

About the Section Editors

Peter L. Hays is an Adjunct Professor at George Washington University's (GWU) Space Policy Institute and a Senior Policy Advisor with Falcon Research. He has been directly involved in helping to develop and implement major national security space policy initiatives since 2004 and serves as a senior advisor on governance, cadre, and strategic messaging issues. Peter served as a Staff Augmentee at the Office of Science and Technology Policy in 1988 and at the National Space Council in 1990. He served as an Air Force Officer from 1979 to 2004, flew C-141 cargo planes, and previously taught at the Air Force Academy, Air Force School of Advanced Airpower Studies, and National Defense University; he now teaches Air- and Spacepower Seminars at the Marine Corps School of Advanced Warfighting and the Space and National Security and the Science, Technology, and National Security Policy graduate seminars at GWU. Peter holds a Ph.D. from the Fletcher School and was an Honor Graduate of the Air Force Academy. Major publications include: *Handbook of Space Security*, *Space and Security*, and *Toward a Theory of Spacepower*.

Dr. Jana Robinson is Space Security Program Director at the Prague Security Studies Institute (PSSI). She previously served as a Space Policy Officer at the European External Action Service (EEAS) in Brussels as well as a Space Security Advisor to the Czech Foreign Ministry. From 2009 to 2013, Dr. Robinson worked at the European Space Policy Institute (ESPI), seconded from the European Space Agency (ESA), leading the Institute's Space Security Research Programme. Dr. Robinson is an elected member of the International Institute of Space Law (IISL) and the International Academy of Astronautics (IAA). She is also a member of the Advisory Board of the George C. Marshall Missile Defense Project of the Center for Strategic and International Studies (CSIS) in Washington, D.C. Dr. Robinson holds a Ph.D. in Space Security from Charles University in Prague and received two M.A. degrees from George Washington University's Elliott School of International Affairs and Palacky University in Olomouc, respectively.

About the Managing Editor

Ntorina Antoni is currently a Ph.D. candidate of strategic management in space security at Eindhoven University of Technology. Her research focuses on strategic decision-making under uncertainty and legitimacy in the context of high reliability organizations. She is also an Attorney-at-Law and member of the Athens Bar Association in Greece. Ntorina previously worked at the Strategy Department of the European Space Agency (ESA). In her role as Strategy and Policy Analyst, she got involved in the research and analysis of regulations, policies, and strategies of the ESA member states. Prior to that, Ntorina served as the Legal Counsel for Swiss Space Systems, a company which planned to provide orbital launches of small satellites and manned sub-orbital spaceflights. She was in charge of the legal and regulatory aspects of aviation and aerospace projects including aircraft purchase, certification of parabolic flights and suborbital flights, as well as regulation of small satellites. Ntorina is a Board Member of the Brainport TechLaw in Eindhoven and an elected member of the International Institute of Space Law (IISL). She holds a Law degree from the University of Athens in Greece, a Master's degree (LL.M) in International and European Law from Tilburg University in the Netherlands, and an Advanced Master's degree (LL.M) in Air and Space Law from Leiden University in the Netherlands.

Contributors

Olufunke Adebola Sam Nunn School of International Affairs, Georgia Institute of Technology, Atlanta, GA, USA

Simon Adebola Capacity Direct LLC, Atlanta, USA

Maarten Adriaensen European Space Agency (ESA), Paris, France

William Ailor Center for Orbital and Reentry Debris Studies, The Aerospace Corporation, El Segundo, CA, USA

Hamda Al Hosani Space Policies and Legislations Department, UAE Space Agency, Masdar City, Abu Dhabi, United Arab Emirates

Naser Al Rashedi Space Policies and Legislations Department, UAE Space Agency, Masdar City, Abu Dhabi, United Arab Emirates

Fatima Al Shamsi Space Policies and Legislations Department, UAE Space Agency, Masdar City, Abu Dhabi, United Arab Emirates

Jesse Andries Solar Terrestrial Center of Excellence (STCE) – Solar Influences Data Analysis Center (SIDC), Royal Observatory of Belgium, Brussels, Belgium

Ntorina Antoni Eindhoven University of Technology, Eindhoven, The Netherlands

Tal Azoulay Yuval Neeman Workshop for Science, Technology and Security, Tel Aviv University, Tel Aviv, Israel

Hanamantray Baluragi Indian Space Research Organization (ISRO), Bangalore, India

David Berghmans Solar Terrestrial Center of Excellence (STCE) – Solar Influences Data Analysis Center (SIDC), Royal Observatory of Belgium, Brussels, Belgium

Shamil Biktimirov Skolkovo Institute of Science and Technology, Moscow, Russia

Olavo de O. Bittencourt Neto Catholic University of Santos, Santos, Brazil

Nickolas J. Boensch Bryce Space and Technology, Alexandria, VA, USA

Ulrike M. Bohlmann European Space Agency (ESA), Paris, France

Bleddyn Bowen University of Leicester, Leicester, UK

Denis Bruckert EU Satellite Centre (EUSC), Madrid, Spain

Jean François Bureau Institutional and International Affairs, Eutelsat Group VP, Paris, France

Dean Cheng The Heritage Foundation, Washington, DC, USA

Jean-Pierre Darnis Université Côte d'Azur, Istituto Affari Internazionali, (IAI), Nice, France

Michael Davis Space Industry Association of Australia, Adelaide, Australia

Helene de Boissezon French Space Agency (CNES), Toulouse, France

Committee on Earth Observation Satellites (CEOS), Haiti Recovery Observatory, Port au Prince, Haiti

Philip De Man University of Sharjah, Sharjah, UAE

University of Leuven, Leuven, Belgium

Marta De Oliviera International Space University (ISU), Illkirch-Graffenstaden, France

Marco Detratti European Defence Agency (EDA), Brussels, Belgium

Davide Di Domizio European Defence Agency (EDA), Brussels, Belgium

Ferdinando Dolce European Defence Agency (EDA), Brussels, Belgium

Everett C. Dolman Air Command and Staff College, Air University, Montgomery, AL, USA

Andrew Eddy Athena Global, Simiane-la-Rotonde, France

Pascal Faucher Defence and Security Office, French Space Agency (CNES), Paris, France

Steven Freeland University of Western Sydney, Sydney, NSW, Australia

Daniel Freire e Almeida Catholic University of Santos, Santos, Brazil

Alexandru Georgescu National Institute for Research and Development in Informatics (ICI), Bucharest, Romania

Christina Giannopapa European Space Agency (ESA), Paris, France

Estelle Godard Institut d'études politiques de Paris, Paris, France

Elise Gruttner United Nations Assistance to the Khmer Rouge Trials, New York, NY, USA

Gaia Guiso Women in Aerospace-Europe (WIA-E), Milan, Italy

Xiaoxi Guo China Academy of Space Technology (CAST), Beijing, People's Republic of China

Peter L. Hays Space Policy Institute, George Washington University, Washington, DC, USA

Stacey Henderson Adelaide Law School, The University of Adelaide, Adelaide, SA, Australia

Yvon Henri OneWeb, London, UK

Per Høyland Department of Political Science, University of Oslo, Oslo, Norway

Irteza Imam Department of Defence and Strategic Studies, Quaid-i-Azam University, Islamabad, Pakistan

Isaac Ben Israel Yuval Neeman Workshop for Science, Technology and Security, Tel Aviv University, Tel Aviv, Israel

Anton Ivanov Skolkovo Institute of Science and Technology, Moscow, Russia

Nicole J. Jackson School for International Studies, Simon Fraser University, Vancouver, BC, Canada

Moriba Jah Department of Aerospace Engineering and Engineering Mechanics, The University of Texas at Austin, Austin, TX, USA

Jan Janssens Solar Terrestrial Center of Excellence (STCE) – Solar Influences Data Analysis Center (SIDC), Royal Observatory of Belgium, Brussels, Belgium

Hamid Kazemi Aerospace Research Institute, Department of Air and Space Law, Tehran, Iran

Ioannis Kechaoglou RHEA Group, Wavre, Belgium

Tanzeela Khalil South Asia Center, Atlantic Council, Washington, DC, USA

Ahmad Khan Department of Strategic Studies, National Defence University, Islamabad, Pakistan

John J. Klein George Washington University's Space Policy Institute, Washington, DC, USA

Alexandros Kolovos Automatic Control, AirSpace Technology, Defence Systems and Operations Section, Hellenic Air Force Academy, Athens, Greece

Tereza B. Kupková Prague Security Studies Institute (PSSI), Prague, Czech Republic

George D. Kyriakopoulos School of Law, National and Kapodistrian University of Athens, Athens, Greece

Ajey Lele Institute for Defence Studies and Analyses, New Delhi, India

Stijn Lemmens Space Debris Office, European Space Agency (ESA) – European Space Operations Centre (ESOC), Darmstadt, Germany

Francesca Letizia Space Debris Office, IMS Space Consultancy at Space Debris Office, European Space Agency (ESA) – European Space Operations Centre (ESOC), Darmstadt, Germany

Hongbo Li China Academy of Launch Vehicle Technology (CALT), Beijing, China

Dan Li China Academy of Launch Vehicle Technology (CALT), Beijing, China

Salvador Llopis Sanchez European Defence Agency (EDA), Brussels, Belgium

Lehao Long Chinese Academy of Engineering, Beijing, China

Zhuoyan Lu International Space University (ISU), Strasbourg, France

Holger Lueschow European Defence Agency (EDA), Brussels, Belgium

Tarlan Mammadzada "Azercosmos" OJSCo, Baku, Azerbaijan

Patrik Martínek The Prague Security Studies Institute (PSSI), Prague, Czech Republic

Jean-Christophe Martin CEO Marency SAS, Paris, France
CEO Marency SAS, Brussels, Belgium

Peter Martinez Secure World Foundation (SWF), Broomfield, CO, USA

Attila Matas Orbit/Spectrum Consulting, Grand-Saconnex, Switzerland

Robert Mazzolin RHEA Group, Wavre, Belgium

Wim Mees Royal Military Academy, Brussels, Belgium

Maria Messina Education Unit, Italian Space Agency (ASI), Rome, Italy

Edmondo Minisci University of Strathclyde, Glasgow, UK

Sebastien Moranta European Space Policy Institute (ESPI), Vienna, Austria

Elina Morozova International Legal Service, Intersputnik International Organization of Space Communications, Moscow, Russian Federation
International Institute of Space Law, Paris, France

Jean Muylaert RHEA Group, Wavre, Belgium

Annamaria Nassisi Strategic Marketing, Thales Alenia Space, Rome, Italy

Pat Norris VAPN Ltd, West Byfleet, UK

Francesco Pagnotta Presidency of the Council of Ministers – Office of the Military Advisor, Rome, Italy

Deganit Paikowsky Yuval Neeman Workshop for Science, Technology and Security, Tel Aviv University, Tel Aviv, Israel

Xavier Pasco Fondation pour la Recherche Stratégique (FRS), Paris, France

Isabella Patatti Strategic Marketing, Thales Alenia Space, Rome, Italy

Andrea Patrono EU Satellite Centre (EUSC), Madrid, Spain

Roberto Pelaez European Defence Agency (EDA), Brussels, Belgium

Regina Peldszus Department of Space Situational Awareness, German Aerospace Center (DLR) Space Administration, Bonn, Germany

Massimo Pellegrino Vienna, Austria

Joe Pelton International Association for the Advancement of Space Safety (IAASS), Arlington, VA, USA

Gina Petrovici German Aerospace Center (DLR), Bonn, Germany

Małgorzata Polkowska University of War Studies, Warsaw, Poland

Rajeswari Pillai Rajagopalan Nuclear and Space Policy Initiative, Observer Research Foundation, New Delhi, India

Marco Reali Presidency of the Council of Ministers – Office of the Military Advisor, Rome, Italy

Jana Robinson Space Security Program, Prague Security Studies Institute (PSSI), Prague, Czech Republic

Alvaro Rodríguez EU Satellite Centre (EUSC), Madrid, Spain

Luis Sanchez University of Strathclyde, Glasgow, UK

Chris Schacht Adelaide, Australia

Tommaso Sgobba International Association for the Advancement of Space Safety (IAASS), Noordwijk, The Netherlands

Lin Shen China Academy of Launch Vehicle Technology (CALT), Beijing, China

Byrana Nagappa Suresh Indian Space Research Organization (ISRO), Bangalore, India

Kazuto Suzuki Hokkaido University, Hokkaido, Japan

Mahshid TalebianKiakalayeh Graduate of Islamic Azad University, Tehran North Branch, Tehran, Iran

Leslie I. Tennen Law Offices of Sterns and Tennen, Glendale, AZ, USA

Jean-Daniel Testé OTA (l'Observation de la Terre Appliquée), Peujard, France

Jean-Jacques Tortora European Space Policy Institute (ESPI), Vienna, Austria

Maite Trujillo European Space Agency (ESA) – European Space Research and Technology Centre (ESTEC), Noordwijk, The Netherlands

Sufian Ullah Department of Defence and Strategic Studies, Quaid-i-Azam University, Islamabad, Pakistan

Cristina Valente Marketing and Sales, Telespazio, Rome, Italy

Petra Vanlommel Solar Terrestrial Center of Excellence (STCE) – Solar Influences Data Analysis Center (SIDC), Royal Observatory of Belgium, Brussels, Belgium

Massimiliano Vasile University of Strathclyde, Glasgow, UK

Nikita Veliev Skolkovo Institute of Science and Technology, Moscow, Russia

Xin Wang China Academy of Launch Vehicle Technology (CALT), Beijing, China

Jessica L. West Project Ploughshares, Waterloo, ON, Canada

Douglas Wiemer RHEA Group, Wavre, Belgium

Stefano Zatti European Space Agency (ESA) – Security Office, Frascati, Italy

Dong Zeng China Academy of Launch Vehicle Technology (CALT), Beijing, China

Vasilis Zervos University of Strasbourg and International Space University (ISU), Strasbourg, France

Shengjun Zhang China Academy of Launch Vehicle Technology (CALT), Beijing, China

---

Stefano Zatti has retired.

Part I
International Space Security Setting

International Space Security Setting: An Introduction

Peter L. Hays

Contents

Foundational Themes 3
International Space Security Focus Areas 5
Conclusions 7

Abstract

This chapter provides an introduction to Part 1 of the second edition of the *Handbook of Space Security* by overviewing major issues and themes that frame discourse about space security. This part contains 17 chapters that include foundational discussions about definitional, governance, theoretical, legal, deterrence, and resilience themes for space security as well as more focused discussions about cooperation, strategic competition, export controls, critical infrastructure, cyber threats, safety, traffic management, and the sustainability of space resources. Together, these themes and issues provide a comprehensive setting for refining and advancing our dialogue about international space security.

Foundational Themes

Defining and scoping space security has been and still probably is the single most important issue for any dialogue about this topic. While some decades ago most distinctions between military and civilian uses of space were rather clear, there was always an area in between called "dual-use." The latter has been ever expanding under the label of "security" and it has thereby been diminishing exclusively military and civilian purposes to just small remnants. One can think of the challenges that arise in the fields of environmental security, cybersecurity, food security, water

---

P. L. Hays (✉)
Space Policy Institute, George Washington University, Washington, DC, USA
e-mail: hayspl@gwu.edu

security, and other areas. Tackling the security issues in these domains requires analysts to take into consideration all dual-use elements that serve both civilian and military uses. It then becomes clear that we must speak of a new paradigm for the use of space; a paradigm that blurs the traditional divide between civilian and military uses. Accordingly, the chapter by Ntorina Antoni explains how the definitional approach used across the chapters of this book can be stated as: "Space security" is the aggregate of all technical, regulatory and political means that aims to achieve unhindered access and use of outer space from any interference as well as aims to use space for achieving security on Earth. This adds to an already existing abundance of definitions for space security. Since it is not likely a consensus can be reached, the definition for the *Handbook of Space Security* maintains the practice of attempting to narrow general applicability and tailor definitions to specific uses. This definition serves as a frame for the focus and the structure of the Handbook.

Governance and theoretical perspectives form other foundational aspects of space security. Effective governance is needed for humanity to derive more benefits from space; space governance also seeks to ensure space is used in stable and sustainable ways. The challenges to international space governance in ensuring space as a safer and more secure environment are enormous. These include the increasing number of actors, growing commercialization, expanding military space programs, the proliferation of anti-satellite weapons, and lack of consensus among states on the need for and realization of a conclusive and universally negotiated treaty to prohibit an arms race in space. Ahmad Khan and Sufian Ullah indicate that any model for space governance should aim to reconcile the inherent competitive tendencies among states by incentivizing further cooperation. The idea of effective global space governance should seek to maximize the prospects for peaceful exploitation of space as *a global commons* by encouraging responsible behavior of states. My chapter asserts that spacepower theory can describe, explain, and predict how individuals, groups, and states can best derive utility, balance investments, and reduce risks in their interactions with the cosmos. Such foundational theory should be more fully developed and become a source for critical insights on finding better ways to generate wealth in space, making trade-offs between space investments and other important goals, reordering terrestrial security dynamics as space becomes increasingly militarized and potentially weaponized, and seizing exploration and survival opportunities that only space can provide.

Chapters exploring the laws of war for space and the role of space in deterrence complete the foundational part of Part 1. Steven Freeland and Elise Gruttner explain how the regulation of space is embedded in international law and explicate the major themes of the 1967 Outer Space Treaty (OST), the main source of space law. As technology advances, space has been increasingly used during the course of armed conflict, notwithstanding the "peaceful purposes" provisions of the OST. Reconciling these seemingly incompatible concepts and developments is difficult and requires an understanding of how and to what extent the international law principles of *jus in bello* – body of legal norms that regulate the conduct of participants in armed conflict – as well as international humanitarian law, apply to the conduct of these activities. Freeland describes how the rising number of "dual-use" satellites further complicates matters and asserts that there is a growing need to reach consensus on additional legal regulation for armed conflict that may involve use of space capabilities. Stacey Henderson's chapter about Arms Control stresses that the current international law regime applicable to outer space does not prohibit the placement or use of all weapons in outer space, prohibiting only nuclear weapons and weapons of mass destruction, and is not capable of preventing a conventional arms race in outer space.

John Klein and Nickolas Boensch assert that even though deterrence has a legitimate role in future space strategy, it is not the panacea for preventing conflict. History shows that deterrence will at times fail due to miscalculation, uncertainty, or chance. They conclude that the enduring nature of war and strategy (and therefore deterrence) as well as the evolving character of war, indicate that the implementation of space deterrence should also be expected to change. This change is currently reflected by the growth of the commercial space sector (particularly in the United States, Europe, China, and Japan) – whether in reusable or responsive launch vehicles or mega-constellations of Earth imaging and communications satellites. In view of the increasing complexity of the space environment, resilience has emerged as a pervasive concept in contemporary space security. Regina Peldszus analyzes resilience from two distinct but complementary approaches: mission assurance and deterrence as well as high reliability and resilience engineering. Drawing on contemporary thinking from civilian and military perspectives, resilience is addressed as a distinct yet malleable notion at the intersection of space security and safety. She concludes that resilience is likely to continue as a key concept in space policy and systems planning. Straddling the fields of space security and reliability, it may inform, enrich, or even galvanize the more traditional security and safety management disciplines.

International Space Security Focus Areas

Maintaining the ability of this domain to support safe, sustainable, and secure access and use for all – the essence of space security – requires cooperation. Jessica West explains that although cooperation is embedded as a core value within the institutions and laws that govern outer space, new uses and users of outer space are changing the dynamics of space security cooperation. This means that while cooperation can enhance security in space for those involved, it may come at cost to the long-term security of space by increasing strategic rivalry and facilitating the escalation of conflict into outer space. Jana Robinson and co-authors address the China's and Russia's global space footprint in the economic and financial (E&F) domain. They describe how the pace and nature of these international space partnerships concluded by China and Russia present a strategic and competitive challenge for Europe, the U. S., and other allies, including the development of global space governance, as well as markets based on transparency, good governance, and disclosure. The authors assert that the more subtle strategy deployed in the developed, democratic countries, to gain influence is conducted on an incremental basis, while the other approach, described as "space sector capture," mostly involves developing countries and consists of offering package deals of capabilities, services, and financing, creating sole-source supplier relationship and long-term dependencies.

Ulrike Bohlmann and Gina Petrovici explain how the Cold War drove both innovation in space technology and imposition of controls on the export of these technologies. Balancing national security and commercial interests has been and remains difficult due to the Janus-faced, "dual-use" nature of space technology, serving scientific and commercial interests on the one hand and strategic, defense-related objectives on the other. Export restrictions play a significant role within the sovereignty of a state, assert Annamaria Nassisi and Isabella Patatti, and therefore keeping the technological edge is perceived as a form of dominant power. As the demand for space-based security is very high, major spacepowers are inclined to protect critical technology, rather than exporting it. Alexandru Georgescu highlights that space systems are a key enabler for a wide variety of applications which have become critical to the functioning of modern societies. He uses the Critical Infrastructure Protection framework to argue that space systems may constitute a new form of Critical Infrastructure, dubbed Critical Space Infrastructure, and traces the positive impact that such a perspective may have on space security governance.

With regard to cybersecurity, Stefano Zatti asserts that the security measures implemented in space-based systems may turn out to be insufficient to guarantee information assurance against possible cyberattacks. Accordingly, security-specific aspects of the European Space Agency's (ESA) space missions, along with specific cyber threats and possible countermeasures are addressed. Space safety is necessary for the sustainable development of space yet, as Joe Pelton and his coauthors describe, safety considerations are too often an afterthought for space security issues. Without improved space safety practices and standards from launch, to on-orbit operations, to reentry, billions of dollars of space assets, many astronaut lives, and even people on Earth could all be increasingly in peril. A related topic of growing importance is the concept of space traffic management. William Ailor's chapter begins by providing an overview of the evolution of the near-Earth space environment, discussing the current situation, and projecting how the addition of large constellations of satellites to low-Earth orbit (LEO) will affect that environment. Just as the growth in air travel led to air traffic management, assuring that future space systems will have minimal interference to their operations requires a system to warn operators of potential collisions and other hazards: a space traffic management system.

Space sustainability is a concept that has emerged within the past 15 years to refer to a set of concerns relating to outer space as an environment for carrying out space activities safely and without interference, as well as to concerns about ensuring continuity of the benefits derived on Earth from the conduct of such space activities. Peter Martinez, as a long-time international space policy expert, is in an ideal position to review the role of the various relevant United Nations (UN) entities in ensuring space sustainability and provide a detailed review of the Working Group on the Long-Term Sustainability of Outer Space Activities within the Scientific and Technical Subcommittee of the United Nations Committee on the Peaceful Uses of

Outer Space (UN COPUOS). In addition, his chapter discusses the relationship of the work in UN COPUOS with related work being done in the Conference on Disarmament, the UN Group of Governmental Experts (GGE) on Transparency- and Confidence-Building Measures (TCBMs) in Outer Space Activities, and the initiative by the European Union to propose a draft international Code of Conduct for outer space activities. Finally, George Kyriakopoulos addresses space security as part of the overall international security, the maintenance of which constitutes the fundamental purpose of the UN Charter. In particular, he asserts that preserving security with respect to the celestial bodies requires the activation of mechanisms able to guarantee that the existing *status quo* will not be compromised by the placement of offensive weapons on them or potential conflict over the exploitation of space resources.

Conclusions

This overview of Part 1 of the second edition of the *Handbook of Space Security* provides a comprehensive introduction to major issues and themes that shape humanity's dialogue about space security. The 17 chapters in Part 1 include foundational discussions about definitional, governance, theoretical, legal, deterrence, and resilience themes for space security as well as more focused discussions about cooperation, strategic competition, export controls, critical infrastructure, cyber threats, safety, traffic management, and sustainability. These chapters provide a comprehensive foundation for the more detailed and focused discussions of space security themes and issues in the remainder of the *Handbook of Space Security.*

Definition and Status of Space Security

2

Ntorina Antoni

Contents

Introduction	10
Definition of Space Security	11
Security Definition	12
Space Security Evolution	13
Space Security Definition	15
Status of Space Security	16
Africa	16
Asia-Pacific	18
Europe	21
The Middle East	23
Latin America	25
North America	26
Russia	28
Key Priorities	29
Concluding Remarks: The Way Forward for Space Security	30
References	31

Abstract

Space security has gained increasing importance over the past decade as space becomes part of our everyday life. Yet, space security is not universally defined. The shift of paradigm and transformation of the space domain through new ways of utilizing space and recent technological advances such as mega-constellations, 5G, Internet of Things, artificial intelligence, and advanced materials have resulted both in major challenges and new opportunities. Over the years, space security has evolved. Since the signature of the Outer Space Treaty in 1967, space security has become a complex, broader, and multilayered concept. The topic dominates space

N. Antoni (✉)
Eindhoven University of Technology, Eindhoven, The Netherlands
e-mail: ntorina.antoni@gmail.com

© Springer Nature Switzerland AG 2020
K.-U. Schrogl (ed.), *Handbook of Space Security*,
https://doi.org/10.1007/978-3-030-23210-8_126

law and space policy agendas at the United Nations General Assembly Committees and the Committee on the Peaceful Uses of Outer Space along with its subsidiary bodies. In this context, this chapter aims to effectively capture the multifaceted concept of space security and provide an overview of its current status.

Introduction

Space security entails the possibility to access and use space for all nations. Although traditionally it has been associated with military engagement, over the past years it has been enriched with safety aspects. The space race between the United States and the former Soviet Union in the 1960s triggered the first concerns regarding space security. The attempt to end an arms race in space was effected with the conclusion of the United Nations (UN) Outer Space Treaty in 1967 (United Nations 1967). The treaty sought to define boundaries for the security of outer space by establishing the principle of peaceful purposes in accordance with the UN Charter and by prohibiting the militarization and weaponization of space. The ratification of the Outer Space Treaty was a remarkable endeavor of resolving the space race tension, ensuring stability, and promoting international cooperation. Thus, space security – although not explicitly defined – was the result of the stabilizing effect of a treaty-based mechanism, and *vice versa* space security meant that activities in outer space ensure stability and peaceful uses of outer space (United Nations 1967). In this context, the interrelatedness between space security and stability was reinforced by the explicit distinction between civil and military uses of outer space.

Five decades later, the scope of space security has changed. As Sheehan notes in the chapter "Defining Space Security," of the previous edition of this Handbook, "space security includes now aside the military dimension, also, economic, societal and environmental dimensions" (Sheehan 2015). These elements are indispensable to space security, in view of the ongoing transformation of the space sector that moves away from the traditional confines of space activities. The so-called New Space encapsulates major changes taking place at unprecedented rate. These are related to the growing participation of private actors, the rising number of space-faring nations, and the emergence of the civil-military paradigm. This means that the dividing line between civilian and military uses of outer space has yet become artificial leading to uncertainty regarding governance of dual-use or hybrid areas. The terms "safety," "security," and "defense" are intertwined and used interchangeably with no clear separation between areas of action. In many languages there is no clear distinction between the words safety and security. The cultural aspects of safety, security, and defense vary from country to country and from region to region. What is more, the understanding of space security has been redefined considering the new often blurred borders between safety – a clearly civilian area – and defense – a clearly military one. Security lies in between and for some countries/regions is closer to safety while for others closer to defense. This debate extends to governance questions as to who has legitimacy to act in space security and for what type of actions. Also, what would the role of the civil and defense actors respectively be and in which area. Accordingly, the various and divergent concepts, approaches, and definitions across the chapters of this Handbook are representative of an evolving space security landscape.

The absence of an internationally agreed definition – combined with the systemic nature of the space sector with multiple strategic objectives – presents challenges when endeavoring to build cooperative approaches among diverse organizational actors. As such, this requires the development of a mechanism that fosters new forms of cooperation among states in the advent of the new space era. Therefore, stability remains of strategic importance to the space sector, as it influences the effectiveness of states to manage the growing challenges and ultimately ensure space security. Accordingly, the remainder of this chapter will address definitional aspects and the current status of space security. It will provide an overview of space security perspectives in Africa, the Asia-Pacific, Europe, the Middle East, Latin-America, North America, and Russia. Such an approach will help to identify the underlying challenges related to space security and advance the understanding of the civil-military paradigm therein. The latter is a main challenge that needs to be taken into consideration for the development of space security enhancing mechanisms.

Definition of Space Security

There is no commonly agreed definition and uniform understanding of space security. Be that as it may, there are myriad definitions adopting either a "soft" or "hard" approach. Often, the concept of "security" is used instead of the term "safety" or the term "defense," or instead of both. This creates ambiguity concerning the content of space security and the set of underlying shared values and principles. As a result, the lack of clear boundaries between these concepts poses a major definitional challenge for space security, as depicted in Fig. 1. In attempt to address this definitional challenge, this section will first take a closer look into the security concept under international relations/law perspective and, then, it will examine the evolution of the security concept in the outer space context.

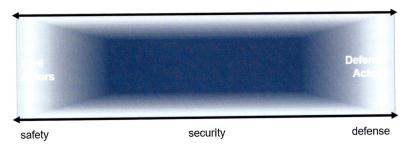

Fig. 1 Definitional challenge for space security

Security Definition

Security is derived from the Latin term *se + cura*, meaning free from care. Security means the quality or state of being secure, such as freedom from danger (safety) and something that secures protection. International relations scholarship has not agreed on a single definition of space security, due to its contested nature in the post-Cold War era as well as its overall subjectivity. In the words of Gallie (1956) security is often referred to as an "essentially contested concept" one for which, by definition, there can be no consensus as to its meaning (Williams 2013). According to Williams, "security is most commonly associated with the alleviation of threats to cherished values; especially those which, left unchecked, threaten the survival of a particular referent object in the near future" (Williams 2013). Maintaining international peace and security is the central mission of the UN as per Preamble of the United Nations Charter. As such, the UN has agreed to: "take effective collective measures for the prevention and removal of threats to the peace, and for the suppression of acts of aggression or other breaches of the peace, and to bring about by peaceful means, and in conformity with the principles of justice and international law, adjustment or settlement of international disputes or situations which might lead to a breach of the peace" (United Nations Charter, Article 1). Thus, although security is not defined under international law, it is perceived as closely influenced by the concept "peace" and "peaceful means."

The lack of clear definition has led to the interpretation of security from several perspectives, namely, individual, national, international, and global (McDonald and Brollowski 2011). Over time, security became intertwined with the concepts of territoriality and sovereignty of states, as reflected in the term "national security" and "defense." *Defense* is derived from the Latin term *defensum* meaning "thing protected or forbidden." In the broad sense it means "the act or action of defending." Defense pertains to the protection of states' territory, including its property and population, via diplomatic channels or by use of force (McDonald and Brollowski 2011). The use of force is stipulated in Article 2, paragraph 4 of the UN Charter while the right of a State to use force in self-defense is long-established in customary international law (Greenwood 2011). Provided that no definition of defense is provided, the main understanding of national security and defense in this context is interrelated with political and military security used by states (McDonald and Brollowski 2011).

Although the traditional concern has been related to security from external military threats and the use of force, the notion of security has evolved to include additional threats to a number of values: environmental security, economic security, physical security, human security, etc. (Baldwin 1997). In 1995, the United Nations Secretary-General called for a "conceptual breakthrough" of security which goes beyond the confines of "armed territorial security" to include also "the security of people in their homes, jobs and communities" (Rothschild 1995). This shifts the focus of security from states to people. This type of security stands closer to the notion of "safety" which is related to the human right perspective of security or the security of the individual as stated in Article 3 – right to life – of the UN

Universal Declaration of Human Rights. *Safety* is derived from the Latin term *salvus* meaning "uninjured, in good health." Safety means "the condition of being safe from undergoing or causing hurt, injury, or loss" and "something that secures protection." Accordingly, this notion of security – mainly associated with safety – has gained attention by international relations scholars as well as the UN, who seek to give a broader interpretation to security that includes economic, food, health, environmental, personal, community, and political security (Baldwin 1997; Osisanya 2015).

Space Security Evolution

The definition of space security is as elusive as the definition of security itself. Similarly, to the ambiguity of the security concept within the frame of international relations, there is no universally agreed definition on space security. As such space security is a multifaceted term that many have attempted to define yet no consensus has been reached. The evolution of the security concept over time combined with the evolution of outer space activities poses unique challenges to the understanding and definition of space security. What is more, a significant challenge remains the dual-use nature of space technology and applications.

The military perspective of space security, closer to the "defense" side, has to a large extent derived from the global agenda on international peace and security. The launch of Sputnik-1 in the 1960s, followed by the first manned spaceflights in the 1970s, marked a technological race between the former Soviet Union and the United States. This created the fear of an arms race in space and profoundly influenced the definition of space security. In this regard, the international community was concerned that space could be used for military purposes. Accordingly, the UN General Assembly adopted in 1958 the Resolution 1348 (XIII) "Question of the peaceful use of outer space," where it expressed the desire to "avoid the extension of rivalries into this new field." The principles set forth in this resolution combined with those of the subsequent resolutions (1961 and 1962) were ultimately embodied in the UN Treaty on Principles Governing the Activities of States in the Exploration and Use of Outer Space, including the Moon and other Celestial Bodies, with the United Nations Committee on the Peaceful Uses of Outer Space (UNCOPUOS) being the most important UN body engaging in the development of international space law (United Nations 1967).

The Outer Space Treaty establishes outer space as a *global commons*, not subject to national appropriation. The States Parties to the treaty recognized that it was in the common interest of all mankind to commit to broad international cooperation in the scientific as well as the legal aspects of the exploration and use of outer space for peaceful purposes. The 1996 UNGA Declaration on International Cooperation further elaborates on the modes of cooperation that are considered most effective and appropriate "including, inter alia, governmental and nongovernmental; commercial and non-commercial; global, multilateral, regional or bilateral and international cooperation among countries in all levels of development." In addition, the treaty makes explicit reference to the applicability of international law and the UN Charter to outer space. Article III stipulates that: "States Parties to the Treaty shall carry on activities in the exploration and use of outer space, including the moon and other celestial bodies, **in accordance with international law**, including the Charter of the United Nations, in the interest of **maintaining international peace and security** and promoting international co-operation and understanding" (bold by the authors). Furthermore, Article IV paragraph 1 prohibits the placement in orbit around the earth of "any objects carrying nuclear weapons or any other kinds of weapons of mass destruction," and it adds that "The moon and other celestial bodies shall be used by all States Parties to the Treaty **exclusively for peaceful purposes**. The establishment of military bases, installations and fortifications, the testing of any type of weapons and the conduct of military manoeuvres on celestial bodies shall be forbidden. The use of military personnel for scientific research or for any other peaceful purposes shall not be prohibited. The use of any equipment or facility necessary for peaceful exploration of the moon and other celestial bodies shall also not be prohibited" (bold by the authors). Despite the premise of peaceful purposes, the explicit prohibition of the weaponization and militarization of outer space, and the application of international law, the boundaries of space security under the body of international space law remain yet dubious.

However, the distinction between civilian and military uses of outer space is not easy to draw due to the strong dual-use nature of space technology. From a technical point of view it is not easy to tell whether certain space technologies in the fields of satellite communications, positioning, navigation and timing, and space situational awareness are used for civilian applications and programs or for military and defense purposes. The dual-use factor is the main reason that the United Nations Group of Governmental Experts (GGE), aiming at exploring legal instruments that prevent the placement of weapons in outer space, failed to reach a consensus on a substantive report in October 2019. During the General Assembly's First Committee on discussing related draft resolutions one of the delegates stated that "Due to the dual-use nature of space objects, it is inherently difficult to define an outer space weapon or to know and verify intentions behind certain related activities." Another one highlighted "the current blurred distinction between civil, commercial and military activities in outer space, saying the international community must bring new ideas to discussions going forward" (United Nations 2019). This blurred dividing line between civilian and military applications is further exacerbated by the increasing commercialization of outer space and the new security paradigm of hybrid threats that are also applicable to outer space. On the one hand, the growing commercial sector enables the pursuit of military authorities to modernize space capabilities. Such a trend is reflected in a recently commissioned study to explore the possibilities and risks of employing commercial systems for the proposed US Space Force (SpaceNews 2020). On the other hand, the availability of civilian or commercial space assets to security- and defense-related missions contribute to the proliferation of hybrid threats to space. These include active operations, such as cyberattacks, jamming or spoofing, dazzling, and passive ones such as hiding or moving assets (Robinson 2018).

What is more, the increasing participation of private commercial actors in space security has raised some concerns related to the "softer" side of space security, namely space safety and sustainability of outer space activities. Space safety regards the use of space technology and applications with societal benefits, namely, water management, marine and coastal ecosystems, health care, climate change, disaster risk reduction and emergency response, energy, navigation, seismic monitoring, natural resource management, biodiversity, agriculture, and food security. Hence, space safety extends to the security of space systems in order to provide for security on Earth, as well as for space sustainability (▶ Chap. 15, "Space Safety"). In February 2018, at the 55th session of the UN COPUOS – Scientific and Technical Subcommittee, agreement was reached on nine additional guidelines on the long-term sustainability of the outer space activities, with the following definition: "The long-term sustainability of outer space activities is defined as the ability to maintain the conduct of space activities indefinitely into the future in a manner that realizes the objectives of equitable access to the benefits of the exploration and use of outer space for peaceful purposes, in order to meet the needs of the present generations while preserving the outer space environment for future generations" (▶ Chap. 17, "Space Sustainability").

Space Security Definition

Based on the above, the different perspectives of space security have led to myriad definitions. For instance, the Space Security Index (SSI) defines space security as "the secure and sustainable access to, and use of, space and; the freedom from space-based threats" (Sheehan 2015). Another example is that the European Union defines space security as "security from space, where Space-based assets and systems are critical to ensuring security on Earth, and security of space, where these assets need to be protected in the difficult environment of outer space." Since it will be difficult to reach consensus among states, and any definition might end up being obsolete provided the rapid transformation of outer space activities, it is imperative to narrow general applicability and tailor-made approaches. To that end, the following definition is provided in this Handbook:

> "Space security" is the aggregate of all technical, regulatory and political means that aims to achieve unhindered access and use of outer space from any interference as well as aims to use space for achieving security on Earth.

This approach helps to structure the myriads of definitions and cover the full spectrum of constitutive elements thereof. Such a definition is thus necessary to reach a common understanding and support cooperative and collective frameworks in order to tackle the inherent complexities of space security in a constantly shifting environment. A foundational challenge arising from space security initiatives at the national, regional, and intergovernmental levels is the need for collaboration and synergies between civilian and military entities as depicted in Fig. 1. Such a challenge leads to uncertainty in governance, strategy, and policymaking aspects of space and security. In this regard, considerations need to be made for mechanisms that can effectively deal therewith and ensure stability and international cooperation for space security.

Status of Space Security

The new paradigm shift in space security has influenced the expectations of countries and regions around the world. This part will address the status of space security in order to identify strategic priorities and needs on space security.

Africa

Space activities in Africa have been increasing over the past decades, with different countries having reached different levels of investment and development. African actors include organizations that operate at the continental level, such as the African Union (AU), regional organizations, typically the Regional Economic Communities (RECs), technical organizations such as specialized agencies and institutes in different areas and the African space agencies (Giannopapa 2011).

Space security engagement in African countries is more visible on the civil side, in particular regarding the benefits that space can bring to Earth. African countries that engage on space can be split into three main categories: passive users, active users, and active developers. Passive users are African countries that do not have any space capabilities. They only receive information already processed by others. Active users are African countries that have the capacity to process the information offered. Active developers are those African countries that themselves have capacity in space activities and typically have also a space agency and more advanced space policy provisions, either contained in self-standing documents or as parts of other policies. Space security is perceived according to the user needs of the countries. Most considerations are related to the use of space security for societal purposes on Earth. Space applications can assist in providing solutions to people's basic needs such as providing food and water security, health care, education, early warning, disaster management, and emergency response. Nevertheless, the benefits of space applications are not sufficiently communicated to decision makers or the wider population, and there is not enough basic education at various levels to perform, manage, and operate space-based assets. Various space projects in different areas have been developed in Africa but very few are sustainable beyond the pilot phase. This is because often the local community of end users is not involved from the beginning and does not have a feeling of ownership. Appropriate bodies have not been identified within the government structure to take up the responsibility for running and maintaining the project. The projects developed in Africa are typically conceived by developed countries, which have not properly captured the societal needs and infrastructure restrictions of the underdeveloped countries they purport to be helping (Giannopapa 2011).

African countries have been looking into the European model for regional cooperation on space activities with the African Union taking the lead, while countries that are active developers engaging in bilateral cooperation with global players. The 2017 African Space Strategy, which sets out the objectives for an African Space Program, is mainly focused on scientific and public good applications (African Union 2017). The strategy was adopted by the African Union heads of states and government, representing 55 member states on the African continent. According to the policy, satellites are enabling tools aimed to tackle challenges in Africa, including water resources, weather monitoring, security assistance in conflict zones, disaster aid planning, infrastructures, and food security. The focus is predominantly on space security for Earth. Food security enabled by space systems is further elaborated in ▶ Chap. 40, "Space-Enabled Systems for Food Security in Africa". The African Union has emphasized the importance of security and defense to the development of Africa (African Union 2015). As such, strategic priorities for the African Space Program focus on earth observation, satellite communication, autonomy and space science, navigation and positioning (UNOOSA 2019). Aside the civil use of space applications, the African Space Strategy outlines that earth observation data are important for military applications "where terrain profiling and mapping is critical for the deployment of ground troops, especially in hostile and remote territories" (African Union 2017). For example, at the national level, the 2018 South Africa National Space Strategy takes into consideration the dual-use of space activities and the interconnectedness of civil and military applications (Republic of South Africa 2018). However, no explicit reference is made in either document regarding the cooperation between civil and military authorities and the potential challenges of discrepancies between the regional and national levels.

Acknowledging the importance of coordinating the African space program, Egypt has been assigned to host the African Space Agency (African Union 2017). Yet, establishing the space agency has encountered many challenges due to governance issues and financial implications. In order to realize these objectives, the African Union convenes regular annual space conferences with African space actors as well as UN representatives, academia, and the private sector (African Union 2020). This inclusive approach is necessary provided the increasing number of countries that have national space programs. While South Africa has established links for space security cooperation at the bilateral and multilateral levels and is also finalizing a new space legislation (Lal et al. 2018), other countries in Africa are lagging. Accordingly, the next steps are focused on mobilizing resources across all African countries to facilitate the launching and implementation of the African Space Agency (African Union 2020). These steps are necessary in order to overcome the political fragmentation and improve the governance of the African space program. It is thus fundamental to form a coordinated regional approach for the development of space activities, which can also tackle safety and security challenges in the continent. Such approach can further facilitate the representation of Africa at the international cooperation *fora* and enhance international cooperation.

Asia-Pacific

Independent space powers coexist in Asia, namely Japan, China, India, South Korea, Pakistan, and Australia. Asia is the world's second-largest defense spender while it is becoming increasingly active in space. The geopolitical and military competition in Asia has an impact on the space efforts of these countries (▶ Chap. 27, "Space Security in the Asia-Pacific"). There are three key drivers to space in Asia: increasing use of space for military purposes; civilian use that could also lead to conflict because of congestion and competition; and investments in military technologies such as those for anti-satellite (ASAT) tests and missile defense. The growing space competition is demonstrated by the rapidly growing development of counterspace capabilities, such as kinetic ASAT missiles, electronic and cyber warfare capabilities, and new efforts at creating specialized military agencies devoted to space utilization. The Asia-Pacific Space Cooperation Organization (APSCO) is an intergovernmental space cooperation organization headquartered in Beijing, China, and its members include China, Bangladesh, Iran, Mongolia, Pakistan, Peru, Thailand, and Turkey. It was founded in 2005. Both Japan and China have been trying to set up regional cooperation under their respective leadership, resulting in two different formats for governance. While Japan established an Asia-Pacific Regional Space Agency Forum (APRSAF) in 1993, China founded the APSCO in 2008. Membership is somewhat overlapping, with institutions from 40 states (including non-Asian) participating in APRSAF, and 8 formal member-states in APSCO. This indicates the basic difference between the two: APRSAF is a coordination mechanism of institutions (space agencies, research establishments, space applications users, etc.), while APSCO is an intergovernmental organization.

China, over the past decades, has been rising as one of the major space powers worldwide toward establishing dominance. In a stepwise approach, it has set out its ambitions and a long-term strategic and programmatic development. Among other achievements, the landing of an unmanned mission on the near side of the Moon using its Chang'e 3 in December 2013 and the Chang'e 4 landing on the far side of the lunar surface in January 2019 have marked China's presence as an international space power. The engagement has been increasing on the civil side using space to provide space security on Earth, while information and communication technologies have gained impetus in overall national power and especially military capability (▶ Chap. 29, "Chinese Concepts of Space Security: Under the New Circumstances"). Many of China's satellites are dual-use, supporting urban planners and agricultural programs as well as the military. China's military-dominated and government-monopolized characteristics of space affairs aim to internationalize and commercialize the space industry (Nie 2020). Under the Belt and Road Initiative – expanding from China and Asia to Europe and Africa – China has been actively engaging in the Space Information Corridor project. The latter "takes communication, remote sensing, navigation satellites as the main body, with space-based information resources and ground information sharing network and aims at realizing co-construction and sharing of space information in the region" (Jiang 2019). Simultaneously, the emerging rise and engagement of China in the commercial market will increase competition among other Asian states as well as with other countries. On the defense side, the relationship of space and national security has been evolving as part of a broader ongoing assessment of the role of information in future warfare (▶ Chap. 29, "Chinese Concepts of Space Security: Under the New Circumstances"). In 2015, China established the People's Liberation Army's (PLA) Strategic Support Force which saw the integration of the PLA space, cyber, and electronic warfare capabilities, which is considered a significant achievement considering the future of warfare that would see the interface between all these different capabilities. Due to the PLA's attention to information and communications technologies, the centrality of space dominance has grown as well. As with other Chinese military activities, the PLA's approach to space operates within the context of guiding thoughts. The guiding thoughts for space are "active defense, all-aspects unified, key point is establishing space dominance" (▶ Chap. 29, "Chinese Concepts of Space Security: Under the New Circumstances").

Japan's space policy has been influenced significantly by its overall foreign and security policy. At the start of its space activities, Japan was reluctant to engage in security-related uses of space, largely due to its pacifist constitution, which is interpreted to prohibit using space for security purposes. The Basic Space Law 2008 urged the government to use space systems "to ensure international peace and security and also to contribute to the nation's security" (▶ Chap. 30, "Historical Evolution of Japanese Space Security Policy"). This has been evolving over the past few years. The country updated in 2013 its Basic Space Plan. The latter aimed at creating new opportunities for the involvement of Japan in international efforts to address the most pressing space security-related challenges of the twenty-first century. As such, this update and its subsequent revision in 2015 marked the reorientation of Japan's space program toward tackling the changes in its surrounding security environment. The latest document reflects the new national security policy 2014 (military use of space) and establishes long-term and concrete public investment plan for the upcoming 10 years (Komiya 2016). In particular, the latest version of the Basic Space Plan aims at responding to the growing threat of ASAT weapons and the increasing quantity of space debris by putting emphasis on space security, through strengthening security capabilities and the Japan-US alliance and ensuring the stable utilization of outer space. The two key projects envisaged in the new plan are the development of a Space Situational Awareness (SSA) system with ground and space segments, and the establishment of self-defense forces. In addition, Japan is preparing a new series of Earth observation satellites to tackle natural disasters and limited natural resources. Accordingly, a new space security budget is developed with contributions of Japan Aerospace Exploration Agency (JAXA) and the Ministry of Defence (Euroconsult 2019). Therefore, the revised Basic Plan represents a completely new direction of Japanese space policy with increasing role for the military. On April 19, 2019, the United States and Japan reiterated their commitment to military space activities and highlighted that space, cyberspace, and the electromagnetic spectrum are priority areas to better prepare for cross-domain operations (Spacewatch.global 2019).

India over the past year has also been rising to a world space power. It has acquired multifaceted space capabilities with dual-use applications – both civilian and military – and focuses on achieving autonomy in space including launchers, satellite communications, Earth observation, and navigation. Overall, India has followed the policy of the use of space for socioeconomic development. Over the years, the Indian Space Research Organisation (ISRO)'s program has matured significantly, and, at present, Indian space program is regarded as one of the important space programs in the world. From launching small satellites to undertaking successful missions to the Moon and Mars, India has excelled in almost all areas of space experimentations (▶ Chap. 31, "India in Space: A Strategic Overview"). Additionally, India has made significant investments toward establishing its military architecture owing to its strategic needs. Space technologies have become central to strengthening this architecture, essentially as a force multiplier. The March 2019 ASAT test clearly communicated India's intention and capability to use space for military purposes. Soon after the test, India has announced plans to establish a Defence Space Agency along with a Defence Cyber Agency marking a shift in the evolution of the Indian space strategy (Euroconsult 2019). The launch program remains one of India's main objectives to ensure independent access to space. Starting with the development of the Satellite Launch Vehicle (SLV-3) during the 1970s, it has progressed through the Augmented Satellite Launch Vehicle (ASLV), Polar Satellite Launch Vehicle (PSLV), and Geosynchronous Satellite Launch Vehicle (GSLV). Recently, the development of the next-generation launch vehicle, the Geosynchronous Satellite Launch Vehicle Mark III (GSLV MkIII), has been completed and has become operational. India's next milestone mission of Human Spaceflight has been initiated, and the first crewed flight is expected by 2022 (▶ Chap. 70, "Indian Space Program: Evolution, Dimensions, and Initiatives").

The space sector in *Australia* is experiencing an unprecedented level of public interest and government support. National security considerations and the economic benefits of a fast-growing world market for space products and services are inextricably linked as drivers for a range of government and industry initiatives (▶ Chap. 71, "Australia's Space Security Program"). The Government's clear intention is to enhance Australian Defence sovereign space capabilities progressively through dedicated Intelligence, Surveillance, Reconnaissance (ISR); space and cyber programs; SSA systems; and military satellite capability (Euroconsult 2019). In May 2016 the Australian Department of Defence announced that it is preparing a roadmap for a $2.3 billion next-generation satellite communications investment on mixing commercial and military capability. Australia is now treating space as an integral component of its role in the protection of its national security interests and in the advancement of its international responsibilities. In September 2017, the Australian Space Agency was established to take over the operational and regulatory activities. Regarding the development of space policy, the 2017 Space Industry Association of Australia (SIAA) White Paper suggests that ensuring long-term access to space, for strategic purposes requires both civil and military capabilities. Along the same line, the Australian Civil Space Strategy 2019–2028 focuses on seven national priorities: position, navigation, and timing; earth observation; communication technologies, and services; "leapfrog" research and development; space situational awareness; robotics and automation; and access to space (▶ Chap. 71, "Australia's Space Security Program").

South Korea is implementing a pro-active space program in order to create autonomous operational capabilities in areas such as Earth Observation, as well as to develop domestic industrial capabilities for satellites and launch vehicles, and to build the associated infrastructure. Space development in South Korea is driven by the National Space Development Promotion Basic Plan, with the third pillar including exploration and navigation as two key elements for 2018–2022. South Korea has historically focused on satellite development, while more recently expanding into space launch vehicles. Furthermore, South Korea is trying to meet national user needs, to serve external needs commercially, and to increase its participation in international programs. The next challenge for the country is to complete the development of the launcher KSLV-2 (Euroconsult 2019).

Pakistan, despite political, technological, and economic constraints, is considered an aspiring space power, although with a relatively modest space program compared to the larger, more successful ones of China and India. The country aims to utilize available resources to improve its nascent space infrastructure through collaborative efforts to gain eventual self-sufficiency for socioeconomic and strategic purposes in the South Asian region (▶ Chap. 72, "Pakistan's Space Activities"). The Space Development Program 2040 approved by the National Command Authority aims to ensure space-based benefits for the country and focuses mainly on telecommunications and Earth observation. In addition, Pakistan has close ties to China via the China-Pakistan Economic Corridor (CPEC). It considers a manned space mission in 2022. Yet, SUPARCO's (Space and Upper Atmosphere Research Commission) lack of funding impedes technological advancement and innovation (Euroconsult 2019).

Europe

In Europe, the space sector is a particularly interesting and dynamic field, mainly because Europe includes several space faring nations with varying capabilities and priorities. Due to the inherent dual-use nature of space activities, responsibility for space has traditionally resorted under a State's sovereign competences. Traditionally, security- or defense-related space programs have been established and maintained at the national level or dealt with bilaterally or multilaterally in *ad hoc* cooperative programs. Only civilian space activities, including Earth observation, telecommunications, human spaceflight, space transportation science, and technology development, were the subject of cooperation at the regional and intergovernmental levels. However, the past years/decade the security dimension of space activities has increasingly been coming to the attention of European countries, as well as the European Union (EU) and the European Space Agency (ESA). "Space and security," both in its security from space and security in space aspects, is progressively contributing to the further integration of space activities in sectorial policies (Giannopapa et al. 2018). Today, space security constitutes the second pillar of activities of the ESA as agreed by the Ministers of ESA's Member States in November 2019. Additionally, the European Commission in its new organizational structure creates a dedicated Directorate General Defense Industry and Space, with the new president Ursula von der Leyen viewing it as one of the EU's priorities to reinforce the European defense capacities. This directorate aims to exploit the growing possibilities that space offers for the security of European citizens, including the capitalization of synergies between the civil and defense sectors.

Over the past several years, the EU has formulated a space security strategy, including in the 2018 Proposal for a Regulation for a Space Programme for the EU which is based on the 2016 Space Strategy for Europe. One of the main goals of the 2016 Space Strategy for Europe is to "Reinforce Europe's autonomy in accessing and using space in a secure and safe environment." The Regulation for a Space Programme proposes the development of Governmental Satellite Communications (GOVSATCOM) and SSA programs to accompany the satellite navigation program Galileo and the earth observation program Copernicus. The European family of launchers includes the Ariane 5, Vega, and Soyuz that secure Europe's independent access to space and are launched from the Guiana Space Centre. At the same time, ESA has more explicitly formulated its space security policy, as reflected in the "Elements of ESA's Policy on Space and Security" and the safety and security program adopted at the Ministerial Council in 2019. During the Ministerial Council 2019, ESA adopted a safety and security program and also secured the transition to the next generation of launchers: Ariane 6 and Vega-C, as well as the Space Rider, ESA's new reusable spaceship. European institutional programs are intertwined with the national and multilateral programs of the European countries based on their national budgets and contributions to organizations such as and the EU. In addition to ESA and the EU policy and programmatic developments, in 2019 NATO Defence Ministers approved its first ever space policy.

The different space security policies of the various European countries are to a large extent determined by national needs and priorities as brought forward through their participation in relevant space and security organizations, including ESA, EU, the European Defence Agency (EDA), and the North Atlantic Treaty Organization (NATO). The largest groups of European countries are currently members to all four organizations (ESA, EU, EDA, and NATO): Belgium, Czech Republic, France, Germany, Greece, Estonia, Hungary, Luxembourg, The Netherlands, Poland, Portugal, Romania, Spain, and the United Kingdom (due to Brexit the UK is no longer an EU Member State). A few countries belong to ESA, EU, and EDA, but they do not belong to NATO: Austria, Finland, Ireland, and Sweden. Norway is part of NATO and ESA. Even though Norway is not an EU Member State, it still participates in EU space programs and to EDA programs. Denmark is a NATO, EU, and ESA Member State but is not an EDA Member State. It also opted out of the EU Common Security and Defence Policy (CSDP). Slovenia is an EU and EDA Member State, and an ESA Associate Member State. Switzerland is an ESA Member State. Overall, the current priorities and trends in space and security are reflected on the space and security elements stipulated in national strategic documents. Depending on each European state, either a dedicated space security strategy is in place or space and security aspects are included in strategy documents covering other policy areas. For example, space and security aspects can be found in maritime strategies and arctic strategies that also stress the importance of space-based assets and applications in these domains. The space activities and programs of European countries are centered on the fields of Earth observation, satellite communication, Global Navigation Satellite System (GNSS), SSA, space transportation, satellite operations, and detection, tracking, and warning. The institutional space and security policy developments in Europe have been developing in parallel with the policies of the European countries. All in all, European space and security governance is multifaceted, thereby posing a major challenge to effective cooperation among the EU, ESA, and the European States (▶ Chap. 23, "Strategic Overview of European Space and Security Governance").

The Middle East

The United Arab Emirates (UAE), Saudi Arabia, and Iran have emerged as regional leaders in the Middle East (Euroconsult 2019). In the recent years, Middle Eastern countries have come together to collaborate on satellite programs. Namely, in March 2019 the UAE Space Agency launched the Arab Space Coordination Group to build the first pan-Arab Earth Observation satellite via cooperation among its 11 member nations: Algeria, Bahrain, Egypt, Jordan, Kuwait, Lebanon, Morocco, Oman, Saudi Arabia, Sudan, and the UAE (National Defense 2020).

The UAE, a federation of six emirates with the world's sixth largest oil reserves and the Middle East's primary trading center, had essentially no involvement in space activities – other than shareholdings in communication satellites – until 2006, when the government of Dubai established the Emirates Institution for Advanced Science and Technology (EIAST). The latter was incorporated into Mohamed Bin Rashid Space Center (MBRSC) in 2015 which focuses on space research, satellite manufacturing, systems development, and Earth observation (Euroconsult 2019). In 2014, the UAE Space Agency was established with the mandate of overseeing and promoting the country's space sector and activities. With this mandate the agency is responsible for the development of the so-called National Space Framework consisting of four main components, namely, Space Policy, Space Strategy, Space Law, and Space Regulations (▶ Chap. 34, "UAE Approach to Space and Security"). The UAE Space Agency actively cooperates with several international and regional space agencies such as those in the United States, France, China, India, Japan, South Korea, Italy, Germany, Kazakhstan, Bahrain, among others. The UAE is in the process of establishing its regulatory framework for space activities. In December 2019, the National Space law was approved and came into effect, hence setting the regulatory basis for space activities by covering the organization and objectives of space projects undertaken by the country, including peaceful space exploration and the safe use of space technologies. In the field of satellite communications, Al Yah Satellite aims to provide commercial, governmental, and military services, while at the same time focusing on further growth, empowerment of human capital, and quality enhancement (▶ Chap. 34, "UAE Approach to Space and Security").

Israel's space industry aims at achieving independence and national defense goals. The launch program – Shavit rocket – plays an important role in Israel's vision, making it one of the few nations with the ability to launch unmanned missions to space. The Shavit launch vehicle, operated by the Israel Defense Forces, was developed to enable Israel to launch its military reconnaissance satellites, the Ofeq series (Euroconsult 2019). In the past 30 years, Israel developed an indigenous space capability to develop, launch, operate, and maintain satellites in two main niche areas: Earth observation and communications, including the ground segment of communications satellites. Israel's focus continues to rely on a broad space infrastructure for defense and civilian applications under the auspices of the Ministry of Defence and the Israel Space Agency. The space agency aims at implementing a new space program, geared toward research and development while supporting multiple private and academic initiatives. In addition, the agency has forged bilateral cooperation with the United States and European countries (Euroconsult 2019). While security in the region has been Israel's key concern throughout its history, unsurprisingly security has also been the key driver of the country's space activities. It has, however, also resulted in the growth of the commercial space sector. Israel has expanded, in recent years, its cooperation with international partners, as well as established a civilian space policy backed by modest government funding. Within the context of protecting and encouraging this nationally important ecosystem, Israel considers international space security, safety, and sustainability to be of importance (▶ Chap. 32, "Israel's Approach Towards Space Security and Sustainability").

Iran is a member of the Asia-Pacific Space Cooperation Organization (APSCO) since 2004 (Spacewatch.global 2016). Iran planned to build and launch satellites in 1996, but made little headway until 2004, when a broad review of plans and policies led to the creation of the Iranian Space Agency (ISA) and the allocation of a sizable budget under the sixth Five-Year Development Plan. ISA, which falls under the Ministry of Information and Communications Technology, along with the Ministry of Defence has cooperation agreements in place with Russia, Bolivia, Azerbaijan, and Kazakhstan. The Iranian space program is sustained by substantial research and development capabilities in its universities and defense industry, robust funding, and high-level political support. As such, the space program aims to fulfill both civilian and military objectives (Euroconsult 2019). The "Comprehensive Document of Aerospace Development," which was adopted in 2012 and emphasizes the capabilities of Iran for space activities extending to both civilian and military entities. In February 2007, Iran tested a Sounding Rocket Vehicle (SRV) for research purposes which was followed by SRV 1. SRV 2, which was successfully launched into space, provided the opportunity for Safir SLV to launch the first national satellite, Omid (▶ Chap. 33, "Policies and Programs of Iran's Space Activities"). Iran's space program up to that point had been based on ground stations that relayed Intelsat communications and received Landsat data. In recent years, Iran has made steps in space science, space technologies, and space applications for civilian purposes mainly through communication satellites (Tarikhi 2015).

Latin America

Currently, at the forefront of Latin America's space ambitions are two of Latin America's largest and most technologically advanced countries, Argentina and Brazil (Harding 2015). The Argentine National Space Activities Commission (CONAE) is responsible for Argentina's national civilian space activities, which are free from military control and entirely promote the peaceful uses of outer space. Its program is focused on Earth observation (Euroconsult 2019). In Chile and Peru, the current Earth observation satellite systems provide imagery for both military and civilian applications including disaster management (Euroconsult 2019). The Union of South American Nations (UNASUR) is an intergovernmental union established in 2008 (came into effect in 2011) to encompass all South American Countries. UNASUR previously discussed establishing a South American Space Agency; however, this has not yet been created (Sarli et al. 2018). In addition, the Space Conference of the Americas (CEA) is a continental forum of regional and international cooperation, created in the early 1990s by the United Nations General Assembly to achieve a convergence of positions on issues of common interest related to the peaceful use of outer space by its Member States. The objective is to agree on strategies to promote the practical use of space applications to support programs with a high degree of social content for the region, to encourage progress in and development of space law, and to strengthen educational programs and training in space science and technology (UNSPIDER 2010). Notwithstanding the value of these organizations, there is still no regional understanding or approach about space and security.

Despite the status to the economy in the region, *Brazil* has managed to sustain growth since the end of 2017 with an industrial production growing slowly. The Brazilian Space Agency (AEB), created in 1999, is responsible for the coordination of Brazilian space activities, with significant effort undertaken in Earth Observation and launcher development. AEB oversees implementing and coordinating Brazil's space policy in cooperation with the Ministry of Science, Technology, Innovations and Communication and the Ministry of Defense (Euroconsult 2019). National launching facilities were developed in Brazilian territory, including the Alcantara Launching Center (*Centro de Lançamento de Alcântara*), designed in 1983. Due to its geographic position, launchings benefit from the Earth's rotation in order to achieve greater speed, allowing fuel economy and increased payload capacity. Brazil's space-related objectives are described in the *Programa Nacional de Atividades Espaciais (PNAE) Planejamento* 2012–2020. The final segment of the Brazilian space program revolves around the development of a national launching vehicle, thus securing independent access to outer space. Named VLS (for Satellites Launching Vehicle, "*Veiculo Lançador de Satélites*" in Portuguese), the program has faced budgetary and technical burdens since its conception operations (▶ Chap. 35, "Space Security in Brazil"). In 2012–2014, Brazil set out to indigenously develop a geostationary communications satellite, continued to support the joint China-Brazil Earth Resources Satellite (CBERS) program, developed two indigenous Brazilian space launch vehicles, supported its joint Brazil-Ukraine

Cyclone launch vehicle program referenced above, and established a science and technology research satellite program. Recently, the Brazilian Ministry of Defense signed, in December 12, 2018, a Space Situational Awareness agreement with US Department of Defense, as part of a larger effort to increase safety of space operations (▶ Chap. 35, "Space Security in Brazil").

North America

The *United States* remains the world's leading space program, both when it comes to civil and defense space components. The space program has been further expanded by the Trump administration, for example through civil and defense budget increases, through policy and legislative initiatives, and through a proposed Space Force. In the United States, space policy has remained relatively consistent over the last 60 years with a focus on international cooperation, peaceful uses, and development of outer space for the common good. Throughout this time, the right of self-defense in space has been linked to military activity. Yet, the 2017 National Security Strategy made a notable shift regarding the security aspects, while at the same time the National Space Council was revived (▶ Chap. 20, "War, Policy, and Spacepower: US Space Security Priorities"). Accordingly, US Space Policy Directives 1 and 2 aim at fostering commercial activities through an appropriate regulatory framework, while Space Policy Directives 3 and 4 address the creation of space traffic management and the establishment of a space force respectively. The Space Policy Directive 1 calls for the United States "to lead an innovative and sustainable program of exploration with commercial and international partners," while the Space Policy Directive 2 calls for the streamlining of regulations on commercial use of space. The Space Policy Directive 3 on Space Traffic Management (STM) aims for US leadership in space by stipulating the need to "set priorities for space situational awareness (SSA) and STM innovation in science and technology (S&T), incorporate national security considerations, encourage growth of the U.S. commercial space sector, establish an updated STM architecture, and promote space safety standards and best practices across the international community." The Space Policy Directive 4 establishes the US Space Force as a sixth military branch of the United States Armed Forces within the Department of the Air Force.

Under the Space Policy Directive 3 responsibility for providing SSA data for civil use is assigned to the Department of Commerce (DoC), while the Department of Defense (DoD) will focus on maintaining access to and freedom of action in space. In particular, the Department of Commerce becomes the agency responsible for SSA data sharing and timely warning of collision avoidance, including conjunction assessments and maneuver plans, available to the public through the publicly available portion of DoD authoritative catalogue. The availability of the data is and will remain to be free of direct user fees. The Department of Defense, therefore, shifts the civilian part of its responsibilities to DoC and will oversee the military part of authoritative catalogue of space objects (U.S.FR 2018). Shall the Space Policy Directive 3 proposal be approved it does raise the following fundamental question:

first, how distinctive can the military SSA activities be from the civilian SSA activities and, second, will it be possible for them to integrate under one comprehensive regime? This has implications not only for the governance of the safety of operations and national security but also for the exchanges and coordination with other national and international organizations (Hitchens 2019). Managing STM ultimately boils down to balancing between to seemingly contradictory objectives; one being the safety and sustainability of outer space activities, and the other one being the national security concerns of the government as further depicted in the proposal for the creation of a space force.

Regarding space programs, the US military and intelligence organizations' programs combined constitute by far the world's largest space program. The services provided by these programs include telecommunications, surveillance, missile early warning, meteorology, positioning/timing, radio interception, nuclear detonation detection, and data relay. Space systems provide both tactical and strategic services to the US military and intelligence agencies and in some cases to those of its allies. Strategic functions include monitoring international security treaties, analyzing the security forces of current and potential adversaries, and providing information to the President and the Secretary of State. Tactical functions include supporting US military and intelligence forces around the world. Overall, the US military space program continues to dwarf (a) the military space programs of all other countries combined and (b) of US civilian agencies such as NASA. The USA is unique in deploying military satellites of all types and on a global basis, and there is little sign that this will change in the next decade (▶ Chap. 59, "Satellite Programs in the USA").

The *Canadian* new Space Strategy 2019 issued by the Ministry of Innovation, Science and Economic Development Canada recognizes the importance of space as "a strategic national asset which underpins everything from national security to the ability to connect Canadians living in rural and remote communities." Since 2016, the Ministry has committed to new investments worth over $2.6 billion. Space systems are also considered vital to the Canadian Armed Forces, which rely on them to effectively conduct operations for the defense of Canada and North America and to contribute to global peace, safety, and security. One of the most important objectives of the strategy is Canada's future mission to the Moon by joining the US-led Lunar Gateway mission (Government of Canada 2019). The Canadian Space Agency (CSA) focuses on accelerating space business and modernizing investments through the Space Technology Development Program. The CSA also participates to the European Space Agency ARTES program. Concerning satellite communications, the Department of National Defense (DND) has ties with the United States in the context of the Wideband Global Satellite (WGS) System and the US Advanced Extremely High Frequency program with protected military satellite communications. The DND has also contributed to the Maritime Monitoring and Messaging Microsatellite in the field of automatic identification system (Euroconsult 2019). The DND and the Canadian Armed Forces are seeking to develop a common operating picture of space assets, based on the program Innovation for Defence Excellence and Security (IDEaS). As such, new space-based technologies will enable them to maintain space situational awareness for informed, expedited decision making in support of space system operations (Government of Canada 2019).

Russia

Outer space has become an important area for Russia which aims to rebuild its global status and prestige as a space power by intensifying the links between space and defense. Russia considers outer space predominantly as a strategic region to enhance its military capabilities on Earth, provide intelligence and communication functions, and achieve international esteem. Russia is reactive to US strategy and counterspace technologies The latter, including electronic weapons that can jam satellites, have been developing to provide Russia with an asymmetrical edge to offset US military advantages. Hence, military efforts are but one part of a complex set of tools, employed to navigate what Russia perceives as an increasingly hostile world. Already in 2011, Russia brought about certain institutional modernizations creating the Russian Aerospace Defence Forces which are meant for space security-related activities (▶ Chap. 21, "Russia's Space Security Policy"). The Federal Space Program 2016–2025 places emphasis on telecommunication satellites and the need for space technology to generate direct socioeconomic benefits. In March 2018, the Russian Defense Minister Sergey Shoigu stated that Russia must deploy a modern fleet of military satellites to support its army and navy. To quote him: "only with support from space will it be possible for the Armed Forces to reach maximum effectiveness" (DIA 2019). However, Russia's economic, military, and technological weaknesses compared to the United States and NATO have led Russia to pursue asymmetrical tactics which include working through bilateral bodies and those affiliated through the UN on space policy (▶ Chap. 21, "Russia's Space Security Policy").

Russia can be considered today as having the most complete launch program in the world. Russia currently operates four types of launch vehicles, the Rockot, Soyuz, Zenit, and Proton. The "Russian Space Launch Program" chapter explains how Russia has been successfully engaged in space activities for more than 60 years, having entered the space age as part of the Soviet Union and striding on as a separate state. On the one hand, after the dissolution of the USSR, Russia inherited the large scientific and technical potential and technological developments of one of the two most powerful space nations of that time. But on the other hand, Russia was deprived of a large part of technologies and infrastructure put in place earlier. The launch vehicles that used to be Soviet became foreign, and the key launch site turned out to be located outside Russia's national territory. Also, it proved to be difficult for Russia to use remnants of its own technologies. For Russia, space is thus not only a question of national defense and security or its position in the market of commercial launch services but also, and more importantly, a question of the status of Russia as a highly developed nation in terms of science and technology (▶ Chap. 60, "Russian Space Launch Program").

Key Priorities

The priorities and trends in space security as seen in the countries and regions presented above can be grouped in Fig. 2 below. The identified space, security, and defense priorities areas are related to "Security from Space" and "Security in Space." The "Security from Space" priorities constitute: (1) disaster management, (2) resource management, (3) transport and communications, (4) environment, climate change, and sustainable development, (5) external security including foreign policy and border surveillance, (6) internal security including support to justice and home affairs, (7) military, and (8) financial. The "Security in Space" priorities constitute: (1) defensive space security and control, (2) offensive space security and control, (3) space surveillance and tracking, (4) space weather, (5) near earth objects, (6) orbital debris mitigation, (7) space traffic management, (8) active debris removal, and (9) access to space. These trends demonstrate an evolution of European countries priorities from strictly civil-oriented applications to also encompassing security and defense ones. The grouping of priorities allows for a clear overview of the status of space security. However, the lack of explicit boundaries between "safety," "security," and "defense" makes it rather difficult to clearly distinguish among the different positions of countries and regions. In some countries it seems that space security is closer to the safety side (i.e., Africa, UAE), while in others it is closer to the defense side (i.e., the United States, Russia). Several countries in the regions presented have demonstrated a clear shift of their space policy and programs

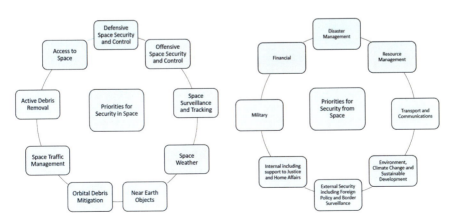

Fig. 2 Priorities for space security in space and from space across countries/regions

from safety to defense, notably Japan. In some cases, it is even more ambiguous to identify an approach due to the absence of a regional understanding such as in Africa and Latin America.

Concluding Remarks: The Way Forward for Space Security

The significance of outer space as a strategic focal area with geopolitical consequences is widely recognized. Outer space, which was perceived during the Cold War as another theatre of operations for the United States and the former Soviet Union, has now become a common strategic medium for governmental and nongovernmental activities around the world. Accordingly, the concept of space security has been changing over time. Traditional defense strategic concepts remain relevant in the face of hybrid threats, counterspace, and proposed Space Forces rendering thereby space as a warfare domain. Even though space security used to consist of exclusively military and defense elements in states' relations, it has evolved to encompass increasing activities of private and commercial actors and their implications for safety and sustainability. In this regard, space security cooperation ensuring the peaceful uses of outer space is absolutely necessary.

The wide range of space security perceptions has been further intensified by the technological and organizational transformation of the outer space environment. The increasing technological advances; the growing interdependencies between governmental, civilian, and commercial actors; and the emergence of the civil-military paradigm have created diversified interests across countries and regions in the world. In this context, multifaceted and interactive space security perspectives with ranging safety, security, and defense elements present a rising concern for cooperation mechanisms in place. Hence, the debated and different understanding of the concept of security is reflected in the various policies and programs across the world, emphasizing the civil-military nexus and the associated challenges. Definitions and concepts of what is encompassed by the term "space security" are diverse, imprecise, and evolving along the strategic priorities and needs of countries and regions. This presents problems when endeavoring to build cooperative approaches among diverse actors. The definition provided in this Handbook allows to structure the uncertainty created by the multitude of definitions, while allowing to reach a common understanding fostering cooperative and collective frameworks in space security. This definition allows to approach space security in its full spectrum capturing all elements of space security instead of differentiating among the various conditions. Hence, the definition manages to resolve the civil-military challenge by incorporating it in a flexible manner instead of turning it into a fixed and obsolete concept. In line with the principles of international cooperation and peaceful uses of outer space, this definition ultimately enables common understanding which is a starting point aiming to ensure strategic stability at the international level.

Based on this definition, the way forward for space security calls for its operationalization through the development of stability-enhancing mechanisms that tackle the challenges of the evolving security notion. Such mechanisms should be underpinned by mutually understood concepts that are translated in a comprehensive regulatory solution at the international level. In this direction, recent developments at the UNISPACE+50 process, prepared by the United Nations Office of Outer Space Affairs in 2018, have considered the development of international legal mechanisms that cope with the broader concept of space security – safety, security, and sustainability. In this regard, Space Traffic Management (STM) is considered as a solution that can ensure space security in the broad sense by resolving practical concerns of the international community, such as in-orbit collisions and interferences. The definition of the STM in the 2006 International Academy of Astronautics Cosmic Study as "the set of technical and regulatory provisions for promoting safe access into outer space, operations in outer space and, return from outer space to Earth free from physical and radio-frequency interference" complements the operational side of the space security definition in this chapter. Hence, STM can serve as the basis of ensuring space security while safeguarding the principles of international cooperation and peaceful uses of outer space.

References

African Union (2015) An overview of the Agenda 2063, First Ten Year Implementation Plan. Retrieved from https://au.int/sites/default/files/documents/33126-doc-11_an_overview_of_agenda.pdf

African Union (2017) African space strategy – Towards Social, Political and Economic Integration, Second Ordinary Session for the Specialized Technical Committee Meeting on Education, Science And Technology (Stc-Est) 21 October To 23 October 2017, Cairo, Egypt. Retrieved from https://au.int/sites/default/files/newsevents/workingdocuments/33178-wd-african_space_strategy_-_st20445_e_original.pdf

African Union (2020) First continental report on the implementation of Agenda 2063, African Union Development Agency, 8 February 2020. Retrieved from https://au.int/sites/default/files/documents/38060-doc-agenda_2063_implementation_report_en_web_version.pdf

Baldwin DA (1997) The concept of security. Rev Int Stud 23:5–26

DIA (2019) Challenges to space security. Defense Intelligence Agency, Washington, DC

Euroconsult (2019) Government space programs – benchmarks, profiles, and forecasts to 2028. Euroconsult, Paris

Giannopapa C (2011) Improving Africa's benefit from space applications: the Europeane-African partnership. Space Policy 27:99–106. Elsevier

Giannopapa C, Adriaensen M, Antoni N, Schrogl K-U (2018) ESA's policy on space and security. Acta Astronautica 147:346–349

Government of Canada (2019) Exploration imagination innovation – a new space strategy for Canada, Ministry of Innovation, Science and Economic Development. Retrieved from https://www.asc-csa.gc.ca/pdf/eng/publications/space-strategy-for-canada.pdf

Greenwood C (2011) Self-Defence. In: Max Planck Encyclopedia of Public International Law. Oxford University Press, Oxford, April 2011

Harding RC (2015) Emerging space powers of Latin America: Argentina and Brazil. In: Al-Ekabi C, Baranes B, Hulsroj P, Lahcen A (eds) Yearbook on space policy 2012/2013. Springer, Vienna

Hitchens T (2019) Space traffic management: U.S. military considerations for the future. J Space Saf Eng 6:108–112

Jiang H (2019) Programme and development of the "Belt and Road" Space Information Corridor, CNSA at 2019 United Nations/China Forum on Space Solutions. Retrieved from https://www.unoosa.org/documents/pdf/psa/activities/2019/UNChinaSymSDGs/Presentations/Programme_and_Development_of_the_Belt_and_Road_Space_Information_Corridor_V5.1.pdf

Komiya Y (2016) Basic plan on space policy, implementation schedule (Revised FY2015). Retrieved from Japan Space Forum 2016: http://www.jsforum.or.jp/stableuse/2016/pdf/6.%20DG_Komiya.pdf

Lal B, Balakrishnan A, Caldwell BM, Buenconsejo RS, Carioscia SA (2018) Global trends in space situational awareness (SSA) and space traffic management (STM). IDA Science and Technology Policy Institute, Washington, DC

McDonald A, Brollowski H (2011) Security. In: Max Planck encyclopedia of public international law. Oxford University Press, Oxford

National Defense (2020) Middle East Allies look to expand space capabilities. Retrieved from https://www.nationaldefensemagazine.org/articles/2020/2/3/middle-east-allies-look-to-expand-space-capabilities

Nie M (2020) Space privatization in China's national strategy of military-civilian integration: an appraisal of critical legal challenges. Space Policy. In press, 2020,101372, ISSN 0265-9646

Osisanya S (2015) National security versus global security. United Nations Chronicle. Retrieved from https://www.un.org/en/chronicle/article/national-security-versus-global-security

Republic of South Africa (2018) National Space Science and Technology Programme Strategy, To leverage the benefits of space science and technology for socio-economic growth and sustainable development. Retrieved from https://www.sansa.org.za/wp-content/uploads/2018/05/National-Space-Strategy.pdf

Robinson J (2018) Cross-domain responses to space hybrid provocations via economic and financial statecraft. The Prague Security Studies Institute, Prague, Czech Republic

Rothschild E (1995) What is security? The MIT Press on behalf of American Academy of Arts & Sciences, MIT Press 124:53–98

Sarli BV, Cabero Zabalaga M, Telgie A (2018) Review of space activities in South America. J Aeronaut Hist 8:208–232

Sheehan M (2015) Defining space security. In: Schrogl K-U et al (eds) Handbook of space security: policies, applications and programs. Springer, New York

SpaceNews (2020) Space force-sponsored study to assess capabilities of commercial industry. Retrieved from https://spacenews.com/space-force-sponsored-study-to-assess-capabilities-of-commercial-space-industry/

Spacewatch.global (2016) Iranian satellites to share information system of the Asia Pacific Space Cooperation Organisation. Retrieved from Spacewatch.global: https://spacewatch.global/2016/09/iranian-satellites-share-information-system-asia-pacific-space-cooperation-organisation/

Spacewatch.global (2019) Military space: Japan and United States pledge mutual defence cooperation in space, cyber. Retrieved from https://spacewatch.global/2019/04/military-space-japan-and-united-states-pledge-mutual-defence-cooperation-in-space-cyber/

Tarikhi P (2015) The Iranian Space Endeavor. Springer International Publishing, Cham

U.S.FR (2018) Space policy directive-3, National Space Traffic Management Policy, US Federal Register, 18 June 2018, 83 FR 28969, 2018-13521. Retrieved from https://www.federalregister.gov/documents/2018/06/21/2018-13521/national-space-traffic-management-policy

United Nations (1967) Treaty on Principles Governing the Activities of States in the Exploration and Use of Outer Space, Including the Moon and other Celestial Bodies, Article I, 27 January 1967, 18 UST 2410; 610 UNTS 205; 6 ILM 386 (1967)

United Nations (2019) General Assembly, First Committee, First committee delegates exchange divergent views on how best to prevent weaponizing outer space, cyberspace amid eroding global trust, GA/DIS 3636, Seventy-forth session, 17th & 18th meetings, 29 October 2019. Retrieved from https://www.un.org/press/en/2019/gadis3636.doc.htm

UNOOSA (2019) The African space policy and strategy, towards social, political and economic integration of Africa, Dr. SALEY Mahaman Bachir, African Union Commission, United Nations China Forum on Space solutions, 24–27 April 2019 - Changsha – China. United Nations Office for Outer Space Affairs

UNSPIDER (2010) VI Space conference of the Americas. United Nations Platform for Space-based Information for Disaster Management and Emergency Response (UN-SPIDER), November 2010. Retrieved from http://www.un-spider.org/news-and-events/events/vi-space-conference-americas

Williams PD (2013) Security Studies: An Introduction. Routledge, Oxford University Press

Challenges to International Space Governance

Ahmad Khan and Sufian Ullah

Contents

Background	36
Challenges to International Space Governance	37
Increasing Competition in Space	37
No Consensus on Space Arms Control	38
Increasing Reliance on Space Assets	38
Security-Driven Self-Interests of States	39
Dual-use of Space Assets	39
The USA, Russia, and China in Space	41
Proliferation of ASAT Weapons	42
Stalemate on Arms Control Treaty Negotiations	43
Conclusion	46
References	47

Abstract

Space is now a congested, contested, and competitive domain. Space technologies and programs have become vital ingredients of major spacefaring nations' national power. In the past three decades, space has gained importance for security and socioeconomic development of spacefaring nations. However, most of the activities in this domain are unchecked primarily due to lack of an internationally agreed treaty in space. In addition, the challenges to international space governance in ensuring space as a safer and more secure environment are

---

A. Khan (✉)
Department of Strategic Studies, National Defence University, Islamabad, Pakistan
e-mail: ahmad_ishsq669@yahoo.com

S. Ullah
Department of Defence and Strategic Studies, Quaid-i-Azam University, Islamabad, Pakistan
e-mail: sufianullah@gmail.com

© Springer Nature Switzerland AG 2020
K.-U. Schrogl (ed.), *Handbook of Space Security*,
https://doi.org/10.1007/978-3-030-23210-8_116

enormous. These include the increasing pool of actors, growing commercialization, expanding military space programs, the proliferation of anti-satellite weapons, and lack of consensus among states on a conclusive and universally negotiated treaty to prohibit an arms race in space. The expanding number of spacefaring actors and dual-use technologies have made the skies and space more competitive. On top of that, the competing states are following a path from space militarization to weaponization which has aggravated the threat of space war.

Background

Space is a strategic domain in which both major and emerging space powers seek a place. States are shifting gears to maintain their position in space in twenty-first century primarily because of both military and economic benefits. After the end of Cold War, the world is witnessing a gradual increase in the military and peaceful space assets and technologies of states other than USA and Russia. States like India, China, Japan, South Korea, etc. are expanding their international partnership to send other states' satellites in outer space. This has made space a competitive domain; in fact, China and India have sent provided alternative platforms for developing countries to send their satellites more cheaply than from the USA and Russian launching pads. Key challenges to making space a more safe and secure environment include the increasing pool of actors, growing commercialization, expanding military space programs, and lack of agreement by states on a conclusive and universally negotiated treaty to prohibit an arms race in space. The expanding number of spacefaring actors has made the skies and space more competitive. Likewise, these factors are increasingly contributing toward key challenges to international space governance and ensuring space remains a safe and secure domain for commercialization and scientific exploration.

To pursue scientific endeavors and to meet commercial as well as security needs, the reliance of states on space-based capabilities has increased significantly. Along with the commercial and economic uses of space, states have been ambitiously investing in military exploitation of this *global commons* that has now emerged as an arena of potential confrontation. While space has a lot to offer to human growth and prosperity, this potential is hijacked by risks of conflict escalation. The recent trends in technological developments suggest that the space is being rapidly weaponized with little prospect of establishing an agreeable and verifiable framework for global space governance. Divergent policy approaches, coupled with competing strategic aspirations, constrain pursuit of collective action toward achieving this goal. Space activities are generally categorized into three sectors including civilian, military, and commercial. This chapter focuses on the military sector to explicate what drives the arms competition in outer space and argues that cooperation and space governance may enhance mutual security and decrease possibilities of conflict. However, the primary challenge to this is that states tend to see the domain of space as a field of competition whereby one's considerable presence equates to enhanced national stature among the comity of nations. One such example is recently witnessed euphoria among Indian leaders after the country successfully test fired its first Anti-Satellite (ASAT) weapon system.

Challenges to International Space Governance

Currently, spacefaring nations seek to establish decisive dominance at least in the military aspect. This approach consequently heightens competition by stimulating tensions and encouraging military countermeasures. Given the ongoing geopolitical competition in space, whereby this domain has emerged as the fourth medium of warfare, any model for space governance must not overlook the inherent competitive tendencies among states, and should rather aim to promise more incentives through cooperation (Khan and Khan 2019). The rationale behind global space governance revolves around the notion that despite having competing interests, the irrational behavior of spacefaring nations would be equally disadvantageous for their pursuits in this domain. Recognizing the states' inherent instinct to pursue competitive objectives, this chapter suggests cooperation toward space governance in a manner that does not overlook their grounded self-interest but aims to reduce threats in a cooperative manner. The idea of global space governance, therefore, seeks to maximize the prospects of peaceful exploitation of this *global commons* by encouraging responsible behavior.

Increasing Competition in Space

Spacefaring nations are developing their space programs at a fast pace. What really drives this unabated exploitation of space is a question that largely remains unaddressed. While some argue that these programs are driven by "techno-nationalism" to demonstrate technological prowess, a state's natural instinct to hedge against the others in a *global commons* also cannot be ignored (Sheehan 2010). Likewise, since an advanced space program is an indication of nation's military and industrial strength, states tend to see progress in this domain as a token of national prestige (Mission Shakti ... 2019) (One such example is Prime Minister Narendra Modi's statement after India conducted a successful Anti-Satellite Missile Test on 27 March 2019. Notwithstanding that the experts raised concerns regarding possible negative implications that ASAT's debris may have, Prime Minister Modi asserted that this capability placed India among the handful of space superpowers).

While the USA and Soviet Union were the only competitors for supremacy in space during the Cold War, the post-Cold War era has witnessed new entrants in this competition – all aiming to secure advantageous position through dual-use capabilities. The ongoing trends of commingling space and counterspace capabilities give rise to new warfare strategies that not only implicate national securities of other states but also undermine the prospects of peaceful exploration of outer space.

No Consensus on Space Arms Control

One reason behind states' reluctance to enter into any space-related arms control agreements can be the urge to amass maximum technological capabilities. It appears that the major actors intend to buy maximum possible time to develop enough capabilities before fully agreeing to put limits to numbers and types of their capabilities. Nevertheless, it must be realized that the future of global arms control considerably depends upon the arms buildup in the domain of space. If an arms race in space continues to go unchecked, it would seriously jeopardize other efforts to arrest arms competition. The primary reason behind this is the growing diversity of space-based military assets and their unfolding roles in different military strategies. One such example is nuclear deterrence.

Increasing Reliance on Space Assets

Ever since the launch of first satellite in outer space, reliance on space has only increased. The rapid scientific and technological advancements have surged the utilization of outer space for wide range of purposes. Also, the exploitation of this *global commons* no longer remains the prerogative of a handful of great powers; rather even the developing countries are now the beneficiaries of this domain. The enhanced reliance on space-based assets for a wide range of operations – like remote sensing, communication, and so on – further necessitates effective global space governance that could ensure the favorable environment for peaceful ventures. While our dependence on space has immensely increased in the past few decades, the international community has failed to adequately respond to the rising threats and vulnerabilities. A state's reliance on space assets is directly proportional to the increasing risks, vulnerabilities, and challenges to its operational space assets. The more the numbers of space assets are operationalized, the more vulnerable they are to attacks in space from the state adversary that also includes non-state actors. The superpowers in Cold War era took a promising start in the form of new treaties and norms to develop consensus in achieving a peaceful outer-space environment. In the 1960s, both the rivals carried out nuclear explosions above the atmosphere but later agreed on non-testing of nuclear weapons in atmosphere and have complied with this restriction so far. However, with passing times, the international community has largely failed to develop a concrete mechanism to avoid evolving threats confronting a peaceful environment in space.

The competitive or cooperative engagement among states dates back to the Cold War period when the two superpowers were engaged in the Moon race and sought to develop space-based capabilities to support their military and intelligence operations. The space endeavors of the two adversaries seemed to be driven more by the struggle for political and military advantages, instead of pursuit of scientific explorations. Although the two states agreed on peaceful cooperation in space after the Apollo-Soyuz Test Project emerged as a symbol of détente between the competing superpowers in the 1970s, their continued fight for predominance in this domain converted space into a battleground of the arms race (Apollo-Soyuz Test Project).

3 Challenges to International Space Governance

Security-Driven Self-Interests of States

The key challenge to global space governance remains in the very nature of states' behavior. As argued earlier, the self-interested members of international community generally tend to overlook the negative consequences of their behavior and only seek to maximize their benefits. Particularly in the absence of shared strategic goals, states find it difficult to compromise their ambitions and accept limitations on the behavior that otherwise promises relative advantages. In this context, what may encourage states to enter into arms control arrangements is the realization of their mutual vulnerabilities in a destabilizing environment. Likewise, putting "high value on collective benefits" may also encourage them to avoid any irresponsible behavior that could threaten the international peace and security. Therefore, as rightly identified by Thomas Schelling and Morton Halperin, the threefold objectives of any arms control arrangement include reducing the likelihood of occurrence of war, costs of preparations, and destructive consequences of war (Larsen 2002).

At present, states joining hands to develop and operate the International Space Station, joint production and launch of satellites, and greater integration in the use of space-based services such as communication networks are testament of ongoing international cooperation in this domain. While it hints at the huge prospects of any mutually agreeable governance framework, the same is also marred by significant challenges owing to differing aspirations of states toward this domain. Besides peaceful uses of these capabilities, the security implications of space-based technologies are driving interstate competition. Along with jointly pursued peaceful uses, great powers have been exploiting space for their vested military advantages, thus depicting the emerging face of interstate relations where strategic rivalries predominantly define the geopolitical environment. The diverse use of space applications, where the distinguishing line between peaceful and military use of space-based capabilities is too blurry to clearly reflect the difference, further exacerbates one's urge to misinterpret other's actions and respond accordingly. One such example is Global Positioning System (GPS) that, despite it enabling a huge array of peaceful uses in different activities, is considered as a military capability (Ohlandt 2014). Likewise, even the high-resolution commercial remote sensing satellites, which can strengthen stability by providing a real-time picture of an adversary's military operations, may also be used for missile warning, target identification, and other military roles.

Dual-use of Space Assets

The political consequences of the employment of dual-usable assets in space are undesirable. If one were to see the role of military capabilities in geopolitical competition among great powers, it would be plausible to assume that the military values that competing powers attach to their space capabilities would be of paramount significance in shaping the evolving great power competition.

Since the international anarchic system offers no guarantee that a state with considerable military power would not use it in an offensive manner to subdue others, it is only natural that the rival states view each other's technological advancements with skepticism and may even resolve to respond in kind. This is particularly significant in an environment that lacks institutional structures that could assure general security by regularizing the behavior of states and ensuring transparency. It consequently generates a security dilemma that continues to fuel the mutual mistrust, intensify the prevailing arms race, and diminish the prospects for global space governance (Jervis 1978). (Robert Jervis defined security dilemma as a state of affair in which security of one state decreases the security of another state.) Since one state's accumulation of power, as put by John Herz, makes others feel insecure and compel them to take remedial measures, this consequently leads to chain reaction that undermines general security (Herz 1950).

The inherent "ambiguous symbolism" of technology, as referred to by Ken Booth and Nicholas J. Wheeler, makes the security dilemma more intense as other states may see one's weapons as offensive or defensive depending on their own presupposition and threat perceptions (Booth and Wheeler 2008). The ambiguity regarding a dual-use capability is particularly more intense in the domain of space. The claimed roles of space-based assets may alter dramatically during peace and the times of crises. The dilemma of interpretation and dilemma of response are therefore the two most significant factors that constrain the pursuit of agreeable governance structures in the domain of space (Booth and Wheeler 2008).

When a state develops a certain military capability, the leadership of a rival state confronts the dilemma of interpretation in ascertaining whether that development is for offensive purpose or it is just a measure to bolster defensive capability. Likewise, the dilemma of response also constrains one's options to respond to any such development by adjusting force posture. The traditional response of a state in an uncertain environment may include disengagement from other actors and arms development and deployment. If the state goes for such an action that generates military confrontation, it leads to mutual hostility and further diminishes the prospects of reducing threats through cooperation. In such an environment of uncertainties, the best course of action that a state can think for itself, as put by John J. Mearsheimer, is to maximize its relative power and be a hegemon in the system by ensuring that there is no peer competitor with equally offensive or overwhelming capabilities (Mearsheimer 2013). After land, air, and sea, space has now emerged as a domain where competing powers vehemently pursue predominance through technological advancements and refrain, at least at the moment, from committing to limitations on this pursuit.

Such an understanding of security dilemma in international politics is evident from Chinese and Russian perceptions of growing US presence in space. The USA has long maintained unequivocal supremacy in space by virtue of its unmatched civilian as well as military space capabilities. While it successfully superseded Soviet Union in this rivalry for hegemony, China has recently emerged as the most potent competitor to US dominance in this domain. As put by Defense Intelligence Agency, adversary's integration of space and counterspace capabilities into military operations pose a challenge to US space dominance (Challenges to Security in Space 2019). The notion of "congested, contested, and competitive" space domain also signifies how the USA views the growing adversarial capabilities that may erode its strategic advantage. Thus, the USA considers it imperative to deny space to its rivals (Outer Space Increasingly 2012).

The USA, Russia, and China in Space

The USA, Russia, and China are currently the major space powers – with India trying to steadily catch up – which have developed significant space-related weapon systems. These states possess the most sophisticated capabilities and have also shown the intent to test and launch space-based weapon systems.

Russia sees the US space-based capabilities in connection with the growing asymmetries in the military equilibrium between the two states and seeks to contain the USA by modernizing its own assets (Zervos 2011). The USA believes that China and Russia seek to challenge the US position in space, exploit its dependence on space-based assets, and reduce its military effectiveness through counterspace capabilities (Challenges to Security in Space 2019). Given that there are no serious efforts to address disagreements over a desirable code of conduct in space, the strategic distrust among them continues to grow. For China, the advanced space program is a means to project its soft power in terms of technological development and also to accrue strategic advantage by challenging the US supremacy.

With China surpassing Russia as the second leading space power, the US-China rivalry in space may have grave implications for global peace and security. China has been emerging as a rising global power, and its increasing footprint in space now challenges decades-long US supremacy. The two strategic rivals are skeptical of each other's technological developments and space capabilities. This situation is further exacerbated in the absence of any concrete dialogue mechanism over a prospective agreeable code of conduct in space. There have been perceptions within China that the US-led space weaponization not only implicates peaceful uses of outer space but also is also driven by an intent to neutralize China's nuclear deterrent capability (Zhang 2008). This growing fear has shaped the Chinese behavior on two different accounts: at strategic level, it resolves to take the countermeasures and enhance its own space capabilities to neutralize any such threats, and at diplomatic level, it strongly advocates the Prevention of Placement of Weapons in Outer Space Treaty (PPWT) to prevent weaponization of outer space. CD first proposed PPWT in CD in 2008. In addition to these concerns, projecting its own space power has also become paramount for China to register itself as a rising power that could position itself among the other advanced spacefaring nations (Khan and Khan 2019). While dominating space, these developments also exacerbate US fears of a "Space Pearl Harbor," thus contributing to unending security spiral where action-reaction dynamics result in unending arms race (Commission to Assess United States National Security Space Management 2001). (The term "Pearl Harbor" has gained the salience of a metaphor for being caught unaware. The 2001 Space Commission of the USA cautioned of a Space Pearl Harbor stating that the USA needed to be better prepared as its assets in space were vulnerable against a surprise attack.) This is particularly evident in the development of a more destabilizing capability, i.e., the anti-satellite weapons system. China tested its ASAT missile capability in 2007 when it launched SC-19 interceptor to hit a defunct satellite at an altitude of 865 kilometers. A year later, the USA also demonstrated its capability to neutralize a failed satellite with an SM-3 missile. Though an ASAT capability is a Cold War technological development, this development suggests that states are moving on a path from space militarization to weaponization.

The USA emphasizes on sustaining its supremacy in space while ensuring its "freedom of action," by dissuading or deterring others from developing similar capabilities (National Space Traffic Management Policy 2018). Such hegemonic aspirations, combined with Chinese and Russian resolve to challenge US predominance, set the stage for potential conflict.

Proliferation of ASAT Weapons

Besides its claimed deterrent imperatives, India's testing of ASAT weapon system appears to signify a symbol of national pride and prestige. However, the destabilizing consequences of this technology far outweigh such potential benefits, thus posing a question on the strategic and military rationale behind its development and operational deployment. Introduction of ASAT weapon systems only exacerbates the risks of crisis escalation and eruption of conflict at lower levels. By reducing the early warning capability or disrupting the satellite-dependent communication channels, the ASAT weapon directly undermines one state's ability to effectively retaliate against a possible strike that implicates the strategic stability (Oznobishchec 1989). Likewise, ASATs are also inherently aggressive weapon systems that increase the vulnerabilities of other state's command and control & intelligence, information, surveillance, and reconnaissance (I^2SR) systems against a possible attack. This particular attribute of neutralizing some of an adversary's critical infrastructures makes ASATs a potent counterforce weapon. It consequently not only raises nuclear alert levels but also emboldens response options that may not remain within the confines of the law of proportionality (Ullah and Imam 2019). This scenario only contributes to an uncertain strategic environment and intensifies the dilemmas of interpretation and response by increasing the element of unpredictability between the two adversaries (Lele 2019). In a nuclear environment, the possessors of this capability may argue for the deterrent role of ASAT weapon systems and consider this capability as a contributor to strategic stability (U.S. Anti-Satellite (ASAT) Program 1987). (For instance, the USA has since long considered its ASAT program as an essential component of its deterrence.) However, the aforementioned aspects suggest that ASAT weapon systems may only have some role in fighting a war, but not in averting one, and thus contribute very little deterrent value. ASAT weapons can put at the risk the very satellites – including early warning systems, communication satellites, and so on – which are believed to be vital for strategic stability.

There is also a possibility that given the growing vulnerabilities to space-based assets, particularly in context of kinetic or other threats to revolving satellites, states may eventually resolve to invest more in ground-based and aerial systems to ensure survivability of their critical infrastructures. Major powers need to realize that any use of ASAT weapons in counterforce roles may lead to crisis instability and possible nuclear escalation. Besides these growing risks of escalation, the kinetic use of weapons in space, as argued by Michael Krepon, would be self-defeating (Krepon 2013). For instance, the use of ASAT offers limited military advantage as the resulting debris would pose an equally dangerous threats to one's own space assets. By recognizing this inherent character of ASAT weapon, major stakeholders may attempt to develop consensus on considering this tool of warfare as a destabilizing weapon. This would not only reinforce strategic equilibrium in space but also pave way toward more encompassing arms control mechanisms.

Stalemate on Arms Control Treaty Negotiations

Given that the military and peaceful uses of space-based technologies generate competition among spacefaring nations, the undesirable political and strategic consequences of this arms race require that concerted efforts be made to reach a legal framework to govern space. As the advancing technologies offer more potent and survivable space weapon systems and the distinguishing lines between civilian and military use of space capabilities become more blurred, the chances of accidental use of weapons would increase the dangers of crisis instability. The continued dismissive attitude of concerned states would steadily bring us closer to a point where it will be difficult to arrest the buildup of space-related weapons systems. Further delays in reaching an agreeable legal framework of governance will result in increasingly irresponsible behavior involving ambitious use of military capabilities in this global common.

Unlike other domains of warfare including air and sea, space is more prone to risks of collateral damage as any disruption of infrastructure would also directly implicate international commercial ventures. The debris caused by any kinetic-energy warfare would not only implicate military satellites of competing states but also result in unstable space environment for all stakeholders. It is, therefore, important that all spacefaring nations push for collective action to reach an agreeable mechanism to put limits on weaponization of space and work toward creating a more transparent environment in space. Nevertheless, collective action is a function of mutual trust and transparency. The lesser the security dilemma, the greater are the prospects of developing consensus. It is the uncertain environment in space that has thus far prevented states from achieving that level of mutual trust. The Treaty on Principles Governing the Activities of States in the Exploration and Use of Outer Space, including the Moon and Other Celestial Bodies, commonly known as the Outer Space Treaty (OST) serves as a baseline for a collective action. It recognizes space as a potential "province of all mankind" and emphasizes the peaceful exploration of this *global commons* (Treaty on Principles Governing the Activities of States 1967). The major provisions of the Treaty call for adherence to international law to ensure international peace and security, agreement on not to place nuclear weapons in Earth orbit or on any celestial body, and establish liability for damages. (Article III of the Treaty on Principles Governing the Activities of States in the Exploration and Use of Outer Space, Including the Moon and Other Celestial Bodies, deals with the provision related to adherence to international law. Article IV provides for a commitment on not to place weapons of mass destruction in outer space. Articles VI and VII deal with the liability clauses and specify that the state parties to the Treaty shall be responsible for their national activities. Any activities in space carried out by nongovernmental entities would require authorization by the concerned national authority, and the latter shall be held responsible for any violations to the principles of the Treaty.) Though the Treaty offers a foundation for further global cooperation through more treaties, a number of shortcomings in the scope of the Treaty serve as obstacles in achieving this broader goal. These include, first, the anarchic nature of international system whereby there is no central authority to regulate the behavior of states. Second, there is no effective verification mechanism to check the member states' compliance with agreed framework, and the OST regime has largely failed to generate collective action toward ensuring well specified and rules-based global space governance. Third, as pointed out by Nancy Gallagher and John D. Steinbruner, the states are unwilling to offer transparency about their technical ventures and tend to classify these developments under secrecy (Gallagher and Steinbruner 2008). It only exacerbates the dilemma of interpretation for adversarial states in assessing the true motivations of other states. Fourth, to arrest the weaponization of space, the Treaty only banned placement of weapons of mass destruction in space. The limited scope of the Treaty thus could not address the post-Cold War evolving trends in the development of space weapons including the Conventional Prompt Global Strike capabilities, ASAT weapons, and their technological links with the antiballistic missile systems.

Effective space governance requires that in absence of a central authority that could verify compliance, spacefaring nations take on the responsibility and observe restraint though voluntary or less formal means. Eligar Sadeh suggests that these may include establishing global norms, codes of conduct, confidence building measures, and diplomacy (Sadeh 2015). Nevertheless, the greatest constraint in driving a collective action in this regard remains to be the prevailing mistrust among states and absence of any verification mechanism. Strategic distrust among major spacefaring nations demonstrates how growing mistrust may put existing legally binding commitment in jeopardy. This challenge is further magnified in an environment where there are no verification mechanisms to ensure member states' compliance to the agreed framework.

The divergent approaches of major spacefaring nations toward global space governance further add complexities to the problem. While the USA and its allies, including the European Union (EU) and Japan, argue for an international code of conduct that could enhance rules-based space order, China and Russia emphasize negotiating a treaty to prevent an arms race in outer space (Krepon 2013). The divergent approaches of different states are partially shaped by how they viewed the defunct Anti-Ballistic Missile (ABM) Treaty between the USA and Russia. Many legal experts held the opinion that this bilateral treaty directly dealt with the weaponization of space as it included the space-based ABM systems (Johnson-Freese 2007). For three decades, ABM Treaty proved to be an effective tool to prevent deployment of space weapons, but US withdrawal from the Treaty in June 2003 was a serious blow to the norms against weaponization of space. Withdrawal from this Treaty created incentives for the USA to deploy space-based element of its multilayered missile defense system and also other war-fighting capabilities as part of military uses of space (Outer Space and Global Security 2003). The vacuum created by US decision to withdraw encouraged states not parties to the treaty to press for a treaty-based mechanism to ban use of weapons in outer space. Furthermore, in February 2008, Russia and China jointly drafted and submitted a Prevention of the Placement of Weapons in Outer Space (PPWT) Treaty in Conference on Disarmament to prohibit the weaponization of space. However, the USA opposes the proposed PPWT as it continues to vote against the UN Resolutions prohibiting an arms race in outer space.

The key difference among states remains what constitutes a weapon in space. Also, states tend to justify their space programs while highlighting the deterrent value of their infrastructure. However, even if a weapon is placed in space for defensive purposes, its ambitious and aggressive use would have irreversible consequences for strategic stability as well as peaceful exploration of space. This necessitates that efforts be made to develop consensus on destabilizing implications of certain types of technologies and outlaw the development and placement of those weapon systems in outer space. After having successfully developed a consensus on placement of nuclear weapons in space, global peace and stability requires that similar restrictions be placed on the testing and placement of ASAT weapon systems. By pursuing arms control to constrain the growth of ASATs, the dangers of escalation and preemptive first strikes may be averted. The first requirement for a peaceful or rules-based governance structure for space is to ensure the survival of satellites. The proposals that emphasize only regulating behavior in space tend to overlook the anarchic nature of the system and states' urge to maximize their relative gains, which often lead to an irresponsible behavior without any regard for the negative consequences for other states (Defrieze 2014) (For instance, it is emphasized that regulating and punishing behavior is the best approach to control the weaponization of space).

Through cooperative engagements, transparency may be ensured to minimize the security dilemma. At present, it appears that the ongoing technological developments and innovations in space programs also cause the dismissive attitude toward arms control. Major powers are seemingly buying time to accumulate maximum capabilities as they are also not sure about the status of research and development in adversary's space programs. This raises the level of uncertainty.

Conclusion

We conclude that states are following a path from space militarization to weaponization. This indicates that states have the capability to weaponize outer space depending on the political, economic, and strategic circumstances at the global and regional levels. The growing number of actors in outer space has increased space traffic, makes space assets more vulnerable, and creates incentives for major spacefarers to take offensive countermeasures to protect their space assets. Their reliance on space assets has been increased as well as the future requirements to conduct combat missions on the ground. Therefore, outer space has become a strategic field for the major powers. There is a possibility of deployment of weapons in the outer space, which could create conflicts of interest among the nations. Therefore, space is now a strategic domain that could play a determining factor in the future engagements. The US unilateral withdrawal from the ABM Treaty in June 2002 and its initiation of Ballistic Missile Defence (BMD) have prompted Russia and China to revisit their strategic force posture. Both countries are continuously increasing their military capabilities at a moderate level, because of the US missile defense system in Europe and in the Far East. Because of this, there is a chance of horizontal proliferation of space weapons in the world. Now, there is a possibility that emerging space powers in view of their security objectives will move toward building ASATs in upcoming years, which will raise the chances of the vertical proliferation of space weapons.

In the absence of multilateral international agreements to prohibit the weaponization of outer space, there is an urgent need for an international agreement or treaty, which could bar the proliferation of space weapons both horizontally and vertically. In this regard, there is a need to re-evaluate and strengthen prevention of an arms race in outer space (PAROS) issue in CD. The world is facing a looming threat of space warfare. And this threat has emerged as reality due to non-agreement of major powers on framing any legal instrument/treaty which prohibits the weaponization of outer space. Russia and China have proposed several drafts to prevent the weaponization of space, like PPWT and No First Placement of Weapons in Space, but the USA and its allies are not agreeing on such draft proposals. These efforts suggest that Russia and China are not willing to fight the war in space and more inclined toward framing legal instruments to avoid space war. On the other hand, it is the USA which is not interested in having any treaty which restricts its freedom of action in space. The chapter concludes that there is a real threat in space that could germinate the seed of armed conflict. The USA and China don't trust each other, and there is strategic distrust between two adversaries, which could be the triggering point of an armed conflict in outer space. The USA's concerns are emanating from China military and economic rise, and the USA wants to contain China in its region of influence. China considers space to be a building block of its national power. This fosters strategic distrust between two space powers, diminishing chances of cooperation between two states. This also provides incentives for both states to build their military space programs to counter each other's space power. Both countries have developed kinetic and non-kinetic means to disrupt, degrade, and damage each other's space assets. All these efforts are a perfect recipe for a competition in outer space. Space is one of the *global commons* providing crucial resources for humanity. The competition is relatively new and different from what the world has observed in nuclear and conventional military buildups. The military buildup in outer space has brought a paradigm shift in international security as the fear of war in space would cause catastrophe on the Earth. The problem of space debris in the aftermath of a future space war may last for centuries. Therefore, the development of space weapons on the Earth by major spacefarers and the doctrines and strategy to fight a war in space has brought a paradigm shift in international security.

References

Apollo-Soyuz Test Project. National Aeronautics and Space Administration. https://history.nasa.gov/astp/index.html. Accessed 28 Mar 2019

Barnes E, William J (2019) Russia has restarted low-yield nuclear tests, U.S. believes. New York Times, 29 May. https://www.nytimes.com/2019/05/29/us/politics/russia-nuclear-tests.html. Accessed 30 July 2019

Booth K, Wheeler J (2008) The security dilemma: fear, cooperation and trust in world politics. Palgrave Macmillan, Basingstoke, p 43

Buchan A (1965) The dilemma of India's security. Survival 7(5):204–207

Challenges to Security in Space (2019) U.S. Defence Intelligence Agency, Washington, DC, pp 7, 33

Defrieze C (2014) Defining and regulating the weaponization of space. Joint Force Q 74:110–115. https://ndupress.ndu.edu/Portals/68/Documents/jfq/jfq-74/jfq-74_110-115_DeFrieze.pdf. Accessed 10 July 2019

Gallagher N, Steinbruner D (2008) Reconsidering the rules for space security. American Academy of Arts and Sciences, Cambridge, p 33

Herz H (1950) Idealist internationalism and the security dilemma. World Polit 2(2):157

International Space Station. National Aeronautics and Space Administration. https://www.nasa.gov/mission_pages/station/cooperation/index.html. Accessed 28 Mar 2019

Jervis R (1978) Cooperation under the security dilemma. World Polit 30(2):169

Johnson-Freese J (2007) Space as a strategic asset. Columbia University Press, New York, p 107

Khan Z, Khan A (2019) Space security trilemma in South Asia. Astropolitics 17(1):4

Krepon M (2013) Space and nuclear deterrence. In: Krepon M, Thompson J (eds) Anti-satellite weapons, deterrence and Sino-American space relations. Stimson Center, Washington, DC, p 33

Larsen JA (2002) An introduction to arms control. In: Larsen JA (ed) Arms control: cooperative security in changing environment. Lynne Reinner Publishers, Boulder, p 2

Lele A (2019) The implications of India's ASAT test. The Space Review, 1 April. http://www.thespacereview.com/article/3686/1. Accessed 15 Apr 2019

Mearsheimer J (2013) Structural realism. In: Dunne T, Kurki M, Smith S (eds) International relations theories: discipline and diversity, 3rd edn. Oxford University Press, Oxford

Mission Shakti Successful, Announces PM Modi. Know All About It (2019) NDTV, 27 March. https://www.ndtv.com/india-news/mission-shakti-successful-announces-pm-narendra-modi-know-all-about-anti-satellite-missile-test-2013581. Accessed 31 Apr 2019

Mowthorpe M (2004) The militarization and weaponization of space. Lexington Books, Lanham, p 110

Ohlandt R (2014) Competition and collaboration in space between the U.S., China, and Australia: Woomera to WGS and the impact of changing U.S. national space security policy. Asian Surv 54(2):399

Outer Space and Global Security (2003) United Nations Publications, Geneva, p vii

Outer Space Increasingly 'Congested, Contested and Competitive', First Committee Told, As Speakers Urge Legally Binding Document to Prevent Its Militarization (2012) Press release no. GA/DIS/3464, October 23. https://www.un.org/press/en/2013/gadis3487.doc.htm. Accessed 20 Apr 2019

Oznobishchec S (1989) Vulnerability of satellites and ASAT weapons. In: Hassard J et al (eds) Ways out of the arms race: from the nuclear threat to mutual security – proceedings of the second international scientists' congress. World Scientific, London, p 143

Report of the Commission to Assess United States National Security Space Management and Organization (2001) Committee on armed services of the U.S. House of representatives, January 11, p 13. https://fas.org/spp/military/commission/executive_summary.pdf. Accessed 2 Apr 2019

Sadeh E (2015) Obstacles in international space governance. In: Schrogl K-U et al (eds) Handbook of space security: policies, applications and programs. Springer, New York

Sheehan M (2010) Rising powers: competition and cooperation in the New Asian space race. RUSI J 155(60):44

Space Policy Directive-3, National Space Traffic Management Policy (2019) White House. https://www.whitehouse.gov/presidential-actions/space-policy-directive-3-national-space-traffic-management-policy. Accessed 10 Apr 2019

The U.S. Anti-Satellite (ASAT) Program (1987) The U.S. Anti-Satellite (ASAT) program: a key element in the national strategy of deterrence. The White House, May 11. https://fas.org/spp/military/program/asat/reag87.html. Accessed 1 Apr 2019

Treaty on Principles Governing the Activities of States in the Exploration and Use of Outer Space, Including the Moon and Other Celestial Bodies (1967) Opened for signature January 27. http://www.oosa.unvienna.org/pdf/publications/STSPACE11E.pdf. Accessed 1 Apr 2019

Ullah S, Imam I (2019) What to make of India's ASAT test. South Asian Voices, April 1. https://southasianvoices.org/what-to-make-of-indias-asat-test/. Accessed 2 Apr 2019

Zervos V (2011) Conflict in space. In: Braddon L, Hartley K (eds) Handbook on the economics of conflict. Edward Elgar Publishing Limited, Northampton, p 215

Zhang H (2008) Chinese perspectives on space weapons. In: Russian and Chinese responses to U.S. military plans in space. American Academy of Arts and Sciences, Cambridge, MA, p 31

Spacepower Theory and Organizational Structures

Peter L. Hays

Contents

Introduction	50
Noteworthy Efforts to Develop Spacepower Theory	51
Spacepower Theory and Current US Space Policy	56
Spacepower Theory, Hard Power, and the Quest for Sustainable Security	59
Spacepower Theory, Harvesting Energy, and Creating Wealth in and from Space	65
Spacepower Theory, Environmental Sustainability, and Survival	69
Conclusions	71
References	71

Abstract

Spacepower theory is useful in describing, explaining, and predicting how individuals, groups, and states can best derive utility, balance investments, and reduce risks in their interactions with the cosmos. Spacepower theory should be more fully developed and become a source for critical insights as humanity wrestles with our most difficult and fundamental space challenges. This theory can help to guide us toward better ways to generate wealth in space, make tradeoffs between space investments and other important goals, reorder terrestrial security dynamics as space becomes increasingly militarized and potentially weaponized, and seize exploration and survival opportunities that only space can provide. This chapter reviews noteworthy efforts to develop spacepower theory and overviews recent changes in US organizational structures for spacepower. It then considers ways

---

The opinions expressed in this chapter are mine and do not imply endorsement by the Space Policy Institute, Falcon Research, or Department of Defense.

P. L. Hays (✉)
Space Policy Institute, George Washington University, Washington, DC, USA
e-mail: hayspl@gwu.edu

theory and structures could help to refine current US space policy and address some of the most significant challenges and issues surrounding space security, space commercialization, and environmental sustainability and survival.

Introduction

The goal of spacepower theory is to describe, explain, and predict how individuals, groups, and states can best derive utility, balance investments, and reduce risks in their interactions with the cosmos. These are long-term, broad, indeterminate, and ambitious goals – it is hardly surprising that more than 60 years into the space age humanity has yet to develop spacepower theory able to address these goals in comprehensive and accepted ways. Incomplete and immature theory inhibits our ability to identify, pursue, and sustain major space objectives. More mature spacepower theory would provide critical insights as humanity wrestles with our most difficult and fundamental space challenges and guide us toward better ways to generate wealth in space, make tradeoffs between space investments and other important goals, reorder terrestrial security dynamics as space becomes increasingly militarized and potentially weaponized, and seize exploration and survival opportunities that only space can provide. This chapter reviews noteworthy efforts to develop spacepower theory and overviews recent changes in US organizational structures for spacepower. It then considers ways theory and structure could help to refine current US space policy and address some of the most significant challenges and issues surrounding space security, space commercialization, and environmental sustainability and survival.

Current perceptions that more robust spacepower theory is needed are undoubtedly most acute in the United States, but they are also growing worldwide as space becomes an increasingly contested and important domain. For decades, space capabilities gave the United States important asymmetric advantages that provided foundational elements of America's strength in the information age. These advantages are now being undermined by many factors including the reemergence of great power competition, the rise of China as a near-peer competitor with significant space and counterspace capabilities, continuing growth in the numbers and capabilities of space actors, and US uncertainties and missteps in determining and implementing its best strategy for developing and employing space capabilities. The trajectory of spacepower development has reached an inflection point where business as usual will no longer improve or even maintain US advantages – a point where the United States must implement different approaches or face diminishing returns from its space investments and the loss of space leadership. Attempting to identify and act upon inflection points is associated with strategic thinking and concepts of operations for terrestrial military operations; these approaches now hold obvious appeal to Americans pondering their space future. More mature and robust spacepower theory could help provide a more broad and stable foundation for the United States to develop a more deliberate, comprehensive, long-term, and consistent space strategy that would draw on all instruments of power from all levels of government, foster unity of effort in national space activities, improve the viability of the US space industrial base, and, in particular, craft better ways to leverage state-of-the-world commercial and international space capabilities.

Despite its importance, movement toward developing better spacepower theory is likely to be slowed by discouraging attributes associated with spacepower that include lack of acceptance that such theory is needed, very large investments and long timelines, requirements for sustained popular and political support, and prospects for only potential or intangible benefits. These factors can erode acceptance of and support for improving spacepower theory at both the personal and political levels, but they also point to the need for an incremental approach and reinforce the long-term benefits of theory in providing guidance, stability, and predictability. Indeed, more robust spacepower theory could provide an essential foundation for improving the structure and predictability of humanity's interactions with the cosmos. Perhaps more than any other approach, the issues spacepower theory addresses, the precedents from which it is drawn, and the pathways ahead it helps to illuminate could help guide the future development of spacepower.

Noteworthy Efforts to Develop Spacepower Theory

Many studies touch on aspects of spacepower theory, but few focus solely on this topic, and fewer still address the topic comprehensively and have widespread acceptance. This section briefly considers some of the most noteworthy efforts as well as elements of major and enduring themes and analogies any robust spacepower theory would need to address. The first major, comprehensive, and focused effort to develop spacepower theory began in 1997 when the Commander of US Space Command, General Howell M. Estes III, commissioned Dr. Brian R. Sullivan to write a book on this topic. James Oberg then became the leader of the effort and he published *Space Power Theory* in 1999 (Oberg 1999). Oberg draws on his academic background in astrodynamics and computer science as well as more than 20 years' experience with the National Aeronautics and Space Administration (NASA) Space Shuttle program to present a cogent narrative about the importance of spacepower that is particularly strong on the technical underpinnings of spacepower and emphasizes the need for space control. The book provides a strong foundation for spacepower theory, details the range of elements that contribute to its development, reviews how major space-faring states developed and used spacepower, and discusses several significant technical and political impediments to its development. Unfortunately, the political dimension of Oberg's spacepower theory is less well developed, primarily because his analysis does not provide much focus on the ways the attributes of spacepower relate to strategy or the development and employment of power in other domains.

By contrast, Everett Dolman, a professor in the Schriever Scholars program and at the School of Advanced Air and Space Studies at Air University, provides a spacepower theory that is focused almost entirely on the political rather than technical aspects of spacepower. *Astropolitik: Classical Geopolitics in the Space*

Age (Dolman 2001) explains how the physical attributes of outer space and the characteristics of space systems shape the application of spacepower and then uses this *astropolitical* analysis to develop a compelling vision for the United States to reject the Outer Space Treaty (OST) regime, promote free-market capitalism in space, and use space to help provide global security as a public good. His book is intellectually grounded in the best traditions of geopolitics, has something genuinely new to say, makes vital contributions to the dialogue about the interrelationships between space and national security, and is easily the most important book on space and security since the publication of Walter A. McDougall's Pulitzer prize-winning...*the Heavens and the Earth: A Political History of the Space Age* in 1985. *Astropolitik* is a stunning intellectual achievement and the first book that can legitimately claim to present a comprehensive theory of spacepower. It challenges conventional thinking about the status quo for space and has generated a great deal of controversy and provoked many responses. To be sure, many of the major points Dolman asserts are open to debate, such as whether space will actually become a virtually limitless source of wealth, what technologies and strategies the United States might employ to assert dominance over low-Earth orbit (LEO), and how and why domestic and international political forces might come to align with his astropolitical prescriptions. But one mark of a great book is that it helps to define and structure subsequent debate; *Astropolitik* has clearly advanced the study of spacepower theory by providing the language and lines of argumentation for future discourse.

There are several other noteworthy additions to this field: M. V. Smith's *Ten Propositions Regarding Spacepower*, John J. Klein's *Space Warfare: Strategy, Principles and Policy*, the National Defense University's (NDU) edited volume *Toward a Theory of Spacepower: Selected Essays*, and Klein's *Understanding Space Strategy: The Art of War in Space* (Smith 2001; Klein 2006, 2019; Lutes et al. 2011; Klein 2019). *Ten Propositions Regarding Spacepower* is written from the perspective of an Air Force officer who spent several years integrating space-related capabilities into numerous exercises and real-world combat; the study seeks to answer the philosophical question "what is the nature of spacepower?" Smith describes the nature of spacepower by presenting ten propositions, supporting each with historical evidence: Space is a distinct operational medium; the essence of spacepower is global access and global presence; spacepower is composed of a state's total space activity; spacepower must be centrally controlled by a space professional; spacepower is a coercive force; commercial space assets make all actors space powers; spacepower assets form a national center of gravity; space control is not optional; space professionals require career-long specialization; and weaponizing space is inevitable. Smith's propositions build from and are consistent with main themes in Oberg's and Dolman's works, but they independently advance spacepower theory by providing a more comprehensive and thorough exposition of the attributes of spacepower and its employment.

John J. Klein is a naval aviator, his books build from Carl von Clausewitz and other classic military theorists. In particular, Klein modifies for space the classic maritime theory presented by Julian Corbett in *Some Principles of Maritime Strategy* and first published in 1911. Corbett's theory about maritime activity is among the best developed and comprehensive of all theories designed to explain military operations in terrestrial domains. Klein assesses airpower, seapower, and maritime strategies, finding that maritime strategy is most suitable for application to space; builds from Alfred Thayer Mahan's and Corbett's ideas about sea lines of communications to discuss the importance of celestial lines of communications; and asserts that there is an overemphasis on power and offensive space operations in current American spacepower thought. Klein's work advances spacepower theory by creating tight linkages with Corbett's well-developed maritime theory and providing a firm foundation for further refining spacepower theory.

The NDU spacepower theory study was commissioned by the Department of Defense (DoD) as the result of deliberations during preparation of the 2005 Quadrennial Defense Review. The study was a team effort to produce an edited volume and does not attempt to present a single point of view about spacepower theory. Instead, the study published 30 chapters written by national and international space experts and organized into six sections: introduction to spacepower theory; economics and commercial space perspectives; civil space perspectives; national security space perspectives; international perspectives; and evolving futures for spacepower. The strength of this approach is that it presents the most broad and wide-ranging perspectives ever assembled about spacepower theory, but weaknesses also stem from this approach, because there is no unified perspective or even many major common themes that emerge from the work. The overarching goal of the study was to foster dialogue and incubate further development of spacepower theory; it is hoped that the study's broad and wide-ranging perspectives will encourage advancement of spacepower theory along multiple paths.

Major and enduring themes and analogies any robust spacepower theory should address include perspectives on the growing use and importance of space, debates about the economic potential of space, debates over the need for and inevitability of space weaponization, perspectives of space as a frontier to be tamed, and perspectives that link space to humanity's purpose and destiny. Another set of factors shaping spacepower theory are the oft-invoked analogies between spacepower and seapower or airpower. Seminal theorists who developed important perspectives on sea and air operations include Mahan, Corbett, Giulio Douhet, William "Billy" Mitchell, and John Warden. (Several of these individuals were quite prolific; the following list represents their best known works: Mahan (1890), Corbett (1988), Douhet (1983), Mitchell (1988), and Warden (1988). On the importance of these works see Sumida (1997), Meilinger (1997), and Mets (1999).) Some of the key concepts that these theorists developed or applied to the air and sea mediums are command of the sea, command of the air, sea lines of communication, common routes, choke points, harbor access, concentration and dispersal, and parallel attack. Several of these concepts have been appropriated directly into various strands of embryonic space theory; others have been modified slightly and then applied. For example, Mahan's and Corbett's ideas about lines of communications, common routes, and choke points have been applied quite directly onto the space medium. Seapower and airpower concepts that have been modified to help provide starting points for thinking about spacepower include harbor access and access to space and command of the sea or air and space control. But, of course, to date, no holistic spacepower theory has yet emerged that is fully worthy of claiming a place alongside the seminal seapower and airpower theories listed above. There are also many fundamental questions concerning the basic attributes of the space medium and how appropriate it is to analogize directly from seapower or airpower theory when attempting to build spacepower theory. Few concepts from seapower theory translate directly into airpower theory, and it is not reasonable to expect either seapower or airpower theory to apply directly for the distinct space domain.

Organizational structures also play a critical role in shaping US spacepower. Creation of the US Space Force (USSF) in December 2019 marks a momentous change in the structure of the US military and a significant shift in American strategic thought about the military utility of space (National Defense Authorization Act 2019). While this development alone cannot resolve all spacepower theory and space strategy issues, it may end an era of more than 30 years when the US was not satisfied with how it organized its national security space activities and churned through several different structures. Yet, because organizational structure is only a second-order issue, deeper questions remain related to the lack of consensus and direction on US objectives and priorities for its spacepower theory. The United States has yet to focus enough or reach consensus even domestically, let alone internationally, on first-order issues such as the long-term viability of the Outer Space Treaty (OST) regime, space weaponization, options for exploiting space resources and creating wealth in and from space, or other overarching issues related to the objectives it seeks from space, why these are important, and what the best strategies are to pursue these objectives. Worse, far too much current attention has been diverted toward third-order issues such as in which congressional districts Space Force units should be located or even what the Space Force patch and uniforms should look like. These are all issues and decisions that deserve some level of attention and hold some importance, but, as it stands up its first new military branch in more than 70 years, it is critical for the United States to focus initially on the first-order issue of the spacepower theory that will prioritize what the Space Force should do. Focusing on the organizational structure of the Space Force and first-order priorities for space can help the United States ask the right questions and move toward doing the right things, at the right times, and for the right reasons.

For the past several years, the United States was not able to reach consensus on the need for a Space Force, Space Corps, or other potential major reorganizations. Between 2017 and 2019, the House of Representatives and Senate Armed Services Committees were unable to reach a compromise on the Space Force. Elevating the issue above these committees in the fall of 2019 during budget negotiations allowed broader compromises between the parental leave provisions some Democrats wanted and Space Force provisions the President and some Republicans wanted. Unfortunately, expansion of the scope for compromises on a new organization did not extend beyond considering only a very narrow slice of military structure options or even the name of the new organization – the President insisted it be called the Space Force. The new Space Force resulting from these uncertainties and compromises holds significant potential but faces continuing disagreements about its most important and appropriate near-term priorities and, as a military organization, is limited in its ability to effectively address the full range of first-order strategic space issues the United States currently faces. Another reflection of congressional concerns about the Space Force is the unprecedented level of oversight and reporting requirements Congress has levied on the new organization, including bi-monthly reporting on progress in establishing the Space Force.

There are several important concepts that can help us examine the Space Force and determine where it fits in relation to previous models and structures. Aristotle originated the idea that "form follows function," a broad philosophical construct that includes deliberations on how organizational functions ought to determine organizational structure. This concept, along with Clausewitz's assessments about inflection and culminating points, may be helpful as we consider the development of spacepower and how organizational structures may need to evolve. Another consideration for framing discussions on the Space Force is the adage that "when all you have is a hammer, everything looks like a nail" and the potential consequences of the United States choosing this military organization model and discussing the need for space dominance. Unlike most other national security mission areas, during the past 30 years, the Pentagon was directed to or chose itself to make several significant changes in its national security space organizational structures including the Deputy Under Secretary of Defense for Space (1994–1997); National Security Space Architect (1998–2004); National Security Space Office (2004–2010); Department of Defense (DoD) Executive Agent for Space (2003–2015); and Principal DoD Space Advisor (2015–2018). In addition, Air Force Space Command was established in 1982 and redesignated as the Space Force under the 2019 Space Force Act; and US Space Command was established in 1985, merged underneath US Strategic Command in 2002, and reestablished as an independent geographic combatant command in 2019.

The Space Force Act gives DoD 18 months to establish the initial operational capability of the Space Force and implement several other key provisions. The Act established the Space Force, a distinct armed force within the Department of the Air Force under the Secretary of the Air Force, and created a new General Officer position, the Chief of Space Operations (CSO). On 14 January 2020, Vice President Mike Pence administered the oath of office to Air Force General John "Jay" Raymond, making Raymond the first CSO and first member of the Space Force. The CSO is already attending meetings of the Joint Chief of Staff (JCS) and under the Act becomes a member of the JCS on 20 December 2021. The Space Force must establish its headquarters along with determining its subordinate units and their basing locations. All Space Force units will initially come from the Air Force, but, over time, it is expected that some Army and Navy units, along with appropriate Guard and Reserve units, will also transfer to the Space Force. Likewise, officers and enlisted personnel, initially from the Air Force, can voluntarily transfer into the Space Force. It is expected that some officers and enlisted personnel from the other Services, along with new accessions to the military, will also volunteer for the Space Force. Other key provisions in the Space Force Act establish a new Assistant

Secretary of the Air Force for Space Acquisition and Integration (ASecAF SA&I) position and a new Space Force Acquisition Council (SFAC). The ASecAF SA&I is a Senate-confirmed position that serves as the senior architect for space systems and programs across the Department of the Air Force, chairs the SFAC, is to become the Air Force Service Acquisition Executive (SAE) for space systems and programs as of 1 October 2022, and provides fiscal and strategic guidance by overseeing and directing the Space Rapid Capabilities Office, the Space and Missile Systems Center, and the Space Development Agency. The SFAC is to meet monthly, and its membership includes the Under Secretary of the Air Force, the Assistant Secretary of Defense for Space Policy, the Director of the National Reconnaissance Office (NRO), the CSO, and the Commander of the US Space Command.

Spacepower Theory and Current US Space Policy

The United States has the most developed, open, and mature process for promulgating national space policy, and these space policies contain many elements that would be needed in robust and comprehensive spacepower theory. This is not to suggest that US space policy is the same as or can substitute for spacepower theory, but it does mean that attempts to develop spacepower theory need to be aware of and interact with these elements of US space policy. Widely accepted and comprehensive spacepower theory could help the United States refine its space policy, provide a stronger and more sustainable and consistent foundation for its implementation, and also improve its strategic-level management and organizational structures for implementing goals from the National Security Strategy, National Defense Strategy (NDS), and National Strategy for Space (NSfS).

The Trump Administration's National Security Strategy, released in December 2017, established America's vital national interest in space, reemphasized the importance of space for US security, and provided several overarching yet demanding objectives that will require focused attention and considerable effort to pursue:

> The United States must maintain our leadership and freedom of action in space. Communications and financial networks, military and intelligence systems, weather monitoring, navigation, and more have components in the space domain. As U.S. dependence on space has increased, other actors have gained access to space-based systems and information. Governments and private sector firms have the ability to launch satellites into space at increasingly lower costs. The fusion of data from imagery, communications, and geolocation services allows motivated actors to access previously unavailable information. This "democratization of space" has an impact on military operations and on America's ability to prevail in conflict.
> Many countries are purchasing satellites to support their own strategic military activities. Others believe that the ability to attack space assets offers an asymmetric advantage and as a result, are pursuing a range of anti-satellite (ASAT) weapons. The United States considers unfettered access to and freedom to operate in space to be a vital interest. Any harmful interference with or an attack upon critical components of our space architecture that directly affects this vital U.S. interest will be met with a deliberate response at a time, place, manner, and domain of our choosing.

4 Spacepower Theory and Organizational Structures

Priority Actions
ADVANCE SPACE AS A PRIORITY DOMAIN: America's newly re-established National Space Council, chaired by the Vice President, will review America's long-range space goals and develop a strategy that integrates all space sectors to support innovation and American leadership in space.
PROMOTE SPACE COMMERCE: The United States will simplify and update regulations for commercial space activity to strengthen competitiveness. As the U.S. Government partners with U.S. commercial space capabilities to improve the resiliency of our space architecture, we will also consider extending national security protections to our private sector partners as needed.
MAINTAIN LEAD IN EXPLORATION: To enable human exploration across the solar system and to bring back to Earth new knowledge and opportunities, we will increase public-private partnerships and promote ventures beyond low Earth orbit with allies and friends. (National Security Strategy 2017)

Then Secretary of Defense James Mattis released an unclassified summary of the NDS in January 2018. The strategy sets three overarching objectives for DoD to address the reemergence of great power competition: rebuilding military readiness to develop a more lethal Joint Force, strengthening alliances and attracting new partners, and reforming DoD's business practices for greater performance and affordability. The NDS designates space as a "warfighting domain" and indicates DoD will "prioritize investments in resilience, reconstitution, and operations to assure our space capabilities" (National Defense Strategy 2018).

In March 2018, the White House released the National Strategy for Space. The strategy established four pillars for a unified approach:

- Transform to more resilient space architectures: We will accelerate the transformation of our space architecture to enhance resiliency, defenses, and our ability to reconstitute impaired capabilities.
- Strengthen deterrence and warfighting options: We will strengthen U.S. and allied options to deter potential adversaries from extending conflict into space and, if deterrence fails, to counter threats used by adversaries for hostile purposes.
- Improve foundational capabilities, structures, and processes: We will ensure effective space operations through improved situational awareness, intelligence, and acquisition processes.
- Foster conducive domestic and international environments: We will streamline regulatory frameworks, policies, and processes to better leverage and support U.S. commercial industry, and we will pursue bilateral and multilateral engagements to enable human exploration, promote burden sharing and marshal cooperative threat responses. (National Strategy for Space 2018, p. 2)

Cumulatively, these documents move the Trump Administration considerably beyond the space policy of the Obama Administration. They reflect the "America First" more unilateral tone of many Trump Administration policies and move away from the stress on cooperation and responsible behavior in space in the 2010 National Space Policy. More specifically, the Trump Administration rejected the Obama Administration's categorization of space stability and sustainability as vital national interests and returned to the previous approach that categorized just national security considerations as vital national interests in space as found in the 2006 and previous National Space Policies.

Additionally, the United States must continue to implement the many approaches and comprehensive actions detailed in the National Security Space Strategy (NSSS). The NSSS was signed by the Secretary of Defense and Director of National Intelligence and released in February 2011 (Secretary of Defense and Director of National Intelligence 2011). The NSSS publicly substantiated that space is growing increasingly congested, contested, and competitive: DoD is tracking over 22,000 man-made objects in space (including 1,100 active satellites), there are hundreds of thousands of additional debris pieces too small to track with current sensors but that could still damage satellites in orbit, and there is also increasing congestion in the radiofrequency spectrum due to satellite operations by more than 60 states and consortia and as many as 9,000 satellite communications transponders expected to be in orbit by 2012 (Secretary of Defense and Director of National Intelligence 2011, pp. 1–2).

Space is increasingly *contested* in all orbits. Today space systems and their supporting infrastructure face a range of man-made threats that may deny, degrade, deceive, disrupt, or destroy assets. Potential adversaries are seeking to exploit perceived space vulnerabilities. As more nations and non-state actors develop counterspace capabilities over the next decade, threats to US space systems and challenges to the stability and security of the space environment will increase. Irresponsible acts against space systems could have implications beyond the space domain, disrupting worldwide services upon which the civil and commercial sectors depend (Secretary of Defense and Director of National Intelligence 2011, p. 3).

And with respect to increasing competition, while the United States "maintains an overall edge in space capabilities," its "competitive advantage has decreased as market-entry barriers have lowered"; its "technological lead is eroding in several areas"; "US suppliers, especially those in the second and third tiers, are at risk due to inconsistent acquisition and production rates, long development cycles, consolidation of suppliers under first-tier prime contractors, and a more competitive foreign market"; and the US share of world satellite manufacturing revenue has dropped from an average of more than 60% during the 1990s to 40% or less during the 2000s (Secretary of Defense and Director of National Intelligence 2011).

To address these challenges, the NSSS seeks three strategic objectives: strengthening safety, stability, and security in space; maintaining and enhancing the strategic national security advantages afforded to the United States by space; and energizing the space industrial base that supports US national security (Secretary of Defense and Director of National Intelligence 2011, p. 4). The strategy advocates five strategic approaches to pursue these objectives: promoting responsible, peaceful, and safe use of space; providing improved US space capabilities; partnering with responsible nations, international organizations, and commercial firms; preventing and deterring aggression against space infrastructure that supports US National Security; and preparing to defeat attacks and to operate in a degraded environment (Secretary of Defense and Director of National Intelligence 2011, pp. 5–11). Pursuit and implementation of these strategic objectives has proven challenging, but the NSSS correctly assesses the most significant changes in the space strategic environment and presents a responsible way for the United States to address these changes that begins to approach the comprehensive advances needed for spacepower theory.

Spacepower Theory, Hard Power, and the Quest for Sustainable Security

There are several of hard power issue areas where spacepower theory might provide insights on space security including the OST regime and other transparency- and confidence-building measures (TCBMs), space situational awareness (SSA), space weaponization, and the rise of China as a major factor in space security. The OST regime is by far the most important and comprehensive mechanism in shaping space security. Although there is some substance to arguments that the OST only precludes those military activities that were of little interest to the superpowers and does not bring much clarity or direction to many of the most important potential space activities, the treaty nonetheless provides a solid and comprehensive starting point for spacepower theory and is an important foundation for thinking about additional theoretical structures needed to advance spacepower. Moreover, there is broad consensus on the merits and overall value of the OST regime; space-faring actors are much more interested in building upon this foundation than in developing new structures.

Spacepower theory should provide guidance on the most effective ways to confront the OST regime. Some theories would advocate abandoning this regime; most others would seek ways to improve and build upon the OST regime including working toward achieving more universal adherence by all space-faring actors to the regime's foundational norms and expanding the regime beyond just states to include all important space-faring actors. Beginning work to include major non-state space actors in the OST would be a significant step that would require substantial expansion of the regime and probably would need to be accomplished incrementally. The security dimensions of the regime have opened windows of opportunity, and important precedents have been set by expanding participation in the United Nations Committee on the Peaceful Uses of Outer Space (UN COPUOS) and the World Radio Conferences of the International Telecommunications Union (ITU) to include non-state actors as observers or associate members. Some form of two-tiered participation structure within the OST regime might be appropriate for a number of years, and it could prove impractical to include non-state actors in a formal treaty, but steps toward expanded participation should be carefully considered, both to capture the growing spacepower of non-state actors and to harness their energy in helping achieve more universal adherence to the regime. Perhaps most importantly, these initial steps would help promote a sense of stewardship for space among more actors and increase attention on those parties that fail to join or comply with these norms. Other particular areas within the OST regime that spacepower theory should address, perhaps through creation of a standing body with specific implementation responsibilities, include the Article VI obligations for signatories to authorize and exercise continuing supervision over space activities and the Article IX responsibilities for signatories to undertake or request appropriate international consultations before proceeding with any activity or experiment that would cause potentially harmful interference.

Another key area for security and spacepower theory for the United States and other leading space-faring actors that would help better define OST implementation obligations and demonstrate leadership in fostering cooperative spacepower would be improvements in how SSA data is developed and shared globally. Due to increasing use of space by more actors, the growing number of active satellites, and, especially, recent deliberate and accidental debris creating events caused by the Chinese ASAT test in January 2007, the February 2009 collision between Iridium and Cosmos satellites, and the Indian ASAT test in March 2019, there is now more worldwide interest in spaceflight safety and considerable motivation for improvements in developing and sharing SSA data with more users in more timely and consistent ways. As a result of the 11 January 2007 Chinese ASAT test, the US Space Surveillance Network has cataloged 2,378 pieces of debris with diameters greater than 5 cm, is tracking 400 additional debris objects that are not yet cataloged, and estimates the test created more than 150,000 pieces of debris larger than 1 cm^2. Unfortunately, less than 2% of this debris has reentered the atmosphere so far, and it is estimated that many pieces will remain in orbit for decades and some for more than a century (NASA Orbital Debris Proogram Office 2009).

Spacepower theory should provide guidance on the most effective approaches toward achieving these objectives. One approach would be continuation and improvements in US Government efforts to create a data center for sharing SSA data globally including ephemeris, propagation data, and pre-maneuver notifications for all active satellites. SSA issues are framed by specialized concepts and jargon. Conjunctions are close approaches, or potential collisions, between objects in orbit. Propagators are complex modeling tools used to predict the future location of orbital objects. Satellite operators currently use a number of different propagators and have different standards for evaluating and potentially maneuvering away from conjunctions. Maneuvering requires fuel and shortens the operational life of satellites. Orbital paths are described by a set of variables known as ephemeris data; two-line element sets (TLEs) are the most commonly used ephemeris data. Much of this data is contained in the form of a satellite catalog. The United States maintains a public catalog at space-track.org. Other entities maintain their own catalogs. Orbital paths constantly change, or are perturbed, by a number a factors including Earth's inconsistent gravity gradient, solar activity, and the gravitational pull of other orbital objects. Perturbations cause propagation of orbital paths to become increasingly inaccurate over time; beyond approximately four days into the future, predictions about the location of orbital objects can be significantly inaccurate (Weeden 2009; McGlade 2007). Under Space Policy Directive-3, the Trump Administration is working to improve space traffic management and is planning to move the center for sharing SSA data globally from DoD to the Department of Commerce. Another approach would be to transition and operate such a data center under international auspices and perhaps create an international space traffic management organization that would be somewhat analogous to the International Civil Aviation Organization (ICAO). A final approach would build from commercial efforts such as the Space Data Association and Commercial Space Operations Center to encourage the commercial sector, rather than governments, to play the leading role in providing SSA data globally. In each case, processes would need to be developed and refined for users to voluntarily contribute data to the center, perhaps through a Global

Positioning System (GPS) transponder on each satellite, and for spaceflight safety data to be constantly updated, freely available, and readily accessible so that it could be used by satellite operators to plan for and avoid conjunctions.

Spacepower theory should also address difficult legal, technical, and policy issues that inhibit progress on sharing SSA data that include bureaucratic inertia, liability, and proprietary concerns; nonuniform data formatting standards and incompatibility between propagators and other cataloging tools; and security concerns over exclusion of certain satellites from any public data. Some of these concerns could be addressed by working toward better cradle-to-grave tracking of all cataloged objects to help establish the launching state and liability; using opaque processes to exclude proprietary information from public databases to the maximum extent feasible; and indemnifying program operators, even if they provide faulty data that results in a collision, so long as they operate in good faith, exercise reasonable care, and follow established procedures.

Theories for operating in other domains and history suggest there are very important roles for militaries both in setting the stage for the emergence of international legal regimes and in enforcing the norms of those regimes once they emerge. Development of any TCBMs for space, such as rules of the road or codes of conduct, should draw closely from the development and operation of such measures in other domains such as sea or air. The international community should consider the most appropriate means of separating military activities from civil and commercial activities in the building of these measures, because advocating a single standard for how all space activities ought to be regulated or controlled is inappropriately ambitious and not likely to be helpful. DoD requires safe and responsible operations by warships and military aircraft, but they are not legally required to follow all the same rules as commercial traffic and sometimes operate within specially protected zones that separate them from other traffic. Moreover, operational security considerations dictate that these military forces often do not provide public information about their location and planned operations. More robust spacepower theory as well as full and open dialogue about these issues will help us develop space rules that draw from years of experience in operating in other domains and make the most sense for the unique operational characteristics of space.

Other concerns surround the implications of various organizational structures and rules of engagement for potential military operations in space. Spacepower theory should help us address key questions such as whether military space forces should operate under national or only international authority, who should decide when certain activities constitute a threat, and how such forces should be authorized to engage threats, especially if such engagements might create other threats or potentially cause harm to humans or space systems. Clearly, these and several other questions are very difficult to address and require careful international vetting well before the actual operation of such forces in space. In addition, we should consider the historic role of the Royal and US Navies in fighting piracy, promoting free trade, and enforcing global norms against slave trading, as well as the current international effort to combat piracy off the Horn of Africa. What would be analogous roles in space for the US military and other military forces today and in the future and how might the United States and others encourage like-minded actors to cooperate on such initiatives? Attempts to create legal regimes or enforcement norms that do not specifically include and build upon military capabilities are likely to be divorced from pragmatic realities and ultimately frustrating efforts (Joseph DeSutter 2006).

Robust and comprehensive spacepower theory should also address the viability and utility of various top-down and bottom-up approaches to TCBMs. The OST regime was developed through top-down methods, but since that success, many factors have made this approach increasingly difficult. The most serious of these problems include disagreements over the proper forum, scope, and object for negotiations; basic definitional issues about what is a space "weapon" and how they might be categorized as offensive or defensive and stabilizing or destabilizing; and daunting concerns about whether adequate verification mechanisms can be found for any comprehensive and formalized TCBMs that would likely prohibit certain space activities while seeking to encourage others. These problems relate to a number of very thorny, specific issues such as whether the negotiations should be primarily among only major space-faring actors or more multilateral, what satellites and other terrestrial systems should be covered, and whether the object should be control of space weapons or TCBMs for space; the types of TCBMs which might be most useful (e.g., rules of the road or keep out zones) and how these might be reconciled with the existing space law regime; and verification problems such as how to address the latent or residual ASAT capabilities possessed by many dual-use or military systems or how to deal with the significant military potential of even a small number of covert ASAT systems.

New space system technologies, continuing growth of the commercial space sector, and new verification technologies interact with these existing problems in complex ways. Some of the changes would seem to favor TCBMs, such as better radars and optical systems for improved SSA, attribution, and verification capabilities; technologies for better space system diagnostics; and the stabilizing potential of redundant and distributed space architectures that create many nodes by employing larger numbers of hosted payloads and less expensive satellites. Many other trends, however, would seem to make space arms control and regulation even more difficult. For example, very small satellites are becoming increasingly capable and might be used as virtually undetectable active ASATs or passive space mines; proliferation of space technology has radically increased the number of significant space actors to include a number of non-state actors that have developed or are developing sophisticated dual-use technologies such as autonomous rendezvous and docking capabilities; satellite communications technology can easily be used to jam rather than communicate; and growth in the commercial space sector raises issues such as how quasi-military systems could be protected or negated and the unclear security implications of global markets for dual-use space capabilities and products.

There is disagreement about the relative utility of top-down versus bottom-up approaches to developing space TCBMs and formal arms control, but, following creation of the OST regime, the United States and many other major space-faring actors have tended to favor bottom-up approaches, a point strongly emphasized by US Ambassador Donald Mahley in February 2008: "Since the 1970s, five consecutive U.S. administrations have concluded it is impossible to achieve an effectively verifiable and militarily meaningful space arms control agreement" (Ambassador Mahley 2008). Yet this assessment may be somewhat myopic since strategists need to consider not only the well-known difficulties with top-down approaches but also the potential opportunity costs of inaction and recognize when they may need to trade some loss of sovereignty and flexibility for stability and restraints on others. Because the United States has not tested a kinetic energy ASAT since September 1985 and has no program to develop a dedicated ASAT system, would it have been better to exchange the option to maintain this capability for pursuit of a global ban on testing kinetic energy ASATs, and would such a norm have produced a restraining effect on development and testing of the Chinese ASAT? This may have been a lost opportunity to pursue TCBMs but is a complex, multi-dimensional, and interdependent issue shaped by a variety of other factors such as inabilities to distinguish between ballistic missile defense and ASAT technologies, reluctance to limit technical options after the end of the Cold War, the emergence of new and less easily deterred threats, and the demise of the Anti-Ballistic Missile (ABM) Treaty.

To circumvent significant challenges with top-down approaches, there have been several attempts to make progress through primarily incremental, pragmatic, technical, and bottom-up steps. Examples of this approach include the December 2007 adoption by the United Nations General Assembly of the Inter-Agency Debris Coordination Committee (IADC) voluntary guidelines for mitigating space debris, work initiated by the European Union toward an International Code of Conduct for outer space activities, the Long-Term Sustainability of Space Activities effort at UN COPUOS, and the United Nations Group of Governmental Experts on TCBMs (Council of the European Union 2008; United Nations General Assembly Resolution 62/217 2008).

Moreover, the Chinese, in particular, apparently disagree with pursuing only bottom-up approaches and, in ways that seem both shrewd and hypocritical, are currently developing significant counterspace capabilities while simultaneously advancing various top-down proposals in support of prevention of an arms race in outer space (PAROS) initiatives and moving ahead with the joint Chinese-Russian draft treaty on Prevention of Placement of Weapons in Outer Space (PPWT) introduced at the Conference on Disarmament in 2008 and updated in 2014. Thus far, the Chinese have seemed quite disinterested in pursuing space TCBMs; they are moving further and faster than any previous spacefaring actor and in 2013 tested a dedicated high-altitude ASAT system able to hold geostationary satellites at risk, a capability never pursued by the superpowers at the height of the Cold War. With respect to the PPWT in particular, while it goes to considerable lengths in attempting to define space, space objects, weapons in space, placement in space, and the use or threat of force, there are still very considerable definitional issues with respect to how specific capabilities would be addressed. An even more significant problem relates to all the terrestrial capabilities that could eliminate, damage, or disrupt normal functioning of objects in outer space such as the Chinese direct ascent ASAT. One must question the utility of a proposed agreement that does not address the significant security implications of current space system support of network-enabled terrestrial warfare, does not deal with dual-use space capabilities, seems to be focused on a class of weapons that does not exist or at least is not deployed in space, is silent about all the terrestrial capabilities that are able to produce weapons effects in space, and would not even ban development and testing of space weapons, only their use (Reaching Critical Will, "Preventing the Placement of Weapons in Outer Space: A Backgrounder on the draft treaty by Russia and China"). Given these glaring weaknesses in the PPWT, it seems plausible that it is designed as much to continue political pressure on the United States and derail US missile defense efforts as it is to promote sustainable space security.

Since Sino-American relations in general and space relations in particular are likely to play a dominant role in shaping spacepower theory and the quest for sustainable security during this century, proposed Sino-American cooperative space ventures or TCBMs are worthy of special consideration. For example, the United States could make more specific and public invitations for the Chinese to become involved with the International Space Station program and join other major cooperative international space efforts. The United States and China could also work toward developing non-offensive defenses of the type advocated by Philip Baines (2003). Kevin Pollpeter explains how China and the United States could cooperate in promoting the safety of human spaceflight and "coordinate space science missions to derive scientific benefits and to share costs. Coordinating space science missions with separately developed, but complementary space assets, removes the chance of sensitive technology transfer and allows the two countries to combine their resources to achieve the same effects as jointly developed missions" (Pollpeter 2008). Michael Pillsbury outlined six other areas where US experts could profitably exchange views with Chinese specialists in a dialogue about space weapons issues: "reducing Chinese misperceptions of U.S. Space Policy, increasing Chinese transparency on space weapons, probing Chinese interest in verifiable agreements, multilateral versus bilateral approaches, economic consequences of use of space weapons, and reconsideration of U.S. high-tech exports to China" (Pillsbury 2007). Finally, Bruce MacDonald's report on *China, Space Weapons, and U.S. Security* for the Council on Foreign Relations offers several noteworthy additional specific recommendations for both the United States and China. For the United States, MacDonald recommends: assessing the impact of different US and Chinese offensive space postures and policies through intensified analysis and "crisis games," in addition to wargames; evaluating the desirability of a "no first use" pledge for offensive counterspace weapons that have irreversible effects; pursuing selected offensive capabilities meeting important criteria – including effectiveness, reversible effects, and survivability – in a deterrence context to be able to negate adversary space capabilities on a temporary and reversible basis, refraining from further direct ascent ASAT tests and demonstrations as long as China does, unless there is a substantial risk to human health and safety from uncontrolled space object reentry; and entering negotiations on a kinetic energy ASAT testing ban. MacDonald's recommendations for China include providing more transparency into its military space programs; refraining from further direct ascent ASAT tests as long as the United States does;

establishing a senior national security coordinating body, equivalent to a Chinese National Security Council; strengthening its leadership's foreign policy understanding by increasing the international affairs training of senior officer candidates and establishing an international security affairs office within the People's Liberation Army; providing a clear and credible policy and doctrinal context for its 2007 ASAT test and counterspace programs more generally and addressing foreign concerns over China's ASAT test; and offering to engage in dialogue with the United States on mutual space concerns and become actively involved in discussions on establishing international space codes of conduct and confidence-building measures (MacDonald 2008).

Spacepower Theory, Harvesting Energy, and Creating Wealth in and from Space

Moving from hard to soft power considerations, spacepower theory can help to guide spacefaring actors in a number of important areas including further developing and refining the OST regime, adapting the most useful parts of analogous regimes such as the Law of the Sea and Seabed Authority mechanisms, and rejecting standards that stifle innovation, inadequately address threats to humanity's survival, or do not provide opportunities for rewards commensurate with risks undertaken. Revising and further developing the OST regime could be a key first step toward seeking better ways to harvest energy and create wealth in and from space. Expanding participation in the OST as discussed above might also be helpful, but other steps such as reducing liability concerns and improving legal incentives for harvesting energy and generating wealth are likely to be even more effective in pursuit of further commercial development of space. Of course, as with security, more comprehensive and robust spacepower theory would be helpful in considering a range of objectives and values that are in tension and require considerable effort to change or keep properly balanced. The OST has been extremely successful thus far with respect to its primary objective of precluding replication of the colonial exploitation that plagued much of Earth's history. The international community should now consider whether the dangers posed by potential cosmic land grabs continue to warrant OST restrictions that stifle development of spacepower, and, if these values are found to have become imbalanced, how these restrictions might best be changed. Space-faring actors should use an expansive approach to consider how perceived OST restrictions and the commercial space sector have evolved and might be further advanced in a variety of ways including reinterpreting the OST regime itself, becoming more intentional about developing spacepower, creating space-based solar power capabilities, and improving export controls.

While the OST has thus far been unambiguous and successful in foreclosing sovereignty claims and the ills of colonization, it has been less clear and effective with respect to pragmatic property rights and commercialization issues. Part of the problem in this regard stems from the fact that OST is not linked to robust and mature spacepower theory; the regime is also embedded within a broader body of international law and that regime is evolving, sometimes in unclear ways and under different interpretations. Elements within the regime are of unclear and unequal weight: the Moon Agreement with its Common Heritage of Mankind (CHM) approach to communal property rights and equally shared rewards has some effect but more limited standing as customary international law due to its lack of signatories, especially among major spacefaring states; moreover, it falls well short of the OST, a treaty that has been signed by 109 states and in force for over 50 years. Most fundamentally, however, the lack of clarity within space law about property rights and commercial interests is the result of the regime still being underdeveloped and immature. There is also a "Catch-22" factor at work since actors are discouraged from undertaking the test cases needed to develop and mature the regime because of the immaturity of the regime and their unwillingness to be guinea pigs in whatever legal processes would be used to resolve property rights and reward structures. The most effective way to move past this significant hurdle would be to create clear mechanisms for establishing property rights and processes by which all actors, especially commercial actors, can receive rewards commensurate with the risks they undertake. In addition, consideration should be given to reevaluating liability standards by assessing factors, including how much of a disincentive toward appropriate risk taking they may create and whether use of graduated or reduced liability standards might be more suitable in advancing positive incentives for more commercial space activity. Although Art. VII of the OST discusses liability, that article was further implemented in the Convention on International Liability for Damage Caused by Space Objects, commonly referred to as the Liability Convention. Under the Liability Convention, Article II, a launching state is absolutely liable to pay compensation for damage caused by its space object on the surface of the Earth or to aircraft in flight. However, under Articles III and IV, in the event of damage being caused elsewhere than on the surface of the Earth by a space object, the launching state is liable only if the damage is due to its fault or the fault of persons for whom it is responsible (i.e., commercial companies), under a negligence standard. The challenge is how best to evolve the existing space law regime with its two-tiered liability system based on either absolute liability or fault/negligence, depending upon the location of the incident, into a structure that might provide more incentives for commercial development of space (Convention on International Liability for Damage Caused by Space Objects (resolution 2777 (XXVI) annex)). In the Commercial Space Launch Act of 2015 and subsequent legislation, the US Congress indicated that US citizens "engaged in commercial recovery of an asteroid resource or a space resource under this chapter shall be entitled to any asteroid resource or space resource obtained, including to possess, own, transport, use, and sell the asteroid resource or space resource obtained in accordance with applicable law, including the international obligations of the United States." Other states including Luxembourg and the United Arab Emirates have enacted similar legislation, but in all cases the details of how these domestic laws will be implemented and remain compliant with state obligations under the OST remain to be seen. Finally, any comprehensive reevaluation of space property rights and liability concerns should also consider how these factors are addressed in analogous regimes such as the

4 Spacepower Theory and Organizational Structures

Seabed Authority in the Law of the Sea Treaty. Unfortunately, however, several of the analogous regimes like the Law of the Sea are largely premised on CHM approaches and may be somewhat better developed than the OST but are also currently underdeveloped and immature with respect to actual commercial operations, limiting the utility of attempting to draw from these precedents.

Provisions of the OST regime are probably the most important factors in shaping commercial space activity, but they are clearly not the only noteworthy legal and policy factors at work influencing developments within this sector. Commercial space activity was not that significant during the Cold War, but that has changed radically. In the 1960s, the United States was first to begin developing space services such as communications, remote sensing, and launch capabilities but did so within the government sector. This approach began to change in the 1980s, first with the November 1984 Presidential Determination to allow some commercial communication services to compete with Intelsat, and continued with subsequent policies designed to foster development of a commercial space sector. By the late 1990s, commercial space activity worldwide was outpacing government activity, and although government space investments remain very important, they are likely to become increasingly overshadowed by commercial activity. Other clear commercial and economic distinctions with the Cold War era have even more significant implications for the future of spacepower: the Soviet Union was only a military superpower, whereas China is a major US trading partner and an economic superpower that recently passed Germany and Japan to become the world's second largest economy and, if current growth projections hold, is on a path to become larger than the US economy by 2030. Because of its economic muscle, China can afford to devote commensurately more resources to its military, including a wide range of increasingly capable space and counterspace capabilities.

The United States and other major spacefaring actors lack, but undoubtedly need, much more open and comprehensive visions for how to develop spacepower theory and advance spacepower. This study is one attempt to foster more dialogue about space security, but the process should continue, become more formalized, and be supported by enduring organizational structures that include the most important stakeholders in the future of spacepower. Spacepower theory should be a foundational part of creating and implementing spacepower and guide approaches "focused on opening space as a medium for the full spectrum of human activity and commercial enterprise, and those actions which government can take to promote and enable it, through surveys, infrastructure development, pre-competitive technology, and encouraging incentive structures (prizes, anchor-customer contracts, and property/exclusivity rights), regulatory regimes (port authorities, spacecraft licensing, public-private partnerships) and supporting services (open interface standards, RDT&E [research, development, test, and evaluation] facilities, rescue, etc.)" (Garretson 2009). In addition, consideration should be given to using other innovative mechanisms and nontraditional routes to space development, including a much wider range of federal government organizations and the growing number of state spaceport authorities and other organizations developing needed infrastructure. Finally, the United States should make comprehensive and careful exploration of the potential of space-based solar power its leading pathfinder in creating a vision for developing spacepower. Working toward harvesting this unlimited power source in economically viable ways will require development of appropriate supporting structures, particularly with respect to incentives, indemnification, and potential public-private partnerships.

Better spacepower theory should also provide guidance on better ways to implement global licensing and export controls for space technology. It is understandable that many states view space technology as a key strategic resource and are very concerned about developing, protecting, and preventing the proliferation of this technology, but the international community, and the United States in particular, needs to find better legal mechanisms to balance and advance objectives in this area. Many current problems with US export controls began after Hughes and Loral worked with insurance companies to analyze Chinese launch failures in January 1995 and February 1996. A congressional review completed in 1998 (Cox Report) determined these analyses violated the International Traffic in Arms Regulations (ITAR) by communicating technical information to the Chinese. The 1999 National Defense Authorization Act transferred export controls for all satellites and related items from the Commerce Department to the Munitions List administered by the State Department. The January 1995 failure was a Long March 2E rocket carrying Hughes-built Apstar 2 spacecraft, and the February 1996 failure was a Long March 3B rocket carrying Space Systems/Loral-built Intelsat 708 spacecraft. Representative Christopher Cox (R-California) led a 6-month-long House Select Committee investigation that produced the "U.S. National Security and Military/Commercial Concerns with the People's Republic of China" Report released on 25 May 1999. In January of 2002, Loral agreed to pay the US government $20 million to settle the charges of the illegal technology transfer, and in March of 2003, Boeing agreed to pay $32 million for the role of Hughes (which Boeing acquired in 2000). Requirements for transferring controls back to State were in Sections 1513 and 1516 of the Fiscal Year 1999 National Defense Authorization Act. Related items were defined as "satellite fuel, ground support equipment, test equipment, payload adapter or interface hardware, replacement parts, and non-embedded solid propellant orbit transfer engines." The stringent Munitions List controls contributed to a severe downturn in US satellite exports. To avoid these restrictions, foreign satellite manufacturers, beginning in 2002 with Alcatel Space (now Thales) and followed by European Aeronautic Defense and Space Company (EADS), Surrey Satellite Company, and others, replaced all US-built components on their satellites to make them "ITAR-free" (de Selding 2005; Barrie and Taverna 2006).

Following the recommendations for rebalancing overall US export control priorities in the congressionally mandated National Academies of Science (NAS) study (National Research Council 2009), the Center for Strategic and International Studies (CSIS) study on the space industrial base (Briefing of the working group on the health of the U.S. space industrial base and the impact of export controls 2008), and the congressionally mandated section 1248 report completed by the Departments of State and Defense that assessed risks associated with removing satellites and related components from the US Munitions List, both the Obama Administration and Congress moved to reform US export controls in significant ways. The administration's proposal was advanced in August 2009 and called for "four singles": a single export control licensing agency for both dual-use and munitions exports, a unified control list, a single enforcement coordination agency, and a single integrated information technology (IT) system supporting the export control process. Following the significant space export control reforms enacted in 2014, the Trump Administration is looking toward additional export control reforms as well as streamlining efforts for other regulatory and licensing procedures under Space Policy Directive-2. These changes will help the United States avoid two major problems with an overly restrictive export control regime: First, an overly broad approach that tries to protect too many things dilutes resources and actually results in less protection for "crown jewels" than does a focused approach; and second, a more open approach is more likely to foster innovation, spur development of sectors of comparative advantage, and improve efficiency and overall economic growth.

Spacepower Theory, Environmental Sustainability, and Survival

The area where insights from spacepower theory undoubtedly could help provide the most significant contributions would be in improving environmental sustainability and humanity's odds for survival. More mature and robust spacepower theory is needed, because advancements in these areas face many daunting challenges, including a high "giggle factor," very long timelines that can be beyond our political and personal awareness, and potential returns that are uncertain and intangible. While difficult, work in these areas is absolutely critical, since it may hold the key to humanity's very survival, and it must be pursued with all the resources, consistency, and seriousness it deserves. The quest to improve the ways spacepower theory can support environmental and survival objectives should focus in three areas: space debris, environmental monitoring, and planetary defense.

Human space activity produces many orbital objects; when these objects no longer serve a useful function, they are classified as space debris. Over time, human activity has generated an increasing amount of debris from a variety of causes; the number of cataloged debris objects has gone from about 8,000 to over 22,000 over the past 20 years. The most serious cause of debris is deliberate hypervelocity impacts between large objects at high orbital altitudes such as the Chinese direct ascent kinetic energy ASAT weapon test of January 2007. If current trends continue, there is growing risk that space, and LEO in particular, will become increasingly unusable. Fortunately, there is also growing awareness and earnestness across the international community in addressing this threat. Overall goals for spacefaring actors with respect to space debris include minimizing its creation while mitigating and remediating its effects – spacepower theory can play an important role in raising awareness and providing guidance in all these areas. Key approaches to minimizing creation of debris and mitigating against its effects are commercial best practices and evolving regimes such as the IADC voluntary guidelines. Space-faring actors need to consider mechanisms to transition these voluntary guidelines into more binding standards and ways to impose specific costs such as sanctions or fines on actors that deliberately or negligently create long-lived debris. Fines could be applied toward efforts to further develop and educate spacefaring actors about the debris mitigation regime as well as to create, implement, and improve remediation techniques. An additional potential source of funding for mitigation and remediation would be establishing auctions for the radiofrequency spectrum controlled by the ITU that would be analogous to the spectrum auctions conducted at the national level by organizations like the Federal Communications Commission. Finally, it must be emphasized that techniques for remediating debris using lasers or other methods are likely to have significant potential as ASAT weapons, and very careful consideration should be given to how and by whom such systems are operated.

Space provides a unique location to monitor and potentially remediate Earth's climate. It is the only location from which simultaneous in situ observations of Earth's climate activity can be conducted and such observations are essential to develop a long-term understanding of potential changes in our biosphere. Because so much is riding on our understanding of the global climate and our potential responses to perceived changes, spacepower theory could play a particularly important role in helping us apply apolitical standards in getting the science right and controlling for known space effects such as solar cycles when making these observations and building climate models. If alarming models about global warming are correct and the global community must implement active measures to remediate these effects, space also provides a unique location to operate remediation options such as orbital solar shades, and space-based solar power has the potential to replace the use of some fossil fuels on Earth.

It is also imperative that the United States and all spacefaring actors use insights from spacepower theory and other sources to be more proactive, think more creatively, and transcend traditional approaches toward emerging threats to our survival. Spacepower theory can help to illuminate paths toward and develop incentives to create a better future. Space, perhaps more than any other medium, is inherently linked to humanity's future and very survival. We need to link these ideas together and better articulate ways spacepower can light a path toward genuinely cooperative approaches for protecting the Earth and space environments from cataclysmic events such as large objects that may collide with Earth or gamma ray bursts that have the potential to extinguish all life on Earth if we are unlucky enough to be in their path. Better knowledge about known threats such as near-Earth objects (NEOs) is being developed, but more urgency is required. The predicted near approach of the asteroid Apophis on 13 April 2029 ought to serve as a critical real-world test for our ability to be proactive in developing effective precision tracking and NEO mitigation capabilities. In the near term, it is most important for national and international organizations to be specifically charged with developing better understanding of NEO threats and developing avoidance techniques that can be effectively applied against likely impacts. Ultimately, however, as any robust and comprehensive spacepower theory would tell us, we cannot know of or effectively plan for all potential threats but should pursue multidimensional approaches to develop capabilities to improve our odds for survival and one day perhaps become a multi-planetary species.

Conclusions

This chapter reviewed noteworthy efforts to develop spacepower theory, considered ways it could help to refine current US space policy, and used it to address some of humanity's most significant space challenges including space security, space commercialization, and environmental sustainability and survival. Spacepower theory can describe, explain, and predict how individuals, groups, and states can best derive utility, balance investments, and reduce risks in their interactions with the cosmos; it should be more fully developed and become a source for critical insights. It could help to guide us toward better ways to generate wealth in space, make tradeoffs between space investments and other important goals, reorder terrestrial security dynamics as space becomes increasingly militarized and potentially weaponized, and seize exploration and survival opportunities that only space can provide.

There will be inevitable missteps, setbacks, and unintended consequences, but the inexorable laws of physics and of human interaction indicate that we will create the best opportunities for success in advancing spacepower by beginning long-term, patient work now rather than a crash program later. Spacepower theory should provide an essential foundation for this progress.

References

Ambassador Mahley DA (2008) Remarks on the state of space security. In: The state of space security workshop, Space Policy Institute, George Washington University, Washington, DC

Baines PJ (2003) The prospects for 'non-offensive' defenses in space. In: Moltz JC (ed) New challenges in missile proliferation, missile defense, and space security. Center for Nonproliferation Studies occasional paper no 12. Monterey Institute of International Studies, Monterey, pp 31–48

Barrie D, Taverna MA (2006) Specious relationship. Aviat Week Space Technol 17:93–96

Center for Strategic and International Studies (2008) Briefing of the working group on the health of the U.S. space industrial base and the impact of export controls. Center for Strategic and International Studies, Washington, DC

Corbett JS (1988) In: Grove EJ (ed) Some principles of maritime strategy. Naval Institute Press, Annapolis (First published 1911)

Council of the European Union (2008) Council conclusions and draft Code of Conduct for outer space activity. Council of the European Union, Brussels

de Selding PB (2005) European satellite component maker says it is dropping U.S. components because of ITAR. Space News Business report, 13 June 2005

Dolman EC (2001) Astropolitik: classical geopolitics in the space age. Routledge, New York

Douhet G (1983) In: Kohn RH, Harahan JP (eds) The command of the air. Office of Air Force History, Washington, DC (First published 1921)

Garretson P (2009) Elements of a 21st century space policy. The Space Review, 3 August 2009. Downloaded from http://www.thespacereview.com/article/1433/1

Joseph DeSutter R (2006) Space control, diplomacy, and strategic integration. Space Def 1(1): 29–51

Klein JJ (2006) Space warfare: strategy, principles, and policy. Routledge, New York

Klein JJ (2019) Understanding space strategy: the art of war in space. Routledge, New York

Lutes CD, Hays PL, Manzo VA, Yambrick LM, Bunn ME (eds) (2011) Toward a theory of spacepower: selected essays. National Defense University Press, Washington, DC

MacDonald BW (2008) China, space weapons, and U.S. security. Council on Foreign Relations, New York, pp 34–38

Mahan AT (1890) The influence of sea power upon history, 1660–1783. Little, Brown, Boston

McGlade D (2007) Commentary: preserving the orbital environment. Space News, 19 February 2007, p 27

Meilinger PS (ed) (1997) The paths of heaven: the evolution of airpower theory. Air University Press, Maxwell

Mets DR (1999) The air campaign: John Warden and the classical airpower theorists. Air University Press, Maxwell

Mitchell W (1988) Winged defense: the development and possibilities of modern airpower – economic and military. Dover, New York (First published 1925)

NASA Orbital Debris Program Office (2009) Fengyun 1-C Debris: Two Years Later. Orbital Debris Quarterly News 13(1):2. Johnson Spaceflight Center: NASA Orbital Debris Program Office

National Defense Authorization Act (NDAA) (2019) Conference report to accompany National Defense Authorization Act for fiscal year 2020. The Space Force Act is Sections 951–61 of the fiscal year (FY) 2020. US Congress. Washington, DC. https://docs.house.gov/billsthisweek/20191209/CRPT-116hrpt333.pdf

National Defense Strategy (2018) Summary of the 2018 National Defense Strategy. Department of Defense, Washington, DC, p 6. https://dod.defense.gov/Portals/1/Documents/pubs/2018-National-Defense-Strategy-Summary.pdf

National Research Council (2009) Beyond "Fortress America": national security controls on science and technology in a globalized world. National Academies Press, Washington, DC

National Security Strategy (2017) The White House, Washington, DC, p 31. https://www.whitehouse.gov/wp-content/uploads/2017/12/NSS-Final-12-18-2017-0905.pdf

National Strategy for Space (2018) The White House, Washington, DC, p 2. https://www.whitehouse.gov/briefings-statements/president-donald-j-trump-unveiling-america-first-national-space-strategy/

Oberg JE (1999) Space power theory. United States Space Command, Colorado Springs

Pillsbury MP (2007) An assessment of China's anti-satellite and space warfare programs, policies, and doctrines. Report prepared for the U.S.-China Economic and Security Review Commission, p 48

Pollpeter K (2008) Building for the future: China's progress in space technology during the tenth 5-year plan and the US response. Strategic Studies Institute, Carlisle Barracks, pp 48–50

Secretary of Defense and Director of National Intelligence (2011) National security space strategy: unclassified summary. Office of the Secretary of Defense and Office of the Director of National Intelligence, Washington, DC

Smith MV (2001) Ten propositions regarding spacepower. Air University Press, Maxwell

Sumida JT (1997) Inventing grand strategy and teaching command: the classic works of Alfred Thayer Mahan reconsidered. Woodrow Wilson Center Press, Washington, DC

United Nations General Assembly Resolution 62/217 (2008) International cooperation in the peaceful uses of outer space. UNGA, New York

Warden JA III (1988) The air campaign: planning for combat. National Defense University Press, Washington, DC

Weeden B (2009) The numbers game. The Space Review, 13 July 2009. Downloaded from http://www.thespacereview.com/article/1417/1

The Laws of War in Outer Space

Steven Freeland and Elise Gruttner

Contents

Introduction	74
General Principles of Space Law	75
Principles Regulating the "Military" Uses of Outer Space	80
The Laws of War: General Principles	82
Distinction	83
Military Objective	84
Proportionality	84
The Relevance of the Laws of War to Outer Space	85
Regulating the Threat of Space Warfare: Some Recent Initiatives	87
Conclusion: Perspectives on the Way Forward	89
References	91

Abstract

With the development of technology accelerating at a rapid pace, there has not been a more crucial time to analyze the international legal framework of outer space. The use of outer space for armed conflict is now a reality for space-faring nations and, as such, attention needs to be placed on the legal implications of this modern (potential) theatre of warfare. The applicability of international law to outer space was confirmed in the United Nations Outer Space Treaty that applies to the use and exploration of outer space. With new technologies such as dual-use satellites emerging, complex international law issues relating to the use of force and understanding how and to what extent the international law principles of *jus in bello* – international humanitarian law – apply to the regulation of these outer

---

S. Freeland (✉)
University of Western Sydney, Sydney, NSW, Australia
e-mail: S.Freeland@westernsydney.edu.au

E. Gruttner (✉)
United Nations Assistance to the Khmer Rouge Trials, New York, NY, USA
e-mail: elise.gruttner@gmail.com

© Springer Nature Switzerland AG 2020
K.-U. Schrogl (ed.), *Handbook of Space Security*,
https://doi.org/10.1007/978-3-030-23210-8_59

space activities. This chapter examines the evolution of outer space technology and the relevant legal frameworks that exist, and looks at certain aspects of the *jus in bello* principles that relate to the use of outer space. Some legal principles that exist in international humanitarian law may apply to activities in outer space; however, it remains unclear whether these principles are specific enough to take into account the increasingly diverse ways in which outer space could be utilized during the course of armed conflict. Consequently, there is a growing need for clarity in this evolving field of law, particularly as it relates to armed conflict in outer space.

Introduction

It is now more than 60 years since humankind began its "adventures" in outer space. On 4 October 1957, a Soviet space object, Sputnik I, was launched and subsequently orbited the Earth. The launch of Sputnik I heralded the dawn of the space age, the beginning of the space race (initially between the USSR and the United States), and the legal regulation of the use and exploration of outer space. Since then, laws regulating outer space technology have developed which significantly improve the standard of living for all humanity, through, for example, the facilitation of public services such as satellite telecommunications, global positioning systems, remote sensing technology for weather forecasting and disaster management, and television broadcast from satellites. Outer space offers the opportunity for immense social, economic, and scientific growth; however, new challenges are being posed to humanity through this advancement, and law will continue to play a crucial role in this regard.

The commencement of the space age began at the height of the Cold War, when both the United States and the USSR strove to flex their respective technological "muscles." At the time, the two leading space superpowers raced to develop their space capabilities, generating tension on a global scale. In October 1962, shortly after the launch of Sputnik I, the "Cuban Missile Crisis" threatened global security. This security-sensitive environment resulted in the international community's first endeavors to regulate this new frontier of outer space. While the purpose was to reduce the build up of weapons in space, the conventional obligations and restrictions that were agreed upon and codified in the major space treaties were neither entirely clear nor sufficiently comprehensive to meet all of these challenges. While a large body of space scholars would interpret the relevant provisions as prohibiting military space activities in outer space, those who had the capability to use space technology did not follow these restrictions. Since that time, it is clear that space has been utilized for military activities almost from the time of the very infancy of space activities.

Since the commencement of outer space activities, the environment has become significantly more complex and the consequences have heightened. Through the use of remote satellite technology and communication technology, information is being gathered and constantly relied upon by all aspects of society. This information is also an element in direct terrestrial military activity and represents an integral part of the major superpowers' respective military hardware. With outer space being increasingly used as part of active engagement in the conduct of armed conflict, it is now within the bounds of possibility to imagine outer space as an emerging theatre of warfare (Ricks 2001).

With the development of these outer space capabilities in mind, this chapter focuses on the potential application of the current laws of war to the use of outer space. While the last 60 years has shown us that outer space is being utilized for military purposes, it is not clear how these activities are treated and regulated at the international level. Through applying an analysis of the existing *jus in bello* principles to the domain of outer space, it would seem that the current capability of space-related technology exists outside the established international framework that has regulated the laws of war to date. The unique environment that outer space poses leaves some instances in which the established *jus in bello* principles application are not suitable as a regulatory system for the distinctive domain that is outer space.

Accordingly, this chapter will briefly outline the fundamental principles governing the international legal regulation of outer space and focus more specifically on those that are most relevant to military and warfare-related activities that utilize space technology. Subsequently, it examines the general principles that govern the laws of war in brief, before discussing these principles' relevance to outer space. Following on from this, a number of initiatives designed to (potentially) fill some of the lacunae that appear to exist within the current legal regime will be outlined, before we make some more general observations regarding the way forward in terms of legal regulation.

Ultimately, notwithstanding that the laws of war do theoretically apply to activities in outer space, the principles may not be specific enough to provide appropriate regulation for the increasingly diverse ways in which outer space could be used during the course of armed conflict. With the world's growing dependence on space technology and the ever-increasing space race among the major super powers, it is more important than ever to examine the existing international principles and reach a consensus on additional legal regulation directly applicable to the conduct of armed conflict in outer space. This will require close cooperation and greater trust between the major space powers, supported by other States and the international community. If this framework can be established, it may provide more certainty among the leading space-faring nations and could reduce the risk of the negative repercussions that are associated with space assets and activities.

General Principles of Space Law

The launch of Sputnik I immediately gave rise to difficult and controversial legal questions, involving previously undetermined concepts. Early academia contemplated the nature and scope of laws that might apply to the exploration and use of outer space, but only at a hypothetical level (Lyall and Larsen 2009). The world has since moved on from mere contemplation and has entered a new age of space technology and reliance. Humanity's ability to adapt and aspire to greater heights has led to the explosion of scientific exploration, the commercialization, and the militarization of outer space. With this rapid change came the international community's need to react to an unprecedented event of an unregulated legal environment.

These changes in space activities were largely driven at the time by the geopolitical situation – predominantly the state of Cold War that prevailed between the United States and the USSR. Time has shown us that the desire for ever-increasing technological prowess was as much motivated by military aspirations as a wish to explore and use space for other (scientific) purposes. It was in this conflicting environment that the international community had to respond. Between the desire of the two superpowers and the greater concerns among the international community, regulation had to somehow strike a balance between contrasting interests.

Accordingly, soon after the launch of Sputnik I, the United Nations established a new committee to take primary responsibility for the development and codification of the fundamental rules relating to the use and exploration of outer space with the name of United Nations Committee on the Peaceful Uses of Outer Space (UNCOPUOS). An ad hoc Committee on the Peaceful Uses of Outer Space, with 18 initial Member States, was established in 1958 by the United Nations General Assembly (UNGA 1958), which subsequently converted it into a permanent body in 1959 (UNGA 1959). UNCOPUOS is now the principal multilateral body involved in the development of international space law. In addition to States, a number of international organizations, including both intergovernmental and nongovernmental organizations, have observer status with UNCOPUOS.

The first question that was posed to UNCOPUOS sought a clarification as to the legal categorization of outer space for the purposes of international law. In order to be in a position to do this, a legal definition of what constituted outer space and where it began was required. While many theories have been put forward since the question was first posed, where air space "ends" and outer space "begins" has thus far remained unanswered from an international legal viewpoint.

Notwithstanding the lack of a clear definition of outer space, a number of fundamental legal principles relating to the exploration and use of outer space quickly emerged – in particular the so-called "common interest," "freedom," and "non-appropriation" principles. These principles were later incorporated into the terms of the United Nations Space Law Treaties, for example Articles I and II of the 1967 Treaty on Principles Governing the Activities of States in the Exploration and Use of Outer Space including the Moon and other Celestial Bodies, with the result that they also constitute binding conventional rules codifying what had already amounted to principles of customary international law. In essence, the international community, including both of the major space-faring States of the time, had accepted that outer space was to be regarded as being similar to a *res communis omnium* (Cassese 2005).

The aforementioned three fundamental rules that underpin the international law of outer space represent a significant departure from the legal rules relating to air space, which is categorized as constituting part of the "territory" of the underlying State. The principal air law treaties reflect the territorial nature of air space.

5 The Laws of War in Outer Space

For example, reaffirming the principle already acknowledged as early as 1919 (ICAN 1919), the 1944 Convention on International Civil Aviation (ICAO 1944) provides that "every State has complete and exclusive sovereignty over the air space above its territory" (ICAO 1944). Even though, a demarcation between air space and outer space has not yet definitively emerged – at least thus far – this has not in practice led to any significant confusion as to "which law" might apply in particular circumstances (Freeland 2010b). However, as the range of activities in outer space becomes ever broader, the issue will become more important in relation not only to the broad principles of international space law but also on a practical level – for example, to the regulation of commercial suborbital space tourism activities, which, at least under current technological constraints, involve paying passengers being taken to an altitude slightly in excess of 100 km above the Earth (Freeland 2010b).

By contrast, Article II of the Outer Space Treaty encompasses the so-called "non-appropriation" principle, which is regarded as one of the most fundamental rules regulating the exploration and use of outer space (Freeland and Jakhu 2010). The provision reads:

> Outer space, including the moon and other celestial bodies, is not subject to national appropriation by claim of sovereignty, by means of use or occupation, or by any other means.

In essence, Article II confirms that outer space (which includes the Moon and other celestial bodies) is not subject to ownership rights and prohibits inter alia any sovereign or territorial claims to outer space. Outer space therefore is not to be regarded as "territorial," a principle that, by the time the treaty was concluded in 1967, was already well accepted in practice. This was evidenced by the fact that although the USSR had not sought the permission of other States to undertake the Sputnik mission, there were no significant protests that this artificial satellite had infringed on any country's sovereignty as it circled the Earth. As was observed by Judge Manfred Lachs of the International Court of Justice (Lachs 1969):

> [t]he first instruments that men sent into outer space traversed the air space of States and circled above them in outer space, yet the launching States sought no permission, nor did the other States protest. This is how the freedom of movement into outer space, and in it, came to be established and recognised as law within a remarkably short period of time.

By the time that the Outer Space Treaty was finalized, both the United States and the USSR had already been engaged in an extensive range of space activities. However, neither had made a claim to sovereignty over any part of outer space, including celestial bodies, notwithstanding the planting by the Apollo 11 astronauts of an American flag on the surface of the Moon. This is to be compared with the situation in Antarctica, which had seen a series of sovereign claims by several States in the period leading up to the finalization in 1959 of the Antarctic Treaty, 402 U.N. T.S. 71. Article IV of the Antarctic Treaty has the effect of suspending all claims to territorial sovereignty in Antarctica for the duration of that instrument, as well as prohibiting any "new claim, or enlargement of an existing claim." The Protocol on

Environmental Protection to the Antarctic Treaty, 30 I.L.M. 1455, which came into force in 1998, augments the Antarctic Treaty by protecting Antarctica from commercial mining for a period of 50 years. As a result, although it was of great importance to formalize this principle of non-appropriation of outer space, the drafting process leading to the finalization of Article II of the Outer Space Treaty was relatively uncontroversial, particularly given its early acceptance as a fundamental concept by these two space faring States.

It is no coincidence that the non-appropriation principle is set out immediately following Article I of the Outer Space Treaty, which elaborates on the "common interest" and "freedom" principles and confirms that the exploration and use of outer space is to be undertaken "for the benefit and in the interests of all countries" and freely "by all States without discrimination of any kind, on a basis of equality and in accordance with international law." In broad terms, the primary intent of Article II was to reinforce these concepts by confirming that principles of territorial sovereignty do not apply to outer space. Not only does this reflect the practice of States from virtually the beginning of the space age but it also helps to protect outer space from the possibility of conflict driven by territorial or colonizing ambitions.

There has, however, been one notable exception in this regard – the Bogota Declaration. In 1976, a number of equatorial States – including Brazil, Colombia, the Congo, Ecuador, Indonesia, Kenya, Uganda, and Zaire – issued the Bogota Declaration, in which they claimed sovereign rights over segments of geostationary synchronous orbit above their respective territories. They asserted their claims principally because of the lack of an accepted delimitation between airspace and outer space. Such assertions were strenuously opposed by other States and have not been successful.

In this regard, the sentiments reflected in Article II of the Outer Space Treaty are fundamental to the regulation of outer space and its exploration and use for peaceful purposes. It is for these reasons that a binding principle of non-appropriation is an essential element of international space law.

Unlike the corresponding provision in United Nations Convention on the Law of the Sea (UNCLOS) (1982) dealing with the high seas, Article II does not expressly limit itself to the purported actions of States; rather, the provision is drafted in more general terms, in that it seeks to prohibit specific actions that constitute a "national appropriation." One should note, however, that the Chinese version of the Outer Space Treaty differs in this respect from all other versions, in that it prohibits appropriation "through the state by asserting sovereignty, use, occupation or any other means." In accordance with Article XVII of the Outer Space Treaty, the Chinese version is "equally authentic" with all other versions. However, it has also been noted that the fact that the other four versions (English, Russian, French and Spanish) all concur on the text of the provision is significant, "the more so if they include the languages which were mostly used in negotiations of the Outer Space Treaty" (Kopal 2006). With the obvious exception of the reference to "by claim of sovereignty," there is no express limitation in Article II *only* to the actions of States. This has, over the years, given rise to frequent debate among commentators as to the

5 The Laws of War in Outer Space

precise scope of the prohibition and, more particularly, the extent (if at all) to which "private property rights" (Harris 2004) may exist in outer space, notwithstanding (or perhaps as a result of) the terms of Article II.

In other aspects, the degree to which international law governs outer space is not entirely clear. The Outer Space Treaty affirms that activities in space are to be carried on "in accordance with international law," (UNGA 1967, Article III) but the fact that most existing international law at the time was developed for "terrestrial" purposes meant that it was not readily or directly applicable in every respect to this new paradigm of human endeavor. Moreover, the non-sovereignty aspect of outer space meant that any then existent national law (which, in any event, did not at that time specifically address space-related issues) would not prima facie apply to this frontier, and would not be the appropriate legal basis upon which to establish the initial framework for regulating the conduct of humankind's activities in outer space. It was clear, therefore, that, at the dawn of the development of "space law," specific international binding rules would be required to address the particular characteristics and legal categorization of outer space.

There is now a substantial body of law dealing with many aspects of the use and exploration of outer space, mainly codified in and evidenced by Treaties, United Nations General Assembly resolutions, national legislation, the decisions of national courts, bilateral arrangements, and determinations by Intergovernmental Organizations.

Five important multilateral treaties have been finalized through the auspices of UNCOPUOS (UNGA 1959). These are:

(i) 1967 Treaty on Principles Governing the Activities of States in the Exploration and Use of Outer Space, including the Moon and other Celestial Bodies (UNGA 1967)
(ii) 1968 Agreement on the Rescue of Astronauts, the Return of Astronauts and the Return of Objects Launched into Outer Space (UNGA 1968)
(iii) 1972 Convention on International Liability for Damage Caused by Space Objects (UNGA 1972)
(iv) 1975 Convention on Registration of Objects Launched into Outer Space (UNGA 1975)
(v) 1979 Agreement Governing the Activities of States on the Moon and other Celestial Bodies (UNGA 1979)

The United Nations Space Treaties were formulated in an era when only a small number of countries had space-faring capability. The international law of outer space thus, at least partially, reflects the political pressures imposed by the superpowers at that time.

The United Nations General Assembly has also adopted a number of space-related Principles, which include:

(i) 1963 Declaration of Legal Principles Governing the Activities of States in the Exploration and Use of Outer Space (UNGA 1963)

(ii) 1982 Principles Governing the Use by States of Artificial Earth Satellites for International Direct Television Broadcasting (UNGA 1982)
(iii) 1986 Principles Relating to Remote Sensing of the Earth from Outer Space (UNGA 1986)
(iv) 1992 Principles Relevant to the Use of Nuclear Power Sources in Outer Space (UNGA 1992)
(v) 1996 Declaration on International Cooperation in the Exploration and Use of Outer Space for the Benefit and in the Interest of All States, Taking into Particular Account the Needs of Developing Countries (UNGA 1996)

These sets of principles provide for the application of international law and the promotion of international cooperation and understanding in space activities, the dissemination and exchange of information through transnational direct television broadcasting via satellites and remote satellite observations of Earth, and general standards regulating the safe use of nuclear power sources necessary for the exploration and use of outer space. More recent "guidelines" have also been agreed relating to various other issues, including the problem of space debris (UNCOPUOS 2007).

In the context of the regulation of the exploration and use of outer space, these five sets of principles have therefore largely been considered as constituting "soft law" (Freeland 2012); however, a number of specific provisions may now represent customary international law.

Yet, despite all of these developments, it is clear that the existing legal and regulatory regime has not kept pace with the remarkable technological and commercial progress of space activities since 1957. This represents a major challenge in relation to the ongoing development of effective legal principles, all the more in view of the strategic and military potential of outer space in an era of globalization.

Principles Regulating the "Military" Uses of Outer Space

The Outer Space Treaty provides a number of general principles that are intended to restrict the military uses of outer space, including the requirement that activities in the exploration and use of outer space shall be carried out "in accordance with international law, including the Charter of the United Nations" (UNGA 1967, Article III). One of the primary drivers behind the inclusion of this provision was the concern among many States that outer space would become a new arena for international conflict. Article 2 of the Moon Agreement extends these sentiments by referring to "the Declaration on Principles of International Law concerning Friendly Relations and Co-operation among States in accordance with the Charter of the United Nations" (UNGA 1970).

As noted, many of the fundamental principles that formed the basis of the Outer Space Treaty were concluded at a time when the world was in the midst of uncertainty and mistrust, largely as a result of the prevailing geopolitical environment of the Cold War. At the time, there was a genuine fear held by the international community that outer space would be utilized for military purposes, as well as concern that it could perhaps ultimately become a theatre of war. In December 1958, the United Nations emphasized the need "to avoid the extension of present national rivalries into this new field" (UNGA 1958).

By 1961, the General Assembly had recommended that international law and the United Nations Charter should apply to "outer space and celestial bodies" (UNGA 1961). This was repeated in General Assembly Resolution 1962, which set out a number of important principles that were ultimately incorporated into the Outer Space Treaty (UNGA 1963). The specific reference to the United Nations Charter was considered to be important, given that the maintenance of international peace and security is the underlying principle of the system established under that instrument (UN 1945).

The sentiments underlying the United Nations Charter were strengthened further by the restrictions imposed in relation to nuclear weapons and weapons of mass destruction by Article IV of the Outer Space Treaty, although, as has been well documented by leading commentators, this provision in and of itself does not represent a complete restriction on the placement of weapons in outer space, nor of their use (Schrogl and Neumann 2009). Indeed, there have been, from time to time, proposals put forward to amend Article IV in order to enhance these restrictions, but this has not (yet) eventuated (Bogomolov 1993).

The "peaceful purposes" provision set out in Article IV of the Outer Space Treaty has been the subject of much analytical discussion as to its scope and meaning. While there exists some consensus among space law commentators – but not complete unanimity – that this provision is directed against "nonmilitary" rather than merely "nonaggressive" activities, the reality has been different. As noted, it is undeniable that, in addition to the many commercial, civilian, and scientific uses, outer space has and continues to be used for an expanding array of military activities. Gone are the days where there existed only two superpowers leading the world in space technology and exploration. The modern space age sees a new space race that exists among existing and emerging powers. It has now been established that competitive capabilities in outer space directly correlate to strategic and military capabilities on Earth.

In this context, if one were to adopt a hard-line pragmatic view, the "nonmilitary vs. nonaggressive" debate relating to the peaceful purposes requirement is a redundant argument, even though it represents an extremely important issue of interpretation of the strict principles of international space law. This then assumes that the militarization of space is a given, as much as it may pain international and space lawyers to admit this.

Moreover, Article 51 of the United Nations Charter – which confirms the "inherent right" of self-defense "if an armed attack occurs" – is also applicable to the legal regulation of outer space. Under the principles of public international law, this right remains subject to express legal limitations – the requirements of necessity and proportionality (ICJ 2003). Even where the right of self-defense is lawfully exercised, the State so acting will remain subject to the laws of war. While this is, in theory, uncontroversial, the difficulty is to determine precisely whether (and how) these fundamental principles can be applied to the unique legal and technological context of outer space.

This is particularly relevant given that the use of satellite technology already represents an integral part of the military strategy and the conduct of many armed conflicts. With space capabilities and technology developing at a rapid pace, any case of armed conflict in the twenty-first century and beyond will continuously involve the utilization of outer space. This is the direction and environment that the current political climate it operates in.

While the weaponization of space flies in the face of the principles of the Outer Space Treaty, it would be naive to ignore the realities of the twenty-first century. It would rather be more pragmatic to understand both what (and how) existing legal principles, including the rules of the laws of war, apply to any military activities involving outer space and to determine what needs to be done to provide, at least from a regulatory perspective, an appropriate framework to protect humankind in the future.

The Laws of War: General Principles

Over time, the international community has gradually agreed upon the principles of the laws of war (also known as international humanitarian law or the *jus in bello*) designed to provide legal constraints applicable to the conduct of armed conflict. It is only relatively recently that these minimum international standards have been developed to regulate *how, with what*, and *against whom* wars could be fought. It is interesting to note that "war" as a concept was declared illegal by the 1928 Pact of Paris (Kellogg–Briand 1929). Facing the realities of the modern world, it is evident that armed conflict still continues and has become more complex over time. This is not only due to the rapidly developing technology that has become increasingly integral to the conduct of armed conflict but also to the increasing role of non-State actors in conflict. As such, the scope for cataclysmic destruction and loss of life has increased due to the development of sophisticated weaponry, which includes the use of space technology.

The origins of the "laws and customs of war" were from the customary practices of armies on the battlefield and has since developed as a principal branch of international law (Henckaerts and Doswald-Beck 2005). The application of these customary practices was not uniform, and it therefore became evident that more formalized standards were required. A major step forward in the development of the rules of war, which inter alia limit the method and means of conducting warfare and also provide for classes of protected persons and protected objects, came with the Brussels Conference of 1874 and, more significantly, The Hague Peace Conferences of 1899 and 1907, which gave rise to some important standard-setting treaties that are still applicable today. The 1899 Conference concluded that "[t]he right of belligerents to adopt means of injuring the enemy is not unlimited" (U.K.T.S. 1899).

As time went on, further treaties followed, each specifying in greater detail the limits of what constituted (un)acceptable behavior in the context of armed conflict.

For instance, those provisions of the Hague Conventions that applied the laws of war to restrict the use of poison or poisoned weapons and asphyxiating gases were further extended by the 1925 Geneva Protocol (L.N.T.S. 1929).

The scars of the Second World War demonstrated the inadequacy of the existing rules, especially in respect to the treatment of civilians and noncombatants. The critical four 1949 Geneva Conventions were concluded to address these issues, (UNGA 1929, 1949a, b, c, d) and these were strengthened by the Additional Protocols of 1977 (UNGA 1949a, b, c, d, 1977). There have also been a growing number of other important treaties that have added to the corpus of international humanitarian law and the rules regulating armed conflict, particularly in relation to restrictions on specific weapons and means of warfare. Among these are several treaties that relate to the use of outer space, including those limiting the testing of nuclear and other weapons (UNGA 1963, 1972, 1996, 1998), as well as the 1977 Convention on the Prohibition of Military or Any Other Hostile Use of Environmental Modification Techniques (ENMOD) (UNGA 1977), which was the first instrument that dealt with deliberate destruction of the environment during warfare, although it also applies in time of peace.

International humanitarian law is now a well-developed area of international law, covering many aspects of terrestrial warfare. The obligations that arise under the fundamental principles of international humanitarian law, particularly those contained in The Hague Conventions and the Geneva Conventions and their Additional Protocols, have been reaffirmed by the United Nations Security Council (UNSC 2006). Moreover, the formation of multiple national, regional, and international enforcement mechanisms of justice – culminating in the International Criminal Court, the world's first permanent court of its kind – clearly indicates that the international community is determined that those senior officials (both military and political) who breach these established standards are to be held to account (Freeland 2006).

While there are many principles that have arisen through the evolution of the *jus in bello*, there are three specific concerns that form the basis of any decision to undertake an act of military engagement. They are the principles of distinction, military objective, and proportionality. Each of these is relevant to a consideration of the applicability of the laws of war to the use of outer space. Many commentators combine issues of distinction and military objective into a broader principle known as "discrimination." However, by differentiating between these two issues, it emphasizes the need to distinguish between civilians and combatants without reference to sometimes subjective considerations as to what constitutes a military target in the context of military advantage.

Distinction

Under the principle of distinction, deliberate attacks against civilians and noncombatants are prohibited. Article 48 of Additional Protocol I provides inter alia that "[i]n order to ensure respect for and protection of the civilian population . . . the Parties to a conflict shall at all times distinguish between the civilian population and combatants" (UNGA 1977). In addition, those engaged in armed conflict must not use weapons that are incapable of distinguishing between combatants and non-combatants. These represent fundamental concepts in the conduct of military activities and illustrate the strong linkages between the scope of international humanitarian law and the development of formal legal principles for the human rights of the individual. In his Dissenting Opinion in *Legality of the Threat or Use of Nuclear Weapons*, Judge Koroma pointed out that "both human rights law and international humanitarian law have as their *raison d'être* the protection of the individual as well as the worth and dignity of the human person, both during peacetime or in an armed conflict" (Koroma 1996).

Military Objective

The principle of military objective asserts that attacks not directed at a legitimate military target are prohibited. The important issue is the need to distinguish between civilian persons or objects and military objectives – comprising the elements of "effective contribution to military action" and "definite military advantage" specified in Article 52 of Additional Protocol I (UNGA 1977). Article 52 of Additional Protocol I provides inter alia that "[i]n so far as objects are concerned, military objectives are limited to those objects which by their nature, location, purpose or use make an effective contribution to military action and whose total or partial destruction, capture or neutralization, in the circumstances ruling at the time, offers a definite military advantage" (UNGA 1977).

Proportionality

The principle of proportionality promulgates that even when attacking a legitimate military objective, the extent of military force used and any injury and damage to civilians and civilian property should not be disproportionate to any expected military advantage. This standard demands an assessment of any potential "collateral damage" in the case of military action. However, in practice, it is often difficult to apply the proportionality principle given that different people and States ascribe differing relative "values" to military advantage *vis-à-vis* civilian injury and damage. One only need recall the Advisory Opinion in the *Legality of the Threat or Use of Nuclear Weapons*, where the International Court of Justice, while noting that the threat or use of a nuclear weapon should comply with the requirements of international law relating to armed conflict, in particular, the principles of international humanitarian law, could not say categorically that the threat or use of nuclear weapons would in every circumstance constitute a violation of international law (ICJ 1996).

The Relevance of the Laws of War to Outer Space

As noted above, the existing principles of international humanitarian law, as an integral part of international law, are, in theory, applicable to the military use of outer space. There is no specific "territorial" limitation to the laws and customs of war, which apply both to the area where the hostilities actually take place, as well as to other areas affected by those hostilities. If, for example, direct military action takes place in one area, but the effects of that action impact on civilians elsewhere, that represents a relevant consideration in determining whether such action is consistent with, for example, the principle of proportionality. As a result, any military activity that takes place in outer space will prima facie be subject to the *jus in bello* in relation not only to that direct action but also as to its effects and consequences elsewhere, including on Earth.

Having reached this conclusion, it is then necessary to determine whether the rules of war are "relevant" to activities in outer space. Looking at past and current events, the answer appears self-evident. During the Gulf War in 1990, the military value of space assets were first utilized for the conduct of warfare. Indeed, "Operation Desert Storm" is regarded as "the first space war" (Maogoto and Freeland 2007). It was recognized that the use of space technology would create an "integrated battle platform" to aid in the implementation of military strategy. Following the attacks of 11 September 2001, the United States Administration embarked on a policy designed to dominate the space dimension of military operations. This necessitates having the ability to protect critical US infrastructure and assets in outer space. The United States' approach to military advantage has now become heavily reliant upon space capabilities. The latest announcement from the Trump Administration establishing a "Space Force" further indicates the pivot of the United States military towards outer space, an area that it has increasingly designated as the newest "war-fighting domain."

Further, the reliance on space assets by other leading space-faring nations such as Russia and China illustrates the global space race that is developing in outer space. China has been rapidly consolidating its status as a space power, adding further to the tensions relating to space-related weapons technology. The first Gulf War demonstrated to China's military leadership the importance of high-tech integrated warfare platforms, and the ability of sophisticated space-based command, control, communications, and intelligence systems to link land, sea, and air forces. While one of the strongest motivations for China's space program appears to be political prestige and scientific exploration, China's space efforts will almost certainly contribute to the development of improved military space systems.

While not the leading space-faring power that it once was at the height of the Cold War, Russia still maintains a strong focus on developing its space capabilities and striving for technical dominance in this domain. The European Union has also identified outer space as "a key component for its European Defence and Security Policy" (Hagen and Scheffran 2005). Even for smaller countries such as Australia, the political landscape of national space policy highlights military and national security concerns (Freeland 2010a).

In this context, several commentators have opined that space warfare is, in fact, inevitable and cannot be avoided (De Angelis 2002). If these assertions turn out to reflect reality, the principles of the laws of war should be applied. However, it is not clear how this will be done in practice and what consequences will follow.

One complicating factor in this analysis is the increasing prevalence of what are referred to as "dual-use" satellites. The concept of a dual-use facility or resource – typically a commercial facility or resource that is also utilized by the military for military purposes – has become a common feature of contemporary technological society. This presents particular difficulties for those conducting armed conflict, since an asset that could prima facie be regarded as a legitimate military target on the basis of military objectives (see further below) might also – even at the same time – be operating for civilian/commercial uses. It is sometimes very difficult, or indeed impossible, to "quarantine" what is the civilian/commercial aspect of a facility from the military component. Additionally, military "customers" are now regularly utilizing commercial satellites to undertake military activities. In this respect, the language of the international law instruments relating to outer space has not hampered the increasing utilization of satellite technology for an expanding array of military activities.

One terrestrial example is illustrative of the difficulties of engaging in a straightforward legal analysis of any attack against such a facility. During the 1999 NATO bombing campaign directed towards forcing the Serbian military to leave Kosovo (known as "Operation Allied Force"), one deliberate target was the RTS Serbian TV and Radio Station in Belgrade. NATO missiles destroyed the station on 23 April 1999, with significant – and only civilian – loss of life. The bombing of the TV studio was part of a planned attack aimed at disrupting and degrading the C3 (Command, Control, and Communications) network of the Government of the Former Yugoslavia.

At a press conference on 27 April 1999, NATO officials justified this attack in terms of the dual military and civilian use to which the communication system was routinely put (NATO 2000; Freeland 2002). In essence, NATO stressed the dual-usage to which such communications systems were put, emphasizing the fact that "military traffic is ... routed through the civilian system" (NATO 2000).

This concept is, as noted above, also a common feature of space technology. A combination of factors – the increasing dependence by military and strategic forces within (the major) powers on the use of satellite technology; the inability of Governments to satisfy such demands for reasons associated either with costs or the lack of technological expertise (or both); and the advent of commercial satellite infrastructure and services that are responsive, technologically advanced, available, and appropriate to meet these demands – means that military "customers" are now regularly utilizing commercial satellites to undertake military activities. Given that such an increasingly important group of space assets used for military purposes are these dual-use satellites, one is also drawn to the question of whether, and in what circumstances, such a satellite can (ever) be regarded as a legitimate target of war.

The answer will depend upon a number of fundamental principles of international law. Clearly, the physical destruction of a satellite constitutes a use of force. Apart from a consideration of the principles in the United Nations Space Treaties, one would have to determine whether such an action represents a legitimate (at law) use of force, with the only possible justification being Article 51 of the United Nations Charter.

For example, assume that a combatant regards a dual-use satellite – in this scenario, a GPS or remote sensing satellite – as representing a legitimate military objective in accordance with the principles of distinction and military advantage. Even if this were a correct assessment, the principle of proportionality would also apply. Moreover, one could argue that, implicit in the principle of distinction is the obligation on the parties to a conflict to take "all feasible precautions" to protect civilians from the effects of an attack (Henckaerts and Doswald-Beck 2005). There would also be adverse environmental consequences (including significant space debris) resulting from the destruction of a satellite, and various international environmental law principles would therefore also be applicable in these circumstances.

One can certainly envisage that the deliberate destruction of such a satellite could, even if it does not result in any immediate civilian casualties, have a devastating impact on communities, countries, or even regions of the world. Millions of lives and livelihoods could, potentially, be affected, economies destroyed, and essential services incapacitated. Naturally, some of the consequences of such an attack may be difficult to foresee, but it would, one could argue, be regarded at the least as reckless. However, there is likely to be some uncertainty as to whether and how a "recklessness" test is to be applied in such a situation (Freeland 2002).

Overall, given the unique nature of outer space, the fundamental principles of the laws of war – developed to regulate *terrestrial* warfare and armed conflict – are probably neither sufficiently specific nor entirely appropriate for military action in outer space. Even though every effort should be made to apply the existing principles as directly as possible, the largely unprecedented nature of such an environment means that more specific rules will almost certainly be required, if they are to provide a comprehensive framework to properly protect humanity from the disastrous consequences of outer space becoming another theatre of warfare.

Regulating the Threat of Space Warfare: Some Recent Initiatives

There is reluctance among the major powers to address the question of international space law regulation through the use of binding treaty instruments. As such, a voluntary soft law approach has been preferred, such as utilizing "transparency and confidence building measures" (TCBMs). A principal TCBM in the area of space regulation had been the (draft) International Code of Conduct for Outer Space Activities (CoC), which was initially developed as a European initiative but has since become broader in scope (European External Action Service 2014). Discussions in 2015 intended to facilitate agreement on this instrument failed, with the instrument now not seriously referred to in relevant fora.

Space debris, an issue that was addressed in the draft CoC, is increasingly becoming a signification risk to operations in outer space as on-orbit collisions can have catastrophic repercussions and leads to further creation of debris. Related to the issue of space debris is, of course, the issue of maintaining the integrity of space assets, both in terms of adhering to measures on debris control and mitigation as well as remediation. Equally it serves to minimize the possibility that a state would destroy another state's satellite (and in the process almost certainly create additional space debris).

Another approach was that of the Draft Treaty on the Prevention of the Placement of Weapons in Outer Space, the Threat or Use of Force Against Outer Space Objects (PPWT). Since the early 1980s, there have been a series of United Nations General Assembly (UNGA) Resolutions on the specific issue of preventing an arms race in outer space. Such Resolutions focused on addressing this issue and drew further attention of the international community to the need to respond to various military initiatives taken by major space powers in their use of outer space. Ostensibly responding to these calls, in February 2008, the then Minister of Foreign Affairs of the Russian Federation, Sergey Lavrov, presented a draft document headed "Treaty on the Prevention of the Placement of Weapons in Outer Space, the Threat or Use of Force Against Outer Space Objects" to the 65 members attending the Plenary Meeting of the United Nations Conference on Disarmament (CD) in Geneva. The PPWT had been developed by Russia and China, two of the major space superpowers in the world.

The formal submission of the PPWT to the CD followed several years of diplomatic discussion, directed towards agreeing the terms of legally binding rules addressing the dangers of an arms race in space. In general terms, the PPWT focused on three primary obligations of States Parties, each of which are specified in Article II: not to place in orbit around the Earth, install on celestial bodies, or station in outer space in any other manner "any objects carrying any kind of weapons"; not to "resort to the threat or use of force against any outer space objects"; and not to encourage another State(s) or Intergovernmental Organization to "participate in activities prohibited" by the PPWT.

In responding to the PPWT, the United States Administration has continually reiterated that it opposes any treaty that seeks "to prohibit or limit access to or use of space," adding that, in any event, such a treaty would be impossible to enforce. Indeed, verification measures in relation to the obligations of State Parties under the PPWT would undoubtedly prove to be difficult and complex – though perhaps not impossible – to implement. Instead, the United States has indicated that it prefers "discussions aimed at promoting transparency and confidence building measures."

Overall however, and despite its shortcomings, the PPWT has raised issues of crucial importance to the future use and exploration of outer space, indeed to the very nature of space activities. It was therefore unfortunate that the document was so quickly rejected out of hand by the United States. Indeed, in February 2008, barely a week after Russia and China submitted the PPWT to the CD, the United States fired an SM-3 missile from USS Lake Erie that destroyed a failed satellite approximately 150 km above the Pacific Ocean. Although the United States argued that this action

5 The Laws of War in Outer Space

was necessary to prevent the fuel tank of the satellite – containing hydrazine – from breaking up and polluting the atmosphere, others have suggested that this was simply a "test" by the United States of its anti-satellite capability.

Yet, despite these setbacks, the formal submission of the PPWT by two of the world's space superpowers has had the effect of generating further momentum in relation to other initiatives to address the impending perils associated with the possible weaponization of space, for instance, the draft CoC.

While there are obvious benefits in developing greater trust between the space powers in issues relating to space security, the danger of the aforementioned TCBMs is that they are nonbinding TCBM. For all practical purposes, they considered as the "end game" on this issue, so that the formalization of binding obligations may *never* emerge.

At its core, the draft CoC provisions, for example, are merely guidelines or recommendations that do not have the force of law, unless they are to be regarded as reflecting rules of customary international law, itself a very difficult assertion to substantiate in the absence of, say, a ruling by the International Court of Justice. This approach appears inadequate to meet the complex risks associated with the continued development of space-related weapons.

Conclusion: Perspectives on the Way Forward

The above brief discussion gives rise to several conclusions and reflections: first, present indications suggest that there is an increasing likelihood that outer space will not only be used to facilitate armed conflict (as it already is) but will ultimately become a theatre of war. The tendency of the major superpowers to unequivocally rely on space technology has spurred a space weapons race, despite the efforts of the international community. Even though the United States may currently claim space superiority, leading nations such as China and Russia closely follow behind and have access to equally sophisticated (and potentially devastating) space weapons technology.

Secondly, the development of such technology and the increasing range of military uses of outer space heighten the dangers of a space war. The proliferation of crucial military space assets means that, from a military and strategic viewpoint, the disabling or destruction of satellites used by another country may be perceived as giving rise to very significant advantages. The fact that it has not happened in the past is no reason to assume that we will never see a space conflict.

Thirdly, all countries in the world are highly dependent on space technology to maintain and improve their livelihood and standard of living. The nonmilitary uses of space have become vital aspects of any community's survival. At the same time, however, many of the satellites providing these commercial and civilian services are dual-use, in that they are also utilized for military and strategic purposes. This raises difficult questions about the "status" of such assets under the rules of war – particularly as to whether they may, under certain circumstances, be regarded as legitimate military objectives.

Fourthly, the Outer Space Treaty, which also reflects customary international law, specifies that the rules of international law apply to the use and exploration of outer space. These include not only the *jus ad bellum* principles regulating the use of force but also the principles of the laws of war. Respect for these rules is absolutely vital for the safety and security of humankind, as well as the interests of future generations. However, with the exception of those treaties that seek to ban the use and testing of certain types of weapons, as discussed there are many uncertainties that arise when one seeks to apply, in particular, the laws of war to a space conflict. The consequences of a space war are potentially so great and unknown that one cannot be sure as to exactly how these existing rules are to apply.

Fifthly, if we are to avoid "gray areas" in the law, it is therefore necessary to develop specific and clear rules and standards that reduce the weaponization of outer space, as well as any form of conflict in the region of outer space and against space assets. The Outer Space Treaty, as well as the other United Nations Space Treaties, do not currently provide stringent rules or incentives to prevent an arms race in outer space, let alone a conflict involving (and perhaps "in") space. This may, therefore, require additional specific legal regulation of outer space that is directly applicable to armed conflict involving the use of space technology. The position is, of course, further complicated by the applicability of the right of self-defense, a right that States will never abandon.

As part of these new rules, clear definitions must be developed for concepts such as "space weapons," "peaceful purposes," and "military uses." Moreover, the fundamental issue of "where space begins" should be definitively resolved, so as to counter any arguments that outer space is, in fact, an area akin to the territory of a State for the purposes of national security.

Sixthly, at the same time, careful consideration must be given to the application of the principles of the laws of war to this new paradigm of potential conflict. While, of course, there already exist very well established fundamental rules regulating terrestrial warfare, it is not clear whether these are entirely sufficient to protect humanity from the consequences of any future "space wars." Ultimately, the legal regulation of outer space is not likely to take the form of binding treaty obligations which supplement the existing laws of war (as they may apply to such activities) in the short-medium term, but rather will be on a voluntary nonbinding basis. This illustrates the sensitivities related to further regulating outer space activities that relate to issues of national security interests, particularly those of the major space powers.

It seems that a "softly, softly" approach involving the development of TCBMs is the preferred strategy, particularly of the United States, but this brings with it much more uncertainty, a lack of formal enforcement capability and enforcement mechanisms, and the possibility of undue flexibility of approach by the main global players. It is imperative on all stakeholders to find a path forward, in order to meet the challenges of the twenty-first century. The existing international regulatory framework, while important, cannot alone stand up to the unknowns that military-related space technology imposes upon us. It is critical that an appropriate and acceptable regulatory regime is found; however, the mechanism and conformity that this might take still remains unclear, particularly considering the distinct complexities that outer space presents.

References

Bogomolov V (1993) Prevention of an arms race in outer space: the deliberations in the Conference on Disarmament in 1993. J Space Law 21(2):141

Cassese A (2005) International law, 2nd edn. Oxford University Press, Oxford/New York, p 95

Comprehensive Nuclear Test-Ban Treaty (not yet in force), 1998, United Nations, Dept. for Disarmament Affairs and Dept. of Public Information

Convention on Civil Aviation ("Chicago Convention"), 7 December 1944, International Civil Aviation Organization (ICAO), 15 U.N.T.S. 295 (1944)

Convention on the Law of the Sea (UNCLOS), 1982, Article 137(1), United Nations Division for Ocean Affairs and the Law of the Sea, United Nations Treaty Series

Convention on the Prohibition of Military or Any Other Hostile Use of Environmental Modification Techniques (ENMOD), 1977, United Nations, 16 I.L.M. 88

Convention on the Prohibition of the Development, Production and Stockpiling of Bacteriological (Biological) and Toxin Weapons and on their Destruction, 1972, United Nations, 1015 U.N.T.S. 163

Convention on the Regulation of Aerial Navigation ("Paris Convention"), 1919, International Commission for Air Navigation (ICAN), 11 L.N.T.S. 173

De Angelis IM (2002) Legal and political implications of offensives actions from and against the space segment. Proc Colloq Law Outer Space 45:197

Declaration of the first meeting of equatorial countries or 'Bogota Declaration', 1976, Heads of Delegations, Japan Aerospace Exploration Agency

Draft International Code of Conduct for Outer Space Activities, 2014, European External Action Service. http://www.eeas.europa.eu/non-proliferation-and-disarmament/pdf/space_code_con duct_draft_vers_31-march-2014_en.pdf. Accessed 18 Feb 2019

Final Report to the Prosecutor by the Committee Established to Review the NATO Bombing Campaign Against the Federal Republic of Yugoslavia, 2000, paragraph 72 (2000) 39 I.L.M. 1257 (NATO Report)

Freeland S (2002) The bombing of Kosovo and the Milosevic Trial: reflections on some legal issues. Aust Int Law J 150:165–168

Freeland S (2006) How open should the door be?: Declarations by non-states parties under Article 12(3) of the Rome Statute of the International Criminal Court. Nord J Int Law 75(2):211

Freeland S (2010a) Sensing a change? The re-launch of Australia's space policy and some possible legal implications. J Space Law 36(2):381

Freeland S (2010b) Fly me to the Moon: How will international law cope with commercial space tourism? Melb J Int Law 11(1):90

Freeland S (2012) The role of 'soft law' in public international law and its relevance to the international legal regulation of outer space. In: Marboe I (ed) Soft law in outer space: the function of non-binding norms in international space law, vol 9, Bohlau Publishing (Austria)

Freeland S, Jakhu R (2010) Article II. In: Hobe S, Schmidt-Tedd B, S. Hobe S, Schmidt-T B, Schrogl K.-U (eds) Cologne Commentary on Space Law, vol. 1. Outer Space Treaty, Carl Heymanns Verlag, Köln

General Treaty for the Renunciation of War ("Kellogg–Briand Pact"), U.K.T.S., 1929, Article I, p 29

Geneva Convention for the Amelioration of the Condition of the Wounded, Sick and Shipwrecked Members of Armed Forces at Sea, 1949, United Nations, 75 U.N.T.S. 85 (UNGA 1949a)

Geneva Convention for the Amelioration of the Condition of the Wounded and Sick in Armed Forces in the Field, 1949, United Nations, 75 U.N.T.S. 31 (UNGA 1949b)

Geneva Convention Relative to the Protection of Civilian Persons in Time of War, 1949, United Nations, 75 U.N.T.S. 287 (UNGA 1949c)

Geneva Convention Relative to the Treatment of Prisoners of War, 1929, United Nations, 75 U.N.T. S. 135 (UNGA 1949d)

Hagen R, Scheffran J (2005) International space law and space security – expectations and criteria for a sustainable and peaceful use of outer space. In: Benkö M, Schrogl K-U (eds) Space law:

current problems and perspectives for future regulation, vol 273, Eleven International Pub. (Utrecht, The Netherlands), pp 281–282

Hague Convention II, 1899, [1907] Supp 1 American Journal of International Law 129, United Nations

Harris DJ (2004) Cases and materials on international law, 6th edn. Sweet & Maxwell, London, p 252

Henckaerts J-M, Doswald-Beck L (2005) Customary international humanitarian law – Cambridge University Press (Cambridge), vol 1: Rules, xxv, p 70

Kopal V (2006) Comments on the issue of "Adequacy of the current legal and regulatory framework relating to the extraction and appropriation of natural resources of the Moon". In: Policy and law relating to outer space resources: examples of the Moon, Mars, and other celestial bodies, (28–30 June 2006) workshop proceedings. McGill Institute of Air & Space Law, Montreal, pp 227–230

Legality of the Threat or Use of Nuclear Weapons, 1996, p 577, 1 ICJ Rep. 245, Judge Koroma Dissenting Opinion

Lyall F, Larsen PB (2009) Space law: a treatise. Ashgate, Farnham, pp 3–9

Maogoto J, Freeland S (2007) Space weaponization and the United Nations Charter: a thick legal fog or a receding mist? Int Lawyer 41(4):1091–1107

Moon Agreement, 1979, United Nations Treaty Collection. 1363 U.N.T.S. 3

North Sea Continental Shelf Cases (Federal Republic of Germany v. Denmark and Federal Republic of Germany v. The Netherlands), Judgment, 1969 p 230, I.C.J. 3, Dissenting Opinion of Judge Lachs

Outer Space Treaty, 1967, United Nations Treaty Collection. 610 U.N.T.S. 205

Protocol for the Prohibition of the Use in War of Asphyxiating, Poisonous or Other Gases, and of Bacteriological Methods of Warfare, 1929, United Nations, xciv L.N.T.S. 65–74

Protocol I Additional to the Geneva Conventions of August 12, 1949, and relating to the Protection of Victims of International Armed Conflicts, (Additional Protocol I), 1977, United Nations, 16 I.L.M. 1391

Protocol II Additional to the Geneva Conventions of 12 August 1949 and relating to the Protection of Victims of Non-International Armed Conflicts, (Additional Protocol II), 1977, United Nations, 16 I.L.M. 1442

Resolution 26/25, United Nations General Assembly, (XXV), on the Declaration on Principles of International Law concerning Friendly Relations and Co-operation among States in accordance with the Charter of the United Nations (1970)

Resolution 1348, United Nations General Assembly, (XIII) on Questions on the Peaceful Uses of Outer Space, preambular paragraph 3, (1958)

Resolution 1472, United Nations General Assembly, (XIV), on International co-operation in the peaceful uses of outer space, (1959)

Resolution 1674, United Nations Security Council, on the Protection of civilians in armed conflict, paragraph 6, (2006)

Resolution 1721, United Nations General Assembly, (XVI), on International co-operation in the peaceful uses of outer space, paragraph 1(a), (1961)

Resolution 1962, United Nations General Assembly, (XVIII), on the Declaration of Legal Principles Governing the Activities of States in the Exploration and Uses of Outer Space, paragraph 4 (1963), United Nations

Resolution No 37/92, United Nations General Assembly, on the Principles Governing the Use by States of Artificial Earth Satellites for International Direct Television Broadcasting, (1982)

Resolution No 41/65, United Nations General Assembly, on the Principles relating to Remote Sensing of the Earth from Outer Space, (1986)

Resolution No 47/68, United Nations General Assembly, on the Principles relevant to the Use of Nuclear Power Sources in Outer Space (1992)

Resolution No 51/122, United Nations General Assembly, on the Declaration on International Cooperation in the Exploration and Use of Outer Space for the Benefit and in the Interest of All States, Taking into Particular Account the Needs of Developing Countries (1996)

Ricks T (2001) Space is playing field for newest war game; air force exercise shows shift in focus. The Washington Post, p A1

Schrogl K-U, Neumann J (2009) Article IV. In: Hobe S, Schmidt-T B, Schrogl K.-U. (eds), Cologne Commentary on Space Law, Volume I: Outer Space Treaty, Carl Heymanns Verlag, Köln

The Caroline Case, ICJ Rep. p 161. (2003). 29 B.F.S.P. 1137–1138; 30 B.F.S.P. 195–196

Treaty Banning Nuclear Weapon Tests in the Atmosphere, in Outer Space and Under Water, 1963, United Nations, 480 U.N.T.S. 43

Treaty on Principles Governing the Activities of States in the Exploration and Use of Outer Space, including the Moon and other Celestial Bodies, 1967, United Nations General Assembly

UNCOPUOS (2007) Report of the Scientific and Technical Subcommittee on its forty-fourth session. A/AC.105/890, Annex 4, 42. http://www.oosa.unvienna.org/pdf/reports/ac105/AC105_890E.pdf. Accessed 18 Feb 2019

United Nations, Charter of the United Nations, 24 October 1945, 1 U.N.T.S. XVI

United Nations Treaty Collection. 672 U.N.T.S. 119 (1968)

United Nations Treaty Collection. 961 U.N.T.S. 187 (1972)

United Nations General Assembly, Convention on Regulation of Objects Launched into Outer Space, United Nations Treaty Collection. 1023 U.N.T.S. 15 (1975)

Further Reading

See Lyall F, Larsen PB (2009) Space law: a treatise. Ashgate, Farnham, pp 3–9 for a summary of the main academic theories relating to 'space law' in the period prior to the launch of Sputnik I

See United Nations General Assembly Resolution 1472 (XIV) on International Cooperation in the Peaceful Uses of Outer Space (12 December 1959). UNCOPUOS currently has 87 Members, which, according to its website, means that it is 'one of the largest Committees in the United Nations.' For more see http://www.unoosa.org/oosa/en/members/index.html. Last accessed 18 Feb 2019

Arms Control and Space Security

Stacey Henderson

Contents

Introduction	96
Outer Space Treaty	96
Moon Agreement	98
Arms Control in Outer Space: Historic and Current Efforts	99
Partial Test Ban Treaty	99
Anti-Ballistic Missile Treaty	99
SALT II	100
United Nations General Assembly	100
Draft Treaty on Prevention of the Placement of Weapons in Outer Space and of the Threat or Use of Force Against Outer Space Objects	102
International Code of Conduct for Outer Space	106
Domestic Arms Control for Outer Space	107
Conclusions	107
References	109

Abstract

With the increasing reliance on space assets for civilian, commercial, and military purposes, there is a need to safeguard their full operability and enhance their security. The uncontrolled use of weapons in outer space has the potential to increase space debris, destroy critical space assets and impair their ongoing functionality, and negatively impact space security. The current international law regime applicable to outer space does not prohibit the placement or use of all weapons in outer space, prohibiting only nuclear weapons and weapons of mass destruction, and is not currently capable of preventing a conventional arms race in outer space.

S. Henderson (✉)
Adelaide Law School, The University of Adelaide, Adelaide, SA, Australia
e-mail: stacey.henderson@adelaide.edu.au

Introduction

At the commencement of the space age, activities in outer space were dominated by two space powers: the United States of America (USA) and the Union of Soviet Socialist Republics (USSR). Today, there are many states, an intergovernmental organization (the European Union), and private actors active in the use and exploration of outer space. As the number of space actors has grown, so too has the reliance on space assets, which are now a critical part of everyday life. The increased dependence on space assets brings with it an increased vulnerability and the need to ensure their security. The continuing growth in the number of space actors and space activities suggests that debates over the placement and use of weapons in outer space will only increase in the future (López 2012).

Arms control has long been discussed between states in the context of space security. Indeed, "the research, development, and testing of aggressive space weapons can be traced back to the 1950s" (Su 2017). Prior to the negotiations on the 1967 Treaty on Principles Governing the Activities of States in the Exploration and Use of Outer Space, including the Moon and Other Celestial Bodies (Outer Space Treaty), the USA and the USSR had agreed "not to station in outer space any objects carrying nuclear weapons or other kinds of weapons of mass destruction" (General Assembly Resolution 1884, 1963). Yet the current international legal regime governing outer space does not prohibit all weapons in outer space, prohibiting only nuclear weapons and other weapons of mass destruction. This chapter outlines the international law applicable to arms control in outer space and examines historic and recent efforts to limit the placement and use of weapons in outer space and to prevent an arms race in outer space.

Outer Space Treaty

The Outer Space Treaty is the fundamental international law treaty regulating activities in outer space. All space-faring states are party to the Outer Space Treaty; the Outer Space Treaty has 109 States Parties, with a further 23 states having signed the treaty, but not yet ratified it (as at 19 December 2019).

The Outer Space Treaty is the only one of the four main international space law treaties to specifically address arms control; the Rescue and Return Agreement (1968 *Agreement on the Rescue of Astronauts, the Return of Astronauts and the Return of Objects Launched into Outer Space*), the Liability Convention (1972 *Convention on International Liability for Damage Caused by Space Objects*), and the Registration Convention (1976 *Convention on Registration of Objects Launched into Outer Space*) are all silent on this issue.

Article IV of the Outer Space Treaty directly addresses the issue of weapons in outer space. Article IV provides:

> States Parties to the Treaty undertake not to place in orbit around the Earth any objects carrying nuclear weapons or any other kinds of weapons of mass destruction, install such weapons on celestial bodies, or station such weapons in outer space in any other manner.

> The Moon and other celestial bodies shall be used by all States Parties to the Treaty exclusively for peaceful purposes. The establishment of military bases, installations and fortifications, the testing of any type of weapons and the conduct of military manoeuvres on celestial bodies shall be forbidden. The use of military personnel for scientific research or for any other peaceful purposes shall not be prohibited. The use of any equipment or facility necessary for peaceful exploration of the Moon and other celestial bodies shall also not be prohibited.

However, Article IV contains ambiguities, "partly due to the lack of clear terminological definitions, and partly due to subsequent technical developments" (Su 2017).

What would amount to "placement" in outer space is not defined or clarified in the Outer Space Treaty. Some are of the view that a full rotation in orbit is not required in order to amount to placement in orbit (Hobe et al. 2010), whereas others take the view that nuclear armed "pop-up" anti-satellite (ASAT) interceptors that ascend directly to their targets without entering into orbit are not placed into orbit and therefore do not fall within the Article IV prohibition (Bhat and Mohan 2009).

The term "peaceful purposes" is also not defined in the Outer Space Treaty. During negotiations for the Outer Space Treaty, it became clear that neither the USA not the USSR wanted the Treaty to include a definition of peaceful purposes that might have imposed limitations on their future uses of outer space (Sullivan 1990). As such, the term was left undefined. There are ongoing debates about whether "exclusively for peaceful purposes" means that celestial bodies are to be completely nonmilitarized in the sense that they cannot be used for any military activities, or whether peaceful purposes means that they may be used for non-aggressive military activities (Su 2010). The latter view is more consistent with the terms of the Outer Space Treaty, given that paragraph 2 of Article IV details prohibited and permissible military activities and uses of celestial bodies. If all military activities were prohibited in outer space, there would be no need for paragraph 2 of Article IV.

The Outer Space Treaty prohibits the placement of nuclear weapons and other weapons of mass destruction (generally accepted as including chemical and biological weapons) in outer space or on celestial bodies. It "does not explicitly restrict (and so allows) other military-related activities in outer space, such as the deployment of military satellites and conventional weapons in outer space" (Tronchetti and Hao 2015). The Outer Space Treaty does not prohibit "the deployment of offensive devices in orbit," provided they are not nuclear or other weapons of mass destruction, nor does it prohibit "the development, storage, and testing of ground-based ASAT devices" (Tronchetti and Hao 2015). The Outer Space Treaty is silent on the use of Earth-based weapons against space objects. It is also unclear whether nuclear devices used to deflect near-Earth objects threatening the Earth should be characterized as weapons and fall within the Article IV prohibition (Su 2017). The ease with which nuclear devices for near-Earth object diversion could be re-purposed for use as nuclear weapons, and the consequential risks of such re-purposing, suggests that they should be covered by the prohibition.

Article IV, paragraph 2, of the Outer Space Treaty imposes restrictions on the military use of celestial bodies and, in the context of arms control, prohibits the testing of any type of weapon on celestial bodies. However, the prohibition on weapons testing does not extend to outer void space and is limited to a prohibition on weapons testing on celestial bodies.

Article IX of the Outer Space Treaty obliges States Parties to "conduct all their activities in outer space...with due regard to the corresponding interests of all other States Parties to the Treaty." This Article may have incidental application to the testing of Earth-based ASAT weapons, as it imposes obligations on States Parties to avoid harmful contamination, harmful interference, and to undertake appropriate international consultations (Su 2017). However, the standard for avoiding harmful contamination and what is required under the obligation to undertake international consultations are both highly subjective, and Article IX has not been invoked by States in practice when ASAT weapons have been tested, even in situations where large amount of debris were created (Su 2017). Based on subsequent state practice, Article IX must therefore be interpreted as not prohibiting or restricting the testing of Earth-based ASAT weapons.

The Outer Space Treaty regulates space activities and sets the foundation for international space law. The Outer Space Treaty prohibits States Parties from deploying or using nuclear weapons or other weapons of mass destruction in outer space. Additionally, the Outer Space Treaty prohibits States Parties from conducting weapons tests of any kind of weapon on celestial bodies. However, the Outer Space Treaty does not prohibit States Parties from deploying or using conventional weapons in outer space, nor does it prohibit the use of Earth-based weapons against space objects. Additionally, the verification mechanisms in the Outer Space Treaty are very weak, being limited to the opportunity to observe the flight of space objects (Article X, Outer Space Treaty) and a reciprocal right of visitation (Article XII, Outer Space Treaty), falling far short of the verification mechanisms in most arms control agreements.

Moon Agreement

The Moon Agreement (1984 *Agreement Governing the Actvities of States on the Moon and Other Celestial Bodies*) mirrors the provisions of the Outer Space Treaty in relation to arms control and is similarly silent in relation to conventional weapons. The Moon Agreement has only 18 States Parties (as at 19 December 2019): Armenia, Australia, Austria, Belgium, Chile, Kazakhstan, Kuwait, Lebanon, Mexico, Morocco, the Netherlands, Pakistan, Peru, Philippines, Saudi Arabia, Turkey, Uruguay, and Venezuela. A further four states – France, Guatemala, India, and Romania – have signed, but not ratified, the Moon Agreement. The Moon Agreement remains unratified by major space-faring states, such as China, Russia, and the USA.

Article 3, paragraph 3 of the Moon Agreement provides:

> States Parties shall not place in orbit around or other trajectory to or around the Moon objects carrying nuclear weapons or any other kinds of weapons of mass destruction or place or use such weapons on or in the Moon.

Further, Article 3, paragraph 4 of the Moon Agreement prohibits the "establishment of military bases, installations and fortifications, the testing of any type of weapons and the conduct of military manoeuvres on the Moon." The Moon Agreement does not impose any additional obligations in relation to arms control on its States Parties beyond those specified in the Outer Space Treaty.

Arms Control in Outer Space: Historic and Current Efforts

Partial Test Ban Treaty

The 1963 Treaty Banning Nuclear Weapon Tests in the Atmosphere, in Outer Space, and under Water (Partial Test Ban Treaty) entered into force on 10 October 1963. It is one of the most widely accepted international arms control treaties, with 125 States Parties as at 19 December 2019. After signing the Partial Test Ban Treaty, the then two space faring states, the USA and the USSR, reaffirmed "that they did not intend to station any objects carrying WMDs [weapons of mass destruction] in outer space" (Su 2013).

Article I of the Partial Test Ban Treaty provides that

> Each of the Parties to this Treaty undertakes to prohibit, to prevent, and not to carry out any nuclear weapon test explosion, or any other nuclear explosion, at any place under its jurisdiction or control:
> (a) in the atmosphere; beyond its limits, including outer space; or under water, including territorial waters or high seas

The Partial Test Ban Treaty is limited to a prohibition of nuclear explosions in outer space, including nuclear weapon test explosions. It does not contain any restrictions on conventional arms in outer space.

Anti-Ballistic Missile Treaty

The 1972 Anti-Ballistic Missile Treaty (*Treaty between the United States of America and the Union of Soviet Socialist Republics on the limitation of anti-ballistic missile systems*) was a bilateral agreement negotiated between the USA and the USSR as part of the Strategic Arms Limitation Talks (SALT I). The Treaty expressly prohibited the development, testing, and deployment of sea-based, air-based, space-based, and mobile land-based anti-ballistic missile systems (Article V, ABM Treaty). To provide assurance of compliance with the Treaty's provisions, each Party was to use national technical means of verification, such as satellites, and was not to use deliberate concealment measures or interfere with such verification (Article XII, ABM Treaty). The USA withdrew from the Anti-Ballistic Missile Treaty in June 2002, bringing the treaty to an end.

SALT II

The 1979 Strategic Arms Limitation Talks II Treaty (SALT II) prohibited "systems for placing into Earth orbit nuclear weapons or any other kind of weapons of mass destruction, including fractional orbital missiles" (Larsen 2017). The Treaty also restricted the USA and the USSR from developing more than one new type of intercontinental ballistic missile each, enforcing that limitation by allowing each party to conduct flight testing of only one new such weapon (Koplow 2018). Verification of the SALTII Treaty obligations was again by national technical means of verification, including reconnaissance satellites. The SALT II Treaty expired on 31 December 1985.

United Nations General Assembly

Prevention of an Arms Race in Outer Space

Since 1981, the General Assembly has adopted annual resolutions on the Prevention of an Arms Race in Outer Space (PAROS). The 1988 resolution reaffirmed "that general and complete disarmament under effective international control warrants that outer space shall be used exclusively for peaceful purposes and that it shall not become an arena for an arms race" and called for complete disarmament and effective verification measures (A/RES/43/70, 1988). The resolution passed with a vote of 154 to 1, with only the USA opposing the resolution, which it has repeatedly done on an annual basis.

The 2019 resolution "Further practical measures for the prevention of an arms race in outer space" echoed previous PAROS resolutions and was passed with a vote of 124 in favor to 41 against, with 10 abstentions:

> *Recognizing* the catastrophic consequences of the weaponization of outer space or any military conflicts in outer space and that the prevention of an arms race in outer space would avert a grave danger for international peace and security,
>
> . . .
>
> *Recognizing* that, while the existing international treaties related to outer space and the legal regime provided for therein play a positive role in regulating outer space activities, they are unable to fully prevent the placement of weapons in outer space and therefore avert an arms race there, and that there is a need to consolidate and reinforce this regime,*Expressing serious concern* over the plans declared by certain States that include the placement of weapons, in particular strike combat systems, in outer space,
>
> . . .
>
> 6. *Urges* the international community to continue its efforts aimed at preventing an arms race, including the placement of weapons, in outer space, with a view to maintaining international peace and strengthening global security. (UN Doc A/C.1/74/L.58/Rev.1, 2019)

The PAROS resolutions "typically recognize that the legal regime applicable to outer space is insufficient to guarantee the prevention of an arms race in outer space and there is an urgent need to consolidate and reinforce it, since prevention of an arms race in outer space would avert a grave danger for international peace and security" (Su 2013). Although recognizing the gaps in the existing international space law regime, and the inability of the current treaty regime to prevent an arms race in outer space, the annual PAROS resolutions do not attempt to establish any normative obligations in relation to arms control and space security. The voting pattern on the PAROS resolutions has significantly changed over the years, indicating that states are now markedly less committed to this ideal than in the past.

No First Placement of Weapons in Outer Space

In October 2004, Russia announced a policy of "no first deployment of weapons in outer space" at the First Committee of the United Nations General Assembly, calling on all other space-faring and space-using states to join them in this pledge (Su 2017). The initiative was supported by other Member States of the Collective Security Treaty Organisation (Armenia, Belarus, Kazakhstan, Kyrgystan, and Tajikistan), Brazil, Indonesia, Sri Lanka, Argentina, and Cuba. These states have declared that they "will not in any way be the first to place weapons of any kind in Outer Space, that they will make all possible efforts to prevent Outer Space from becoming an arena for military confrontation and to ensure security in Outer Space activities" (Su 2017).

In 2014, Russia submitted a draft resolution on no first placement of weapons in outer space to the First Committee of the 69th session of the United Nations General Assembly (A/C.1/69/L.14, 2014a). The resolution was adopted by the General Assembly on 2 December 2014 by a vote of 126 in favor and 4 against (Georgia, Israel, Ukraine, and the USA), with 46 abstentions. The US Permanent Representative to the Conference on Disarmament stated that the resolution "does not adequately define space weapons, leaving the nonbinding resolution difficult to enforce, or for compliance with the agreed-upon measures to be verified" (Chow 2018). In explaining its vote, the USA stated that "Russia's military actions do not match their diplomatic rhetoric," observed that any space "object with maneuvering capabilities can in theory be used for offensive purposes," and noted that the lack of a "common understanding of what we mean by a space weapon [in the resolution]…would increase mistrust or misunderstanding with regard to the activities and intentions of States" (Plath 2018).

The "No First Placement of Weapons in Outer Space" resolution (GA Res 69/32, 2014b) "encourages all States, especially space-faring nations, to consider the possibility of upholding as appropriate a political commitment not to be the first to place weapons in outer space" (para 5). The resolution relates only to deployment of weapons in outer space and does not restrict research or development of such weapons. Further, the commitment on pledging states is conditional – merely not to be the first to place weapons in outer space; the deployment of weapons in outer space by any other entity, another state or private actor, would free the pledging states from their commitment not to place weapons in outer space. Since the adoption of the No First Placement of Weapons in Outer Space resolution, 22 States have made such a commitment (as at 31 October 2019): Argentina, Armenia, Belarus, Bolivia, Brazil, Cambodia, Cuba, Ecuador, Guatemala, Indonesia, Kazakhstan, Kyrgyzstan, Nicaragua, Pakistan, Russia, Sri Lanka, Suriname, Tajikistan, Uruguay, Uzbekistan, Venezuela, and Vietnam (UN Doc A/C.1/74/L.58/Rev.1, 2019).

The no first placement of weapons in outer space has remained an annual agenda item at the General Assembly since 2014. The most recent resolution in 2019 was passed with a vote of 123 in favor to 14 against, with 40 abstentions. While not imposing any normative obligations on states, these resolutions encourage states to make the political commitment not to be the first to place weapons of any kind in outer space.

Draft Treaty on Prevention of the Placement of Weapons in Outer Space and of the Threat or Use of Force Against Outer Space Objects

The Draft Treaty on Prevention of the Placement of Weapons in Outer Space and of the Threat or Use of Force Against Outer Space Objects (PPWT) is a proposed arms control treaty that would impose new definitions, terms, conditions, and obligations on States Parties and fill the gaps in the Outer Space Treaty in relation to arms control. The initial draft PPWT was submitted to the Conference on Disarmament by Russia and China in 2008 (CD/1839, 2008). The draft treaty drew upon a 2002 working paper that had been jointly submitted to the Conference on Disarmament by Russia, China, Indonesia, Belarus, Vietnam, Zimbabwe, and Syria (Su 2010), which confirmed the preference of those states to a hard law approach to the prevention of the weaponization of outer space (Tronchetti and Hao 2015). China has affirmed that it "opposes any weaponization of outer space and any arms race in outer space... [and] believes that the best way for the international community to prevent any weaponization of or arms race in outer space is to negotiate and conclude a relevant international legally-binding instrument" (Shimabukuro 2014).

The 2008 initial draft PPWT contains a preamble and 14 Articles. The preamble specifically highlights "the risk that the placement of weapons and an arms race in space pose to the security of space objects as well as to international peace and security as a whole" (Tronchetti and Hao 2015). The draft treaty provides a definition for a weapon in outer space, which is defined as

> Any device placed in outer space, based on any physical principle, which has been specially produced or converted to destroy, damage or disrupt the normal functioning of objects in outer space, on the Earth or in the Earth's atmosphere, or to eliminate a population or components of the biosphere which are important to human existence or inflict damage on them. (Article I(c), CD/1839)

A weapon is considered to be "placed in outer space" if "it orbits the Earth at least once, or follows a section of such an orbit before leaving this orbit, or is permanently located somewhere in outer space" (Article I(d), CD/1839). The draft treaty provides a comprehensive ban on any hostile action against objects in space, defining the use of force and threat of force broadly as

> any hostile actions against outer space objects including, inter alia, actions aimed at destroying them, damaging them, temporarily or permanently disrupting their normal

functioning or deliberately changing their orbit parameters, or the threat of such actions. (Article I(e), CD/1839)

The core provision of the 2008 draft PPWT is contained in Article II, which sets out a series of prohibitions:

> The States Parties undertake not to place in orbit around the Earth any objects carrying any kinds of weapons, not to install such weapons on celestial bodies and not to place such weapons in outer space in any other manner; not to resort to the threat or use of force against outer space objects; and not to assist or induce other States, groups of States or international organizations to participate in activities prohibited by this Treaty. (Article II, CD/1839)

Article II must be read in conjunction with Articles IV and V, which confirm that nothing in the draft PPWT should be seen as interfering with the right to self-defense or the right to peacefully explore and use outer space. Finally, States Parties are encouraged to implement transparency and confidence building measures (Article VI, CD/1839), although the draft treaty does not establish or recommend any such measures, leaving that to subsequent agreement.

While it was "praised for taking up the challenge of increasing the security of space assets by legal means," the initial draft PPWT received criticism "for not addressing the most serious threats to space objects, strategically favoring the interests of its co-sponsors and lacking a reliable means of verification" (Tronchetti and Hao 2015). Criticisms leveled against the draft PPWT included its failure to specifically ban the development and testing of Earth-based ASATs (Jaramillo 2009), as well as the overly restrictive definition of "weapon in space" and failure to address "the possible utilization of dual-use satellites and ballistic missiles as means to damage or destroy active space objects" (Tronchetti and Hao 2015). There was also criticism of the broad definition of "use of force" and "threat of force" and the inclusion of a definition of "outer space," which specified 100 km altitude demarcation; no international agreement on where outer space begins has ever been reached and the delimitation between airspace and outer space has been one of the most controversial topics since the beginning of the space age. The USA staunchly opposed the 2008 draft PPWT and indeed remains opposed to the development of arms control agreements or other restrictions that would restrict its ability to conduct research, development, testing, operations or other activities, or to operate freely in outer space (Jaramillo 2009).

In June 2014, an updated version of the PPWT was presented to the Conference on Disarmament by Russia and China (CD/1985, 2014). The updated PPWT contains a preamble and 13 Articles. The updated preamble emphasizes the commitment "not to threaten international peace and security by placing weapons in space or by transforming outer space in an area of military confrontation" (preamble, CD/1985). However, the updated preamble has been criticized for creating uncertainty regarding scope as any references to the right to explore and use outer space for peaceful purposes, as well as any direct references to arms control and disarmament agreements relating to outer space, have been removed (Tronchetti and Hao 2015).

The 2014 draft PPWT contains several notable changes from the 2008 version, including the removal of any definition of outer space. The definition of weapon in space has been slightly amended to mean:

> any outer space object or its component produced or converted to eliminate, damage or disrupt normal functioning of objects in outer space, on the Earth's surface or in the air, as well as to eliminate population, components of biosphere important to human existence, or to inflict damage to them by using any principles of physics. (Article 1(b), CD/1985)

However, the definition of weapon in space remains a restrictive one. The challenge of properly defining weapon in space where so many space objects are dual-use is not effectively addressed in the 2014 draft PPWT. While the definition prohibits the deployment of any space object produced or converted for hostile purposes, the testing, deployment, and use of space objects for purposes such as active debris removal and near-Earth object diversion (which could easily be subsequently used as weapons) are not prohibited.

A space object is considered to have been placed in outer space if it "orbits the Earth at least once, or follows a section of such an orbit before leaving that orbit, or is permanently located in outer space or on any celestial bodies other than the Earth" (Article 1(c), CD/1985).

The definition of use of force and threat of force has been narrowed to:

> any intended action to inflict damage to outer space object under the jurisdiction and/or control of other States, or clearly expressed in written, oral or any other form intention of such action. Actions subject to special agreements with those States providing for actions, upon request, to discontinue uncontrolled flight of outer space objects under the jurisdiction and/or control of the requesting States shall not be regarded as use of force or threat of force. (Article I(d), CD/1985)

This definition makes it clear that under the 2014 draft PPWT it would still be permissible for States Parties to test Earth-based ASAT technology on their own space objects. Article II remains the core provision of the 2014 draft PPWT and has been amended to read as follows:

> States Parties to this Treaty shall:
>
> – not place any weapons in outer space
> – not resort to the threat or use of force against outer space objects of States Parties
> – not engage in outer space activities, as part of international cooperation, inconsistent with the subject matter and the purpose of this Treaty
> – not assist or incite other States, groups of States, international, intergovernmental, and any nongovernmental organizations, including nongovernmental legal entities established, registered, or located in the territory under their jurisdiction and/or control to participate in activities inconsistent with the subject matter and the purpose of this Treaty (Article II, CD/1985)

While explicitly recognizing the need for measures to verify compliance, Article V of the 2014 draft provides:

> The States Parties recognize the need for measures to verify compliance with the Treaty, which may form the subject of an additional protocol.
> With a view to promoting confidence in compliance with the provisions of the Treaty, States Parties may implement agreed transparency and confidence-building measures, on a voluntary basis, unless agreed otherwise. (Article V, CD/1985)

The USA opposed the 2014 draft PPWT, stating that "it does not address the significant flaws in its previous version, such as including an effective verification regime or dealing with terrestrially-based anti-satellite systems" (Rose 2014). As with the 2008 draft PPWT, the 2014 draft does not indicate what kind of transparency and confidence building measures states could take to ensure compliance with the treaty (Tronchetti and Hao 2015). The USA stated that there "is no integral verification regime to help monitor or verify the limitation on the placement of weapons in space... Moreover,....it is not possible with existing technologies or cooperative measures to effectively verify an agreement banning space-based weapons" (CD/1998, 2014). The lack of a verification regime stands as a major impediment to the progress of the PPWT initiative (Su 2017). Russia and China responded that the Outer Space Treaty also does not provide any mechanisms for verification but that the development of verification mechanisms, to the extent possible, would be desirable for the full implementation of the PPWT (CD/2042, 2015). The USA also based its objections to the 2014 draft PPWT on the lack of a prohibition on "possession, testing, production, and stockpiling of such weapons" to prevent a Party from developing "a readily deployable space-based weapons break-out capability" (CD/1998, 2014). Russia and China responded that effective "monitoring of 'research, development, production, and terrestrial storage of space-based weapons' – on which there is no prohibition... – is not feasible in practical terms for objective reasons" (CD/2042, 2015). Additionally, the USA stated that the draft PPWT "does not address the most pressing, existing threat to outer space systems: terrestrially-based anti-satellite weapon systems. There is no prohibition on the research, development, testing, production, storage, or deployment of terrestrially-based anti-satellite weapons; thus, such capabilities could be used to substitute for, and perform the functions of, space-based weapons" (CD/1998, 2014). Russia and China responded that

> While anti-satellite weapons as a class of weapons are not prohibited under the draft PPWT, the proliferation of such weapons is restricted through a comprehensive ban on the placement in outer space of weapons of any kind, including anti-satellite weapons. A ban on ground-based anti-satellite (ASAT) weapon systems has been introduced into PPWT through the ban on the use of force, regardless of its source, against space objects. (CD/2024, 2015)

Russia and China have further argued that ground-based ASATs are covered in the draft PPWT through the prohibition on the use of force and that the placement of weapons of any kind in outer space is prohibited (Chow 2018). Additionally, that there is a need to clarify the issue of the use of force in outer space on the grounds provided for in the Charter and to reach a common understanding of the right of self-defense under the Charter as regards outer space (CD/2042, 2015).

Although its provisions were substantially reworded, the 2014 draft PPWT retained the most criticized aspects of the 2008 draft PPWT (Tronchetti and Hao 2015). As the Conference on Disarmament requires consensus, it is unlikely that the PPWT will progress further given the ongoing opposition by the USA.

International Code of Conduct for Outer Space

In 2008, the European Union (EU) released its proposal for an International Code of Conduct for Outer Space Activities (Code). While not legally binding, the Code was an attempt to establish "rules of the road" to enhance the safety, security, and sustainability of space operations (Koplow 2018). The initial draft of the Code was met with criticisms about the lack of consultation with other non-EU States. Informal bilateral consultations were then undertaken with Brazil, Canada, China, India, Indonesia, Israel, Russia, South Africa, the Republic of Korea, Ukraine, and the USA. This was followed by a series of multilateral consultations between June 2012 and May 2014, with meetings held in Vienna, Kiev, Bangkok, and Luxembourg (Johnson 2014). Several revised drafts of the Code were then released, with the latest version being released on 31 March 2014 (European Union, 2014 draft Code).

The purpose of the Code is to "enhance the safety, security, and sustainability of all outer space activities pertaining to space objects, as well as the space environment" (Article 1.1, 2014 draft Code). The Code "addresses outer space activities involving all space objects launched into Earth orbit or beyond, conducted by a Subscribing State, or jointly with other States, or by non-governmental entities under the jurisdiction of a Subscribing State, including those activities conducted within the framework of intergovernmental organisations" (Article 1.2, 2014 draft Code). The focus of the Code is directed less at the legality of weapons in outer space, and more at establishing rules of behavior and ensuring "transparency so that states can avoid accidental war in outer space" (Larsen 2017).

In the context of arms control and weapons in space, Article 4.2 provides that

> The Subscribing States resolve, in conducting outer space activities, to:
> - refrain from any action which brings about, directly or indirectly, damage, or destruction, of space objects unless such action is justified:
>   - By imperative safety considerations, in particular if human life or health is at risk or
>   - In order to reduce the creation of space debris or
>   - By the Charter of the United Nations, including the inherent right of individual or collection self-defense
>
> and where such exceptional action is necessary, that it be undertaken in a manner so as to minimize, to the greatest extent practicable, the creation of space debris

Article 4.2 does not expressly prohibit the deployment, research, or development of weapons in outer space and permits the use of such weapons in the specific circumstances identified in the Article. Article 4.2 could be interpreted as permitting the use of Earth-based ASAT weapons in certain circumstances, where the use of such weapons is justified by imperative safety considerations, in order to reduce the creation of space debris, or in self-defense. The Code does not explicitly address space-to-Earth weapons in any way, which is potentially a significant shortcoming of the Code from a security perspective (Su 2017). The Code also does not address the challenges posed by dual-use space objects, nor does it specify what measures could be taken to verify compliance with the Code. Both Russia and China oppose the proposed Code, strongly preferring the treaty approach to arms control and the prevention of an arms race in outer space (Larsen 2017).

Domestic Arms Control for Outer Space

In December 2019, the New Zealand government announced four policy principles which had been approved as an enhanced approach to payload permit assessments under the Outer Space and High-altitude Activities Act (Bradley 2019). In the context of arms control, under the updated policy the following payloads are banned from being launched into space from New Zealand:

- Payloads that contribute to nuclear weapons programs or capabilities
- Payloads with the intended end use of harming, interfering with, or destroying other spacecraft, or space systems on Earth
- Payloads with the intended end use of supporting or enabling specific defense, security or intelligence operations that are contrary to government policy
- Payloads where the intended end use is likely to cause serious or irreversible harm to the environment

These domestic restrictions on the launch of certain payloads may cover a broader range of weapons than the international space law regime.

Conclusions

The Outer Space Treaty and the Moon Agreement both prohibit States Parties from placing nuclear weapons and other weapons of mass destruction in orbit, installing such weapons on celestial bodies, or stationing such weapons in outer space in any other manner. In addition, States Parties are prohibited from carrying out tests of weapons of any kind on celestial bodies. However, the current international space law regime does not prohibit (and so permits) states to develop conventional orbital weapons and Earth-based ASAT weapons.

The existing international space law regime addressing the potential weaponization of outer space is outdated, inadequate, and insufficient to adequately address arms control in outer space (Jaramillo 2009). An arms race in outer space, with states competing for control of space could "deter optimal patterns of peaceful exploitation of space, and even the current amplifying rhetoric about weaponization of space may make some wary potential space actors hesitant about investing in new space capabilities" (Koplow 2018). The rapidity with which space technologies are being developed and the increasing prevalence of dual-use space technologies which could also be used for hostile purposes, such as those used for active debris removal and near-Earth object diversion, challenge the existing space arms control regime, and remain an unresolved challenge for space arms control initiatives. One of the biggest challenges still to be overcome is definitional; the need for States to agree on a definition of what amounts to a weapon in space.

Arms control in outer space is not inherently more complicated or difficult than arms control in other domains (Koplow 2018), with the exception of verification challenges and the prevalence of dual-use objects in space compared to other domains. Any effective arms control regime for outer space must, by necessity, include all states (four as at the time of writing – the USA, Russia, China, India) that possess significant military arms technology that can be deployed in outer space (Larsen 2017). The problem is one of political will; "the resistance to effective arms control in outer space seems especially entrenched; even modest measures provoke a strong allergic reaction" (Koplow 2018). The USA has repeatedly stated that

> The United States will oppose the development of new legal regimes or other restrictions that seek to prohibit or limit US access to or use of space. Proposed arms control agreements or restrictions must not impair the rights of the United States to conduct research, development, testing, and operations or other activities in space for US national interests. (Peoples 2011)

In contrast, Russia and China have repeatedly expressed a preference for treaty-based arms control measures to prevent an arms race in outer space. However, the international political and legal conditions for the acceptance of a new treaty regulating arms control and space security appear to be currently lacking, given that the international community has been unable to agree on a new treaty regulating space activities for more than 40 years (Tronchetti and Hao 2015).

As most states appear unwilling to accept binding restrictions on their ability to protect their space assets, any treaty-based arms control initiatives, such as the PPWT, appear unlikely to succeed. Consequently, nonbinding, soft law instruments may have the highest chance of advancing space security through states voluntarily adopting the best practice behaviors established. Arms control measures for space must be well-crafted, with adequate verification mechanisms and robust enforcement (Koplow 2018). Any arms control regime for space, whether hard or soft law, must be able to fill the normative void in the current space security treaty regime and reduce the risk of miscalculation and misinterpretation. Only in this way can space security be enhanced.

References

Agreement governing the activities of states on the moon and other celestial bodies. 11 July 1984. GA resolution 34/68, annex

Bhat S, Mohan V (2009) Anti-satellite missile testing: a challenge to article IV of the outer space treaty. NUJS Law Rev 2(2):205–212

Bradley G (17 December 2019) Government updates rules on what can be sent into space. NZ Herald. https://www.nzherald.co.nz/business/news/article.cfm?c_id=3&objectid=12294704

Chow B (2018) Space arms control: a hybrid approach. Strateg Stud Q 12(2):107–132

Delegation of the United States of America to the Conference on Disarmament. Analysis of the 2014 Russian-Chinese draft treaty on the prevention of the placement of weapons in outer space, the threat or use of force against outer space objects; conference on disarmament. CD/1998. 3 Sept 2014. https://documents-dds-ny.un.org/doc/UNDOC/GEN/G15/007/57/PDF/G1500757.pdf?OpenElement

Draft treaty on the prevention of the placement of weapons in outer space and of the threat or use of force against outer space objects (2008) CD/1839 and (2014) CD/1985

European Union (2014) Draft international code of conduct for outer space activities. http://www.eeas.europa.eu/non-proliferation-and-disarmament/pdf/space_code_conduct_draft_vers_31-march-2014_en.pdf

Hobe S, Schmidt-T B, Schrogl K.-U (eds) (2010) Cologne Commentary on Space Law, vol 1, Outer Space Treaty, Carl Heymanns Verlag, Köln

Jaramillo S (2009) In defence of the PPWT treaty: toward a space weapons ban. Ploughshares Monit 30(4):11–14. https://ploughshares.ca/pl_publications/in-defence-of-the-ppwt-treaty-toward-a-space-weapons-ban/. Accessed 26 Dec 2019

Johnson C (2014) Draft international code of conduct for outer space activities fact sheet. Secure World Foundation, Washington, DC

Koplow D (2018) The fault is not in our stars: avoiding an arms race in outer space. Harv Int Law J 59(2):331–388

Larsen P (2017) Outer space arms control: can the USA, Russia and China make this happen. J Confl Secur Law 23(1):137–159

López L (2012) Predicting an arms race in space: problematic assumptions for space arms control. Astropolitics 10:49–67

Peoples C (2011) The securitization of outer space: challenges for arms control. Contemp Secur Policy 32(1):76–98

Permanent Representative of the Russian Federation and the Permanent Representative of China to the Conference on Disarmament. Follow-up comments by the Russian Federation and China on the analysis submitted by the United States of America of the updated Russian-China draft PPWT. CD/2042. 14 Sept 2015. https://documents-dds-ny.un.org/doc/UNDOC/GEN/G15/208/38/PDF/G1520838.pdf?OpenElement

Plath C (2018) Explanation of vote in the First Committee on resolution: L.50, "no first placement of weapons in outer space." United Nations, 5 Nov 2018

Rose F (2014) Statement by Deputy Assistant Secretary for Space and Defence Policy, Bureau of arms control, verification and compliance at conference on disarmament plenary session, 10 June 2014

Shimabukuro A (2014) No deal in space: a bargaining model analysis of U.S. resistance to space arms control. Space Policy 30:13–32

Su J (2010) The "peaceful purposes" principle in outer space and the Russia-China PPWT proposal. Space Policy 26:81–90

Su J (2013) The environmental dimension of space arms control. Space Policy 29:58–66

Su J (2017) Space arms control: lex lata and currently active proposals. Asian J Int Law 7(1):61–93

Sullivan C (1990) The prevention of an arms race in outer space: an emerging principle of international law. Temp Int Comp Law J 4(2):211–237

Treaty between the United States of America and the Union of Soviet Socialist Republics on the limitation of anti-ballistic missile systems, signed at Moscow 26 May 1972, entered into force 30 Oct 1972

Treaty on principles governing the activities of states in the exploration and use of outer space, including the moon and other celestial bodies (10 Oct 1967) GA resolution 2222 (XXI), annex

Tronchetti F, Hao L (2015) The 2014 updated draft PPWT: hitting the spot or missing the mark? Space Policy 33:38–49

United Nations General Assembly (1963) Question of general and complete disarmament. GA Res 1884 (XVIII)

United Nations General Assembly (1988) Prevention of an arms race in outer space. UN Doc A/RES/43/70

United Nations General Assembly (2014a) Draft resolution on "no first placement of weapons in outer space." UN Doc A/C.1/69/L.14

United Nations General Assembly (2014b) No first placement of weapons in outer space. GA resolution 69/32

United Nations General Assembly (2019) Further practical measures for the prevention of an arms race in outer space. UN Doc A/C.1/74/L.58/Rev.1

Role of Space in Deterrence

John J. Klein and Nickolas J. Boensch

Contents

Introduction	112
Space Deterrence	113
Deterrence by Punishment	114
Deterrence by Denial	116
Principles of Space Deterrence	119
Conclusions	125
References	125

Abstract

A proper space strategy agrees with the universal and overarching logic of strategy. Therefore, the concept of deterrence has applicability in the space domain. Space activities and policies are relevant for deterring conflict, as well as maintaining international peace and stability. Although deterrence has a legitimate role in future space strategy, it is not the panacea for preventing conflict, because history teaches that deterrence will at times fail due to miscalculation, uncertainty, or chance. This is also true for deterring acts of aggression in space.

The views expressed in this article are solely those of the authors and do not necessarily reflect those of Falcon Research, the George Washington University, Bryce Space and Technology, or the US government.

J. J. Klein (✉)
George Washington University's Space Policy Institute, Washington, DC, USA
e-mail: kleinjj@gwu.edu

N. J. Boensch
Bryce Space and Technology, Alexandria, VA, USA
e-mail: nick.boensch@brycetech.com

Introduction

Secure access to space is a critical national security interest for many countries. Space-reliant technologies enable vital activities – including commerce, trade, environmental monitoring, intelligence collection, and governmental actions. Consequently, many countries will seek to ensure access to and use of space through diplomatic, informational, military, and economic instruments of national power. The strategic importance of space means that in a potential conflict, an adversary will be incentivized to blunt this advantage and challenge command of the domain. To discourage this behavior and protect one's national interests in space, the concept of deterrence is salient. The recent return of discourse focused on great power competition – fueled by Chinese belligerency in its territorial disputes and Russian adventurism on its borders and abroad – has a distinct space element and highlights that space deterrence will be a principal theme in space strategy for many space powers.

Since the beginning of the Space Age, some polities considered operations and activities in space as a means of supporting the ends of national policy. Thorough analysis of the nexus between space and deterrence, however, remained unexplored for decades largely for two reasons (Thomson 1995). First, space deterrence was considered to be closely coupled with nuclear deterrence thinking from the Cold War because space systems enabled nuclear command and control, supported early warning of ballistic missile launch, and served as national technical means of verification in arms control measures. Any interference against national security space systems was thought to be a potential precursor to a nuclear war. Second, in the immediate post-Cold War world, there was not a significant or explicit space threat to be deterred. China's 2007 direct ascent anti-satellite (ASAT) weapon test and subsequent military posture in space has been a catalyst that has prompted policy makers and strategists within the United States to more fully consider the role of deterrence in space strategy.

When developing space deterrence strategies, strategists should acknowledge the unity in strategic experience. The fundamental concept of deterrence is enduring and has been studied in depth. Just as Carl von Clausewitz identified the universal nature and changing character of war, the nature of deterrence is enduring, while its implementation differs between different domains of warfare and each geopolitical context (Clausewitz 1989). Consequently, when considering the role of deterrence in space, the strategist may use historical experience and lessons – from antiquity to the present day – to better understand the relationship between the space domain and deterrence theory. This chapter presents many of the most fundamental topics of space deterrence. Admittedly, much of the current literature on space deterrence focuses on the strategic challenges facing the United States; however, the lessons to be gleaned are often relevant to other countries. The concepts presented in this chapter are meant to guide readers and future strategists, thereby aiding them in thinking about deterrence in space and allowing them to identify sound arguments, train their judgment, and avoid pitfalls when crafting strategy (Clausewitz 1989).

Space Deterrence

Fundamentally, deterrence efforts seek to affect the decision-making of others and influence their behavior. This is reflected in a commonly used definition that *deterrence* refers to persuading a potential enemy that it is in its own interest to avoid certain courses of activity (Schelling 1966). When a potential adversary forgoes certain actions or some forms of behavior that they would otherwise have carried out due to intolerable costs, this is commensurate with deterrence – through either denial or punishment. In its most simple form, deterrence involves persuading an adversary that the risks or costs of an action exceed any perceived benefit or gain.

Because states derive strategic advantages from satellites and potential rivals may seek to deny a state this advantage, the concept of "space deterrence" is a relevant concept for space powers. Deterrence by punishment, compellence, deterrence by denial, and dissuasion are important ideas in the formulation of a sound space strategy. Taking the commonly accepted definition mentioned previously, *space deterrence* refers to persuading a potential enemy that it is in its own interests to avoid certain courses of activity in, through, or from space. Regardless of the chosen terminology or definition, what is ultimately important is that there are actions that can be taken relative to space that affect the decision-making of others.

One of the most essential distinctions in deterrence theory is between deterrence by punishment and by denial (Snyder 1961). *Deterrence by punishment* concerns the threat of credible and potentially overwhelming force or other retaliatory action against any would-be adversary to discourage potential aggressors from conducting hostile actions. *Deterrence by denial* refers to the capability to deny the other party any gains from the behavior that is to be deterred (Snyder 1961). This concept refers similarly to deterring an adversary by presenting a credible capability to prevent it from achieving the potential gains adequate to motivate the action (Krepinevich and Martinage 2008).

A related but distinct concept is *compellence*, which involves convincing an adversary to cease some current undesired action. Compellence is often described as a direct action that persuades an opponent to give up something that is desired (Schelling 1966). While deterrence has a negative object – it discourages unwanted actions – the object of compellence is positive. Effort is expended to force or convince an actor to conform to one's will.

Both military and nonmilitary means are applicable when seeking to affect the thinking of others to enable deterrence by punishment, deterrence by denial, and compellence. Nonmilitary means equate to the *soft power*, or the diplomatic, informational, and economic instruments of national power (Nye 2005). Nonmilitary means can be used to affect another state leader's thought processes – whether reinforcing a currently held view that is beneficial to the deterring state or changing the view of another state's leadership or polities. Consequently, a practical implementation of space deterrence may entail political and diplomatic efforts, such as new international treaties or agreements; multimedia stories presenting news in a favorable perspective; and commerce and trade activities that increase one's own economic influence or affect negatively a potential adversary or opposing alliance (Klein 2019).

Deterrence by Punishment

A deterrence by punishment approach in space is underpinned in the belief that the threat of credible and potentially overwhelming force or other retaliatory action against any would-be adversary is sufficient to deter most potential aggressors from conducting hostile actions in space. Such a strategy should clearly convey the capability and credibility behind the threat and communicate the specific behavior sought to be discouraged (Morgan 1977). As part of its broader space strategy, the United States seeks to deter attacks against its satellites. The 2017 US National Security Strategy conveys that harmful interference or attacks targeting US satellites will be met with a deliberate response in the "time, place, manner, and domain" of its choosing (The White House 2017). A US joint doctrine describes that, consistent with the right to self-defense, the United States may utilize its space assets to target the space capabilities of an adversary to deter potential threats (Joint Chiefs of Staff 2018). Some security experts view that the punishment portion of the US space deterrence strategy has been pursued and emphasized extensively, perhaps to the detriment of other approaches to secure US interests in space (Johnson-Freese 2017).

Many analysts have identified challenges associated with implementing a deterrence by punishment approach in the space domain. These include establishing appropriate thresholds for retaliation for both non-kinetic and reversible attacks on satellites, differences in severity due to no loss of life when compared to terrestrial action, and having the requisite attribution capabilities and processes.

The absence of explicit threshold that a state would retaliate against complicates efforts to deter adversaries. Some policy makers question whether non-kinetic and reversible actions are hostile acts or armed attacks that warrant a military response (Harrison et al. 2017). Reversible and non-kinetic actions on satellites supporting tactical operations may be treated differently from large-scale kinetic attacks on satellites supporting nuclear command or control or early warning missions. However, between these extremes, there is still a highly uncertain boundary that complicates deterrence efforts (MacDonald et al. 2016). Ultimately, what is considered an armed attack or hostile act in space necessitating a retaliatory response will depend on the broader geopolitical context.

Another challenge for a space deterrence strategy is that attacks on satellites typically are unlikely to result in loss of human life. Consequently, hostile actions in space may be considered by some polities to be less escalatory or grave as conflict on Earth. The frequently used adage that captures this thinking is "satellites don't have mothers." This view may cause decision-makers to view aggression in space as never rising to levels that would warrant a military response, whether terrestrially or in space. Moreover, because military actions in space are unlikely to produce direct casualties, there may be an appeal to turn to these activities as tensions between competing states escalate (MacDonald et al. 2016). Perceptions that hostilities in space are less severe than terrestrial conflict can be discouraging to those hoping to deter attacks against one's satellites.

Regardless, the thought that the non-casualty-generating effects of space actions preclude a deterrent threat is unfounded. Article 2(4) of the United Nations Charter describes the need to refrain from the threat or use of force against a state's territorial integrity – which may be interpreted as a state's physical property. Self-defense and retaliatory threats to deter a potential armed attack against a state's satellites are then appropriate and justified. Upon further examination, one may dispel a historical challenge surrounding space deterrence by punishment: a hostile action against a state's space systems may still be deterred by threat of retaliation, even if there is no loss of life.

Yet another challenge to effective deterrence in space lies in the difficulty of attributing who or what caused a satellite to cease to function normally. Military actions in space can produce various effects, may be non-kinetic and reversible, and in some cases these effects may be difficult to identify and attribute. An effective deterrent requires timely assessment of the event to orient and respond appropriately. Operating at hundreds to more than 30,000 km above the Earth's surface makes it difficult to physically inspect and track satellites, thereby making determining and assessing damage an onerous endeavor. The hostile space environment – where satellites face solar activity, scorching and frigid temperatures, radiation, electromagnetic activity, and an increasing amount of debris – further complicates efforts (Wright et al. 2005). Operators must distinguish between intentional interference from adversaries and interference arising from normal operation in a hostile environment.

Some authors argue that the difficulties associated with attribution may be less worrisome than originally thought (Harrison et al. 2009). An attack on a state's satellites unconnected to a terrestrial strategic event is thought to be highly unlikely. Attacks on satellites will occur following the terrestrial breakdown of general deterrence between states. The source of an attack may be less nebulous than space deterrence literature has declared, particularly if the attacking state launches a coordinated attack on many satellites to try to gain command of space early in the conflict. Drawing from this example, intelligence gleaned prior to the attack may be a more meaningful method of attribution than enabled by postattack space situational awareness (SSA) assessment.

Regardless of this assessment, in the current context of the global proliferation of counterspace capabilities, there will likely be ample room for misperception and miscalculation in a state's leadership. This necessitates robust SSA capability to address issues of identifying, assessing, and attributing activities that occur in orbit. Greater SSA capabilities allow a state to differentiate between intentional attacks and malfunctions due to the satellite itself or the hostile environment it inhabits, thereby reducing the potential for misinterpretations and miscalculations (Sheldon 2008). Effective SSA capabilities will necessitate knowing what on-orbit systems are present, along with their location, capabilities, historical anomalies, operating patterns, and intended use. Such information will facilitate those preparatory measures needed to pit one's strengths against a potential adversary's weakness. Because SSA is a global endeavor, information sharing architectures must be designed to include the international community and commercial industry. This means that much of the data and resulting information provided through SSA systems should be releasable and disseminated to many of those participating in the global effort.

Today's security challenges can complicate the implementation of a deterrence by punishment strategy. While some security analysts assess that thresholds for retaliation, differences in severity for space actions, and ensuring a sufficient attribution capability may be less problematic than many think, it remains to be seen whether this is confirmed in practice.

Deterrence by Denial

Among many security professionals, *deterrence by denial* is often associated with the concept of *dissuasion* – activities that seek to influence the decision calculus of potential adversaries to discourage the initiation of military competition. A strategy incorporating dissuasion seeks to convey the futility of conducting a hostile act, affecting the confidence of a potential adversary's leadership and causing decision-makers to not pursue a military confrontation in the first place. To be most effective, dissuasion activities occur before a threat manifests itself. Some national security professionals note that dissuasion works outside the potential threat of military action as a kind of "pre-deterrence," because those states dissuaded will not require to be deterred by punishment (Krepinevich and Martinage 2008). While a deterrent that seeks to punish an adversary is tailored to distinct actions by specific actors at definite times, deterrence by denial commonly lacks this specificity and exists as a general deterrent, one that shapes the security environment through a broad, latent deterrent effect originating from one's reputation and capabilities (Morgan 1977).

A deterrence by denial strategy for space seeks to frustrate or complicate the adversary's plans by introducing greater costs and reducing associated benefit. Over the past several years, there has been a greater emphasis on the role of deterrence by denial in the broader US space deterrence strategy. Rather than threatening retaliation against the aggressor's satellites or terrestrial targets of value, a US space deterrence by denial strategy emphasizes reducing an adversary's incentive to attack US satellites (Vedda and Hays 2018). A potential adversary may be deterred if it concludes that an attack in space will be ineffectual in achieving the desired effect. Much of deterrence by denial and dissuasion necessitates preparing for potential conflict during peacetime. Because dissuasion involves discouraging the initiation of military competition, for the space domain the requisite peacetime preparedness is included within the contexts of *space mission assurance* and *resilience*.

According the US joint literature, mission assurance entails a process to protect or ensure the continued function and resilience of capabilities and assets – including personnel, equipment, facilities, networks, information and information systems, infrastructure, and supply chains – critical to the performance of the Department of Defense mission essential functions in any operating environment or condition (Office of the Assistant Secretary of Defense for Homeland Defense 2015). Similar to mission assurance but with a different focus, *resilience* is an architecture's ability to support mission success with higher probability; shorter periods of reduced capability; and across a wider range of scenarios, conditions, and threats, despite hostile action or adverse conditions (Joint Chiefs of Staff 2018). Resilience may leverage cross domain solutions, along with commercial and international capabilities. By definition, space mission assurance and resilience efforts can prevent a potential adversary from achieving its objectives or realizing any benefit from aggressive action. Space mission assurance and resiliency help convey the futility of conducting a hostile act and, consequently, enhance deterrence by denial and dissuasion efforts.

Space mission assurance efforts consist of *defensive operations*, which include off-board protection elements; *reconstitution*, which includes launching replacement satellites or activating new ground stations; and *resilience*, which includes on-board protection elements (Joint Chiefs of Staff 2018). Resilience in capabilities includes disaggregation, distribution, diversification, deception, protection, and proliferation. *Disaggregation* is the separation of dissimilar capabilities into separate platforms or payloads. *Distribution* utilizes a number of nodes, working together, to perform the same mission or functions as a single node. *Diversification* is contributing to the same mission in multiple ways, using different platforms, different orbits, or systems and capabilities of commercial, civil, or international partners. *Deception* is hiding one's strengths and weaknesses from one's adversaries. *Protection* is utilizing active and passive measures to ensure space systems provide mission support in any operating environment or condition. *Proliferation* is deploying larger numbers of the same platform, payloads, or systems of the same type to perform the same mission.

Space mission assurance may be supported by a number of preparations preceding a potential conflict. These preparations may include hardening against cyber threats and signal jamming, incorporating shutters for remote sensing satellites to minimize the effects of dazzling by lasers, or increasing the mobility of satellites through novel propulsion technologies (Kueter and Sheldon 2013). Preparations taken in peacetime may include employment of proliferated constellations of small satellites to complicate an adversary's space ambitions. Furthermore, the conduct and training of one's space and terrestrial forces may grant an ample deterrent effect, even if no ancillary preparations have been made. One method of frustrating an adversary's plans may be to train forces to fight under degraded conditions where military forces lose access to space-enabled capabilities, thereby depriving potential aggressors some of the appeal of attacking satellites (Harrison et al. 2009). Consequently, a potential aggressor may be convinced that the prospects for success are too costly, with little benefit.

Another method of frustrating an adversary's space control plans is to reduce one's vulnerability by transitioning traditional space-derived services to terrestrial alternatives, a concept termed *space avoidance*. Its advocates seek to increase space deterrence by minimizing one's presence in space, thereby diminishing an adversary's perceived benefits of attacking one's satellites (Coletta 2009). For example, some space avoidance advocates suggest this may be achieved by using unmanned aerial systems (UAS) for tactical reconnaissance systems instead of remote sensing satellites. Creating redundancy through terrestrial alternatives is prudent, but one should not be misled when judging whether reliance on space can be abated entirely. UAS are a valuable supplementary resource to space-derived intelligence, surveillance, and reconnaissance (ISR); however, most UAS still require space-derived positioning information and communications to operate. Many forms of military power – sea power and airpower, for example – cannot easily reduce reliance on space-derived services. While states should seek to increase terrestrial redundancy to complicate an adversary's plans, a strategy of space avoidance intending to greatly reduce reliance on space is not feasible in modern warfare.

Alliances, international cooperation, and the global proliferation of space power also play a significant role in deterrence by denial. This international dimension influences deterrence in several ways. First, the proliferation of states operating or deriving benefits from satellites creates stakeholders who would likely prefer that their satellites were not put in jeopardy. States outside of the deterrence relationship may have their satellites affected negatively if deterrence fails and conflict ensues, such as by orbital debris from kinetic attacks or the indiscriminate effects of broad radio-frequency jamming. Second, the deterring state may provide a global or multinational space-derived service, such as the US Global Positioning System satellites, which if attacked could potentially draw countries reliant on this service into the conflict on the side of the non-aggressor (Harrison et al. 2009). In these situations, an aggressor may be hesitant to attack space systems if it will have to potentially contend with an international response (Sheldon 2008). Third, allied or partner states may assist the deterring state when a conflict breaks out. The space systems of friendly countries can complement and supplement the deterrer's own capabilities, such as through data sharing agreements, interoperability, or even by assisting in the reconstitution of lost space capabilities. Adversary leadership may be deterred from targeting US satellites if they perceive that the United States could leverage the capabilities of its allies to nullify any anticipated benefit (Sheldon 2008).

Some security experts consider the North Atlantic Treaty Organization (NATO) as being uniquely positioned to bolster deterrence in space through its cooperative alliance. The alliance is increasingly reliant on space for its collective defense and economic prosperity, and an attack on the space assets of any one ally impacts the security of all allies (Schulte 2012). Security experts assert that while NATO is dependent on space-enabled capabilities, its space doctrine and planning have not kept up. Presently, NATO officials are considering how the alliance should address the growing military capabilities of Russia and China, to include issuing NATO's first strategy for space. The strategy is expected to make space an official domain of operation, giving structure to discourse on military developments in space and NATO's response. The alliance may also decide that attacks in space would trigger the organization's Article 5 provisions on collective defense, although internal differences on the subject remain. Analysts have long held that NATO should continue to build the expertise and capacity to conduct operations enabled by space; ensure that doctrine, requirements, and planning account for the operational advantages provided by space; and adapt exercises and training to ensure forces can effectively exploit space-based capabilities (Schulte 2012). It is still uncertain whether NATO's space strategy will implement these recommendations.

A deterrence by denial strategy presents its own challenges. The cost of fielding and launching the most robust, defendable space systems can become a financial burden (Coletta 2009). Hardened, dispersed, disaggregated, or diversified capabilities may cost more to develop, launch, integrate, and operate. Also, resilient architectures may not be able to match the performance of those exquisite space systems. In most cases, smaller, proliferated constellations of satellites will augment, rather than replace these exquisite systems. The space strategist then must consider the benefits of defensive approaches, along with associated time and fiscal procurement costs, when finally deciding upon the best approach. Another challenge of deterrence by denial is that one's space mission assurance and resilience efforts must be widely publicized to be effective in dissuading others.

Both deterrence by punishment and deterrence by denial are fundamental to an understanding of deterrence theory in space. Though deterrence can be valuable in one's attempts to prevent attacks and dissuade aggression, deterrence is not a panacea that will always prevent conflict. Clausewitz's wisdom is insightful. An adversary may strive to have the greatest likelihood of success by expanding its relative superiority, but even without this advantage, an adversary may find war attractive if there is no better option (Clausewitz 1989).

Principles of Space Deterrence

Because deterrence is a strategic behavior, its fundamental nature is enduring, even though its implementation changes with time and for each geopolitical situation. To understand the role of deterrence in space, it is important to identify deterrence's most enduring concepts along with their relation to space strategy.

Primacy of the Adversary's Decision-Making

A deterrence strategy is not a game of solitaire. All too often, policy makers and warfighters forget that those to be deterred may be unwilling or even unable to be deterred. Because war and deterrence are both within the realms of strategy, one must recognize that deterrence is a contest between two independent wills (Clausewitz 1989). The adversary's perceptions and decision-making are the paramount variables determining whether deterrence succeeds or fails. Regardless of the potential credibility of deterrence efforts, an adversary has an independent will and may not necessarily comply. Those to be deterred may fail to comprehend the threat or costs before them, doubt the credibility of the deterrent, or find that their policy ends are significant enough to warrant the costs and risks associated with ignoring the deterrence attempt (Sheldon 2008). Even if the deterring state has increased the costs and minimized adversary benefit through its defensive capabilities or with the demonstrated ability to respond, the decision to be deterred rests with the potential adversary.

Polities and their leadership cannot always be deterred, and they may decline to be coerced, even when heavily physically damaged, hoping for a change in strategic fortune (Gray 2007). "Fools," as some may call them, are far more likely to commit errors of a kind that result in wars or at least a high measure of regional disorder, because they will not be swayed in their decision for violence regardless of the threat of a severe military response to a hostile attack (Gray 2007). In such situations, deterrence could be irrelevant, because the foolish foreign leader may not believe in the latent or explicit threats issued or may not care whether or not the threat of retaliation is honored. Sometimes, if the enemy has nothing to lose, even a very risky action may be preferable to maintaining the status quo. Ultimately, it does not a matter whether one thinks a potential adversary should be deterred given an action or situation; it only matters how the adversary's leadership and decision-makers interpret any action within their worldview and mental constructs. Regardless of the amount of political will and military strength behind a deterrent message, the potential adversary's perceptions are what decide the success of a deterrence strategy.

Deterrence Cannot Be Guaranteed

Strategic history demonstrates that one should be less than confident in the certainty of deterrence. It is possible, and perhaps even probable at times, that deterrence will fail. Ambiguity, miscalculation, incompetence, friction, and chance are all prevalent in deterrence and serve to ensure that deterrence is a highly uncertain venture. The primacy of the psychological aspects in this manifestation of strategy further adds to the uncertainty of deterrence (Sheldon 2008). There is the fundamental, persistent threat that the countries in a deterrence relationship will trip, accidentally or inadvertently, into war (Gray 1991). Some may be quick to forget that, much like war, deterrence is a strategic behavior and accordingly suffers from complexity, nonlinearity, and unforeseeable occurrences that can thwart even the most careful and comprehensive planning. This complexity and nonlinearity should be considered and addressed when developing national strategies and operational plans.

One of the significant drawbacks of deterrence theory is identifying when it actually succeeds in causing an adversary not to proceed with an undesired behavior. Assessing the efficacy of deterrence is onerous because successful deterrence must be tested negatively with events that do not occur (Gray 1991). Because of the inability to draw convincing conclusions from events which did not happen, both the policy maker and strategist will likely be left with ambiguous lessons for the development of future strategies.

Credibility and Political Will Are Required

Even with a sufficient capability to support affective deterrence, this capability can be rendered inconsequential if the deterring state lacks the will and credibility to carry out the deterring action (Sheldon 2008). Credibility is the perceived likelihood that the deterring state will follow through with its threat, if its terms are not obliged (Snyder 1961). There is a fundamental tension between credibility and prospective pain. Because of a rational fear of retaliation, the more painful or extreme a potential action is, the less likely it is to be taken, and the less likely it is that anyone will believe it will be taken (Gray 1993). Credibility is dependent on the specific context of the security relationship, and effective credibility relies heavily on the political will of the deterring state to carry out its punitive actions.

While the possession of capability is essential, projecting the willingness to use punitive military force is paramount for deterrence to succeed (Schelling 1966). For this reason, there needs to be a belief that the political will exists to respond with severe military response if attacked in order for deterrence to work. One of the most dangerous scenarios is one in which the deterring state's determination to fight is underappreciated (Gray 1991). Having the requisite political will in using punitive action should deterrence fail is easily subject to misperception, and communicating political will does not inherently make it true or believed (Schelling 1966). Measuring the efficacy of projecting capability and political will, therefore, lies with those to be deterred.

Effective Communication Is Required

Any effort to affect an adversary's decision-making is best served by clearly communicating one's desire, intent, capability, credibility, and rationale for military response. This requisite communication may be achieved through official statements or policy documents or more importantly through a demonstrated history of consistent actions (Klein 2019). If the deterring state is not clear in identifying the specific behaviors that it is trying to deter and conveying the threat of what will transpire if an aggressor chooses not to be deterred – along with the defensive capabilities mobilized to discourage them – then the prospects for successful deterrence are diminished. If one's deterrent message is not received or comprehended, then it will be difficult for deterrence to succeed (Schelling 1966).

In addition to the impediments in communicating deterrence in general, deterrence in space presents its own unique challenges that further complicate it potential success. The remoteness of space, highly classified nature of many of these systems, and perpetual concerns regarding dual-use technology all contribute to an environment where both sides of the deterrence relationship have limited awareness of or insight on the others behavior and conduct (Todd Harrison et al. 2017). Indeed, the dual-use nature of space systems can be particularly troublesome when attempting to clearly comprehend or communicate intent, because motive and intent are made more ambiguous when a state fields dual-use capabilities that can be used for civil, commercial, or military purposes.

Often in analysis of high-technology systems, capability is considered equal to intent (Gray 1993). While China is often at the center of debates over capability and intent, some security experts note the United States fields many of the same dual-use systems that elicit concern among its rivals (Johnson-Freese 2017). Intent is a frequently subjective matter and dependent upon one's worldview. For example, the Soviet Union viewed the US Space Shuttle program as a potential ASAT weapon because Soviet military leadership thought the Shuttle was capable of retrieving satellites and de-orbiting them (Wright et al. 2005). Assuming a worst case of intent based solely on an enemy's capabilities can raise the possibility of miscalculation and increase tensions among states when potentially none may be warranted.

To avoid any potential breakdowns in a deterrence strategy, clear communication of intent, credibility, capability, and what behaviors are sought to be deterred is paramount. While this would be difficult within the other domains, it is particularly important for deterrence in space.

Managing Escalation May Be Problematic

Escalation is an especially complicated issue for the space domain, where an absence of historical experience of military conflict leaves the strategist with little empirical evidence to draw upon. Clausewitz explains the challenge of managing escalation, describing how the interaction of forces tends to drive war to the extreme (Clausewitz 1989). Schelling agrees, writing that escalation sets a pace that cannot be directly controlled (Schelling 1966). The propensity for conflict to escalate means that the space strategist should not act first without considering the potential repercussions of military action (Clausewitz 1989). As a result, prudence is necessary in the formation of a space strategy centered on deterrence.

Escalating horizontally into a different domain could result in much greater escalation than previously anticipated. For instance, a state's response to an attack on its satellites could involve terrestrial targets, thereby potentially causing causalities or violating another state's sovereignty. This horizontal escalation may be politically provocative and could drive further escalation. In many cases, militaries rely heavily on commercially procured and provided satellite services. Attacks against these commercial services could be seen as an inappropriate action that is escalatory to the international community.

Prospects for Strategic Misperception

Strategic theory shapes how states prepare for and conduct strategy. This dimension then is pertinent when considering the execution of strategy. Understanding the strategic theoretical dimension calls for an appreciation, or at the very minimum recognition, of potential differences in interpretation of strategic theory of the adversary. Strategy mismatches – in which there are different cultural and social understandings in the theories of deterrence and escalation control – are some of the most dangerous situations between states. This danger arises because states, whose leaders may consider themselves to be rational and reasonable in not seeking direct military confrontation, may find themselves in such a war, despite their intent or desire. Because of the different understandings of deterrence in preventing war or deterrence's ability to control escalation during conflict, it is important to underscore the differences between American and many Western countries' views and the perceptions of Russia and China. The Russian military's strategy of "escalate to de-escalate" and the Chinese view of using "compellence" through military actions to avoid conflict are two strategic approaches to deterrence that are not emphasized in Western views on deterrence but must be well understood by policy makers and strategists.

In describing *strategic deterrence*, Russian military writings describe the term as an approach seeking to induce fear in opponents, whether in peace or war.

Therefore, the concept includes elements of what others may call *deterrence*, *containment*, and *coercion* (Fink 2017). Russia's approach to deterrence is grounded in its understanding of internal and external threats, including a sense of military asymmetry compared to the West. Russian military doctrine describes perceived dangers from the United States and NATO readiness to use military force, instability and terrorism that could challenge Russia's sovereignty, and a local conflict on its vast borders that could escalate into hostilities, which could include the use of nuclear weapons (Klein 2019). In the Russian perspective, strategic deterrence is not entirely defensive. Within US security circles, some may consider Russia's view of strategic deterrence as an "escalate to de-escalate" strategy – even though that term is not used within Russian military doctrine or strategies – because the strategy comprises the use of military force and actions to potentially de-escalate hostilities or tensions (Schneider 2017). The Russian concept transcends a traditional perception of deterrence having failed if conflict erupts. Therefore, deterrence can continue to work in times of war to prevent escalation, to ensure de-escalation, or for the swift termination of conflict on terms acceptable to Russia.

As with Russia, the Chinese concept of deterrence is fundamentally different than American and Western thinking. Analysis of Chinese writings notes the Chinese concept of deterrence includes a significant element of compellence and coercion; therefore, Chinese deterrence goals may include actions seeking to intimidate the opponent through economic, diplomatic, or military coercion in a way that directly affects an opponent's interests in order to compel the foe to submit to Beijing's will (Kaufman and Hartnett 2016). As a result, the Chinese see deterrence as having a positive object for achieving political ends, whereas the West typically places emphasis on the negative object of deterrence: the discouragement of actions. There are nuances in the Chinese terms used, especially those with more coercive connotations.

Chinese strategists view escalation not as a risk to be avoided but a means to manipulate an adversary (Lewis 2018). China also places special emphasis on overwhelming an opponent through rapid escalation, an approach that – when coupled with manipulation of an opponent's perceptions of the costs of a conflict through coercive measures – increases the chances for dangerous misperceptions. Chinese writings note that along a continuum of conflict, there may be scenarios where militaries are involved but war has not yet formally broken out (Kaufman and Hartnett 2016). Differences between Chinese and American views of deterrence include the Chinese focus on compellence, including coercion, rather than solely on dissuasion. Therefore, the Chinese idea of deterrence manifests itself in both coercive and dissuasive terms (Cheng 2018).

Differences in deterrence theory among Russia, China, and the United States are significant in how they may manifest in practice. Russia's emphasis on harnessing escalation to its advantage and Chinese views on compellence, military activities short of war, and rapid escalation create opportunities for misperceptions and potentially an irreversible slide into conflict that no state desires. These strategy mismatches also have implications for space deterrence because they demonstrate how states may approach deterrence in this domain differently, potentially in ways that make it more likely that conflict will occur.

Space Deterrence Has a Terrestrial Aspect

While this chapter focuses on space deterrence, it is essential to emphasize that deterrence and prospective conflict in space do not occur in isolation from a political and terrestrial context. This has two implications for space deterrence. First, even when there are times of significant instability in the space domain between two powers, the broader strategic landscape of inherent stability within other domains may restrain the initiation of conflict in space (MacDonald et al. 2016). Alternatively, a stable space environment could witness conflict if instability in other domains caused deterrence efforts to break down between rival space powers. Second, deterrence by punishment strategies attempting to prevent aggressive behavior in space is not limited in their targeting to the space environment. A strategy of space deterrence could succeed if it threatened terrestrial assets valued by the potential aggressor, and not necessarily just their space systems, a view that corresponds with the 2017 National Security Strategy. Therefore, there are complex interdependencies that exist between space and terrestrial domains in the effective implementation of a space deterrence strategy.

Space strategy indirectly influences general deterrence by enhancing the lethality of terrestrial forces and by increasing transparency in a deterrence relationship between competing states.

Space-based or space-enabled communications, surveillance, early warning, and navigation services can enable better coordination, communication, logistics, and superior situational awareness to terrestrial forces, thereby enhancing these terrestrial forces' response time, tempo, and operating efficiency. Space-enabled forces can typically engage an adversary with greater speed, precision, and coordination when compared to forces that lack sufficient command of space. Some analysts assess that the strategic effect accorded by space capabilities shifted the basis of US deterrence strategy from the threat of nuclear punishment to denial of the adversary's conventional offensive success (Coletta 2009). Space-enabled capability is thought to give a state's military the ability to increase the lethality and efficacy of its forces, which can in turn create a powerful deterrent to a would-be adversary. Many within China and Russia believe that US space-enabled conventional forces can cripple the command and control of their forces, even without the use of nuclear weapons (Lewis 2018).

Space also contributes to deterrence by creating transparency between adversary states. Space-based systems' global and nearly ubiquitous nature allows satellites to peer into the normally opaque actions of states and provide greater insight to decision-makers (Smith 2016). Satellites' freedom of overflight creates transparency between states, which is essential for deterrence to succeed. This information and knowledge help alleviate some of the unfounded fears between states and may aid in preventing strategic miscalculations. It must be emphasized that space-based capabilities do not allow one to be privy to thoughts and intentions of an adversary, and consequently, uncertainty will persist, even if mitigated to some degree (Smith 2016).

Conclusions

Space imparts many strategic benefits that enable the military and nonmilitary activities of states. The strategic effect derived from space-based capabilities will not remain unchallenged when states drift toward war. Consequently, space powers will likely seek to implement a practical space deterrence strategy to protect their national interest and achieve political ends. Even though deterrence has a legitimate role in future space strategy, it is not the panacea for preventing conflict. Strategic history teaches that deterrence will at times fail due to miscalculation, uncertainty, or chance – ideas incorporating the concept of Clausewitzian friction. This may also be the case for deterring acts of aggression in space, especially considering China, Russia, and the United States have different perspectives on deterrence and escalation control. Facing recent nefarious activities of China and Russia, security commentators in the United States now emphasize a return to great power competition. Space has a unique role in this competition because all three great powers are also great space powers that seek to broaden their use of space while also fielding capabilities to contest command of this domain. Space deterrence will then play an important role within the global community in the future.

Albeit this chapter has emphasized the enduring nature of war and strategy (and therefore deterrence), the character of war changes with time. The implementation of space deterrence should also be expected to change. This change is currently reflected by the growth of the commercial space sector (particularly in the United States, Europe, China, and Japan) – whether in reusable or responsive launch vehicles or mega-constellations of Earth imaging and communications satellites. Studies of space deterrence often omit the potential role of the burgeoning commercial space sector. The exponential growth in commercial capabilities means that denying space services or degrading another's access to or use of space will become even more challenging for great space powers. The commercial space industry can help convey the futility of conducting a hostile act in space, because it will be difficult to deny products or services through a hostile action. This fact may cause a potential adversary's leadership to avoid military confrontation in the first place. Therefore, deterrence by denial may play a greater role than deterrence by punishment during future strategic deliberations than it has to date. This situation is an advantageous development, because governments can focus less time and resources on fielding military-related programs for use in times of conflict, instead giving more support to those commercial services and capabilities that can be used for the benefit of all.

References

Cheng D (2018) Evolving Chinese thinking about deterrence: what the United States must understand about China and space. The Heritage Foundation
Clausewitz C von (1989) On war (trans and ed: Michael Howard and Peter Paret). Princeton University Press, Princeton
Coletta D (2009) Space and deterrence. Astropolitics 7(3):171–192

Fink A (2017) The evolving Russian concept of strategic deterrence: risks and responses. Arms Control Today 47
Gray C (1991) Deterrence resurrected: revisiting some fundamentals. Parameters 21:40–47
Gray C (1993) Weapons don't make war: policy, strategy, and military technology. University Press of Kansas, Lawrence
Gray C (2007) Fighting talk: forty maxims on war, peace, and strategy. Greenwood Publishing, Westport
Harrison R, Jackson D, Shackelford C (2009) Space deterrence: the delicate balance of risk. Space Def 3(1):1–30
Harrison T, Cooper Z, Johnson K, Roberts T (2017) Escalation and deterrence in the second space age. Center for Strategic and International Studies, Washington, DC
Johnson-Freese J (2017) Space warfare in the 21st century: arming the heavens. Routledge, Abingdon
Joint Chiefs of Staff (2018) Space operations. Joint Publication 3–14
Kaufman A, Hartnett D (2016) Managing conflict: examining recent PLA writings on escalation control. CNA, Arlington
Klein J (2019) Understanding space strategy: the art of war in space. Routledge, Abingdon
Krepinevich A, Martinage R (2008) Dissuasion strategy. Center for Strategic and Budgetary Assessments. https://csbaonline.org/research/publications/dissuasion-strategy
Kueter J, Sheldon J (2013) An investment strategy for national security space. Special report no. 129. The Heritage Foundation
Lewis J (2018) "Bottom line thinking" about the "Commanding Heights". In: Wright N (ed) Outer space; earthly escalation? Chinese perspectives on space operations and escalation. NSI, Boston
MacDonald B, Blair D, Cheng D, Mueller K, Samson V (2016) Crisis stability in space: China and other challenges. Foreign Policy Institute. Washington, DC
Morgan P (1977) Deterrence: a conceptual analysis. Sage Publications, Beverly Hills
Nye J (2005) Softpower: the means to success in world politics. Public Affairs, New York
Office of the Assistant Secretary of Defense for Homeland Defense (2015) Space domain mission assurance: a resilience taxonomy. http://policy.defense.gov/Portals/11/Space%20Policy/Resilience TaxonomyWhitePaperFinal.pdf?ver=2016-12-27-131828-623
Schelling T (1966) Arms and influence. Yale University Press, New Haven
Schneider M (2017) Escalate to De-escalate. U.S. Naval Institute Proceedings. https://www.usni.org/magazines/proceedings/2017-02/escalate-de-escalate
Schulte G (2012) Protecting NATO's advantage in space. Transatlantic Current (5). National Defense University. www.dtic.mil/dtic/tr/fulltext/u2/a577645.pdf
Sheldon J (2008) Space power and deterrence: are we serious? The George C. Marshall Institute. Policy Outlook
Smith MV (2016) Space power and the strategist. In: Bailey R Jr, Forsyth J Jr, Yeisley M (eds) Strategy: contest and adaptation from Archidamus to airpower. Naval Institute Press, Annapolis
Snyder G (1961) Deterrence and defense. Princeton University Press, Princeton
The White House (2017) The national security strategy of the United States of America
Thomson A (1995) Satellite vulnerability: a post-cold war issue? Space Policy 11(1):19–30
Vedda J, Hays P (2018) Major policy issues in evolving global space operations. The Mitchell Institute for Aerospace Studies
Wright D, Grego L, Gronlund L (2005) The physics of space security a reference manual. American Academy of Arts and Sciences, Cambridge

Resilience of Space Systems: Principles and Practice

Regina Peldszus

Contents

Introduction	128
Resilience as Concept in Space Security Policy	129
Resilience for Deterrence in an Emerging Threat Environment: US Perspective	129
Resilience for Critical Infrastructure Protection and Non-dependence: European Perspective	131
Resilient Architecture and Infrastructure: The Mission Assurance and Deterrence Perspective	132
Resilience as Key Quality of Functional Architecture	132
Functional Elements of Resilient Architecture	133
Practical Measures	134
Trading Off Resilience and Capability in Architecture	134
Resilient Operations and Organizations: The High Reliability and Resilience Engineering Perspective	135
Resilience Through Sensemaking	136
Resilience Through Performance Variability	138
Practical Measures	138
Assessing Resilience in Operations and Organizations	139
Disciplines Contributing to Resilience	139
Resilience Through Space Situational Awareness	140
Resilience Through Transparency Measures and Partnerships	140
Resilience Through Foresight	141
Conclusion	141
References	142

Abstract

In view of the increasing complexity of the space environment, resilience has emerged as a pervasive concept in contemporary space security. This chapter provides an overview on the principles and practice of resilience of space systems

R. Peldszus (✉)
Department of Space Situational Awareness, German Aerospace Center (DLR) Space Administration, Bonn, Germany
e-mail: regina.peldszus@dlr.de

and operations. It frames the emerging field from two distinct but complementary approaches: mission assurance and deterrence and high reliability and resilience engineering. Drawing on contemporary thinking from civilian and military perspectives, the chapter posits resilience as a distinct yet malleable notion at the intersection of space security and safety and highlights specific areas meriting further engagement for policy makers, systems analysts, and operators.

Introduction

In the past decade, the concept of resilience has come to the fore in contemporary space policy and systems development as a critical quality of space infrastructure and a prerequisite for space security (Pace 2015). As the utilization of the orbital environment transforms, space assets are becoming increasingly exposed to the hazards and dynamics of ever more heterogeneous activity.

Situated within a complex operational domain, space systems are highly complex themselves. Characterized by nonlinear, interdependent interactions and tight coupling, they are prone to incidents or failure (Perrow 2007). Specifically, the bespoke exquisiteness of spacecraft and fleet of spacecraft – the current paradigm of communications, positioning, navigation, timing (PNT), and Earth observation systems – makes them susceptible to internal and external disturbance that includes a hostile physical environment with extreme temperature changes and radiation but may also be posed by operational constraints, mishaps, and, as it is increasingly asserted, the potential of adversary threat.

In view of this "brittleness" or fragility, strategic planners increasingly advocate rendering systems more resilient. (The terms "resilience" and "resiliency" are used synonymously in the relevant body of thought and practice across the sector; for the purpose of this chapter, "resilience" will be used to denote both.) Depending on domain and methodological vantage point, resilience in highly complex large-scale sociotechnical systems refers to the ability to withstand disturbance, bounce back from failure, and continue operations under varying conditions through qualities such as robustness, redundancy, resourcefulness, flexibility, survivability, and contingency planning (Haimes 2009, here 496; Air Force Space Command 2016). Whether articulated as a property or process (cf. sections "Resilient Architecture and Infrastructure: The Mission Assurance and Deterrence Perspective" and "Resilient Operations and Organizations: The High Reliability and Resilience Engineering Perspective"), resilience essentially manifests the *state* of a given system and its subsystems to respond to specific threats, and addressing through physical configuration, operation, and organization is understood as integral to risk management processes (Haimes 2009).

This chapter maps out the conceptual notions of resilience specifically for the space domain. In adopting a distinction parallel to that of security in and from space, focus is placed on resilience *in* space, or of space systems, which here refers to both the space and ground segment.

There is currently no generally agreed nomenclature in the community of practice for defining, describing, implementing, and assessing resilience in space. At the same time, emerging frameworks for the resilience of space systems are not yet widely shared, and their applicability is often tailored to the specific sub-domain they have been conceived to address, i.e., military or civilian system. Charting the notion of "resilience" in the space domain therefore requires casting a wider net in the safety and security sciences and relating already consolidated concepts to emerging formulation in the context of space security.

Two major perspectives on resilience in space currently inform policy making and implementation: on the one hand deterrence and mission assurance and on the other resilience engineering and high reliability organizing. The following first outlines how these strands feature in discussions on resilience in contemporary space security policy and then addresses them in detail with regard to how they theoretically and practically relate to space architecture and infrastructure and organization and operations. Disciplines contributing to resilience in space are then highlighted, and a conclusion suggests issues meriting further attention in the immediate future.

Resilience as Concept in Space Security Policy

As space actors explore future directions in a changing and complex domain, the notion of resilience has lodged itself firmly as an important element in developing space policy. However, despite its pervasive invocation as an end of policy efforts – both to render space systems more resilient and ensure societal resilience through space infrastructure and services – the terminology and approaches to resilience are little consolidated in theory, deconstructed for practical application, or subject of a sustained discourse akin to that for resilience in other domains such as aviation or offshore operations. In the past years, two major conceptual directions have begun to emerge in the USA, Five Eyes community, and Europe, which represent discrete approaches that are mutually complementary.

Resilience for Deterrence in an Emerging Threat Environment: US Perspective

Resilience first surfaced prominently in the US Space Policy of 2010, whose objectives included "increas[ing] assurance and resilience of mission-essential functions" (Arnold and Hays 2013, here 121). The idea was broken down into the development of instruments, structures, and capabilities required for the continuity of space-based services in view of a "degraded, disrupted, or denied space environment," and mechanisms to ensure that requirements for mission assurance and space system resilience would be addressed during acquisition processes for future space capabilities (US National Space Policy 2010, here 9).

In response to disruptive changes in the orbital environment – specifically with regard to new state actors with capabilities that increasingly included a repertoire of technology that could be used for offensive actions – resilience was posited as one of the key approaches to maintaining superiority in space (Pawlikowski et al. 2012): if it were evident to a potential adversary that a system would bounce back from attack, or the damage inflicted would either be recovered swiftly or have limited repercussion on the overall capability afforded through a system or architecture, this would change the calculus of an adversary to attack. The aim was hence a resilient architecture that would be able to "support the functions necessary for mission success with higher probability, shorter periods of reduced capability, and across a wider range of scenarios, [environmental] conditions, and threats, in spite of hostile action or adverse conditions" (DOD 2012, here 4 and 14). Designing and maintaining a resilient system that would withstand and recover would thus mitigate or deny an adversary the benefit of attack (ibid.). In this sense, resilience formed a fundamental element of layered deterrence (Johnson-Freese 2016).

The US approach to framing and creating taxonomies for resilience in space to this date draws predominantly on the defense vantage point. This is also reflected in the space policies of the Five Eyes community. Here, the focus on architecture and capability subtly shifts to missions or infrastructure but in a similar context of sovereignty, emerging threats, and the terminology of a "congested and contested" domain, whereby resilience would serve as countermeasure against adversarial activities to "disrupt, degrade, or damage" (A New Space Strategy for Canada 2019, here 16; National Space Policy 2015). While not all space policies of the Five Eyes explicitly echo the US concept of resilience in these specific terms, their strategies and reasoning are supported by language to that effect: in placing emphasis on the need for space security, they focus on pillars regarded as contributors to resilience in US space policy, such as international partnerships and Space Situational Awareness (see also section "Disciplines Contributing to Resilience") (cf. the space strategies, respectively policies, of New Zealand and Australia).

The dedicated UK Space Security Policy (2014) further draws on resilience as element of its definition of space security per se. It highlights a dedicated approach with regard to becoming resilient in view of concrete disruptions, both human-made and natural. To this end, the overall goal of resilience of space capabilities and services is diversified into several objectives (ibid., here 4). These include the pursuit of a "proportionate approach to investing in resilience, balancing protective measures with other means [...] such as alternative or fallback capabilities" that allow for continued availability of services, with the aim of "enhance[ing] the resilience of essential services [...] to the disruption of satellite operations"; and the commitment to "work with [other partners including the US, EU, EU Member States, and ESA] on an integrated approach to security in European space programs [...] including infrastructure and systems resilience" (ibid.).

By emphasizing disruption rather than threat and deconstructing resilience in business continuity terms that dovetail with the nomenclature used by a wide range of stakeholders including commercial industry, the UK policy already uses a number of concepts that resonate with a civilian-rooted approach more prevalent in Europe.

Resilience for Critical Infrastructure Protection and Non-dependence: European Perspective

Recent European perspectives on resilience in space reflect US policy to a large degree, albeit on a higher level of abstraction and by situating resilience at the intersection of a multifaceted interpretation of space security, reliability, and safety. They place greater emphasis on space-based systems as critical infrastructure, rather than prominently as means of force projection, space control, and dominance or superiority. Instead, resilience is presented and proposed through a prism of views including civil protection, continuity of services, strategic non-dependence, and autonomy, as necessary measures in response to asymmetric and hybrid threats, as means to foster the synergy of civil and military capabilities or a robust industrial base, and as societal protection from natural hazards (cf. Robinson et al. 2018; ESPI 2018; Pellegrino and Stang 2016, here 8; cf. also ESA's proposed Space Safety program).

National space policies in Europe can be positioned on this spectrum. The French space defense strategy explicitly situates resilience as a central consideration of strategic, operational, and space systems development efforts in view of adversaries' offensive counterspace capabilities (Ministère des Armées 2019). The current German space strategy does not refer to resilience, but the government's overarching security policy deliberations identify resilience of space systems – as critical infrastructure – as fundamental need in the context of hybrid threats (BMVg 2016, pp. 58, 60). Other European space policies do not employ the concept of resilience but draw on compatible concepts, such as reliability, continuity, and access to space, and dedicate considerable attention to laying out specific elements conducive to resilience, such as international partnerships and Space Situational Awareness (cf. the space policies or strategies of Italy, Norway, and Sweden).

On a supranational level, the distinct element of European non-dependence in view of third-party capabilities surfaces, which echoes sovereignty concepts in the US deterrence context. As part of its four strategic goals, the European Union's Space Strategy (European Commission 2016) ascribes to "resilience of critical European space infrastructure" a central role as catalyst to reinforce European autonomy in space access and utilization, by ensuring the "protection" and "integrity" of the flagship programs for navigation and Earth observation, Galileo and Copernicus (here 8–9, cf. also European Defense Action Plan). To this end, and analogous to the USA, the specific measure of consolidating diverse European Space Situational Awareness capabilities is proposed; their current transformation into a dedicated program articulates distinctly civilian terms but accommodates considerable leeway in extending both operational and research and development efforts to a wider range of hazards and threats.

Beyond the specific reference to Space Situational Awareness (SSA) as a driver – and apparent placeholder – for resilience, the concept for resilience is neither fleshed out theoretically in European space policy in further detail nor translated into concrete activities or instruments for assessment and evaluation.

Indeed, the development of a dedicated methodological framework has been pointed out as a necessary element for further discussion (Pellegrino and Stang 2016). Yet, on a higher level in the context of EU foreign and security policy, the idea of resilience has become an overall *leitmotif* and guiding principle in the evolution of shared structures that allow averting external risks and threats (Bendiek 2017, here 14). Similarly, the upcoming NATO space policy can be expected to employ language on resilience, given its increasing awareness of the space domain and an overall stance on addressing hybrid warfare (cf. Prior 2017).

Regardless of the idiosyncratic nuances that are being emphasized in the integration of resilience into space security policy and the level of maturity of the related discourse, there is nascent understanding of the various elements for application in space systems, both in architecture and operations.

Resilient Architecture and Infrastructure: The Mission Assurance and Deterrence Perspective

Extending the scope of policy, a limited but growing body of work exists on deconstructing the concept of resilience into applicable elements for space. One approach, departing from the perspective of assurance and deterrence as outlined previously for US space policy, involves describing discrete measures that can be practically applied in space systems and guide their development.

Resilience as Key Quality of Functional Architecture

In response to the National Security Space Strategy from 2011, and in explicit recognition of the lack of a commonly shared taxonomy to facilitate a discussion on resilience and an approach to measuring it, resilience was fleshed out further through a dedicated taxonomy (OSD 2015). The concept was structured specifically though the lens of mission assurance in the warfighting rather than systems engineering domain, focusing on space-based and ground-based infrastructure. (This was contextualized by highlighting the overall assurance afforded by being able to switch to an alternative domain outside space, which was, however, not subject of the taxonomy effort.) Thus, "Space Domain Mission Assurance" was defined as distinct pillars that flanked resilience between defensive operations and reconstitution (ibid., here 3 and following):

– *Defensive Operations* (disrupting an adversary's ability to target; direct intercept; systematic maneuvering to avoid, confuse, or overwhelm a targeting system; active measures to deceive, degrade, or destroy a targeting system)
– *Reconstitution* (providing backup capacity by launching additional satellites or providing additional ground stations; replenishing parts of a constellations; add new signals or spectrum)

- *Resilience* (an to support "functions necessary for mission success with higher probability, shorter periods of reduced capability ... in spite of hostile action or adverse conditions")

The taxonomy emphasizes that, rather than resilience per se being the primary goal of an assurance effort, "it is the warfighting mission assurance benefit, derived from resilience, which [it] seek[s] to assure" (ibid., here 2). As a *means* for assurance rather than an inherent overall goal of a system, resilience is understood as an "internally focused characteristic" – a critical quality or property of a capability that helps ensure its continued availability, reliability, and integrity.

Functional Elements of Resilient Architecture

Much of the discussion on mission assurance was initially centered on the concept of disaggregation as one approach to achieve or improve resilience (Air Force Space Command 2016). Disaggregation here meant the "dispersion of space-based missions, functions or sensors across multiple systems" (ibid., here 3). This would entail five disaggregation approaches, including *modular decomposition* within a single system to allow, e.g., individual subcomponent replacement; *functional* disaggregation by distributing sub-missions on separate platforms, including *hosted payloads* on assets of different missions or agencies; *multi-orbit* disaggregation that employs multiple orbital planes; and *multi-domain* disaggregation, whereby space-based and ground-based systems cooperatively or complementarily perform a mission. From a general resilience viewpoint, disaggregation resonates with the fundamental notion of deconcentration of critical or exposed capabilities for target reduction, which constitutes a paramount approach to addressing vulnerabilities of large-scale sociotechnical systems (Perrow 2007, here 6 and 261).

However, for the space domain, additional – and partially overlapping – dimensions beyond disaggregation were being explored. Distribution, dispersion, and diversity by leveraging the capabilities of government and commercial stakeholders, increasing the number of platforms, and focusing on hosted payloads and mixed architectures were advocated early on as architectural – and hence also acquisitions-related – responses to the contemporary challenges faced by the traditional class of aggregated, highly integrated assets with long lifetimes (Pawlikowski et al. 2012).

Eventually, the focus of policy makers shifted from disaggregation toward a wider context of space protection after the 2014 Space Strategic Portfolio Review (cf. Johnson-Freese 2016, here 171; McLeod et al. 2016, here xii). Resilience itself was broken down into six characteristic architectural "sub-elements," systematically defining several concepts that had been shown as partially interrelated elements previously (OSD 2015) (Here slightly changed in order of appearance for easier comparison):

 - *Disaggregation*, the separation of dissimilar capabilities into separate platforms and payloads, thereby in cases also reducing overall complexity of the system

- *Distribution*, by employing a number of nodes that jointly perform the same function of mission as a single node, allowing for graceful degradation despite failure of a single node
- *Diversification*, by employing different platforms, orbits, systems or actors' capabilities to contribute to the same mission; flexible or adaptable systems
- *Proliferation*, by deploying larger numbers of the same [similar] platforms, payloads, or systems of the same types to perform the same mission
- *Protection*, through active and passive measures including protection from jamming, nuclear hardening, extended maneuverability, internal hosted decoys, onboard countermeasures, onboard/operational event characterization or attribution efforts/instruments
- *Deception*, through measures to confuse or mislead regarding location, capability, operational status, mission type, robustness of system/platform or payload; measures at architectural, operational, or organizational level

While deception was identified explicitly as "a critical element of any space system resilience effort," all elements need not necessarily be present in a single architecture but rather enhance resilience cumulatively or in combination (ibid., here 8).

Practical Measures

A wide range of detailed practical measures can be mapped onto these elements for different levels of space systems. (Note that this selection of practical aspects dovetails with the concepts outlined in the next section.) On the platform level, this may include the hardening and shielding against radiation and kinetic and non-kinetic manipulation; fitting bimodal receivers for different navigation systems or equipping the spacecraft with measures for easier tracking; increased onboard autonomy for measures such as passively safe trajectories during proximity operations; reactive maneuvering in view of another approaching object; and cyber protection to safeguard commanding and telemetry. In the ground segment, there are a number of measures ranging from situating facilities in remote areas for limiting discovery, access, or interference; installing backup facilities; ensuring interoperability with legacy, novel, and partner infrastructure; and putting physical and information security of command and control infrastructure in place (e.g., protecting from mishaps such as severed cables of ground stations during off-site building works, damage or wear of critical equipment through climatic conditions, or compromised mission control software).

Trading Off Resilience and Capability in Architecture

Resilience has been included as a key criterion in the evaluation of alternative space architectures (National Security Space Strategy 2011) and ought to be taken into account at the beginning of the systems planning process as a "critical component to define at system level" (OSD 2015, here 8). Aside from weighing the different elements of resilience against each other, also resilience itself must be traded off with other characteristics of a capability – indeed, since resilience is not understood as a capability itself, resilience and capability must be treated as distinct concepts (Jakhu and Pelton 2017, here 296).

Since in view of resource constraints the benefits of architectural resilience come at a cost, affordability is a key driver in this capability-resilience trade-off (cf. Pawlikowski et al. 2012, here 47). Implementing elements such as distribution or diversification means that other performance aspects of a system or architecture – e.g., sensor coverage, integration times, and procurement cost – may be constrained. These dependencies must be traded against their benefits across the system life cycle and the complete system hierarchy with regard to different threat scenarios (Aerospace 2018).

For want of extensive dedicated metrics, five tentative criteria to assess resilience of a functional architecture include the anticipated level of adversity, functional capability goals of the architecture itself, the risks of not achieving these goals in view of adversities, the severity of functional shortfall, and the duration of downtime that can be tolerated by the mission (DOD 2011).

Resilient Operations and Organizations: The High Reliability and Resilience Engineering Perspective

Once hardware on the ground or in space has been commissioned, changes and modifications to increase the quality of resilience are either infeasible or involve considerable resources. Other system elements are, however, more malleable and may be actively adapted across the life cycle to different extent, including human operators, procedures, or mission rules (McLeod et al. 2016). Next to mission assurance for architecture, a second perspective on resilience hence focuses on operations and organizations rather than infrastructure. Rather than property of a system, resilience here means a continuous pursuit or process, not a characteristic that can be instilled in a system, rather, something that a system is enabled to *perform*.

Normal accident theory (NAT) posits that failure of complex sociotechnical systems in high-risk domains such as space is both inevitable and rare (Perrow 2007). In response to NAT, two proactive fields have formed in safety management across the past decades: high reliability organizing (HRO) and resilience engineering (Haavik et al. 2016). They propose that in view of inherently unsafe systems – or systems exposed to continuous risk – it is in fact the performance of human operators that contributes to safety under varying conditions (Dekker 2012). As they share fundamental terminology with the field of RAMS (Reliability, Availability, Maintainability, Safety) and dependability, and specifically include use cases in both civil and military domains, resilience engineering and high reliability organizing offer an important contemporary lens on aspect of resilience and its context that have not yet matured for the space domain in the mission assurance context.

Resilience Through Sensemaking

Sectors and organizations that are understood through the HRO lens or operate according to HRO principles (i.e., air traffic management, aircraft carriers, utility grids) share fundamental characteristics with space systems. They are highly complex on all levels of the system hierarchy – respectively, nested into systems of systems – with interdependent elements, components, and parts that are tightly coupled and integrated and feature nonlinear interactions. In view of the constraints in the operational environment, they are governed by a high degree of causality, i.e., the laws of physics more so than purely organizational intent, and predominantly face either a physically hostile operating domain (i.e., submarine) or the handling and control of highly hazardous assets or processes (i.e., nuclear power plant). In operations, they rely on the collaboration of distributed actors that may be situated at a distance from the process in a control room environment; in several instances, the system in operation is either highly bespoke or of international significance and sophistication (i.e., a fusion experiment, sample return mission). (In comparison, automotive manufacturing and healthcare environments are also understood as complex undertakings and require a high degree of reliability and continuity but are characterized less by the constraints of physical causality, i.e., when processes come to a halt through a disturbance, the system does not necessarily fail (despite the cost incurred). In contrast, processes in domains such as nuclear power or missile operations require immediate attention and intervention both in routine operations and in view of anomalies in order to avoid irreparable damage or loss.)

Yet, despite tightly coupled processes and constant hazard, highly reliable organizations are able to maintain "continuously safe operations" (Weick and Sutcliffe 2001, here 9). To this end, HROs employ the principle of "collective mindfulness" (ibid., here 9–14). This describes an awareness of ongoing processes by all organizational constituents, combined with an acute understanding of the dependencies and implications of an individual operation or element. Specifically, and in contrast to other types of organizations, HROs operate according to five concrete principles:

- *Preoccupation with failure*, i.e., they cultivate an awareness of small lapses, disturbances, and weak signals; they encourage error reporting and analysis of near misses and foster a culture that challenges complacency and hubris in view of past success.
- *Reluctance to simplify interpretations*, i.e., they deliberately strive for nuanced pictures of a situation and do not rely purely on key indicators; they challenge received wisdom and hear diverse viewpoints.
- *Sensitivity to operations*, i.e., they maintain a situational picture of the "sharp end" (or front line) of operations, which allows continuous adjustments to be made in order to cope with external disturbances.
- *Deference to expertise*, i.e., they foster diverse thinking and an encouragement for decision-making beyond rigid hierarchies by those that are best placed to judge a situation based on their command of the subject matter rather than status or rank.

– *Commitment to resilience*, i.e., maintaining dynamic capabilities for recovery and containment of situations, including flexibility and creative solutions for unexpected problems, but also ensuring continuous supply of fresh resources (i.e., shift personnel) during a crisis or incident.

The final principle can be further understood through the field of resilience engineering. In extending the vantage point and analytical repertoire of HRO, particularly with regard to sensemaking of past, ongoing, and future events, Resilience engineering specifically places emphasis on operator interaction with, and as part of, a system (Leveson 2011; Hollnagel et al. 2008).

Similar to HRO, resilience engineering seeks to understand and leverage the significant part of operations where and how in the face of disturbance the system manages to remain available or "bounces back." (Cf., when looking at a reliability metric of, e.g., 98% in traditional safety management approaches, focus would be directed not only at understanding the 2% of failure cases or incidents, i.e., through failure mode or error analysis, but on the considerable amount of time where "things go right" either in routine or during recovery of anomalies.) Organizational resilience is hence described as the "capability to recognize the boundaries of safe operations ... to steer back from them in a controlled manner" (Dekker 2005) and as the "ability to anticipate and adapt to potential for surprise and error" (Reason 2008, here 8).

Practically, four key abilities contribute to this adaptation and control, i.e., the overall ability of predicting, planning, and executing (Hollnagel et al. 2008):

– *Factual*, learning from past experience such as incidents in view of devising practical measures to address resilience, knowing what has happened
– *Actual*, responding to actual disturbances and regular and irregular threats, i.e., know what to do, being capable of doing it
– *Critical*, monitoring the system's own performance in order to respond to critical events, i.e., know what to look for and direct attention to the right areas
– *Potential*, anticipating potential disruptions, pressures, and their consequences in the near future, finding out and knowing what to expect

This collective "anticipation of the potential" is a key feature of resilient organizations (Hollnagel et al. 2006). Anticipation focuses on both past and future manifestations and pathways toward failure (Dekker 2012) and aims to make sense of events (i.e., incidents, successful operations). Crucially, it also involves investing resources in the anticipation, adaptation, and growth in response to disruption – both in view of negative stressors and novelty (Reason 2008). The latter part of "growth" represents a critical distinction to other assurance concepts: resilience engineering explicitly includes the possibility that a system is strengthened through meaningfully responding to a continuous barrage of internal and external disturbances. While this potential is usually not formally foreseen in systems planning (e.g., as a performance indicator) and its assessment is not afforded by current safety management tools, in space operations it has been anecdotally evident and crucial to functioning in routine or contingency operations (i.e., creating automated protocols to work around frequently occurring ground station time constraints or outperforming nominal mission life time by ingeniously handling the fuel budget of a spacecraft) or even been assumed as an underlying necessity for missions with high degrees of uncertainty (i.e., devising cutting edge trajectories "on the fly" in response to gradual discovery of targets during special missions).

Resilience Through Performance Variability

Personnel contribute to this quality of adaptation on organizational, team, and individual levels. Practically, they create resilience by adjusting their tasks, inserting buffers or automated routines, using heuristics and double checks, or devising decision-making aids (Dekker 2005, here 12–13) whenever the demands of a situation eclipse predefined rules in an otherwise highly proceduralized domain. This habitual or intentional adjustment is called "performance variability" (Hollnagel et al. 2008).

Rather than being understood as deviation from the norm (i.e., violation of a procedure), performance variability is "normal and necessary" and can be identified by determining the discrepancy of normative and descriptive models of work (ibid.). In space operations, the normative – or designer's – model is described in mission rules, system specifications, formal training manuals, or flight plans as aggregate of procedures. The operator's or "actual" model, however, incorporates also experience after commissioning, as operators "continually test their model against reality," often under time and productivity pressures (Leveson 2011, here 42). Operators may thus determine the change or evolution of the system and its state and the need for subsequent updating of their mental model through varying their performance. These practices require what is termed efficiency-thoroughness trade-off, i.e., where operations have to be compromised in view of resource constraints and increasing demands (Hollnagel et al. 2008).

Practical Measures

Specifically for space operations, some work has explored practical measures for resilience, for both the defense and civilian contexts, by transferring practice from external safety-critical sectors (McLeod et al. 2016; Peldszus 2015). These measures include activities and processes that can be implemented or integrated seamlessly in running operations or carry a comparatively modest cost when juxtaposed with changes in infrastructure, such as:

- *Operational simulation and exercises* using different degrees of fidelity for routine, contingency, and special operations
- *Human performance training* (i.e., communication, situation awareness skills for critical operations)
- *Actionable information and appropriate decision aids*, capturing of shared mental models (i.e., subsystems of an asset)
- Structured, standardized *anomaly resolution* (i.e., according to predefined protocols)

- Centralized, non-punitive *anomaly and near-miss reporting*
- Centralized *operational practice reporting* (incl. recovery and what went "well" in routine)
- Encouraging *smart tacit practice* (i.e., double checks, four-eye principle, informal communication)
- Fostering *culture of openness,* ensuring *availability of resources* to handle serious events
- Building in *slack* or backup plans in processes (including shift planning)

Implementing these measures can be achieved on a spectrum of resource intensity, from updating a rule or procedure, scheduling a short regular review forum, or distributing a familiarization resource to assigning a new position, commissioning a software tool, or rolling out a training campaign. Incentivizing the use of these measures, however, requires an organizational and operational culture that regards resilience as priority (McLeod et al. 2016) and recognizes the critical role of human operators in maintaining resilient operations.

Assessing Resilience in Operations and Organizations

Whether the prerequisites for resilient performance are in fact in place can be evaluated by verifying the deployment and implementation of measures such as those highlighted above. However, whether a complex system really behaves in a resilient manner may arguably only be assessed through a case-by-case appraisal of concrete responses, in view of a specific threat and the particular state of the system at a given time (Haimes 2009). Describing and understanding the functioning of complex systems in routine and contingency may necessitate formal modeling techniques. (These would, for instance, be utilized for architectural trade-offs, cf. previous section.) In order to evaluate resilience, there are, however, additional approaches that include natural language and visual tools.

A structured evaluation of whether and how resilience measures are actually implemented can be performed through methods such as the resilience analysis grid. The method diversifies, in fine granularity, the four key abilities of organizations described earlier (factual, actual, critical, potential) and their requisite resources and processes. As to how fare these measures then impact on operations – and are successful – can be analyzed through modeling methods such as the functional resonance analysis method, which considers the specific conditions, resources, input, and resulting states and can be applied in various operational domains and at various levels of a system.

Disciplines Contributing to Resilience

In addition to the specific measures taken in design and operations to achieve resilience in the mission assurance and reliability context, there are a range of stand-alone fields that contribute to resilience in their own right (see Table 1).

Table 1 High-level elements of resilience in the space domain at a glance

Mission assurance and deterrence	Resilience engineering	High reliability organizing	Contributing disciplines
Disaggregation Distribution Diversification Deception Proliferation Protection	Learning from factual Responding to actual Monitoring critical Anticipating potential	Preoccupation with failure Reluctance to simplify Sensitivity to operation Deference to expertise Commitment to resilience	Space situational awareness Partnerships Information sharing Foresight Transparency and confidence building measures
Architecture	*Operations*	*Organization*	*Governance*

Resilience Through Space Situational Awareness

A key element to remaining resilient from a systems and operational perspective is the ability to understand and act upon risks in the orbital environment in real or near real time through Space Situational Awareness (SSA) (Pellegrino and Stang 2016, here 9). By producing actionable information on the location and behavior of space objects and natural hazards through a general recognized space picture and related services (e.g., collision avoidance), SSA constitutes a fundamental background function that enables the protection of critical services such as navigation and Earth observations. Furthermore, both in the operational and deterrence context, SSA is a prerequisite for resolving certain types of anomalies and for verifying activities that occur in the vicinity of a spacecraft (i.e., rendezvous and proximity operations). Finally, as one of the approaches to mitigating the proliferation of space debris, SSA links directly to the effort of resilience and sustainability of the various orbital regimes per se (McCormick 2013).

A comprehensive understanding of the overall operational environment through SSA benefits considerably from burden sharing. In its reliance on distributed sensor networks for surveillance and tracking, SSA is today viewed as a global undertaking. Efforts to share and fuse information and data from various different sources are currently gaining momentum.

Resilience Through Transparency Measures and Partnerships

The growing heterogeneity and granularity of actors in the space domain both lend itself to – and indeed necessitates – cooperation and transparency. For recovery in operation but also to achieve redundancy already during architectural development, information sharing and cooperation constitute essential means (Jakhu and Pelton 2017, here 269). Resilience considerations specifically encompass the strategic engagement in partnerships with stakeholders in international and domestic government agencies, industry, and academia (Defense Science Board 2017). Allied or partner systems are the subjects of protection efforts, but the forging of closer architectural, operational, and diplomatic ties with allies and partner also constitutes a key resilience measure as such (DOD 2012, here 14). Leveraging a wide range of capabilities facilitates directly the resilience concepts of diversification and distribution (i.e., through payloads hosted on allied or commercial platforms).

Resilience Through Foresight

Enlarging the scale from operational anticipation that is characteristic of highly reliable and resilient enterprises, resilience must build on foresight in order to anticipate wider ranging future challenges (Pawlikowski et al. 2012). Foresight methods are used to explore uncertainties and chart various possible futures (Healey and Hodgkinson 2008). For resilience in space operations, they may range from the systematic cross-disciplinary scanning of risks, developments, and change drivers to the appraisal of low-probability-high-impact events (cf. rare but inevitable failure in normal accident theory) and the crafting of possible scenarios, to the in-depth exploration of specific potential event and the rehearsing of protocols in large-scale tabletop exercises and red teaming (Peldszus 2018). These activities are most frequently undertaken in collaboration with different actors and are employed to inform both operations and strategy. They thus contribute to the facilitation of collaborative decision-making and good governance for space as a resilient domain and global commons.

Conclusion

Despite the current lack of globally shared nomenclature, two salient perspectives have emerged for the principles and practice of resilience for the space domain. They focus on maximizing the continuation and reliability of operations in various conditions or seek to imbue an architecture with qualities that minimize incentives for adversary actions in an evolving threat environment.

Resilience is likely to continue to feature as a key concept in space policy and systems planning. Straddling the fields of space security and reliability, it may inform, enrich, or even galvanize the more traditional security and safety management disciplines. Its incorporation in European policy may, on the one hand, be influenced by US thinking and its current narrow but very applicable focus; on the other hand, the assurance and deterrence context will be enriched by wider use of complementary insights from the civilian domain.

Quite certainly, the onset of the deployment of unprecedentedly large constellations will both exacerbate the dynamics of the operational environment of orbit and offer new challenges and avenues for the notion of resilience. Its apparent ubiquity and perseverance call for deepened engagement in further developing the nascent field. Specifically, there is a need for the cultivation of a broad discourse to facilitate shared nomenclatures, detailed taxonomies, and the development of assessment methods. Here, much insight can be drawn from other high reliability domains: the scholarly and industrial communities of practice hailing from the nuclear, transport, and offshore sectors have been prolific – if not conclusive – in their quest for shared theory and application on resilience. Finally, it will be crucial to examine how notions of resilience are interpreted and addressed in the programs and strategies of other major spacefaring actors (Russia, China, India), whose advanced capabilities may be viewed in the context of both deterrence and high reliability.

References

A New Space Strategy for Canada (2019) Ministry of Innovation, Science and Economic Development, Alberta
Aerospace (2018) Resilience for space systems: concepts, tools and approaches (ATR-2017-02226). Aerospace Corporation, Washington, DC
Air Force Space Command (2016) Resiliency and disaggregated space architectures: a white paper. US Air Force, Colorado Springs
Arnold DC, Hays PL (2013) Strategy and the security space enterprise. In: Sadeh E (ed) (2012) Space strategy in the 21st century: theory and policy. Routledge, London, pp 120–158
Bendiek A (2017) A paradigm shift in the EU's common foreign and security policy: from transformation to resilience. German Institute for International and Security Affairs, Berlin
BMVg (2016) Weissbuch zur Sicherheitspolitik und Zukunft der Bundeswehr. German Federal Ministry of Defence, Berlin
Defense Science Board (2017) Task Force on Defense Strategies for Ensuring the Resilience of National Space Capabilities, March 2017, Office of the Secretary of Defense for Acquisition, Technology and Logistics, Washington D.C.
Dekker S (2005) Ten questions about human error: a new view of human factors and system safety. CRC Press, New York
Dekker S (2012) Just culture: balancing safety and accountability, 2nd edn. Ashgate, Burlington
DOD (2011) Fact sheet: resilience of space capabilities. US Department of Defense, Washington, DC
DOD (2012) Directive 3100.10: space policy (update 2016). US Department of Defense, Washington, DC
ESPI (2018) Security in outer space: rising stakes in Europe. European Space Policy Institute, Vienna
European Commission (2016) A space strategy for Europe, COM (2016) 705. European Commission, Brussels
Haavik TK, Antonsen S, Rosness R, Hale A (2016) HRO and RE: a pragmatic perspective. Saf Sci. https://doi.org/10.1016/j.ssci.2016.08.010
Haimes YY (2009) On the definition of resilience in systems. Risk Anal 29(4):498–501
Healey MP, Hodgkinson GP (2008) Troubling futures: scenarios and scenario planning for organizational decision making. In: Hodgkinson GP, Starbuck WH (eds) The Oxford handbook of organizational decision making. Oxford University Press, Oxford, pp 565–585
Hollnagel E, Woods DD, Leveson NG (2006) Resilience engineering: concepts and precepts. Ashgate, Burlington
Hollnagel E, Nemeth CP, Dekker S (eds) (2008) Remaining sensitive to the possibility of failure. Resilience engineering perspectives, vol 1. Ashgate, Burlington
Jakhu RS, Pelton JN (2017) Global space governance: an international study. Springer, New York
Johnson-Freese J (2016) Space warfare in the 21st century: arming the heavens. Routledge, New York
Leveson NG (2011) Engineering a safer world: systems thinking applied to safety. MIT Press, Cambridge, MA

McCormick PK (2013) Space debris: conjunction opportunities and opportunities for international cooperation. Sci Public Policy 40(6):801–813

McLeod G, Nacouzi G, Dreyer P, Eisman M, Hura M, Langeland KS, Manheim D (2016) Enhancing space resilience through non-material means. RAND Corporation, Santa Monica

Ministère des Armées (2019) Stratégie Spatiale de Défense: Rapport du groupe de travail ≪ Espace ≫, Ministere des Armees, Paris

NSSS (2011) National security space strategy: unclassified summary. US Department of Defense and Office of the Director of National Intelligence, Washington, DC

OSD (2015) Space domain mission assurance: a resilience taxonomy. Office of the Assistance Secretary of Defense for Homeland Defense & Global Security, Washington, DC

Pace S (2015) Security in space. Space Policy 33(2):51–55

Parly F (2018) Intervention de Florence Parly, ministre des Armées: Espace et défense, 7 September 2018, CNES, Toulouse. https://www.defense.gouv.fr/actualites/articles/direct-florence-parly-s-exprime-sur-les-enjeux-de-l-espace-pour-la-defense. Accessed 22 Jan 2018

Pawlikowski E, Loverro D, Cristler T (2012) Space: disruptive challenges, new opportunities and new strategies. Strateg Stud Q 6(1):27–54

Peldszus R (2015) The Human Element and System Resilience at the European Space Operations Centre, AMCO-TN-0006 and AMCO-TN-00011 (Internal Reports), ESOC, Darmstadt

Peldszus R (2018) Foresight methods for multilateral collaboration in space situational awareness (SSA) policy and operations. J Space Saf Eng 5(2):115–120

Pellegrino M, Stang G (2016) Space security for Europe. EUISS report no. 29, July 2016. EU Institute for Security Studies, Paris

Perrow C (2007) The next catastrophe: reducing our vulnerabilities to natural, industrial, and terrorist disasters. Princeton University Press, Princeton/Oxford

Prior T (2017) NATO: pushing the boundaries for resilience. CSS analyses in security policy, CSS report no. 213. Center for Security Studies, Zurich

Reason J (2008) The human contribution: unsafe acts, accidents and heroic recoveries. Ashgate, Farnham

Robinson J, Šmuclerová M, Degl'Innocenti L, Perrichon L, Pražák J (2018) Europe's preparedness to respond to space hybrid operations. PSSI report July 2018. Prague Security Studies Institute, Prague

UK National Space Policy (2015) HM Government, London

UK Space Security Policy (2014) HM Government, London

US National Space Policy (2010) Office of the President of the United States, Washington, DC

Weick KE, Sutcliffe KM (2001) Managing the unexpected: assuring high performance in an age of complexity. Wiley, New York

Space Security Cooperation: Changing Dynamics

Jessica L. West

Contents

Introduction	146
The Case for Cooperative Approaches to Space Security	146
Moderating Strategic Rivalry: Technical and Utilitarian Modes of Cooperation	148
From Practical to Symbolic: Cooperation in Space Exploration	149
Expanding Access to Space: Cooperation and Capacity-Building	151
Cooperation for Safety and Sustainability	153
New Patterns of Cooperation: Space Security Versus National Security	155
New Issues: The Moon and Space Resources	158
Conclusion: The Future of Space Security Cooperation	159
References	160

Abstract

The security of outer space is a cooperative endeavor to achieve a shared benefit. Yet, while cooperation is essential for space security, it is often fraught. This chapter examines the logic for cooperation as an approach to space security, including supportive governance mechanisms, and traces the impetus and evolution of such efforts over time, marked by struggle to overcome strategic competition. Increasingly, competition is giving way to new patterns of cooperation focused on military alliances and new strategic interests. In this context, it is not clear that cooperation will be maintained as a core value and principle of space activities.

J. L. West (✉)
Project Ploughshares, Waterloo, ON, Canada
e-mail: jwest@ploughshares.ca

© Springer Nature Switzerland AG 2020
K.-U. Schrogl (ed.), *Handbook of Space Security*,
https://doi.org/10.1007/978-3-030-23210-8_123

Introduction

The security of outer space is focused on the security and sustainability of outer space as a global environment that can be used safely by all, rather than the narrow interests of individual actors. At heart, this is a cooperative endeavor to achieve a shared benefit. Yet, while cooperation is both an individual and collective interest, it is rarely straightforward. Like geopolitical relationships on Earth, outer space is subject to not only cooperative impulses, but also competition, self-interest, power disparities, and fear. Sustaining the security of the outer space environment thus involves extensive coordination, but like a dance, it is also marked by missteps.

This chapter begins by examining the logic for international cooperation as an approach to space security, including supportive governance mechanisms. It then traces the impetus and evolution of cooperative efforts in outer space from technical coordination at the dawn of the space age, through large-scale exploration efforts symbolized by the International Space Station (ISS), capacity-building, and contemporary governance initiatives aimed at safety and sustainability. It is clear that international cooperation is a core value and pursuit of national space activities, and over time it has both widened and deepened. Yet, cooperation is at times stymied by competing values, particularly concerns for national security in outer space, reflecting the ups and downs of broader geopolitical relations and tensions. But cooperation is not merely a reflection of politics. As a mode of governing the security of outer space, cooperative relationships and practices contribute to trust, transparency, and interdependencies capable of transcending political pressures elsewhere.

Space security cooperation has thus been marked by an ongoing struggle to overcome strategic competition. Recently, however, such competition is giving way to new patterns of cooperation. Focused on national security *in* outer space rather than the security *of* outer space, the expansion of military alliances and security partnerships into the space domain – increasingly viewed as one of warfare – raises questions about the strategic stability of the outer space environment and the implications for collective wellbeing in outer space. Looking forward, the chapter also considers how heightened geopolitical competition and shifting strategic interests in outer space might influence emerging activities in outer space including lunar and human space exploration and possible resource extraction.

Cooperation is an essential and persistent feature of activities in outer space and necessary to achieve the long-term security of the outer space environment. But such cooperation is also fraught, striving, and sometimes failing to overcome strategic competition. As the nature of this competition changes alongside new actors and ventures, the continued value placed on cooperation is at risk of diminishing.

The Case for Cooperative Approaches to Space Security

The 1967 Outer Space Treaty recognizes the "common interest of all mankind in the progress of the exploration and use of outer space for peaceful purposes" (OST). But like all global commons, the use of outer space is subject to competing – even conflicting – interests. The natural resources of outer space, such as radiofrequency and orbital positions are limited and shared. The environment is fragile and vulnerable to contamination from the accumulation of debris. Growing use of outer space means that it is becoming more congested, especially in popular orbits where, in the next five years alone, the number of satellites in low Earth orbit (LEO) could grow tenfold if proposals for large-scale constellations advance. As a shared environment, threats to safety, security, and sustainability – be they manmade contamination and interference, or natural hazards such as space weather – are mutually harmful.

Despite the declaration of outer space as a province of all mankind, it is also a place of inequality. Long a domain of the powerful, technology and cost barriers limit both access to and use of space and by extension, the tremendous benefits that it supports including remarkable tools for communication, navigation, and vast data collection enabled by Earth observation. And while threats within the space environment are indiscriminate, the ability to mitigate harm is not equally shared.

Outer space is also a place of strategic competition and tensions. Initially marked by existential competition for military, scientific, technological, and economic supremacy exemplified by the space race, today the strategic use of outer space has evolved into dependency and intense military vulnerability.

From a governance perspective, outer space is thus inherently vulnerable to numerous challenges including a tragedy of the commons, persistent inequality, and security dilemmas. It is a place of mutual interests, but also competition, suspicion, and fear.

The concept of space security is a response to these challenges. Defined here as the "secure, and sustainable access to and use of space, and freedom from space-based threats" (West 2019), this approach to space security promotes a secure and sustainable space environment to assure safe and responsible access to and use for all, as promoted in the 1967 Outer Space Treaty. Reflecting a collective approach to the security of – and in – a global commons, the security of outer space depends on international cooperation.

Along with peaceful purposes, such cooperation is the bedrock of the international governance framework for outer space. Institutionally, the United Nations Committee on the Peaceful Uses of Outer Space (UN COPUOS) is the focal point of cooperation. Spurred by the confluence of scientific and military interests in outer space alongside Cold War competition, the Committee was established in 1959 by UN Resolution 1472 (XIV) "International cooperation in the peaceful uses of outer space." Core to its mandate is to facilitate information exchange related to outer space activities, and to promote and support international cooperation as a means to expand the peaceful use of outer space and to avoid extending national rivalries into this domain. Today, with 92 Member States and growing, it maintains a prominent role in the governance of outer space.

International cooperation is also a key principle of space activities enshrined in the 1967 Outer Space Treaty, which, in addition to the United Nations Charter, provides the fundamental basis for legal order in outer space. Specifically, Article III of the treaty mandates that states pursue outer space activities "…in the interest of maintaining international peace and security and promoting international co-operation and understanding" (1967).

Cooperation in outer space is thus essential, mandated, but often fraught. Security in and of outer space, where threats and vulnerabilities are shared and individual actions have collective consequences, mean that there is clearly a mutual benefit to cooperation. But fostering this cooperation requires overcoming strategic rivalry, national security concerns, and competing interests. Efforts to navigate these tensions in pursuit of shared safety, security, and sustainability benefits in outer space have been ongoing since the early days of space activities, built largely on the basis of technical and utilitarian modes of cooperation.

Moderating Strategic Rivalry: Technical and Utilitarian Modes of Cooperation

The first space age is synonymous with the existential competition of the space race. But even amid deep, strategic rivalry, there were efforts to temper competition with cooperative impulses. The promise of cooperation was held out by U.S. President John F. Kennedy in his inaugural address where he declared "Let both sides seek to invoke the wonders of science instead of its terrors. Together let us explore the stars" (1961). Concretely, following his landmark declaration that the United States would land a man on the Moon within the decade, Kennedy is reported to have reached out to the Soviet Union on several occasions to foster cooperation (Kay 1998). However, terror ultimately overwhelmed cooperation. At a time of heated nuclear confrontation, Soviet reciprocity was foregone in favor of focused attention to the negotiation of a nuclear test ban treaty; later it was stymied by Kennedy's death.

Cooperation was nonetheless established on more technical areas. A 1962 agreement facilitated cooperation in the exchange of weather data and the launching of meteorological satellites, as well as efforts to map the geomagnetic field of Earth, and in the experimental relay of satellite communications (Sagdeev and Eisenhower 2008). Such functional approaches to cooperation in outer space remain a core feature today, having evolved into what are considered global utilities. This includes the sharing of meteorological and climate data, open access to and interoperability of civilian positioning, navigation and timing services, and the increasing public availability of Earth observation (EO) data.

This coordination and sharing of data and services has been formalized through organizations such as the International Committee on Global Navigation Satellite Systems (ICG) established in 2005 under the umbrella of the United Nations to facilitate compatibility, interoperability, and transparency between systems. The Coordination Group for Meteorological Satellites provides a forum for the exchange of technical information on geostationary and polar-orbiting meteorological satellite systems. Collected data is made available to the World Meteorological Organization, which distributes it to more than 3,000 weather-forecast outlets in 187 member states and 6 territories. Efforts to share and expand access to Earth observation data include the Committee on Earth Observation Satellites, which has 62 member agencies from around the world that work to coordinate and harmonize civil EO programs and data exchange from 170 satellites. Similarly, there is an international effort to create a Global Earth Observation System of Systems (GEOSS) that includes government agencies, academia, and the private sector, to enhance the sharing and integration of EO data worldwide. States also cooperate extensively for the use of satellite data to support disaster response and search and rescue through programs such as the International Charter on Space and Major Disasters and the Cospas-Sarsat international satellite system for search and rescue.

Such cooperation contributes to space security by providing essential global services that not only enrich lives, but also save them. This is the primary way in which most people on Earth access and enjoy the benefits of outer space. And, like in the early days of the space age, it remains critical to fostering cooperative relationships and reciprocity across diverse space actors. Indeed, on this basis of narrow, technical cooperation, cooperative relationships in outer space have extended much further, encompassing space exploration as both a practical and symbolic endeavor bridging self-interest and shared goals.

From Practical to Symbolic: Cooperation in Space Exploration

Exploration beyond Earth is at the heart of efforts to access outer space. And perhaps more so than any other activity, exploration bridges the enduring tension between national interest and collective aspiration in outer space. This was evident with the landing of the Apollo 11 mission on the Moon in 1969, which marked both a national achievement and an historic moment for all of humanity. The astronaut remains an enduring symbol of such unity. Taking their place among national icons, astronauts are also global figures, assigned a special status as "envoys of mankind in outer space" under the Article V of the Outer Space Treaty, which affords them the right for assistance, rescue, and return by all states.

Indeed, despite the competitive nature of space activities during the Cold War, the pursuit of space exploration gradually enabled a critical precedent of cooperation, starting with the 1975 Apollo-Soyuz Test Project. Marking the first ever international human spaceflight, the Test Project symbolized growing détente between the United States and Soviet Union; but it was also practical and self-interested. Involving a nine-day spaceflight during which an Apollo spacecraft carrying three American astronauts docked with a Russian Soyuz spacecraft with a crew of two, the mission allowed both parties to test the feasibility of international space rescue through compatible rendezvous and docking systems. Critically, the mission also demonstrated the viability of cooperation on more sensitive areas of technology which continued throughout the Cold War, namely through the exchange of scientific data related to ongoing space probes and robotic missions (Launius 2016). And it laid a foundation for the cooperative spirit that has been a hallmark of space exploration since the end of the Cold War.

In space, the end of the Cold War was marked by a 1992 agreement between the United States and Russia that led to astronaut exchanges and docking of NASA's

Space Shuttle with the Russian Mir space station. This process led to the creation of the International Space Station (ISS), an enduring symbol of space cooperation for the last two decades. Estimated to cost $150 billion to date, the ISS is the single largest, and most expensive space venture ever undertaken. Featuring a permanent human presence in outer space, it is made possible through collaboration among core partners, namely NASA in the United States, Roscosmos in Russia, the European Space Agency, the Japan Aerospace Exploration Agency (JAXA), and the Canadian Space Agency. In all, the ISS has received contributions from 15 states and hosted 236 astronauts from 18 different countries and counting (NASA 2019b).

From a space security perspective, such cooperation is critical to expanding access to outer space. Indeed, the significant expense and technical challenges associated with space exploration means that it is almost impossible without the pooling of financial resources and technical expertise, which in turn helps to expand both individual and collective capacity and participation in outer space. Cooperation on space cooperation marks a meeting of self-interest and shared achievement.

Space exploration reflects both the security of outer space and the international cooperation necessary to sustain it. It is also a means to this end. While the ups and downs of cooperative ventures are influenced by geopolitical and national security interests, over time such cooperation and shared interests in space has transcended these dynamics. Collaboration provides a critical mode of transparency and promotes a shared understanding of space activities. Mutual dependency in such a challenging environment builds trust in a field of activity that overlaps with strategic competition. Working and living together in outer space demands not only language training, but cultural understanding. Much like the iconic Earthrise image instills a sense of shared humanity, cooperation and co-existence in outer space introduces a shared vulnerability and mutual dependency.

Indeed, today the United States and Russia remain bound together on the ISS, mutually dependent on one another for access to and use of it. Since the retirement of the Space Shuttle, NASA has been dependent on Russia for access via Soyuz, while Russia depends on the United States for satellite communication. Although both parties strive to end such dependency, this entrenched cooperation has largely transcended geopolitical tensions on Earth, including political fallout related to recent interventions in Ukraine. Joint activities on the ISS have been largely exempt from rising hostilities and sanctions elsewhere.

Nonetheless, there are exceptions to the spirit of cooperative space exploration, most notably between the United States and China. China is not a member of the ISS and cooperation between the United States and China is extremely limited. This is largely a reflection of security concerns, which escalated following the Chinese anti-satellite (ASAT) demonstration that successfully destroyed one of its own ageing weather satellites in 2007. In 2011, the U.S. Congress adopted legislation barring any scientific activity between the United States and China involving either the National Aeronautics and Space Administration (NASA) or the White House Office of Science and Technology Policy (United States Congress 2012). However, American law does not ban private sector agreements with China, and in 2017 SpaceX carried the first experiment independently designed and fabricated in China to the

ISS. Further, a 2015 inaugural Civil Space Dialogue initiated tentative efforts to improve cooperation and transparency between the two states (U.S. Department of State 2015). This dialogue is tepid but ongoing, a testament to both the importance and challenges of cooperative relationships in a strategic environment.

More recently, cooperative efforts related to space exploration have been expanding beyond advanced spacefaring states to include emerging ones. Led by UN COPUOS, the 50th anniversary meeting of the first United Nations Conference on the Exploration and Peaceful Uses of Outer Space (UNISPACE+50) took place in June 2018. First among seven thematic priorities was to expand global partnerships on space exploration and innovation, specifically to "promote cooperation between spacefaring states and emerging space states," so that exploration becomes "open and inclusive on a global scale" (UN Office of Outer Space Affairs (OOSA) 2017). Reinforcing this goal, China marked the occasion by inviting all members of UN COPUOS to participate in its upcoming Tiangong-3 space station and intends to train astronauts from developing countries. In this way, cooperation in space exploration is a means to bridge not only strategic divides in outer space, but also varying abilities to access outer space. Indeed, capacity-building to expand access to outer space is another core feature of space security cooperation.

Expanding Access to Space: Cooperation and Capacity-Building

The central tenet of space security is the ability for all to be able to access and use space for peaceful purposes. Today, in addition to the European Space Agency (ESA), eight countries have direct access to space through national space launch capabilities; more than 70 operate national satellites (Union of Concerned Scientists 2019). International cooperation has been essential to this growth in access to space.

Like space exploration, some initiatives are international. For example, the KiboCUBE joint project between UNOOSA and the Japan Aerospace Exploration Agency (JAXA) makes use of Japan's Kibo module on the ISS to launch CubeSats on behalf of educational and research institutions from developing countries. But most cooperation is bilateral. NASA currently has over 700 agreements with international organizations (NASA 2019a), China has 120 (Xinhua 2018). And the Indian Space Research Organisation (ISRO) cooperates with at least 50 states (ISRO 2017). The essential role of bilateral relationships in expanding national capabilities is evident using the example of the United Arab Emirates (UAE). Established in 2014, its national space agency signed more than 16 cooperative agreements with international space agencies within the first 3 years of operation. (Permanent Mission of the United Arab Emirates to the United Nations 2017). Cooperative endeavors include advanced capabilities such as space exploration and human spaceflight.

Regional cooperation is also a critical tool for increasing access to outer space and its benefits. It is most developed in Europe, where the European Space Agency (ESA) facilitates space activities among its 22 Member States. A similar approach is being adopted in Africa, where progress on an African space strategy and African Space Agency is spurring greater cooperation. Likewise, in 2019 the Arab Space Coordination Group was initiated by the UAE and ten other countries (Algeria, Bahrain, Egypt, Jordan, Kuwait, Lebanon, Morocco, Oman, Saudi Arabia, and Sudan); its first collective project will be an Earth observation satellite used to monitor the environment and climate.

And yet regional cooperation also illustrates the enduring tensions between cooperation and strategic competition. This is clear in Asia, where two competing organizations foster cooperation: the Asia-Pacific Regional Space Agency Forum (APRSAF) and the Asia-Pacific Space Cooperation Organization (APSCO). The APRSAF was established by Japan in 1993; it currently includes participation by public and private entities from 40 counties. Modest achievements include the Sentinel Asia collaborate initiative to apply remote-sensing capabilities to support disaster management in the region. APSCO, established by China in 2005, includes Bangladesh, China, Iran, Mongolia, Pakistan, Peru, Thailand, and Turkey. Its activities have focused on training and data-sharing, disaster monitoring, and an Asia-Pacific Ground-Based Space Object Observation System (APOSOS) for monitoring objects in Earth orbit.

Indeed, security tensions and competition mean that cooperative efforts are rarely straightforward. The example of India's GSAT-9 communications satellite is a case in point. Described as a "gift" for the South Asian Association for Regional Cooperation (SAARC), Pakistan nonetheless opted out of participation (Set 2017). Likewise, the BRICS (Brazil, Russia, India, China, and South Africa) economic association, with its goal of decreasing dependency on the West, also provides a vehicle for space cooperation, but struggles with internal competition. Nonetheless, it has agreed to a first substantive project, namely the creation of a "virtual" remote sensing satellite constellation through a data-sharing system.

It is also clear that cooperative efforts can reinforce rather than transcend strategic interests. Although still taking shape, China's ambitious Belt and Road development and infrastructure initiative may be a case in point. Intended to integrate China into a network of global trade, the Belt and Road includes a Spatial Information Corridor to bring participants into China's space-based infrastructure services, including the BeiDou satellite navigation system, satellite communications, meteorology, remote sensing, and space-based broadband Internet service (Hui 2018). Including 65 national participants as of 2018, it is described as a cooperative initiative aimed at capacity-building and common development across members.

Some have questioned the long-term aims of such deep integration (Robinson 2019). More concretely, however, it speaks to the presence of underlying strategic undertones that can influence space security cooperation and capacity building. Specifically, Pakistan's participation in 2018 was expanded to include access to the BeiDou's military service (Abi-Habib 2018). Indeed, the persistence and even growth of national security uses – and corresponding geopolitical tensions – in outer space can impede other areas of cooperation related to safety and sustainability, which are needed to mitigate the challenges associated with more extensive uses of outer space.

Cooperation for Safety and Sustainability

While indicative of space security, growing access to and use of space is not without challenges. In particular, the natural environment of outer space, while seemingly vast, is also fragile. As a global commons, it is open to everyone, and almost everything, from satellites to Tesla Roadsters, giant disco balls, and advertising. Most of what we put into space never returns, contaminating the environment for future use. To avoid a tragedy of the commons – and to enhance the safety of operations for everyone – cooperation is essential. And it is increasingly taking place. Indeed, it is noteworthy that the limited dialogue between the United States and China is focused largely on safety including "space and terrestrial weather; space debris and spaceflight safety; and the long-term sustainability of outer space activities" (U.S. Department of State 2016). But here too, there are limits, largely imposed by national security interests.

The mitigation of space debris is one of the most significant examples of cooperative efforts to enhance the security of outer space. The Inter-Agency Space Debris Coordination Committee (IADC) evolved from cooperation between NASA and ESA following the creation of a large debris cloud in low Earth orbit caused by an Ariane 1 second stage explosion in 1986. It now includes 13 of the leading civil space agencies from around the world, including Roscosmos and the China National Space Administration (CNSA). The Committee published the first set of international guidelines related to space debris mitigation, a version of which was adopted by the UN General Assembly in 2008 as "voluntary measures to which all space actors should comply" (UNOOSA 2010). While implementation is uneven, collective efforts to limit the production of new debris in orbit have significantly reduced the rate of debris accumulation and contributed to enhanced sustainability of the environment.

Cooperation on safety is another core contribution to space security, primarily through efforts to mitigate natural threats including Near Earth Objects (NEOs) and space weather. Depending on size, a NEO that enters Earth's orbit can damage or destroy populated areas such as cities, or even the planet itself. Cooperation is emerging to mitigate this risk. In 2013 members of UN COPUOS created two international networks to coordinate detection, early warning, and future planetary defense measures: the International Asteroid Warning Network (IAWN), and the Space Mission Planning Advisory Group (SMPAG). The goal of each network is to ensure that all countries – including those with limited space capabilities – are aware of the threats – and to enable global warning, mitigation, and response processes. Space weather is another focus of safety cooperation. Space weather refers to changes in the space environment and geomagnetic storms that stem from flares and electromagnetic radiation emitted from the sun, which threatens security of objects both in outer space and on Earth by causing radiofrequency blackouts, orbital drag on satellites, and powerful power surges. In 2017 the expert group first convened by UN COPUOS in 2014 laid out a roadmap for greater international cooperation and information exchange on space weather events aimed at developing global modelling and forecasting capabilities (UN COPUOS 2017). Separately, the

World Meteorological Organization is wrapping up a 4-year plan that includes similar aims.

Key to these efforts is the role of UN COPUOS in coordinating cooperation for improved safety and sustainability. One of its most significant achievements in this regard is the identification of, and agreement to, a set of 21 voluntary guidelines for the long-term sustainability of outer space activities. Adopted by the Scientific and Technical Subcommittee in 2018 and referred to the UN General Assembly in 2019 along with a comprehensive preamble, the guidelines are indicative of the intersection of space security and cooperation. As stated in the preamble, they are "premised on the understanding that outer space should remain an operationally stable and safe environment that is maintained for peaceful purposes and open for exploration, use and international cooperation by current and future generations, in the interest of all countries...." (UN COPUOS 2018). The aim of the guidelines is to assist both individual and collective mitigation of risks; moreover, the guidelines emphasize that international cooperation is *required* to implement and monitor their effectiveness and impact.

Adoption of these guidelines is significant. In addition to articulating the link between cooperation, safety, and sustainability, they lend further impetus to the efforts on which states are already pursuing cooperation, such as space weather and debris mitigation. However, there are clear omissions. Beyond noting that they should be compatible with the "defense or national security" interests of states, the guidelines exclude activities more closely related to these interests. This includes issues that involve dual-use capabilities such as active debris removal and advanced rendezvous and proximity operations, as well as issues that approach arms control, such as restraints on intentional interference or harm of satellites. Also absent is an effort to create a more global or inclusive approach to space situational awareness. This issue lends insight into the tension between the security of outer space as a global commons that requires cooperation, and national security interests that drive strategic competition.

Indeed, while debris mitigation has emerged as a focal point of international cooperation for the security of outer space, safety from debris – largely a function of space situational awareness (SSA) – reflects much more cartelized modes of cooperation. An extension of space surveillance, SSA refers to the ability to generate actionable knowledge from surveillance data in order to identify, track, and catalog objects in orbit. This focus on action means that it is a critical capability for both safety *and* security in outer space. And, because no single actor has an absolute capability to precisely monitor every object on orbit, SSA depends on cooperation. But despite its widespread utility, there is no global system for monitoring objects and activities in outer space. Neither is there a global system to manage space traffic and safety.

This does not mean that there is no cooperation; indeed, cooperation on SSA is extensive, but also selective, and largely military (Lal et al. 2018). The most prominent measures are supported by the United States. The U.S. Department of Defense, which has by far the most advanced capabilities through its Space Surveillance Network of global terrestrial and space-based telescopes. It shares significant information on a public-platform, free of charge, through the Spacetrack.org website as part of the SSA Sharing Program run by the Combined Space Operations Center under the U.S. Strategic Command (USSTRATCOM). The U.S. Department of Defense also supports general space traffic management by providing conjunction warnings to other operators.

However deeper cooperation to share classified data that supports more advanced safety and security needs on orbit is restricted to bilateral agreements between USSTRATCOM and key allies and security partners. As of early 2019, these included agreements with 19 states (the Netherlands, Brazil, the United Kingdom, the Republic of Korea, France, Canada, Italy, Japan, Israel, Spain, Germany, Australia, Belgium, the United Arab Emirates, Norway, Denmark, Thailand, and New Zealand), in addition to ESA and the European Organization for the Exploitation of Meteorological Satellites, and more than 77 commercial space companies (US Strategic Command Public Affairs 2019).

Other actors are in turn developing their own, independent SSA capabilities. This includes European states, who are pooling national capabilities under a Space Surveillance and Tracking Support Framework. Russia and China also maintain extensive national capabilities, but do not widely share data; China is working narrowly with APSCO partners to develop the Ground-Based Space Object Observation Network. Several private companies also have commercial SSA capabilities and services. Such duplication would be beneficial to space security if data were pooled or otherwise used for verification and corroboration, but it is not. Instead, the persistent lack of *global* collaboration and cooperation on SSA and corresponding efforts to manage traffic in space reflects the ongoing difficulty of balancing the security *of* space as a common interest and national security concerns linked to the growing use of outer space.

New Patterns of Cooperation: Space Security Versus National Security

The physical security of objects in outer space is a core element of space security, entwining the objectives of national security with common security interests. In addition to natural threats such as space weather or impacts from debris, physical harm to satellites can include intentional efforts to interfere with space systems. From a space security perspective, core challenges include not only how to protect individual systems from harm, but also how to maintain strategic stability and prevent escalation of conflict into the space environment. This is a key function and goal of early efforts to foster cooperative space exploration activities and remains a feature of the ISS. However, the ability to adopt cooperative approaches on strategic issues closely related to national security such as restrictions on the deployment of weapons or the use of force in outer space remains the most intractable challenge to the security of outer space.

To be sure, there are mutual interests in preventing the use of military force in outer space, including overwhelming dependency on space assets for national security as well as the indiscriminate and long-lasting harm that violent conflict could inflict on the space environment. These concerns coalesced following the 2007 ASAT demonstration by China, which both threatened assured access to critical space systems in low Earth orbit and created the largest ever debris cloud in space. The event also marked a turning point in strategic relations in outer space from self-restraint to a simmering arms race.

The OST includes some provisions to prevent the worst of foreseeable conflict in outer space, including a ban on the orbiting of weapons of mass destruction and all military installments on the Moon. Other restrictions on armed conflict in outer space are scant, and mostly bilateral. Evidence of nascent protections for strategically sensitive satellites can be glimpsed in the Anti-Ballistic Missile Treaty, the Strategic Arms Limitation Talks, the Intermediate-Range Nuclear Forces Treaty, the Threshold Test Ban treaty, the Peaceful Nuclear Explosions Treaty, the Strategic Arms Reduction Treaty, the Conventional Forces in Europe Treaty, and the second Strategic Arms Reduction Treaty, which all included measures barring interference with "national technical means of verification," widely understood to mean satellites used to monitor treaty compliance (Black 2008). Although narrowly applied and eventually abrogated, the Anti-Ballistic Missile Treaty involved a restriction against the placement of ballistic missile interceptors in outer space. To be sure, this era also coincided with rampant and sometimes outlandish development of anti-satellite weapons. But self-restraint avoided the operational deployment of such weapons. The general belief was that space is too important to risk becoming a domain of military conflict.

This tacit cooperation to maintain the strategic stability of the outer space environment has eroded. Beginning with the abrogation of the ABM treaty by the United States in 2002, and including renewed interest and demonstration of ASAT capabilities including by China in 2007, the United States in 2008, and India in 2019, as well as the revival of Soviet-era weapons systems by Russia, there is now a simmering arms race in outer space. Insecurity generated by these activities is exacerbated by new on-orbit capabilities such as advanced rendezvous and proximity operations. These capabilities can support a range of both legitimate and more nefarious activities in outer space, blurring safety and security issues.

Efforts to agree to additional arms control measures in the Conference on Disarmament have stalled for over 30 years. So have efforts to develop additional voluntary measures related to behavior in outer space – for example through a code of conduct. A cooperative approach to support additional transparency and confidence-building measures (TCBMs) has also eroded (West 2018). In place of a shared belief in the need to avoid armed conflict in outer space, and international cooperation to restrict it, a growing number of states including China, India, Russia, France, Japan, the United Kingdom, and the United States now see space as a likely domain of armed conflict in the near future. From a strategic perspective, this shift introduces significant vulnerabilities for national security because of dependency on space systems for almost all military and security operations. Ongoing military developments such as a new United States Space Force are symptomatic of this growing sense of insecurity.

Like SSA, this vulnerability is leading to new patterns of selective cooperation based on deepening military alliances and strategic partnerships. Most cooperation involves the sharing of space-based capabilities and data for terrestrial military purposes. Examples include the participation of Canada, the Netherlands, and the United Kingdom in the U.S. Advanced Extremely High Frequency (AEHF) satellite program, and the shared use of the U.S. Wideband Global Satcom communications service by Canada, Denmark, Luxembourg, the Netherlands, New Zealand, and Australia. But such cooperation is expanding to include more formal alliance structures based on defense interests in outer space. This includes cooperation within the Five Eyes intelligence alliance (Australia, Britain, Canada, New Zealand, and the United States) such as the sharing of signals intelligence. Five Eyes partners also participate in the annual U.S. Air Force Space Command Wargames (Schriever wargames) which in recent years has expanded to include France, Germany, and Japan. Expanded cooperation is the focus of the newly renamed Combined Space Operations Center, which provides command and control of space forces and features greater cooperation with U.S. allies and partners including the Five Eyes, Germany, and Japan. The NATO (North Atlantic Treaty Organization) alliance is also making moves to recognize space as a domain of warfare, and military cooperation.

The number of security partnerships in outer space is growing, particularly in Asia. The long-standing US–Japan alliance now firmly includes defense cooperation in space. The United States has also increased defense-related cooperation in space with India, now a major defense partner. Japan and India are also coordinating bilaterally; in 2018, the Japan–India Space Dialogue included a focus on security, namely sharing satellite data and surveillance technology (Hayashi 2018). India and France, which had long cooperated on civil space programs, have also extended cooperation to security applications (Rajagopalan 2018). Likewise, India and Vietnam have expanded their strategic relationship to include defense cooperation in space, primarily through satellite imagery (Parameswaram 2018). China's ongoing cooperative endeavors also have strategic undertones, particularly the Belt and Road Initiative, which includes military cooperation with Pakistan and could expand to include additional partners.

There are positive aspects to such cooperation. The pursuit of objectives such as inter-operability and shared capabilities builds capacity and is a key mode of resilience in outer space: the ability to withstand interference with a satellite's capabilities and maintain core functions. As a technical ability, resilience can enhance security to both deliberate and natural threats, bridging safety and security concerns in outer space. This has been a clear benefit of global cooperation related to satellite services for positioning, timing and navigation. Some argue that resilience could also deter aggression in space and stabilize the strategic environment (Air Force Space Command 2016). But the extension of strategic partnerships into space could also further escalate military tensions and even conflict in outer space, particularly in the absence of broader cooperative efforts to restrict the most damaging forms of conflict and protect strategic assets. This is particularly concerning in the face of rising geopolitical competition and acute vulnerability in outer space.

Further, it is also unclear how rising strategic competition and deep but narrow security cooperation in outer space will affect emerging areas of space activities such as lunar exploration and resource extraction.

New Issues: The Moon and Space Resources

China's historic robotic landing on the far side of the Moon in 2019 heralded a new focus of human activity in outer space defined by lunar exploration and the possible exploitation of space-based resources. Other missions – either underway or planned – include India's Chandrayaan 2 robotic mission to the lunar South Pole and NASA's new Artemis program to build a lunar Gateway in orbit around the Moon and return American astronauts to the lunar surface. China has long-term plans to send astronauts to the Moon and develop a research base there. The European Space Agency also has a robotic lunar program and interest in resource extraction, as does Japan. The collective focus is on the lunar south pole, where resources critical for human survival and sources of power – including water ice and helium-3 – are known to exist. Unlike in the past, the goal is not merely to touch the Moon, but to leave a permanent mark: to establish bases and even human settlements, and to extract resources. Non-state actors are also participating. In 2019 SpaceIL launched the first private robotic lander to the Moon. Commercial ventures such as Moon Express – which focus on extracting the Moon's resources – are also set to arrive. Several companies are setting up businesses to shuttle items between Earth and the Moon. Billionaires Elon Musk (SpaceX) and Jeff Bezos (Blue Origin) aim to establish private exploration programs and human colonies. Whether or not individual missions advance, the long-term trend is toward a more expansive and possible exploitive human presence in outer space. Implications for the cooperative security of outer space are unclear.

Although colonization and the search for resources are long-standing themes of human history, they introduce new questions in relation to the security of outer space. These include issues related to contamination and the environmental integrity of the Moon; processes for – and the implications of – claiming locations for research and settlement; the mingling of scientific, commercial, and military interests; and how to extend the benefits of lunar access and extractive resources in space to the global community. Critically, these new activities reinforce established tensions between cooperation and strategic competition that drive dynamics related to sustainability, security, and equity in the global commons of outer space.

Thus far, signs of cooperation are strong. The return to the Moon is a global pursuit. The 2018 Global Exploration Roadmap published by the International Space Exploration Coordination Group describes "an emerging international consensus to proceed with lunar exploration using a cislunar platform as the initial step in space exploration beyond low Earth orbit" (International Space Exploration Coordination Group 2018). The Group of 15 space agencies, including NASA, Roscosmos, and CNSA, participate in this nonbinding initiative, discussing common interests and identifying potential areas of cooperation. It has also adopted new terms of reference as a basis to foster international space cooperation and dialogue. Other cooperative initiatives include the nonprofit Moon Village Association, which is working to foster an international collaborative approach to lunar exploration and For All Moonkind, which seeks to protect and preserve human heritage including individual landing sites on the Moon.

There is also considerable bilateral cooperation emerging. Significantly, NASA received Congressional approval to collaborate with China on lunar landing research and transmitted images of the lunar landing site for the Chang'e 4 mission in 2019 (David 2019). China has invited additional international partnerships for its planned Chang'e 6 lunar sample return mission. The United States is engaging both international and private sector partners for the Artemis human exploration program. India and Japan are pursuing joint projects; China has also reached out to India.

Some efforts to cooperate on the governance of resource-use are taking shape. The Hague International Space Resources Governance Working Group is formulating governance recommendations and guidelines; the *Draft Building Blocks for the Development of an International Framework on Space Resource Activities* was published in 2017 (Universiteit Leiden 2017). There are discussions within UN COPUOS to potentially create a working group to further explore legal considerations. Bilaterally, Luxembourg – one of the greatest proponents of private sector resource extraction in outer space – is cooperating with like-minded countries including the UAE and Japan.

These are all good signs, but there are few agreed upon rules to put inspiration into practice. Efforts to operationalize peaceful uses and cooperative approaches of the OST in the 1979 Moon Agreement failed. And despite a global focus, the sense of a new race to the Moon and underlying strategic interests – including a possible scramble for resources – cannot be ignored. Beyond a focus on national security in outer space, the United States aims for pre-eminence in the space domain (The White House 2018), while China seeks to be a "space power in all respects" (The State Council Information Office of the People's Republic of China 2016). Private and commercial interests introduce yet another competitive component. How these tensions will interact with lunar exploration – and resource ambitions – is not clear. Neither is it clear that the spirit of cooperation that informs the principles of peaceful and equitable use of outer space in the OST will endure. As U.S. Vice President Pence has asserted, those who get there first – and stay – will write the "rules and values of space" (The White House 2019). The future of cooperation in outer space may depend on who gets there first.

Conclusion: The Future of Space Security Cooperation

Outer space is a fragile environment, a critical resource, and a focus of strategic competition. Maintaining the ability of this domain to support safe, sustainable, and secure access and use for all – the essence of space security – requires cooperation. Further, cooperation is embedded as a core value within the institutions and laws that govern outer space, a *raison d'être* of both the UN Committee on the Peaceful uses of Outer Space, and the Outer Space Treaty. Over time, cooperative efforts to improve the safety and sustainability of space operations, and to expand global access to outer space, have widened, increasing both individual and collective capacity and well-being in a challenging environment.

But while cooperation is the norm in outer space, it is not straightforward. National security interests present the most persistent impediment. At times cooperation has provided a way to transcend relationships by developing trust and transparency. The ISS is a key example. Other times, cooperation trails strategic and geopolitical interests, marked most strongly by the ongoing absence of international cooperation to limit the use of force in outer space. Combined with intense dependency on vulnerable space-based systems for military and national security objectives, this void is giving way to new, narrow patterns of cooperation among national security allies and partners. While such cooperation can enhance security *in* space for those involved, it may come at cost to the long-term security *of* space by increasing strategic rivalry and facilitating the escalation of conflict into outer space.

New uses and users of outer space are also changing the dynamics of space security cooperation. Examining the revival of lunar and human exploration alongside interest in the exploitation of space-based resources indicates a shift toward a more intense, long-term, and strategic human activities in presence that will leave a fundamental mark on worlds beyond our planet. This shift is being undertaken with considerable international cooperation. And yet underlying strategic rivalry as well as commercial and private interests may well impede efforts to implement the values of the Outer Space Treaty, including peaceful uses, cooperation, and global benefit. A cooperative approach to the security of outer space remains a prudent way to ensure that these values are upheld.

References

Abi-Habib M (2018) China's 'Belt and Road' plan in Pakistan takes a military turn. New York Times, 19 December. https://www.nytimes.com/2018/12/19/world/asia/pakistan-china-belt-road-military.html. Accessed 11 Sept 2019

Air Force Space Command (2016) Resiliency and disaggregated space architectures. White paper. https://www.afspc.af.mil/Portals/3/documents/AFD-130821-034.pdf?ver=2016-04-14-154819-347. Accessed 11 Sept 2019

Black S (2008) No harmful interference with space objects: the key to confidence-building. In: Security in space: the next generation, United Nations Institute for Disarmament Research, 31 Mar–1 Apr

David L (2019) Farside politics: The West eyes Moon cooperation with China. Scientific American, 7 February. https://www.scientificamerican.com/article/farside-politics-the-west-eyes-moon-cooperation-with-china/. Accessed 11 Sept 2019

Hayashi S (2018) Japan-India 'Space-Dialogue' to include surveillance sharing. Nikki Asian Review, 9 December 2018. https://asia.nikkei.com/Politics/International-relations/Japan-India-Space-Dialogue-to-include-surveillance-sharing. Accessed 11 Sept 2019

Hui J (2018) The spatial information corridor contributes to UNISPACE+50. China National Space Administration technical presentation to the scientific and technical subcommittee of the UN Committee on the Peaceful Uses of Outer Space http://www.unoosa.org/documents/pdf/copuos/stsc/2018/tech-08E.pdf. Accessed 11 Sept 2019

Indian Space Research Organization (ISRO) (2017) International cooperation. https://www.isro.gov.in/international-cooperation. Accessed 9 Sept 2019

International Space Exploration Coordination Group (2018) Government representatives from 45 countries and international organisations meet at the 2nd International Space Exploration Forum (ISEF2), 26 April. https://www.globalspaceexploration.org/wordpress/?p=792. Accessed 11 Sept 2019

Kay WD (1998) John F. Kennedy and the two faces of the U.S. space program, 1961–63. Pres Stud Q 28(3):573–586

Kennedy JFK (1961) Inaugural address. Special message to Congress on Urgent National Needs, 25 May 1961, Washington, DC. John F. Kennedy Library and Museum website: https://www.jfklibrary.org/asset-viewer/archives/JFKWHA/1961/JFKWHA-032/JFKWHA-032

Lal B et al (2018) Global trends in space situational awareness (SSA) and space traffic management (STM). Science and Technology Policy Institute, 10 October. https://csis-prod.s3.amazonaws.com/s3fs-public/event/181010_SSA_CSIS.PDF. Accessed 11 Sept 2019

Launius R (2016) Key developments in USA/USSR space cooperation during the Cold War. https://launiusr.wordpress.com/2016/08/15/some-key-developments-in-usaussr-space-cooperation-during-the-cold-war/. Accessed 9 Sept 2019

NASA (2019a) Active international agreements by signature date (as of 30 June 2019). https://www.nasa.gov/sites/default/files/atoms/files/house_approps_action_international_saas_active_as_of_6-30-2019.pdf. Accessed 9 Sept 2019

NASA (2019b) International cooperation. https://www.nasa.gov/mission_pages/station/cooperation/index.html. Accessed 9 Sept 2019

Parameswaram P (2018) India-Vietnam defense relations in the spotlight with bilateral visit. The Diplomat, 18 June. https://thediplomat.com/2018/06/india-vietnam-defense-relations-in-the-spotlight-with-bilateral-visit/. Accessed 11 Sept 2019

Permanent Mission of the United Arab Emirates to the United Nations (2017) UAE statement to the fourth committee on international cooperation in the peaceful uses of outer space. 17 October. https://www.un.int/uae/statements_speeches/uae-statement-fourth-committee-%E2%80%9Cinternational-cooperation-peaceful-uses-outer. Accessed 9 Sept 2019

Rajagopalan RP (2018) From sea to space: India and France deepen security cooperation. The Diplomat, 15 March. https://thediplomat.com/2018/03/from-sea-to-space-india-and-france-deepen-security-cooperation/. Accessed 11 Sept 2019

Robinson J (2019) State actor strategies in attracting space sector partnerships: Chinese and Russian economic and financial footprints. Prague Security Studies Institute. http://www.pssi.cz/download/docs/686_executive-summary.pdf. Accessed 11 Sept 2019

Sagdeev R, Eisenhower S (2008) United States-Soviet space cooperation during the Cold War. NASA. https://www.nasa.gov/50th/50th_magazine/coldWarCoOp.html. Accessed 9 Sept 2019

Set S (2017) India's regional diplomacy reaches outer space. Carnegie India https://carnegieendowment.org/files/7-3-2017_Set_IndiaRegionalDiplomacy_Web.pdf. Accessed 11 Sept 2019

The State Council Information Office of the People's Republic of China (2016) China's space activities in 2016. White Paper 27 December. http://www.scio.gov.cn/wz/Document/1537091/1537091.htm. Accessed 11 Sept 2019

The White House (2018) President Donald J. Trump is unveiling an America first national space strategy. 23 March. https://www.whitehouse.gov/briefings-statements/president-donald-j-trump-unveiling-america-first-national-space-strategy/. Accessed 11 Sept 2019

The White House (2019) Remarks by Vice President Pence at the fifth meeting of the National Space Council, Huntsville Alabama, 26 March. https://www.whitehouse.gov/briefings-statements/remarks-vice-president-pence-fifth-meeting-national-space-council-huntsville-al/. Accessed 11 Sept 2019

Treaty on Principles Governing the Activities of States in the Exploration and Use of Outer Space, Including the Moon and Other Celestial Bodies (Outer Space Treaty) (1967). https://2009-2017.state.gov/t/isn/5181.htm. Accessed 11 Sept 2019

UN COPUOS (2017) Report of the scientific and technical subcommittee on its fifty-fourth session, held in Vienna from 30 January to 10 February 2017. http://www.unoosa.org/oosa/oosadoc/data/documents/2017/aac.105/aac.1051138_0.html. Accessed 11 Sept 2019

UN COPUOS (2018) Guidelines for the long-term sustainability of outer space activities. Conference room paper by the Chair of the working group on the long-term sustainability of space activities (June). http://www.unoosa.org/res/oosadoc/data/documents/2018/aac_1052018crp/aac_1052018crp_20_0_html/AC105_2018_CRP20E.pdf. Accessed 11 Sept 2019

Union of Concerned Scientists (2019) Satellite database. https://www.ucsusa.org/nuclear-weapons/space-weapons/satellite-database. Accessed 9 Sept 2019

United States Congress (2012) H.R.2112 – consolidated and further continuing appropriations act. Sec. 539. https://www.congress.gov/bill/112th-congress/house-bill/2112/text. Accessed 9 Sept 2019

United States Department of State (2015) The first meeting of the U.S.-China space dialogue. Media note. www.state.gov/r/pa/prs/ps/2015/09/247394.htm. Accessed 9 Sept 2019

United States Department of State (2016) The second meeting of the U.S.-China space dialogue. Media note, October 24. https://2009-2017.state.gov/r/pa/prs/ps/2016/10/263499.htm. Accessed 11 Sept 2019

Universiteit Leiden (2017) Draft building blocks for the development of an international framework on space resource activities, September. https://www.universiteitleiden.nl/binaries/content/assets/rechtsgeleerdheid/instituut-voor-publiekrecht/lucht%2D%2Den-ruimterecht/space-resources/draft-building-blocks.pdf. Accessed 11 Sept 2019

UNOOSA (2010) Space debris mitigation guidelines of the committee on the peaceful uses of outer space. United Nations. http://www.unoosa.org/pdf/publications/st_space_49E.pdf. Accessed 11 Sept 2019

UNOOSA (2017) UNISPACE +50 thematic priorities, p 3. http://www.unoosa.org/documents/pdf/unispace/plus50/thematic_priorities_booklet.pdf. Accessed 9 Sept 2019

US Strategic Command Public Affairs (2019) USSTRATCOM, Polish space agency sign agreement to share space services, data. USSTRATCOM, 11 April. https://www.stratcom.mil/Media/News/News-Article-View/Article/1811729/usstratcom-polish-space-agency-sign-agreement-to-share-space-services-data/. Accessed 11 Sept 2019

West J (2018) Why the chances of conflict in outer space are going up. Ploughshares Monit 39(4), winter. https://ploughshares.ca/pl_publications/why-the-chances-of-conflict-in-outer-space-are-going-up/. Accessed 11 Sept 2019

West J (ed) (2019) Space security 2019. Project Ploughshares, Waterloo

Xinhua (2018) China strengthens international space cooperation. China Daily, 19 April. http://www.chinadaily.com.cn/a/201804/19/WS5ad899eea3105cdcf65195a1.html. Accessed 9 Sept 2019

Strategic Competition for Space Partnerships and Markets

Jana Robinson, Tereza B. Kupková, and Patrik Martínek

Contents

Introduction	164
Global Chinese and Russian Economic and Financial Space Activities	165
Africa	169
Latin America	169
Europe	171
The Arctic	173
Antarctica	174
The Middle East	175
South and Southeast Asia	175
Western, Central, and Eastern Asia	176
Top Space Sector Capture Trends	177
Key Findings	178
Conclusion	180
References	180

Abstract

China and Russia's global space footprint in the economic and financial (E&F) domain is not well understood today. This chapter, through analyses of space-related transactions of China and Russia globally, describes the pro-active approach to international space partnerships by these two state actors. It concludes that these partnerships are often skewed, exposing recipient countries to

---

J. Robinson (✉)
Space Security Program, Prague Security Studies Institute (PSSI), Prague, Czech Republic
e-mail: jrobinson@pssi.cz

T. B. Kupková
Prague Security Studies Institute (PSSI), Prague, Czech Republic
e-mail: kupkova@pssi.cz

P. Martínek
The Prague Security Studies Institute (PSSI), Prague, Czech Republic
e-mail: martinek@pssi.cz

© Springer Nature Switzerland AG 2020
K.-U. Schrogl (ed.), *Handbook of Space Security*,
https://doi.org/10.1007/978-3-030-23210-8_141

vulnerabilities and dependencies on the benefactor(s). The more subtle strategy deployed in the developed, democratic countries to gain influence is conducted on an incremental basis (e.g., through commercial contracts, academic exchanges, scientific research). The other approach, described as "space sector capture," mostly involves developing countries and consists of offering package deals of capabilities, services, and financing, creating sore-source supplier relationship and long-term dependencies. The chapter argues that the pace and nature of these international space partnerships concluded by China and Russia present a strategic and competitive challenge for Europe, the USA, and other allies, including the development of global space governance, as well as market based on transparency, good governance, and disclosure.

Introduction

The actions and conduct of Russian and Chinese state-owned and -controlled enterprises (SOEs) are often driven by both commercial *and* strategic considerations. This has become increasingly evident with the emergence of new brands of soft power projection adopted by both countries. In the economic and financial (E&F) domain, this means seeking to gain influence and strategic advantage over targeted states via ostensibly commercial, legal transactions, and projects (e.g., acquisitions, partnerships, loans, joint ventures, minority investments).

There is a dearth of understanding concerning how China and Russia are using the legitimate E&F domain to compromise the integrity of the space sectors of various countries that lack space programs, adequate funding, operating personnel, and technical expertise. These include various levels of "space sector capture" achieved through the offer of end-to-end capabilities (i.e., vertically integrated packages of design/manufacturing of satellite(s), launch services/launch insurance, ground segment construction/equipment, provision of operating personnel, the training of local staff, and financial assistance) (Robinson et al. 2019).

These offers often involve the use of nonmarket trade and financial practices for the purpose of expanding their global space footprint at a strategic level (with a number of associated operational, political, geographic, and military benefits). This is being accomplished through securing desired foreign projects/assets, beachheads in priority regions, the acceptance of subsidized loans often to non-creditworthy state borrowers to acquire political leverage and/or secure the collateralized assets in default scenarios. China's and Russia's economic and financial activities in the developed, democratic countries involve more incremental approach, often through seemingly benign scientific research/development, academic exchanges, individual commercial contracts, or broader funding commitments beyond the space sector (Robinson 2018a).

As the economies of China and Russia are inextricably linked with their governments, its companies do not operate as traditional commercial enterprises. Many decisions pertaining to overseas investments are subject to approval by government authorities. State-controlled enterprises, including quasi-private companies, usually have some level of state involvement in their management structures and are obliged to comply with government policies and directives.

The activities of these companies in the international trading and financial systems are designed to appear benign and commercial, providing space aspirants with capabilities they crave, ostensibly to advance the prosperity and security of these targeted countries. Countries lacking a space program, adequate funding, and technical expertise are generally open to such seemingly magnanimous offers, even if it means their countries could well become perilously dependent on these outside benefactors (Robinson 2018b).

Such international partnerships result in dependency, even control, over the space sectors of the recipient countries (e.g., Belarus, Bolivia, Nigeria, Pakistan, Sri Lanka, and Venezuela). The transactions, including offers of large-scale financing at below-market terms, are primarily for the purpose of expanding China and/or Russia's global space footprint at a strategic level (with a number of associated operational, political, geographic, and military benefits) (Robinson 2018c).

This article provides an overview concerning how China and Russia are using the legitimate E&F domain to compromise the integrity of the space sectors of various countries that lack space programs, adequate funding, operating personnel, and technical expertise. Through granular analysis of space-related transactions of Chinese and Russian state-controlled enterprises, the article demonstrates that this trend represents a material risks to the targeted countries from the perspectives of national security and sovereignty. In short, this chapter introduces a new risk category within the space security portfolio, namely the economic and financial (E&F) operations of nondemocratic state actors. It brings forward the above-referenced "space sector capture" concept and delineates its elements. It then describes the transactional approach of China and Russia to international space partnerships. Finally, it offers key findings and recommendations.

Global Chinese and Russian Economic and Financial Space Activities

The analysis provided in this section draws from the research, including an open-source database, of the Prague Security Studies Institute (Robinson et al. 2019) concerning the strategic and commercial dimensions of space-related Chinese and Russian economic and financial activities globally. It provides an overview of prominent space-related partnering arrangements, which often involve partial or full dependency of the recipient countries on Chinese or Russian financing, technology, equipment, services, and/or expertise.

As of February 2020, China and Russia have been actively engaged in space partnerships in at least 78 countries (see Fig. 1). These two state actors have been especially active in Latin America, Europe, South/Southeast Asia, and Africa. China has also increased substantially its outreach to the Arctic countries in the past decade (Figs. 2 and 3).

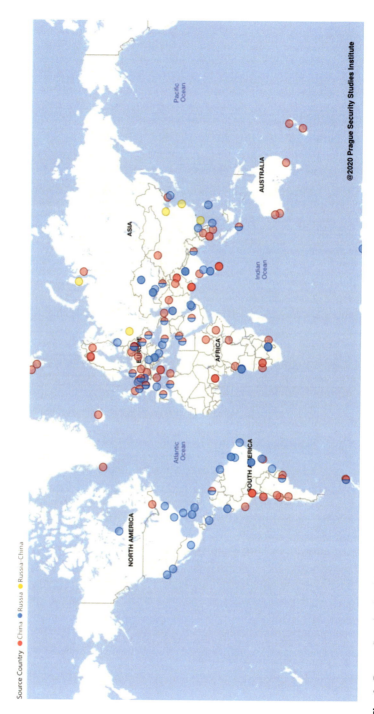

Fig. 1 Prague Security Studies Institute (PSSI) global map of space transactions of China and Russia as of February 2020 (does not include the Arctic and Antarctica)

10 Strategic Competition for Space Partnerships and Markets

Fig. 2 PSSI map of space transactions of China and Russia in the Arctic as of February 2020 (does not include the Arctic and Antarctica)

Fig. 3 PSSI map of space transactions of China and Russia in Antarctica as of February 2020 (China in red, Russia in blue)

Africa

Both Russia and China have assisted a number of African countries in establishing, or expanding, their space activities. Russia has worked with Angola, Algeria, Egypt, and South Africa and China has reached out to Algeria, Democratic Republic of Congo, Egypt, Ethiopia, Mozambique, Namibia, Nigeria, Sudan, Tunisia, and Uganda. In all of these cases, space ground infrastructure was provided, often through subsidized lending arrangements or even taking stakes in local companies.

Russia, for example, launched a satellite for Egypt in 2014 and also assisted in the development of their third Earth observation (EO) satellite that was launched in 2019 from Baikonur. It has offered engineering and other assistance to Ethiopia and helped South Africa in developing a satellite surveillance program.

Notable examples of China's activities in Africa are those in Nigeria. China entered Nigeria's space sector in 2004, manufacturing Nigeria's first communications satellite, NigComSat-1 (based on the Sinosat-2), which was launched in 2007. It failed after 18 months in orbit. Its replacement, NigComSat-1R, was launched in 2011 (Krebs 2017). Both satellites were built and launched under a contract with the state-controlled China Great Wall Industry Corporation (CGWIC), and the second satellite was subsidized 100% by China. It was the first time that China had reached out to a foreign country in this fashion and the first time that CGWIC provided all aspects of in-orbit delivery of a satellite for an international customer. This means China also provided two ground stations (one tracking station in Kashi, China, and one fully operational station in Abuja, Nigeria), training of personnel, financing, and insurance. In 2018, China agreed to finance (through the Export-Import Bank of China) the building of two new communication satellites, but in exchange for CGWIC's stake in state-owned NigComSat Ltd. (Nigeria's satellite communications operator and service provider). China also offered to possibly send a Nigerian astronaut to space in the 2030s.

The Nigeria example demonstrates that a recipient country can become largely dependent on its benefactor (China in this case) for its space program through this strategy of offering a complete "package deal." In a similar fashion, China has offered a satellite "package deal" to Ethiopia, Algeria, Sudan, and Congo. As for Russia, it offered a package deal to Angola and Egypt.

Latin America

Both China and Russia successfully built and operate space infrastructure in this region. There is a robust increase in China's influence in Latin America, including through promises and agreements to provide space-related technology, expertise, and services. China has worked with Argentina, Bolivia, Brazil, Chile, Cuba, and Venezuela.

A controversial project was built by the China Harbour Engineering Company (CHEC), a subsidiary of People's Liberation Army (PLA)-affiliated China Communications Construction Company (CCCC), in Argentina – a satellite tracking, telemetry, and command station in the Patagonia region, operational since 2017. CHEC is also involved in illegal island-building, and the militarization of same, in the South China Sea. China was able to secure a 50-year lease agreement of the land and facility and does not permit the entry of local officials. It also does not employ local personnel (Londoño 2018). The Neuquen TT&C station in Argentina is in the proximity of a number of infrastructure assets constructed by China in South America, the Fibre Optic Austral in Chile and the China-funded multi-billion-dollar turnkey railway and infrastructure projects in Argentina (Giri 2018). China also built a 60 cm-diameter telescope which resides in the Observatorio Astronomico "Felix Aguilar" (OAFA) of the National University of San Juan (UNSJ). It provides data to the International Laser Ranging Service (ILRS). Another China-Argentina Radio Telescope (CART) is planned to be completed by June 2020.

More recently, an Argentinean company, Satellogic (developing a constellation of EO satellites with panchromatic, hyperspectral, multispectral and infrared capabilities), announced that it would launch its fleet of spacecraft on Chinese rockets under a contract with CGWIC. It is the largest single contract for Chinese launch industry in the international commercial market in more than 20 years (Clark 2019). The first two satellites were launched in January 2020.

Brazil is subject to US export control-related restrictions and indirectly affected by the US ban on space commerce with China. It has viewed collaboration with China and Russia in space as fundamental to its efforts to overcome bottlenecks related to the development of its space program (without any apparent concern over the country's dependency on China and Russia in this strategic sector).

The flagship project of Beijing's space collaboration with Brazil is the China-Brazil Earth Resources Satellites CBERS program. It is a collaboration between CAST (China Academy of Space Technology) and INPE (Instituto de Pesquisas Espaciais). China provided the technology, launch service, and subsidized financing. Data from the project have been shared since 2004, with third parties (ESA 2019).

Russia was the first to offer collaboration to Brazil in the area of global navigation. As of February 2020, Brazil hosted four GLONASS ground stations on its territory (the largest number outside Russia). Besides Brazil, Russia established GLONASS stations in Argentina, Ecuador, Nicaragua, and Venezuela. China is also seeking to launch collaboration with Brazil on its global satellite navigation system, BeiDou (which it asserts will be global by 2020) (Selding 2015). Russia also partnered with Brazil on its Russian telescope at Pico dos Dias Observatory in Brazopolis, Minas Gerais, operational since 2018. This electro-optical facility is designed to help fill the observation gaps in the geostationary orbit. Russia is reportedly planning to deploy an optical-electronic monitoring station in Chile, Mexico, and South Africa (Ibeh 2019).

Cuba, not surprisingly, hosted a large Russian signals intelligence facility between 1962 and 2002. It is said to have reopened in 2014 (Kelley 2014). Interestingly, that same year Russia wrote off some 90% of Cuba's $32 billion Soviet-era debt. Currently, Cuba's largest international creditor and trading partner is, perhaps not surprisingly, China. In May 2018, satellite images revealed a newly constructed radome within the signals intelligence base in Bejucal, Cuba. It is believed that Beijing financed this new facility (Lee 2018).

China also seeks to expand its space observation capabilities through its collaboration with Chile, establishing an astronomical research center at the Catholic University of Chile in 2013 and a plan to build its own observatory some 30 km from Chile's Paranal Observatory. As Chile has relatively good governance and strong institutions, however, China may not have gotten a preferential deal, typically involving public procurement regulations beneficial to Chinese companies, or Chinese loans and investments (Ellis 2017).

When it comes to the telecommunications sector in Latin America, China delivered complete packages (involving construction, financing, delivery, ground stations, and operations) to Bolivia, Venezuela, Cuba, and Uruguay. Argentina, Bolivia, Brazil, and Venezuela are all laboring under a troubling degree of space sector dependency with a common thread of China financing their projects through direct investments.

Europe

Unlike in Africa and Latin America, where China, and to a lesser degree Russia, have been frequently offering vertically integrated space sector packages (partial or complete), including large-scale subsidized financing, their approach in the developed, democratic countries is more subtle and incremental, often involving seemingly benign scientific research/development initiatives, academic exchanges, individual commercial contracts, or broader funding commitments beyond the space sector.

Russia has benefited from its established ties in Europe, especially in the launch subsector. Germany, France, Spain, and the Netherlands have all been recipients of Russia's launch services. For instance, in February 2014, TsSKB-Progress signed a $400 million supply agreement with Arianespace to provide a batch of seven Soyuz-ST rockets for launch from Kourou in French Guiana. This agreement was built on a previous arrangement between Roscosmos and the French company to launch "mid-class Soyuz-ST rockets over 15 years" (Nowakowski 2016). Russian VNIIEM Corporation also managed to secure an agreement in the EO subsector when it signed a cooperation protocol with a British company Surrey Satellite Technology in 2015 "for the creation of a small Earth remote-sensing (ERS) satellite" (Glavkosmos 2019).

There are interesting cases of corporate acquisitions by China in Germany. For example, on June 19, 2018, it was announced that Fosun International had agreed to acquire FFT Produktionssysteme GmbH & Co. from ATON GmbH for an undisclosed sum. Subsequently in August 2018, Fosun won regulatory approval from the European Commission for the acquisition, as it was deemed that no competitive concerns would arise (European Commission 2018). Similarly, in 2018, Changzhou QFAT Composite Material, a subsidiary of China Iron and Steel Research Institute, acquired aerospace firm Cotesa after the approval of German regulators in April 2018 (Xinhua 2018). Another acquisition of the controlling 94.55% stake of the robotics firm Kuka by Midea in 2016 demonstrates that such acquisition can quickly turn into an opening for other Chinese entities to enter the local market (Taylor 2016). Within a year from the acquisition, Kuka signed a memorandum of understanding with a Chinese company Huawei to "deepen their global partnership" (Williamson 2017). Many experts believe that Huawei represents a serious national security risk.

A similar effort was discovered in Ukraine. The Beijing-based Skyrizon acquired the majority stake in Ukraine's aerospace company Motor Sich in 2017, following a PLA contract with Motor Sich for 250 jet engines for JL-10/L-15 jets, a deal worth $380 million concluded in 2016 (Wang 2018). As the Pentagon, NATO officials, and G7 diplomats raised concern over Skyrizon interest in Motor Sich, Ukraine's antitrust authority eventually launched an investigation into the case to potentially block the deal. As of late February 2020, the antitrust authority says it might have a strong case against the deal and it appears that it could be canceled (Gorchinskaya 2020).

As a backdrop to these developments, in November 2017 Ukrainian representatives signed a long-term cooperation program with their Chinese counterparts that included some 70 projects that involve, among other areas, "implementation of China's Lunar Exploration Program, a mission to study the planets of the solar system, new materials development, and remote sensing" (Interfax-Ukraine 2017). Russia's complex involvement in Ukraine, following the 2014 invasion of its southeastern territory, involved the reintegration of some of its Soviet-built infrastructure into its ground station network (Foust and Bodner 2016) in an extraordinary breach of international law.

Belarus is a prime example of a country's space sector being shaped by both China and Russia. As a former Soviet republic, its space sector has been closely tied to that of Russia. China made its foray into Belarus through its state-owned enterprises, monitored by a special Committee established by Belarus called the Belarusian-Chinese Intergovernmental Committee on Cooperation. The 91.5 km^2 "Great Stone Industrial Park" located outside Minsk was stood up in 2010 under an agreement between Belarus Economy Ministry and China CAMC Engineering Co. Ltd. (CAMCE) and is overseen by the China-Belarus Industrial Park Development Company (JSC), owned 60/40 by the Chinese and Belarusian governments, respectively. Chinese telecom giants ZTE and Huawei have been heavily involved in the development of this industrial park. Both companies have a history of assisting China's intelligence operations and the alleged theft of intellectual property (Pai 2019).

PLA-affiliated China Aerospace Science and Technology Corporation (CASC) signed a letter of intent in March 2018 to become an anchor company in the industrial complex (Belta 2018a). In addition, the China-Belarus Cooperation Center for Science and Technology Achievement is to be built by another Great Stone resident, China No. 15 Metallurgical Construction Group (15MCC), and "funded using an economic and technical assistance grant from the Chinese government" (Belta 2018b).

The same Chinese entity (CAMCE) overseeing Chinese investment into Great Stone in Belarus signed a memorandum of understanding with Lithuanian Kaunas Free Enterprise Zone regarding the development of a pilot free trade zone, which should involve funding for research and development in biotechnology, information technology, space research, and photonics. While the current status of the joint project is not available in the open source, the Chinese side has made a connection between the zones in Lithuania and Belarus saying: "Minsk and Kaunas will be developed as inland terminals for the North and Baltic Seaports (especially Klaipeda Seaport)" (Rail Working Group 2016).

Belarus is also a recipient of a comprehensive package deal involving the construction, launch, training, ground station, and temporary management of a communications and broadcasting satellite, Belintersat-1 offered through the China Great Wall Industry Corporation (CGWIC). The satellite's launch (from Xichang) in January 2016 coincided with a sharp increase in exports (75.5% in the first 9 months) from the Belarusian military-industrial complex to China. This was the first CGWIC contract with a European client. (CGWIC previously launched six other satellites for international clients). It was also the first time that CGWIC got involved in the satellite's operations (Hill 2011).

Belarus stands as a post-Soviet state that is largely compelled to remain Moscow's strategic ally through economic, political, and diplomatic leverage, as well as hybrid operations. Interestingly, Belarus has sought to attract China's investment possibly to help offset Russia's inordinate influence. In 2018, Minsk was the destination of Chinese Defense Minister Wei Fenghe's first foreign visit (together with Moscow), demonstrating Beijing's strategic interest in this region. Although it may seem as a smart move to reduce traditional Russian domination, Minsk seems to be on the path of full space sector dependency on Beijing and Moscow.

The Arctic

As an Arctic state, Russia's space-related activity in the Arctic largely remains confined to its vast amount of its claimed Arctic territory, and, as such, is not analyzed in this article. That said, Russian companies have been identified as having other business ties in certain Arctic states, including Svalbard.

China's space activities in the region have been expanding. In December 2018, the relatively new Ministry of Natural Resources (MNR), which now oversees the Chinese Arctic and Antarctic Administration (CAA), launched the "Arctic Environment Satellite and Numerical Weather Forecasting Project." According to MNR, it is to assist China's role in the governance of the Arctic and in the building of the Polar Silk Road (Eiterjord 2019). China currently has its stations in Kiruna (Sweden), Karholl (Iceland), Ny-Ålesund (Svalbard), and Longyearbyen (Svalbard), and plans to establish ones in Finland (Sodankyla) and Greenland (Nuuk).

The China Remote Sensing Satellite North Polar Ground Station (CNPGS) in Kiruna, Sweden, is the first Chinese overseas EO satellite data receive station. CAS declared CNPGS in

Kiruna to be an important part of China's Gaofen project (launched in 2010) – a global EO satellite network to be completed in 2020. Concerns have been raised about its potential dual-use purpose. In January 2019, the Swedish Defense Ministry's Defence Research Agency (FOI) publicly expressed a concern that the ostensibly civilian cooperation with China could, in fact, be controlled by the PLA and used to supplement military surveillance of the Arctic region with implications for Sweden's national security (Hinshaw and Page 2019).

The China-Iceland Arctic Science Observatory (CIAO) in Karholl, Iceland, is jointly operated by PRIC and the Icelandic Centre for Research (Rannis). The facility, not far from the Icelandic port town Akureyri, has been operational since October 2018. China's Ny-Ålesund Yellow River Station on Svalbard Island, operational since 2004, has the world's largest space physics observatory and is able to accommodate 37 personnel in summer and 4 in winter (the highest occupancy of any other country with facilities there).

China's interest in Finland spiked in the period of its Chairmanship of the Arctic Council from 2017–2019, resulting in, among other developments, a China-Finland Joint Action Plan (2019–2023), which laid the groundwork for additional Chinese investment in the country going forward (MFA Finland 2019). Implementation of this action plan has included an agreement between Chinese RADI and Finnish Meteorological Institute to establish a joint Research Center for Arctic Space Observations and Data Sharing to be built in Sodankyla, Lapland.

With regard to Greenland, although it has resisted, to date, Chinese demarches, it remains a target for Chinese investment. In 2017, rather discreetly, a Chinese-funded satellite ground station and a research facility were launched in Greenland, a collaboration between a local Greenland Institute of Natural Resources and Global Change and Earth System Science Research Institute of the Beijing Normal University (BNU) (CAS 2018).

Antarctica

With regard to space-related activities in Antarctica, Russia's GLONASS stations and Chinese BeiDou stations have been installed in this region and both countries have research stations there some of which have space capabilities. China's research stations include the Changcheng (Great Wall), Zhongshan (established in 2010), Kunlun (since 2013), and Taishan (since 2014). The Polar Research Institute plans to build China's fifth station on the Inexpressible Island with construction to be completed by 2022. This research station would be close to the world's largest Antarctic station – McMurdo Station of the United States as well as New Zealand's Scott Base (Liu 2018).

Russia reopened in 2006 its Molodyozhnaya Research Station, the Soviet Union's largest station in Antarctica. Russia's most important location in Antarctica is its newer Progress station. Its other stations include the Vostok Research Station, Novolazarevskaya Research Station, Bellingshausen Research Station, and the oldest Mirny Research Station. The activities of both countries in Antarctica demonstrate their determination to bolster their presence in this strategic outpost.

The Middle East

Russia and China's space presence in this region has sharply increased in the past few years with the signing several state-to-state Memorandum of Understanding (MoUs) on future collaboration. Russia, for example, has used human space flight as a diplomatic tool. The United Arab Emirates (UAE) has already had their astronaut sent to the ISS in September 2019 with Russian assistance. Saudi Arabia and Bahrain are in talks with Roscosmos, and Egypt and Iran are also believed to be interested in having their astronauts sent into orbit by Russia.

Russian activity in this region was recorded in the UAE, Israel, Iran, and Turkey. Roscosmos and other Russian companies (i.e., VNIIEM and Barl) are assisting Iran as it seeks to create its own remote sensing capabilities. The assistance comes in many forms, including the supply of parts and technology, ground network equipment, and potential launches of spacecraft which could be provided by Roscosmos.

Iran's first satellite, Sina-1, was built and launched in 2005 by Russia (NPO Polyot). At that time, information concerning the satellite's payload was not disclosed. Roscosmos is to potentially provide future launches as Iran struggles to build its own reliable launch infrastructure, despite being ostracized from the international community for its nuclear program. Russia also signed a deal with Iran (2005) promising to build and launch two telecommunications satellites, Zohreh 1 and Zohreh 2. The deal was terminated for unknown reasons but could possibly be resurrected.

Russia's collaboration with the UAE takes advantage of the UAE's strategy of funding high-tech projects to promote the country's capabilities. Moscow can offer its space hardware and expertise, including in the Space Situational Awareness (SSA) SSA and GNSS subsectors. Research and development projects include, for example, the "Martian Town" project which is to be built in Dubai by 2023.

Chinese involvement in the Middle East has included collaboration with Israel, Saudi Arabia, and Turkey, largely through CGWIC. The company facilitated China's collaboration with the King Abdulaziz City for Science and Technology (KACST) of Saudi Arabia. KACST developed an optical micro-camera that was launched in May 2018 on one of China's Longjiang (Dragon River) microsatellites for lunar orbit operations. It was part of China's Queqiao Chang'e-4 relay satellite mission in May 2018. China also launched the Saudi-made SaudiSat 5A and SaudiSat 5B EO satellites in December 2018, demonstrating its reliable low-Earth orbit (LEO) launch services.

Interestingly, the development of the two SaudiSat-5 satellites was based on the Saudi Arabia-Belarus agreement from May 2016, in an effort to bolster Riyadh's own manufacturing capabilities (Barbosa 2018). In December 2012, CGWIC also assisted Turkey in the launch of its EO satellite (GÖKTÜRK-2). China has also sought to expand its GNSS presence in the Middle East and North Africa (MENA) region through its "BeiDou Center of Excellence" in Tunis, Tunisia.

South and Southeast Asia

Both Beijing and Moscow have a formidable network of partnerships in this region. China has provided assistance to Afghanistan, Bangladesh, Cambodia, Indonesia,

Laos, Pakistan, Sri Lanka, and Thailand. Overall, most of China's activities involved construction and launch of telecommunications satellites (Afghanistan, Pakistan, Sri Lanka, Cambodia, and Indonesia), often through some form of package deal. Other transactions focused on sole launches (e.g., Palapa-D, SupremeSAT-I, PakTES-1A) and the use of BeiDou system (e.g., construction of BeiDou stations in Pakistan or lease of BeiDou for Thailand).

Space sectors dependent on China include those of Sri Lanka (SupremeSAT, Pallekele Space Academy, etc.), Indonesia (Palapala-N1), Cambodia (Techo-1), and Pakistan (PakSat-1R). Pakistan, Thailand, and Sri Lanka also facilitate BeiDou's coverage, hosting stations on their territories. Bangladesh is the only country in which CGWIC lost a bid for the manufacturing and launch of Bangladesh's first telecommunications satellite, Bangabandhu Satellite-1, to Thales Alenia Space (Shamrat 2018).

Russia has worked with Bangladesh, Indonesia, India, the Philippines, and Vietnam. Relations with India are most extensive. In 2010, Russia and India signed an intergovernmental agreement on granting India access to the encrypted high-accuracy military GLONASS signal (Sputnik 2018). In May 2015, Roscosmos signed an MoU with the Indian Space Research Organization (ISRO) to increase their cooperation in a series of subsectors (e.g., satellite navigation, launch vehicle development, remote sensing, the use of ground infrastructure) and provide India's Space Program with GLONASS technology (TASS 2018). Out of these agreements emerged several potential projects that are in their initial stages. These include a joint communications satellite, creating a remote sensing constellation, training of Indian personnel to send them to the International Space Station on board a Russian spacecraft, monitoring of Indian Railways, and construction of ground stations in India (including Russian global navigation satellite system (GLONASS) stations) (Cozzens 2019).

Western, Central, and Eastern Asia

Although China has sought to establish ties in this region, and reached out to Armenia, Azerbaijan, Turkmenistan, and Armenia, Russia remains the more prominent space player in this region, working with Armenia, Azerbaijan, South Korea, Kazakhstan, and Uzbekistan.

Russia has been historically active in Kazakhstan mainly because of the existing launch infrastructure located in Baikonur, the lease of which expires in 2050. Kazakhstan was granted access to GLONASS military signal in 2018. (India (in 2010) and Algeria (in 2019) also received access to the GLONASS military signal.) A GLONASS station was also opened at the Byurakan Observatory in Armenia, which included some modernization of the existing station. Following this transaction, Russia, aiming to further strengthen its ties with Armenia, promised to train an Armenian astronaut for a mission to the International Space Station.

Russia also recently reached out to Uzbekistan. It was reported on May 1, 2018, that Roscosmos offered the Uzbek State Space Research Agency (Uzbekcosmos) to finance a program that would enable the country to launch its first satellites. Shortly thereafter, a proposal on a trilateral (Russia, Uzbekistan, and Kazakhstan) satellite launch was tabled (SpaceWatchGlobal 2018).

Top Space Sector Capture Trends

When analyzing partnering arrangements described in section two of this article, there seem to be four prominent space sector capture trends. They include vertically integrated package deals (e.g., in Angola, Belarus, Brazil, Congo, Ethiopia, Bolivia, Pakistan, Nigeria, Algeria, Venezuela); the active involvement of China and Russia in the Arctic and Antarctica; China's global space power projection through EO and space observation partnerships; and expanding the number of GNSS (GLONASS and BeiDou) ground stations abroad enhancing the capabilities of the respective systems.

The package deals are a hallmark of Chinese and Russian influence attempts and have been most evident in the telecommunications subsector (but also found, for example, in the EO subsector). While the general pattern of the package deals is similar across the board, the number and type of components in each offer varies.

China has been a driver of multilateral Space and Earth observation partnerships. China leads the international Asia-Pacific Ground-based Optical Space Objects Observation System (APOSOS) initiative, launched through the Asia-Pacific Space Cooperation Organization (APSCO). The stated goal is to build a global optical observation network with at least one facility in each of the APSCO Member States and elsewhere. China has sensors in all eight APSCO member countries, as well as in Brazil and Ukraine. Mexico joined the APOSOS network in 2017 (IDA 2018).

The primary objective of this organization is to build a data sharing platform and use existing infrastructures from Member States. A second objective is to bring new capabilities to the table and extend the ability to observe MEO and GEO. It is to be operated under APSCO observation mission management department. For example, new telescopes (manufactured by the Changchun Institute of Optics, Fine Mechanics and Physics of CAS) have been completed and declared functional in Iran, Pakistan, and Peru.

China has also tied its economic and strategic interests to the grouping of Brazil, Russia India, China, and South Africa (BRICS) nations, providing financial backing to a sizeable portion of their activities. This includes the first BRICS space project – sharing of EO data with an intention to eventually build a remote sensing satellite constellation. While a specific timeline has not been made available, the current plans are for the project to have two phases: (1) the creation of a remote sensing data-sharing system, making the data from each of the member countries' existing EO satellites available to all the other members; and (2) the creation of a new EO satellite constellation (Campbell 2017). There is also an EO collaboration between Europe and China through the Dragon program (currently in its 4th iteration) between MOST's National Remote Sensing Centre of China and the European Space Agency (ESA).

Spearheaded by China and Russia, BRICS was configured as a platform for engagement of emerging market economies and developing countries. Initially, the initiative struggled to fund itself, but the creation of the BRICS Development Bank, now called the New Development Bank (NDB), based in Shanghai, represented an important pivot operationally. In April 2017, Brazil was the recipient of the bank's first development loan. Together with the Asian Infrastructure Investment Bank and the Silk Road Fund, NDB is a key investment tool for China's power projection strategy abroad (Reuters 2018).

With regard to the GNSS sector, both Russia and China continuously work on global expansion of their respective systems. Russian President Vladimir Putin prioritized the GLONASS system's restoration in the early 2000s. The full constellation of satellites was reestablished by 2011. GLONASS satellites have undergone several upgrades over the years, the latest being the GLONASS-K. The GLONASS stations have been (or are planned to be) placed in Antarctica, Argentina (planned), Armenia, Cuba (planned), Ecuador (planned), India (two stations, both planned), Kazakhstan (one existing and one planned), Nicaragua, South Africa, South Korea (envisioned), UAE (planned), and Venezuela (planned).

With regard to China, on December 27, 2018, China's BeiDou Navigation Satellite System (BDS) was declared to be providing global service as the construction of the BDS-3 primary system had been completed. China wants BeiDou to become an alternative to America's Global Positioning System (GPS). The BeiDou applications are also promoted through the Belt and Road Initiative (BRI). The more widespread use of Russia's and China's GNSS will help them integrate these countries into their respective economic and military orbits.

Key Findings

Although it is not possible to accurately assess the precise number and status of all existing space partnerships that China and Russia have concluded globally with the use of open source materials, it has been possible to determine that the number of these partnerships has expanded substantially, especially on the part of China. A Chinese media source asserts that, as of April 2018, China had signed 121 space cooperation agreements with 37 countries and four international organizations (MFA PRC 2018). Back in 2016, according to the State Council Information Office on China's space activities, China had signed 43 space cooperation agreements with 29 countries, space agencies, and international organizations (SCIO PRC 2016). If accurate, this would mean an increase of 78 international cooperation agreements over about 15 months. PSSI has identified Chinese space relationships with 60 countries as of February 2019.

Russia has been less transparent when it comes to declaring publicly the number of international space partnerships it has concluded. It has mentioned international cooperation in relation to its technological development goals in its "Russian Federal Space Program 2016–2025" (Roscosmos 2016). Russia has mostly focused on reviving, or maintaining, its post-Soviet ties. CIS states are important targets (especially Belarus, Kazakhstan, and Armenia, but also Azerbaijan, Moldova, Tajikistan, Turkmenistan, and Uzbekistan) (Roscosmos 2020). PSSI unearthed Russian space partnerships with 44 countries.

Overall, Chinese and Russian economic and financial activities in various regions indicate that space sector partnerships that create various levels of dependencies (and even full-scale space sector capture) have been vigorously pursued with respect to both developing and developed countries. As democratic countries have more rigorous requirements for transparency, accountability, and the rule of law, a more subtle, incremental approach is evident (e.g., projects related to scientific research/development, academic exchanges, or individual commercial contracts).

The research suggests that China and Russia enter into space partnerships globally for two main reasons: (1) because the recipient state is in an important, or even strategic, geographic location for enhancing Chinese/Russian space capabilities (e.g., for GLONASS/BeiDou, SSA, EO); and (2) a country has strategic importance that a space partnership helps leverage (e.g., a country's energy resources, mineral wealth, supportive geopolitical policy positions). In some cases, these motivations are co-mingled, such as the case of the Arctic countries (e.g., Greenland).

The second pattern of behavior described above (where space considerations are not the prime mover) most often occurred in less economically successful, but resource-rich, countries (e.g., Bolivia, Nigeria, Venezuela), or countries that are geographically and/or geopolitically strategic for China or Russia (e.g., Pakistan for China, Iran for Russia, and Cuba for both countries). In the case of China, some of these recipient countries (e.g., Belarus, Cambodia, Laos, and Pakistan) are valued clients of its BRI, without a direct linkage to space.

Incremental space sector dependencies in the developed countries are more difficult to detect and guard against (e.g., academic exchanges, scientific and research projects, broader funding commitments beyond the space sector). More visible space sector capture largely takes place via the offering of vertically integrated "package deals." On a number of occasions, China and Russia have been able to construct successfully dual-use space infrastructure and services due to hospitable political relations, corruption, and internal economic and social strife in the targeted countries (e.g., Argentina, Brazil, Cuba, Colombia, Nicaragua, Nigeria, Pakistan, Sri Lanka, Venezuela).

China and Russia generally use their state-controlled enterprises (e.g., China Great Wall Industry, Roscosmos) to penetrate a specific country's space sector. These state companies position themselves as preferred "go-to" entities. This is, in no small part, because of generous financing that often does not reflect the targeted country's creditworthiness.

The pattern of Chinese and Russian space-related transactions reveals a global approach, signaling that both countries are determined to expand their space stature and competitiveness, and close the gap with Europe and the United States. Some targeted countries receive financial backing well beyond the space sectors. Influencing/capturing the space sector is just part of a broader strategic outreach (e.g., in Bolivia).

Conclusion

Tracking and visually mapping the international transactions of Chinese and Russian state-controlled enterprises in the space sectors of various countries revealed that Beijing and Moscow offer assistance to the nascent space programs that, in many cases, create dependencies, including a full-scale space sector capture.

Today, the implications of the active pursuit of international space partnerships globally by China and Russia to increase their influence over the space domain through such means as the offer of vertically integrated "package deals" of capabilities and services are not well-understood. The economic and financial largesse provided by these state actors, including subsidized financing, are accepted, and often even welcomed, by the recipient countries which lack their own space funding, technical expertise, and human resources, even if it exposes them to partial or complete dependency on these outside "benefactors."

One of the risks stemming from these global networks of space dependencies is the opening for China and Russia to increasingly shape applicable rules, norms, and standards for access to, and operations in, space. This asymmetric threat to global space governance stemming from these hybrid economic and financial maneuvers needs to be better appreciated, including the underlying rationales for these activities.

The United States, Europe, and other allies, including Japan, need to also review their space partnership approaches and configure more effective and attractive offers to their nontraditional space partners to counter Chinese and Russian predatory economic and financial practices. It is clear that countries of all economic performance levels are intent on benefiting from the value provided by space. If upgraded space-related engagement and an enhanced level of support do not occur from Western countries, the void is likely to be filled – as we are already witnessing – from these nondemocratic actors, with negative consequences.

As in other domains, a sustainable model of international partnerships cannot be established without transparency, good governance, accountability, respect for national sovereignty, and the rule of law. Western nations will have to be able to demonstrate that there is a clear benefit in collaboration with countries that respect free and fair market principles and behavior versus state-controlled economies that often show little regard for such principles.

References

Barbosa R (2018) Long March 2D launches SaudiSAT-5A/B NASA space flight. NASA Spaceflight. https://www.nasaspaceflight.com/2018/12/long-march-2d-saudisat-5a-b/. Accessed 10 Mar 2020

Belta (2018a) China aerospace science and technology corporation to open research center in Belarus. Belarus Telegraph Agency. https://eng.belta.by/society/view/china-aerospace-science-and-technology-corporation-to-open-research-center-in-belarus-113476-2018/. Accessed 10 Mar 2020

Belta (2018b) Plans to build R&D commercialization center in China-Belarus industrial park within one year. Belarus Telegraph Agency. https://eng.belta.by/economics/view/plans-to-build-rd-commercialization-center-in-china-belarus-industrial-park-within-one-year-117427-2018/. Accessed 10 Mar 2020

Campbell K (2017) BRICS bloc agree remote sensing space constellation project. Engineering News. http://www.engineeringnews.co.za/article/brics-bloc-agree-remote-sensing-space-constellation-project-2017-07-04/rep_id:4136. Accessed 10 Mar 2020

CAS (2018) China, Finland to enhance Arctic research cooperation. Chinese Academy of Sciences. http://english.cas.cn/Special_Reports/Belt_of_Science_Road_for_Cooperation/Technology_Cooperation/201810/t20181029_200564.shtml. Accessed 9 Mar 2020

Clark S (2019) Chinese company inks deal to launch 90 commercial smallsats. Spaceflightnow. https://spaceflightnow.com/2019/01/30/chinese-company-inks-deal-to-launch-90-commercial-smallsats/. Accessed 3 Mar 2020

Cozzens T (2019) India to host GLONASS ground station for Russia. GPS World. https://www.gpsworld.com/india-to-host-glonass-ground-station-for-russia/. Accessed 10 Mar 2020

Eiterjord T (2019) China's busy year in the Arctic. The Diplomat. https://thediplomat.com/2019/01/chinas-busy-year-in-the-arctic/. Accessed 8 Mar 2020

Ellis E (2017) China's relationship with Chile: the struggle for the future regime of the Pacific. Jamestown Foundation. https://jamestown.org/program/chinas-relationship-chile-struggle-future-regime-pacific/. Accessed 8 Mar 2020

ESA (2019) CBERS – 1st generation. eoPortal Directory. https://directory.eoportal.org/web/eoportal/satellite-missions/c-missions/cbers-1-2. Accessed 5 Mar 2020

European Commission (2018) Daily news press release 27/08/2018. European Commission.. http://europa.eu/rapid/press-release_MEX-18-5181_en.htm. Accessed 7 Mar 2020

Foust J, Bodner M (2016) How Crimea fractured Ukraine's space program. SpaceNews. https://spacenews.com/from-the-magazine-ukraines-untethered-orbital-manuevers-how-crimea-hurt-ukraines-space-program/. Accessed 7 Mar 2020

Giri C (2018) Version 2: mapping China's global telecom empire. Gateway House. https://www.gatewayhouse.in/china-global-telecom-tentacles/ Accessed 9 Mar 2020

Glavkosmos (2019) Services/earth observation data and solutions. Glavkosmos JFC. http://glavkosmos.com/en/earth-observation/. Accessed 7 Mar 2020

Gorchinskaya K (2020) Ukraine prepares to snub China in aerospace Deal with U.S. help. Forbes. https://www.forbes.com/sites/katyagorchinskaya/2020/02/17/ukraine-prepares-to-snub-china-in-aerospace-deal-with-us-help/#2aa25509328c. Accessed 9 Mar 2020

Hill J (2011) China wins first European satellite construction, Launch Contract. Via Satellite. https://www.satellitetoday.com/telecom/2011/09/21/china-wins-first-european-satellite-construction-launch-contract/. Accessed 10 Mar 2020

Hinshaw D, Page J (2019) How the pentagon countered China's designs on Greenland. Wall Street J. https://www.wsj.com/articles/how-the-pentagon-countered-chinas-designs-on-greenland-11549812296. Accessed 10 Mar 2020

Ibeh J (2019) Russia to deploy space monitoring stations in South Africa, Mexico and Chile. Space in Africa. https://africanews.space/russia-to-deploy-space-monitoring-stations-in-south-africa-mexico-and-chile/. Accessed 5 Mar 2020

IDA (2018) Global trends in space situational awareness (SSA) and space traffic management (STM). Institute for Defense Analyses, p 50. https://www.ida.org/idamedia/Corporate/Files/Publications/STPIPubs/2018/D-9074.pdf. Accessed 10 Mar 2020

Interfax-Ukraine (2017) Ukraine, China approve updated bilateral program of space cooperation until 2020. KyivPost, Interfax-Ukraine News.Agency. https://www.kyivpost.com/ukraine-politics/ukraine-china-approve-updated-bilateral-program-space-cooperation-2020.html. Accessed 7 Mar 2020

Kelley M (2014) Russia is reportedly reopening its Spy Base in Cuba. Business Insider.. https://www.businessinsider.com/russia-is-reportedly-reopening-its-spy-base-in-cuba-2014-7. Accessed 9 Mar 2020

Krebs G (2017) NIGCOMSAT 1 and 1R. Gunter's space page. https://space.skyrocket.de/doc_sdat/nigcomsat-1.htm. Accessed 9 Mar 2020

Lee RV (2018) Satellite images: a (Worrying) Cuban mystery. The Diplomat. https://thediplomat.com/2018/06/satellite-images-a-worrying-cuban-mystery/. Accessed 2 Mar 2020

Liu N (2018) What does China's fifth Research Station mean for Antarctic governance?. The Diplomat. https://thediplomat.com/2018/06/what-does-chinas-fifth-research-station-mean-for-antarctic-governance/. Accessed 10 Mar 2020

Londoño E (2018) From a Space Station in Argentina, China Expands its Reach in Latin America. New York Times. https://www.nytimes.com/2018/07/28/world/americas/china-latin-america.html. Accessed 5 Mar 2020

MFA Finland (2019) Joint action plan between China and Finland. The Ministry of Foreign Affairs of Finland. https://um.fi/documents/35732/0/Joint+Action+Plan+2019-2023+%283%29.pdf/bd639013-a815-12d2-ae44-4bc50ded7a97. Accessed 9 Mar 2020

MFA PRC (2018) China strengthens international space cooperation. Ministry of Foreign Affairs of the People's Republic of China. https://www.fmprc.gov.cn/zflt/eng/jlydh/mtsy/t1555457.htm. Accessed 9 Mar 2020

Nowakowski T (2016) Arianespace to launch Soyuz rocket with European satellites. SpaceFlight Insider. https://www.spaceflightinsider.com/organizations/arianespace/arianespace-launch-soyuz-rocket-european-satellites/. Accessed 7 Mar 2020

Pai Ajit (2019) Protecting National Security and public safety. Federal Communications Commission. https://www.fcc.gov/news-events/blog/2019/10/28/protecting-national-security-and-public-safety. Accessed 16 Mar 2020

Rail Working Group (2016) Kaunas free economic zone joins the rail working group rail working group. Rail Working Group. https://www.railworkinggroup.org/kaunas-free-economic-zone-joins-the-rail-working-group/. Accessed 9 Apr 2019

Reuters (2018) BRICS development bank to expand lending to private sector. Reuters, Business News. https://www.reuters.com/article/us-china-brics-bank/brics-development-bank-to-expand-lending-to-private-sector-idUSKCN1IU0P2. Accessed 10 Mar 2020

Robinson J (2018a) Competition for international space partnerships. Prague Security Studies Institute, Prague, p 6. http://www.pssi.cz/download/docs/617_sagar2018-presentation-robinson.pdf. Accessed 26 Nov 2018

Robinson J (2018b) Cross-domain responses to space hybrid provocations via economic and financial statecraft. In: USSTRATCOM 2018 deterrence and assurance academic Alliance conference, March 2018. https://www.stratcom.mil/Portals/8/Documents/AA_Proceedings/3.pdf?ver=2018-10-04-141147-287. Accessed 9 Mar 2020

Robinson J (2018c) Europe's preparedness to respond to space hybrid operations. Prague Security Studies Institute, Prague, p 4. http://www.pssi.cz/download/docs/590_europe-s-preparedness-to-respond-to-space-hybrid-operations.pdf. Accessed 9 Mar 2020

Robinson J, Robinson R, Davenport A, Kupkova T, Martinek P, Emmerling S, Marzorati A (2019) State actor strategies in attracting space sector partnerships: Chinese and Russian economic and financial footprints. Prague Security Studies Institute, Prague. http://www.pssi.cz/download/docs/686_executive-summary.pdf. Accessed 4 Mar 2020

Roscosmos (2016) The main provisions of the Federal Space Program 2016–2025. State Space Corporation ROSCOSMOS, Moscow. https://www.roscosmos.ru/22347/. Accessed 9 Mar 2020

Roscosmos (2020) Intergovernmental agreements and commissions for economic, scientific and technical cooperation. Roscosmos. https://www.roscosmos.ru/22887/. Accessed 7 Mar 2020

SCIO PRC (2016) White paper on China's Space Activities in 2016. The State Council Information Office of the People's Republic of China. http://www.spaceref.com/news/viewsr.html?pid=49722; http://english.scio.gov.cn/whitepapers/2017-01/10/content_40535777.htm. Accessed 10 Mar 2020

Selding P (2015) Brazilian government bypassing the US as it builds out a space sector. SpaceNews. https://spacenews.com/brazil-bypassing-the-us-as-it-builds-out-a-space-sector/. Accessed 8 Mar 2020

Shamrat AS (2018) Bangladesh joins the space age. YaleGlobal, Yale University. https://yaleglobal.yale.edu/content/bangladesh-joins-space-age. Accessed 10 Mar 2020

SpaceWatchGlobal (2018) Uzbekistan widens partnerships for emerging space programme. SpaceWatch.Global. https://spacewatch.global/2018/11/uzbekistan-widens-partnerships-for-emerging-space-programme/. Accessed 9 Mar 2020

Sputnik (2018) Russia Grants Kazakhstan access to military satellite signal – space agency. Sputnik News. https://sputniknews.com/military/201808041066939986-russia-kazakhstan-military-satellite-signal/. Accessed 9 Mar 2020

TASS (2018) Russia's Roscosmos invites India to cooperate in manned space missions, satellite program. TASS Russian News Agency. http://tass.com/science/1023084. Accessed 9 Mar 2020

Taylor E (2016) China's Midea receives U.S. green light for Kuka takeover. Reuters. https://www.reuters.com/article/us-kuka-m-a-mideamidea-group-idUSKBN14J0SP. Accessed 9 Mar 2020

Wang B (2018) Ukraine helping South Korea, China and Canada With Rockets or Jets or Both. Nextbigfuture. https://www.nextbigfuture.com/2018/12/ukraine-helping-south-korea-china-and-canada-with-rockets-and-jets.html. Accessed 9 Mar 2020

Williamson J (2017) Huawei and KUKA forge new smart manufacturing deal. The Manufacturer. https://www.themanufacturer.com/articles/huawei-and-kuka-forge-new-smart-manufacturing-deal/. Accessed 3 Mar 2020

Xinhua (2018) German government gives Chinese AT&M's takeover of Cotesa green light: report. Xinhua Net. http://www.xinhuanet.com/english/2018-04/26/c_137139362.htm. Accessed 9 Mar 2020

Space Export Control Law and Regulations

Ulrike M. Bohlmann and Gina Petrovici

Contents

Introduction	186
Terminology	186
International Legal Regimes	189
The Melee of International Legal Instruments on Export Control	190
The Specificities of the Outer Space Regime	192
National and Regional Legal Regimes	194
The Export Control Regime of the United States	194
Export Regulations of the European Union	200
The Export Control Regulations of the European Space Agency	203
Conclusions	205
References	206

Abstract

The world continues to face an increasing demand for dedicated space applications. Global partnerships and cooperation provide the baseline for numerous space missions. International cooperation entails peaceful exploration, exploitation, and use of outer space since the early post-Cold War years. Space technology is dual-use by nature as it is serving scientific or even commercial interests on

---

The views expressed are purely personal and do not necessarily reflect the view of any entities with which the authors may be affiliated. Legal developments up to 28 February 2019 have been taken into account.

U. M. Bohlmann (✉)
European Space Agency (ESA), Paris, France
e-mail: ulrike.bohlmann@esa.int

G. Petrovici
German Aerospace Center (DLR), Bonn, Germany
e-mail: gina.e.petrovici@googlemail.com

© Springer Nature Switzerland AG 2020
K.-U. Schrogl (ed.), *Handbook of Space Security*,
https://doi.org/10.1007/978-3-030-23210-8_35

the one hand and strategic, defense-related objectives on the other. A functioning and reliable system for space export controls is a prerequisite for a well-functioning international space industry. This contribution sheds light on the relevant international, regional, and national legal mechanisms and their respective effects on the space industry.

Introduction

Since the inception of space activities, with the launch of Sputnik 1, enduring scientific and technological development ensured the success of outer space activities.

Outer space developed as an enabler and outer space activities, and their applications are tremendously important tools to answer the needs of modern society. The strategic and political objectives of outer space activities have always been intrinsically linked to the economic success and social returns (ESA Space Economy 2019). The Satellite Industry Association detected a global growth of satellite industry revenues of $ 268.6 billion (SIA 2019), and Europe's space sector is facing an extraordinary long series of growing sales, with a growth of sales to European customers of € 703 million (Aerospace and Defence Industries Association of Europe (ASD) EUROSPACE 2018). "Taking together Europe's defense and civil aeronautics, exports are at the same order of magnitude as the US" (ibid.). Export control systems are closely connected to the realm of international cooperation and foreign policy, to national security, and, as for the technical aspects, to space technology and its underlying technical knowledge. Understanding the logic behind the export control systems necessitates considering all four of these areas and their respective interconnections. While space technology determines the object of export controls and its practical scope, national security and international cooperation form the nature of the rules with their political impact. At the same time, it is precisely space technology and its characteristics that influence the national approaches toward its export controls and the willingness to cooperate on an international level. Although these interrelations are apparent and recognized, an analysis revealing their structure and inner functioning is necessarily multifaceted. The tangled connections give rise to rather complex and very heterogeneous systems of export control rules and its further development.

Terminology

Export control regulations are considered to be part of foreign trade law (Wegner et al. 2006, p. 21). More precisely, export controls can be understood as segment of commercial administrative law (Tietje 2009, aup.681). Governments exercise export controls as a means to promote foreign policy and commercial interests and to protect strategic industrial sectors as well as national security and the nonproliferation of sophisticated weapons and weapons of mass destruction (WMD) (Hertzfeld and Jones 2011).

For the sake of this chapter, we define *export controls* as the restriction on the export of goods or services imposed by the exporter's own country. Export controls apply to a variety of items: chemical, biological, nuclear, military, and dual-use in nature (Aubin and Idiart (2011), pp. 1–18). The laws and regulations governing export controls are in fact posing an exception to the free trade principles of international commercial exchanges. Thus, entities involved in trade of controlled goods and services are obliged to follow and comply with the applicable regulations articulated by national and international regimes. The right to exercise export controls belongs to the sovereign rights of states to control and regulate cross-border transfer of goods and services (Pezzullo 2014). And it has to be noted that each state performs export controls of space technologies unilaterally. Export control regulations are in fact a reaction to the need to control items – the hardware, the software, or the technology and services leaving their territory – for the reason of security and technology protection.

In accordance with numerous relevant laws and regulations, *export* may be defined as the transfer (note that the term transfer is used for intra-European Union exports, for more details see below) of an item from one country (the country of exportation) to a foreign country (the country of destination) regardless of the method that is used for this transfer. Laws apply based on the nationality of the item. The term *item* may encompass hardware, software, technology, and even know-how. Export is also considered to have occurred if the item leaves the country only temporarily or as an item which is not for sale. Other forms of temporary exports are a wholly owned subsidiary in a foreign country or the transfer of an item with foreign origin from the exporting country to the country of origin (Aubin and Idiart 2011, p. 4). Also, the release of information regarding controlled technology to a foreign entity while the item itself remains on the soil of the country of exportation is considered an export. National export control regimes, such as in the United States (the US export control regime will be further examined below), have a broader scope of application and include deemed exports and the re-transfer inside the same territory of a product to another consignee. Export controls typically focus on strategic items, but some controls on embargoed countries can apply to the export of any item. Further, it is worth noting that more than one set of rules may apply to a single transaction.

The export control systems are often being implemented in the form of licensing authorization processes. In order to estimate whether a license can be granted, several factors are taken into consideration by the relevant administration. First, the goods and the technology submitted to the licensing process have to be examined with regard to their nature. In addition, the end user, the intended end use, and the destination have to be determined. Classification of goods and their placement on such a list derives from their characteristics. Although no consolidated definition of the term *space technology* can be offered, it can be described as a systematic application of engineering and scientific disciplines to the exploration and utilization of outer space (McGraw-Hill Science & Technology Dictionary). Within the export control laws and regulations, the term often has a very specialized meaning which is applied in the range of the concerned legal or regulatory text. The export control system typically deals not only with the physical item itself but also with the specific information required for the development, production, or use of an item classified commonly as technical data or technical assistance. The term *space technology* refers simultaneously to the physical item as well as to the technical knowledge the item is built upon.

However, the close link between the Fourth Industrial Revolution, also known as Industry 4.0, and new space developments enables the rapidly increasing amount of new space entities to make use of material and technology originally used on Earth in outer space. This kind of technology might not yet fall under the given definitions of *space technology*, which is also varying between the existing export control regimes (see, e.g., EAR and EU Council Regulation 428/2009). In some cases, these items may then be classified as commercial. In the United States, *exempli causa*, these items fall under the jurisdiction of the US Department of Commerce and are designated as EAR99, as the majority of commercial products, which will not require a license to be exported or reexported as long as not exported to an embargoed or sanctioned country, to a party of concern, or in support of a prohibited end use. Nonetheless, the classification of such items has of course to be made on a case-by-case basis and might not be straightforward, where innovative technology is used for the first time in the harsh space environment. Therefore, the expression *technology used in space* might be more suitable in view of the export control classifications than merely *space technology* as the former encompasses a broader range of items than the latter.

The technical characteristics of the space item are a crucial factor in this whole process. If an item intended for export is referenced by a specific export control legal or regulatory text, the prescribed necessary measures in order to obtain a license for its export have to be taken.

The space industry covers a wide range of activities, such as design, construction, and assembling of complete spacecrafts, major subsystems and electronic systems such as earth observation technologies (Rhodes et al. 2015). Consequently, space export-controlled items range from satellites, launching pads to component parts, such as lasers and sensors as well as imagery and high-sensitive data.

Controls are usually conducted on exports of nuclear, chemical, biological, military, or dual-use goods, technology, or services. Virtually, the majority of controlled exports relate to military and dual-use goods and services. *Dual-use goods* are goods that may be used for both nonmilitary and military applications. Dual-use technology can then be described as those products, technologies, and services that can address and serve the needs of the civil and military fields. The military application can be either proven or even potential (Wetter 2009 xv).

As a matter of fact, all space items and technologies, *exempli causa* launch vehicles and satellites, are inherently dual-use since outer space as such is militarily strategic. While some early launch vehicles, such as the launch of Sputnik 1 on the Russian 8K71PS rocket (Konyukhov 2003) or the launch of the first US satellite "Explorer 1" on Army Jupiter-C (Angelo 2006), were converted military ballistic missiles with almost identical technology which may then have undergone further technological developments, it holds true also for contemporary launch vehicles that are most of the time independent system developments and that they still share many characteristic features and technological bases. Therefore, they are, together with ballistic missiles, classified as delivery mechanisms for weapons of mass destruction. Satellites and their components are modified according to their intended use, which means that they can also be adapted to the requirements imposed by a military purpose. Satellites as well as satellite components are therefore generally categorized as dual-use and may be considered militarily sensitive.

The swings of technology in the aerospace industry, from military to civil applications after the era of the Cold War and back, prove that both areas benefit from a cross-fertilization. Items developed strictly for civil or commercial purposes might still be used for military purposes and vice versa.

The possibility to provide support and strategic advantages to military and intelligence operations is regarded as one of the most important export control characteristics of space technologies. Space technologies have already become an essential element of military and intelligence activities and play an important role in their future planning and structures. These characteristics are not based on the technical qualities of the space technologies but result from their potential use and the benefits they bring.

Therefore, the question that needs to be asked is not how to prevent this inherent dual-use nature, as the solution would be clearly beyond the regulatory or even technical possibilities. The question is what the export control regulations concerning these items should look like (Mineiro 2012, pp. 3–12).

In contrast to export controls, sanctions can be defined as restrictions on transactions with other countries, persons, or entities based on security or policy concerns, which do not necessarily involve items. Instead of the nationality of the item, the nationality of the person involved in the transaction gains central importance in the context of sanctions. Sanctions are not only limited to items, software, or technology and can also encompass financial transactions, related agreements, or services (Wolf, How U.S. Export Controls and Sanctions Affect the Work of the European Space Agency, p. 2.).

International Legal Regimes

Internationally coordinated export controls can contribute to the promotion and preservation of global stability and security. Even though there are no binding multilateral agreements focusing *exclusively* on space technology export control yet, it must be highlighted that a number of multilateral agreements concerning space technology used as weapon delivery systems and documents prohibiting their deployment and operation have been agreed upon and entered into (Gerhard and Creydt 2011, p. 191). Furthermore, some bilateral agreements serving as a basis to exempt respective states from the related licensing process can be found (Mineiro 2012).

The following paragraphs will examine the general international export control regimes and their applicability to space technology. Following that, the impact of the traditional international space law, in a narrower sense, to the space export controls will be investigated.

The Melee of International Legal Instruments on Export Control

Enhancing international cooperation is a means to maintain peace and international security. "[...] Growing tensions between states wishing to develop their civilian space programs, on one side, and states willing to prevent the traffic of dangerous military items, on the other side, have developed" (Von der Dunk/Tronchetti, Handbook of Space Law 2017, p. 360). The dialogue between nations resulted in the establishment of a number of international legal instruments intended to balance the international law principle of protection of peace and international security with the right to legitimate self-defense as well as to development and economic freedom (Achilléas 2007, p. 20). Four separate and almost wholly independent functional regimes compose the current export control system, which supplement the provisions of other binding, multilateral treaties primarily focused on the development and possession of weapon technologies, such as the 1968 Nuclear Non-Proliferation Treaty (NPT) (Treaty on the Non-Proliferation of Nuclear Weapons 1968), the 1972 Biological Weapons Convention (BWC) (Convention on the Prohibition of the Development 1972b), and the 1993 Chemical Weapons Convention (CWC) (Convention on the Prohibition of the Development 1993; Joyner 2004).

The four functional supplier state regimes are:

- The system of the Nuclear Suppliers Group (hereinafter "NSG", http://www.nuclearsuppliersgroup.org), which governs the area of nuclear weapons and materials
- The Wassenaar Arrangement on Export Controls 1996 (hereinafter "Wassenaar Arrangement"), which sets the rules in the context of conventional weapons
- The Australia Group (The Australia Group, http://www.australiagroup.net/), which deals with chemical and biological weapons proliferation
- The Missile Technology Control Regime (hereinafter "MTCR" or "The MTCR Guidelines 2012"), which regulates the export of missile and related delivery system technologies

In addition, these regimes are complemented by provisions of the Nuclear Non-Proliferation Treaty (hereinafter "NPT"), the IAEA Comprehensive Safeguards Agreement and Model Additional Protocol (1997), the Zangger Trigger List (1974), the Limited Test Ban Treaty (hereinafter "LTBT") (Treaty Banning Nuclear Weapon Tests 1963), and the Comprehensive Nuclear-Test-Ban Treaty (hereinafter "CTBT 1996") regarding nuclear materials and tests; by the Geneva Protocol (1925), the Biological and Toxin Convention (Convention on the Prohibition of the Development 1972), and the Chemical Weapons Convention with regard to chemical and biological weapons proliferation; by the Hague Code of Conduct (hereinafter "HCoC" or "International Code of Conduct 2002") concerning ballistic missile proliferation; and by United Nations Register of Conventional Arms (hereinafter "UNROCA 1992") in the domain of conventional weapons.

By the conclusion of bilateral or multilateral agreements, states coordinate their domestic regulations but generally do not install a specific authority to enforce these obligations. In addition to these partly soft-law instruments, there are only a few specific cases, in which international law imposes direct obligations to control space technology exports. In accordance with Article 39 of the United Nations Charter, the United Nations Security Council is entitled to take actions "to determine the existence of any threat to the peace and shall make recommendations or decide what measures shall be taken in accordance with Artt. 41 and 42, to maintain international peace and security."

One concrete example is United Nations Security Council Resolution 1540 (UN Doc. S/Res/1540 2004), which was adopted unanimously on 28 April 2004. Security Council Resolution 1540 establishes the obligation under Chap. VII of the UN Charter for all member states of the UN to develop and enforce appropriate legal and regulatory measures against the proliferation of chemical, biological, radiological, and nuclear weapons and their means of delivery, in particular, to prevent the spread of weapons of mass destruction to non-state actors (on an implementation strategy for Resolution 1540, see Heupel 2007). Thus, the UN Security Council may pass on behalf of its 15 member states a resolution, which then establishes the need for member states to adopt and enforce controls over exports insofar as materials and technologies might have the abovementioned impact so that it poses a security threat. This explicit international obligation also impacts the right of states to export space technologies, to the extent that they could be used as a means of delivery for weapons of mass destruction (Mineiro 2012, p. 21). However, so far it never passed a legally binding resolution concerning the export of satellite technology (ibid., p. 20).

In the field of soft-law agreements (on the notion of soft law, see Freeland 2012, with further references), the Wassenaar Arrangement on Export Controls for Conventional Arms and Dual-Use Goods and Technologies plays a major role. Numerous space faring nations are parties to it. It is a nonbinding export control agreement aiming at "contributing to regional and international security, by promoting transparency and greater responsibility in transfers of conventional arms and dual-use goods and technologies" (Wassenaar Arrangement, paragraph 1). Certain space and satellite technologies are listed in the categories of sensitive and very sensitive dual-use goods in the annexed List of Dual-Use Goods and Technologies, and their transfer or denial must be therefore duly notified according to the rules set up by the arrangement (Mineiro 2012; Achilléas 2007, p. 53). Following increasing cybersecurity vulnerabilities, the state parties to the Wassenaar Arrangement decided already in December 2017 to include also computer network intrusion software into Category 4 of its Dual-Use List. According to the respective definition the term "Intrusion software" means software specially designed or modified to avoid detection by "monitoring tools," or to defeat "protective countermeasures," of a computer or network-capable device, and performing any of the following:

(a) The extraction of data or information, from a computer or network-capable device, or the modification of system or user data
(b) The modification of the standard execution path of a program or process in order to allow the execution of externally provided instructions (Wassenaar Arrangement Secretariat 2018, pp. 221–222)

Moreover, launch vehicle space technologies are addressed in two other non-binding instruments – in the Guidelines for Sensitive Missile-Relevant Transfers and the International Code of Conduct against ballistic missile proliferation. The MTCR is a voluntary regime (Jakhu and Wilson 2000, pp. 165–7), applicable to rocket systems, including space launch vehicles and sounding rockets whose transfer could possibly make a contribution to deliver systems other than manned aircraft for weapons of mass destruction. According to the provisions of HCOC, states exporting launch vehicle technology must promote the nonproliferation of ballistic missiles capable of delivering weapons of mass destruction and be vigilant in consideration of assistance to space launch vehicle program in any other country. MTCR and HCOC concern the export of satellite technology only indirectly through the control of items enumerated on their lists. When a satellite technology is used together with these items, these two arrangements have to be taken into consideration and complied with. The MTCR control list contains, e.g., complete rocket systems (suborbital and space launch vehicles) with a payload of 500 kg at a distance of minimum 300 km, its component parts and production facilities, etc. (MTCR, Equipment, Software and Technology Annex 2014, para 1(a); Category I 1.A.1; Category I 1. B.1.). Items listed in Category I underlie a strong presumption of denial (MTCR, Guidelines for Sensitive Missile-relevant Transfer 2003). Category II contains rocket systems with a range of minimum 300 km and dual-use satellite technologies, like flight control systems and hydrazine (ibid., Category II 10.A.1–10. A.3; Category II 4. C.2). An export is authorized if its end use is in line with the MTCR Guidelines and if this is assured by the recipient state (§§ 2,5 MTCR Guidelines; Von der Dunk 2011, p. 194). The MTCR is a nonlegally binding export control regime. To ensure its impact, the HCoC came into force in 2002. In November 2018, 139 states are signatory states of the HCoC, which is a politically binding document (HCoC 2019). It aims certainty and transparency through publication of annual reports.

The above-enumerated international legal regimes apply to export controls of space technology because of its inherent dual-use characteristics and the potential to be used or incorporated in a system of weapons, missiles, or other applications destined for non-peaceful purposes. However, the trade with space technologies falls also under the restrictions posed by specific regimes applying to outer space.

The Specificities of the Outer Space Regime

Outer space as an area beyond national control is subjected to special rules of international public law characterized by principles such as peaceful use and non-armament (Achilléas 2007, p. 60; for an analysis of outer space law and space security, see Freeland 2012). Article III of the Treaty on Principles Governing the Activities of States in the Exploration and Use of Outer Space, including the moon and other celestial bodies (hereinafter "OST" or "Outer Space Treaty 1967"), stipulates that "States Parties to the Treaty shall carry on activities in the exploration and use of outer space, including the Moon and other celestial bodies, in accordance with international law, including the Charter of the United Nations, in the interest of maintaining international peace and security and promoting international cooperation and understanding." As a consequence, Article IV paragraph 1 regulates that no objects carrying nuclear weapons or any other kinds of weapons of mass destruction shall be placed in orbit around the Earth and no such weapons shall be installed on celestial bodies or stationed in outer space in any other manner. This Article has been considered as one of the first provisions on arms control (Schrogl and Neumann 2010). As may be observed, placing of other than nuclear weapons and weapons of mass destruction in outer space is not expressly excluded; also, a transfer of weapons is not explicitly prohibited. With regard to the fact that the Earth's orbits are of an immense strategic and also military value, this loophole may serve states future interests.

According to Article IV paragraph 2, the moon and other celestial bodies shall be used "exclusively for peaceful purposes." Since the Outer Space Treaty itself does not provide a definition of the term "peaceful," its interpretation has over the years given rise to much debate centered on the question as to whether it is to be understood as meaning "nonmilitary" or "nonaggressive," with the latter meaning seeming to have gained general acceptance (e.g., Markoff 1967; Hobe and Hedman 2010). Irrespective of the meaning attributed to the term in other provisions of the Outer Space Treaty, its combination with the term "exclusively" in Article IV paragraph 2 leads to an all-embracing prohibition of military use of the moon and other celestial bodies even if of nonaggressive nature (Schrogl and Neumann 2010). This is supported by a parallel to other international treaties and legal instruments: Article I of the Antarctic Treaty providing "for peaceful purposes only" is commonly understood as complete demilitarization of Antarctica. In addition, the United Nations Convention for the Law of the Seas (hereinafter "UNCLOS") distinguishes between "exclusively for peaceful purposes," referring to nonmilitary use of the seabed (Art. 141) and "shall be reserved for peaceful purposes" (Art. 88), allowing for military but nonaggressive use.

The establishment of military bases, installations and fortifications, the testing of any type of weapons, and the conduct of military maneuvers on celestial bodies shall be forbidden. Only scientific research and peaceful exploration may be conducted, also by military personnel.

Article III OST provides that "States Parties to the Treaty shall carry on activities in the exploration and use of outer space, including the Moon and other celestial bodies, in accordance with international law, including the Charter of the United Nations, in the interest of maintaining international peace and security and promoting international cooperation and understanding."

Accordingly, it may be concluded that all disarmament and nonproliferation treaties that are part of international law are also applicable to activities carried out in outer space. This is further supported by Article 38 (1) of the Statute of ICJ; the recognized source of international law includes apart from international conventions also international custom, as evidence of a general practice accepted as law and the general principles of law recognized by civilized nations.

National and Regional Legal Regimes

In order to provide an overview of the complexities related to the legal export control regimes of space technologies, the following sub-chapters look at two prominent examples of national and regional legal regimes: the US regime on the one hand and the European Union system on the other. Following that, a sub-chapter sheds light on the additional particularities of dealing with export control issues in an international organization dedicated to the promotion of cooperation among its member states in space research and technology and their space applications, the European Space Agency (hereinafter "ESA").

The Export Control Regime of the United States

One unique characteristic of the US export control regime is the extraterritorial jurisdiction it establishes over US goods and technology. US export controls apply also when these items or technologies are located outside the territory of the United States, based on the fact that these are of US-origin or contain significant US content (for further details, Gerhard and Creydt 2011). This extraterritorial application of jurisdiction, which has given rise to criticism and debate over the years (e.g., Ress 2000; Clement 1988), is, however, a fact of law (see for more details Little et al.) and has a great impact not just on US American firms but on exporters all around the world since any foreign company seeking to re-export a product of US origin or with a certain percentage of US technology (details can be found in Part 734.4 EAR) will always have to comply with the applicable US regulations, regardless of whether they export from US territory or from any other place in the world. More precisely, § 734.4 and Supplement No. 2 to part 734 Export Administration Regulation (hereinafter "EAR") set out the scope of application of the US *De minimis* Rules and Guidelines. It applies if non-US-made commodity or software incorporates or is bundled with controlled US-origin commodities or software or in case that non-US-made technology is commingled with or drawn from controlled US-origin technology. If the abovementioned circumstances apply, the non-US-made item is subject to the EAR if the US-origin-controlled content exceeds a dedicated percentage based on destination.

In the *Nottebohm Case* (Liechtenstein v. Guatemala, I.C.J. 4 1955), the extraterritorial applicability of US export controls was fiercely discussed and finally considered to be in accordance with international law through the channel of national sovereignty under compliance with the "genuine link requirement." Failure to comply may give rise to prosecution, blacklisting, or other forms of punishment. It is worth pointing out in this context that re-export has been elaborated as follows: "In addition, for purposes of *satellites* controlled by the Department of Commerce, the term "re-export" also includes the *transfer of registration of a* satellite *or operational control* over a satellite from a party resident in one country to a party resident in one country to a another country" (Part 77. Part 772 EAR). From a US perspective and understanding, there is no general freedom or right to export goods. Rather, an export license is considered to be a privilege which can be revoked in case of noncompliance with applicable regulations. Article I Sect.8 US Constitution authorizes the US congress to enact norms to ensure security and foreign trade. The United States is involved in international agreements, e.g., WA and MTCR, and implements them into national law (von der Dunk and Tronchetti 2017, p. 364).

Despite their importance for national security and for the overall economy, the US regime of export regulations has not been consolidated into a single and unified text. Instead, there are different rules applied by different departments depending on the individual item and recipient in question:

- *Dual-use goods*, i.e., goods that may be used for both nonmilitary and military applications, are regulated by the Export Administration Regulations (EAR).
- *Military goods* are governed by the International Traffic in Arms Regulations (ITAR).
- Finally, the Office of Foreign Assets Control Regulations (OFAC) relates to *particular countries*, *organizations*, and *persons* for the protection of national security interests.

With regard to OFAC, it must be pointed out that, unlike the EAR and ITAR, the main focus of these regulations is not on particular items and services but on targeted countries and end users: OFAC administers, upholds, and enforces economic trade sanctions based on US foreign policy and national security goals against targeted foreign states, organizations, and individuals. OFAC derives its authority from a variety of US federal laws regarding economic sanctions. The effect on the American space industry of these regulations is, however, only marginal given that the targeted countries are not important trading partners in the field of space technology.

Further, on 30 June 2017, US President Trump signed the Executive Order on Reviving the National Space Council. The Council's recommendations to the US President for regulatory reforms and other actions aim to "unleash the economic potential of the U.S. commercial space industry" (Office of Space Commerce 2019). The Aerospace Industries Association (AIA) announced "While the U.S. experienced the third largest trade deficit on record, our industry generated $143 billion in exports and a positive trade balance of $86 billion, effectively reducing the U.S. trade deficit by 10 percent" (Aerospace Industries Association (AIA) 2018, p. 1.).

Dual-Use Goods: The Export Administration Regulations (EAR)

The US Department of Commerce is empowered to administer and to enforce rules for the export of dual-use items under the Export Administration Act of 1979 (Export Administration Act of 1979). More specifically, its Bureau of

Industry and Security is charged with the development, implementation, and interpretation of US export control policy for dual-use items under the EAR. § 730.5 EAR refers to a broad meaning of the term "export," which applies to transactions outside of the United States or to activities other than exports. For instance, re-exports are defined by § 734.2 (b) and include in regard of space items also the transfer of registrations or monitoring/control over satellites (Part 772.1 EAR), which plays a decisive role in the monitoring of space debris. In addition, EAR applies to deemed re-exports, which are defined as releases of technology or software source code subject to EAR, within a state outside the United States, to persons who are not citizens or lawful permanent residents of that state (third-country nationals). Exceptions are imposed for the releases to nationals of certain countries (A:5 Country Group) and to certain bona fide regular and permanent employees.

Moreover, EAR encompasses in-country transfers of commodities, software, and technology that remain "subject to the EAR" once outside of the United States. Such items include:

- US-origin items, wherever located, if in their original form
- Non-US items incorporating more than de minimis amount of US-origin-controlled content
- Non-US items that are direct product of certain US-origin technology or software

The EAR establishes a number of general prohibitions (Part 736.2 EAR) relating to certain exports, re-exports, and other conduct, subject to the scope of the EAR, which necessitate a license from the Bureau of Industry and Security or the qualification for a license exception. Facts that determine the applicability of the general prohibitions are:

- The classification of the item on the Commerce Control List
- The country of ultimate destination for an export or re-export
- The ultimate end user
- The ultimate end use
- Conduct such as contracting, financing, and freight forwarding in support of a proliferation project

The relevant details are provided in accompanying lists and schedules. It is interesting to note that the otherwise existing possibility to obtain a license exception is specifically excluded for a number of items, among which feature certain "space-qualified items" (Part 740.2(a)).

Once established that an envisaged export activity does not fall under a general prohibition, the exporter needs to examine the specific license requirements, if any. This examination is accomplished by means of the CCL and the Country Chart, as well as the established reason for the control of the item (Gerhard and Creydt 2011, p. 198). The space relevant categories are Category 6, dealing with lasers and sensors and Category 9, which is listing propulsion systems, space vehicles and related equipment. The Commerce Control List (hereinafter "CCL") includes dual-use items such as listed on the WA and MTCR list and also items under the "catch-all clause" (U.S. Department of State, A Resource on Strategic Trade Management and Export Controls 2016). Criminal sanctions, in addition to administrative and civil penalties, can be imposed where the EAR regulations have been violated (Part 764.3 EAR).

As a result of the 2013/2014 Export Control Reform, numerous items, such as transitioned military items (600 series) and transitioned spacecraft items (ECCNs 9A515), were moved from ITAR to the EAR. As of end 2014, the majority of commercial spacecraft items were moved from ITAR to EAR, where they are classified under:

- ECCN 9A515: Commercial communication satellites and spacecraft including lower-performing remote sensing satellites, planetary rovers and (inter-)planetary probes, related systems for these and parts/components of satellite bus, and payloads not listed on the USML
- ECCN 9A004: International Space Station (hereinafter "ISS"), James Webb Space Telescope, and parts or components thereof.

The revised United States Munitions List (hereinafter "USML") Category XV covers now satellites and spacecraft providing unique military and intelligence functions, human-rated habitats, and certain ground control equipment and parts/components. The latter includes 18 specific technologies critical to military functions, any payload performing the military functions described above, and US DoD funded payloads.

Restrictions on certain "military end uses" in the People's Republic of China and Russia or "military end users" in Russia are imposed. Exceptions are defined in EAR 744.21 for export, re-exports, transfers, and launch within Russia for the ISS. Stricter requirements for military end users in the People's Republic of China are expected in May 2019.

License exceptions can result from the *de minimis* rule, which is based on:

- The percentage, by value, of US-origin-controlled content in a foreign-made item
- The intended destination of the re-export or deemed re-export
- *De minimis* percentage: 0% threshold for 9x515 (The 9x515 Export Control Classification Number (hereinafter "ECCN") describes "spacecraft," related items, and some radiation-hardened microelectronic circuits that were previously subject to the International Traffic in Arms Regulations (hereinafter "ITAR")) to China, 10% threshold for Iran, North Korea, Syria, and Sudan, as well as 25% threshold applicable to all other countries.

Secondly, it can result from a Strategic Trade Authorization (hereinafter "STA"), requiring a prior consignee statement for transfers and re-exports in the A:5 country group (37 US partners and allies including all ESA member states). Lastly, license exemptions apply for US government programs (hereinafter "GOV").

Military Goods: The International Traffic in Arms Regulations

"[...] For everyone active in the space sector, knowledge about how to deal with International Traffic in Arms Regulations (hereinafter "ITAR") is absolutely necessary" (Gerhard and Creydt 2011, p. 203; current version http://pmddtc.state.gov/regulations_laws/itar_official.html). ITAR governs the export, re-export, re-transfer, and temporary import of defense items and services. Under the Arms Export Control Act of 1976, the State Department has the delegated power to control, enforce, and administrate the regulations. These defense articles and services are listed in the United States Munitions List (hereinafter "USML") (Part 121 ITAR). USML categories of particular interest in the context of space activities are:

- Category XV: Satellites and Spacecraft, providing unique military and intelligence functions
- Category IV: Launch Vehicle, including any interface between any spacecraft and a launch vehicle
- Category XII: Fore Control. Laser, Imaging and Guidance Equipment, such as Global Navigation Satellite System (hereinafter "GNSS") receiving equipment exceeding the "CoCom rule"
- Category XIII: Materials and Miscellaneous Articles

Following the passing of the Strom Thurmond National Defense Authorization Act for fiscal year 1999 (Public Law 105–261 1998), the export control competence concerning commercial satellites was shifted from the Department of Commerce to the Department of State following concerns about the proliferation of sensitive satellite technology (for the background and details, van Fenema 1999, p. 332). Exports under the ITAR are broadly defined (Annex 5 reflects a detailed scope of application), also covering:

- Exports of defense articles outside the United States.
- Transfer of registration, control, or ownership to a foreign person of any aircraft, vessel, or satellite covered by the USML, whether in the United States or abroad.
- Disclosing/transferring in the United States of any defense article to foreign government and disclosing technical data to a foreign person
- Performing a defense service for foreign person

ITAR consists of 11 parts (§§120–130) and aims to enforce the Arms Export Control Act (hereinafter "AECA"). Administrating body of the AECA is primarily the Department of State under the Directorate of Defense Trade Controls (hereinafter "DDTC").

An authorization is the prerequisite for conducting an export, re-export, or transfer of an ITAR-controlled item, regardless of its destination. Different license requirements exist for the permanent or the temporary export of any defense article or technical data (§§ 123 and 125 ITAR). The performance of defense services also requires the prior approval by the Directorate of Defense Trade Controls of the State

Department, which requires the conclusion of specific agreements between the performer of such services and the respective international partner (§§ 124.1 ITAR). Conclusively, authorization can be obtained in form of a license (DSP-5. DSP-83, DSP-73) or an agreement (Technical Assistance Agreement (hereinafter "TAA") (§ 120.22 defines the Technical Assistance Agreement as an agreement for the performance of a defense service and/or the disclosure of technical data (including the assembly of defense articles, providing that production rights or manufacturing know-how are not conveyed, which are to be covered by a MLA), Manufacturing License Agreement (hereinafter "MLA") (§ 120.21 defines a Manufacturing license agreement as an agreement granted by a US national to a foreign person to manufacture defense articles abroad, which involve either the export of technical data (§120.10 ITAR), defense articles, the performance of a defense service, or the use by the foreign person of technical data or defense articles previously exported by the US national (§124 ITAR))).

The "see-through" rule applies to commodities, software, and technical data. According to this rule, any non-US-made item incorporating US-origin ITAR content will require DDTC approval to re-transfer the ITAR content, regardless of the type of content or destination. Compared to the EAR, ITAR "sees through" the item and continues to control the original ITAR content. Further, there is no *de minimis* threshold in the ITAR. However, some ITAR-controlled items are exempted from the "see-through" rule when incorporated into an EAR controlled system, e.g., USML spacecraft (Category XV) items.

There are only a few exemptions from ITAR authorization requirements, such as the Defense Trade Cooperation Treaties with Australia and the United Kingdom for certain governmental end users (Long 2013, p. 58 ff.). In addition, ITAR contains exemptions for transfers to regular employees (including loan-employment contractors) from NATO and EU member states, Switzerland, Australia, Japan, and New Zealand. It follows that this exemption applies to all ESA member states. A transfer of items or services to dual/third-country nationals outside these countries can be made in the framework of:

- Foreign vetting
- DDTC vetting

The ITAR regulations have been criticized as too cumbersome, complex, and time-consuming (e.g., Abbey and Lane 2009), which in turn is said to have led to a competitive disadvantage for the US space industry and a decline of exports, as competitors around the globe started to invest in the non-dependence in certain fields of technology (see also Landry 2010). However, with Space Policy Directive – 2 "Streamlining Regulations on Commercial Use of Space," which was signed by President Donald J. Trump in May 2018, the White House is eager to ensure the promotion of economic growth and the minimization of uncertainties for taxpayers, investors, and private industry while protecting national security, public safety, and foreign policy interests.

The overall objective is the encouragement of American leadership in space commerce (the White House. President Donald J. Trump is Reforming and Modernizing American Commercial Space Policy 2018).

Space Policy Directive 2 addresses five areas:

1. Commercial launch and licensing
2. Commercial remote sensing
3. Creation of an Office of Space Commerce within the Department of Commerce
4. Radio frequency spectrum management
5. Export licensing

The latter requests the National Space Council to review of export licensing affecting commercial space flight as well as commercial remote sensing activities with strict consideration of US commercial space policy objectives. As a result, the NASA Advisory Council Regulatory and Policy Committee issued recommendations to relieve the Lunar Gateway from such licensing restrictions as are applicable to the ISS (NASA Advisory Council Regulatory and Policy Committee Observations, Findings, and Recommendations of 12 October 2018).

In October 2018, the US State Department issued minor ITAR revisions, including two space relevant changes. As result of a previous revision conducted by the Department of Commerce's Bureau of Industry and Security (BIS), the State Department added a note clarifying that the rocket engine controls in Category IV(d) of the USML do not apply to satellite and spacecraft thrusters. Such thrusters fall under USML Category XV(e)(12) or ECCN 9A515 of the CCL. Further, the ITAR update also amended the definition of spacecraft-related defense services in USML Category XV(f), but without changing the actual scope of the controls (Office of Space Commerce. ITAR Clarification on Satellite Thrusters 2018). After the recent update of the USML, Category XV now includes letter (f) "the furnishing of assistance (including training) to a foreign person in the launch failure analysis of a satellite or spacecraft, regardless of the jurisdiction, ownership, or origin of the satellite of spacecraft, or whether technical data is used."

Export Regulations of the European Union

The European Union (EU) has for many years played a leading role in arms export control, both regionally and internationally. Nowadays, export control regulations on dual-use goods and defense equipment have been harmonized within the EU under a common regime. Similar to the US approach, products are categorized into either dual-use goods or military products. However, unlike to the situation in the United States, none of the regulations or national laws has extraterritorial effect on third countries outside the EU.

However, the categorization of items in dual-use and defense equipment caused conflicts of competences. EU competences are laid down in the EU treaties. Article 3 (1.e) of the Treaty on the Functioning of the European Union (hereinafter "TFEU") provides for the exclusive competence of the EU over Common Commercial Policy (hereinafter "CCP"). Article 207 defines the scope of CCP, which can be summarized as unified basis for trade affairs, e.g., liberalization and export control. At the same time, Art. 4(1) of the Treaty of the European Union (hereinafter "TEU") sets out the exclusive member state authority over national security matters since member states prefer to exercise their sovereignty in this area. Moreover, Art. 346 of the Lisbon Treaty provides for member state's sovereignty on "public security."

Prior to 2009, national rules regulating the transfer of defense equipment did not necessarily distinguish between exports to third countries and transfers between EU member states.

Space activities are business activities. However, they cover the transfer of dual-use items, which has to be assigned to the objectives of national security and foreign policy as it refers to nonproliferation. Accordingly, the EU set up a Council Regulation (Council Regulation (EC) No. 3381/94, OJ L 367/1) based on CCP and a Council Decision (Council Decision 94/942/CFSP, OJ L 367/8) based on the Common Foreign and Security Policy (hereinafter "CFSP") to create a common export control regime for dual-use items. The underlying competence of CCP was defined by the European Court of Justice (hereinafter "ECJ"), resulting in the exclusive competence over EU external trade relations (Opinion of the Court of 11 November 1975 given pursuant to Article 228 of the EEC Treaty – Avis 1/75). Thus, according to Art. 2(1) TFEU only the EU might legislate and adopt legally binding acts in CCP.

The currently applicable Council Regulation (EC) No. 428/2009 (hereinafter referred to as EU Dual-Use Regulation) was established after the ECJ decided to assign dual-use exports to the scope of the CCP and, by doing so, to the exclusive EU competence (Fritz Werner Industrie-Ausrüstungen GmbH v Federal Republic of Germany; Criminal proceedings against Leifer and others). Council Regulation 428/2009 of May 2009 (Council Regulation No. 428/2009 of 5 May 2009 setting up a community regime for the control of exports, transfer, brokering, and transit of dual-use items, OJ L 134) governs the export of dual-use goods from a country within the EU to a third country. In addition, the export of military goods and defense-related products is regulated by Directive 2009/43 (http://eur-lex.europa.eu/LexUriServ/LexUriServ.do?uri¼OJ:L:2009:146:FULL:EN:PDF.), the European Union Code of Conduct on Arms and Exports, and the Common Military List of the European Union. Under the regime established by Directive 2009/43, member states can issue general licenses for those exports throughout the EU where the risk of re-exportation to foreign countries is under control. At the same time, the nation states retain their discretion to determine the eligibility of products for the different types of license and to fix their terms and conditions. In 2008, the European Council also agreed on the Common Position 2008/944/CFSP defining common rules governing control of export of military technology and equipment to *third* countries. It should be noted, however, that the impact of these rules on the European space industry is not particularly important, since most space-related items are considered dual-use and not military items. There are also council regulations restricting the export against specific countries, e.g.,

Myanmar (Council Decision 2010/232/CFSP). Moreover, Council Regulation (EU) No 833/2014 of 31 July 2014 introduces certain restrictive measures as applicable for exports of items and services under Council Regulation (EC) No 428/2009 and on certain service related to the supply of arms and military equipment, if an embargo on such goods is applied by the member states. However, the prohibition should not affect the exports of dual-use goods and technology, including for aeronautics and for the space industry, for nonmilitary use, or for a nonmilitary end user (Council Regulation (EU) No 833/2014, pp. 1–11). The aim of this harmonization is to promote European and international security as well as to allow for comparable conditions for all economic entities within the EU's common market. Despite this development, some differences do, however, remain among EU member states with regard to export control licensing (Gerhard and Creydt 2011, p. 210).

Dual-Use Items

§12 of the preamble of Council Regulation 428/2009 provides that "[...] Member States retain the right to carry out controls on transfers of certain dual-use items within the Community in order to safeguard public policy or public security." As a consequence, member states implement the Regulation by issuing licenses and enforcing export control in accordance with Art. 9 of this Regulation (Art. 9, (EC) No.428/2009.; Wetter 2009, p. 49). It was adopted as a single document abandoning the "cross-pillar" approach (Aubin and Idiart 2011, p. 109). Free trade of dual-use items within the EU is one of the general principles (§ 4 Preamble, European Union Regulation (EC) 428/2009). The Council Regulation 428/2009 establishes a common community export licensing system, a common control list, and a common export authorization. Dual-use items which require a license to be exported from the EU are listed in Annex I. Most space assets are regarded as dual-use items under Annex I: Category 9, aerospace and propulsion, lists the different systems, equipment, components, materials, software, and technology subject to the regulation. Authorizations, which are valid throughout the community, are granted by the competent authorities of the member state where the exporter is established. Pursuant to Article 22.1 of Regulation 428/2009, dual-use goods items in Annex IV of the regulation are considered to be particularly sensitive and require a license even to be traded *within* the European market. This procedure, which concerns also quite a number of space-related items, constitutes an exception to the principle of free movements of goods (Wetter 2009, p. 54). It is interesting to note that Annex IV establishes some explicit exemptions (OJL 134, p. 264) and as such does not control the following items of the MTCR technology:

1. That are transferred on the basis of orders pursuant to a contractual relationship placed by ESA or that are transferred by ESA to accomplish its official tasks
2. That are transferred on the basis of orders pursuant to a contractual relationship placed by a member state's national space organization or that are transferred by it to accomplish its official tasks

3. That are transferred on the basis of orders pursuant to a contractual relationship placed in connection with a community space launch development and production program signed by two or more European governments
4. That are transferred to a state-controlled space launching site in the territory of a member state, unless that member state controls such transfers within the terms of this regulation

On 10 October 2018, Commission Delegated Regulation (EU) 2018/1922 amended Council Regulation 428/2009. This delegated regulation revises the EU dual-use list in Annex I in accordance with the decisions taken within the framework of the international nonproliferation regimes and export control arrangements in 2017. The majority of changes results from amendments of the 2017 Plenary of the Wassenaar Agreement. Space relevant among them is the increase of controls for ground-based spacecraft control equipment (9A004). In addition, Section 4E001 now covers new decontrol for technology for "vulnerability disclosure" and "cyber incident response."

The Missile Technology Control Regime agreed in 2017 to amend the control on satellite navigation systems (7A105) to include regional and global systems.

Another renewal in European Union Export Controls is the EU foreign investment screening regulation, which entered into force on 10 April 2019. It is based on a proposal scheduled by the European Commission in September 2017 and intends to ensure Europe's security and policy order in relation to foreign direct investment within the EU.

The EU is the main destination for foreign direct investment in the world with an amount of € 6.295 billion at the end of 2017. The new framework aims to create a cooperation mechanism between member states and EU Commission to enable exchange on the matter and "[...]allow the Commission to issue opinions when an investment poses a threat to the security or public order of more than one Member State, or when an investment could undermine a project or programme of interest to the whole EU, such as Horizon 2020 or Galileo[...]" (European Commission. EU foreign investment screening regulation enters into force 2019). Moreover, it intends to foster international cooperation on investment screening and to establish requirements for national screening mechanisms.

The Export Control Regulations of the European Space Agency

The European Space Agency (ESA) is an international organization created by the Convention for the establishment of a European Space Agency, which was opened for signature in Paris on 30 May 1975 and entered into force on 30 October 1980. According to Article II ESA Convention, its member states entrusted ESA "[...] to provide for and to promote, for exclusively peaceful purposes, cooperation among European States in space research and technology and their space applications, with a view to their being used for scientific purposes and for operational space applications systems [...]".

While Article VI of Annex I to the ESA Convention provides that "Goods imported or exported by the Agency or on its behalf, and strictly necessary for the exercise of its official activities, shall be exempt from all import and export duties and taxes and from all import or export prohibitions and restrictions," in drawing up the Convention, ESA member states also paid heed to the potentially highly sensitive nature of the technology developed and introduced with Article XI.5(j) of the ESA Convention, a starting point for ESA's own rules and procedures with regard to export control issues.

It provides that:

> The ESA Council shall adopt, by a two-thirds majority of all Member States, rules under which authorisation will be given, bearing in mind the peaceful purposes of ESA, for the transfer outside the territories of the Member States of technology and products developed under the activities of ESA or with its assistance.

These rules supplement the regular national export control procedures. Exports remain in their essence governed by national laws and regulations, given their foreign policy and security implications, even in the case of an export of technologies or products developed under the activities or with the help of ESA.

This basic provision of Article XI.5(j) of the ESA Convention is implemented by Chap. IV of the ESA Rules on Information, Data, and Intellectual Property, adopted by the ESA Council on 19 December 2001 (rules), ESA/CCLV/Res. 4 (final). These rules are intended to ensure close liaison between the agency and the national export control authorities of the member states. In line with general ESA policy considerations, the rules contribute to promoting the maximum exploitation of ownership rights by drawing a clear distinction between technology and products that are owned by ESA, on the one hand, and those which are owned by contractors of the agency, on the other hand: the transfer of technology or products owned by ESA necessitates the *authorization* by the Agency Technology and Product Transfer Board (hereinafter "ATB"), whereas the transfer of technology or products owned by contractors only needs to be subject of a *recommendation* by the ATB.

The ATB's authorization or recommendation, which is – again – not a substitute for the national-level authorization process but rather an additional procedure, is not necessary when the transfer of technology or products is made pursuant to a cooperative agreement between ESA and a government agency of the country of destination. In such case, it is assumed that the ESA Council, when approving the cooperative agreement, has given an overall authorization for the transfer of data and goods in accordance with the relevant provisions of the agreement.

In a first instance, the ATB functions according to a written procedure: when the technology or products are owned by ESA, the transfer proposal is *rejected* when one-third or more of all member states have communicated their opposition. When the technology or products are owned by a contractor, a transfer is *not recommended* where one-third or more of all ATB delegations have communicated their opposition. If, however, within a given time frame, one or more ATB delegations request a meeting to discuss the matter, the ATB Chair convenes such a meeting. The decision shall then be taken at that meeting, with a two-thirds majority of all ATB delegations present (resolution amending the terms of reference for the Agency Technology and Product Transfer Board (ATB), adopted on 11 October 2006, ESA/C/CLXXXIX/Res. 2 (Final), which amends the terms of reference adopted by the resolution on the creation of an Agency Technology and Product Transfer Board, ESA/C/CLXVII/Res. 1 (Final), adopted on 8 October 2003).

In considering its authorizations and recommendations, the ATB takes into account several factors, such as:

- That the purpose of the Agency is to provide for and to promote, for exclusively peaceful purposes, cooperation among European states in space research and technology and their space applications
- The competitive position of the member states industrial entities as a whole and the competitive edge and technical lead for technology and products
- The relevant provisions of the member states export control laws and regulations
- The requirement for timely implementation of the Agency's programs and activities
- The requirement for restrictions on re-export and/or the existence of any relevant technology transfer agreements

ESA's rules do not prejudice the fact that export control is a national competence, governed by the national laws and regulations of the member states and, in numerous instances, subject to those international agreements by which the member states are bound. This implies also that a certain technology or product that has been developed under an ESA program may still need to be submitted to the regular national export control procedure in case it is planned to be used in another ESA program by another member state than the member state of the originator. In line with that, Article XXIII of Annex I to the Convention provides, "Each Member State shall retain the right to take all precautionary measures in the interests of its security."

In the particular case of an ESA project incorporating US-origin content, software, or information or in when ESA is partnering with NASA and US companies, complying with US export controls is of utmost importance. As a result, ESA may be restricted in its ability to:

- Re-transfer or re-export that US-origin content to other entities within European countries or from one country to another
- Share the US-origin information or software with certain employees, contractors, or other individuals within ESA
- Re-transfer or re-export ESA prototypes or finished products to other countries, including for launch

Conclusions

"Space has a security dimension and security has a space dimension" (European Commission White Paper 2003, p. 17). Space export controls are conducted in an extremely complex set of rules and regulations governing the trade in space items,

items that are not only used in an ultra-hazardous environment but are also dual-use by nature. As the evolution of space activities is closely linked to scientific and technological innovations, new developments in these sectors need to be taken into account. The presented revisions to the existing export control regimes reflect the growing awareness of technological developments and areas of vulnerability.

In addition, with the ever-increasing interest in space activities and related growing participation, diverse policy objectives and varying policy considerations play a central role in this global environment that is evolving at an ever-increasing pace. The diverging interests of governments, international organizations, space agencies, and private actors need to be investigated and balanced while taking into consideration that the international trade in technology is influenced by the interplay between commercial interests, foreign policy objectives, and national security considerations in respect to the proliferation of sensitive technologies. The different approaches and national concepts to regulate the trading in space items provide therefore a characteristic exemplification of how individual nations balance these different and conflicting interests – what degree of importance they attach in their system of political values to one in relation to the others.

References

Abbey G, Lane N (2009) United States space policy: challenges and opportunities gone astray. American Academy of Arts and Sciences. http://carnegie.org/fileadmin/Media/Publications/PDF/spaceUS.pdf

Achilléas P (2007) International regimes. In: Aubin Y, Idiart A (eds) Export control law and regulations handbook – a practical guide to military and dual-use goods trade restrictions and compliance. Kluwer Law International BV, p 20

Aerospace Industries Association (AIA) (2018) Facts & figures U.S. Aerospace & Defense. http://www.aia-aerospace.org/wp-content/uploads/2018/07/2018_-Annual-Report_Web.pdf

Angelo JA Jr (2006) Encyclopedia of space and astronomy. Facts on File Inc, New York

Art 9 (2) Council Regulation No. 428/2009

ASD Europe (2018) The State of the European Space Industry in 2017. Available at https://eurospace.org/wp-content/uploads/2018/06/eurospace-facts-and-figures-2018-press-release-final.pdf. Accessed 30 Apr 2019

Aubin Y, Idiart A (2011) Export control law and regulations handbook – a practical guide to military and dual-use goods trade restrictions and compliance. Kluwer Law International BV

Brussels, 10 April 2019 (2019). http://europa.eu/rapid/press-release_IP-19-2088_en.htm

Clement I (1988) American export controls and extraterritoriality. School of law LLM Theses and Essays, Paper 121, University of Georgia

Comprehensive Nuclear-Test-Ban Treaty, Adopted by the United Nations General Assembly on 10 Sept 1996, but it has not entered into force as of Aug 2012, published in A/50/1027

Convention on the Prohibition of the Development (1972a) Production and stockpiling of bacteriological (biological) and toxin weapons and on their destruction, signed on 10 April 1972, entered into force on 26 March 1975. http://www.opbw.org/convention/conv.html. Accessed 20 Aug 2012

Convention on the Prohibition of the Development (1972b) Production and stockpiling of bacteriological (biological) and toxin weapons and on their destruction, signed on 10 April 1972, entered into force on 26 March 1975. Available at http://www.unog.ch/80256EDD006B8954/

(httpAssets)/C4048678A93B6934C1257188004848D0/$file/BWC-text-English.pdf. Accessed 24 Aug 2012

Convention on the Prohibition of the Development, Production, stockpiling and use of chemical weapons and on their destruction, opened for signature on 13 Jan 1993, entered into force on 29 Apr 1997, A/RES/47/3

Council Decision 94/942/CFSP of 19 December 1994 on the joint action adopted by the Council on the basis of Article J.3 of the Treaty on European Union concerning the control of exports of dual-use goods, OJ L 367/8. Available at https://eur-lex.europa.eu/legal-content/EN/TXT/PDF/?uri=OJ:L:1994:367:FULL&from=DE

Council Decision 2010/232/CFSP (OJ L 105, 27 Apr 2010, p 22). Available at http://eur-lex.europa.eu/LexUriServ/LexUriServ.do?uri¼OJ:L:2010:105:0022:0108:EN:PDF

Council Regulation (EC) No. 3381/94, of 19 December 1994 setting up a Community regime for the control of exports of dual-use goods, OJ L 367/1. Available at https://eur-lex.europa.eu/legal-content/EN/TXT/?uri=CELEX%3A31994R3381

Council Regulation (EU) No 833/2014 of 31 July 2014 concerning restrictive measures in view of Russia's actions destabilising the situation in Ukraine OJ L 229, pp 1–11. Available at https://eur-lex.europa.eu/legal-content/GA/TXT/?uri=CELEX:32014R0833

Council Regulation No. 428/2009 of 5 May 2009 setting up a community regime for the control of exports, transfer, brokering, and transit of dual-use items, OJ L 134. Available at http://trade.ec.europa.eu/doclib/docs/2009/june/tradoc_143390.pdf

Directive 2009/44/EC of the European Parliament and of the Council of 6 May 2009 simplifying terms and conditions of transfers of defense-related products within the community, OJ L 146/1. http://eur-lex.europa.eu/LexUriServ/LexUriServ.do?uri¼OJ:L:2009:146:FULL:EN:PDF

ECJ on 17th October 1995, Case C-70/94 Fritz Werner Industrie-Ausrüstungen GmbH v Federal Republic of Germany [1995] ECR I-03189 [cited as: Fritz Werner Industrie-Ausrüstungen GmbH v Federal Republic of Germany].

ESA Space Economy. Available online at: https://www.esa.int/About_Us/Business_with_ESA/GSEF/Space_Economy/(print). Accessed 07 Feb 2019

European Commission (2003) White paper. Space: a new European frontier for an expanding Union. An action plan for implementing the European Space policy http://www.dlr.de/rd/en/Portaldata/28/Resources/dokumente/WhitePaper_en.pdf

European Commission. EU foreign investment screening regulation enters into force

Freeland S (2012) The role of "soft law" in public international law and its relevance to the international legal regulation of outer space. In: Marboe I (ed) Soft law in outer space, the function of non-binding norms in international space law. Böhlau, Wien, p 9

Gerhard M, Creydt M (2011) Safeguarding national security and Foreign policy interests – aspects of export control of space material and technology and remote sensing activities in outer space. In: von der Dunk FG (ed) National space legislation in Europe, issues of authorisation of private space activities in the light of developments in European Space Cooperation. Brill, Leiden, pp 189–224

Guidelines for Sensitive Missile-Relevant Transfers, adopted 16 Apr 1987, 26 ILM 599

Hertzfeld H, Jones R (2011) International aspects of technology controls. In: Brünner C, Soucek A (eds) Outer space in society, politics and law. Springer, Vienna

Heupel M (2007) Implementing UN Security Council Resolution 1540: a division of labour strategy, Carnegie papers, nonproliferation program, number 87

Hobe S, Hedman N (2010) On the preamble of the 1967 Outer Space Treaty. In: Hobe S, Schmidt-T B, Schrogl K.-U (eds) Cologne Commentary on Space Law, Vol 1. Outer Space Treaty, Carl Heymanns Verlag, Köln

IAEA Comprehensive Safeguards Agreement and Model Additional Protocol, entered into force on 12 Sept 1997, published in INFCIRC/540

International Code of Conduct against ballistic missile proliferation established on 25 Nov 2002. Available at http://www.unhcr.org/refworld/docid/3de488204.html. Accessed 19 Aug 2012

International Court of Justice on 6th April 1955, Nottenbohm Case, Liechtenstein v Guatemala, Judgement 1955 I.C.J., 4 [cited as: Liechtenstein v Guatemala, I.C.J., 4.]

International Emergency Economics Powers Act, as amended (Public Law 95–223; 50 U.S.C. §§1701–1706; 91 Stat. 1628). Available at https://www.govinfo.gov/content/pkg/STATUTE-91/pdf/STATUTE-91-Pg1625.pdf

Jakhu R, Wilson J (2000) The new United States export control regime: its impact on the Communications Satellite Industry. In: XXV annals of air and space law, 2000, pp 157–181

Joyner D (2004) Restructuring the multilateral export control regime system. In: Journal of conflict and security law, Oxford University Press, Oxford, United Kingdom, 9(2):181–211

Konyukhov SN (2003) Conversion of Missiles into Space Launch Vehicles. In: Mark H (ed) Encyclopaedia of space science & technology, vol 1. 441 ff

Landry KL (2010) Exploring the effects of international traffic in arms regulations restrictions on innovations in the U.S. Space Industrial Base. Air Force Institute of Technology, Wright-Patterson Air Force Base, Ohio, p 63. http://www.dtic.mil/cgi-bin/GetTRDoc?AD¼ADA535245

Little KC, Reifman SD, Dietrick AJ (2015) U.S. export controls apply extraterritorially – circumstances in which foreign persons are subject to U.S. export laws and regulations. http://www.cailaw.org/media/files/SWIICL/ConferenceMaterial/2015/bootcamp/ear-ita-article1.pdf

Long C (2013) An imperfect balance: ITAR exemptions, national security, and U.S. competitiveness. 2 Natl Secur Law J 43:58. Arlington (US)

Marboe I (2012) Soft law in outer space, the function of non-binding norms in international space law. Böhlau, Vienna

Markoff MG (1976) Disarmament and "Peaceful Purposes" Provision in the 1967 Outer Space Treaty: In Journal of Space Law. University of Mississippi School of Law, Vol. 4, p 6

Markoff MG (2005) Disarmament and "Peaceful Purposes" provisions in the 1967 outer space treaty. J Space Law 1967:3. Institute of Air and Space Law, McGill University, "Peaceful" and Military Uses of Outer Space: Law and Policy, Feb 2005. Available at http://www.e-parl.net/pages/space_hearing_images/BackgroundPaper%20McGill%20Outer%20Space%20Uses.pdf

McGraw-Hill Science & Technology Dictionary, accessible at http://www.answers.com/library/Sci%252DTech+Encyclopedia-cid-3566557

Mineiro MC (2012) Space technology export controls and international cooperation in outer space. Springer, Dordrecht/Heidelberg/London/New York

MTCR, Equipment, software and technology annex 2014, para 1(a);, Category I 1.A.1; Category I 1. B.1. Available online at http://mtcr.info/wordpress/wp-content/uploads/2016/10/MTCR-TEM-Technical_Annex_2016-10-20.pdf. Accessed 30 Apr 2019

NASA Advisory Council Regulatory and Policy Committee Observations, Findings, and Recommendations of 12 October 2018. https://www.nasa.gov/sites/default/files/atoms/files/regulatory_policy_committee_report_dec2018_tagged.pdf

Office of Space Commerce (2018) ITAR clarification on satellite thrusters. https://www.space.commerce.gov/itar-clarification-on-satellite-thrusters/

Office of Space Commerce (2019) Space Council focuses on regulatory reform. https://www.space.commerce.gov/space-council-focuses-on-regulatory-reform/

OJL 134 (2009) p 264. Available at http://trade.ec.europa.eu/doclib/docs/2009/june/tradoc_143390.pdf

Opinion of the Court of 11 November 1975 given pursuant to Article 228 of the EEC Treaty. – Avis 1/75

Part 121 of the International Traffic in Arms Regulations (ITAR). Available at http://pmddtc.state.gov/regulations_laws/documents/official_itar/ITAR_Part_121.pdf

Part 734.4 of the EAR. http://www.bis.doc.gov/policiesandregulations/ear/734.pdf

Part 736.2 of the EAR. http://www.bis.doc.gov/policiesandregulations/ear/736.pdf

Part 740.2(a) 7 of the EAR

Part 764.3 of the EAR. Available at http://www.bis.doc.gov/policiesandregulations/ear/764.pdf

Part 772 EAR. Available at http://www.bis.doc.gov/policiesandregulations/ear/772.pdf

Pezzullo M (2014) Sovereignty in an age of global interdependency: the role of borders. Australian Strategic Policy Institute. Canberra, Australia

Public Law 105–261 – 17 Oct 1998. Available at http://www.opbw.org/nat_imp/leg_reg/US/nat_def_auth_act.pdf

Regulation (EU) No. 388/2012 of the European Parliament and of the Council of 19 Apr 2012 amending Council Regulation (EC) No. 428/2009 setting up a community regime for the control of exports, transfer brokering, and transit of dual-use items; OJ L 129, 16 May 2012. Available at http://eur-lex.europa.eu/LexUriServ/LexUriServ.do?□uri¼OJ:L:2012:129:0012:0280:EN:PDF

Ress H-K (2000) Das Handelsembargo, völker-, europa- und außenwirtschaftsrechtliche Rahmenbedingungen. Praxis und Entschädigung. Springer, Berlin

Rhodes Ch, Hough D, Ward M (eds) (2015) The aerospace industry: statistics and policy. Available at https://researchbriefings.parliament.uk/ResearchBriefing/Summary/SN00928

Satellite Industry Association (SIA) – represented by the Tauri Group. 2018 State of the Satellite Industry. Available online at: https://www.sia.org/wp-content/uploads/2018/06/2018-SSIR-2-Pager-.pdf. Accessed 07 February 2019

Schrogl K-U, Neumann J (2010) Article IV. In: Hobe S, Schmidt-T B, Schrogl K.-U (eds) Cologne Commentary on Space Law, vol 1, Outer Space Treaty, Carl Heymanns Verlag, Köln

The Australia Group. http://www.australiagroup.net/

The current version is Available at http://pmddtc.state.gov/regulations_laws/itar_official.html

The Export Administration Act of 1979. Public Law 96–72; Enacted September 29, 1979. https://legcounsel.house.gov/Comps/The%20Export%20Administration%20Act%20Of%201979.pdf

The Hague Code of Conduct against ballistic missile proliferation, adopted 25 Nov 2002

The Hague Code of Conduct against ballistic missile proliferation, List of Subscribing States (2019). Available at https://www.hcoc.at/?tab=subscribing_states&page=subscribing_states

The International Traffic in Arms Regulations (ITAR), §§ 123 and 125, respectively. Available at http://www.pmddtc.state.gov/regulations_laws/itar_consolidated.html

The International Traffic in Arms Regulations (ITAR), §§ 124.1

The MTCR Guidelines. Available online at http://www.mtcr.info. Accessed 20 Aug 2012

The Nuclear Supplier Group. http://www.nuclearsuppliersgroup.org

The report is available at. http://www.defense.gov/home/features/2011/0111_nsss/docs/1248_Report_Space_Export_Control.pdf

The Wassenaar Arrangement on Export Controls for Conventional Arms and Dual-Use Goods and Technologies, opened for signature on 12 July 1996 and entered into force on 1 Nov 1996, guidelines and procedures. Available online at https://www.wassenaar.org/app/uploads/2015/06/WA-DOC-17-PUB-001-Public-Docs-Vol-I-Founding-Documents.pdf. Accessed 30 Apr 2019

The Wassenaar Arrangement Secretariat, List of dual-use goods and technologies and munitions list – public documents volume II of 06 December 2018, pp 221–222. Available online at https://www.wassenaar.org/app/uploads/2018/01/WA-DOC-17-PUB-006-Public-Docs-Vol.II-2017-List-of-DU-Goods-and-Technologies-and-Munitions-List.pdf Accessed 30 April 2019

The White House. President Donald J. Trump is Reforming and Modernizing American Commercial Space Policy (2018). Available online at: https://www.whitehouse.gov/briefings-statements/president-donald-j-trump-reforming-modernizing-american-commercial-space-policy/. Accessed 10 Feb 2019

Tietje C (2009) In: Herdegen M (ed) Internationales Wirtschaftsrecht. Ein Studienbuch. Beck, Munich, p 681

Treaty Banning Nuclear Weapon Tests in the atmosphere, in outer space, and underwater, signed 5 Aug 1963 and entered into force on 10 Oct 1963. Available online at http://www.un.org/disarmament/WMD/Nuclear/pdf/Partial_Ban_Treaty.pdf. Accessed 20 Aug 2012

Treaty on Principles Governing the Activities of States in the Exploration and Use of Outer Space, including the Moon and other celestial bodies (signed 27 Jan 1967, entered into force on 10 Oct 1967) 610 UNTS 205

Treaty on the Non-Proliferation of Nuclear Weapons, signed on 12 June 1968, entered into force on 22 Apr 1970, reproduced in IAEA INFCIRC/140

U.S. Department of State (2016) A resource on strategic trade management and export controls. https://www.state.gov/strategictrade/

UN Doc. S/Res/1540 (2004) Text available at http://www.un.org/News/Press/docs/2004/sc8076.doc.htm

United Nations Register of Conventional Arms (UNROCA). Established on 1 Jan 1992 under General Assembly Resolution 46/36 L of 9 Dec 1991

van Fenema PH (1999) The international trade in launch services: the effects of U.S. laws, policies and practices on its development, Leiden

von der Dunk FG (2011) National space legislation in Europe, issues of authorisation of private space activities in the light of developments in European space cooperation. Brill, Leiden

von der Dunk FG, Tronchetti F (eds) (2017) Handbook of Space Law. Edward Elgar Publishing, Cheteham/Northhampton

Wassenaar Arrangement, Initial elements: statement of purposes, paragraph 1

Wegner C, Weith N, Ehrlich W (eds) (2006) Grundzüge der Exportkontrolle. Bundesanzeiger Verlag. Cologne, Germany

Wetter A (2009) Enforcing European Union law on exports of dual-use goods. Oxford University Press, Oxford

Space Systems and Space Sovereignty as a Security Issue

12

Annamaria Nassisi and Isabella Patatti

Contents

Introduction	211
Notion of Sovereignty and Jurisdiction	212
Why Does Space Security Matter?	214
Space Systems and Security from Space	215
State Sovereignty and Homeland Security	215
State Sovereignty and the Military Domain of a state	218
Space Systems and Security in Space	220
Vertical Territorial Sovereignty	221
Space Systems and Economic Sovereignty	223
Conclusion	224
References	225

Abstract

Since the beginning of space activities, the global community speculated about the relation between planet Earth and the space environment, and on the potential offered by the space enabled services to safeguard a country's political, economic, and social sovereignty. Through the decades, space technologies progressively enhanced global safety, by improving domestic and international coordination and strategies. In particular, this chapter will focus on the relation between space systems and the security issues linked to a state's sovereignty.

A. Nassisi (✉) · I. Patatti
Strategic Marketing, Thales Alenia Space, Rome, Italy
e-mail: Annamaria.Nassisi@thalesaleniaspace.com; Isabella.Patatti@external.thalesaleniaspace.com

© Springer Nature Switzerland AG 2020
K.-U. Schrogl (ed.), *Handbook of Space Security*,
https://doi.org/10.1007/978-3-030-23210-8_144

Introduction

Space has always been the stage of humankind's greater achievements and the focus of inspiring collaboration among nations, mainly within the scientific domain. The space sector has always been considered a strategic resource, able to contribute to the pursuit of a multitude of political, social, and economic objectives of a country (Darnis et al. 2016). The ability to access to satellite capabilities and operate them has always been critical to both the major powers on the international geopolitical stage and also the global community as a whole. The space environment has acquired greater importance, not only for the institutional actors but also for non-state actors, scientific and academic institutions, international organizations, and all other players that use space technologies and services in order to improve their activities. Outer space resources embrace all kind of applications, ranging from global communications to farming, from weather forecasting to environmental monitoring and climate change, from navigation to surveillance and disaster management. Being critical to the well-being of all countries and people, it becomes imperative that all humankind can access and enjoy its many benefits. Therefore, given its importance and practical utilization, space has become a particularly challenging conundrum of public policy.

Space is also interlinked to the concept of security, safety, and defense. Within the security and safety context, space infrastructures and services are key elements to the political and strategic dimension. Whether we take into consideration international agreements and policies or situation of crisis and disasters, space-based capabilities appear to be strategic and effective instruments, critical to the well-being as well as safeguarding the sovereignty of states. Furthermore, the increase of dual-use space systems, which are blurring the line between military and civil space-based missions, is linked to public returns of investments in proprietary assets. The progress in technological advancements seems today to be inadequately regulated by the 1967 Outer Space Treaty (OST), and an update should be envisaged. In the last years, the concept of space security is being linked to the idea of having a dedicated space force, but this proposal is still being object of discussion among the spacefaring nations. This chapter will focus on the security and defense concerns of space-enabled capabilities linked to the sovereignty and interests of a state, by examining the issue from different angles, in particular those connected to the security, defense, political, and economic dimensions. The objective of this discussion is dual: we want to highlight the importance of space assets as strategic elements for the security and well-being of the sovereignty of a state.

Notion of Sovereignty and Jurisdiction

The increasing awareness of vulnerabilities has led to a debate about state sovereignty. Given that this term carries some weight, it is worth defining it within the context of this discussion. Sovereignty, once a relatively uncontested concept, lately had become a question of rivalry within the national and international relations theory (Alshdaifat 2018). In the context of contemporary public international law, we can define sovereignty as the basic international legal status of a state that, within its territorial jurisdiction, is not subject to governmental, executive, judicial, or legislative jurisdiction of a foreign state or to foreign laws other than public international law (Steinberger 2013).

Sovereignty is a legal principle by which each state is entitled to exercise exclusive control and supreme authority within its boundaries. Article 1 of the Montevideo Convention of Rights and Duties of states of 1993 indicates *"the state, as a person of international law, should possess the following qualifications: Permanent population, a defined territory, government, capacity to enter into relations with other states"* (Montevideo Convention 1933). Furthermore, Article 2 of the Charter of the United Nations recognizes that all states are equal and sovereign because they all are politically independent. Sovereignty can therefore be considered the benchmark for the doctrines of responsibility, jurisdiction, and nationality. The concept of jurisdiction refers to the power of the states to affect its nationals, property, and circumstances, and therefore reflects the basic principles of sovereignty, equality of states, and non-interference in national affairs. The competence of states in respect to their territories is generally attributed to their sovereignty and jurisdiction, but a distinction in the two terminologies should be noted: while sovereignty can be intended as the legal personality of a state, jurisdiction refers to the rights, claims, powers, and freedoms of a state and therefore refers to its regulatory authority to make and enforce rules upon people.

The notion of sovereignty applied to outer space has been introduced as an object of discussion, following the launches of the first satellites in 1957, and then further developed, with the creation of the OST. in 1967. Articles I and II of the treaty affirm that all space activities shall be undertaken in the sovereignty-free outer space, including the Moon and other celestial bodies. Furthermore, outer space is recognized to be *res communis*, which according to Roman law is the *"property of all,"* that is outer space is not subject to private ownership. However, the exclusion of sovereignty in outer space laws does not exclude the exercise of certain sovereign rights by states in space (Zhang 2019). Article VI of the Outer Space Treaty, as *lex specialis*, recognizes the concept of jurisdiction to be applicable to a state's activities in outer space, and asserts that states are responsible for all *governmental* and *nongovernmental* space activities. Article VI does not make a distinction as to whether the activities at issue are the state's own activities or those of private actors. Given that space activities are undertaken by a state (and/or nongovernmental body) by means of objects and infrastructures, a state's supervision over the said activities invites concurrent jurisdiction over it: this *quasi-territorial jurisdiction* provides space objects with a nationality and converts them into pieces of quasi-territory of a particular state (Von der Dunk 2011). The concerns encompassing the concept of sovereignty, however, have become more critical in recent decades, as the growing lack of natural resources and the need for national security are major issues of the twenty-first century.

In this perspective, also non-spacefaring countries are going to procure or develop their own space infrastructures in order to be independent and to strengthen their own sovereign jurisdiction. As a consequence, the protection of space infrastructures becomes fundamental to guarantee continuity of the space services. The sovereignty of a state could have implications also in the removal of satellites, or part thereof, when it is classified at the end of its life as a debris. New Active Debris Removal (ADR) technology and investments from the private sector have voiced doubts on how to deal with these activities being the satellite, or part thereof, a sensitive element for a state. Analogous considerations can be done for other types of technologies, such as On-Orbit Services (OOS), where the private sector is investing to supply services. There is no clearly defined legal framework that reconciles the sovereignty needs and the return on investment for the private sector.

Why Does Space Security Matter?

At the dawn of the space program, civilian and military space systems were developed by the Soviet Union and the United states according to their respective competitive strategies. In particular, during the years of the Cold War and the nuclear buildup, the two nations wanted to detect the construction of the nuclear arsenals from afar and find storage and preparation sites for the missiles through the use of observation and early warning satellites, which later became one of the benchmarks of the strategic dialogue that opened in the late 1960s. In order to keep outer space a safe environment and prevent its weaponization, the 1967 the Outer Space Treaty was drafted and signed by 132 countries. The idea behind the treaty was to have a dedicated document clearly indicating that the use of space is a privilege for the whole humanity and, hence, state sovereignty cannot be extended to outer space. However, countries started to progressively understand the strategic value offered by the ownership of space assets, and they started to invest more heavily in space activities for competitive, defensive, political, and economic reasons.

Space applications have quickly become a powerful asset in the new geopolitical strategic arena, as governments have started to integrate and use space systems for various purposes. The new role of space activities as a component of state power has opened up new debates nationally and internationally that could radically change the world scene at a political, economic, and industrial level. Although the threat of the Cold War is now over, countries should still be prepared to address a multitude of security problems that could arise without warning and in unpredictable ways.

We can identify two dimensions of security offered by the space environment:

1. *Security from space* that entails the contribution of space systems in achieving enhanced security on Earth and encompasses Earth observation satellites, early-warning systems, navigation satellites, and electronic intelligence systems to guarantee security for the country and for international cooperation, such as food and water security, study of climate change, management of natural resources and disasters, migration, border control, environmental protection.

2. *Security in space* that is focused on the protection of the assets in the outer space environment against natural and human threats.

Over the past decade, international fora have pursued legal frameworks for responsible conduct in space, but as of today, the international community has not reached a general consensus on new laws or regulatory norms. The first step in developing new legal frameworks must be based on a realistic, and holistic, assessment of risks and threats (Hitchens and Johnson-Freese 2016).

Space Systems and Security from Space

Nowadays, the link between space and sovereignty appears to be stronger than ever, and a lot of non-spacefaring countries are in the process of acquiring these capabilities. Governments (civil and military) act as both facilitator and regulator to support national development in order to guarantee independence and autonomy. In the spacefaring nations, government funding supports the technological advancements of their national industries to maintain competitiveness and to support high-performance programs development. This approach has been followed also by the non-spacefaring countries that are motivated by self-sufficiency to serve national policy interests (sovereignty). Accordingly, they procure Earth Observation (EO) space systems equipped with cybersecurity capabilities in order to fulfill a more immediate dual-use role, or they develop their own national manufacturing capabilities motivated by the growth opportunity of qualified labor, and an increase in local industry's competencies. Within the context of homeland security, space assets contribute to strengthen both external and internal security of states together with other platforms (e.g., ground-, air-, and sea-based ones) (Directorate-General for External Policies of the Union 2014). The demand of satellites for security objectives has increased in the last few years motivated also by the growing number of threats, expeditionary missions in remote environments, as well as an increased number in humanitarian relief missions. These operations, and in particular the civilian and humanitarian ones, are likely to characterize the states' security efforts in the years ahead.

State Sovereignty and Homeland Security

National governments have intensified their commitment to homeland security, increasing their operational activities in domains such as border control and maritime surveillance missions. When contextualizing the development and management of space capabilities for security and defense of a state's sovereignty, it must be borne in mind that a state constantly features both public and private actors. states worldwide are progressively shaping and implementing an inclusive approach to security, one that takes into account synergies among different technologies and services, and tries to make the best of existing resources and capabilities. The space sector represents a strategic resource, able to contribute to the pursuit of a multitude of political and socioeconomic objectives. As mentioned before, space is also naturally tied to the constantly evolving concept of security, which is not always tied to the offense-defense dimension. In this framework, space assets, encompassing a wide spectrum of performances, can answer to both civil and military needs, originating from the growing number of global challenges (natural and man-made) as well as non-state actors that are present on the international scene. In the face of these new security challenges, a state requires timely and reliable information, either when it is operating on its own territory or when it is involved in international matters. Besides the daily sovereign affairs within its territory, a state must also keep into consideration its commitments towards the global community, borne to safeguard its own interests or necessary to maintain the stability of the international landscape. The production of information relevant to security in the shape of satellite based information, if coupled with in-situ, aerial, and other source of intelligence, represents a strategic tool able to influence decision-making processes at both national and international level.

At national level, security relies on governments, represented by institutional actors (e.g., space agencies, ministers of defense, minister of interior, and minister of foreign affairs, etc.) and is hence related to national sovereignty: in particular, homeland security is an especially critical element in the overall security of a country, as it does not only protect the state from attacks but also ensures the safety of people by helping government bodies to prepare for and mitigate damage from various security threats (Wu and Wang 2018). Effective homeland security operations rely on information collection, integration, and analysis. Hence, a secure and integrated intelligence network is required. The dual-use tied to space-based data and information is particularly useful in making sure that the sovereignty of a state is protected *internally*, through the creation of a stable environment for its people, and *externally*, through the protection of national sovereignty and interests against foreign interference and violation. Remote sensing intelligence can help government bodies in establishing border security and ensure territorial protection by monitoring national borders and territorial seas.

The satellite infrastructure contributes also to the homeland security department in the fight against attacks perpetrated from non-state actors. Institutions and public structures, which are related to the image of a state or are symbolic of a state's power, can quickly become targets of attacks. For example, the attacks of the 11 September 2001 targeted symbols of American power. The term *non-state actor* is a very broad one and can refer to any entity or force that is not directly controlled, integrated into, or legally part of a sovereign state (Boyce 2013). Non-state actors can range from terrorists to ruthless guerrillas, or even to private and commercial entities. One of the top priorities of a country's homeland security is to protect its people from groups or individuals that, for political, religious, or economic motives, engage in terrorist attacks, criminal acts, or actions that threaten national safety and security. In this context, one of the main problems of the protection of national sovereignty is the possibility of incurring in asymmetric warfare against non-state players engaged in terrorist attacks.

There are many different definitions of terrorism. Some of these would suggest that an act only counts as terrorism if it directly causes death or injury to innocent people, while other definitions are much broader. For the purposes of this chapter, we will define terrorism as the *unlawful use of violence and intimidation (...) in the pursuit of political aims* (Oxford English Dictionary). Terrorist acts will differ based on the behavior and characteristic of the criminal group, and how it would respond to different types of government actions. Groups could attack governments or military targets to gain autonomy from their existing regimes, or could attack civilian moved by political or religious motives. Attacks could target a state's technologies: one example was the use of jamming during Operation Iraqi Freedom, in which insurgents deliberately jammed commercial satellite communications used by the US military. In case of an attack against space systems, terrorist groups would more likely engage in cyberattacks (Coleman and Coleman 2017), or in practices to degrade an orbit, or to disable communication links, or blind surveillance satellites to reduce a state's military advantage. For example, the Liberation Tigers of Tamil Eelam (LTTE) frequently hacked government networks and websites to engage in propaganda and, in 2007, pirated a US satellite to broadcast to other countries.

To enhance national security, the market trend highlights major requests to protect the space assets from cyberattacks. Cybersecurity encompasses aspects related to computing and network that will include the satellite and having an impact on all elements within the network topology and connected computers. As a consequence, this capacity extends to data delivery and cloud systems. The protection of space-based assets enables secure data integrity, data availability, data confidentiality, and resilience. A set of regulations, stemming from spacefaring countries, has been put in place to address the business practices with the aim to preserve national security and to comply with international obligations. These regulations are applicable to both institutional and commercial programs, and could help to prevent future non-state actors to acquire their own capabilities in space, with the intent of using them to launch direct attacks. A 2016 research paper stated that *"cyber threats against space-based systems include... well-resourced organized criminal elements seeking financial gain; (and) terrorist groups wishing to promote their causes, even up to the catastrophic level of cascading satellite collisions"* (Livingstone and Lewis 2016). It is important to not underestimate non-state actors that carry out asymmetric attacks to influence states, and this is true for the space segment as it is for the ground one. According to Miller (2019), *"current technology makes space an offense-dominant domain. Despite the cost and technological difficulty of reaching space, it is relatively easy to carry out attacks, at least compared to the cost of defending capabilities in space."* Therefore, it is important to develop defense capabilities in the space domain in order to reduce the chances of an attack and be prepared in case of one.

State Sovereignty and the Military Domain of a state

With respect to the relationship between state sovereignty and military space operations, the potential of space capabilities for military operations represent a key element in the analysis of space as a strategic resource of a state. In particular, military reconnaissance came to be viewed as a staple in a state's exercise of territorial sovereignty, whereas having knowledge of the adversary's military and industrial abilities was considered essential to receive an accurate situational awareness and to prevent foreign intervention. Even today, despite the proliferation of scientific and commercial satellite data, the technology's military roots continue to be evident: for example, defense departments still control the lion's share of high-resolution satellite imagery.

The prominence of the defense domain continues to be of critical importance in safeguarding the sovereignty of a state. As it is often remarked, the security of its own citizens, who gave their allegiance to the sovereign entity, is the first duty of a state. Hence, with this consideration in mind, the use of the defense industry to assure the safety of its citizens and security of its territory against internal and external threats can often become a necessary function (Yeo 2014). Overarching goals, both civilian as well as military, have been defined and adapted to match the changing security environment. Space assets can provide strategic help to support the operative theaters in case of international cooperation for crisis management operations. Significant changes feature not only the miniaturization of technologies for small satellites but also the launch services encompassing new generation low-cost launchers that offer speedy rocket launches in short time

Military Activity in Space

It is worth mentioning that the legality of military activities in space is tied to the 1967 OST (de Gouyon Matignon 2019). Most significant from a military perspective is Article I, which stipulates that space is *"free for exploration and use by all states without discrimination of any kind, on a basis of equality and in accordance with international law and there shall be free access to all areas of celestial bodies,"* hence not explicitly prohibiting the use of satellites to perform surveillance, reconnaissance, communications, and other functions without authorization of other states, even during peacetime. Other articles of the OST bear on the military use of outer space, such as Article IV, which calls for the de-weaponization of space: *"States Parties to the Treaty undertake not to place in orbit around the Earth any objects carrying nuclear weapons or any other kinds of weapons of mass destruction, install such weapons on celestial bodies, or station such weapons in outer space in any other manner. The Moon and other celestial bodies shall be used by all states Parties to the Treaty exclusively for peaceful purposes. The establishment of military bases, installations and fortifications, the testing of any type of weapons and the conduct of military maneuvers on celestial bodies shall be forbidden. The use of military personnel for scientific research or for any other peaceful purposes shall not be prohibited. The use of any equipment or facility necessary for peaceful exploration of the Moon and other celestial bodies shall also not be prohibited.*" Historically, military space operations have been nonthreatening in character and raised very few contentious legal issues. They have typically consisted of space control (passive defensive counter space missions) and space support, while space warfare has always remained purely notional. The legal architecture governing military operations in space was originally designed for space exploration and commercial applications, and the resilience of the applicable law in the face of the challenges arising from the changes in the global political landscape has yet to be determined (Schmitt 2006).

Peaceful Use in the Defense Domain

There are questions about the interpretation of the term *peaceful* as it can be intended as either *non-military* (broad interpretation) or *non-aggressive* (narrow interpretation). In particular, the narrow interpretation could explain how it is imperative for a nation to retain its right of self-defense, as expressed both in customary law and in Article 51 of the Charter of the United Nations. For example, the United states provide that the term "peaceful purposes" allows for "*intelligence-related activities in pursuit of national defense*". Through this, it is possible to adopt a "battlefield awareness" model, thanks to which the programs that have defense purposes can focus on information collection for tactical applications. In this sense, space systems are viewed as strategic enablers that offer better knowledge to a state's military operations through value-added information that increase the ability of a state to apply precision military force. National military bodies can rely on space support provided by numerous kind of satellites, such as Satellite Communications (SatComs), Intelligence, Surveillance and Reconnaissance (ISR), Position Navigation and Timing (PNT), although potential adversaries could develop anti-satellite skills, that, supported by an array of sensors, would be able to attack space systems through multiple manners (e.g., cyber, electronic, missiles, directed energy weapons, jamming) (Defense Intelligence Agency 2019). The level of modernization of technologies has completely reinvented satellite utilization in modern warfare, where battles can be won or lost depending on who has the most sophisticated, secure, and specialized infrastructure. Space offers persistent coverage, and, unlike ground vehicles or aircrafts, satellites are unfettered by earthly features such as atmospheric drag or terrain. Through remote sensing real-time intelligence, military targets can be detected and battlefield features (e.g., ground terrain, weapon equipment, enemy location) can be unveiled and a strategic support within the operative theater can be provided.

Satellites offer capabilities to monitor and, in addition, provide warning messages against the transportation and launch of ballistic missiles and enemy movements inside and outside a country's frontier. Navigation systems deliver guiding information to accurately strike targets. Space-based sensors provide the first indication of attacks and terrestrial sensors provides follow-up information useful for countries to deliver the appropriate defensive and/or offensive response. The space dimension becomes indispensable to answer to the prerequisites of precision, efficacy, and promptness that are essential to military operations.

The Dawn of New Regulations and Space Policy Directives

As we have seen, nowadays, space systems represent a significant constituent of national defense by means of aiding government bodies in the creation of an active and dependable defense strategy. In recent years, a number of countries have started to recognize space as a distinct location or concept where conflict can take place, such as on land, sea, air, or space, or within digital systems (Liptak 2019). Indicative of the likelihood of the space domain to become more and more interlinked with the defense industry are, indeed, the 2019 recent events, which saw the re-establishment by the Trump's Administration of the U.S. Space Force, followed closely by the creation of the French Space Command by President Macron. Born as a way of "proactive prevention," the newest branch of the American and French forces will protect the interests of their respective countries in space. In the wake of these events, also the North Atlantic Treaty Organization (NATO) turned its attention to space as an *"operational domain"* over concerns that enemies of the Western military alliance could cause chaos by jamming satellites. NATO's Secretary General Jens Stoltenberg reportedly said that there was no question of weapons being deployed, but the alliance had to protect civilian and military interests (Boffey 2019). There will always be a case of discrepancy between the non-territorial nature of space and the principle of state sovereignty, whereas the notion that the jurisdiction, affairs, and entities within a territory are solely business of that territory becomes more complex when there are no physical lines. However, given the defense department role and use of resources in space for the purpose of national defense, and for the purpose of this discussion, reference should be made to Article VI of the OST that attributes the responsibility of the activities of the governmental entities and their contractors to states.

Space Systems and Security in Space

One of the fundamental principles of the concept of sovereignty of a state is autonomy. In the modern international system, countries continue to perceive themselves as independent units and strive to preserve their autonomy and decision-making ability. The space systems are classified as critical infrastructures on which states rely for their well-being. As technological and cost barriers to space lower, more countries and private entities partake in space infrastructure construction and rely on ownership of space assets.

Today, a significant proportion of the economies and infrastructures of modern states depend on such technologies. In this framework, it is easy to see how space-based service interruption would severely affect a large number of activities. Thus, protecting them by reducing their vulnerability is becoming critical for the sovereignty of a state.

There are several cases of space threats that would put at risk the safety of infrastructures in space, such as space-debris collisions, or the uncontrolled reentry of a spacecraft. Other than uncontrolled disasters, with the rapid increase in space technologies comes also the risk of utilizing space systems for direct attack purposes (e.g., North Korea recently tested ICBMs missiles to ascertain they can use these weapons against Japan and the United states). Space debris around the Earth constitute a considerable hazard to both crewed and unmanned space operations. Objects in LEO could impact or be impacted by pieces of debris, and the force generated by the impact could be so powerful to damage or render inoperable the satellite or even create more debris, causing a collisional cascade effect known as Kessler Syndrome.

Another threat to space assets could be represented by adversaries jamming communication and navigation systems, or blinding imagery satellites or other strategic sensors. Physical or cyberattacks against ground infrastructures can also threaten space assets capabilities. The outer space and cybernetic environments have been intertwined. Hence, they find themselves facing common threats that they would need to be addressed by common strategies. In particular, it is worth reminding the attack against NASA in 2010–2011, in which NASA's computers experienced more than 5,400 incidents of unauthorized access and attacks by malicious software (Protalinski 2012). According to the investigations carried out, the attacks may have come from individuals wanting to test their abilities, foreign intelligence services and criminal enterprises wanting to profit from the information gained.

The international community needs to recognize the level of dependence modern societies have on space assets and capabilities. Many institutional and private actors rely on the space sector to create a set of strategies, initiatives, and programs at a national and international level. The last few years have seen a rekindle of the strategic great-power competition for the conquest of the space environment, which has become object of interest of the major global actors. China and the United states, as in other dimensional domains, are first in line in the newest space race, especially when it comes to strengthening their position in an environment that has numerous implications, particularly economic and strategic. Compared to the historical space race of the United states versus the Soviet Union, the newest race sees the involvement of numerous countries, other than the aforementioned China and the United states: even though Russia remains a great power in the space segment, other countries like France, Israel, India, and Japan are making their voices heard. As the new space race reaches the heart of the competition, it could have a strong impact on the balance of power in the world. How this will affect to the concept of state sovereignty remains to be seen and should be the object of investigation by policy makers.

Vertical Territorial Sovereignty

The debate over the delineation of the boundary between outer space and state sovereignty precedes the beginning of the space race; however, following the launch of Sputnik in 1957, two legal concepts concerning spaceflight started to be the object of discussion of policy makers. Originally, when the United states and the Soviet Union started their expansion towards outer space, they tacitly assumed that international law did not prohibit it, and the other countries did not protest as well. Today, following the ever-changing space technology, it is easy to notice how the understanding and implementation of a state's sovereignty in outer space needs to be addressed legally. A state's sovereignty remains important for the security of the state. As of today, the delimitation of outer space and airspace is still not regulated and therefore sovereignty cannot be presumed. It has happened that disagreements arose in this matter, like in the case of the 1976 Bogotá declarations regarding the supremacy of equatorial states over geostationary orbit (Polkowska 2018).

Concerning vertical extension of sovereignty, it should be recalled that according to the Chicago Convention of 1944, states hold absolute and exclusive jurisdiction in relation to their respective air space. However, the 1967 OST establishes that outer space cannot be subjected to national claims of appropriation (Bittencourt Neto 2012). The problem of defining a state's extension of vertical sovereignty is primarily based on the lack of a natural boundary separating air and space. In the years following the Chicago Convention, states have taken different positions on the matter, but as of today, there is still no general consensus. For example, after the launch of the Sputnik, the Soviet Union claimed vertical sovereignty without a defined upper limit. South Africa, on the other hand, pinned down outer space as *"the space above the surface of the Earth from a height at which it is in practice possible to operate an object in an orbit around the Earth."* The United states' position in the matter changed repeatedly between the 1950s and 1960s, yet with a 2003 regulation, and with the purpose of defining the qualifications of an astronaut, the U.S. Air Force defined *space* as the area of 50 miles (80.4 km) above the Earth's surface (Reinhardt 2007). The attitude of states generally varies depending on the current political and economic situation. However, the exercise by the state of unlimited control and power in the air is also a condition for the security of the state and its citizens (Shrewsbury 2003). The security issue is therefore an essential argument in favor of the concept of territorial authority.

Delimitation is also important to ensure equal access to space for all states. In the words of John Cobb Cooper, *"unless [the upper boundary of national airspace] is fairly close to the Earth's surface, few states will be able to put a satellite into orbit... without passing through the national airspace of other states. In other words, few states will be free of a political veto by other states in planning orbital flights."* As more and more states are developing their own domestic space launch capability, only few of these new space powers will be able to freely access space, or utilize the most efficient launch azimuths, if neighboring states can claim sovereignty up to even 62 miles (100 km). Setting a low vertical limit on state sovereignty will ensure all states have equal access to space (Reinhardt 2007).

Another issue raised by the absence of an international definition of the space boundary is liability for space activities. The *Liability Convention* imposes absolute liability on the launching state for damage caused on the surface of the Earth or to aircraft in flight by the state's *"space object"* or the *"launch vehicle and parts thereof,"* as in the case of the uncontrolled reentry of the Soviet Union satellite Kosmos 954 over northern Canada.

As space becomes more and more economically and politically important due to the inexorable progress of science, the issue of vertical sovereignty will continue to grow. Defining the limit between a state's sovereign territory and free outer space could also add clarity to all the treaties that are written in a functional manner without defining where space begins.

Space Systems and Economic Sovereignty

In a world of mutual dependence, economic sovereignty hinges on the ability to protect economic power. In order to safeguard its sovereignty, the aim of a state should be to become a player in all fields that are vital for the resilience of the economic system, and that could contribute to shape the global community's future in a critical way. Today, economic sovereignty becomes a geopolitical power and economic relationships can be used as broader geostrategic goals.

The economic sovereignty agenda of a country should hold several objectives, such as boosting a state's research, scientific and technological base, protecting assets critical to national security, promoting a level playing field in national and international competition, and employing policies to strengthen a state's monetary and financial autonomy (Leonard et al. 2019). The concept of economic sovereignty inspired major initiatives in fields such as energy, geopositioning, artificial intelligence, computing and, of course, aerospace. Space is considered by states a strategic economic domain as it is a major enabler and multiplier (Zervos 2017), while being borderless and virtually unregulated by existing treaties. This characteristic makes outer space one of the main tactical elements of a country and offers states numerous opportunities for leadership and partnerships.

In the consideration of space applications within the economic dimension, it is important to remember that space applications are considered *public goods*, and hence, since the dawn of the space race, they have been mainly funded by public investments. The underlying rationale for public space investments is the concept of *market failure*. Space is an *externality-inducing* industry, and thus governments are needed to manage the externalities into a socially optimum outcome. Furthermore, there is a risk linked to the *underinvestment* within the upstream segment of the space value chain and the long development of programs, commonly regarded as high-risk (Return from Public Space Investments 2015). As a result of these market failures, together with the security considerations associated with the space technologies, the responsibility for production and control of space assets has been historically laid on government institutions. The latter in turn reap the benefits in the form of direct revenues, of territory and disaster management, and indirect revenues, in the form of education and qualified employment. Space activities stimulate the development of new technologies – as an innovation factor, as a competitiveness factor, and as a key to the consolidation of national industrial capabilities and internationally recognized economic power on the world stage. In a world where international alliances are of particular importance for national and worldwide security, the inter-alliance specialization offered by space becomes critical for stability and economic profile of states. A state's strategic autonomy is strengthened by national ownership of assets for the defense and security applications. It is worth mentioning also that the space industry is subject to economies of scale and scope. In this context, it is possible to notice that space technologies can boost the economic growth of a nation and establish the technological advantage of countries. This is illustrated in the case of the military satellites, which the spacefaring nations domestically produce, due to both the demand side of the country (countries tend to select their home industries to enhance their economies of scale and scope, other than for security reasons) and also the supply side (on which trading restrictions are applied in order to safeguard technologies as trade can rapidly diminish technological gaps).

In particular, the export restrictions play a significant role within the sovereignty of a state (Noble 2008). Keeping the technological edge is perceived as a form of dominant power from countries. As the demand for space-based security is very high, when it comes to exporting major space powers are inclined to protect critical technology. As national security issues are gaining prominence everywhere, so is the relationship between national security and economics. As economics is becoming again an area of great-power competition, economic tools are employed to secure geopolitical advantage. The strategic intent of export control is to keep sensitive technologies out of the hands of potential adversaries and guarantee to a particular state a larger market share. Export control of space technologies in a particular state are, usually, more concerned with the relative performance of foreign systems to those of the state of origin. It is important to protect a country's technological lead and strategic independence through the prevention of the proliferation of technologies and systems to potential adversaries. This is compatible with space systems, as well as the strategic nature of space security, and the fact that industries subject to economies of scale have been long considered strategic to the sovereignty of a country.

However, it is important to keep in mind that economic sovereignty does not mean containing the spread of technology at all costs. In the current interconnected world, technological leadership also depends on continuous innovation and investments benefits stemming from cooperation. While it is important for a state's sovereignty to protect its core assets – especially when security interests are at stake – economic sovereignty does not mean resisting to globalization. A state's competitive advantage also helps to increase the qualified human capital at a remarkable rate. In every business environment – and undoubtedly in the space sector – it is vital to have access to individuals with technical training. An educated workforce fuels the economy of innovation of a state. Accordingly, innovation creates competition and competition creates jobs that, in turn, create growth. Space activities also impact the economic sovereignty of a state through their ability to increase dramatically the capacity of humans to act and to interact with other people or countries with increasing strength (European Space Agency 2005).

Conclusion

Through this discussion, we have seen what important effects space has on modern sovereignty. Space systems reinforce the exclusive structure of sovereignty and its potentiality to foster decisions within its territory and on the world stage. To prosper and preserve their independence in a world of geopolitical competition, states must address globally the space and security challenges. This could involve creating a new idea of sovereignty that sees the space environment as part of their identity, power, and bureaucratic interests. Creating an environment tailored to such incentives requires work to be done at the legal and policy levels.

References

Alshdaifat SA (2018) Who owns what in outer space? Dilemmas regarding the Common Heritage of Mankind. Pécs J Int Euro Law 2018/II:24

Bittencourt Neto O (2012) The elusive frontier: revisiting the delimitation of outer space. IAC-12. E7.1.9. https://iislweb.org/docs/Diederiks2012.pdf

Boffey D (2019) Nato leader identifies space as the next 'operational domain, The Guardian. https://www.theguardian.com/world/2019/nov/20/nato-identifies-space-as-next-operational-domain

Boyce J (2013) Surrendering sovereignty: the private military industry, the state, and the ideology of outsourcing. https://skemman.is/bitstream/1946/14746/2/snidmat_ma_i_stjornmalafraedi_0.pdf

Coleman N, Coleman S (2017) Terrorism in space. Examining the issues and mitigating the risks. In: 68th International Astronautical Congress (IAC), Adelaide, Australia, 25–29 Sept 2017

Darnis J, Sartori N, Scalia A (2016) Il futuro delle capacità satellitari ai fini della sicurezza in Europa: quale ruolo per l'Italia? IAI, Edizioni Nuova Cultura, Quaderni

De Gouyon Matignon L (2019) The legality of military activities in outer space. https://www.spacelegalissues.com/space-law-the-legality-of-military-activities-in-outer-space/

Defense Intelligence Agency (2019) Challenges to security in space. Defense Intelligence Agency, Washington, DC. https://www.dia.mil/Portals/27/Documents/News/Military%20Power%20Publications/Space_Threat_V14_020119_sm.pdf

Directorate-General for External Policies of the Union (2014) Space, sovereignty and European security: building European capabilities in an advanced institutional framework. Publications Office, Luxembourg. http://www.europarl.europa.eu/RegData/etudes/etudes/join/2014/433750/EXPO-SEDE_ET(2014)433750_EN.pdf

European Space Agency (2005) The impact of space activities upon society. ESA, Noordwijk. http://www.esa.int/esapub/br/br237/br237.pdf

Hitchens T, Johnson-Freese J (2016) Toward a new national security space strategy. Atlantic Council Strategy paper no. 5. https://espas.secure.europarl.europa.eu/orbis/sites/default/files/generated/document/en/AC_StrategyPapers_No5_Space_WEB1.pdf

Leonard M, Pisani-Ferry J, Ribakova E, Shapiro J, Wolff G (2019) Redefining Europe's economic sovereignty. European Council on Foreign Relations, Policy Contribution 31321 Issue n°9, Bruegel. https://www.bruegel.org/wp-content/uploads/2019/06/PC-09_2019_final-1.pdf

Liptak A (2019) France's air force is getting a space command. https://www.theverge.com/2019/7/13/20693087/france-military-air-force-space-command-president-emmanuel-macron

Livingstone D, Lewis P (2016) Space, the final frontier for cybersecurity? Research paper, Chatham House International Security Department, London

Miller GD (2019) Space pirates, geosynchronous guerrillas, and nonterrestrial terrorists. Air Space Power J 33:33

Montevideo Convention on the Rights and Duties of states, Art. 1, 1933, 165 LNTS 19

Noble MJ (2008) Export controls and United states space power. Astropolitics 6(3):251–312. https://www.tandfonline.com/doi/full/10.1080/14777620802469798

Polkowska M (2018) Limitations in the airspace sovereignty of states in connection with space activity. https://pdfs.semanticscholar.org/cbbb/ebd8da928ea965653bbd6708b8bbfc0cafa0.pdf

Protalinski E (2012) NASA: hackers had "full functional control, ZDNet. https://www.zdnet.com/article/nasa-hackers-had-full-functional-control/

Reinhardt DN (2007) The vertical limit of state sovereignty. J Air Law Com 72:65. https://scholar.smu.edu/jalc/vol72/iss1/4

Return from Public Space Investments – an initial analysis of evidence on the returns from public space investments. (2015). London Economics IV. https://londoneconomics.co.uk/wp-content/uploads/2015/11/LE-UKSA-Return-from-Public-Space-Investments-FINAL-PUBLIC.pdf

Schmitt MN (2006) International law and military operations in space, Max Planck Yearbook of United Nations Law Online 10:89–125. https://www.mpil.de/files/pdf3/04_schmittii1.pdf

Shrewsbury MSM (2003) September 11th and the Single European Sky: developing concepts of airspace sovereignty. J Air Law Com 68:115

Steinberger H (2013) Max Planck Institute for comparative public law and international law. In: Encyclopedia for public international law, vol. 414. Oxford University Press, Oxford

Von der Dunk FG (2011) The origins of authorisation: article VI of the outer space treaty and international space law and international space law. Space, Cyber, and Telecommunications Law Program Faculty Publications. 69. https://digitalcommons.unl.edu/cgi/viewcontent.cgi?article=1068&context=spacelaw

Wu X, Wang J (2018) Space system as critical infrastructure. Springer, Boston

Yeo AI (2014) Security, sovereignty, and justice in U.S. overseas military presence. Int J Peace Stud 19(2):43–67

Zervos V (2017) The European space-industrial complex: new myths, old realities. Econ Peace and Sec J EPS Pub 12(1):28–36

Zhang W (2019) Extraterritorial jurisdiction on celestial bodies. Space Policy 47:148–157

13
Critical Space Infrastructures

Alexandru Georgescu

Contents

Introduction	228
Critical Infrastructure Protection	228
Critical Space Infrastructures	231
Distinguishing Characteristics of SI and CSI	233
Critical Space Infrastructure Protection	234
Results from Framework Application	236
Principles of Resilience	237
Complex System Governance	240
Conclusions	242
References	242

Abstract

Space systems are a key enabler for a wide variety of applications which have become critical to the functioning of modern societies. This chapter uses the Critical Infrastructure Protection framework to argue that space systems may constitute a new form of critical infrastructure, dubbed Critical Space Infrastructure, and traces the positive impact that such a perspective may have on space security governance. Critical Infrastructure Protection has developed a conceptual toolbox, as well as practical policy prescriptions, which may be of use to policy and decision-makers to increase resilience and meet future space security challenges.

Keywords

Critical infrastructure · Resilience · Space systems · Governance · Complex system

---

A. Georgescu (✉)
National Institute for Research and Development in Informatics (ICI), Bucharest, Romania
e-mail: alexandru.georgescu@ici.ro

© Springer Nature Switzerland AG 2020
K.-U. Schrogl (ed.), *Handbook of Space Security*,
https://doi.org/10.1007/978-3-030-23210-8_129

Introduction

Space systems have become a key enabler for a wide variety of applications related to command, control, coordination, data gathering, and communications. With their growing capabilities and numbers, the quantity and quality of applications also increase, while lowering access barriers, thereby improving usability and leading to an increase in the number of beneficiaries.

The Organisation for Economic Co-operation and Development (OECD) (2016) notes that the world is entering a fifth stage of space development, one in which we are witnesses to "growing uses of satellite infrastructure outputs (signals, data) in mass-market products and possibly for global monitoring of treaties (land, ocean, climate), third generation of space stations, extensive mapping of solar system and beyond thanks to new telescopes and robotic missions, new space activities coming of age (e.g. new human-rated space launchers, in-orbit servicing)." Space inputs permeate many of the products (tangible and intangible) that we consume, which are the result of extensive global supply and production chains or of the processing of information and the combining of symbols within globalized networks.

Therefore, space services may be consumed directly or indirectly through their role in the functioning of other systems on which we are dependent. The use of space capabilities in energy, transport, financial markets, agriculture, weather forecasting, and other fields is well-known. These latter systems represent a small cross-section of critical infrastructures (CI), sociotechnical systems whose disruption or destruction would generate significant economic damage, casualties, and loss of confidence (Gheorghe et al. 2018, p. 3). Their security is paramount and, therefore, we must consider the question of their governance. While government deals with decision-making, governance encompasses mechanisms, norms, and organizations that mediate the decision-making and implementation process.

The governance of the aforementioned infrastructures like energy and transport relies on Critical Infrastructure Protection (CIP), a comprehensive framework for managing the risk to the key infrastructures, assets, and resources on which our societies are critically dependent, which has been developed for the past two decades.

This chapter aims to introduce space systems into the CIP framework and define them as Critical Space Infrastructures (CSI), arguing that CIP can close some of the gaps that have manifested in the governance of space security and which are creating significant troubles from the perspective of sustainable exploitation of space. With the articulation of the existence of CSI, we follow up with a discussion on Critical Space Infrastructure Protection (CSIP) from the perspective of the specialty literature.

Critical Infrastructure Protection

CIP was first conceived during the Clinton Administration, but only came to the fore after the September 11 attacks, when the systemic impact of the attacks was noted and provided ample argument in favor of the defining trait of CIP, the interdependencies between components, infrastructures, and systems which lead to the transmission of risk and the cascading disruption of critical infrastructures. Presidential Decision Directive (PDD)-63 (1998) identified critical infrastructures as being "those physical and cyber-based systems essential to the minimum operations of the economy and government" (The White House 1998). CIP did not stay confined at the national level. Later, the EU would create its European Program for Critical Infrastructure Protection (EPCIP) through which it set guidelines for improvement of national CIP governance and the identification, designation, and protection of European Critical Infrastructures. Directive 114/2008 established that European Critical Infrastructures are "essential for the maintenance of vital societal functions, health, safety, security, economic or social well-being of people" and are distinguished from national CI through their impact on two or more Member States "as a result of the failure to maintain those functions" (European Commission 2008).

CIP works most often with the concept of resilience, which is the capacity or quality of a system to retain or rapidly regain an adequate level of functioning in the face of a crisis event, with minimal disruption, material damage, or loss of human life. Linkov et al. (2014) argued that resilience should be a priority from the design phase of new systems and of their regulatory frameworks, because resilience is the only consistent answer to the issues of uncertainty and complexity. CIP also works with numerous other concepts, which encompass different aspects of CI qualities, behavior, and interactions during crisis events (Table 1).

Interdependencies are a key feature of CIP systemic thought. Gheorghe and Schläpfer (2006) define interdependencies as bidirectional relationships wherein the status of one infrastructure affects the status of others and is affected in its turn by others. The topology of critical infrastructure risk is built also with the mapping of interdependencies. These are varied, being physical, sectoral, geographic, logical, social/political, cybernetic, or informational and with many taxonomies in existence. The other key features are the dynamics of cascading disruptions – "cascading disasters are extreme events in which cascading effects increase in progression over time and generate unexpected secondary events of strong impacts. These tend to be at least as serious as the original event and contribute significantly to the overall duration of the disaster's effect" (Pescaroli and Alexander 2016). They result from the vulnerabilities and rigidities that accumulate within a system-of-systems across multiple domains until a trigger event or mechanism manifests alongside the alignment of key breaking points. The absence of these alignments prevents the actual cascading disruption event, as it interrupts the vector for the transmission and escalation of the disruption. Pescaroli and Alexander (2016) distinguish between cascading effects and cascading disasters. The former are the multidimensional and complex dynamics which produce the latter.

Regionalization and globalization have brought these issues to the fore, as cooperation becomes a key facet of CIP efforts when global supply and production chains as well as globally synchronized databases and markets produce new risks, vulnerabilities, and threats which are beyond the ability of single jurisdiction authorities to tackle. If chains are only as strong as their weakest links, decision-makers and CIP practitioners cannot count on localized resilience and CIP success to

Table 1 An overview of concepts related to resilience. (Source: author compilation)

Concepts related to resilience in CIP specialty literature	
Vugrin et al. (2011)	
Absorptive capacity	An internal quality of the system that allows it to absorb the effects of systemic or environmental disruption with little degradation in functioning. It is associated with robustness and the presence of redundancies
Restorative capacity	The system recovers easily from the effects of a disruption and also experiences permanent modifications as a result of the episode (adoption of new technology, reorganization, etc.)
Adaptive capacity	The system reorganizes itself in order to maintain functionality, reduce disruptive impact, and rapidly recover full function levels. For instance, a factory may switch suppliers or modify its designs to limit the impact of resource scarcity
Jonkeren et al. (2012)	
Static resilience	The ability of a system to continue functioning after suffering a major shock
Dynamic resilience	The rapidity with which a system recovers from a disruptive event. It is related to repair and reconstruction times
Rockefeller and Arup (2014)	
Reflectivity	Such a system is conscious of the uncertainties and of the changes in the security environment
Robustness	The system actively eschews designs which render it vulnerable to cascading disruptions, catastrophic malfunctions, and overdependence on certain assets
Redundancy	Is found in the diversity of pathways and options for fulfilling system tasks. A system with redundancies can weather significant increases in pressure, upstream shocks, or the malfunctioning of individual assets and system components
Flexibility	The quality that a system possesses to change as a result of shocks and to even find benefits in those changes
Adaptability	The capacity to mobilize systems and resources during temporary stresses or shocks in order to attenuate the impact of the negative events
Inclusivity	The system seeks out and accepts inputs from all categories of stakeholders and includes it in the process of developing strategies, plans, priorities, and resource distribution patterns
Integration	An integrated system responds efficiently to challenges and features short and rapid feedback channels. The governance mechanisms transcend sectorial and other limitations in order to adequately reflect the complexity of the system and to adequately implement policies and decisions

maintain systemic integrity. Helbing (2013) emphasized that global critical infrastructure networks facilitate the propagation of risks and generate the potential for cascading disruption stemming not just from external factors (such as attacks, sabotage, natural disaster) but also from internal ones resulting from system errors, attrition, malinvestment, lack of maintenance, and, most important of all, the complexity of the system-of-systems (SoS). These disruptions within the SoS may lead to the contagion and the escalation of the effects, sometimes in a mutually reinforcing pattern. Perrow (1999) discussed the "normal accidents" or spontaneous malfunctions that arise from the complexity and tight couplings of a system, sometimes without the possibility having ever been foreseen. Eusgeld et al. (2011)

argue that a SoS perspective acknowledges that the components of infrastructure systems may be large-scale systems as well, sometimes operating autonomously from a legal, administrative, or governance standpoint, but linked to the wider system through dependencies and interactions which assign systemic consequences to localized disruptions through the propagation of risks and disruptions.

There was a tacit acceptance of these risks, to the extent to which they were anticipated, in exchange for the efficiencies and gain in well-being that accompanied them. If one were to describe the evolution of CI in the past hundred years, it is that formerly autonomous and vertically integrated infrastructure systems separated by geography, information lag, and risk aversion suddenly found themselves in much greater contact (Bucovețchi et al. 2019), a situation which Setola et al. (2017) called "rapid change in the organizational, operational and technical aspects of infrastructures." Cyber infrastructures and now space infrastructures are some of the initiators and facilitators of systemic changes which result in the increase in CI SoS surface contact and in the tightening of the couplings within the system that accelerate the transmission of risk. The systemic transformations give rise to new sources of added value, new functionalities, and also punctual increases in safety and security through higher governance capacity, but also new risks, vulnerabilities, and threats.

The field is constantly evolving to keep pace with the demands of a SoS beset by and in thrall to growing complexity. One of the recent evolutions, for instance, is complex system governance (CSG), which emphasizes complexity as a source of emergent and sometimes unanticipated behaviors and properties in the system not found in its individual components. A later section of the chapter will elaborate on this idea.

Critical Space Infrastructures

OECD (2019) defines space economy as the "the full range of activities and the use of resources that create and provide value and benefits to human beings in the course of exploring, understanding, managing and utilizing space. Hence, it includes all public and private actors involved in developing, providing and using space-related products and services, ranging from research and development, the manufacture and use of space infrastructure (ground stations, launch vehicles and satellites) to space-enabled applications (navigation equipment, satellite phones, meteorological services, etc.) and the scientific knowledge generated by such activities. It follows that the space economy goes well beyond the space sector itself, since it also comprises the increasingly pervasive and continually changing impacts (both quantitative and qualitative) of space-derived products, services and knowledge on economy and society."

Infrastructure serves not only the economy but also society, and we may draw on this definition to define a space infrastructure (SI) as a sociotechnical system whose main functional component is located beyond the arbitrary line separating the Earth's atmosphere from outer space. Critical Space Infrastructures have the added trait of criticality – their disruption or destruction would cause significant casualties, economic damage, or loss of confidence. CSI have components that are also intra-atmospheric – for instance, ground stations and communication links (Fig. 1).

The identification and designation of CI serves an important role in CIP processes, but this tendency may be muted in space. A country may have tens of thousands of miles of roads, serving a large number of settlements of all sizes, which makes the designation of the critical ones for national functioning and continuity in the face of attacks all the more important. However, SI do not have an especially large inventory, given the high number of functions they serve and the number of beneficiaries. For the rest of the chapter, we will discuss CSI primarily as a function of orbiting assets, or satellites. As the field develops, we will one day be able to talk about CSI composed of probes, research bases on other planets, and interplanetary transport networks. As of yet, a theoretical threshold of criticality will likely only be met by SI containing satellite components. The only other likely candidates are the various probes which measure the activity of the Sun and are part of early warning systems regarding solar flares that give CI operators opportunities to enter conservation states or initiate measures to safeguard system integrity. If we were to speculate on the criticality of early warning systems for another high-impact, low-frequency event – the collision of the Earth with asteroids – we would find that operational assets are also located on Earth or in orbit.

According to the frequently updated open-source database of the Union of Concerned Scientists, by 31 March 2019, there were 2062 satellites in orbit (UCS 2019). Table 2 breaks down that number.

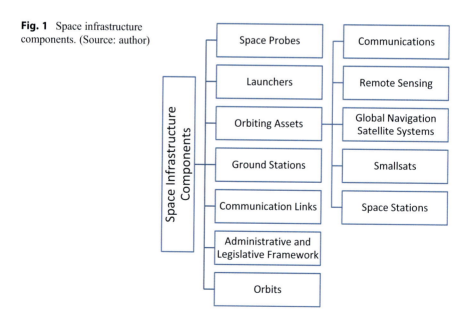

Fig. 1 Space infrastructure components. (Source: author)

Table 2 Breakdown of orbital asset inventory, compiled by author with data from UCS (2019)

Total number of satellites = 2218 by 30 September 2019				
By country	United States 1007	China 323	Russia 164	Other 724
By orbit	Low 1468	Medium 132	Geostationary 562	Elliptical 56
Total estimated number of US satellites = 1007				
By character of owner	Civil 35	Commercial 620	Government 163	Military 189

Distinguishing Characteristics of SI and CSI

The extreme technical, operational, and financial constraints under which satellite operators labor have significant results on the characteristics of CSI and the subsequent emerging risk profile, as opposed to terrestrial CI.

Firstly, their main distinguishing characteristic is, as mentioned, their low numbers. This is compounded by the size factor. While the correlation is waning with the advances in miniaturization, the larger satellites (by mass) will tend to be the most critical, because the expense needed to develop and launch them must be justified by their function and capabilities. However, according to Ambassador Sorin Ducaru, Director of the European Union Satellite Centre, speaking on 4 October 2019 in Bucharest during the International Eurodefense Conference, only half of the existing assets weigh more than 1000 kg (wet mass). Niederstrasser (2018) pegs this as the upper bounds of the smallsat category of system size, and Bryce Space and Technology (2019) confirms that around half of launched systems were in this range in 2018.

Secondly, the larger systems are more likely to be one of a kind, designed specifically for the respective mission, resulting in little interoperability with other systems or opportunities for stakeholders reliant on them to substitute for any lost capacity.

Thirdly, the cost structure and technical barriers make system replacement an expensive, long-term, and uncertain proposition, one that governments are more likely to find palatable. The CSI are also less likely to feature intermediate thresholds of functioning, where partial utility may be maintained in the event of the materialization of a risk, unlike many terrestrial CI, for whom total disruption is just one end of a long spectrum of partial disruption states (transport network carrying capacity, processing power, public services delivery rate and area coverage, partial production of energy and of goods, etc.).

These barriers are being subverted by the wider application of new technologies, like CubeSat architecture and by miniaturization, which are also an initial enabler of convergence with terrestrial CI in another important way – preponderance of ownership or operation by private entities. In Europe and the United States, the estimates of this rate are very high and vary between 60% and 85%, depending on source, and where the upper bound is set by the United States (Cellucci 2018). Bryce's annual State of the Satellite Industry reports have noted the increase of CubeSat numbers among launches, as well as the preponderance of nongovernmental entities (commercial, academic) among the owners. Bryce (2019) noted that 1300 smallsats (according to Bryce definitions, up to 600 kg) were launched between 2012 and 2018, of which 961 were CubeSats. And half of all of these provide commercial services, especially in remote sensing. This trend accounts for the high use of low Earth orbit, and the differences in inventory breakdown would be even starker, if orbital dynamics and atmospheric drag did not lower the mission time of such satellites.

CSI are expected to function in a very challenging security environment, featuring both natural and man-made threats, the latter divided into unintentional and deliberate. The Royal Academy of Engineering (2013) notes the high likelihood of spontaneous malfunction from environmental pressures, derived also from mass and cost constraints for the engineering of more robust systems, though this is also proceeding apace. Terrestrial CI do not generally feature such high background risk levels, since existing regulations regarding resilience from the design phase, on a sectorial and national basis, attempt to reduce the impact of background factors (geology, hydrography, etc.) or include them in decision-making regarding critical infrastructure commissioning. The space-specific threats of debris collisions and extreme space weather phenomena also significantly impact CI, while the former also presents a significant collective action failure that any emerging governance framework will have to address. Moreover, from the beginning, the US National Security Space Strategy has defined space as a "congested, competitive and contested" environment (DoD and ODNI 2011).

CSI are also interesting for their limited current range of interdependency types. Until we have arrived at the level of space industrialization and active resource exploitation, there will be very few cases of physical interdependencies, or of bidirectional dependencies.

Critical Space Infrastructure Protection

Heino et al. (2019) argue, with regard to traditional CIP efforts, that "a severe disruption in the system can go beyond geographical, organizational, and administrative boundaries, thus activating a multifaceted set of actors whose ability to collaborate is required to restore the situation." Stakeholders will have to engage in Critical Space Infrastructure Protection (CSIP) efforts in order to improve the resilience of CSI or of the system-of-systems in which it is operating. Starting from a common definition of CIP, CSIP efforts comprise all of the programs and activities in which stakeholders will engage in order to maintain the level of functioning of Critical Space Infrastructures above a predefined threshold in case of the materialization of a threat and to minimize the casualties, material damage, and systemic impact on other CI. The stakeholders run the gamut from manufacturers of components and providers of services to the owners/operators of CSI, the competent national authorities (civilian, military), and various international or supranational organizations. The latter is the case of the EU, which administers the EU space program, comprising the Galileo/European Geostationary Navigation Overlay Service (EGNOS) global navigation satellite system, the Copernicus remote sensing program, and the future governmental satellite communications network for EU and EU Member State government communications.

While CSIP is a subset of CIP efforts, it is important to acknowledge the differences in approach which CSIP will entail, deriving from the specific CSI characteristics, the space security environment, and the actual ability of a respective stakeholder to govern and positively impact space security outcomes. CSIP is, from the start, a very international activity which will find distinct advantages stemming from the application of global solutions to persistent issues such as space debris, frequency fratricide (already governed globally), and hardening against space weather phenomena. Countries may not have any SI and still be critically dependent on those operated under the jurisdiction of others. Few countries have full spectrum capabilities, and none of them have total space autonomy so that their society and economy are exclusively provisioned with critical space services (and, in the future, goods) by systems under their jurisdiction. However, CSIP efforts start at the national level, which is the ideal starting point for relations with the lower orders of stakeholders involved in the process. What CSIP cannot do is solve the persistent and intentional gaps in the legal and administrative framework of space, as those are contingent on the political will of major spacefaring nations to agree to bind themselves with rules, as opposed to implementing their own programs for space security governance. CSIP can offer tools, mechanisms, and activities short of political action that can lead to an understanding of security issues, their proper communication to important decision-makers and other stakeholders, and their gradual amelioration through resilience-building measures.

It is important to note that designing CSIP governance mechanisms must also take into account the possible future state of the CSI system-of-systems and its environment. This estimate is not only technological but also social, legal, political, geopolitical, and economic. Exercises in strategic foresight like the one described in The Future of Space 2060 report released by Air Force Space Command in 2019 may hold particular significance for each organization, but they are ultimately developing scenarios for system contexts, infrastructure, and interrelationships that will determine the governance solutions necessary to maintain system viability (Air Force Space Command 2019), which is an almost exact description of complex system governance under CIP efforts.

What we can say with near certainty about the future is that we will have many more space systems in place, as well as space actors (especially private ones), that our dependence on them will have increased and new dependents will have materialized from the developing world. The potential results of the materialization of a negative event will have increased as well, maybe to the point of becoming an existential threat at systemic levels. Significant uncertainties will persist in the legal and administrative realm, as the best positioned countries to profit materially and strategically from this ambiguity will head off most efforts aimed at collective action toward positive security transformations. Just as in the steady globalization of CIP, CSIP will be an important vector for the coalescing of common concepts, definitions, toolboxes, practices, and standards, because of the previously stated need for practitioners to cooperate because of interdependencies. The increasingly globalized private sector will act as a vector for the spread of these elements, as they will be able to cooperate with CSIP efforts and CIP efforts aimed at managing the exposure to CSI risk in each country in which individual companies are present and are regulated from a CIP perspective by the national authorities.

Results from Framework Application

If we were to apply the CIP framework to space systems, we would have to be careful to keep in mind the specificities of these systems, of their operating environment, and of their threat matrix, as mentioned in the prior sections. It would also require acknowledgment that some of the elements of a CIP framework may already be under development or in place, such as a space situational awareness program.

It is important to note that CIP also influences considerations on future SI design in favor of resilience and the achievement of acceptable levels of the qualitative and quantitative indicators of resilience listed in the CIP theory section.

The application of CIP would require competent national authorities to begin an identification and designation process for CSI, based on a methodology they have developed for this purpose, in order to allocate scarce security resources to where they would have the highest impact. Owners/operators (OO) of CSI would come under the purview of the competent sectoral authority and would have to conform to regulations regarding protection measures and coordination with the other stakeholders in the national CIP system. This varies on a state-by-state basis, but it generally entails the development of an operator's security plan (OPS) for approval by the competent authority, its regular update (on a set schedule or whenever the situation calls for it), and the implementation of a communication system or structure with the liaising authority. In the European system, this is achieved through security liaison officers embedded within the infrastructure operator and the competent authority, as well as mechanisms for the sharing of relevant information.

CI operators/owners in energy, transport, and other domains would also have to take note of their dependence on CSI, apply adequate methodologies to estimate it, and factor it into their respective OPS.

A generic OPS identifies vulnerabilities, describes existing security programs, and details the ones that will be implemented, as well as the gradual and permanent measures that are instituted with every increase in the alert level.

For a CSI OO, the OPS would include not only references to background security levels but also readiness levels for expected threats such as space weather. The OO is tasked to articulate a security vision for its CSI, in order for the OPS to meet with approval from the regulator, such as plans for new satellite design policies that stress resilience through physical and electromagnetic shielding, for security through obscurity by using specially designed software and other systems, or for redundancy in the form of higher satellite counts, lower replacement times, and so on.

For an OO of an infrastructure dependent on CSI, the OPS would include its plans to mitigate the risk of disruption of critical space services. The possibilities are varied – it may have contracts in place with alternate providers or even alternate system providers, who are not subject to cascading disruptions in the space environment (e.g., ground-based data collection as opposed to space-based remote sensing). The OO may also commit to a reduction in its dependence on CSI, though many entities have limited ability to negotiate and limit CSI risk from third parties (such as OOs of other CI on which they are dependent).

Depending on the country in question, it is possible that there are no CSI OOs within the national jurisdiction. These countries must manage their dependence on CSI without having the possibility of exerting influence to ameliorate their vulnerabilities and therefore face a slew of other challenges and uncertainties (including political) when compared to countries with partial or full spectrum space capabilities. These countries may become active in international fora (such as UN COPUOS – the Committee on the Peaceful Uses of Outer Space) and intergovernmental arrangements in order to pursue a part in collective governance efforts.

Georgescu et al. (2019, p. 272) speculated that one answer to the issue of space governance may lie in the change of incentive structure resulting from the formalization of the space governance issue, for instance, through the CIP framework. Whereas a collision in space or some other disruptive or destructive event may currently be written off as an "Act of God," the average CI operator on Earth finds himself under greater scrutiny when disruptions occur. The CI designation system formalizes a responsibility on the part of the OO, as well as the competent authority and the ultimate coordinating authority at higher levels (the Ministry of Interior in the European model, the Department of Homeland Security in the American one). A failure of due diligence, such as inadequate security measures, exposes the OO to liability issues, whose potentially significant costs may eventually make them more accepting of the higher costs of greater security. A market-centric governance model may also emerge, where unsustainable behavior in space, with impact on the security outcomes, may also be sanctioned in an emergent manner through market mechanisms – interest rates for funding new investment or insurance premia may be lower for security-conscious actors whose systems are less prone to disruption or destruction by design or through other factors.

Given the nature of the space environment, CIP efforts must also focus on cooperation between nations, as they are starting to do also on the ground, with the emerging coordination for the protection of transborder or global infrastructure chains in energy or transport (EPCIP's efforts have been mainly in the energy and transport areas, if one looks at the list of designated ECI). By stressing the mutual dependence of countries and the security gains from having a resilient CI system-of-systems, cooperation under a CIP framework, especially between different orders of stakeholders below the political level, may have far-reaching impact.

Principles of Resilience

As mentioned before, the application of a CIP framework seeks to increase the resilience of the system in question or of the wider CI system-of-systems. CSI would also strive for resilience and thereby utilize the conceptual toolbox described in the previous chapter to plan for an increase in resilience. In this regard, Johnsen (2010) describes seven principles of resilience whose application to space systems may clarify the results of a CSIP framework.

Chief among these is the graceful or controlled decline, which is the result of competencies and capacities which arrest the quick decline of a system, for instance, through the prompt and efficient intervention of emergency response teams or other contingency measures. The system is also set up in such a way that the chance of catastrophic system degradation, such as a facility exploding, is minimized. Solutions may vary depending on the type of space system in question (single or constellation, for instance), but they generally rely on operators having identified key issues in the CSI functioning which may accelerate service degradation.

Key for CI resilience is also the management of margins, where operators do not just evaluate risks, but also acknowledge them when they occur, as they erode the system margins which allow a still acceptable level of functioning above the critical threshold for rapid system degradation. Systems are tested for their capacity to remain within safe operational margins, and operators use proactive indicators to measure the state of the system's margins. Such a principle may appear less applicable to space systems but one may find examples, such as the management of fuel for station keeping in the backdrop of the need to maneuver to avoid impact with debris, but also to maximize system lifespan and minimize debris creation upon mission end.

For CSIP efforts to work as a collective endeavor, they require common mental modes among the various categories of active stakeholders and *system governors*. The OOs must be able to communicate not only among themselves but also with the OOs of CI dependent on CSI and with the sectoral authorities and overall authorities, as well as the growing layer of global stakeholders, such as international institutions. This is an effective way to prevent accidents, mitigate their effects, and assimilate lessons from various disruptions.

Resilient systems are also flexible and redundant. The former is less applicable to space systems, which are generally path dependent on the specific architecture of the system in question, but flexible systems are also open to incremental improvements and improvisations, which are possible in the realm of cybersecurity, among others. The latter principle of resilience, redundancy, is also problematic with regard to the space component of a CSI, though it may be applicable elsewhere. Having reserves and multiple systems running in parallel are workable ideas, but one should remember that redundancy is also another source of complexity, which is a source of risks such as common cause failures. Diversity of systems, as a subset of flexibility, is also an option.

The issue of complexity is a permanent concern for CIP efforts, as its rise obscures interactions between systems that may result in new risks, vulnerabilities, and threats or paths for cascading disruptions. The reduction of complexity is, therefore, a common security concern, though it often develops into the management of the growth rate of complexity, since the prioritization of economic growth, efficiency, and development is a leading cause of complexity buildup. Space systems are also subject to this "iron law," with systemic trends indicating that complexity will increase in the CSI operating environment through the rise in new systems, the rise in orbital crowdedness, the growing number of interactions between systems and system components, and, last, but not the least, the growing complexity of the individual assets themselves, which are on their way to integrating AI, blockchain, and other developments. An often-overlooked source of complexity when it comes to technological assets is also the organization behind its operation. Generally, a system becomes less complex as it reduces the possibility of feedback loops, as it segregates functions and creates direct lines of information with unique pathways for each.

Under these conditions, the most that CSIP practitioners may hope for is the reduction in system couplings, or the rate of transmission of disruption from one system or system component to another. For instance, fossil fuel-based power generation facilities may reduce their coupling with the mining asset or the transport infrastructure by having reserves on hand, granting autonomy during the initial phases of a crisis. With regard to CSI, we may consider reducing the rate of propagation of cyber threats, but other examples are possible. Overall, systems that reduce couplings are flexible in their operating manner and in the resources they use, and they have delayed or non-sequential functioning compared to their upstream CI influencers.

Johnson and Gheorghe (2013) added two more principles for resilience. The first is the reduction in system fragility, which is an endogenous factor in the system, and the opposite of vulnerability, which is an exogenous factor affecting the system. One may find many instances in the functioning of a CSI which may be assimilated to a state of fragility, but an often-overlooked factor is the organization of the OOs and, for instance, their financial vulnerabilities, cost structures, openness to subversion from abroad, etc.

The second, drawing from the financial sector, is the concept of anti-fragility, which is the quality of a system to be strengthened by the repeated application of small stressors. The classic example is of a forest experiencing regular small fires and then ceasing to do so, following human intervention. The accumulation of plant matter makes the inevitable future fire much more dangerous and stronger when compared to the strength of the smaller, regular fires. We may argue that this quality is also present in space systems, whose challenging security environment has led prospective developers toward increased robustness, within the financial and mass constraints of the launch systems. Baker et al. (2008) have noted that the various coronal mass ejections and other space weather phenomena that have been analyzed do not compare to the potential of the largest solar storm ever recorded (the Carrington Event – 1859), but they are sufficient to produce damage both in orbital and ground-based assets. However, no episode has, so far, been an existential threat, and this has also spurred research into hardening systems and into the development of early warning systems and mitigation measures. The system, overall, becomes stronger than it would have been under ideal conditions, when a Carrington Event-level solar storm would have much graver consequences (Hapgood and Thomson 2010).

Complex System Governance

As mentioned in the description of the CIP framework, CSG is an emerging field at the intersection of several disciplines that has the potential to resolve some of the issues inherent in the manner in which the governance of space systems has developed over time –piecemeal and self-organized, based on gradual accumulation. This organic development is sometimes satisfactory, but it often leaves important gaps in the space governance framework which CSG is uniquely positioned to address. According to Keating et al. (2014), CSG is the design, execution, and evolution of the metasystem functions necessary to provide control, communication, coordination, and integration of a complex system. Metasystems "are sets of related functions which only specify 'what' must be achieved for continuing system viability (existence), not specifying 'how' those functions are to be achieved" (Keating and Katina 2016). These include system identity, system context, strategic system monitoring, system development, learning and transformation, environmental scanning, system operations, operational performance, information, and communications. It is beyond the scope of this chapter to engage in detailed explanations of the underpinning of CSG, but it suffices to say that CSG focuses on system viability and purposeful design in that direction, through a loop of system analysis (*initialization*), readiness levels assessment, and governance development (Keating and Bradley 2015).

In the absence of a purposeful design, Georgescu et al. (2019, p. 323) diagnose the existence of *system drift*, a state in which the system accrues unintended consequences. These consequences are also the result of the emergent behaviors and phenomena of the system which could not necessarily have been anticipated from the analysis of its individual components. The deviations from healthy system conditions are termed "pathologies" (Keating and Katina (2016) mentioned 53 identified pathologies), and they result from a violation of one or more of the metasystem functions. They degrade system functioning to the point where viability becomes in doubt. Identifying these pathologies and resolving them is necessary for system health.

From a CSG perspective, the governance of CSIP has four key issues. The first is the increased complexity in design, execution, and development. The second and third are the importance of including a wide range of considerations from multiple fields in the development process while maintaining a design which offers direction, oversight, and accountability. Lastly, the stakeholders involved should have different worldviews and must participate voluntarily.

Many of the challenges in Fig. 2 will be present in the complex system represented by the space infrastructures and the interactions with their environment or the wider system-of-systems in which space infrastructures are embedded.

From the perspective of CSG, its application to CSI starts with the clarification and structuring of the problem matter around critical systemic issues from across the spectrum of relevant problems, from political to economic and strategic. It continues by mapping the CSI governance metasystem, with its contexts and the various interrelationships, which will allow for the discovery of profound systemic problems

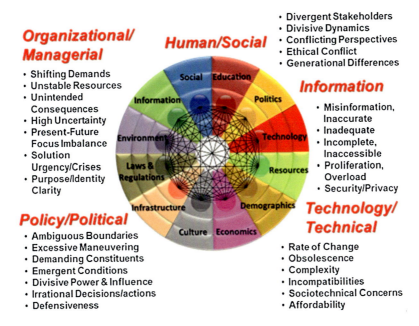

Fig. 2 The complex system problem domain and its five challenge fields for practitioners (Georgescu et al. 2019, p. 322)

once the practitioner has applied different CSG-specific methods and tools. These steps are done through extensive modeling and simulation, which relies on the systematic disassembly of the issues in the initial stage. Since "the map is not the territory," we must be careful with the biases and mistakes of our base definitions and functions. Following this, governance options can be formulated, designed, modeled, and tested before execution in the real world where they will hopefully increase system viability.

For instance, one important issue that affects the viability of the CSI complex system is the issue of space debris. The inadequate development of space governance in this field has led to a "tragedy of the commons" type situation, where a critical asset and resource (orbital bands) is steadily deteriorating. The risks of collision in these areas become ever higher (Salter 2015) and inflict steadily higher damages on the collective of users and possibly triggering also a cascading collision event once past a certain critical threshold (the so-called Kessler syndrome). This is a governance failure because system viability is imperiled and the makeup of the governance structure does not incentivize self-restraint in the creation of debris; does not punish the act of polluting, even deliberate pollution as part of anti-satellite weapon tests; and does not foster the financial preconditions for designing and deploying debris cleanup measures.

Following the CSG process outlined above, we might find something similar to the system briefly outlined in the previous section, where financial incentives are created for sustainable behavior, or something else entirely.

Conclusions

Over the course of this chapter, we have argued in favor of space systems as a new CI category and the potentially significant security benefits stemming from the application of the CIP framework in order to increase the resilience of CSI and of the societies which are critically dependent on them.

The significant advantage of the CIP framework is that it is already a debated and developed field, with significant impact on security policy and the legislative/administrative frameworks for security in the United States, the European Union Member States (as well as the EU itself), and other countries. Extending it to the space environment has the potential to improve security outcomes and to address some of the gaps stemming from the organic development of the space security governance framework under the unique conditions of the space security environment. Critical Infrastructure Protection provides a comprehensive framework for the management of key infrastructures, assets, and resources on which individual countries or the global community depends. Since space systems are already an acknowledged component of the existing critical infrastructure domains, the CSI concept is a natural outgrowth.

To sum up, the CIP framework provides tools and concepts with which to analyze space security and describe the relationships formed with terrestrial infrastructures. It is also a gateway to an extant governance framework which is in use at American and European levels, both for individual nations and collectively. It also provides a coherent vision for a holistic understanding of security, in which space security is not cordoned off into its own field, but is integrated in the wider security domain as befits the reality of the complexity of the critical infrastructure system-of-systems. The CIP framework has been working toward alleviating the impact of trends in terrestrial CI which are also present among space systems, most notably the growing rate of ownership by private entities of prospective CSI, but also the potential of counterspace operations within the hybrid warfare becoming the new normal among rival states (Robinson et al. 2019). In addition, the recent developments in the CIP field, such as complex system governance, are also applicable to CSI.

For these reasons, the concept of CSI and all that derives from it provides a useful perspective and roadmap for improving space security or at least mitigating the security impact of current space sector dynamics.

References

Air Force Space Command (2019) The Future of Space 2060 and Implications for U.S. Strategy: Report on the Space Futures Workshop, 5th, Sept 2019. http://www.spaceref.com/news/viewsr.html?pid=52822

Baker D, Balstad R, Bodeau JM, Cameron E, Fennell JF, Fisher GM, Forbes K, Kintner P, Leffler L, Lewis W, Reagan J, Small A, Stansell T, Strachan LS (2008) Space Weather Events – Understanding Societal and Economic Impacts: A Workshop Report, Space Studies Board,

National Research Council, ISBN 10: 978–0–309-13811-6, available online at http://lasp.colorado.edu/home/wp-content/uploads/2011/07/lowres-Severe-Space-Weather-FINAL.pdf

Bryce Space and Technology (2019) Smallsats by the numbers 2019, [Online] Available at https://brycetech.com/downloads/Bryce_Smallsats_2019.pdf

Bucoveţchi O, Georgescu A, Badea D, Stanciu RD (2019) Agent-based Modeling (ABM): support for emphasizing the air transport infrastructure dependence of Space systems. Sustainability 11 (19):5331

Cellucci T (2018) Perspective: Innovative Public-Private Partnerships Help Secure Critical Infrastructure, published by the National Security Today, 27 Nov 2018. https://www.hstoday.us/subject-matter-areas/infrastructure-security/perspective-innovative-public-private-partnerships-accelerate-technology-and-secure-critical-infrastructure/

Department of Defense, Office of the Director of National Intelligence (2011) National Security Space Strategy Unclassified Summary. Washington DC, https://www.dni.gov/index.php/newsroom/reports-publications/reports-publications-2011/item/620-national-security-space-strategy

European Commission (2008) Council Directive 2008/114/EC of 8 December 2008 on the identification and designation of European critical infrastructures and the assessment of the need to improve their protection. https://eur-lex.europa.eu/LexUriServ/LexUriServ.do?uri=OJ:L:2008:345:0075:0082:EN:PDF

Eusgeld I, Nan C, Dietz S (2011) "System-of-systems" approach for interdependent critical infrastructures. Reliab Eng Syst Saf 96(6):679–686. https://doi.org/10.1016/j.ress.2010.12.010

Georgescu A, Gheorghe A, Piso M-I, Katina PF (2019) "Critical Space infrastructures: risk, resilience and complexity", topics in safety, risk, reliability and quality, series 36, eBook ISBN 978-3-030-12604-9. https://doi.org/10.1007/978-3-030-12604-9. Springer International Publishing

Gheorghe A, Schläpfer M (2006) Critical infrastructures: ubiquity of digitalization and risks of interdependent critical infrastructures, *Systems Man and Cybernetics 2006*. SMC '06. IEEE international conference, vol 1, pp 580–584

Gheorghe AV, Vamanu DV, Katina PF, Pulfer R (2018) Critical infrastructures, key resources, key assets. Risk, vulnerability, resilience, fragility, and perception governance, topics in safety, risk, reliability and quality, series 34, eBook ISBN 978-3-319-69224-1. https://doi.org/10.1007/978-3-319-69224-1. Springer International Publishing

Hapgood M, Thomson A (2010) Space weather. Its impact on earth and implications for business, Lloyd's 360 risk insight briefing, available online at https://www.lloyds.com/~/media/lloyds/reports/360/360-space-weather/7311_lloyds_360_space-weather_03.pdf

Heino O, Takala A, Jukarainen P, Kalalahti J, Kekki T, Verho P (2019) Critical infrastructures: the operational environment in cases of severe disruption. Sustainability 11(3):838. https://doi.org/10.3390/su11030838

Helbing D (2013) Globally networked risks and how to respond. Nature 497(7447):51–59. https://doi.org/10.1038/nature12047

Johnsen S (2010) Resilience in risk analysis and risk assessment. In: Moore T, Shenoi S (eds) Critical infrastructure protection IV – fourth annual IFIP WG 11.10 international conference on critical infrastructure protection. IFIP Advances in Information and Communication Technology series (311), p. 215–227. Washington DC, SUA: Springer, ISBN 978-3-642-16806-2

Johnson J, Gheorghe G (2013) Antifragility Analysis and measurement framework for systems of systems. Int J Disaster Risk Sci 4(4):159–168

Jonkeren O, Ward D, Dorneanu B, Giannopoulos G (2012) Economic impact assessment of critical infrastructure failure in the EU: a combined systems engineering – inoperability input-output model, Joint Research Centre, 20th international input-output conference, Bratislava. https://www.semanticscholar.org/paper/Economic-impact-assessment-of-Critical-failure-in-A-Jonkeren-Ward/8682d0ef992243115196a66651f3d86483e37b3d

Keating CB, Bradley JM (2015) Complex system governance reference model. Int J Syst Syst Eng 6 (1/2):33. https://doi.org/10.1504/ijsse.2015.068811

Keating CB, Katina PF (2016) Complex system governance development: a first generation methodology. Int J Syst Syst Eng 7(1/2/3):43–74. https://doi.org/10.1504/ijsse.2016.076127

Keating CB, Katina PF, Bradley JM (2014) Complex system governance: concept, challenges, and emerging research. Int J Syst Syst Eng 5(3):263–288

Linkov I, Bridges T, Creutzig F, Decker J, Fox-lent C et al (2014) Changing the resilience paradigm. Nat Clim Chang 4:407–409

Niederstrasser C (2018) Small launch vehicles – a 2018 state of the industry survey, SSC18-IX-01, 32nd Annual AIAA/USU conference on small satellites, Northrop Grumman Corporation, publisher: Utah State University Research Foundation (USURF). https://digitalcommons.usu.edu/cgi/viewcontent.cgi?article=4118&context=smallsat

OECD (2016) Space and innovation. OECD Publishing, Paris. https://doi.org/10.1787/9789264264014-en

OECD (2019) The Space economy in figures: how Space contributes to the global economy. OECD Publishing, Paris, available at https://doi.org/10.1787/c5996201-en

Perrow C (1999) Normal accidents: living with high-risk technologies. Princeton University Press. isbn:9781400828494

Pescaroli G, Alexander D (2016) Critical infrastructure, panarchies and the vulnerability paths of cascading disasters. Nat Hazards 82(1):175–192. https://doi.org/10.1007/s11069-016-2186-3

Robinson J, Robinson R, Davenport A, Kupkova T, Martinek P, Emmerling S, Marzorati A (2019) State Actor Strategies in Attracting Space Sector Partnerships: Chinese and Russian Economic and Financial Footprints, Prague Security Studies Institute, Prague, available online at: http://www.pssi.cz/download/docs/686_executive-summary.pdf

Rockefeller Foundation, Arup Development Group (2014) The City Resilience Index. https://www.arup.com/perspectives/themes/cities/city-resilience-index

Royal Academy of Engineering (2013) Extreme space weather: impacts on engineered systems and infrastructure, ISBN 1-903496-95-0, designated (RAENG, 2013), available at http://www.raeng.org.uk/publications/reports/space-weather-full-report

Salter AW (2015) Space debris – a law and economics analysis of the orbital commons, Mercatus Center, George Mason University, USA, available at https://www.mercatus.org/system/files/Salter-Space-Debris.pdf

Setola R, Luiijf E, Teocharidou M (2017) Critical infrastructures, protection and resilience. In: Setola R, Rosato V, Kyriakides E, Rome E (eds) Managing the complexity of critical infrastructures: a modelling and simulation approach. Studies in systems, decision and control, vol 90. Springer open, ISBN: 978-3-319-51042-2/978-3-319-51043-9, https://doi.org/10.1007/978-3-319-51043-9

The White House (1998) Presidential Decision Directive/NSC-63 (as PDD-63), Washington DC. https://clinton.presidentiallibraries.us/items/show/12762

Union of Concerned Scientists (2019) UCS Satellite Database, accessed 12 Oct 2019. https://www.ucsusa.org/resources/satellite-database

Vugrin E, Wahren D, Ehlen M (2011) A resilience assessment framework for infrastructure and economic systems: quantitative and qualitative resilience analysis of petrochemical supply chains to a hurricane. Process Saf Prog 30(3):280–290. https://doi.org/10.1002/prs.10437

Space and Cyber Threats

Stefano Zatti

Contents

Introduction: The European Space Agency and Its Missions	246
A Security-Flavored Space	246
Hacking in Space: Astro-Hackers?	247
Motivations of Attackers	250
Threats and Countermeasures	251
End-to-End Cybersecurity	254
Countermeasures Related to the Information Assurance Properties	255
Tele-Commands	255
Telemetry	255
Payload Data	256
ESA's Own Approach to Mission Security	257
Mission Categories and Security Profiles	259
Conclusions: New Space, New Cyber Threats!	262
References	262

Abstract

Space-based systems play an important role in our daily life and business. The evolution of our well-being is likely to rely on the use of space-based systems in a growing number of services or applications that can be either safety-of-life critical or business and mission-critical. The security measures implemented in space-based systems may turn out to be insufficient to guarantee the information assurance properties, namely, confidentiality (if required by the data policy), availability and integrity of these services/applications, as well as authenticity and non-

---

Stefano Zatti has retired.

S. Zatti (✉)
European Space Agency (ESA) - Security Office, Frascati, Italy
e-mail: Stefano.Zatti@gmail.com

© Springer Nature Switzerland AG 2020
K.-U. Schrogl (ed.), *Handbook of Space Security*,
https://doi.org/10.1007/978-3-030-23210-8_92

repudiation. The various types of possible cyberattacks on space segments, ground stations, and control segments are getting increasingly visible and have been indeed frequent. What to do in order to counter such occurrences is less obvious and needs to be addressed with priority and a whole new family of countermeasures. This paper will first introduce ESA and its constituency, and then it will address the security-specific aspects of its space missions. Threats specific to the different types of missions from the cyberspace will be presented, and possible countermeasures will be analyzed. The motivations that may induce some offenders to cause damage to space missions will be then examined. A categorization of the different types of space missions will then lead to the proposal of creating different protection profiles, to be respectively implemented at increasing degrees of sophistication for the different mission categories.

Introduction: The European Space Agency and Its Missions

The European Space Agency, ESA, was founded in 1975 by merging two existing launch and space research organizations, with the aim expressed in Article 2 of the ESA Convention: "To provide for and promote, for exclusively peaceful purposes, cooperation among European states in space research and technology and their space applications." Composed of 22 member states, with eight major sites/facilities in Europe and a large number of ground stations around the earth, providing attractive jobs to about 2200 staff, ESA has in the course of its lifetime designed, tested, and operated in flight over 80 satellites.

A Security-Flavored Space

Although they are conceived, designed, and operated for peaceful purposes, the space missions of ESA can indeed present security aspects and address security elements. The different critical elements of space missions can influence the level of sensitivity and consequently the level of threats that each one of those space missions has to face. In particular, the following aspects of "security from space" have been highlighted by the ESA Council as critical for the benefit of European citizens, leading to the development of specific missions to address those.

Security on Earth:

- **Disaster management**: the use of space systems to support natural disaster management, such as climate change or earthquakes
- **Critical infrastructures protection**: space-based applications and technologies used for the safety and security of critical infrastructures
- **Transportation and logistics**: space-based solutions to support transportation on land, air, and water

- **Energy:** space-based technical contributions to the management of energy production and supply
- **Agriculture and water:** space-based applications in support of the 2030-agenda formalized as the Sustainable Development Goals, in particular no. 6 (clean water), 2 (zero hunger), and 15 (life on land)
- **Surveillance (air, sea, land, space):** space surveillance in support of national and international programs on air, sea, land, and space
- **Humanitarian crisis support** and emergency rescue tasks
- **Migration and border control:** space-based applications in support of grand challenges such as migration and border control
- **Public safety (including civil protection)**

Security in Space:

- **Space situational awareness:** real-time information of the status (position, direction, speed) of specific objects in space
- **Near-earth objects:** asteroids, meteorites, in the vicinity of our planet
- **Space weather:** phenomena outside the atmosphere that can affect terrestrial infrastructures, like solar winds and electric power grids
- **Satellite tracking:** knowledge of position and trajectory and speed of man-made objects through active and passive observation and tracking

It is clear that when dealing with such subjects that the space missions can create interest in malevolent individuals or organizations to have an impact on them in different ways (see section later). Having such sensitive scope, the space missions require therefore that the responsible authorities establish a set of measures to protect them from adverse effects, encompassing both the assets composing them (space and ground) and the data they provide.

Hacking in Space: Astro-Hackers?

We introduce first the definition of a few basic concepts that are necessary in the broad context of addressing threats and analyzing risks. ISO 27005 (Information Technology—Security Techniques—Information Security Risk Management 2018) defines "threat" as "A potential cause of an incident that may result in harm of systems and organization," whereas the Consultative Committee on Space Data Systems (CCSDS) 350.8-G-1 (Information Security Glossary of Terms 2012)) defines "risk" as "Possibility that a particular threat will adversely impact a system by exploiting a particular vulnerability."

A threat can be generated by human or nonhuman sources and can be intentional or unintentional. All threat agents attempt to do harm against a physical or virtual resource or asset. In case that the resource has one or more vulnerabilities, it can potentially be exploited by a threat agent, resulting in a compromise of the properties of information assurance (IA) of the system, namely, confidentiality, integrity, or availability (traditionally, C-I-A), plus authenticity or non-repudiation. In particular, loss of confidentiality will result in unauthorized disclosure of information; loss of integrity will result in unauthorized modification or destruction of information; loss of availability will result in a loss of access to critical resources; lack of authenticity will result is doubts of the sources and the genuine nature of the available information. Lack of non-repudiation will result in uncertainty of transmission and/or reception of particular pieces of information. Overall, the loss of information assurance might result in harm to an agency's operations, assets, or individuals.

Computer systems are typical targets of threats given the attractive nature of the information they contain and make offer to the majority of legitimate users, while they often suffer from a number of vulnerabilities. The National Institute of Standards and Technology's (NIST) Special Publication 800-12 (An Introduction to Computer Security—The NIST Handbook 1995) states that:

> Computer systems are vulnerable to many threats that can inflict various types of damage resulting in significant losses. This damage can range from errors harming database integrity to fires destroying entire computer centers. Losses can stem, for example, from the actions of supposedly trusted employees defrauding a system, from outside hackers, or from careless data entry clerks.

The pervasive interconnectivity of networks and the ever-increasing dependency of industrial, commercial, and scientific activities, including space missions, on information technology and communications, create a whole new dimension to the threats that can affect a space mission. In the past, in order to reach a satellite in orbit to threaten its functions, in fact, it would have been necessary for the adversary to build or possess an infrastructure to generate and send tele-commands to the spacecraft, an expensive and massively complex endeavor. In the most brutal case, the attacker would have to gain physical access to the ground infrastructures to raise havoc to the people and infrastructures located therein, thus gaining the capability to affect the spacecraft. Nowadays, via the pervasive nature of the access networks all interconnected through the Internet, it is sufficient for a hacker to tamper with and bypass the existing protection measures simply while sitting in their own offices or homes. This is not just science fiction: there have been several publicly documented events including the cases described below:

- In 1998, the German-US ROSAT space telescope inexplicably turned toward the Sun, irreversibly damaging a critical optical sensor, following a cyber-intrusion at the Goddard Space Flight Center of NASA in the USA (HTTPS://EN.WIKIPEDIA.ORG/WIKI/ROSAT).
- On October 20, 2007, Landsat 7 experienced 12 or more minutes of interference. Again, on July 23, 2008, it experienced other 12 min of interference. The responsible party did not achieve all steps required to command the satellite, but the service was disturbed.(Satellite Services and Interference-the Current Situation).
- In 2008, NASA EOS AM–1 satellite experienced two events of disrupted control: in both cases, the attacker achieved all steps required to command the satellite but did not issue commands.

These cases made the news, as shown in Fig. 1. (HTTPS://WWW.DAILYMAIL.CO.UK/NEWS/ARTICLE-2055311/HACKERS-INFILTRATE-US-SATELLITES-TAKEN-COMPLETE-CONTROL-ACHIEVING-STEPS-REQUIRED-COMMAND-SATELLITE.HTML).

ESA itself has not been immune to attacks, of course. As recently as December 2015, an attack by Anonymous has compromised a large number of accounts of external users who were accessing the targeted service to gather ESA mission data that was already openly available to the public but required preregistration (Fig. 2). The fact that the service was entrusted to a third party did not actually mitigate the damage, as proper protection measures should have been imposed on the provider by the ESA customer, according to the prevalent rules. Although the data compromised was not highly relevant, due to the open data policy, the reputation of the organization was damaged by the occurrence. Moreover, all affected identities had to be urgently contacted and informed, to avoid that the compromised identities could be used by the attackers to access other sites, increasing the impact and potential damage. "A group of hackers operating under the Anonymous banner hacked the European Space Agency (ESA) and leaked the data for no reason other than for "lulz." Over 8000 people will not find anything amusing about the breach since their names, email addresses and passwords were posted in one of three data dumps on JustPaste.it" (Anonymous Hacks The European Space Agency; Attackers Hack European Space Agency, Leak Thousands Of Credentials 'For The Lulz).

Some more cases are presented below, organized by the categories of the missions that were affected (Fig. 3).

Fig. 1 Press report on US satellite hacking

Fig. 2 Anonymous hacks ESA data servers

Motivations of Attackers

What are motivations for potential attackers to expend effort and undertake risks to damage space systems? The financial gain in the cases we have presented is nonexistent. Other possibilities for the motivations, as well as what could be the characteristics of such perpetrators in a broader spectrum of cases, are addressed in the following.

One possible motivation can be the search of technological information by commercial or institutional competitors, possibly by means of third parties: the knowledge gained by hacking equipment or data could be used to bridge technological gaps and gain competitive advantages in the space arena.

Cybercriminals could insert themselves in this race by gathering information and technical details that they could sell to interested parties for some sort of financial advantage. This would require considerable technical skills including ability to do reverse engineering, to make sense out of the gathered information.

Employees of the organization could be the sources of additional threats, seeking some sort of revenge for perceived mistreatment or simply by unwittingly creating havoc with their negligent behavior. Insider threats are indeed often referred to in the literature as major sources of hacking problems.

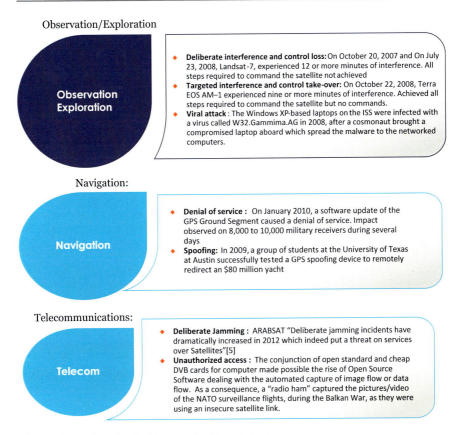

Fig. 3. Cases of satellite hacking organized by mission types

Given the very high resonance of space-related news, some hacktivists could turn out to be politically or socially motivated to hamper space mission, thus gaining a high visibility, like the case of the Anonymous attack to ESA mentioned before. Even worse, some terrorists, driven by motivations of political or religious nature, may want to penetrate space-based infrastructures and impact the services they provide to the broad user community, to cause cascaded damage to any other critical infrastructures of different nature, with potentially disastrous impact (e.g., health, energy, water, transportation, telecommunications).

Threats and Countermeasures

The infrastructures supporting all space missions can be in general characterized by a *ground segment, a space segment,* and *a control segment.* The control segment is used by the mission controllers, who issue tele-commands (TC) that via the ground segment can be uploaded all the way up to the space segment, a set of spacecraft that circles in space at possibly different altitudes, depending on the mission type. In the other direction, the spacecraft send back to Earth messages containing housekeeping telemetry (HKTM) to indicate the status of the various instruments and on-board parameters, and the payload, that is the *raison d'etre* of the mission and can be constituted by very large amounts of data (like in the case of Earth observation missions).

The threats that can affect a spacecraft in orbit can be characterized by the different ways that adversaries can use to tamper with the tele-commands that are normally sent from the control center to the spacecraft to perform specific mission-related actions, as well as with the data that returns to Earth, be it either payload related to the mission or housekeeping telemetry that informs ground control of the status of the instruments on board (Security Threats against Space Missions, 2015).

In a recent talk at the RSA Conference on cybersecurity, for example, Bill Malik, VP of Infrastructure Strategies at Trend Micro, calls the range of vulnerabilities exposed on satellites "astonishing" (HTTPS://WWW.EXTREMETECH.COM/EXTREME/287284-HACKING-SATELLITES-IS-PROBABLY-EASIER-THAN-YOU-THINK). Malik stresses the need for satellite design to incorporate security at the most basic levels, by implementing an appropriate set of countermeasures, as we will see in the following.

When addressing the need to protect space missions and develop the appropriate set of countermeasures, a systematic approach based on risk analysis methodology must be followed, as described by the author in a paper that summarizes the outcome of two studies commissioned by ESA in 2014 to address the specific situation of its current missions (Del Monte and Zatti 2015). The risk assessment methodology described in the paper is structured in four phases:

- *The first phase: Cyber threat analysis* to define the context of the analysis, by defining and modeling the space missions and considering all the possible threats they can be subject to, with particular attention to the new generation of cyber threats, as described in the following
- *The second phase: Identification and assessment of vulnerabilities* to identify the potential existing vulnerabilities of the assets of the space mission classes and to define elementary threat scenario
- *The third phase: Identification and assessment of risks* to assess identified vulnerabilities, to evaluate the related risks, and to build attack trees based on the space mission architectures
- *The fourth and final phase: Definition of the necessary measures to counteract the threats* and address the risk as defined in the previous phases, resulting in a set of recommendations and a mitigation plan

There are events that, when happening, could have a significant impact on the mission, performance, economics, human safety, data loss or compromise, or even the total loss of the mission. These events can be classified in a hierarchy based on what are the most feared events among them, on which a mission planner has to put extra effort by devising appropriate countermeasures in order to decrease the chances of suffering the resulting damage.

14 Space and Cyber Threats

Each category of missions has its own most feared events; we will consider in this paper only those most frequent or likely and potentially harmful, among all the feared ones.

- *Unauthorized operation of satellite/launcher/spacecraft/ground facilities*

This is one of the most feared threats among the flight control team members and possibly one of the easiest ones to achieve.

Satellites are usually designed with an on-board and autonomous failure detection and recovery system, which upon detection of any failure on-board they will configure themselves into a safe position, awaiting for ground to investigate. So, if someone managed to send commands to cause any harm, the satellite will go into a safe mode by itself before the damage was irreversible. In this sense, the damage could be the loss of science, service, and money. However, there are circumstances where the mission could be lost. Those cases could occur:

- If someone knows the spacecraft design and its operations very well, it could generate a sequence of commands that would really cause the end of the mission, like de-orbiting the satellite or causing a failure of the on-board communications system. The places where someone could introduce commands are via the mission planning system, the flight dynamics system, the mission control system, or directly at the telemetry and command system located at the ground stations.
- If someone manages to command the satellite for a longer period, even after the satellite has gone to safe mode, it could also create an on-board failure to end of the mission.
- The receivers that live in spacecraft which are flying in orbits below geostationary (GEO) are sensitive to the received radiofrequency signal power. If the ground station uplink amplifier was manipulated to uplink with a higher power than the allowed one, and then the receiver could be broken, meaning the end of the mission. Usually, the spacecraft have two receivers, but as they are working in hot redundancy, if this event occurred, it could end with both receivers at once. Manipulation of the uplink amplifiers can be done through the hardware itself, via the ground station jobs generated at the control center and sent to the station via ftp, via the ground station jobs once stored at the station computer of the ground station.

The launch is a very critical phase of the mission. If someone managed to insert a wrong command into the launch-automated sequence to inject a failure on any of the critical activities, like the separation sequence, or the burns needed, then it could cause the loss of the launcher and of the spacecraft carried inside.

- *Unavailability of the communications for tele-commands and telemetry during critical maneuvers*

There are different activities that are considered critical, those ones that have an impact on the success of the overall mission. Typically, these activities consist on different type of maneuvers: launch and early orbit phase (LEOP) maneuvers, planetary swing/fly-by, trajectory correction maneuvers, in-orbit insertion maneuvers, and collision avoidance maneuvers, among others. During the execution of any of these maneuvers, it is very important to maintain contact with the spacecraft or the launcher in order to get real- or near real-time ranging and Doppler shift measurements.

- *Unavailability of critical services for medium or large periods of time*

The HVAC systems (heating, ventilation, and air conditioning system) provide the air quality to the human spaceflight vehicles. The maintenance of the atmosphere aboard the spacecraft is critical, not only for its habitability but also for its function. If this system failed, the spacecraft cabin will get contaminated with so many chemical contaminants that it will be impossible to live there.

- *Equipment destruction or theft*

The most feared and disruptive of all events is the loss of the whole equipment, by destruction or theft, with the prospect of a total prejudice to the mission.

Analyzing more in depth the ways such events can be determined, it must be realized that the syntax of the tele-commands is known and common to many missions. An intruder can thus prepare her own attack by counterfeiting the tele-commands in proper syntax, but she has then to upload them to the spacecraft for execution. In the past, such action was not simple to perform, requiring access to space via complex and expensive ground infrastructures, not easily available and not easily accessible. However, nowadays, given the very widespread and ever-increasing connectivity of operational networks with public networks, the intruder can gain access to the ground infrastructures by hacking their access control at the network boundaries, and they find a way to issue commands to the spacecraft, affecting its functionality.

Likewise, the downward stream can be affected by capturing the telemetry (that can provide important indications on the state of the spacecraft and its instruments, as well as the payload, affecting the confidentiality and the availability of the data to the intruder and further on to other parties).

End-to-End Cybersecurity

In order to ensure the proper protection of all the assets related to space missions, including the segments described above, material and human, as well the mission data, it is necessary to tackle the various aspects of security as a process that spans all components and all phases, in an end-to-end fashion.

This implies the consideration of the security pillars and the respective countermeasures, as follows.

- Physical: zoning, access control for data centers, perimeter and internal fencing
- Personnel: vetting, clearances, building trust in employees, peer control
- Information protection: classified vs unclassified data and parameters
- Information assurance (IA) properties and respective ways to ensure them:
Confidentiality – encryption
Integrity – media access control (MAC)
Availability – redundancy
Authenticity – identity management, cross check, access control, signature of data
Non-repudiation – notarization, certificates

Countermeasures Related to the Information Assurance Properties

The following analysis will provide insight of the specific countermeasures that must be enacted to counteract the respective threats, to be applied to tele-commands (ground to space), telemetry (space to ground), and payload data (space to ground too). The quoted algorithms are introduced and well explained in reference (CCSDS Cryptographic Algorithms 2014).

Tele-Commands

TC Availability: a combination of spread spectrum, firewall and autonomy techniques, high-power uplink margins, and diversity of the sites of the telemetry and tele-command stations seem appropriate to reduce the risk down to an acceptable level.

TC Integrity: integrity of individual commands can be achieved by appending an integrity check value (ICV) in a manner similar to the way a digital signature is appended. To ensure that it is not modified or corrupted, the ICV is often keyed (e.g., a keyed hash), or the ICV value can be encrypted using a symmetric encryption algorithm (e.g., AES).

TC Authentication: use encryption of the messages at the segment level, with a block cipher-based message authentication code (MAC).

TC Confidentiality: at packet level, use Advanced Encryption Standard (AES) algorithms in cipher feedback mode (CFB), output feedback mode (OFB), or in counter mode of operations (CTR), the latter recommended by CCSDS.

TC Sequencing (Anti-replay): insert in the command an integrity-checked progressive sequencing number based on a counter.

Telemetry

To ensure *housekeeping telemetry confidentiality* at virtual channel level, use Advanced Encryption Standard (AES) algorithms in cipher feedback mode (CFB), output feedback mode (OFB), or in counter mode of operations (CTR), the latter recommended by CCSDS.

Housekeeping telemetry Integrity can be obtained by appending an ICV in a similar manner to that used for tele-commands (see above).

Housekeeping telemetry (anti-replay): Sequencing based on the introduction of a sequence number generated by a counter on board.

Payload Data

At virtual channel level, use Advanced Encryption Standard (AES) algorithms in cipher feedback mode (CFB), output feedback mode (OFB), or in counter mode of operations (CTR), the latter recommended by CCSDS.

As far as the key management is concerned, it is necessary to be able to change the keys after a certain amount of data had been sent, to avoid the possibility of the intruder to accumulate large amounts of encrypted data and to exercise cryptanalysis on them. This can be done with *over the air rekeying* techniques (OTAR), based on preloaded master keys.

In any case, in order to be able to apply the abovementioned security measures on the ground-to-orbit link, it is fundamental that a set of cryptographic functionalities is installed on board, implemented in bespoke hardware that normally goes simply under the term "Crypto-chip," before the launch (Fig. 4). This function is able to perform on the tele-commands and on the telemetry all the functions necessary to implement the information assurance properties, as required by the mission designers on the basis of the risk assessment specific to that mission (see next section).

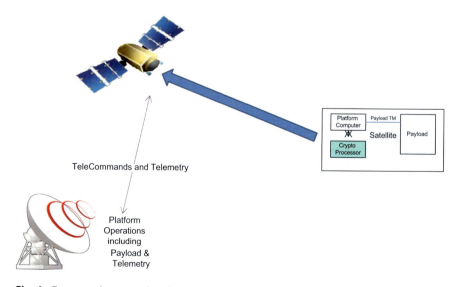

Fig. 4 Crypto-equipment on board

In addition, some missions may require that also the payload should be protected (mainly for reasons of the property of confidentiality). This implies the encryption of the whole payload, with the consequent need to renew the keys used for the encryption on a periodic basis, determined by the amount of the data to be transmitted. The same equipment can be used to that end, with the use of appropriate encryption primitives (via an application programming interface, API) to be invoked by the on-board software on each data package or on the whole data stream, to achieve the required level of protection.

ESA's Own Approach to Mission Security

In order to address specifically the security risks for space missions that concern specifically the missions designed and operated by ESA, the ESA security office held a workshop with participation from ESA mission planners from different directorates and member states' delegations. The methodology suggested by the ISO 27005 standard (Information Technology – Security Techniques – Information Security Risk Management 2018) was applied, in order to perform the assessment and develop appropriate recommendations. The ISO 27001 is the main driver of the whole exercise on ESA systems and missions aimed at analyzing the risks and establishing an information security management system (Information Technology – Security Techniques – Information Security Management Systems – Requirements 2013).

The following recommendations stood out among those presented:

- Increase a risk awareness culture throughout the agency, by identifying the key people with the directorates and programs involved
- Mitigation of risks by improving the detection and reaction process
- Mitigation of risks coming from the supply chain when building a satellite or a ground station
- Need to create synergies between different directorates, systems, infrastructures, and actors within the agency
- Need to create a permanent security risk assessment, management, and maintenance process, based on the loop of continuous improvement (PDCA, plan-do-check-act)

It has been agreed that the goal of the exercise is to develop an "ESA policy on security risks in space programmes and mitigation measures," to induce mission planners to address security at the moment they start their work of the definition of all the mission aspects and characteristics. At that point, the owner of the mission must decide on the mission-specific security measures, based on a risk assessment methodology, and of course must ensure the necessary funding.

The following elements stood out as elements to be included in the to-be-developed ESA policy:

- Provision of horizontal functions, building a cross-mission infrastructure providing a catalogue of basic common services to all missions, e.g., crypto-chips including standardized primitives to invoke the respective functions, software components, key management, etc.
- Selection of mission-specific functions, to be decided on the basis of the risk assessment by the mission planner, depending on types of missions, categories, and data policy (this will be developed in the following section)

In addition, the policy shall address, in a modular way:

- The different categories and the different phases of the missions (A-to-F), based on a critical assessment of the suggestions from two studies
- The cost aspects in the risk assessment model, to enable a better decision by management, proposing a way to assess and quantify the intangible costs for loss of image/reputation (reputation index)

Concerning the general aspects to be addressed, the policy shall:

- Focus on public-private partnership (PPP) for definition of missions, by involving industry in planning and support
- Ensure the involvement of small and medium enterprises (SMEs), as they increase the choice of providers, and they are often the expert in the specific fields, e. g., in software development

The policy will also have to include:

- Provisions to ensure the supply chain to avoid from any risks stemming from the use of untrusted vendors, to be applied to future missions
- A common mandatory base for all missions to address the minimum risks (loss of spacecraft, image, etc.), whereas all other specific decisions on risk mitigation shall remain with the mission owner
- Involvement of ESA mission partners including the European Defence Agency, the European Commission, Eumetsat, to be decided on a case-by-case basis.

As a complement to the policy document, as customary, at a lower level, an implementation document shall be developed, containing specific technical/commercial details and tools, to evolve over time alongside the development of technology.

Different security profiles would then have to be defined to address an increasingly stronger hierarchy of security measures, similarly to what was done by the working group on security of the Sentinels, the space components of the Copernicus program, which had defined a number of security profiles of increasing coverage and complexity and recommended to each mission the adoption of the most suitable one of them.

Secure ground-to-space connectivity is in particular being standardized at worldwide level by the CCSDS, in the data link layer security standard.

Mission Categories and Security Profiles

There is a difference in the threats that different categories of mission can be subject to. Different mission types have actually different security requirements, based on the need to protect one or more of the five property of information assurance, plus and with priority the survival and well-being of humans (safety of life applications and manned spaceflight). Figure 5 illustrates the hierarchy of space missions based on such priorities.

The different space missions can be categorized by different categories of risks, with increasing depth and level of concerns:

- Scientific
- Earth observation
- Navigation
- Communications
- Space situational awareness
- Manned spaceflight and exploration

In order to approach the cybersecurity of missions in a systematic way leading to a streamlined engineering, corresponding to different levels of security implementation, five different protection profiles have been developed, to be applied to the different mission categories (0 to 4). They cover tele-commands, housekeeping telemetry, as well as payload data, as detailed in the following.

Profile 0: no specific security
No tele-command authentication or encryption
No encryption of housekeeping telemetry or payload data
Standard terrestrial links security (firewalls, intrusion detection/prevention, security information and event management (SIEM), etc.)
Implemented, for example, in the historic ESA missions ERS/ENVISAT and the Earth Explorers, as a matter of fact, in most of the space missions until recently

Profile 1: static tele-command protection
Tele-command authentication and anti-replay
Authentication key(s) preloaded on board before launch
Tele-command authentication can be enabled/disabled automatically (via a timeout mechanism) or by ground via specific protected tele-commands
Currently implemented on the ESA missions MetOp and Automated Transfer Vehicle (ATV)

Profile 2: dynamic tele-command protection
Tele-command authentication and anti-replay.
Authentication keys are loaded by ground using preinstalled master keys for the encryption of the related tele-commands.
Tele-command authentication can be enabled/disabled automatically (via a timeout mechanism) or by ground via specific protected tele-commands.
Implemented in the Sentinels of the Copernicus program.

Profile 3: dynamic tele-command + payload data protection

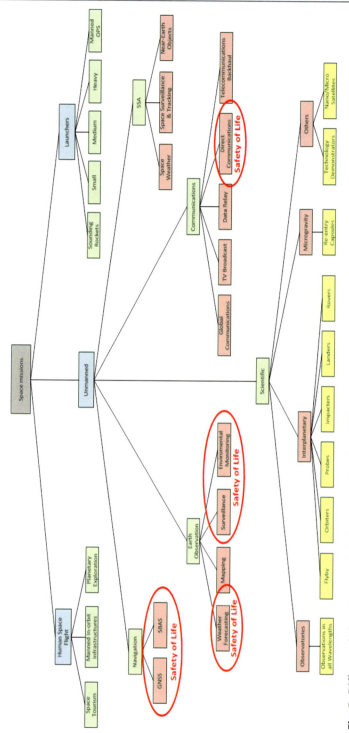

Fig. 5 Different mission categories, including those with safety of life characteristics

Payload data is also encrypted, catering to a specific data policy that requires protection and access control of the data.

Four types of keys: master keys, tele-command authentication keys, payload data encryption keys, and tele-command encryption keys.

Payload data encryption can be enabled/disabled automatically (via a timeout mechanism) or by ground via specific protected tele-commands.

Profile 4: dynamic tele-command + payload + telemetry data protection

Telemetry data is also encrypted.

Five types of keys: master key, tele-command authentication key, data encryption key, telemetry data encryption key, and tele-command encryption key.

Some of the data encryption can be enabled/disabled automatically (via a timeout mechanism) or by ground via specific protected tele-commands.

When a new mission is planned and designed, it is thus important that the specific threats, according to the specific mission category to streamline efforts, are analyzed and that the necessary security measures are put in place accordingly. Stitching up security upon an existing system would be, in fact, a very complex and expensive endeavor, not recommended from the managerial and financial points of view.

As a first example, we consider the Galileo program. As far as the protection of the Galileo infrastructure is concerned, the following measures are applied:

- Physical protection of the ground segment elements
- Strong interaction with host countries: critical infrastructure plan and arrangements for efficient intervention
- Location of ground stations only in EU member states, with full control of the territory

With respect to the protection of the Galileo signals in space, Profile no. 4 is selected, providing dynamic tele-command + housekeeping telemetry data protection:

- Payload protected only for particular, specifically defined services.
- End-to-end signal protection, from spacecraft to control center.
- Signal modulation features allow robustness against low-level interference, uplink diversity, and anti-jamming antennas.

As a second example, the Copernicus program has selected Scenario 2, with dynamic tele-command protection as baseline. The following security functions are performed on the Sentinel spacecraft:

- Tele-command authentication + anti-replay.
- Tele-command "encryption" limited to security-related tele-commands, like the uploading of new keys.
- "Encryption" affects *only* the tele-command "data field."
- No housekeeping telemetry and payload data encryption.
- Preinstalled fixed master keys: used as key encryption keys.
- Session keys: used for authentication, uploaded by ground using master keys.

- Keys are referenced by meta-information avoiding the need to encrypt housekeeping telemetry.
- Tele-command authentication can be bypassed automatically upon critical mission failure.
- Tele-command authentication bypass can be enabled/disabled by ground via authenticated tele-command or by a watchdog based on timeout.

Conclusions: New Space, New Cyber Threats!

It has emerged clearly in this discussion that the cybersecurity of space missions is a matter of competiveness for the European space industry and that, at the same time, is a vital interest of the European Union that is the owner of the Copernicus and Galileo programs.

Moreover, the need to guarantee high production rates (e.g., four satellites per day in the case of the densest constellations) requires the system integrators to stretch globally the existing supply chain while including new components' providers. This introduces the additional need to ensure that the whole chain is trusted through all its components and thus subjected to strict controls (that must dive all the way down to the silicon).

The globalization of manufacturing capabilities and the increased reliance upon commodity software and hardware for space and ground segments have expanded the opportunities for malicious modification in a manner that could compromise critical functionality. This is introducing a whole family of additional risks, challenging the frontier of technology.

References

An Introduction to Computer Security—The NIST Handbook (1995) National Institute of Standards and Technology special publication 800–12. NIST, Gaithersburg, Maryland
Anonymous Hacks The European Space Agency. https://www.inverse.com/article/9152-anonymous-hacks-european-space-agency-releases-data-online
Attackers Hack European Space Agency, Leak Thousands Of Credentials 'For The Lulz'. https://www.computerworld.com/article/3014539/attackers-hack-european-space-agency-leak-thousands-of-credentials-for-the-lulz.html
CCSDS Cryptographic Algorithms (2014) Informational report (green book), CCSDS 350.9-G-1, Issue 1. CCSDS, Washington, DC
Del Monte L, Zatti S (2015) Preliminary reflections about the establishment of a cyber-security policy for a sustainable, secure and safe space environment. In: IAC-15,D5,4,3,x27636, 66th international Astronautical congress, Jerusalem, pp 12–16
https://en.wikipedia.org/wiki/rosat
https://www.dailymail.co.uk/news/article-2055311/hackers-infiltrate-us-satellites-taken-complete-control-achieving-steps-required-command-satellite.html
https://www.extremetech.com/extreme/287284-hacking-satellites-is-probably-easier-than-you-think

Information Security Glossary of Terms (2012) Report concerning space data system standards (green book), CCSDS 350.8-G-1, Issue 1. CCSDS, Washington, DC

Information Technology—Security Techniques—Information Security Management Systems—Requirements (2013) International standard, ISO/IEC 27001:2013, 2nd edn. ISO, Geneva

Information Technology—Security Techniques—Information Security Risk Management (2018) International standard, ISO/IEC 27005:2018, 3rd edn. ISO, Geneva

Satellite Services And Interference-The Current Situation., https://www.itu.int/en/ITU-R/space/workshops/2013-interference-geneva/presentations/Yasir%20Hassan-%20Arabsat.pdf

Security Threats against Space Missions (2015) Report concerning space data system standards (green book), CCSDS 350.1-G-2, Issue 2. CCSDS, Washington DC

Space Safety

15

Joe Pelton, Tommaso Sgobba, and Maite Trujillo

Contents

Introduction	266
The Many Facets of Space Safety	268
Acceptable Safety Level	269
Safety Standards and Compliance Verification	270
Launch Safety	271
Launch Site Ground Safety Risk	271
Launch Flight Safety Risk	272
Launch Risk for Maritime and Air Transportation	274
Air-Launch Safety	274
On-Orbit Safety	276
Orbital Debris	276
Collision Risk with Orbital Debris	277
Controlling Orbital Debris Risk	279
Orbital Debris Remediation: Active Debris Removal	280
Reentry Safety Risk	282
Environmental Risk	283
Risk for Aviation	285
Existing Regulations and Standards	286

J. Pelton (✉)
International Association for the Advancement of Space Safety (IAASS), Arlington, VA, USA
e-mail: joepelton@verizon.net

T. Sgobba
International Association for the Advancement of Space Safety (IAASS), Noordwijk, The Netherlands
e-mail: tommaso.sgobba@esa.int

M. Trujillo
European Space Agency (ESA) – European Space Research and Technology Centre (ESTEC), Noordwijk, The Netherlands
e-mail: trujillo_m@hotmail.com; maite.trujillo@esa.int

© Springer Nature Switzerland AG 2020
K.-U. Schrogl (ed.), *Handbook of Space Security*,
https://doi.org/10.1007/978-3-030-23210-8_50

Human Spaceflight Safety 287
   System Safety 287
   Commercial Suborbital Regulatory Safety Framework: A Case Study 288
   Human Rating: A Historical Perspective 292
   Human Spaceflight Safety Risks 294
Conclusions 297
References 297

Abstract

Space safety is necessary for the sustainable development and performance of space missions. Space safety best practices are important in almost any kind of space mission, manned or unmanned, commercial, scientific, exploratory, and even military. Space safety aims to protect human lives relevant infrastructures in space and on ground, and affected Earth, orbital, and planetary environments. As the pace of space exploration and commercial exploitation increases, without improved space safety practices and standards, billions of dollars (US) of space assets, many astronaut lives, and even people here on Earth could all be increasingly in peril.

This chapter introduces the many facets of safety that must be addressed by spacefaring nations around the world. This chapter assesses a wide range of space safety risks in relation to the various flight phases, from launch, to on-orbit operations, to reentry while stressing the importance of new forms of international agreement to coordinate areas such as space traffic management.

Introduction

Space safety includes the protection of human life, the safeguard of critical and/or high-value space systems and infrastructures, as well as the protection of Earth, orbital, and planetary environments.

Space safety is necessary for the sustainable development of space activities. Space safety actually covers many diverse areas that are discussed in this chapter. Space safety can be defined as freedom from or mitigation of human or natural harmful conditions. These conditions can cause death, injury, illness, damage to or loss of systems, facilities, equipment or property, or damage to the environment. The term "safety" refers to threats that are nonvoluntary in nature (design errors, malfunctions, human errors, natural hazards, etc.), while "security" refers to threats which are voluntary (i.e., of aggressive nature such as use of anti-satellite weapons). In some languages, a single term is used for both, which may sometimes lead to confusion. Space safety thus covers many different areas as shown in Fig. 1. This figure shows the various fields of space safety, the relevant interest scope (national, international, or global), and the preferred processes used for risks mitigation, risk-based design, and operational hazard controls, although a mixture of the two is generally used.

Fig. 1 Space safety fields – Credit: IAASS (International Association for the Advancement of Space Safety)

Space safety can refer to human crew and passengers, personnel directly involved in system integration and operation, personnel not directly involved but co-located, as well as the general public – whether on land, the oceans, or aloft. In the case of unmanned systems such as robotic satellites or high altitude platforms, etc., space safety refers also to (non-malicious) external causes that lead to degradation or loss of mission objectives. For example, it could include such matters as the collision between two operational satellites, or the collision of an operational satellite with space debris.

Absolute freedom from harmful conditions (i.e., safety) is impossible to achieve. To be absolutely safe, a system, product, device, material, or environment should never cause or have the potential to cause an accident. In the realization and operation of systems, the term safety is generally used to mean acceptable, or mitigated, risk levels, not absolute safety. The increasing level of activities in the stratosphere, subspace, or what is sometimes called the proto-zone (i.e., 21–160 km), is an area that will increasingly be considered as an area of concern under space safety as well.

Acceptable risk level is not the same as personal acceptance of risk, but it refers to risk acceptability by stakeholders' community or by society in a broad sense. Acceptable risk levels vary from system to system and evolve with the passing of time due to socioeconomic changes and technological advancement. Implementing proven best practices at status-of-art is a prerequisite for achieving an acceptable risk level, or in other words to make a system "safe" or "safer." Safety best-practices are usually established by government regulations and norms, or by industrial standards, and enforced through authoritative organizations (e.g., government bodies or delegated independent organizations). Without enforced regulations, norms, or standards, the term safety (i.e., acceptable risk) becomes meaningless. In other words, compliance with regulations, norms, and standards represents the safety yardstick of a system.

Firstly, the chapter introduces the many facets of safety and discusses acceptable safety levels. Then, we address the risks inherent to each flight phase, from launch to on-orbit and reentry safety risks. Finally, human spaceflight safety considerations are described.

The Many Facets of Space Safety

A total of 23 astronauts and cosmonauts have lost their lives since the beginning of human spaceflight. The count includes four casualties on ground during training and the most recent casualty during flight test on the part of Michael Alsbury. Alsbury died on October 32, 2014, in the crash of the suborbital Virgin Galactic SpaceShipTwo "Enterprise." The first casualty on ground was the soviet cosmonaut trainee Valentin Bondarenko who died in a pressure chamber fire during training in March 1961. Few years later, three American astronauts were killed by a fire during training inside an Apollo capsule. There have been in total three accidents during reentry: Soyuz 1 in April 1967, Soyuz 11 in June 1971, and Shuttle Columbia in February 2003. In the latter case, in addition to the loss of crew, the public on ground and the passengers travelling by air were subjected to an unprecedented level of risk due to the US continental-wide path of falling debris, with a projected 1% chance that a fatal collision with aircraft would occur (Helton-Ingram et al. 2005).

Although a rare occurrence, space accidents are not perceived by the public simply as caused by random or unfortunate circumstances, but as the dramatic demonstration that human spaceflight programs and the entire organizations behind them failed their core mission. The risk of loss of life in human spaceflight is currently around 1 in 100 flights! Enormous, if we compare with accident rates in commercial aviation which is around 1 in 10 million flights in the USA. Although high in percentage, the loss of life in human spaceflight is still very low in absolute terms due to the low number of flights per year. Nevertheless, the entire human spaceflight program is putted in question following an accident. The reason is that the ultimate purpose of the program is the achievement of safe (and routine) physical human access to space. An accident makes clear to the public (and to political representatives) that such objective is far from being achieved. Furthermore, there is no sign of convinced effort by the cost-conscious emerging commercial human spaceflight industry in learning the lessons from 60 years of government program. They seem to prefer pursuing the development of obsolete rule-based consensus standards based on future data instead of modern techniques of risk-based design based on proven performance requirements.

Improving the safety culture of the companies, building robust safety organizations, and promoting safety education and research is paramount to maintain and expand public support for human spaceflight.

Space safety is not only about astronaut safety. Unmanned space access has become increasingly important to the great majority of countries worldwide. Upon achieving the status of a spacefaring nation, however, a key responsibility that devolves is to establish the technology and processes to protect (national and foreign) life and property against the consequences of malfunctioning rockets and reentry space systems (e.g., satellites, rockets upper stages). Safety risk in space missions also includes general public safety (on ground, on air, and at sea) and safety of launch site personnel. Space safety in a wider sense also encompasses the safeguarding of strategic and costly systems on orbit (i.e., satellites, international space station, and global utilities), valuable facilities on ground (e.g., launch pads), as well as the protection of the orbital space and of the Earth environment.

Acceptable Safety Level

The safety level achieved by a system can be objectively determined by data; however, defining an acceptable level of safety is not a simple job. The definition of acceptable safety level is based not only on technical state-of-the-art considerations but also on a number of nontechnical factors such as cultural, economic, market, or political assessments. For such reason, the safety acceptability level in any field, from drinkable water to toys or nuclear power plants, is generally established by national regulations. They may differ from country to country and also evolves with time and public expectation. When international commerce is involved, such rules need necessarily to be harmonized. An example are the international safety standards for air navigation issued by the International Civil Aviation Organization (ICAO), which represent one of the most clear-cut successes in the field of international safety cooperation.

Due to the fact that there is nothing as "absolute safety" and given that the "acceptable risk" is usually a critical balance of industrial interests and public rights, the lack of national or international regulations as currently the case in the USA for the commercial human spaceflight industry represents a business survival risk, beyond the legal protection afforded by the "informed consent" approach. This is to say that without a government-defined safety standard (i.e., acceptable level of safety), an operator would have a hard time defending the vehicle actual risk level after an accident. Indeed following a fatal accident, it seems likely that the operator's fleet would be grounded and perhaps made obsolete by newly issued (and likely strict) standards in the emotional wake of the accident. We can say that obtaining a certification of compliance with safety regulations serves the interests of the customer but at the same time also protect industry from (unbounded) tort liability by implicitly or explicitly defining the acceptable risk level at the current state of art.

For instance, in 2008, the US Supreme Court ruled in favor of a manufacturer of a balloon catheter that burst and severely injured a patient during an angioplasty. The Court wrote that the Food and Drug Administration (FDA) spent an average of 1,200 h reviewing each device application and granted approval only if found there was a "reasonable assurance" of its "safety and effectiveness." The manufacturer argued that the device design and manufacturing had been in accordance with FDA's regulations and that FDA and not the courts was the right forum on imposing requirements on cutting edge medical devices, arguing that "nothing is perfectly safe."

Safety Standards and Compliance Verification

In the safety field, it is an axiom that safety rules and compliance verification are under a single authority. There is no industry in which this principle is not applied. Usually such single authority is vested into a dedicated government organization (e.g., the Federal Aviation Administration (FAA) in the USA or the European Aviation Safety Agency (EASA) in Europe). The rapid advancement of technologies, however, pose a problem of staying abreast of technical advances. Government agencies have difficulty in keeping up with the pace of accelerating technical and scientific knowledge. One concept that is taking hold is the establishing of intermediate organizations, funded and supported by industry, i.e., so-called safety institutes, which are being tasked to develop and maintain safety standards and to verify compliance. However, care must be exercised to prevent that under the pressure of economic interests, such "self-controlling" safety-standards processes become ineffective or unduly influenced by industry interests.

The Presidential Commission that investigated the Deep Horizon oil drilling platform disaster in the Gulf of Mexico in 2011 clearly made the point about what a safety Institute should be, and how to ensure that it can effectively contribute to the continuous improvement of safety for the benefit of industry and society. The Commission identified in particular three elements: (1) commitment of companies CEOs; (2) involvement and cooperation among the best technical experts from industry (on the basis of the principle that "safety is not proprietary"); and (3) distancing these Institutes from industrial advocacy/lobbyist organizations. In this respect, the Commission noted that an organization that works as the industry's principal lobbyist and public policy advocate cannot serve as a reliable standard-setter, because it would regularly resist anything that could make industry operations potentially more costly. Such organization would fail to reflect "best industry practices" and would express instead the *"lowest common denominator." "In other words, a standard that almost all operators could readily achieve."*

As we will see later, the risks related to space activities (e.g., launch and reentry) are often of international nature; however, currently there are no international regulations but only few national regulations, which are often scattered among different government agencies and organizations or not applied in a uniform manner by key players. To explore the different facets of space safety, the next sections will discuss the safety risks associated to the different flight phases, from launch to on-orbit safety (e.g., space debris-related hazards) and reentry. Then, the risks inherent to human spaceflight will be introduced, both for orbital and suborbital flights.

Launch Safety

Launch Site Ground Safety Risk

On August 22, 2003, at 13:30 h (local time), a massive explosion destroyed a Brazilian Space Agency VLS-1 rocket as it stood on its launch pad at the Alcantara Launching Center in northern Brazil. Twenty-one technicians close to the launch pad died when one of the rocket's four first-stage motors ignited accidentally. The investigation report established that an electrical flaw triggered one of the four solid fuel motors while it was undergoing final launch preparations. The report said that certain decisions made by managers long before the accident occurred led to a breakdown in safety procedures, routine maintenance, and training. In particular, the investigation committee observed a lack of formal, detailed risk management procedures, especially in the conduct of operations involving preparations for launch. As of today, there have been nearly 200 people killed on ground by rocket explosions during processing, launch preparations, and launch. In the last 10 years, there have been also at least six launches which have been terminated by explosion commanded by the launch range safety officer to prevent risk for the public. There have been also several more cases of launchers that did not make to orbit, exploded on the pad, or came back prematurely to Earth in an uncontrolled fashion.

The main ground hazards during launch are explosive, toxic, or radioactive hazards. Explosive hazards (overpressure and fragments thrown by an explosion) are an important component in the launch area. Toxic hazards from the rocket's exhaust products and meteorological conditions are often an additional consideration in defining so-called exclusion areas. Additional sections of the launch complex may be restricted to protect against the kinetic energy of inert debris (i.e., spent stages) or radiation from radars and other support instrumentation. During preparation for launch, a very common issue that the ground processing safety community encounters is lack of recognition of the need for detailed ground safety documentation and rigorous technical safety reviews. Many hardware and mission designers assume that if the hardware is safe to fly, it will also be safe during ground processing. Some also assume that the industrial safety processes used during development and manufacture are sufficient for use at the launch and landing sites.

When an exclusion area is defined, each country has its own procedures for communicating the boundaries of the area. On land, this is commonly through sign postings and guards. Formal notices are frequently used to communicate with operators of ships and aircraft. Moreover, the degree of compliance varies with location and time. When the exclusion area is near the launch complex, ranges frequently employ different forms of surveillance to determine whether any vessels have intruded into the hazardous area. When intruders can be identified, the ranges may request them to depart, passively wait for their departure, or proceed with the launch, based on the decision that the risk to the vessel is sufficiently small.

Nowadays, international commercial spaceports are proliferating, and the growing need is felt to equally and uniformly protect worldwide the local personnel as well as the foreign teams which participate to launch campaigns. When in October 2002 a Russian Soyuz exploded at launch killing a young Russian soldier who was watching the launch from the first floor of a building, it was by pure luck that no one was injured of the large international support team on site which was watching the launch from a closer location.

Launch Flight Safety Risk

Knowledge of best practices and techniques in launch safety and risk assessment are not widespread and may vary greatly from country to country. Furthermore, currently during launch, a country may take risks on the population of a foreign country that even if equal to that for their own population is a unilateral decision and not the outcome of consultation. Space treaties define liabilities, but they neither define nor require uniform risk assessment and management methods and standards.

The way of achieving public protection from launch activities is by isolating the hazardous condition from populations at risk. When this is not feasible, launch vehicle performance and health is monitored for automatic or manual flight termination. Flight termination strategies are meant to limit rocket excursions from planned trajectory. The residual risk is evaluated with reference to where people may be at risk because of debris generated by flight termination.

Identification of high-hazard areas may range from simplistic rules of thumb to sophisticated analyses. When simple rules are applied, they commonly specify a hazard radius about a launch point and planned impact points for stages, connected by some simple corridor. More sophisticated analyses attempt to identify credible rocket malfunctions, model the resulting trajectories, and determine the conditions that will result in debris such as exceeding the structural capacity of the rocket or a flight termination action by a range safety officer. These analyses typically include failure analyses to identify how a launch vehicle will respond under various failure scenarios. This will include failure response analyses to define the types of malfunction trajectories the vehicle will fly. The vehicle loads are assessed along the malfunction trajectory to determine whether structural limits will be exceeded. Vehicle position and velocity may be compared against abort criteria to assess whether the vehicle should be allowed to continue flight, terminate thrust, or be destroyed. Debris-generating events then become the basis for assessing the flux of debris falling through the atmosphere and the impact probability densities. The debris involved may be screened by size, impact kinetic energy, or other criteria to assess which fragments pose a threat to unsheltered people, people inside various types of buildings, people on ships, and people in aircraft. The resulting debris impact zones are then commonly used as part of the basis for defining exclusion areas.

Although full hazard containment is considered to be the preferred protection policy, it is not always possible. The next line of protection after defining exclusion areas is real-time tracking and control of the rockets. Range safety systems are used for this purpose. They include a means of tracking a launch vehicle's position and velocity (tracking system) and a means of terminating the flight of a malfunctioning vehicle (flight termination system).

Flight termination criteria are customarily designed based on the capability of the range safety system to limit the risk from a malfunctioning launch vehicle. Frequently, ranges assume that they can reliably detect a malfunctioning launch vehicle and terminate its flight whenever good quality tracking data is available. This assumption is based on high-reliability designs customarily used for range safety systems. At present, however, there are no international design standards for range safety systems. Moreover, efforts to assure that the design standard does, in fact, achieve the intended reliability levels are rare.

The final tiers of protection are risk analysis and risk management. Residual risks from the launch are quantified and assessed to determine if they are acceptable. This step involves an extension of the model outlined above for assessing hazardous areas. It is common to perform these protection steps in an iterative manner, using the results of each step to adjust the approach to the others until the desired level of safety is achieved with acceptable impacts on the proposed launch. The current practice is to assess risks for each launch and to approve the launch only when risk levels are acceptable. Unlike most other activities, annual risk levels are addressed by exception.

A proper risk analysis addresses the credible risks from all launch-related hazards. These may include inert debris, firebrands, overpressure from exploding fragments, and toxic substances generated by normal combustion as well as toxic releases from malfunctions. When assessing launch risks, as it occurs for reentry, it is important to account for all exposed populations: people on land, people in boats, and people in aircrafts. Proper consideration must be given to the effect of sheltering (i.e., type of construction and materials of houses, buildings) on the risks. It is often assumed that neglecting sheltering will overstate the risk. When sheltering is adequate to preclude fragment penetration, this assumption is valid. When fragments are capable of penetrating a structure, debris from the structure increases the threat to its occupants. As launch vehicles proceed downrange, they typically leave the territorial domain of the launching country and begin to overfly international waters and the territory of other countries.

Tolerable risks for a launch are commonly expressed in terms of a collective or societal risk level and risk to the maximally exposed individual (individual risk). Collective risk is commonly expressed as the number of individuals statistically expected to be exposed to a specified injury level. Individual risk is commonly expressed as the probability that the maximally exposed individual will suffer the specified injury level. The two most commonly used levels of injury are fatality and serious injury. When it is difficult to quantify risk directly, impact probability for specified classes of debris is often used as a proxy measure. Thus, for example, it is customary to protect people on ships or people on airplanes by creating exclusion zones based on impact probabilities.

Outside of the immediate launch area, surveillance is more difficult and more costly. Consequently, most ranges use surveillance very selectively outside of the immediate launch area, typically restricting surveillance to planned impact areas for spent stages and other planned jettisons. As a result, publishing exclusion areas at these distances is much less effective. More efficient tools for surveying these remote locations and communicating with intruders would enhance the effectiveness of protecting ships and aircraft in these areas.

Launch Risk for Maritime and Air Transportation

Controlling risks to seafaring vessels from space launch activities is most successful when mariners are notified about hazard areas and when the responsible launching agency surveys the potentially affected areas to detect intruders and to warn them to leave the exclusion area. Following a mishap, communication with these vessels to proceed at maximum speed in a prescribed direction to minimize impact probability is essential to control undue risks. Currently, costs and technology limit surveillance and communication to locations near land.

For launch preparations, the management of airspace must also consider aircraft traffic. At present, there are limited capabilities for addressing this issue. The Federal Aviation Administration (FAA) has begun an initiative to address these concerns for US operations. It should be noted that the current practice is for each launch range to manage risks on a mission-by-mission basis through Launch Collision Avoidance (LCOLA) processes. Minimal attention is paid to annual risks generated by the range's launch operations. There is no agency – national or international – that monitors and controls risk posed to overflown populations. A city may be placed at risk by launches from multiple launch sites without the performance by involved launching nations of any coordinated assessment to assure that the risk levels are acceptable.

Citizens of all countries should be equally protected from the risk posed from overflying by launch vehicles and returning spacecraft(s). The common practice is to make these determinations on a launch-by-launch basis with no consideration of previous, planned, or future launches.

Air-Launch Safety

Launching from ground means that the first stage of the rocket will be traveling through the denser layers of the atmosphere where drag is a significant issue. The launchpad is at a fixed location, selected to meet logistics, safety, and environmental constraints, not always the most favorable one to reach orbit. What if one could do away with the first stage and replace it with a high-altitude platform, such as an aircraft or a balloon? Compared with ground launch, air-launch provides flexible and reusable "first stage" and a lighter expendable rocket.

To place a satellite on low Earth orbit, let us say at 300 km, you need to reach a velocity of about 9.4 km/s. Because the starting velocity is zero, we can also talk of it as difference of velocity, or delta-V, of 9.4 km/s. The delta-V depends on several parameters, like initial altitude, speed, and angle of attack. If we launch, for example, from a balloon at 15 km, with zero launch velocity and 0° angle of attack, the required delta-V is 8.8 km/s. If we launch from an aircraft flying at 1.200 km/h, the delta-V drops to 8.5 km/s. It further drops to 8.3 km/s if the launch angle of attack is 30°. Delta-V reductions translate into lighter rocket, less fuel, and lower cost.

On 13 June 1990, Pegasus of Orbital Sciences Corporation became the first commercial air-launch vehicle. Released from a modified Lockheed L-1011 airliner, Pegasus can put a satellite of 450 kg in low Earth orbit.

In the past, the market of small satellites (100–500 kg) and miniature satellites, which comprises microsatellites (10–100 kg) and nanosatellites (1–10 kg), was small. They were usually launched as secondary payloads on larger launch vehicles. Nowadays, the commercial and military market of small and miniature satellites is the fastest growing segment of the space launch business, and several companies are developing air-launch vehicles specifically targeted to such market. Ability to "launch on demand" at low cost will allow launch operators to offer unprecedented flexibility for schedule and orbital placement. Several air-launch systems are under development as adaptation of commercial and military airplanes or as dedicated systems. Any country or operator with experience of military supersonic aircraft has the potential to develop its homegrown micro- and nanosatellites air-launch service.

In July 2012, Virgin Galactic announced that their rocket called LauncherOne would be air-launched from the same WhiteKnightTwo aircraft carrier they developed for suborbital human spaceflight. Later a Boeing 747, called Cosmic Girl, has been adapted for the purpose. Air-launch operations will start in 2019.

Stratolaunch Systems is developing a gigantic new air-launch system. The project comprises three main components: the carrier aircraft being built by Scaled Composites, a multistage rocket, and a mating and integration system. Stratolaunch's carrier aircraft has wingspan of 117 m and a weight of over 540.000 kg including the fully fueled launch vehicle. The test flights of the carrier aircraft are planned in 2019.

The diffusion worldwide of air-launches raises the issue of safety. The record of current spacefaring countries in performing traditional ground-based launches is not uniform. It may seriously worsen for air-launches because of lack of experience of newcomers. Most of what we know about air-launch safety comes from Orbital ATK Pegasus and from supersonic air-launches performed by USAF and NASA. As of 2018, Orbital ATK Pegasus was launched 44 times safely, but in two cases the rocket veered off course and was destroyed by command sent by range officer.

Orbital ATK Pegasus launches are performed under safety oversight of one of the US launch ranges. Some key points:

(a) Rocket assembly and payload integration processes are identical to those followed for traditional expendable rockets. Safety rules are identical.
(b) Integration of rocket on aircraft carrier is performed on an isolate runway section, subjected to safety rules similar to pad operations. Safety requirements for barriers to prevent inadvertent rocket ignition are identical with those for ground launches, but arming is done in flight close to launch time.
(c) Aircraft flight to reach the launch location is subjected to constraints. For example, it is forbidden to overfly populated areas.
(d) Launch is performed following a countdown process with teams and equipment - on-board the aircraft carrier and remotely at the range.
(e) Rocket flight is redundantly tracked. Manual commands are sent for flight termination, if it leaves the planned trajectory (use of mobile range equipment). Special procedures are defined for the case of launch abort.
(f) Air traffic is cleared in advanced from launch location, but procedures are not tightly enforceable, as for launches from ground. In the latter case, the spaceport and the airspace overhead can be "sealed." For air-launches, which usually take place from the international airspace therefore outside national authority, only advisory NOTAMs can be sent to alert the air traffic.

On-Orbit Safety

Orbital Debris

Space is not an empty vacuum but contains both natural debris (i.e., micrometeoroids, interplanetary dust) and human-made space debris. Humans generally have no involvement in natural debris; thus, here we will concentrate exclusively on human-made debris. Orbital debris generally refers to any human-made material on orbit which is no longer serving its intended function. There are many sources of debris. One source is discarded hardware such as upper stages of launch vehicles or satellites which have been abandoned at the end of their operational life. Another source is spacecraft items released in the course of mission operations. Typically, these items include launch vehicle fairings, separation bolts, clamp bands, adapter shrouds, and lens caps. Various shapes and sizes of debris are also produced as a result of the degradation of hardware due to atomic oxygen, solar heating, and solar radiation and also from combustion of solid rocket motors. Examples of such products are paint flakes, aluminum oxide exhaust particles, and solid motor-liner residuals.

Fifty years of spaceflight have cluttered the space around the Earth with an enormous quantity of human-made debris. Scientists assume that there are approximately 500,000 objects in orbit whose sizes are above 1 cm. Currently, about 22,000 of such objects (i.e., 10 cm in diameter or larger) are being tracked by the US Space Surveillance Network (including about 1,000 objects representing functional satellites). The number is expected to rise significantly as new large-scale satellite constellations are deployed in the future to provide communications, remote sensing, and other services such as frequency monitoring.

When the new S-band radar "Space Fence" is fully operational, this number will increase to a much larger number and will be able to track objects down to the size of a marble in Low Earth Orbit (LEO). Among the tracked pieces of debris, there are about 200 satellites abandoned in Geostationary Earth Orbits (GEO) occupying or drifting through valuable orbital positions and posing a collision hazard for functional spacecraft(s). The survival time of the debris can be very long. Objects in 1,000 km orbits can exist for hundreds of years. At 1,500 km, the lifetime can go up to thousands of years. Objects in geosynchronous orbit can presumably survive for one million years.

The future population of orbital debris will depend upon whether the creation or removal rate dominates. Currently, the only mechanism for removal of debris is orbital decay through atmospheric drag, which ultimately leads to atmospheric reentry. This mechanism is only effective in a restricted range of low Earth orbits (LEO). At higher orbits, it takes hundreds to thousands of years for objects to reenter the Earth's atmosphere. Consequently, there is no effective removal mechanism. Historically, the creation rate of debris has outpaced the removal rate, leading to a net growth in the debris population in low Earth orbit at an average rate of approximately 5% per year. A major contributor to the current debris population has been fragment generation via explosions. As the debris mitigation measure of passivation (e.g., depletion of residual fuel) comes to be implemented more commonly, it is expected that explosions will decrease in frequency. It may take a few decades for the practice to become implemented widely enough to reduce the explosion rate, which currently stands at about four per year.

Several environment projection studies conducted in recent years indicate that, with various assumed future launch rates, the debris populations at some altitudes in LEO will become unstable. Collisions will take over as the dominant debris generation mechanism, and the debris generated will feed back into the environment and induce more collisions. The most active orbital region is between the altitudes of 900 and 1,000 km, and even without any new launches, this region is highly unstable (Liou and Johnson 2006). It is projected that the debris population (i.e., objects 10 cm and larger) in this "red zone" will approximately triple in the next 200 years, leading to an increase in collision probability among objects in this region by a factor of ten. In reality, the future debris environment is likely to be worse than was suggested, as satellites continue to be launched into space.

Collision Risk with Orbital Debris

Orbital debris generally moves at very high speeds relative to operational satellites. In LEO (i.e., altitudes lower than 2,000 km), the average relative impact velocity is 10 km/s (36,000 km/h). In the geostationary orbits, the relative velocity is lower, approximately 2 km/s, because most objects move in an eastward direction orbit.

At these hypervelocities, pieces of debris have a tremendous amount of kinetic energy. A 1 kg object at a speed of 10 km/s has the same amount of kinetic energy that a fully loaded truck, weighing 35,000 kg, has at 190 km/h. A 1-cm-sized aluminum sphere at orbital speed has the energy equivalent of an exploding hand grenade. A 10 cm fragment in geosynchronous orbit has roughly the same damage potential as a 1 cm fragment in low Earth orbit.

Pieces or particles of debris smaller than 1 mm in size do not generally pose a hazard to spacecraft functionality. Debris fragments from 1 mm to 1 cm in size may or may not penetrate a spacecraft, depending on the material composition of the debris and whether or not shielding is used by the spacecraft. Penetration through a critical component, such as the flight computer or propellant tank, can result in loss of the spacecraft. NASA considers pieces of debris 3 mm in size and above as potentially lethal to the retired Space Shuttle and the International Space Station. Debris fragments between 1 and 10 cm in size will penetrate and damage most spacecraft. If the spacecraft is impacted, satellite function will be terminated, and at the same time, a significant amount of small debris will be created. If a 10 cm debris fragment weighing 1 kg collides with a typical 1,200 kg spacecraft, over one million fragments ranging in size from about 1 mm and larger could be created. Such collisions result in the formation of a debris cloud which poses a magnified impact risk to any other spacecraft in the orbital vicinity (e.g., other members of a constellation of satellites).

Certain regions of the debris cloud are constricted to one or two dimensions. Such constrictions do not move with the debris cloud around its orbit. They remain fixed in inertial space while the debris cloud repeatedly circulates through them. In many satellite constellations, there are multiple satellites in each orbital ring. If one of these satellites breaks up, the remaining satellites in the ring will all repeatedly fly through the constrictions. If many fragments are produced by the breakup, the risk of damaging another satellite in the ring may be significant. If satellites from two orbital rings collide, two debris clouds will be formed with one in each ring. The constrictions of each cloud will then pose a hazard to the remaining satellites in both rings.

In February 2009, a nonoperational Russian satellite, Cosmos 2251, collided with Iridium 33, a US commercial telecommunication satellite, over Siberia at an altitude of 790 km. This collision, the first of its kind, was the worst space debris event since China intentionally destroyed one of its aging weather satellites during an antisatellite missile (ASAT) test, in 2007. The Iridium satellite that was lost in the collision was part of a constellation of 66 low Earth-orbiting satellites providing mobile voice and data communications services globally. As expected, the risk of collision of other Iridium satellites in the same plane dramatically increased with daily announcements of possible collisions (i.e., conjunctions) with Iridium 33 debris. Fig. 2 presents the evolution in time of the number of human-made debris objects, which highlights the increasing problem impacting the sustainability of the space environment.

In general, orbital debris collision is among the top risk for human spaceflight. The 2003 Shuttle risk assessment performed after the Columbia accident, the first

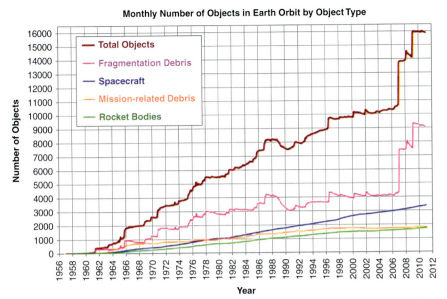

Monthly Number of Cataloged Objects in Earth Orbit by Object Type: This chart displays a summary of all objects in Earth orbit officially cataloged by the U,S, Space Survellance Network, "Fragmentation debris" Includes satelite breakup debris and anomalous event debris, while "mission-related debris" includes all objects dispedsed, separated, or released as part of the planned mission.

Fig. 2 Catalogued human-made space objects in Earth's orbit (Credit: NASA)

one that incorporated the threat posed by orbital debris, determined that the likelihood of orbital debris bringing down the Shuttle was far greater than that of the widely feared failures of main engines, solid rocket boosters, or thermal protection. Orbital debris colliding with different spots of the wing flaps was the most likely catastrophic failure. Damage would have rendered the wing flap (elevon), unable to steer and slow the Shuttle during the reentry phase.

Orbital debris collision is the primary source of risk for the International Space Station (ISS). To minimize such risk for the crew, the ISS is shielded. The ISS is indeed the most heavily shielded spacecraft ever flown. All together there are 100 different shields protecting the ISS. Critical components such as habitable compartments and high-pressure tanks will be able to withstand the impact of debris as large as 1 cm in diameter.

Controlling Orbital Debris Risk

Orbital debris risk is best controlled by limiting creation through a number of design and operational measures, like "passivation," collision avoidance maneuvers, and end-of-life disposal.

Passivation is the term used to describe the prevention of satellite and upper-stages explosions by controlled removal of stored energy at the end of useful life. For example, propellant in upper stages and satellites can be eliminated by either venting or burning to depletion. This process is applied primarily to low earth orbit satellites. Batteries can be also designed to reduce risk of explosion.

Spacecraft maneuvers, when possible, can also mitigate orbital debris risk of collision. The International Space Station has maneuvered on several occasions to avoid collisions with orbital debris. Also, in the case of satellite constellations, because a potential collision will lead to the creation of a debris cloud that may result in damage to other members of the constellation, collision avoidance maneuvers may be necessary. Another means to reduce the risk of collision is to remove satellites and upper stages from mission orbits, the so-called protected orbits, at the end of operational life. Currently, UN guidelines and other internationally agreed standards (e.g., ISO 24113) recommend that a space system should not remain in its mission orbit for more than 25 years. Such objective is met either by lowering the orbit such that residual atmospheric drag is sufficiently strong to cause decay and reentry or by moving the spacecraft to a "graveyard orbit" outside of protected regions. At orbits above 2,000 km, it is not economically feasible to force reentry within 25 years. Spacecraft operating in the geosynchronous orbits are routinely boosted into a higher disposal orbit at the end of their mission life, except in case of malfunction. Propellants need to be reserved to perform the disposal maneuvers. There are penalties in the form of reduced performance and/or mission life linked to the disposal of space systems. Estimates of the amount of "lost" lifetime for geosynchronous satellites vary between 6 months and 2 years. For example, it has been calculated that if a typical commercial communication satellite that has 24 Ku-band and 24 C-band transponders with bandwidths of 36 MHz has to be boosted into a higher disposal orbit at the end of its mission life, this maneuver would cause the satellite operator an average loss (in terms of how much longer the satellite could have continued commercial operations) of as much as 1 year's profit. This problem can be mitigated by employing the so-called inclined orbit operation so as to preserve fuel since North–South station-keeping requires more than ten times more fuel than East–West station-keeping.

Orbital Debris Remediation: Active Debris Removal

In view of the massive amount of debris already in existence in Earth orbits, growing consensus among experts suggests that an active process for the removal of existing debris from space is required, as mitigation is no longer sufficient to ensure the long-term sustainability of outer space activities. Active debris removal (ADR), specifically the removal of nonfunctional spacecraft and spent upper stages, requires the development of advanced technologies and concepts (Fig. 3). Their implementation also raises a number of difficult technical, economic, strategic, institutional, legal, and regulatory challenges that must be addressed at the very outset. For such on-orbit services to become available, the following elements need

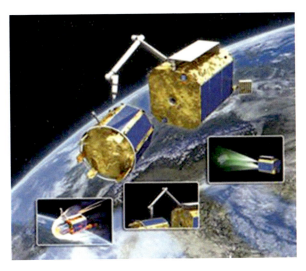

Fig. 3 Active debris removal concept (Credit: DRL)

to be in place: service at the lowest possible cost, spacefaring countries committing to gradually remove their own debris, and new national licensing regulations mandating removal (autonomous or enforced) at the end of mission.

To achieve the lowest possible service costs, international technological cooperation, high rate of missions per year, and possibly multiple service targets per mission would be needed at least for an initial period of operations. The international technological cooperation would serve the purpose of making available all existing technologies and share the cost of new developments. Servicing with a single flexible system multiple (international) customers and perform multiple removals within the same mission would also substantially contribute to lower the operational costs. To a certain extent, the same path used once for the development of the satellite telecommunication industry may be repeated with the establishment of an intergovernmental organization on the model of the early International Telecommunications Satellite Organization (INTELSAT) which would later evolve into full commercial services. Another alternative would be to create an international fund for debris removal. Such a fund could start as national and/or regional cooperative and evolve into an international fund supported by all spacefaring nations. The problem with all such plans or concepts is that the current provisions of Article VII of the Outer Space Treaty of 1967 as well as those of the so-called Liability Convention of 1972 structure "liabilities" associated with such removal activities in such a way as to give little incentives for countries to actively remove debris from orbit.

The development of new technological capabilities, in the area of active debris removal and possible repurposing of the defunct spacecraft, continues apace. There have been a series of projects sponsored by the U.S. Defense Advanced Research Projects Agency (DARPA), by NASA, by DLR of Germany, and by private companies such as McDonnell Detwiler, Vivasat, and Conesat. On June 20, 2018, the "Remove Debris" proof of concept small satellite was launched through the Kaber Launch facility operated by Nanoracks from the International Space Station. This small satellite with a mass of only 100 kg will be able to test a range of possible removal techniques. It was constructed by Surrey Space Technology Ltd. and after some 5 years of development is now testing these removal strategies. ("NanoRacks Deploys Largest Satellite from International Space Station to Date," July 2, 2018. https://mail.aol.com/webmail-std/en-us/suite).

Reentry Safety Risk

As previously mentioned, nonfunctional satellites, spent launch vehicle upper stages, and other orbital debris do not remain in low Earth orbits indefinitely but gradually return to Earth due to residual atmosphere drag. In low Earth orbits, natural orbital decay can take place within few months or requires hundreds or even thousands of years to happen depending on the altitude.

As nonfunctional satellites, spent launch vehicle stages, and other pieces of debris enter denser regions of the atmosphere, they fragment and sometimes explode due to high aerodynamic forces combined with loss of materials strength due to heat caused by friction with air at high velocity. Heat would subsequently cause the demise of major portions of the hardware due to melting and vaporization. However, between 10% and 40% of the original mass will survive and reach Earth's surface. In general, parts and components made of aluminum and similar materials with low melting temperatures do not survive reentry, while those made of materials with high melting temperatures, such as stainless steel, titanium, do survive. Also parts with low mass and large surface area, and therefore large aerodynamic drag, will survive due to slow down and related low heating. The surviving fragments represent a hazard to people and property on the ground. They also represent a potential serious risk to air and maritime traffic.

Due to variability of the atmosphere layers around the Earth, it is difficult to predict the exact reentry time of a randomly reentering satellite or upper stage. As a consequence, it is very difficult to predict where surviving fragments will hit the surface of the Earth. Over the last 50 years, more than 1,400 metric tons of materials are believed to have survived reentries. The largest object to reenter was the Russian Mir Space Station, which weighed 120,000 kg. Reentries are frequent, in particular upper-stages reentries. In 2011, launch vehicles upper stages reentered at a rate of 1 per week with a total mass that was five times that of uncontrolled spacecraft reentries for the same period (Figs. 4 and 5). Many of the reentered parts recovered on ground, including tanks up to 250 kg weight, belonged to rockets.

Currently, a number of countries prescribe that the risk of any personal casualty due to a single reentry event must be less than 1 in 10,000 reentries. France has the most conservative requirement of less than 2 in 100,000 reentries. Of particular concern, although very remote, is the risk for aviation and the emotional and psychological impact on the general public that a single accident with many casualties would cause (Ailor and Wilde 2008), as described in section "Risk for Aviation."

Fig. 4 Stainless steel propellant tank of second-stage Delta 2 reentered launch vehicle (US, 1977) (NASA courtesy)

Fig. 5 Reentered titanium motor casting of third-stage Delta 2 (Saudi Arabia, 2001) (NASA courtesy)

Environmental Risk

There is health risk related to launch ascent failures and reentry of space systems (e.g., rocket bodies and nonfunctional space systems). During normal launches, stages separate sequentially and fall down to Earth. Most launch trajectories and spaceport locations are chosen to ensure that the impact areas are outside populated areas and mainly contiguous to the oceans. Nevertheless, there are inland spaceport locations and land overflying trajectories which lead to stages dropping to ground in sparsely inhabited areas with ensuing soil contamination. Approximately 9% of the propellant from a launch stage remains in the tank once it is dropped. The penetration of contaminants depends on the nature and properties of the soil and can lead to the contamination of groundwater as well as surface water. For example, hydrazine (UDMH) is often used in hypergolic rocket fuels as a bipropellant in combination with the oxidizer nitrogen tetroxide and less frequently with IRFNA (red-fuming nitric acid) or liquid oxygen. UDMH is a toxic carcinogen and can explode in the presence of oxidizers. It can also be absorbed through the skin. A tablespoon of hydrazine in a swimming pool would kill anyone who drank the water. In a study conducted by Vector, the Russian State Research Center of Virology and Biotechnology in Novosibirsk, health records from 1998 to 2000 of about 1,000 children in two areas in southern Siberia polluted due to launches from Baikonur spaceport in Kazakhstan were examined, comparing them with 330 records from a nearby unpolluted control area. Grouping all cases of disease together, the research team concluded that children from the worst-affected area were up to twice as likely to require medical attention for diseases such as endocrine and blood disorders during the 3 years studied and needed to be treated for twice as long. Contamination can be far worse and massive in case of launch failure. In September 2007, the explosion of a Russian Proton M rocket contaminated a vast swath of agricultural land in Kazakhstan with 200 t of toxic fuel.

Reentries may also cause concern because of the toxicity or radioactivity of materials on board. On 21 February 2008, an uncontrolled reentering satellite was shot down on grounds of public safety. The satellite was destroyed at an altitude of 247 km by a ship-launched missile. The malfunctioning spacecraft, a US spy satellite (USA 193), carried 450 kg of highly toxic frozen hydrazine fuel in its titanium fuel tank. In addition, it was expected that about 50% of the satellite's mass of 2,270 kg would survive reentry, thus adding to public risk on ground.

Currently, there are 32 defunct nuclear reactors circling the Earth as well as 13 reactor fuel cores and at least 8 radio-thermal generators (RTGs). RTGs had been used six times in space missions in low Earth orbits up to 1972 and twice in the geostationary geosynchronous orbit up to 1976. Since 1969, another 14 reactors have been used on lunar and interplanetary missions. The total mass of RTG nuclear fuel in Earth orbit today is in the order of 150 kg. Another form of nuclear power source used in space activities is a nuclear reactor. Most of these reactors were deployed on Soviet radar reconnaissance satellites (RORSATs) launched between 1965 and 1988.

Among the space nuclear accidents (i.e., unwanted/unplanned release of radioactive material), two involved orbital debris, and a third was a close call. In 1978, the RORSAT COSMOS 954 failed to separate its nuclear reactor core and to boost it into a disposal orbit as planned. The reactor remained on board the satellite in an orbit that decayed until it reentered the Earth's atmosphere. The satellite crashed near the Great Slave Lake in Canada's Northwest Territories, spreading its radioactive fuel over an area of about 124,000 km^2. Recovery teams swept the area by foot for months. Ultimately, they were able only to recover 12 large pieces which comprised a mere 1% of the estimated quantity of radioactive fuel on board. These pieces emitted radioactivity of up to 1.1 Sv/h. (It should be noted that usually a nuclear emergency is declared on ground at 500 μSv/h.) A few years later, in 1982, another RORSAT, COSMOS 1402, failed to boost the nuclear reactor core into a storage orbit. The ground controller managed to separate the core from the reactor itself to make it more likely that it would burn up in the atmosphere before reaching the ground. The reactor was the last piece of the satellite to return to Earth in February 1983 when its core fell into the South Atlantic Ocean.

Then, in April 1988, yet another Russian spacecraft, COSMOS 1900, failed again to separate and boost the reactor core into a storage orbit. However, later on, the redundant system succeeded in separating and boosting the nuclear core into a storage orbit, although lower than that originally planned.

Risk for Aviation

Many of the practices that apply to launch apply also to reentry, but the latter pose special issues because they are mainly random or related to unique behavior of the reusable vehicle during reentry.

The disintegration during reentry of the Shuttle Columbia on February 1, 2003, was a watershed moment in the history of launch and reentry safety analysis. It highlighted the need to select vehicle reentry trajectories which minimize the risk to ground populations and the need to take measures to keep air traffic away from falling debris if a reentry accident occurs. The Columbia accident initiated a chain of events that demonstrated the need for a deliberate, integrated, and, eventually, international approach to public safety during launch and reentry operations. This is especially true for the management of air traffic and space operations.

Shortly after the breakup of Columbia over a relatively sparsely populated area of Texas, dramatic images of the debris from the breakup were seen around the globe: an intact spherical tank in a school parking lot, an obliterated office rooftop, mangled metal along roadsides, and charred chunks of material in fields. The NASA Administrator testified before the US Senate that it was "amazing that there were no other collateral damage" (i.e., that no members of the public were hurt).

The Columbia Accident Investigation Board (CAIB) raised and answered many questions relevant to public safety during launch and in particular reentry. Given the available data on the debris recovered and the population characteristics in the vicinity, a CAIB study found that the absence of ground casualties was, in fact, the statistically expected result. Specifically, based on census data and modeling methods consistent with US standards and requirements set by other US agencies (e.g., the USAF in the Air Force Space Command Manual and by the FAA in the Federal Register), the study found that "the lack of casualties was the expected event, but there was a reasonable probability (less than 0.5 but greater than 0.05) that casualties could have occurred." However, a similar event over a densely populated area such as Houston would almost certainly have produced multiple casualties among the public on the ground.

At the time of the Columbia accident, NASA had no formal policy regarding public risk during Shuttle reentry. Following the CAIB report, NASA established a new safety policy (NPR8715.5). The NASA public safety policy embraced many of the risk measures and thresholds already in use by other US agencies, such as individual and collective risk limits in terms of casualties. However, NASA's public safety policy also putted forward innovative criteria for risk budgets governing distinct phases of flight which have gained broad acceptance. Therefore, the Columbia accident led to greater consensus and innovation in the management of risk to people on the ground from launch and reentry operations.

The Columbia accident also promoted the development of improved methods and standards for aircraft safety during launch and reentry. Following the release of the final report of the CAIB, the FAA funded a more detailed aircraft risk analysis that used the actual records of aircraft activity at the time of the accident. That study found that the probability of an impact between Columbia debris and commercial aircraft in the vicinity was at least one in a thousand, and the chance of an impact with a general aviation aircraft was at least one in a hundred. The analysis used the current models which assume that any impact anywhere on a commercial transport with debris of mass above 300 g produces a catastrophic accident: all people on board are killed. Current best practices are captured in RCC 321–07 "Common Risk Criteria for the National Ranges," which provides a vulnerability model for the commercial transport class. In 2008, the FAA and USAF sponsored the development of vulnerability models for transoceanic business jets based on the same methods.

After the release of the CAIB report, the FAA investigated the need for new decision support tools to better manage the interface of space and air traffic. The relevant procedures were then developed, and they are currently in use as a real-time tactical tool in the event of a catastrophic event like the Columbia accident to identify how to redirect aircraft around a space vehicle debris hazard area.

Existing Regulations and Standards

The sections above have provided an overview of risks associated to launch, on-orbit operations, and reentry. As it has been illustrated, the issues raised above involve risks that are national and/or international in nature. For launch and reentry activities, national regulations exist in some spacefaring nations; however, no international regulation applicable worldwide has been agreed. There have been ever-increasing safety concerns that are now posed by the pollution of the orbital environment (i.e., orbital debris) to operational spacecraft and the international space station. After many years of debate, international space debris mitigation guidelines have been worked out by the Inter-Agency Space Debris Coordination Committee (IADC), and these have been agreed as voluntary standards within the umbrella of the United Nations Committee of Peaceful Usages of Outer Space Activities (UN COPUOS). In addition, the International Organization for Standardization (ISO) has published a standard on space debris mitigation (i.e., ISO 24113) to put forward design and operational practices for implementation in future space systems to minimize the generation of orbital debris.

However, no remediation activities are yet internationally agreed, and therefore, neither standards nor regulations exist in this field. In addition, although different countries have the observational assets to perform space situational awareness services, currently, there are no agreed space traffic management regulations. Clearly these two issues, i.e., space debris mitigation and space traffic management, will constitute the two most important international space safety standards and regulatory issues to be faced in the next few years.

Human Spaceflight Safety

In the following sections, the risks associated to human spaceflight will be presented. Firstly, the concept of system safety for crewed systems would be introduced. Then, the value of regulations and safety standards would be illustrated by real examples in different fields, and a case study based on the emerging industry of commercial suborbital transportation will be examined. Then, the historical and latest developments for human-rating space systems would be presented, and finally, a selected number of risks associated to human spaceflight will be covered.

System Safety

Prior to the 1940s, flight safety consisted basically of trial and error. The term fly-fix-fly was associated with the approach of building a prototype aircraft, fly it, and repair/modify if broke, and fly it again. For complex and critical systems, such approach is simply impossible. From 1952 to 1966, the US Air Force (USAF) lost 7,715 aircraft in noncombat operations, in which 8,547 persons were killed. As reported by Olsen (2010), "most accidents were blamed on pilots, but many engineers argued that safety had to be designed into aircraft just as any other functional or physical feature related to performance. Seminars were conducted by the Flight Safety Foundation, headed by Jerome Lederer that brought together engineering, operations, and management personnel. At one of those seminars, in 1954, the term 'system safety' was first used in a paper by the aviation safety pioneer C.O. Miller."

In the 1950s, when the Atlas and Titan ICBMs were being initially developed, there was no safety program. Within 18 months after the fleet of 71 Atlas F missiles became operational, 4 blew up in their silos during operational testing. The worst accident occurred in Searcy, Arkansans, on August 9, 1965, when a fire in a Titan II silo killed 53. The US Air Force then developed system safety assessment and management concepts. Such efforts eventually resulted into the establishment of a major standard, MIL-STD-882D, and System Safety Engineering as a discipline (Leveson 2003).

Commercial Suborbital Regulatory Safety Framework: A Case Study

One of the areas of space safety regulation that has received the most attention in recent years has been with regard to overseeing the safety of commercial spaceflights – particularly in the form of suborbital flights.

A suborbital flight is defined as a flight up to a very high altitude beyond 100 km above sea level but in which the vehicle involved does not go into orbit (i.e., does not attain an orbital speed exceeding 11.2 km/s). A suborbital trajectory is defined under US law as "The intentional flight path of a launch vehicle, re-entry vehicle, or any portion thereof, whose vacuum instantaneous impact point does not leave the surface of the Earth." Unmanned suborbital flights have been common since the very beginning of the space age. Sounding rockets covering a wide range of apogees even well above the altitude of the Shuttle and ISS orbits have been routinely launched. Nowadays, suborbital human spaceflight is gaining popularity as demonstrated by the increased interest in space tourism. Still in its nascent phase, the space tourism industry proposes new commercial vehicles which have configurations and operational mode very similar to some early government programs, namely, capsules (e.g., Mercury Redstone) or winged-rocket system (e.g., X-15 aircraft). It should be noted that the two configurations drive very different safety requirements. Safety requirements for the launcher/capsule configuration have been in place for more than 40 years and have been successfully proven, mainly during the performance of (more challenging) orbital flights. The safety requirements for the aircraft-type configuration have a well-established technological basis in the aeronautical engineering field, although they are not reflected in any current civil aviation-type regulation. The experimental aircraft X-15 flew 199 times flights before program cancellation in 1968. The X-15 suffered four major accidents (Fig. 6).

In 2004 and then in 2011, the USA passed the so-called Commercial Launch Amendments Act (CSLAA) and the "Commercial Space Launch Activities" Act that was signed into law in January 2012.

Most recently, the USA enacted H.R.2262 – U.S. Commercial Space Launch Competitiveness Act which has become Public Law No: 114-90 (11/25/2015. This law has four parts that include Title 1 (Spurring Private Aerospace Competitiveness and Entrepreneurship), Title II (Commercial Remote Sensing), Title III (Office of Space Commerce), and Title IV (Space Resource Exploration and Utilization). This new legislation has a variety of shorter term and longer term space safety implications and oversight and licensing arrangements for commercial space launches for manned and unmanned flights. (H.R.2262 – U.S. Commercial Space Launch Competitiveness Act).

The issuance of the new Space Policy Directive-3 in August 2018 that assigned to the new Office of Commercial Space the responsibilities of addressing for commercial space activities both responsibilities for improved space situational awareness and addressing space traffic management issues will the U.S. Department of Defense will address these issues for security and strategic purposes creates a new

Fig. 6 X-15 crash (Credit: NASA)

path forward for the USA. This Directive set for the problem and the way forward to achieve greater space safety in the following manner.

"The future space operating environment will also be shaped by a significant increase in the volume and diversity of commercial activity in space. Emerging commercial ventures such as satellite servicing, debris removal, in-space manufacturing, and tourism, as well as new technologies enabling small satellites and very large constellations of satellites, are increasingly outpacing efforts to develop and implement government policies and processes to address these new activities."

"To maintain U.S. leadership in space, we must develop a new approach to space traffic management (STM) that addresses current and future operational risks. This new approach must set priorities for space situational awareness (SSA) and STM innovation in science and technology (S&T), incorporate national security considerations, encourage growth of the U.S. commercial space sector, establish an updated STM architecture, and promote space safety standards and best practices across the international community."

"The United States recognizes that spaceflight safety is a global challenge and will continue to encourage safe and responsible behavior in space while emphasizing the need for international transparency and STM data sharing. Through this national policy for STM and other national space strategies and policies, the United States will enhance safety and ensure continued leadership, preeminence, and freedom of action in space." (Space Policy Directive-3, National Space Traffic Management Policy, June 18, 2018. https://www.whitehouse.gov/presidential-actions/space-policy-directive-3-national-space-traffic-management-policy/).

It was suggested at the European Space Policy Institute Autumnal Conference in September 2018 that it might be possible for spacefaring nations to undertake a parallel approach to improved space situational awareness and space traffic management as outline in the U.S. Space Policy Directive-3 so that common and coordinated approach to these critical areas of space safety might be undertaken. The proposal made suggested the following: "U.S. Space directive 3 sets new U.S. objectives for 'space situational awareness', the need for an improved registry of space objects in Earth orbit, the need for better sharing of information with regard to operational spacecraft and space debris, the need for an improved space traffic management system and support for collision avoidance services." If other spacefaring nations were to do the same, then a cooperative framework to cooperate in all these areas might be agreed. This could lead to improved security of space operations or space infrastructure security. (Joseph N. Pelton A Path Forward to Improved Space Security: Better Information Sharing, Space Situational Awareness, Space Traffic Management, and More" European Space Policy Institute Autumn Conference, Sept 28, 2018, Vienna, Austria)

The IAASS (International Association for the Advancement of Space Safety) has been particularly concerned with the safety of commercial space systems carrying passengers into space and has developed a safety certification standard that has now been published by the SAE as of July 2018. This standard addresses responsibilities, implementation, and mission safety risk. (SAE International to Publish New IAASS Standard for Commercial Space Travel and Exploration 2018-08-06 WARRENDALE, Pa. August 2018. https://www.sae.org/news/press-room/2018/08/sae-international-to-publish-new-iaass-standard-for-commercial-space-travel-and-exploration)

Self-Regulations: Safety as Business Case

An alternative to government regulations is self-regulations. They are essentially meant to promote a higher level of safety as a business case. Take the example of Formula 1 car racing. In the first three decades of the Formula 1 World Championship, inaugurated in 1950, a racing driver's life expectancy could often be measured in fewer than two seasons. It was accepted that total risk was something that went with the badge. It was the Imola Grand Prix of 1994 with the deaths of Roland Ratzenberger and Ayrton Senna (as shown on direct broadcast TV) that forced the car racing industry to look seriously at safety or risk to be banned forever. In the days after the Imola crashes, the FIA (Fédération Internationale de l'Automobile) established the safety Advisory Expert Group to identify innovative technologies to improve car and circuit safety and mandated their implementation and certification testing. Nowadays, Formula 1 car racing is a very safe multibillion dollar business of sponsorships and global television rights, an entertainment for families that can be enjoyed without risking shocking sights.

Another example comes from the oil industry. The Presidential Commission that investigated the "Deepwater Horizon" disaster in the Gulf of Mexico in April 2010 (11 workers killed plus an oil spill that caused an environmental catastrophe) recommended the establishment of an independent safety agency within the Department of the Interior and that "the gas and oil industry must move towards developing a notion of safety as a collective responsibility. Industry should establish a 'Safety Institute' [...] this would be an-industry created, self-policing entity aimed at developing, adopting, and enforcing standards of excellence to ensure continuous improvement in safety and operational integrity offshore."

Nowadays, sophisticated techniques are available to remove or control hazards in new systems such to minimize the safety risk of new systems before they enter into operation. Such techniques go generally under the name of "safety case."

Prescriptive Requirements Versus Safety Case

The RMS Titanic struck an iceberg on her maiden voyage from Southampton, England, to New York and sank in the early hours of 15 April 1912. A total of 1,517 people died in the disaster because there were not enough lifeboats available. During the Titanic construction, Alexander Carlisle, one of the managing directors of the shipyard that built it, had suggested using a new type of larger davit, which could handle more boats giving Titanic the potential of carrying 48 lifeboats providing more than enough seats for everybody on board. But in a cost-cutting exercise, the customer (White Star Line) decided that only 20 would be carried aboard thus providing lifeboat capacity for only about 50% of the passengers (Titanic 1912). This may seem as a carefree way to treat passengers and crew on board, but as a matter of fact, the Board of Trade regulations stated that all British vessels over 10,000 t had to carry 16 lifeboats. Obviously, the regulations were out of date in an era which had seen the size of ships reaching the 46,000 t of the Titanic.

The above accident illustrates at the same time what is a prescriptive requirement (i.e., an explicitly required design solution for an implicit safety goal) and how it can sometimes dramatically fail. Instead the safety case regime is based on the principle that the regulatory authority sets the broad safety criteria and goals to be attained while the system developer proposes the most appropriate technical requirements, design solutions, and verification methods for their fulfillment. In other words, the safety case regime recognizes that it is the regulatory authority's role and responsibility to define where the limit lies between "safe" and "unsafe" design (i.e., the safety policy in a technical sense), but it is the developer/operator that has the greatest in-depth knowledge of the system design and operations.

A safety case is documented in the Safety Case Report that typically includes the following: (a) the summary description of the system and relevant environment and operations; (b) identified hazards and risks, their level of seriousness, and applicable regulatory criteria/requirements; (c) identified causes of hazards and risks; (d) description of how causes (of hazards and risks) are controlled; and (e) description of relevant verification plans, procedures, and methods.

The safety of the entire International Space Station (ISS) program is based on a process of incremental safety reviews by independent panels of safety case reports (called safety data packages) prepared by systems developers/operators in response to the (generic) safety requirements (NASA SSP 30599 2009). In the course of the operations, further submittals are made to account for configuration changes, previously unforeseen operations, and corrective actions from on-orbit anomalies.

Human Rating: A Historical Perspective

Since the first space programs that achieved human access to space, the identification of system requirements for crewed space systems has been a complex exercise. In the 1950s, the engineering efforts to maximize safety were built on the experience gained about the space environment from unmanned vehicles and experimental platforms with chimpanzees on board, which contributed to gather data for planned crew missions. The concept of human rating (also known previously as manned rated) was used to refer to systems designed to carry humans into space. However, a formal common process designated to grant human-rating certification did not exist at the time, as it is being used in current programs. In the past, the methods for implementing human rating varied as a function of program, across system and subsystems and sometimes across mission phases within a program.

In 1995, 14 years after the Shuttle had entered operations, an agency-wide committee was tasked to develop a human-rating requirements definition for launch vehicles based on conventional (historical) methods. After the revision of past programs both for launchers and spacecrafts such as Gemini, Apollo, and the Space Shuttle, the committee recommended the following definition of human-rating process, that is, "a process that satisfies the constraints of cost, schedule, performance, risk and benefit while addressing the three requirements of human safety, human performance, and human health management and care" in a document reviewing the historical perspective of human rating of US spacecraft (Zupp 1995). Historically, the human-rating process for Mercury, Gemini, and Apollo programs had been centered on human safety. The Skylab and Shuttle programs added to this an emphasis on human performance and health management. Further details on the history of these programs can be found in Logsdon and Launius (2008).

For Gemini as well as for other vehicles since then, an important part of assuring crew safety was the development of a crew escape system in case of abort scenarios. The escape system test program was also quite extensive, leading to the identification of improved designs throughout the testing phase and spanned a 3-year period, which lead to the development of a crew escape system, with an ejection seat qualified for flight crew space from pad aborts to 45,000 ft (Ray and Burns 1976). For the Apollo program, launch vehicles (i.e., Saturn IB and V) were designed for human spaceflight (given that no other launcher was able to deliver the required performance). These vehicles had additional redundancy and safety improvements as compared to its predecessors for Mercury and Gemini. Additionally, there was an extensive ground and unmanned flight plan to validate new design features and to certify the launch escape system uniquely developed for Apollo.

For the Space Shuttle, the considerations for crew safety were a tremendous challenge over previous programs mainly because with its configuration (where the Orbiter vehicle and the crew were much closer to the source of explosive yield of fire and overpressure than in the in-line series burn configurations used on the Mercury, Gemini, and Apollo launch systems). The most significant challenge was how to address the issue of abort during first stage. To enable the possible consideration of crew escape, crew ejection, launch pad ejection, or Orbiter separation and fly way, a method for thrust termination of solid rocket boosters (SRBs) had to be developed. It was a technology that was not proven. Various concepts for thrust termination were examined (i.e., pyrotechnically blow out the head end of the booster and neutralize thrust; another concept was to sever the nozzle to accomplish the same result), but all raised major concerns or introduced significant design challenges. Therefore, a decision was made that the additional safety risks and design complexities introduced by thrust termination were of greater concern that the presumed low failure rate of solid motors. For the areas of "high" risk, more stringent design requirements were derived to build in greater reliability for Shuttle SRBs (i.e., structural design factors of safety, case insulation, and segment seals). The Shuttle used a historical performance database to improve safety design and certified the vehicle to be human rated with no first-stage abort capability. The focus was on system-level integrated methodology.

The human-rating process builds upon data and knowledge acquired during development, manufacturing, and operations. The information derived from the evaluation and analysis of this data can only contribute to strengthening the understanding of failure mechanisms and identifying mitigation strategies to address them. Taking into account the lessons learnt from past programs as well as the technological developments of our time, the need for specific requirements for human rating a space system to enhance crew safety and incorporate the knowledge gained through more than 40 years of space activities materialized with the release of the NASA NPR 8705.2A "Human-Rating Requirements and Guidelines for Space Flight Systems" in 2003. In this first standard addressing human-rating certification, NASA proposed the following definition: "a human-rated system is one that accommodates human needs, effectively utilizes human capabilities, controls hazards and manages safety risk associated with human spaceflight, and provides to the maximum extent practical, the capability to safely recover the crew from hazardous situations."

In 2008 and then in 2011, NASA reissued and updated these requirements (i.e., NPR 8705.2B) with slight modifications from its original version, document that was later updated in 2011. This document contains a set of programmatic and technical requirements that establish a benchmark of capabilities for human-rated space systems. It directs programs to perform human error analysis, evaluate crew workload, conduct human-in-the-loop usability evaluations, prove that integrated human-system performance test results are required to validate system designs, and establish a Human System Integration team to evaluate these activities (Hobbs et al. 2008). NASA Constellation Program (i.e., Ares launchers and Orion capsule) was the first program to incorporate these new human-rating requirements. In parallel, activities are undergoing by other agencies (e.g., ESA and JAXA) for the refinement of safety technical requirements for human-rated space systems (Trujillo and Sgobba 2011). In 2011, the Commercial Crew Program (CCP) issued the CCT-1100 Series that communicates roles and responsibilities, technical management processes supporting certification, crew transportation systems, and ISS-related requirements for potential commercial providers.

Human Spaceflight Safety Risks

The principal safety issues related to orbital human spaceflight are protection from environmental hazards whether space weather (i.e., ionizing radiation) or space debris, the need to provide escape and safe-haven capabilities, and prevention of collision risk. Collision risk may be divided into (1) the risk of collision during proximity operations (i.e., rendezvous and docking) and (2) risk of collision with other space traffic.

Environmental Risk: Ionizing Radiation
The Earth's magnetic field traps electrically charged radiation particles in two belts high above the Earth. The highest extends out to about 40,000 km, and the lowest belt begins at about 600 km above the surface. The intensity of radiation in these belts can be more than a million times higher than on the Earth. For several decades to come, commercial orbital human spaceflight will most probably be limited to low Earth orbit flights where the radiation level is small or negligible. Based on the experience of several decades of human spaceflight in low Earth orbit (Vetter et al. 2002), a safe level of radiation exposure has been defined as that which would increase the lifetime risk of cancer by 3%, and this translates into a total dose of 100–400 rem depending on age and gender (Cucinotta et al. 2011). For comparison, a maximum of 10 rem is the annual dose allowed for workers in occupations involving radiation. Since health risk increases with the total dose, it is important to monitor the dose and to establish norms for the retirement of (commercial) astronauts who reach that level (NRC 2012).

Space Safe and Rescue: Past, Present, and Future
The 1912 Titanic disaster, with a distress message telegraphed in Morse code, was a defining moment in starting the organization of search-and-rescue on a global scale. The shock of the disaster led to the establishment of means for constant distress surveillance on land and aboard ships. In 1914, the first International Convention for the Safety of the Life at Sea (SOLAS) made it an obligation for ships to go to the assistance of other vessels in distress. The system developed and matured gradually in the following decades, and in the early 1950s, it was extended to aviation, but it was only in 1985 that a well-organized international search-and-rescue (SAR) system came into force under the International Convention on Maritime Search and Rescue of 1979. The current international SAR system is based on close coordination between international maritime and aviation organizations and relies on uniform worldwide coverage and use of global space-based monitoring and tracking resources available on board GEO and LEO spacecraft (COSPAS-SARSAT Programme).

As with any comparable system, the safety of crew and passengers on board future suborbital and orbital commercial space vehicles will not depend only on design adequacy, robustness of construction, and the capability to tolerate failures and environmental risks but also upon special provisions which would allow escape, search, and timely rescue in case of emergencies. During a suborbital commercial human spaceflight, an emergency may lead to search and rescue operations at sea or on land not dissimilar from those of an aviation accident. The case of an on-orbit emergency is different, and for that special cooperation, provisions and interoperable means need to be developed. Here, the closest parallel is that of submarine emergencies. Many nations now regularly practice multilateral rescue exercises and coordinate their rescue means and capabilities through the International Submarine Escape and Rescue Liaison Office (ISMERLO).

Ascent Emergencies

During the ascent phase, a so-called abort scenario needs to be considered in order to safeguard the life of the crew and passengers on board a commercial space vehicle. Such scenarios apply to any type of space vehicle and would require also planning and cooperation with foreign countries.

Taking the experience of the Shuttle program as an example, depending on the time a malfunction would have occurred, there were Shuttle international launch abort sites at Halifax, Stephenville, St. Johns, Gander, and Goose Bay (all in Canada). There were also Shuttle transoceanic abort landing sites (TAL) at Ben Guerir Air Base, Morocco; Yundum International Airport, Banjul, The Gambia; Moron Air Base, Spain; Zaragoza Air Base, Spain; and Istres, France. Finally, there were 18 designated Shuttle emergency landing sites spread among Germany, Sweden, Turkey, Australia, and Polynesia, several of which are active international airports. For the purpose of providing the Shuttle program with the necessary assistance, access, and dedicated capabilities at those foreign landing sites worldwide, the US government had to negotiate a large number of specific bilateral agreements. In the future, when commercial human suborbital and orbital spaceflights become common, commercial entities will not be able to gain the same level of assistance on land or at sea and access to foreign facilities unless the necessary international civil space agreements and regulations are put in place by some sort of international space regulatory body similar to ICAO for aviation (Jakhu et al. 2010).

Crashworthiness

Additionally, from the lessons learned of the Columbia accident and based on the findings of the Columbia Accident Investigation Board (CAIB), tasked by NASA to conduct a thorough review of both the technical and organizational causes of the loss of the Space Shuttle Columbia. The CAIB recommended that future vehicles should incorporate the following: (a) a design analysis for breakup to help guide design toward the most graceful degradation of the integrated vehicle system and structure to maximize crew survival; (b) crashworthy, locatable data recorders for accident/incident flight reconstruction; (c) improvements in seat restraint systems to incorporate the state-of-the art technology to minimize crew injury and maximize crew survival in off-nominal acceleration environments; and (d) advanced crew survival suites (including conformal helmets with head and neck restrain devices similar to the ones used in professional automobile racing) and avoidance of materials with low resistance to chemicals, heat, and flames among others.

Orbital Rescue

In 1990, an International Spacecraft Rendezvous and Docking conference was held at the NASA Johnson Space Center. The purpose was to explore the need and international consensus to establish a set of common space systems design and operational standards which would allow docking and on-orbit interoperability in case of emergency. The attributes for such international standards were summarized as follows: (a) each party could implement them with their own systems and resources; (b) cooperation in such standards does not require subordination (i.e., one party does not have to buy parts of the system from another); (c) success of one project or project element is not required to insure success of the other; (d) no one standard requires subordination to another standard; and (e) the functional requirements of the standard can be implemented with a number of alternative technologies. Definition of the standards does not require the transfer of technology.

In 2008, the objective of developing orbital rescue capabilities was restated by the US Congress in the NASA Authorization Act of that year (H.R. 6063). In fact, Sect. 406, EXPLORATION CREW RESCUE, stated that: "In order to maximize the ability to rescue astronauts whose space vehicles have become disabled, the Administrator shall enter into discussions with the appropriate representatives of space-faring nations who have or plan to have crew transportation systems capable of orbital flight or flight beyond low Earth orbit for the purpose of agreeing on a common docking system standard."

In 2010, the international docking system standard (IDSS), based on the original androgynous docking system (APAS) developed in the seventies as part of the Apollo-Soyuz Project, became finally a reality through the initiative of the countries participating to the International Space Station program. Although China was not involved in such standardization effort, the Chinese had already chosen as docking system for their Shenzhou vehicle and for the Tiangong-1 space station a docking system variant called APAS-89, which is the same used on the International Space Station (ISS) and is compatible with the new international docking standard. The Chinese docking system was successfully demonstrated on-orbit in 2011 with a robotic mission. In 2012, further dockings were performed by two Shenzhou (9 and 10), both of with crew board. Following Tiangong 1, a more advanced space laboratory, dubbed Tiangong 2, was launched in 2013 followed by Tiangong 3 in 2015. In the coming years, at least two space stations will be orbiting Earth, the ISS and the Chinese Tiangong, thus making possible for the first time an orbital rescue system. Even private space stations are now envisioned by Bigelow Aerospace with prototypes now in orbit.

In 2004, a cooperative program was launched to implement such capability on the model of the International Submarine Escape and Rescue Liaison Office (ISMERLO) to "establish endorsed procedures as the international standard for submarine escape and rescue using consultation and consensus among submarine operating nations." As for submarines, also in space, the delay between an accident and rescue attempt must be short. Furthermore, the institutionalized contacts and increased transparency engendered by such cooperation orbital rescue would fit with broader trends toward increasing openness and could constitute an important confidence-building mechanism for wider cooperation in making space operations safe and sustainable.

Conclusions

This review has presented a wide variety of space risks. It has explored the safety risks that experienced space organizations and new spacefaring nations are facing. An in-depth understanding of these risks is important to fully comprehend the scope of the safety challenges ahead. Without such an understanding, it will be difficult if not impossible to mitigate them in an effective manner. Both unmanned orbital space systems and crewed vehicles are adversely affected by the growing amount of orbital debris. The cascading effect produced by space objects is a mounting concern. We must seek to minimize the impact of uncontrolled reentering objects that affect the safety of those on land, air, and sea. In addition, the proliferation of new commercial ventures indicates the need to promote space safety in the area of orbital and suborbital tourism and raises the question as how space traffic management might be addressed in future years. The complexity of space safety issues and the scope and nature of future safety challenges may well need to be tackled through an expanded international regulatory framework – one expanded to address the space safety risks that have been described in this chapter.

References

Ailor W, Wilde P (2008) Requirements for warning aircraft of re-entering debris. In: Proceedings of IAASS conference on space safety (2010) Huntsville, Alabama

Cucinotta F, Kim M-HY, Chappell LJ (2011) Space radiation cancer risk projections and uncertainties – 2010. National Aeronautical and Space Administration, Washington, DC. (publisher)/TP-2011-216155

Helton-Ingram S et al. (2005) Federal Aviation Administration report to the space and air traffic executive board on the plan to mitigate air traffic hazards posed by the space shuttle return to flight re-entry. FAA, Washington, DC. (publisher)

Hobbs A, Adelstein B, O'hara J, Null C (2008) Three principles of human-system integration. In: Proceedings of the 8th Australian aviation psychology symposium, Melbourne

Jakhu R, Sgobba T, Trujillo M (2010) An international civil aviation and space organization. In: Proceedings of IAASS conference, 2010, Huntsville, Alabama

Leveson N (2003) White paper on approaches to safety engineering. MIT

Liou LC, Johnson NL (2006) Instability of the present LEO satellite populations. COSPAR

Logsdon M, Launius R (2008) Human spaceflight: projects Mercury, Gemini, and Apollo. NASA Smithsonian Press, Washington, DC, pp 2008–4407

NRC – Committee for Evaluation of Space Radiation Cancer Risk Model (2012) Technical evaluation of the NASA model for cancer risk to astronauts due to space radiation. National Academies Press, Washington, DC

Olsen JA (2010) A History of Air Warfare. Potomac Books, Washington

Ray HA, Burns FT (1976) Development and qualification of Gemini Escape System. NASA TN D-4031

Safety Review Process (2009) International space station program, NASA SSP 30599, Rev. E. Resource document. http://kscsma.ksc.nasa.gov/GSRP/Document/ssp%2030599%20reve.pdf. Accessed 1 June 2012

"Titanic" Inquiry project. Resource Document (1912). http://www.titanicinquiry.org/BOTInq/BOTReport/BOTRep01.php. Accessed 1 June 2012

Trujillo M, Sgobba T (2011) ESA human rating requirements: status. In: Proceedings of the IAASS Conference on Space Safety, Versailles

Vetter R, Baker ES, Bartlett DT, Borak TB, Langhorst SM, McKeever SWS, Miller J (2002) Operational radiation safety program for astronauts in low-earth orbit: a basic framework, Report No. 142

Zupp G (1995) A perspective on the human-rating process of U.S. spacecraft: both past and present. National Aeronautics and Space Administration, Washington, DC. (publisher)

Evolution of Space Traffic and Space Traffic Management

16

William Ailor

Contents

Introduction	300
Objects in Orbit	300
Protected Regions	300
GEO Protected Region	302
LEO Protected Region	303
Space Debris	304
Space Situational Awareness Services	306
Space Situational Awareness Data	306
Best Practices and Standards	307
Changes Coming	308
Large LEO Constellations	309
Environmental Effects on Satellite Lifetime	312
Reentry Disposal of Satellites from Large Constellations	313
Active Debris Removal (ADR)	314
Effect of Large Constellations on SSA Service Requirements	314
Space Situational Awareness and Traffic Management Service Providers	315
Conclusions	315
References	317

Abstract

In 2016, the notion that the local space environment would continue to evolve at rates defined by previous satellite launch histories changed dramatically when SpaceX requested permission to place over 4400 satellites in low Earth orbits. Prior to that time, predictions were that the space environment could be stabilized by space operators abiding by rules and guidelines designed to limit the growth of debris. While all proposed satellites might not be realized, major changes in the

W. Ailor (✉)
Center for Orbital and Reentry Debris Studies, The Aerospace Corporation, El Segundo, CA, USA
e-mail: william.h.ailor@aero.org

near-space environment are coming and may come quickly. This chapter provides background, discusses potential changes, and highlights new polices and services that could arise.

Introduction

Objects in Orbit

Figure 1 shows the growth in the number of operating satellites and debris since the beginning of the space age by object type, Fig. 2 shows orbiting objects sorted by orbit class, and Fig. 3 shows the number of objects per cubic kilometer as a function of altitude.

Protected Regions

Several years ago, spacefaring nations recognized that two regions of near-Earth space are particularly important to space-based services and set these regions aside as "protected regions." The first is the low Earth orbit (LEO) protected region shown in Fig. 4 (Region 4). This region extends to 2000 km above the Earth's surface and is heavily used by satellites that provide communication, Earth monitoring, and other services.

The second protected region is the ring of space surrounding Earth where satellites in geosynchronous equatorial orbits (GEO) operate (Region 3). This is

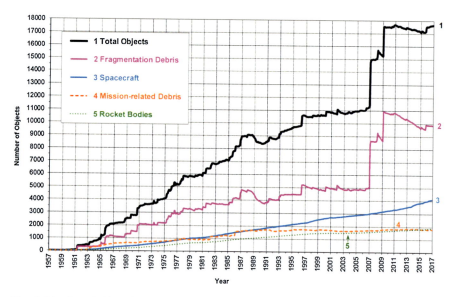

Fig. 1 Number of objects in Earth orbit by year and object type (courtesy NASA)

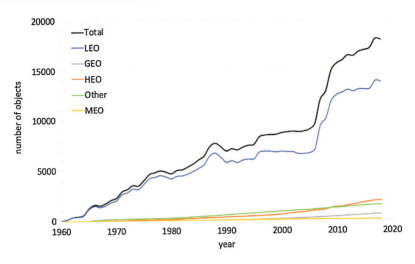

Fig. 2 Number of objects by year and orbit class (LEO, MEO, GEO, HEO)

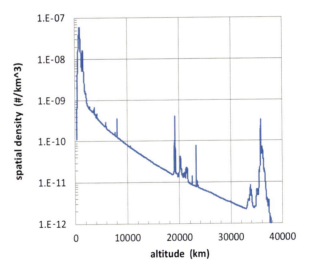

Fig. 3 Number of objects/cubic km as a function of altitude in 2018

where many large communication and weather monitoring satellites operate at fixed locations above specific regions of Earth.

A region not formally protected but should be avoided when disposing of satellites is the medium Earth orbit (MEO) region between 19,700 and 20,700 km above the Earth's surface. This region is home to Global Positioning System (GPS), Glonass, Compass, and Galileo constellations. Due to the critical nature of the navigation and other services these constellations support, current end-of-mission satellite disposal guidelines recommend that orbits for space hardware being disposed avoid passing through this region.

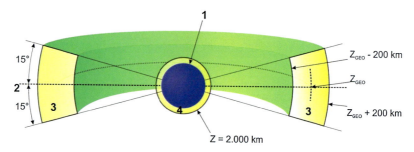

Fig. 4 Protected regions (Key: 1 Earth; 2 equatorial plane; 3 GEO protected region; 4 LEO protected region; Z altitude measured with respect to a spherical Earth whose radius is 6378 km; ZGEO altitude of the geostationary orbit with respect to a spherical Earth whose radius is 6378 km) (from ISO 24113:2019)

Highly elliptical orbits (HEOs) are orbits with high eccentricity, and vehicles in these orbits can pass through LEO, MEO, and GEO regions.

GEO Protected Region

In addition to approximately 470 operating satellites in the GEO protected regime, there are more than 1300 tracked objects (objects larger than 1 meter) and also a number of untracked objects, as well. Based on recent research (Oltrogge et al. 2018), there could be over 1600 objects between 10 cm and 1 m in size and 33,000 objects larger than 1 cm that cross the GEO protected region. The reference argues that a collision of a small, untracked 1 cm-class object with an operating satellite might occur as frequently as every 4 years and every 50 years for an object larger than 20 cm. In the GEO region, collision velocities could reach as high as 4 km/sec with objects in highly eccentric orbits piercing the GEO protected region. More common collisions would be at relative velocities of less than 1 km/sec.

One of the largest and oldest GEO operators is Intelsat, now a commercial company but originally established as an intergovernmental consortium in the 1970s. Intelsat currently operates a fleet of 52 communications satellites. The operating GEO satellites are large, and they have long lifetimes, some exceeding 30 years. Their locations are known well, and many of the satellites operate in orbits very near to Earth's equatorial plane, so they move slowly relative to one another. As a result, operators of GEO satellites generally have more time to make adjustments to avoid close approaches with known objects and maintain their satellites' orbital positions than those in LEO.

Satellites in GEO can provide services to a relatively large area of the Earth's surface, and their 24-hour orbits essentially fix them in place over specific points on the equator, so ground antennas can be pointed at predefined locations to send and receive signals. Satellites in GEO are designed to be very reliable and to operate within predefined slots for long periods. In the past, very large ground antennas were required for communicating with these satellites. The size requirements have been reduced substantially over time. An antenna on a home or business that is pointing to the sky is generally pointing to a satellite in a GEO orbit.

For the near-term, major changes in the design and operational characteristics of GEO satellites or the GEO environment are not expected. An area where we might expect advancements is in the maintenance and disposal of GEO satellites, and some commercial companies are developing satellite servicing and potentially active debris removal (ADR) services as well. These services would perform satellite refueling and disposal services, potentially extending the life of operational satellites by replenishing station-keeping propellant and offering disposal options when satellites fail or end their missions. Since longer lifetimes have economic benefits to the owners, satellite operators could pay for servicing and disposal services. In the future, GEO operators might create a fund that would cover costs of removal of existing derelict satellites or satellites that fail prematurely.

For the GEO region, satellites are to be disposed in orbits above the GEO protected region where they will not reenter that region for at least 100 years. Design of disposal orbits must account for the long-term effects of solar wind, the gravitational attraction of the sun and moon, and other small forces.

LEO Protected Region

Satellites in LEO are at lower altitudes (altitudes less than 2000 km), so their coverage on the Earth's surface is less. They do not operate over fixed regions of the Earth surface but make a complete orbit around Earth in 90 to 120 min. If services are desired by a customer at a fixed location, orbits must be designed to pass within range of that spot at prescribed intervals; if continuous communications is required, a constellation of LEO satellites is necessary, and multiple satellites in that constellation must provide coverage and make and hand off connections with that customer as services are provided. The communications pathway is completed by ground antennas that receive data from linked satellites and pass that data to ground-based users.

Since they are at lower altitudes, LEO satellites have the advantages that they require much less power to communicate with customers on the ground than satellites in GEO. That makes the individual satellites much less expensive to fabricate and put into orbit and enables a customer on the ground to use a low-power, hand-held instrument for communications. The communications time lag is also substantially lower, which is important for machine-to-machine communications.

These features were incorporated in the design of the Iridium satellite system, which includes a relatively large constellation of 66 satellites in low Earth orbit plus several on-orbit spares and provides voice and limited data communication to people located anywhere on Earth. All satellites in the Iridium constellation are essentially of the same design, enabling production on an assembly line. Iridium's first constellation became operational in 1998 and is now being replaced with more capable satellites.

Following Iridium's lead, a revolution is on the horizon as new operators propose very large constellations in the LEO regime. But given the realities illustrated in

Figs. 2 and 3, the environment in LEO – and on the ground – must be considered carefully as this revolution moves forward.

Over 1800 satellites currently operate in the LEO regime, joined by over 13,000 debris objects large enough to be tracked (i.e., larger than 10 square centimeters) plus possibly over 500,000 objects between 1 and 10 cm in size that are harmful to operating satellites but can't currently be tracked. In addition to the large population of orbiting objects, LEO is a more challenging environment than GEO for space operators for several reasons:

- Satellite orbits in LEO are affected by aerodynamic forces that are amplified by environmental factors such as the day/night cycle, solar storms, carbon dioxide in the environment, and other factors that make accurate predictions of where the satellites will be in the future much more challenging.
- For an object to be tracked, it must pass over tracking radar sites on the ground, which will limit the frequency of updates available of the object's orbit. These updates are required to correct for the effects of the atmosphere and other forces on the object's orbit. Without frequent updates, the uncertainty in an object's position at a given time increases, affecting the quality of a collision warning.
- While there are exceptions, operating satellites in the GEO protected region are taking advantage of circular orbits that are at very low inclinations, and positions are well known and well controlled, so most operating satellites move slowly relative to each other compared to those in LEO, where operating satellites are in multiple orbit planes and approach velocities for objects in different planes can be 10 km/sec or higher. As a result, a collision in LEO, even with a small debris object, can destroy an operating satellite and add large numbers of debris objects to the orbital environment.
- There are lots of dead satellites and other debris in LEO that must be frequently tracked, can't maneuver, and must be avoided.

As noted, orbits of satellites in LEO are affected by very small aerodynamic forces that increase as altitude decreases. While these forces help remove debris in low orbits, the lifetime of orbits increases substantially with altitude. For example, the orbit of a dead satellite in a 250-km circular orbit will decay, and the object will reenter in less than a year; a dead satellite in a 1000-km orbit will remain in orbit and be a hazard to other satellites for over a thousand years. Lifetimes of circular orbits continue to increase as altitude increases. Estimates of the lifetime for a particular satellite depend on the satellite's mass, physical dimensions, orientation, and environmental factors such as solar activity as it descends.

Space Debris

Inoperable, human-made objects orbiting Earth are space debris, and space debris can be created by collisions and explosions, by release of objects during satellite deployments and normal operations, by strikes by micrometeoroids and small space debris fragments, and by long-term exposure to the orbital environment. A collision or explosion involving a large object such as a satellite or launch stage creates an expanding cloud of debris objects that eventually blends into the background of debris objects in the orbital environment, slightly increasing the background risk to all satellites in the region. Traveling at orbital speeds, even a flecks of paint can gradually degrade solar panels and damage optical sensors, reducing the mission lifetime of the damaged vehicle. It should be noted that satellites with no maneuver capability are effectively space debris to an operating satellite, which must move to avoid a collision.)

Current guidelines and standards require that satellites in both the LEO and GEO protected regions remove themselves at end of mission to minimize the growth of space debris and minimize the possibility of interfering with other operating satellites in those regions.

At end of life, satellites in LEO orbits should be reentered into the atmosphere for disposal. The reentry process does not necessarily completely "burn up" a reentering object, and some fragments that are potentially hazardous to people on the ground can survive. For this reason, requirements state that if the casualty expectation per reentry exceeds 1 in 10,000 (i.e., surviving fragments might injure or kill one person on Earth should the object reenter 10,000 times), the object should be directed to reenter into a region where there is minimal hazard to people (e.g., an open ocean area). Otherwise, the object may be allowed to simply reenter as its orbit decays – as long as that process takes fewer than 25 years. (As will be discussed later, requirements to minimize hazards due to reentries of large numbers of satellites from large constellations may be a possibility for the future.)

The approximately 20,000 objects included in the figures are only those tracked by ground and optical sensors, and these generally range from 10 cm and larger in the LEO regime to 1 m and larger in the vicinity of the GEO regime. As noted, the orbital environment includes many thousands of objects smaller than 10 cm in size that have resulted from satellite explosions, collisions, debris expelled during normal operations, and other sources over the years. As a result, the total population of orbiting objects is actually considerably higher than the figures indicate, with estimates that the actual total includes as many as 500,000 small, currently untrackable (due to their size) objects in LEO, each of which is large enough to seriously damage a satellite on impact.

The jumps in the population of LEO objects shown in Fig. 1 are due to debris created by the 2007 Chinese anti-satellite (ASAT) test, where a ground-launched vehicle impacted an aging Chinese weather satellite, and collision of the active Iridium 33 satellite and inoperative Russian Cosmos 2251 satellite in 2009. The latter collision ended the operations of the Iridium satellite and also created a debris cloud that included several thousand additional objects, many of which will remain in orbit for decades to come. While some debris from these events has subsequent reentered and is no longer in orbit, Fig. 1 shows that many of the larger tracked objects (and likely many smaller, currently untrackable objects) remain in orbit. And given the altitude where the events occurred, some of these objects will remain hazards to other objects for centuries to come.

The Iridium-Cosmos collision changed the perspective on orbital risks and resulted in a new focus on providing better information to satellite operators on possible threats to their space assets.

Space Situational Awareness Services

Prior to the Iridium-Cosmos collision, the "big sky" perspective, which said that space was so vast that collisions would be very rare, prevailed. Only a few satellite operators felt the need for real collision avoidance and other space situation awareness (SSA) services.

At that time, services were provided by the US Air Force using tracking data collected using government-owned sensor systems. And many operators felt that the available services were not "actionable," meaning that predictions informed a satellite operator when an object might pierce the physical space surrounding a satellite, but that space could be kilometers in size. Given that level of uncertainty, operators did not feel there was sufficient information to warrant moving a satellite, since moving requires expenditure of propellant, with each maneuver fractionally reducing a satellite's lifetime. And a move based on inaccurate data might actually increase the risk of a collision with the approaching object or possibly with another at a later time. Moving a satellite can also affect its ability to fulfil its basic mission objectives. For example, if a satellite must maneuver to avoid a possible collision, its ground coverage area will also move, potentially requiring operator actions to avoid loss of valuable data or service interruptions.

Over the years, new sensors and analysis techniques emerged. Shortly after the Iridium-Cosmos collision, the USA provided close approach distances and probabilities of collision to satellite operators. While this is an improvement over earlier formats for conjunction assessments, some satellite operators still felt that the information from the USA was insufficient for their needs.

In 2009, several operators of satellites in GEO orbits formed the Space Data Association (SDA) to "improve the accuracy and timeliness of collision warning notifications...via sharing of operational data." The SDA, through its Space Data Center, supplements catalog data from the US government with information provided by operators of GEO satellites, who generally know very accurately where their satellites are and also know when satellite maneuvers will occur. The SDA also provides radio frequency interference and other support and assists operator efforts to coordinate maneuvers to avoid interference with other objects.

Today, in addition to the SDA and the US Air Force, several governments provide similar services for operators of their own satellites. But most use the US catalog of resident space objects, supplemented with information from their own sensors and satellite operators, as the basis for their services.

Space Situational Awareness Data

Since the beginning of the space age, the primary catalog used for SSA services has been created and maintained by the US Air Force. This Resident Space Object (RSO) catalog is generally considered to be the most complete of any currently available, and unclassified portions of this catalog have been made available for years (Space-Track.org is the current source). Data for this catalog has been collected primarily by ground-based radar and visual telescope systems operated by the US government. Planned enhancements to tracking resources may decrease the minimum size of tracked objects to as small as 2 cm. As a result, the number of objects in the RSO catalog is expected to increase from approximately 20,000 to over 200,000.

At present, new commercial and international entities are adding their data to the mix. For example, one company currently has two phased-array radar sites in operations and is building a third. That company is currently tracking more than 14,000 objects in LEO and expects to track as many as 250,000 objects 2 cm and larger several times a day when its radars are fully deployed and operational. A second company is operating a global SSA telescope network with more than 25 observatories and 250 telescopes that is tracking man-made space objects in GEO, highly elliptical orbits (HEO), and medium Earth orbits (MEO). Both of these companies offer a variety of services based on the data they collect.

Collecting data on most objects several times a day will enable space situational awareness services of unprecedented accuracy. This type of data will be essential as SSA service providers, who will need to provide accurate and timely warnings of collisions and other interference as the number of objects increases over the next 10–20 years.

Best Practices and Standards

As the number of objects in orbit increased, it became apparent that best practices, guidelines, and even regulations were required to prevent the growth of the space debris population. If this was not done, predictions were that the debris population could continue to increase in an uncontrolled manner as objects and fragments of objects impacted other objects, creating more fragments, etc. (the Kessler effect), eventually making space operations much more difficult and expensive.

And even small, untracked debris can cause problems. For example, impacts of a small debris particle on a solar panel can reduce the power output from that panel – and a large number of impacts can drop the power output so low that the satellite is not able to perform its designated mission and must be deorbited. Ailor (2010) concluded there would be a relatively small decrease in the mean satellite lifetime due to operating in the debris environment for the next 30 to 50 years based on the then-projected environment (an environment with no very large constellations in LEO). A primary driver in the lifetime reduction was the solar panel degradation due to impacts by small debris – impacts that gradually lowered the power provided by the solar panels until power dropped below a critical value, ending the satellite's mission. The effect of the addition of large LEO constellations on these projections will be discussed later.

During this period, it was also recognized that when a launch vehicle or spacecraft reentered the atmosphere at end of life, the object would not completely "burn up." Fragments that could injure or kill a human on the ground might survive.

In the late 1990s, the Inter-Agency Space Debris Coordination Committee (IADC) developed guidelines stating that satellites in LEO should be disposed before their end of life, either by moving them to an orbit that would naturally decay in 25 years or less or preferably by direct reentry – controlling the deorbit process so the debris surviving reentry lands in a safe area. Direct reentry into the atmosphere was recommended for objects where surviving debris might have casualty expectations exceeding 1 in 10,000 (i.e., reentries of that object 10,000 times would be expected to cause one casualty somewhere on Earth). Many nations have incorporated these guidelines into regulations.

The IADC guidelines were developed for reentries of individual satellites. Current proposals suggest that several constellations containing many satellites may be in the offing for the next decades. As will be discussed later, this may lead to new guidelines designed to limit the creation of space debris and hazards to people on the ground and in aircraft as we move forward.

Changes Coming

From the beginning of the space age to the early 1990s, satellites were essentially one-of-a-kind items – each was built with specific capabilities designed for specific missions, and each launch carried only one or, at most, two satellites to orbit. The Iridium satellite system changed that paradigm with its factory-built satellites. The release of the iPhone in June 2007 was another paradigm shifter and was described as "revolutionary" and a "game changer" for the mobile phone industry. These events and subsequent releases of smart phones that included small accelerometers, sensors, cameras, microprocessor, and other technologies encouraged innovators to see how these new satellite manufacturing processes and microelectronics capabilities might be used more broadly in space systems.

During this period, a standard size for a new class of small satellites was defined – a cube $10 \times 10 \times 10$ cm in size – and launch service providers included "piggyback" launchers for these CubeSats that could deploy many such satellites per launch. Most of these small satellites were placed in the low Earth orbit regime, and many were low enough that their orbits would decay within a few years – an important feature given the increasing recognition of a growing space debris problem. Very few of these satellites carried significant propulsion capabilities, and most were experimental in nature, so many rapidly became "space junk," joining the population of nonfunctioning, human-made debris circling our planet.

These new technologies led to a decrease in cost and mass of very capable satellites, enabling providers of space-based services to consider providing worldwide communication and internet services via constellations consisting of large numbers of small satellites (satellites less than 500 kg) in LEO. These satellites would follow the approach used by Iridium: a large number of satellites based on a fixed design would be mass-produced on an assembly line.

A benefit of the smaller satellite size and mass is that launch vehicles could carry more than one satellite to orbit, and Fig. 5 shows the number of launches and the number of satellites carried to orbit per launch from 2005 through 2018. While the number of launches per year has been relatively stable, the number of payloads per

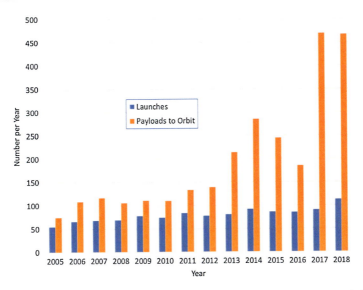

Fig. 5 Number of worldwide launches and payloads carried to orbit from 2005 through 2018

launch has increased substantially (launch failures affected the number of launches and payloads to orbit in 2015 and 2016).

Large LEO Constellations

The lower costs associated with satellite manufacture and launch have led to proposals for placing really large constellations of satellites in LEO, with some exceeding over 1000 satellites in orbit with multiple satellites in each orbit plane. Table 1 lists some organizations that have announced such plans.

The largest proposed constellation to date is SpaceX's StarLink constellation, which would have over 7500 satellites orbiting at 346 km (below the orbit of the International Space Station). If completed, that one constellation would have over four times the total number of satellites that were operating in orbit at the beginning of 2019. SpaceX has proposed locating another 4425 satellites at 1100–1300 km as part of that constellation. While some of the constellations in Table 1 may not materialize, it is evident that the LEO space environment may change dramatically over the next decades.

Clearly the addition of such large numbers of satellites into the LEO regime will raise a number of questions that should be answered. These include:

1. **How will satellites in these constellations be delivered to their operational orbits?** Satellites can be inserted relatively quickly into their orbital positions by launch stages, a common practice today, or they may be inserted into a lower orbit, where they can be deployed, checked out, and then boosted to a higher

Table 1 Proposed constellations

Proposed constellations	Number of satellites
SpaceX K-band (high altitude)	4425
OneWeb	720
LeoSat	120
Theia	112
Telestar	117
Boeing	2956
SpaceX V-band (low altitude)	7518

mission orbit. The advantage of the latter approach is that the checkout orbit can be low enough that the orbit will decay quickly if the satellite is faulty, minimizing the likelihood of subsequent failure and leaving a dead satellite at higher constellation altitudes. If the satellite is found to be healthy during checkout, it can then be put on course to its final orbit.

Some are considering using an electric propulsion system for the transfer from the low orbit to the constellation altitude. While this technique minimizes propellant consumption, it produces a very low thrust, so the satellite will move relatively slowly through altitudes where other satellites operate as it spirals upward, increasing the potential collision threat during transit and the load on operators of space traffic management systems as they seek to minimize the possibility of collisions.

2. **With so many satellites in a large constellation, how will risks to satellites passing through be managed?** In GEO, the International Telecommunications Union (ITU) assigns satellites to operate in specific orbital slots, typically two degrees in size, to minimize frequency interference problems. Since GEO satellites provide services over specific ground areas, these restrictions work well: operators of GEO satellites know whose satellites are operating where and announce moves and activities to avoid interference.

   In LEO, the situation is different: satellites in LEO circle Earth about every 90 minutes and collect data and provide services over relatively small areas as they pass overhead. As a result, an operator wishing to add a satellite to a constellation at a higher altitude must coordinate the passage of that satellite with operators of constellations below that altitude. And if the satellite being moved uses low-thrust propulsion, the transit time will not be quick. The concept of assigning a constellation responsibility for its "shell" and providing best practices for operators wishing to pass through that shell has been suggested. No formal arrangements have been made along these lines.

3. **What is the expected lifetime of satellites in the constellation?** An advantage of the LEO constellations is that it is relatively inexpensive to launch replacement satellites that incorporate the latest systems and capabilities, so some constellation designers might use satellites with 5- to 10-year lifetimes. That could mean a large fraction of each constellation would be moving toward disposal each year as constellations are maintained. For example, assuming a 10-year lifetime, a large

fraction of the satellites could be moving down through lower altitudes toward reentry on a yearly basis. And if electric propulsion is used, that transit could take significant time. All of these satellites in transit to and from the constellations add to the risks of collisions in LEO, so satellite lifetime and constellation disposal plans and strategies will be important considerations as plans for constellations develop.

4. **What happens if a satellite fails while operating in the constellation?** It is inevitable that satellites operating in constellations will fail. These large constellations in low Earth orbit must necessarily have multiple satellites in each orbit plane to provide ground coverage to service their customers (e.g., there are 11 satellites in each of the 6 orbit planes in the 66-satellite Iridium system, and Iridium also maintains several in-orbit spares). A constellation with thousands of satellites could have a hundred or more satellites in each plane, and the operations of each must be well controlled and well-coordinated in orbit. If a satellite fails or communications is lost, it could become a threat to others in the same plane and in the constellation itself. Active support from a space situational awareness service would be required in this case. Clearly, satellites in constellations should be designed with very high reliability for disposal.

5. **How will satellites be disposed at their end of life?** The preferred approach for disposing satellites is to have them leave orbit and reenter the atmosphere in a location where surviving debris will impact in an uninhabited area such as the South Pacific Ocean. In this case, surviving debris has minimal chance of injuring a human or damaging an aircraft. As noted earlier, if the casualty expectation for a reentry is less than 1 in 10,000, current guidelines and regulations say that the satellite can simply be left in an orbit that will decay in less than 25 years. As large constellations emerge, regulators may choose to also limit the hazard posed by cumulative reentries of satellites from these constellations (more on this later). And the 25-year time allotment could be shortened to reduce the transit time to reentry, or might even be eliminated in favor of requiring direct disposal of satellites of larger sizes to minimize ground hazards arising from reentries of large numbers of satellites per year.

6. **How important is satellite disposal?** A key factor in determining the future of debris growth in LEO is the reliability of disposal at end of mission or end of life. Dead satellites in some constellations could remain a threat for hundreds or thousands of years, so managing the LEO environment will require satellites to have disposal systems that can deorbit constellation members with a very high probability of success – some argue the probability of success should be over 90% if we are to maintain some control on the growth of the LEO population.

7. **Since satellites from large constellations will be disposed by reentry into the atmosphere, will risks of hazards to people on the ground or in aircraft increase?** Studies are showing that yes, hazards to people on the ground and in aircraft could increase substantially as a result of hardware associated with large constellations reentering Earth's atmosphere. More on this later.

8. **How will the increase in space traffic be managed?** The current SSA system tracks 20,000 to 30,000 objects, and, as noted earlier, satellite operators in the

LEO regime are calling for services that provide more and more accurate information to help them protect their on-orbit assets. Upcoming improvements in ground-based tracking services may increase the number of objects tracked by a factor of 10. Satellite operators, SSA service providers, and government regulators must work together with regulators to develop best practices, standards, regulations, and policies designed to assure the long-term sustainability of space activities in the LEO environment.

So, given the advent of large LEO constellations, what challenges might be expected in efforts to maintain the long-term sustainability of the space environment and management of space traffic and limit potential hazards to spacecraft in orbit and people on the ground?

Environmental Effects on Satellite Lifetime

The study conducted by Ailor et al. (2010) with no large constellations and assuming business-as-usual satellite operations, disposal, and replenishment activities found that large satellites operating in the LEO regime could experience a mean lifetime reduction over the next 50 years of about 13% (e.g., a satellite with a mean lifetime of 10 years in a no-debris environment would see a reduction of 1.3 years in an environment that included space debris). Much of this reduction would come from degradation of solar panels by small, untrackable debris that would "sandblast" and reduce the power output of solar panels.

An update to that study (Ailor et al. 2017) assumed that over 5000 new small satellites are operating in high LEO orbits, consistent with proposals announced by commercial companies in the 2015–2017 timeframe. As with the 2010 study, this projection included the effects of collisions of both tracked objects, objects greater than 10 cm in LEO, and small debris down to 1 mm and adds changes in the debris environment due to the previously un-modeled new satellites colliding in their constellations and with other objects as they undergo constellation replenishment activities.

Primary results were that satellites being disposed from or added to the new constellations and debris associated with collisions involving these objects could potentially double the reduction in the mean operational lifetime of satellites operating in LEO over that predicted under the business-as-usual approach. As in the 2010 study, the degradation of solar panels due to small debris impacts was a significant factor in this degradation, with the mean lifetime reduction increasing from the 13% predicted in the earlier study to as much as 60% in some cases (i.e., a large satellite with a mean lifetime of 20 years in a no-debris environment could have its mean lifetime reduced by as many as 12 years in an environment that included debris associated with constellation satellite operations). Results suggest that minimizing collisions during replenishment activities and making solar panels and space hardware more resilient to small debris impacts will be increasingly important in the future.

Reentry Disposal of Satellites from Large Constellations

As noted earlier, current regulations limit the casualty expectation for a single satellite reentering in a random, uncontrolled manner to less than 1 in 10,000. If the predicted hazard exceeds that number, the satellite should be commanded to reenter into a safe area such as the South Pacific Ocean. The original Iridium satellites were predicted to each have a casualty expectation of 1 in 17,000, so random reentries were acceptable as the Iridium system evolved.

In 2000, Iridium was having significant financial problems, and as a part of bankruptcy considerations, the government asked that NASA develop an estimate of the cumulative hazard should Iridium be required to dispose all 74 of its satellites. The estimated *cumulative* hazard for reentering all 74 satellites was 1 in 250, well above the 1/10,000 limit for a single satellite. Fortunately, Iridium's financial problems eased, and the constellation has remained in operation. (Notably, satellites from the original constellation are gradually being disposed and replaced by the new Iridium Next satellites, and no fatalities from the disposal operations have been reported to date.)

Similar to the Iridium process, some current proposers of large constellations plan to dispose of their satellites at end of mission by lowering their orbits so random reentries into the Earth's atmosphere would occur within the 25-year timeframe.

A recent study (Ailor 2019) used the past history of debris surviving satellite reentries to develop a first-order estimate of the hazard to people and aircraft from falling debris that could survive reentries of constellation-sized satellites. Projecting to the year 2030, when several proposed constellations were assumed to be operational and were disposing a fraction of their satellites each year, the results show that hazards to people on the ground posed by debris from multiple satellite reentries from a single large LEO constellation could exceed the 1 in 10,000 threshold for a single satellite by 2 to 3 orders of magnitude.

That same study used radar observations of debris falling from a satellite reentry in the 1970s and US departures and destinations of 17 types of large, commercial aircraft to estimate the probability of hazardous debris striking an aircraft given possible masses of satellites proposed for a LEO constellation would be 0.001, corresponding to a maximum yearly casualty expectation for reentries from a single large constellation of nearly 3 in 10 without emergency action by a pilot after such an impact. That estimate would be higher if commercial air traffic was updated to include worldwide flights.

It should be noted that while reentry hazard prediction models can be verified to some degree based on recovered debris (Ailor et al. 2011) and actual data collected during breakup (Feistel et al. 2013), there is very little data on small objects that might survive and be a hazards to aircraft. As large constellations evolve, satellite designs that minimize the number and size of surviving debris should be considered, as should means to verify that the designs are preforming as predicted.

To summarize, there are currently no guidelines or requirements to manage cumulative risks for either ground or aircraft casualties. While new satellite design practices might lower the number of hazardous fragments that result from reentries, the most effective mitigation technique would be to deorbit all satellites into a safe ocean area.

Active Debris Removal (ADR)

Systems are being developed for servicing satellites in GEO orbits, and these systems might also be used to move satellites from the GEO protected region to disposal orbits. There is a potential market for satellite servicing given the "more civilized" operating environment in GEO, where a servicing vehicle could visit several operating satellites travelling in virtually the same orbit. Satellites operating in GEO are generally more expensive to build and launch and are designed with long lifetimes, so servicing that extends a satellite's lifetime could be less expensive and more cost-effective than replacement in some cases.

As noted, the LEO operating environment is different, with satellites operating in multiple orbits with inclinations varying from near zero to over 90 degrees. In addition, satellites are smaller and less expensive to build and launch and offer the advantage that new technologies can be infused quickly by replacing older satellites with newer versions. As a result, while proposals using electrodynamic tether systems and other approaches have been suggested for removal and a new company has announced plans to develop and demonstrate ADR technologies to approach and capture space debris, satellite servicing has not been a potentially attractive market in LEO to date.

But, the addition of new LEO constellations with large numbers of satellites in roughly the same orbits could encourage a new look at LEO satellite servicing and removal, perhaps as part of a constellation's design. For example, given the possibility that new requirements might emerge that would limit the cumulative hazard posed by reentry disposal of satellites from a constellation, constellation designers might include satellites specially designed to remove and dispose of dead satellites or satellites that have reached their end of life via direct disposal into a remote area. And given that a large constellation could have many satellites in the same orbit and orbit plane, a servicer could replace or update components in multiple satellites as an alternative to totally replacing those satellites.

In addition, the LEO environment currently includes several very large, dead satellites and rocket stages that will remain in orbit for many years. Should any of these be impacted by another debris object or an operating satellite, the number of debris objects created could be much larger than that from the Iridium-Cosmos and the Chinese ASAT tests. These objects should be priority targets for an ADR system. To encourage development of ADR capabilities, the International Association for the Advancement of Space Safety (IAASS) has proposed an international grand challenge for ADR (Position Paper 2017).

Effect of Large Constellations on SSA Service Requirements

As noted earlier, space situational awareness services (e.g., predicting close approaches that might threaten an operating satellite) are currently based on databases of some 20,000 objects. New tracking services will increase the size of these catalogs by a factor of at least ten as currently existing small objects, objects from 2 to 10 cm in size, are added. Objects in this size range are capable of seriously damaging or ending the operational lifetime of a satellite.

Recent studies (Peterson et al. 2018) are showing that, using current practices, the addition of large constellations plus the large numbers of satellites transiting the LEO regime as they dispose and replace satellites in these constellations are likely to increase the number of conjunction alerts sent to satellite operators to possibly over 1000 alerts per day. Many of these would not be alerts if tracking uncertainties on all orbiting objects, both operating satellites and hazardous debris objects, were reduced. Installing tracking aids such as transponders on operating satellites and including satellite owner/operator data in conjunction assessments will improve the situation somewhat, but given that debris dominates the population of orbiting objects in LEO, only basic improvements in the tracking accuracy for all objects and in information provided to operators by space situational awareness services will improve the situation substantially.

Space Situational Awareness and Traffic Management Service Providers

As noted earlier, the major provider of SSA services and space tracking data to date has been the US government via the US military. The military manages the collection of data from US-owned ground-based and optical sensors and processes this data to provide conjunction assessment services to satellite operators around the world and basic data to the public. But changes are coming.

The growth in the number of nongovernment satellites and satellite operators is placing an increasing demand on the current military operations, and in response, responsibilities for providing these services may be transferred to a civil agency within the US government. A new civil space traffic management (CSTM) system would likely maximize the use of commercial capabilities and data sources to provide enhanced SSA services as a public good (i.e., at no cost) and could make data available to support the development of specialized, higher level support by commercial entities.

The migration of basic space situational awareness capabilities from the Air Force to a new CSTM entity will likely be complete in the early to mid-2020s. Given its responsibilities, the new organization, likely using data that includes data from commercial data providers, will play an increasingly important role in protecting the near-space environment and shaping the standards, best practices, and regulations that will guide space operations for decades to come.

Conclusions

Specific areas where new services, best practices, standards, and regulations are likely to evolve are:

- **Minimizing the growth of space debris:** Unfortunately, the LEO regime already has a large population of uncontrolled debris objects that can collide with other objects and generate more long-lasting debris. The addition of large numbers of new satellites as large constellations are inserted and maintained has the potential to make operations in LEO much more expensive, forcing satellite designs that are more robust to small debris impacts and have very reliable systems for deorbit at end of mission or after an on-orbit failure. In addition, it is likely that guidelines will be developed encouraging operators of large constellations to deorbit satellites at end of mission quickly, possibly within 5 years or even via direct disposal, to minimize the possibility of collisions. It should be noted that active debris removal could also play an important role: removal of one large, dead satellite or launch stage could eliminate a potential source of debris objects equivalent in number to that possible by decades of mitigation.
- **Providing services to satellite operators:** Services that alert satellite operators of an approaching threat will have a critical role as the number of tracked objects increases due to addition of hundreds of thousands of existing small and hazardous debris objects and satellites associated with large constellations are added to SSA databases. Satellite operators need predictions that are accurate and timely to avoid unnecessary moves. Fortunately, commercial data providers are emerging that promise data that meet necessary requirements, and it is expected that this data will reduce the overall number of alerts to satellite operators. At the same time, the nature of the service providers will also change as providers for SSA and space traffic management services enhance their predictions by incorporating data acquired from commercial entities. Services provided by the US government are sure to evolve as they move from the US military to a civil agency. Active debris removal and satellite servicing may evolve as cost-effective ways to maintain GEO satellites. Similar systems could emerge to maintain and update satellites in large LEO constellations.
- **Addressing needs of governments and regulators:** There will be increased attention to the space environment as proposals for large constellations are realized over the coming years. And if a serious collision occurs, governments will be more likely to collect and analyze data to assure that operators are abiding by agreed restrictions. Governments are also more likely to support regulations to assure that current and future space operators have reasonable access to space and minimal interference with space operations.
- **Assuring safe on-orbit operations and disposal of space systems:** Collection, maintenance, and sharing of accurate data required for assessing possible conjunctions involving all orbiting objects larger than 2 cm will be increasingly important. Included will be sharing of data among satellite operators and space traffic management service providers. This data will be essential for services that provide timely, accurate services to prevent conjunctions among operating satellites and debris and assist with anomaly resolution and mission planning. In addition, the spacefaring community will need to verify proposals that would limit casualties on the ground and in aircraft via "design for demise" techniques and potentially develop cost-effective techniques that will enable direct disposal

of space hardware into safe areas. Finally, practices that minimize interference to satellites and launching systems transiting shells where large constellations are operating need to be developed.

The addition of large constellations of satellites to the LEO environment presents the opportunity to, in effect, re-architect LEO operations for the future. For the last 50 years, satellite operators have operated essentially unfettered by regulations on where they can operate, how they can maneuver, or concepts of "ownership" of a particular region of space. The introduction of thousands of new satellites to this region will make the development of best practices and possibly imposition of restrictions and limitations on where and how satellites can operate necessary if humanity is to preserve that environment for the future.

References

Ailor W (2019) Hazards of reentry disposal of satellites from large constellations. J Space Safe Eng 6:113–121

Ailor W, Womack J, Peterson G, Lao N (2010) Effects of space debris on the cost of space operations. In: IAC-10.A6.2.10, 61st International astronautical congress. Czech Republic, Prague, 27 Sept–1 Oct 2010

Ailor WH, Hallman WP, Steckel GL, Weaver MA (2011) Test cases for reentry survivability modeling. In: 5th IAASS conference, Versailles, 17–19 Oct 2011

Ailor W, Peterson GE, Womack J, Youngs M (2017) Effect of large constellations on lifetime of satellites in low earth orbits. J Space Safe Eng 4:117–123

Feistel AS, Weaver MW, Ailor WH (2013) Comparison of reentry breakup measurements for three atmospheric reentries In: 6th International association for the advancement of space safety conference, Montreal, May 2013

ISO 24113 (2019) Space Systems - Space debris mitigation requirements. Reproduced with permission of International Organization for Standardization, ISO. Standard available from any ISO member or see http://www.iso.org. Copyright remains with ISO

Oltrogge DL et al (2018) A comprehensive assessment of collision likelihood in geosynchronous earth orbit. Acta Astronaut 147:316–345

Peterson GE, Jenkin AB, Sorge ME, McVey JP (2018) Tracking requirements for space traffic management in the presence of proposed small satellite constellations. In: IAC-18,A6,7,6, x43991, 69th international astronautical conference, Bremen, 1–5 Oct 2018

Position Paper (2017) A grand challenge for active removal of space debris In: International association for the advancement of space safety, IAASS-PP-00117-WA, 17 May 2017. See http://iaass.space-safety.org/wp-content/uploads/sites/24/2018/03/Orig.-Position-Paper-on-ADR-v6-1.pdf

Space Sustainability

17

Peter Martinez

Contents

Space Security and Space Sustainability	320
Space Security	320
Space Sustainability	321
The United Nations and Space Sustainability	322
Space in the UN System	322
The United Nations Committee on the Peaceful Uses of Outer Space	323
COPUOS and Space Sustainability	325
Introduction of the Long-Term Sustainability of Outer Space Activities on the Agenda of COPUOS	325
COPUOS Working Group on the Long-Term Sustainability of Outer Space Activities	326
The Guidelines	331
Implementation and Updating of the Guidelines	333
Other Multilateral Initiatives with a Connection to Space Sustainability	334
Conference on Disarmament	334
UN Group of Governmental Experts on Transparency and Confidence-Building Measures (TCBMs) in Outer Space Activities	335
The EU Proposal for an International Code of Conduct for Outer Space Activities	336
Group of Governmental Experts on Further Practical Measures for the Prevention of an Arms Race in Outer Space	337
Concluding Remarks	338
References	339

Abstract

Space sustainability is a concept that has emerged within the past 15 years to refer to a set of concerns relating to outer space as an environment for carrying out space activities safely and without interference, as well as to concerns about ensuring continuity of the benefits derived on Earth from the conduct of such

P. Martinez (✉)
Secure World Foundation (SWF), Broomfield, CO, USA
e-mail: pmartinez@swfound.org

© Springer Nature Switzerland AG 2020
K.-U. Schrogl (ed.), *Handbook of Space Security*,
https://doi.org/10.1007/978-3-030-23210-8_51

space activities. As such, it encompasses the concerns of both space actors and those who are not space actors but who nevertheless benefit from space activities. This chapter reviews the role of the various relevant United Nations entities in ensuring space sustainability and provides a detailed review of the process and discussions held in the Working Group on the Long-Term Sustainability of Outer Space Activities within the Scientific and Technical Subcommittee of the United Nations (UN) Committee on the Peaceful Uses of Outer Space (COPUOS). Finally, the chapter discusses the relationship of the work in UN COPUOS with related work done in the Conference on Disarmament, the UN Group of Governmental Experts (GGE) on Transparency and Confidence-Building Measures in Outer Space Activities, and the initiative by the European Union to propose a Draft International Code of Conduct for outer space activities.

Space Security and Space Sustainability

The terms *space security* and *space sustainability* are sometimes used interchangeably to encompass a set of largely overlapping concerns as seen from two somewhat different perspectives. Underlying both of these perspectives is the acknowledgment that space systems underpin the modern information society and now form part of the critical infrastructure of most nations, whether they are spacefaring or not, and that this infrastructure is exposed to a series of risks of natural and anthropogenic origin. Regardless of the perspective from which one sees the problem, the point is that coordinated global action will be required to address these concerns. Acknowledging and addressing these different perspectives is one of the challenges that will be faced by multilateral initiatives to promote either space security or space sustainability. Hence it is instructive in the context of this chapter on space sustainability in a book devoted to space security to elaborate on this issue of the two perspectives.

Space Security

Security is, in general terms, about being free from danger or threat. In practical terms, this means freedom from doubt, anxiety, or fear based on well-founded confidence that there are mechanisms and processes in place to ensure security as a condition.

However, attempts to pin down exactly what is encompassed by the word *security* prove to be elusive as there is no single universally accepted definition of the concept of "security." In some countries the understanding of the term encompasses human security, environmental security, food security, and so on, while in others the term has a narrower meaning, referring primarily to military and defense-related issues.

Space security is a term that is used among space actors to refer to preserving order, predictability, and safety in space and avoiding courses of action that would ultimately undermine mission assurance, operational safety, and freedom of action in outer space. Another key dimension of this dialogue is the notion that, because of growing reliance on space systems in every facet of modern life, security on Earth (regardless of how one defines it) is increasingly underpinned by security in outer space. Hence one of the key aims of the space security dialogue is to ensure freedom from threats (either ground-based or space-based) to the effective access to and utilization of outer space. For some actors this is closely coupled to concerns about the potential weaponization of outer space, although it is difficult to progress beyond a general acknowledgment of the potential problem to practical measures to avoid it, because of disagreements around the definition of what constitutes a space weapon.

An important point to note is that the space security discourse has, up until recently, been dictated by the national interests and concerns of the major space powers, who are the ones who most heavily invested in space-based infrastructure to support their national security. For some sitting on the sidelines of the debate, space security has sometimes been perceived to be predominantly the preoccupation of the advanced space actors and thus far-removed from the day-to-day concerns of the non-space nations. Others, particularly those from emerging or aspiring space nations, have seen the promotion of multilateral space security discussions as an attempt by the leading space actors to advance and preserve their national space interests and advantages by raising entry barriers to aspiring newcomers on the pretext that the space environment is already "saturated" with actors. Neither of these perceptions has helped to build multilateral consensus on normative rules of behavior for all space actors. However, there are promising signs of middle space powers beginning to play a more active role in promoting multilateral space security dialogues in the future and hence helping to bridge the gap between these different perceptions of space security.

Space Sustainability

The word *sustainability* is derived from the Latin verb *sustinere* (*tenere*, "to hold"; *sus*, "up") and is usually used in the context of being able to maintain an activity at a certain rate or level. Since the 1980s the concept of sustainability has been applied to human habitation and utilization of planet Earth and its resources. This has given rise to the widely used term *sustainable development.* This term was coined in the book *Our Common Future*, which contains the report published by the Brundtland Commission in 1987 (UN GA 1987). The definition for sustainable development given in that book is worth quoting here:

> *development that meets the needs of the present without compromising the ability of future generations to meet their own needs.*

Notice the emphasis on "needs" in this definition. The Brundtland Commission's report placed emphasis in particular on meeting the essential needs of the world's poor, rather than satisfying the nonessential desires of the well-to-do.

The connection of *sustainability* with *outer space* arises from the perspective that space systems are now major global utilities that meet various societal needs. When seen in this light, *space sustainability* is understood to be about using outer space in such a way that all humanity will be able to continue to use it in the future for peaceful purposes and for societal benefit. The sustainability concern here is driven by the realization that the Earth's orbital environment and the electromagnetic spectrum are limited natural resources. This realization leads naturally to a concern for how to ensure that the benefits of space activities will continue to be accessible to future generations and to all nations and raises issues about the equitable and responsible access to and use of space resources.

In other words, from this perspective, space sustainability is seen in the context of wider sustainability discussions and is perceived to be the concern of all beneficiaries of space activities. It is thus an intrinsically multilateral issue. This is a significantly and fundamentally different point of departure for addressing a very similar set of issues driving the space security discourse.

The United Nations and Space Sustainability

The space arena today encompasses a much larger and much more diverse group of space actors than was the case in the first few decades of the space age. These include the "traditional" space actors, such as national space agencies and other national civilian agencies and the military, and a growing number of non-state actors, such as private sector commercial entities, academic and research institutions, and civil society organizations. We are also seeing the emergence of new kinds of space activities, many of which involve operations of space objects in close proximity to each other. Since the actions of a single actor can have consequences for all other actors, no single country (or even a group of like-minded countries) can control the space environment by its (or their) behavior or power alone; collective multilateral action is required.

In terms of international space law, states bear international responsibility for all space activities, including the activities of non-state entities (Outer Space Treaty 1967: Article VI). Hence, in spite of the growing number of non-state actors, the United Nations as a forum for states remains the relevant international forum to discuss such issues. Notwithstanding the preeminent role of states in the legal framework for outer space activities, it is worth reflecting on the contribution of civil society to the discussion on space sustainability, since this sector is playing an increasingly prominent and catalytic role in space activities and is in some respects more responsive to the rapidly changing space arena than the "traditional" fora established by states. This sector also has the access to a great deal of expertise, particularly in the conduct of space operations.

Space in the UN System

At present, there are four principal fora at which space issues are discussed multilaterally in the UN system: (i) the United Nations Committee on the Peaceful Uses of

Outer Space (COPUOS) in Vienna; (ii) the Conference on Disarmament (CD) in Geneva; (iii) the UN General Assembly in New York (and two of its committees, the Disarmament and International Security Committee (First Committee) and the Special Political and Decolonization Committee (Fourth Committee)) ; and (iv) the International Telecommunications Union (ITU) in Geneva, which deals with spectrum and geostationary orbital slot assignments. In addition to these, the World Meteorological Organization in Geneva makes use of space systems for monitoring and predicting terrestrial weather and also supports international coordination of space weather activities, an area of growing importance since space weather affects all space systems.

Space is widely used in the UN system and its entities. Each year approximately 20 UN entities and specialized agencies hold the United Nations Inter-Agency Meeting on Outer Space Activities. They discuss matters of mutual interest in the applications of space technologies to address human needs. Considerations include the implementation of the recommendations of the UNISPACE conferences and space-based contributions of the United Nations entities to the achievement of the Sustainable Development Goals as well as to the implementation of the recommendations of various world summits. The meeting issues a report on its deliberations for the consideration of COPUOS.

The United Nations Committee on the Peaceful Uses of Outer Space

The United Nations Committee on the Peaceful Uses of Outer Space (COPUOS) is the principal international forum for the development and codification of laws and principles governing activities in outer space. It is a standing committee of the UN, established in 1959 by 24 member states and given its mandate in UN General Assembly resolution 1472 (XIV). The Committee currently comprises 95 member states and a large number of permanent observers that enrich its work. The technical work of COPUOS is carried out by two subcommittees, the Legal Subcommittee (LSC) and the Scientific and Technical Subcommittee (STSC). Decisions in COPUOS and its subcommittees are reached by consensus. The Secretariat of COPUOS is the UN Office for Outer Space Affairs (UN OOSA), which is situated at the United Nations Office in Vienna.

During the 60 years of its existence, the deliberations in COPUOS have resulted in a number of very positive developments to advance international cooperation in the peaceful uses of outer space. A full discussion of all the activities and outcomes of COPUOS is outside the scope of this chapter, but it may be found in the paper by Hedman and Balogh (2009). Here, we focus on the aspects of COPUOS pertaining specifically to the long-term sustainability of outer space activities.

The International Legal Framework for Space Activities

COPUOS is the only international forum for the development and codification of international space law. Since its inception, the committee has concluded five international treaties and five sets of legal principles governing space-related activities. The five United Nations Treaties are:

- *Treaty on Principles Governing the Activities of States in the Exploration and Use of Outer Space, including the Moon and Other Celestial Bodies* (known as the "Outer Space Treaty"), adopted by the General Assembly in its resolution 2222 (XXI), opened for signature on 27 January 1967, entered into force on 10 October 1967;
- *Agreement on the Rescue of Astronauts, the Return of Astronauts and the Return of Objects Launched into Outer Space* (known as the "Rescue Agreement"), adopted by the General Assembly in its resolution 2345 (XXII), opened for signature on 22 April 1968, entered into force on 3 December 1968;
- *Convention on International Liability for Damage Caused by Space Objects* (known as the "Space Liability Convention"), adopted by the General Assembly in its resolution 2777 (XXVI), opened for signature on 29 March 1972, entered into force on 1 September 1972;
- *Convention on Registration of Objects Launched into Outer Space* (known as the "Registration Convention"), adopted by the General Assembly in its resolution 3235 (XXIX), opened for signature on 14 January 1975, entered into force on 15 September 1976;
- *Agreement Governing the Activities of States on the Moon and Other Celestial Bodies* (known as the "Moon Agreement"), adopted by the General Assembly in its resolution 34/68, opened for signature on 18 December 1979, entered into force on 11 July 1984.

The 1967 Outer Space Treaty laid the general legal foundation for the peaceful uses of outer space and provided a framework for developing the law of outer space. The four other treaties deal more specifically with certain concepts contained within the Outer Space Treaty.

It is instructive to review some of the principles in these treaties that provide the legal context for discussions on space sustainability and space security. These include the non-appropriation of outer space by any country; the freedom of exploration, scientific investigation, and the use (and even exploitation) of natural resources in outer space; state liability for damage caused by space objects; the avoidance of potentially harmful interference with space activities of other states; the sharing of information on space activities; and the registration of space objects.

The treaties affirm the agreement of states that the domain of outer space is a *res communis* and that the activities carried out therein and the benefits arising therefrom should be devoted to enhancing the well-being of all countries and humankind. Article I of the Outer Space Treaty is of particular relevance to the space sustainability discussion:

> *The exploration and use of outer space, including the Moon and other celestial bodies, shall be carried out for the benefit and in the interests of all countries, irrespective of their degree of economic or scientific development, and shall be the province of all mankind.*

Outer space, including the Moon and other celestial bodies, shall be free for exploration and use by all States without discrimination of any kind, on a basis of equality and in accordance with international law, and there shall be free access to all areas of celestial bodies.

There shall be freedom of scientific investigation in outer space, including the Moon and other celestial bodies, and States shall facilitate and encourage international cooperation in such investigation.

These principles provide the reference points for many delegations in COPUOS against which they will judge the relevance and legitimacy of the space sustainability discourse and its outcome.

In addition to the codification of these treaties and principles, progress has also been made in developing a common understanding on other issues related to the exploration and peaceful uses of outer space. All in all, 132 UN General Assembly resolutions or recommendations relating to outer space have been adopted from 1958 to 2018 (UN OOSA 1958–2018). These resolutions have been complemented by additional instruments containing more technically detailed guidance. These instruments include a set of voluntary Space Debris Mitigation Guidelines (UN OOSA 2010) adopted in 2007 and a Safety Framework for Nuclear Power Source Applications in Outer Space, developed jointly by the Scientific and Technical Subcommittee of COPUOS and the International Atomic Energy Agency, which was adopted in 2009 (UN COPUOS and IAEA 2009).

The UN also maintains a Register of Objects Launched into Outer Space that contains information provided by member states and intergovernmental organizations that are party to the Registration Convention (UN Register). As of 1 January 2019, 69 states had acceded to or ratified the Convention, and another four states had signed it. As of 30 August 2019, the Register contained 8737 space objects launched by 87 states or international intergovernmental organizations, for which an international space object designator had been assigned. It is worth noting that only 7859 of those space objects had been registered with the United Nations.

COPUOS and Space Sustainability

Introduction of the Long-Term Sustainability of Outer Space Activities on the Agenda of COPUOS

Although several aspects of the work of COPUOS are directly relevant to space sustainability, prior to 2010 these topics were being addressed in isolation; the emergence of a more holistic view of these issues goes back to 2005, as the Committee was approaching its 50th year. In that year, Mr. Karl Deutsch of Canada (Chair of the STSC from 2001 to 2003) presented a discussion paper to the Committee on the future role of COPUOS in its next 50 years. Deutsch made the connection between the sustainability of life on Earth and the cooperative international use of space systems; the very subject COPUOS was established to address.

In 2006–2007, the Committee was chaired by Mr. Gérard Brachet of France. He highlighted the issue of space sustainability during his term as Chairman of COPUOS. At the 50th session of the Committee in 2007, Brachet presented a working paper by the Chair (UN GA 2007) that identified the long-term sustainability of outer space activities as one of the key challenges facing the future peaceful uses of outer space. The working paper further suggested that a working group could be established within the STSC to produce a technical assessment of the situation and to suggest a way forward.

In their sessions during 2008, the STSC and COPUOS discussed the introduction of an agenda item dealing with the long-term sustainability of outer space activities and what such an agenda item might encompass. Subsequently, in 2009, at the 46th session of the STSC, a proposal was put forward by the delegation of France to include a new agenda item on the long-term sustainability of outer space activities on the agenda of the STSC.

At its 52nd session in 2009, COPUOS agreed that the STSC should include, starting from its 47th session in 2010, a new agenda item titled "long-term sustainability of outer space activities" and it proposed a multi-year work plan that was to culminate in a report on the long-term sustainability of outer space activities and a set of best-practice guidelines for presentation to and review by the Committee.

In 2010 the STSC established the Working Group on the Long-Term Sustainability of Outer Space Activities under the chairmanship of Mr. Peter Martinez of South Africa. The first issue to be addressed was reaching agreement on the terms of reference, scope, and methods of work. These deliberations were concluded at the 54th session of COPUOS in June 2011.

This is a very condensed review of the emergence of the long-term space sustainability work in COPUOS. Readers interested in a more detailed review are referred to the article by Brachet (2012).

COPUOS Working Group on the Long-Term Sustainability of Outer Space Activities

The terms of reference for this Working Group (UN GA 2011) mandated it to examine the long-term sustainability of outer space activities in the wider context of sustainable development on Earth, including the contribution of space activities to the achievement of the Millennium Development Goals, taking into account the concerns and interests of all countries. (Nowadays we would refer to the Sustainable Development Goals (SDGs), but the terms of reference for the Working Group predated the adoption of the SDGs at a special UN Summit of Heads of State in September 2015. The SDGs are in a sense the successors of the Millennium Development Goals, and hence the same importance (if not more) is attached to ensuring continuity of space-derived data and services to meet these developmental goals.)

The Working Group was mandated to consider established practices, operating procedures, technical standards, and policies associated with the long-term sustainability of outer space activities throughout all the phases of a mission life cycle. The

Working Group took as its legal framework the existing UN treaties and principles governing the activities of states in the exploration and use of outer space; it did not consider the development of new legally binding instruments.

The Working Group was tasked to produce a report on the long-term sustainability of outer space activities and a consolidated set of voluntary best-practice guidelines that could be applied by states, international organizations, national nongovernmental organizations, and private sector entities to enhance the long-term sustainability of outer space activities for all space actors and for all beneficiaries of space activities.

It is instructive to quote from the terms of reference regarding the expected character of the guidelines to be produced. These guidelines should:

(a) *Create a framework for possible development and enhancement of national and international practices pertaining to enhancing the long-term sustainability of outer space activities, including, inter alia, the improvement of the safety of space operations and the protection of the space environment, giving consideration to acceptable and reasonable financial and other connotations and taking into account the needs and interests of developing countries.*
(b) *Be consistent with existing international legal frameworks for outer space activities and should be voluntary and not be legally binding.*
(c) *Be consistent with the relevant activities and recommendations of the Committee and its Subcommittees, as well as of other working groups thereof, United Nations intergovernmental organizations and bodies and the Inter-Agency Space Debris Coordination Committee and other relevant international organizations, taking into account their status and competence.*

Consideration of Topics

In developing its terms of reference, the Working Group identified a wide range of topics of relevance to the overall considerations of space sustainability, spanning from developmental issues to operational issues, space debris, space weather, and also regulatory issues.

The topics were clustered to allow more efficient consideration of related matters, and four expert groups were established to consider these related sets of topics. These expert groups were populated with experts nominated by their national governments. However, the experts served in an ad hominem capacity and did not necessarily represent their governments' positions in all matters. The expert groups were tasked to contribute inputs to the report of the Working Group and to propose candidate guidelines for consideration by the Working Group. The Working Group was to consider these inputs from the expert groups and take any necessary decisions. In this way, a clear separation was established between the expert groups as technical deliberative fora and the Working Group as a diplomatic negotiating forum.

Based on the inputs from the individual experts and other external inputs (see the subsequent sections titled "Coordination with Other International Intergovernmental Entities and Processes" and "Contributions by Non-state Actors"), the expert groups were tasked to identify issues for which sufficient international expert consensus could be found to recommend guidelines based on established best practices. Where the experts identified issues pertinent to the long-term sustainability of outer space activities, but for which the state of knowledge was such that the experts were not yet able to recommend consensus guidelines based on any operational experience, those issues were referred to the Working Group for its attention and possible future consideration.

The four expert groups and their scopes were as follows:

(a) *Expert Group A: Sustainable space utilization supporting sustainable development on Earth*

Co-chaired by Mr. Filipe Duarte Santos (Portugal) and Mr. Enrique Pacheco Cabrera (Mexico)

This expert group addressed the societal benefits of space activities and their contribution to sustainable development on Earth. It considered space as a shared natural resource, the equitable access to outer space and to the resources and benefits associated with it, as well as access to the benefits of outer space activities for human development. This expert group also considered the role of international cooperation in ensuring that outer space continues to be used for peaceful purposes for the benefit of all nations. This expert group proposed seven candidate guidelines and four topics for further consideration by the Working Group.

(b) *Expert Group B: Space debris, space operations and tools to support collaborative space situational awareness*

Co-chaired by Mr. Richard Buenneke (United States of America) and Mr. Claudio Portelli (Italy)

This expert group considered the issues that make the space environment unpredictable and unsafe for space actors. This included an analysis of risks from space debris and measures to reduce the creation and proliferation of space debris. The implementation of such measures requires strengthened cooperative space situational awareness, which in turn requires the collection, sharing, and dissemination of data on space objects, such as orbits, pre-launch, and pre-maneuver notifications. This expert group also considered tools to support collaborative space situational awareness, such as registries of operators and contact information and procedures for sharing relevant operational information among space actors. This led to a recognition of the importance of developing common standards and practices for information exchange. This expert group proposed eight candidate guidelines and three topics for further consideration by the Working Group.

(c) *Expert Group C: Space weather*

Co-chaired by Mr. Takahiro Obara (Japan) and Mr. Ian Mann (Canada)

This expert group focused on ways to reduce the risks of detrimental effects of space weather phenomena on operational space systems. Such risks may be reduced through the sharing and dissemination of key data on phenomena related to space weather in real or near-real time, as well as sharing of models and forecasts. This expert group proposed five candidate guidelines and two topics for further consideration by the Working Group.

(d) *Expert Group D: Regulatory regimes and guidance for actors*
Co-chaired by Mr. Anthony Wicht (Australia) and Mr. Sergio Marchisio (Italy)
This expert group considered the contribution of international and national legal instruments and regulatory practices to promote the long-term sustainability of outer space activities. This included considerations of how the existing treaties and principles that define the international legal framework for space activities are being implemented at the national level through legal and regulatory regimes and how such national regulatory frameworks for space activities could be developed or further strengthened to support the long-term sustainability of space activities. This expert group proposed eleven candidate guidelines and five topics for further consideration by the Working Group.

The expert groups did not work in silos. Several issues under consideration by the expert groups were intrinsically multidisciplinary in character and therefore fell within the competence of more than one of the expert groups. For this reason, the expert groups held joint meetings to discuss overlaps and gaps.

The expert groups met during the sessions of COPUOS and its STSC from 2011 to 2014 and also took the opportunity of meeting at the International Astronautical Congress in Cape Town in 2011, Naples in 2012, and Beijing in 2013. The four expert groups concluded their work in 2014 and submitted their reports to the Working Group, containing a total of 33 proposed draft guidelines.

Coordination with Other International Intergovernmental Entities and Processes

The Working Group was mandated to liaise with the UN Group of Governmental Experts on Transparency and Confidence-Building Measures in Outer Space Activities, the Conference on Disarmament, the Commission on Sustainable Development, the International Civil Aviation Organization, the International Telecommunication Union, and the World Meteorological Organization, as well as relevant intergovernmental organizations, such as the European Space Agency, the European Organization for the Exploitation of Meteorological Satellites, the Asia-Pacific Space Cooperation Organization, and the Group on Earth Observations.

The overarching principle behind these interactions was that the Working Group should avoid duplicating the work being done within these international entities while at the same time identifying areas of concern relating to the long-term sustainability of outer space activities that were not being covered by them.

Contributions by Non-state Actors

Although the discussions within the Working Group occurred at the intergovernmental level of COPUOS, states recognized that non-state actors play an important role in the space arena and have much knowledge and experience to contribute to the formulation of guidelines based on the best practices.

A number of international organizations and bodies, such as the Consultative Committee for Space Data Systems, the Inter-Agency Space Debris Coordination

Committee, the International Space Environment Service, the International Organization for Standardization, the International Academy of Astronautics, the International Astronautical Federation, and the Committee on Space Research, also provided inputs into the work of the Working Group and its expert groups.

Commercial operators have extensive experience in running their fleets of spacecraft and in dealing with space weather and other on-orbit operational issues. A case in point was the industry coordination that took place during the Galaxy-15 "zombie sat" episode in 2010 (Weeden 2010). Industry associations and entities such as the International Astronautical Federation provide access to the collective expertise of the space industry and space agencies.

Finally, there are institutional actors focusing on the governance of space activities, such as the European Space Policy Institute or the Secure World Foundation, that analyze certain topics in depth and prepare position papers. These entities also made valuable contributions to the space sustainability dialogue in COPUOS.

The role of non-state actors is at times a contentious issue in COPUOS. Some member states (usually those with a well-established space industry) are comfortable with engaging the private sector in issues on the COPUOS agenda, while others (usually the ones without a space industry) are concerned that the agenda of COPUOS should not be dictated by the interests of commercial entities. Those states are of the view that COPUOS is a forum of states and that states should direct the agenda and discussions in COPUOS.

Because consensus could not be reached on the direct participation of non-state entities in the Working Group, the solution that was agreed upon was to continue with the established practice that states could choose to include in their delegation representatives of their own national non-state entities. In this way, the contributions of experts from non-state entities were made possible. The inputs of national nongovernmental organizations and private sector entities were thus obtained through the member states of COPUOS.

Negotiation of the LTS Guidelines

Following the expert group phase, the Working Group began developing the draft guidelines based on the recommendations of the expert groups. A number of member states also proposed draft guidelines for consideration by the Working Group. By the start of 2016, through a process of consolidation and streamlining, the Working Group had narrowed its focus to 29 draft guidelines, all at various stages of maturity.

From the start of the Working Group in 2010 to the end of its mandate in 2018, the membership of COPUOS grew from 70 to 92 states. Moreover, as the LTS discussions gained momentum in COPUOS, more states became actively engaged in the debates. Since COPUOS takes decisions by absolute consensus of its member states, all member states had to reach agreement on the text of each one of these guidelines, in all six official languages of the UN. Progress was gradual and uneven, but by June 2016 COPUOS reached agreement on the first 12 guidelines (UN COPUOS 2016). In February 2018, at the 55th session of the STSC, agreement was reached on a further nine guidelines and the text of a politically significant context-setting preamble that included the following definition of space sustainability (UN COPUOS 2018a):

> *The long-term sustainability of outer space activities is defined as the ability to maintain the conduct of space activities indefinitely into the future in a manner that realizes the objectives of equitable access to the benefits of the exploration and use of outer space for peaceful purposes, in order to meet the needs of the present generations while preserving the outer space environment for future generations.*

Readers will notice the parallels between this COPUOS definition of space sustainability and the definition of sustainable development mentioned earlier in the section titled "Space Sustainability."

In addition to the 21 agreed guidelines, there were a further seven draft guidelines (UN COPUOS 2018b) for which the Working Group could not reach consensus during its mandate, which expired in June 2018. Discussions at the 61st session of COPUOS in June 2018 were inconclusive because the Committee could not reach consensus on the way forward and some states could not agree to decoupling the already agreed guidelines from those still under discussion. The session ended in a stalemate, with all states however agreeing that the LTS discussions should continue in 2019.

At its 62nd session in June 2019, the Committee adopted the preamble and 21 guidelines for the long-term sustainability of outer space activities (UN GA 2019). The Committee encouraged states and international intergovernmental organizations to voluntarily take measures to ensure that the guidelines were implemented to the greatest extent feasible and practicable.

The Committee also agreed on the establishment of a new Working Group with a 5-year work plan under the STSC to advance the work on LTS. The Committee decided that this new Working Group would agree on its own terms of reference, methods of work, and dedicated work plan at the 57th session of the STSC, in February 2020. This new Working Group is expected to focus on:

(a) Identifying and studying challenges for the long-term sustainability of outer space activities and to consider possible new guidelines, including those proposed but not agreed within in the previous Working Group;
(b) Sharing experiences, practices, and lessons learned from voluntary national implementation of the already adopted guidelines;
(c) Raising awareness and building capacity to implement the adopted LTS guidelines, in particular among emerging space nations and developing countries.

The Guidelines

The 21 agreed guidelines (UN GA 2019) comprise a collection of internationally recognized measures for ensuring the long-term sustainability of outer space activities and for enhancing the safety of space operations. They address the policy, regulatory, operational, safety, scientific, technical, and international cooperation and capacity-building aspects of space activities. They are based on a substantial body of knowledge, as well as the experiences of states, international intergovernmental organizations, and relevant national and international nongovernmental entities. Therefore, the guidelines are relevant to both governmental and nongovernmental entities. They are also relevant to all space activities, whether planned or ongoing, as practicable, and to all phases of a space mission, including launch, operation, and end-of-life disposal.

The purpose of the guidelines is to assist states and international intergovernmental organizations, both individually and collectively, to mitigate the risks associated with the conduct of outer space activities so that present benefits can be sustained and future opportunities realized. Consequently, the implementation of the guidelines should promote international cooperation in the peaceful use and exploration of outer space.

These 21 agreed guidelines represent the low-hanging fruit of the LTS discussions, but they also mark a significant step forward in that they represent the tangible progress that has been made in COPUOS in addressing space sustainability. This first set of agreed guidelines creates a foundation for further consensus building in COPUOS.

The guidelines are intended to support the development of national and international practices and safety frameworks for conducting outer space activities while allowing for flexibility in adapting such practices and frameworks to specific national circumstances. They are also intended to support states and international intergovernmental organizations in developing their space capabilities in a manner that avoids causing harm to the outer space environment and the safety of space operations.

The guidelines are voluntary and not legally binding under international law. The existing UN treaties and principles on outer space provide the fundamental legal framework for these guidelines. However, despite their non-binding status under international law, the guidelines can have a legal character in the sense that states may choose to incorporate elements of the guidelines in their national legislation, as has been the case with the COPUOS Space Debris Mitigation Guidelines.

The titles of the 21 agreed guidelines are indicated below. The full text of the guidelines is available in UN document A/74/20, Annex II. The remaining seven draft guidelines that did not reach consensus during the mandate of the Working Group are contained in UN document A/AC.105/C.1/L.367. The progress made in discussions of those draft guidelines will inform future discussions of space sustainability in COPUOS.

A. *Policy and regulatory framework for space activities*
   Guideline A.1: Adopt, revise and amend, as necessary, national regulatory frameworks for outer space activities
   Guideline A.2: Consider a number of elements when developing, revising or amending, as necessary, national regulatory frameworks for outer space activities

Guideline A.3: Supervise national space activities

Guideline A.4: Ensure the equitable, rational and efficient use of the radio frequency spectrum and the various orbital regions used by satellites

Guideline A.5: Enhance the practice of registering space objects

B. *Safety of space operations*

Guideline B.1: Provide updated contact information and share information on space objects and orbital events

Guideline B.2: Improve accuracy of orbital data on space objects and enhance the practice and utility of sharing orbital information on space objects

Guideline B.3: Promote the collection, sharing and dissemination of space debris monitoring information

Guideline B.4: Perform conjunction assessment during all orbital phases of controlled flight

Guideline B.5: Develop practical approaches for pre-launch conjunction assessment

Guideline B.6: Share operational space weather data and forecasts

Guideline B.7: Develop space weather models and tools and collect established practices on the mitigation of space weather effects

Guideline B.8: Design and operation of space objects regardless of their physical and operational characteristics

Guideline B.9: Take measures to address risks associated with the uncontrolled re-entry of space objects

Guideline B.10: Observe measures of precaution when using sources of laser beams passing through outer space

C. *International cooperation, capacity-building, and awareness*

Guideline C.1: Promote and facilitate international cooperation in support of the long-term sustainability of outer space activities

Guideline C.2: Share experience related to the long-term sustainability of outer space activities and develop new procedures, as appropriate, for information exchange

Guideline C.3: Promote and support capacity-building

Guideline C.4: Raise awareness of space activities

D. *Scientific and technical research and development*

Guideline D.1: Promote and support research into and the development of ways to support sustainable exploration and use of outer space

Guideline D.2: Investigate and consider new measures to manage the space debris population in the long term

Implementation and Updating of the Guidelines

States and international intergovernmental organizations are encouraged to implement these guidelines to the greatest extent feasible and practicable, in accordance with their respective needs, conditions and capabilities, and with their existing obligations under applicable international law.

International cooperation is required to implement the guidelines effectively and to monitor their impact and effectiveness. However, COPUOS recognizes that not all space actors have equal capability or capacity to implement these guidelines. Therefore, the guidelines place a strong emphasis on international cooperation and information sharing. States and international intergovernmental organizations with extensive experience in conducting space activities are encouraged to support developing countries to strengthen their national capacities to implement the guidelines.

COPUOS also recognizes that these guidelines should be a "living document" that is periodically updated to ensure that, as space activities evolve, the guidelines continue to reflect the most current state of knowledge of pertinent factors influencing the long-term sustainability of outer space activities. This "living document" aspect of the guidelines is especially important given that the rapid evolution in space activities makes space sustainability a dynamic, multi-scale problem.

States and international intergovernmental organizations are encouraged to share their practices and experiences with COPUOS regarding the implementation of the guidelines. States are also encouraged to promote and/or conduct research on topics relevant to these guidelines and their implementation.

COPUOS envisages that it may periodically review, revise, or add to these guidelines to ensure that they continue to provide effective guidance to promote the long-term sustainability of outer space activities. Proposals for revising this set of guidelines, or for new guidelines, may be submitted by any COPUOS member state for consideration by the Committee.

Other Multilateral Initiatives with a Connection to Space Sustainability

The COPUOS work on space sustainability did not occur in a vacuum. There were, in fact, several concurrent discussions in other fora that related to space security and space sustainability. Those initiatives were (and some still are) to some extent addressing a set of largely overlapping concerns from the perspectives of different groups of actors and different fora. In this section we briefly consider how the work of COPUOS on space sustainability relates to those other initiatives.

Conference on Disarmament

Given the importance of military and civilian space systems in modern warfare, there is a technical possibility that such systems could be targeted in a conflict situation. The possibility that space-based weapons might be developed and deployed in outer space has given rise to concerns that this could lead to an arms race in outer space. Given that COPUOS focuses exclusively on the peaceful uses of outer space, questions of space weaponization and related security implications are dealt with at the Conference on Disarmament (CD), the sole multilateral body for negotiating arms control issues.

Within the CD, a number of delegations, notably China and Russia, have raised the issue of the Prevention of an Arms Race in Outer Space (PAROS). However, the CD has effectively been stagnant since 1988, since the member states have been unable to agree on the annual program of work. Not only do the members of the CD disagree over its priorities, but also the consensus rule, which served this body well in the past, is now being used to maintain the deadlock. It is against this backdrop that in 2008 China and Russia introduced a Draft Treaty on the Prevention of the Placement of Weapons in Outer Space and of the Threat or Use of Force against Outer Space Objects (PPWT). However, not all countries agree that new legal instruments to prevent space weaponization are warranted or even beneficial. So, for the time being, the PAROS discussions in the CD are making no progress because of differences of opinion on some fundamental issues. However, there is agreement on the urgency to make progress in those areas where there is consensus, even if such progress must be made outside the CD.

This impasse in the CD had an influence in COPUOS in the sense that the countries supporting the PPWT proposals in the CD did not want the LTS discussions in COPUOS to be used as a pretext to circumvent the need for discussions on the prevention of an arms race in outer space and the development of a legally binding framework to prevent the placement of weapons in outer space. Thus, the terms of reference for the LTS Working Group called for "appropriate liaison" with the CD. The mandate of COPUOS covers only the peaceful uses of outer space, but some of the LTS guidelines could be seen as de facto transparency and confidence-building measures to enhance collective space security. In this way the implementation of the COPUOS LTS guidelines could potentially be useful for improving mutual understanding and for reducing misperceptions and mistrust, thereby ultimately promoting a more favorable climate for arms control and nonproliferation discussions in the CD.

UN Group of Governmental Experts on Transparency and Confidence-Building Measures (TCBMs) in Outer Space Activities

In 2010, the UN General Assembly adopted resolution A/Res/65/68 (UN GA 2010), which called for the establishment of a Group of Governmental Experts (GGE) on "Transparency and Confidence-Building Measures in Outer Space Activities." The GGE was to conduct a study on outer space TCBMs, making use of the relevant reports of the UN Secretary-General, and without prejudice to the substantive discussions on the prevention of an arms race in outer space within the framework of the CD, and to submit to the General Assembly at its 68th session a report with an annex containing the study of governmental experts.

The GGE, which comprised 15 experts selected on the basis of their knowledge and geographical representation, began its work in July 2012 and submitted its final consensus report (document A/RES/65/68) to the First Committee of the UN General Assembly in October 2012. The report was adopted as resolution 68/50 by a unanimous vote in the First Committee and on 10 December 2012 by the General Assembly. This resolution welcoming the GGE report and endorsing its content was co-sponsored by China, Russia, and the United States and represented a diplomatic breakthrough since the United States had never before voted in favor of the annual TCBM resolution.

The LTS Working Group was tasked in its terms of reference to consider appropriate linkages with the GGE. This was done by the Chairs of two respective processes providing formal briefings to each other's groups. It is instructive to identify some interlinkages between the LTS guidelines and the recommendations contained in the GGE report (UN GA 2010).

The GGE report refers in paragraph 39 to exchanges of information on orbital parameters of outer space objects and potential orbital conjunctions. Reference is also made to the registration of space objects. The LTS guidelines concerning the exchange of contact information, exchange of data on space objects, and risk assessments relating to space objects address such matters.

The GGE report refers in paragraph 40 to exchanges of information on forecast natural hazards in outer space. The LTS guidelines on sharing of operational space weather data, forecasts, and best practices address this issue.

Paragraph 42 of the GGE report refers to notifications relating to scheduled maneuvers that may result in a risk to the fight safety of space objects of other states. The LTS guidelines on the safety of space operations address such matters.

Section V of the GGE report refers to international cooperation and touches, inter alia, on international cooperation for capacity-building and confidence-building. The LTS guidelines on international cooperation in support of long-term sustainability and capacity-building address such issues.

The EU Proposal for an International Code of Conduct for Outer Space Activities

More or less at the same time as the multilateral discussions in COPUOS on the long-term sustainability of outer space activities started, the European Union began a political initiative to develop a Code of Conduct for Outer Space Activities. This initiative was pursued outside of the existing multilateral fora, motivated at least in part as a means to bypass the stalemate on the PAROS issue in the CD and the difficulties posed by the consensus rule in both COPUOS and the CD. The EU expressed its intent to open the code for signature at an international diplomatic conference, to be convened for this purpose.

Outside of Europe, no other major space powers openly endorsed the initiative until January 2012, when US Secretary of State Hillary Clinton announced that "the United States has decided to join with the European Union and other nations to develop an International Code of Conduct for Outer Space Activities" (US Secretary of State 2012). Australia's Foreign Minister, Kevin Rudd, soon followed with a similar statement. However, the initiative was not embraced by a significant number of non-EU space-capable states (notably Brazil, Russia, India, China, and South Africa, the so-called BRICS countries), largely because of concerns about the process and the intent of the EU in having kept this initiative out of multilateral fora. (During the development of the draft code, the EU held numerous bilateral consultations, but no multilateral consultations in the UN format, until 5 June 2012, when the EU External Action/Information Service held an information session on the margins of COPUOS.) This meant that the Code of Conduct initiative had no formal multilateral mandate, unlike the GGE on space TCBMs and the COPUOS LTS processes. This lack of a formal multilateral mandate ultimately led to the demise of the Code of Conduct initiative, on procedural grounds, at a special meeting held at the United Nations in New York in July 2015.

The failure of the late attempt by the EU to "multilateralize" the code through this special UN meeting had a positive ripple effect on the LTS discussions in the COPUOS. From the start of the LTS discussions in COPUOS, a number of delegations had questioned how the long-term sustainability work related to the EU's efforts to promote a Code of Conduct and whether such a Code of Conduct would in some way "trump" the long-term sustainability discussions in COPUOS. This had caused a number of delegations to hold back from full engagement in the LTS discussions, waiting to see how the Code discussions were going to play out. With the demise of the Code discussions, COPUOS became the only forum holding productive multilateral space sustainability discussions.

It is worth noting that, although some observers saw the Code of Conduct and LTS discussions as competing processes, a closer examination would show that, although the underlying goals were the same, their approaches were diametrically opposed. The COPUOS LTS work was a technically based, bottom-up approach of developing guidelines based on the collected best practices of established space actors. The Code of Conduct initiative was a more political, top-down approach. The two approaches could, in fact, have complemented each other if the 2015 efforts to multilateralize the Code of Conduct had succeeded.

Since July 2015, the EU has not actively promoted the Code of Conduct, but it has not given up on the idea either. In several statements delivered in multilateral fora in the past 2 years, the EU has expressed the view that it still believes there would be value in agreeing on an instrument that encourages States to make a voluntary political commitment not to undertake activities detrimental to the safety, security, and sustainability of outer space activities. Such a voluntary instrument, potentially to be negotiated within the framework of the UN, should, in the EU's view, not duplicate the work of COPUOS as the UN's mandated norm-creating body for the peaceful uses of outer space and should respect its role in the further development of the legal regime governing space activities. Such a voluntary instrument would build upon the COPUOS LTS guidelines and would be complementary to these guidelines. As of this writing (August 2019), it is not yet clear whether or how the EU intends to translate these ideas into diplomatic initiatives.

Group of Governmental Experts on Further Practical Measures for the Prevention of an Arms Race in Outer Space

For completeness, we will mention here the Group of Governmental Experts on further practical measures for the prevention of an arms race in outer space. This GGE was established pursuant to resolution 72/250, adopted by the General

Assembly on 24 December 2017, and was tasked to "consider and make recommendations on substantial elements of an international legally binding instrument on the prevention of an arms race in outer space, including, inter alia, on the prevention of the placement of weapons in outer space." This GGE, which comprised experts from 25 nations, carried out its work in 2018 and early 2019 under the leadership of Brazil's ambassador to the Conference on Disarmament, Guilherme de Aguiar Patriota.

In accordance with its mandate, the GGE considered recommendations on substantial elements of an international legally binding instrument on the prevention of an arms race in outer space, including on the prevention of the placement of weapons in outer space. Pursuant to this mandate, it discussed (a) the international security situation in outer space; (b) the existing legal regime applicable to the prevention of an arms race in outer space; (c) the application of the right to self-defense in outer space; (d) general principles; (e) general obligations; (f) definitions; (g) monitoring, verification, and transparency and confidence-building measures; (h) international cooperation; and (i) final provisions, including institutional arrangements.

The sessions of this GGE took place against a backdrop of elevated political rhetoric around the counterspace developments in recent years, and the Indian antisatellite (ASAT) test of March 2019 took place during the final session of the GGE, further adding to the grim disarmament climate. The GGE considered several drafts of a substantive report. No consensus was reached on a substantive report, so the GGE's final report was simply a procedural report issued as UN document A/74/77. Although this outcome was disappointing, the process itself was important in that the GGE held substantive discussions on space arms control.

Concluding Remarks

The golden thread running through the processes in COPUOS, the GGE on space TCBMs, and the Code of Conduct initiative is that they were all aiming to produce instruments that are voluntary in nature. However, although such instruments may be legally non-binding, they are politically binding. Another important point to appreciate is that *non-binding* does not mean *non-legal*, in the sense that states can choose to domesticate their politically binding agreement to such voluntary frameworks in their domestic regulatory practices.

A number of countries have expressed concern that such voluntary instruments are inherently fragile and would not prove effective in preventing the weaponization of and an arms race in outer space. However, there does not seem to be consensus at this point on the desirability of legally binding instruments banning the placement and use of weapons in outer space, so the development of voluntary frameworks for promoting space sustainability provides some scope for making progress. Voluntary frameworks do not necessarily retard the evolution of binding norms and can in fact pave the way for adoption of binding norms. Historically, many legal rules have resulted from the codification of existing practices adopted by consensus.

Progress will also be made in the sense that the states that choose to participate in these processes do so because they recognize the urgency of addressing the problems of space sustainability and space security. That awareness in and of itself may be enough to convince space actors to take corrective and preventative actions on their own. The COPUOS LTS guidelines, while non-binding, have the advantage of being the result of a multilateral consensus-based process and will therefore have a good chance of being implemented by space actors, in their own interest.

References

Brachet G (2012) The origins of the long-term sustainability of outer space activities initiative at UN COPUOS. Space Policy 28:161–165

Hedman N, Balogh W (2009) The United Nations and outer space: celebrating 50 years of space achievements. In: Schrogl K-U, Mathieu C, Peter N (eds) Yearbook on Space Policy 2007/2008: from policies to programmes. Springer, Wien/New York, pp 237–250. ISBN: 978-3-211-99090-2

Outer Space Treaty (1967) The treaty on principles governing the activities of States in the exploration and use of outer space, including the moon and other celestial bodies, of 27 January 1967. http://www.unoosa.org/oosa/en/ourwork/spacelaw/treaties/introouterspacetreaty.html. Accessed 3 Feb 2020

United Nations Committee on the Peaceful Uses of Outer Space (2016) Report of the fifty-ninth session, The first batch of agreed guidelines is contained in the Annex to the report of the 59th session of COPUOS in 2016, contained in UN document A/71/20. http://www.unoosa.org/oosa/oosadoc/data/documents/2016/a/a7120_0.html. Accessed 3 Feb 2020

United Nations Committee on the Peaceful Uses of Outer Space (2018a) Working Group on the Long-Term Sustainability of Outer Space Activities: preambular text and nine guidelines, conference room paper by the Chair of the Working Group on the Long-Term Sustainability of Outer Space Activities. UN document A/AC.105/C.1/2018/CRP.18/Rev.1. http://www.unoosa.org/res/oosadoc/data/documents/2018/aac_105c_1201 8crp/aac_105c_12018crp_18rev_1_0_html/AC105_C1_2018_CRP18Rev01E.pdf. Accessed 3 Feb 2020

United Nations Committee on the Peaceful Uses of Outer Space (2018b) Draft guidelines for the long-term sustainability of outer space activities, working paper by the Chair of the Working Group on the Long-Term Sustainability of Outer Space Activities. The most mature expression of these draft guidelines is contained in UN document A/AC.105/C.1/L.367. http://www.unoosa.org/res/oosadoc/data/documents/2019/aac_105c_1l/aac_105c_1l_367_0_html/V1804974.pdf. Accessed 3 Feb 2020

United Nations Committee on the Peaceful Uses of Outer Space Scientific and Technical Subcommittee and International Atomic Energy Agency (2009) Safety framework for nuclear power source applications in outer space. International Atomic Energy Agency, Vienna. www.unoosa.org/pdf/publications/iaea-nps-sfrmwrkE.pdf. Accessed 3 Feb 2020

United Nations General Assembly (1987) Report of the World Commission on Environment and Development, Annex Our Common Future, UN document A/42/427. http://www.un-documents.net/wced-ocf.htm. Accessed 3 Feb 2020

United Nations General Assembly (2007) Committee on the Peaceful Uses of Outer Space, future role and activities of the Committee on the Peaceful Uses of Outer Space, working paper submitted by the Chairman, UN document A/AC.105/L.268. http://www.unoosa.org/pdf/limited/l/AC105_L268E.pdf with a corrigendum issued in document http://www.unoosa.org/pdf/limited/l/AC105_L268Corr1E.pdf. Accessed 3 Feb 2020

United Nations General Assembly (2010) Resolution adopted by the General Assembly on 8 December 2010, on the report of the First Committee (A/65/410), transparency and confidence-building measures in outer space activities, A/RES/65/68. https://undocs.org/en/A/RES/65/68. Accessed 3 Feb 2020

United Nations General Assembly (2011) Report of the Committee on the Peaceful Uses of Outer Space, fifty-fourth session (1–10 June 2011), UN General Assembly document A/66/20, Annex II

United Nations General Assembly (2019) Report of the Committee on the Peaceful Uses of Outer Space, sixty-second session (12–21 June 2019), the final combined texts of the preamble and 21 LTS guidelines is contained in UN document A/74/20, Annex II. https://www.unoosa.org/res/oosadoc/data/documents/2019/a/a7420_0_html/V1906077.pdf. Accessed 3 Feb 2020

United Nations Office for Outer Space Affairs (1958–2018) Documents and resolutions database. www.unoosa.org/pdf/publications/st_space_49E.pdf. Accessed 3 Feb 2020

United Nations Office for Outer Space Affairs (2010) Space Debris Mitigation Guidelines of the Committee on the Peaceful Uses of Outer Space. United Nations, Vienna. https://www.unoosa.org/oosa/documents-and-resolutions/search.jspx?lf_id=. Accessed 3 Feb 2020

United Nations Register of Objects Launched into Outer Space. http://www.unoosa.org/oosa/en/spaceobjectregister/index.html. Accessed 3 Feb 2020

United States Secretary of State (2012) Statement of Hillary Rodham Clinton on January 17, 2012. http://www.state.gov/secretary/rm/2012/01/180969.htm. Accessed 3 Feb 2020

Weeden B (2010) Dealing with Galaxy 15: Zombiesats and on-orbit servicing. The Space Review, Edition of 24 May 2010. http://thespacereview.com/article/1634/1. Accessed 3 Feb 2020

Security Issues with Respect to Celestial Bodies

18

George D. Kyriakopoulos

Contents

Introduction	342
Celestial Bodies: The Current Status Quo	344
Freedom of Exploration and Use; Freedom of Access; Freedom of Scientific Investigation; Non-appropriation	344
Applicability of International Law	344
Use "Exclusively for Peaceful Purposes"	344
Harmful Contamination Is Prohibited	345
Right to Visit Facilities and Equipment of Other States	345
Threats Arising from the Weaponization of Outer Space	345
Prevention of an Arms Race in Outer Space	346
Transparency and Confidence-Building Measures in Outer Space Activities (TCBMs)	347
Threats Associated with the Evolution of Space Activities on Celestial Bodies: The Space Resources Exploitation Issue	348
Extraction and Appropriation of Space Resources: Security Issues	349
Applicable International Law	350
National Approaches to Space Resource Utilization	350
The Resources' Issue Inside the United Nations Committee on the Peaceful Uses of Outer Space (UNCOPUOS)	352
The Hague International Space Resources Governance Working Group	353
Conclusion	354
References	355

Abstract

Space security is part of the overall international security, the maintenance of which constitutes the fundamental purpose of the UN Charter. In particular, preserving security with respect to the celestial bodies requires the activation of mechanisms able to guarantee that the existing *status quo* (global commons'

G. D. Kyriakopoulos (✉)
School of Law, National and Kapodistrian University of Athens, Athens, Greece
e-mail: yokygr@gmail.com

© Springer Nature Switzerland AG 2020
K.-U. Schrogl (ed.), *Handbook of Space Security*,
https://doi.org/10.1007/978-3-030-23210-8_137

character, applicability of international law, peaceful use, prohibition of harmful contamination) will not be compromised by the placement of offensive weapons on them or potential disputes concerning the exploitation of space resources. Another concern is the control of cosmic hazards (planetary defense), although, in this case, security on Earth will be at stake.

Introduction

Security in international relations is a value. As such, it refers to the maintenance and protection of an international *status quo* against potential threats that would be able to disrupt the current state of affairs (Jakhu and Pelton 2017: 269). At global level, the fundamental regulatory scheme for the protection of post-World War II international society is contained in the Charter of the United Nations, the main purpose of which is "the maintenance of international peace and security" (UN Charter, Preamble). In this respect, Article 1 of the Charter, in listing the purposes of the Organization, refers in particular to the need "to maintain international peace and security, and to that end: to take effective collective measures for the prevention and removal of threats to the peace...." Further, it is not by chance that Article 2 paragraph 3 of the Charter provides, among the fundamental principles of the United Nations, that "all Members shall settle their international disputes by peaceful means in such a manner that international peace and security, and justice, are not endangered."

Given the global character of international security, it is obvious that any threats against it will come from internal actors, such as states or other entities (e.g., terrorist groups). In most cases, these threats will be of an aggressive character. As it was rightly pointed out:

> International security is... an internal problem for international society as a whole. In this context, the use of armed force is directed at what may in essence be thought as the problem of internal subversion by those who would threaten the plural and cooperative character of international society. Secession, irredentism, aggressive war, conquest, illegal occupation, mass expulsion, genocide and other actions which violate international law all threaten to disrupt the general condition of peace, order and lawfulness within international society....
> (Jackson-Preece 2011: 20)

Space security is construed in the same way. State relations in the space domain are part of the general scheme of international relations and, in many instances, are able to affect state policies on Earth. The fact that space security should be faced in the context of the fundamental obligation "to maintain international peace and security" is clearly demonstrated in Article III of the *Treaty on Principles Governing the Activities of States in the Exploration and Use of Outer Space, including the Moon and Other Celestial Bodies* of 1967 (hereinafter "The Outer Space Treaty" or "OST"), which calls for the applicability, in outer space, of international law, "including the Charter of the United Nations, in the interest of maintaining international peace and security and promoting international cooperation and understanding."

Outer space constitutes a relatively new field of activity for states that started when the first artificial satellite, Sputnik-1, was launched and put into orbit on 4 October 1957 by the Soviet Union. This achievement was just a first step in the so-called Race to the Moon between the two superpowers of that time, the US and USSR. It is therefore hardly surprising that international space law (which was mainly formed in the 50s, the 60s, and the 70s) reflected in essence the international relations of the Cold War Era and the confrontation of the two space-faring powers of this era. It is thus obvious that, in those days, space security was a game for two. However, the security concerns of the two superpowers finally resulted in an optimal balance, which was reflected in a set of fundamental principles initially formed in the context of Resolution 1962 (XVIII)/1963 of the UN General Assembly and finally incorporated in the Outer Space Treaty.

However, things are changing, and the state of space affairs is now substantially different than in the past. New space actors, mainly coming from the private sector, emerge, intending to operate in "exotic" space activities such as the extraction of asteroid mineral resources or the provision of "space tourism" services.

In view of the above, space security is dependent on the establishment of a regime capable of confronting potential threats "to the advantages that accrue for humanity from the use of space" (Jakhu and Pelton 2017: 269). The purpose of this book is to assess the multiple threats that might compromise space security in toto; however, the object of this particular chapter is to shed light on those aspects of space security that are related to the celestial bodies.

Although the space treaties repeatedly refer to "the exploration and use of outer space, including the Moon and other celestial bodies," the term "celestial body" is not clearly defined. According to a generic definition, the celestial bodies constitute natural objects "located outside of Earth's atmosphere, such as the Moon, the Sun, an asteroid, planet, or star." (https://www.yourdictionary.com/celestial-body) Although several theories on the definition of the celestial bodies have emerged, some of them arguing that comets and asteroids cannot be considered as celestial bodies (Pop 2001), there is still no generally accepted definition – at least in the context of the *corpus juris spatialis*. However, it can be argued that the celestial bodies are distinguished from space objects, which are artificial manmade objects (Hobe 2009). For the purposes of this chapter, the celestial bodies are regarded as described in article 1 of the 1979 Moon Agreement: The Moon and "other celestial bodies within the solar system, other than the Earth."

As stated, security is compromised by internal actors. This means that natural threats to the celestial bodies will not be dealt with in this chapter. Besides, the planetary defense issue (how to address the threat that asteroids and comets represent) is a particular problem for Earth. The presentation below will therefore focus on the situation of celestial bodies as it stands at present, in conformity with the applicable international legal framework, before examining the main causes of potential friction: the possible placement of weapons in outer space (and on celestial bodies) as well as the growing desire for the appropriation of space resources.

Celestial Bodies: The Current Status Quo

According to the legal framework in force, the current status quo of the celestial bodies exhibits some particular characteristics, which are specified below (global commons' regime, applicability of international law, peaceful use, avoidance of harmful contamination). The maintenance of this status quo is a crucial factor in order to preserve security, as far as state activities related to the celestial bodies are concerned. However, in view of the fundamental changes that take place in space affairs, it is possible that a new balance point will be needed in the near future.

Freedom of Exploration and Use; Freedom of Access; Freedom of Scientific Investigation; Non-appropriation

The exploration and use of the celestial bodies shall be carried out for the benefit and in the interests of all countries (it constitutes a "province of all mankind"). All states, "without discrimination of any kind, on a basis of equality and in accordance with international law," are free to explore and use the celestial bodies; further they can freely access all areas of them and, last but not least, they enjoy the freedom of scientific investigation (Article I, Outer Space Treaty, Articles 4 & 6, Moon Agreement).

Moreover, according to the space treaties, the celestial bodies are "not subject to national appropriation by claim of sovereignty, by means of use or occupation, or by any other means" (Article II, Outer Space Treaty), which means that they belong to the so-called "global commons" (spaces beyond national jurisdiction) (Ranganathan 2016: 693, UN System task team on the post-2015 UN Development Agenda 2013: 5). In view of this particular character of the celestial bodies, activities on them "must be conducted with due regard to the corresponding interests of all other States" (Article IX, Outer Space Treaty).

Applicability of International Law

Activities on the celestial bodies must take place "in accordance with international law, including the Charter of the United Nations, in the interest of maintaining international peace and security and promoting international cooperation and understanding" (Article III, Outer Space Treaty, Article 2, Moon Agreement). In view of the applicability of the UN Charter in space affairs, fundamental precepts of international law, such as the prohibition of the use of force in international relations (Article 2 para. 4) or the "inherent" right to self-defense (Article 51) equally apply to the activities of States in outer space.

Use "Exclusively for Peaceful Purposes"

It is prohibited to States (parties to the Outer Space Treaty) to install nuclear weapons or any other kinds of weapons of mass destruction on celestial bodies (Article IV para. 1, Outer Space Treaty, Article 3 Moon Agreement). Further, the celestial bodies shall be used "exclusively for peaceful purposes." In this respect, the establishment of military bases, installations, and fortifications; the testing of any type of weapons; and the conduct of military maneuvers on celestial bodies are forbidden. States are entitled to use equipment or facilities "necessary for peaceful exploration of the Moon and other celestial bodies" (Article IV para. 2, Outer Space Treaty, Article 3 paras. 1–2-4, Moon Agreement). According to the prevailing interpretation, based on state practice, "peaceful" means "non-aggressive" (Tronchetti 2015: 338–340).

Harmful Contamination Is Prohibited

Exploration of the celestial bodies must take place "so as to avoid their harmful contamination and also adverse changes in the environment of the Earth resulting from the introduction of extra-terrestrial matter." In case of activities, on the celestial bodies, which may cause "potentially harmful interference with activities of other States," "appropriate international consultations" must be undertaken (Article IX, Outer Space Treaty, Articles 7, 8 para. 3, Moon Agreement).

Right to Visit Facilities and Equipment of Other States

A right of states to visit "all stations, installations, equipment and space vehicles on the Moon and other celestial bodies...on a basis of reciprocity," after "reasonable advance notice" is established in the Outer Space Treaty. Said right must be exercised in such a way as to "assure safety and to avoid interference with normal operations in the facility to be visited" (Article XII).

Threats Arising from the Weaponization of Outer Space

Although it is prohibited to use the celestial bodies for non-peaceful purposes (Outer Space Treaty, Article IV para. 2), the issue is far from resolved. It is true that, up to now, those states (US, Russia, China) that possess the technical infrastructure and means to undertake military activities on the celestial bodies (for the time being, on the Moon) have avoided placing weapons in outer space (Schrogl and Neumann 2009: 87; Space Security Index 2019: 97, 132, 137). However, the current legal framework does not prevent states from placing conventional weapons in space and relevant scenarios have been developed in military circles. Thus, the general debate on the weaponization of outer space is still topical.

However, it must be stressed that military activities in outer space that would be permissible under space law must also be compliant with Article 2 para. 4 of the UN Charter. In other words, they must be of a non-aggressive character (Petras 2002: 1255). Furthermore, the "inherent" right to self-defense, pursuant to the UN Charter and the international customary law, is also valid in outer space activities, notwithstanding the limitations established by Art. IV of the Outer Space Treaty (in the context of the "peaceful purposes").

Military applications of non-offensive character became common since the beginning of the Space Age, through the use of reconnaissance and early warning satellites. However, the term "weaponization of outer space" refers to the placement of offensive weapons in outer space, encompassing also the development of weapon systems on Earth whose mission is to destroy space objects (Tronchetti 2015: 333–334, Mosteshar 2019). Anti-satellite (ASAT) weapons stand as an example of the latter: they have already been used to destroy satellites in orbit, for testing purposes (Kyriakopoulos 2015: 599). Further, according to existing scenarios, space devices can be used in order to destroy targets on Earth (e.g., use of space for missile interception – Space Security Index 2019: 132), while in addition cyberattacks against satellites is a matter of concern. The situation is made yet more complex by the "dual-use" nature (civil/military) of most space systems (Lyall and Larsen 2018: 448).

Prevention of an Arms Race in Outer Space

This item was in the Agenda of the Conference on Disarmament (CD) since the early 80s. In 1985, CD mandated a committee especially set up for this purpose to identify and examine issues relevant to the prevention of an arms race in outer space (PAROS), such as the legal protection of satellites, nuclear power systems in space, and various confidence-building measures. It is worth noting that the United States clearly opposed to the mandate of the committee, opting for direct talks with the Soviet Union.

In 2008, China and Russia jointly introduced to the Conference on Disarmament (CD) a draft convention on the prevention of the installation of weapons systems in outer space. The draft treaty, while it recognized that "prevention of the placement of weapons in outer space and of an arms race in outer space would avert a grave danger for international peace and security," proposed a broad definition of "weapons placed in outer space" by comprising in its scope any such weapon that "orbits the Earth at least once, or follows a section of such an orbit before leaving this orbit, or *is permanently located somewhere in outer space.*" It is thus obvious that weapons that could have been placed on the celestial bodies were brought within the field of application of the draft treaty. This approach was further confirmed in the wording of Article II of the draft, pursuant which states undertook "not to place in orbit around the Earth any objects carrying any kinds of weapons, *not to install such weapons on celestial bodies. . . .*"

On 10 June 2014, Russia submitted to the CD an updated draft of this Convention (Draft Treaty on the Prevention of the Placement of Weapons in Space, the Threat or Use of Force Against Space Objects, PPWT). This revised draft further established the obligation for states "Not to resort to the threat or use of force against outer space objects" of other states (Article II). Such objects can be, inter alia, "permanently located . . .on any celestial bodies other than the Earth" (Article I).

Through recurring resolutions, the General Assembly, recognizing the importance of the Transparency and Confidence-Building Measures (TCBMs), has consistently urged the states members to "contribute actively to the objective of the peaceful use of outer space and of the prevention of an arms race in outer space" (A/RES/74/32, 12.12.2019).

Transparency and Confidence-Building Measures in Outer Space Activities (TCBMs)

In 1990, the General Assembly requested the Secretary-General, with the assistance of a group of governmental experts, to carry out a study on different confidence-building measures in outer space. That group delivered its report in 1993 (A/48/305) (Takaya-Umehara 2010: 1301).

In 2011, the UN General Assembly further asked the Secretary-General to set up a "Group of Government Experts" (GGE) to conduct a TCBM survey (A/RES/65/68). The GGE submitted its Report to the Assembly on 29.7.2013 (A/RES/68/189). In this Report, the GGE proposed a set of TCBMs in outer space, including the proposal to establish coordination between United Nations Office for Disarmament Affairs (UNODA), United Nations Office for Outer Space Affairs (UNOOSA), and other appropriate UN entities. Among the measures proposed, the following could be of particular importance with respect to the subject matter of this chapter:

- Exchanges of information on the principles and goals of a State's outer space policy
- Exchanges of information on major military outer space expenditure and other national security space activities
- Exchanges of information on forecast natural hazards in outer space
- Voluntary familiarization visits
- Expert visits, including visits to space launch sites, invitation of international observers to launch sites, flight command and control centers, and other operations facilities of outer space infrastructure
- International cooperation
- Consultative mechanisms
- Outreach measures
- Measures of coordination

The General Assembly further encouraged the implementation of the TCBMs. On 5 December 2013, with Resolution 68/50, after having reaffirmed that "preventing an arms race in outer space is in the interest of maintaining international peace and security and is an essential condition for the promotion and strengthening of international cooperation in the exploration and use of outer space for peaceful purposes," the Assembly encouraged Member States to review and implement, "to the greatest extent practicable," the proposed TCBMs, "through relevant national mechanisms, on a voluntary basis and in a manner consistent with the national interests of Member States."

Of course, said measures are nonlegally binding voluntary measures. Nevertheless, as it was stated in the GGE Report, "such measures can augment the safety, sustainability and security of day-to-day space operations and can contribute both to the development of mutual understanding and to the strengthening of friendly relations between States and peoples." Further, the importance of the said TCBMs for the maintenance of space security (and, in particular, for preserving security with respect to celestial bodies) is explicitly recognized in the Report, in its paragraph 31: "In general terms, transparency and confidence-building measures for outer space activities should be aimed at increasing the security, safety and sustainability of outer space. Particular attention should be given to the development and implementation of voluntary and pragmatic measures to ensure the security and stability of all aspects of outer space activities."

In 2017, a new GGE ("on further practical measures for the prevention of an arms race in outer space") was established pursuant UNGA Resolution 72/250 (24.12.2017), in order to "consider and make recommendations on substantial elements of an international legally binding instrument on the prevention of an arms race in outer space, including, inter alia, on the prevention of the placement of weapons in outer space." The Group met in two sessions, the first from 6 to 17 August 2018 and the second from 18 to 29 March 2019. However, no consensus was reached on a substantive report (A/74/77, Note by the Secretary-General).

Following the aforementioned resolutions of the General Assembly, a European Union initiative took the form of a Draft International Code of Conduct for Outer Space Activities. Having noted, in its preamble, the importance of preventing an arms race in outer space, said code also proposed TCBMs, in order to "prevent confrontation and foster national, regional and global security and stability." General Principle 28 of the Code emphasized "the responsibility of States, in the conduct of scientific, civil, commercial and military activities, to promote the peaceful exploration and use of outer space for the benefit, and in the interest, of humankind and to take all appropriate measures to prevent outer space from becoming an arena of conflict."

Threats Associated with the Evolution of Space Activities on Celestial Bodies: The Space Resources Exploitation Issue

Over the last few years, there has been a great deal of talk about the development of activities related to the exploration, exploitation, and utilization of outer space resources. This is a space activity for the future, as it requires the development of relevant technology as well as the mobilization of significant financial resources in order for such a challenging adventure to be able to have a sustainable future. Moreover, important security issues are raised.

In the aforementioned context, new space actors emerge, mainly coming from the private sector and intending to operate in areas that a few years ago would be classified as "exotic."

It is generally accepted that the term "space resource utilization" encompasses activities into Earth orbits, in situ resource utilization (ISRU), as well as the commercial appropriation of space natural resources ("space mining"). Earth orbits are outside the thematic area of this chapter. ISRU is the collection, processing, storing, and use of materials encountered in the course of human or robotic space exploration that replace materials that would otherwise be brought from Earth to accomplish a mission's critical need at reduced overall cost and risk (Sackstender and Sanders 2007). ISRU, provided it is compatible with article IX OST, constitutes an activity that is in conformity with article I paragraphs 2 and 3 of the Outer Space Treaty, as it is associated with the freedom of use, exploration, and scientific investigation of outer space and, in principle, does not constitute appropriation. However, in due time an ad hoc regulation of ISRU could help to avoid tensions, if such use is to take place by numerous users in a common area on a celestial body. Thus, it is the commercial exploitation of space resources that will be discussed in further detail.

Extraction and Appropriation of Space Resources: Security Issues

The National Aeronautics and Space Administration (NASA) claims that there are approximately 100,000 near-Earth objects, and a great number of them potentially contain water and important minerals, such as nickel, cobalt, and iron. However, is "space mining" feasible, with today's standards?

On 6 August 2014, ESA's orbiter "Rosetta" arrived at Comet 67P/Churyumov-Gerasimenko and the lander "Philae" landed on the comet on 12 November 2014. On 8 September 2016, NASA launched OSIRIS-REx, in order to reach the asteroid Bennu and return samples in 2023 (https://www.nasa.gov/osiris-rex). These experimental missions clearly pave the way for the commercial exploitation of celestial bodies (comets, asteroids, planets) in the near future.

During the Cold War, the competition of US and USSR for the "conquest of outer space" resulted in a set of fundamental principles, first enshrined in UNGA Resolution 1962 (XVIII) of 13 December 1963 and again in the Outer Space Treaty of 1967. In this context, the balance over potential claims for the existing "wealth" in outer space led to the adoption of the non-appropriation principle. At that time, of course, the technology available did not allow for the exploitation of space resources. However, the situation is now changing drastically, and it is quite possible that in the near future there will be intense competition among states for the appropriation of space resources. It is therefore obvious that, in this area of activity, international security must be strengthened. This obviously requires a global space resources management regime, since any unilateral actions are likely to cause friction between the states concerned. Such a regime should take into account the existing fundamental principles of international space law, given that states do not contest their acceptance.

Applicable International Law

With respect to the exploration and exploitation of outer space, the *lex lata* provisions are Article I, II, and III of the Outer Space Treaty. Article I provides for the freedom of exploration and use of outer space, including the Moon and other celestial bodies, and for the freedom of access to all areas of celestial bodies. Article II further establishes the non-appropriation principle while, pursuant Article III, activities in the exploration and use of outer space, including the Moon and the celestial bodies, shall be carried on "in accordance with international law, including the Charter of the United Nations, in the interest of maintaining international peace and security and promoting international cooperation and understanding."

As has been said, there is no specific, ad hoc legal framework for the administration of the space resources exploration, exploitation, and utilization. Nevertheless, the Moon Agreement, although poorly ratified, establishes such a regime, as laid down in Article 11:

1. The Moon and its natural resources are the common heritage of mankind, which finds its expression in the provisions of this Agreement, in particular in paragraph 5 of this article.
2. The Moon is not subject to national appropriation by any claim of sovereignty, by means of use or occupation, or by any other means.
3. Neither the surface nor the subsurface of the Moon, nor any part thereof or natural resources in place, shall become property of any State, international intergovernmental or non-governmental organization, national organization or non-governmental entity or of any natural person...

. . . . . .

5. States Parties to this Agreement hereby undertake to establish an international regime, including appropriate procedures, to govern the exploitation of the natural resources of the Moon as such exploitation is about to become feasible. This provision shall be implemented in accordance with article 18 of this Agreement.

However, it is obvious that, for the majority of States that have not yet ratified said instrument, the international norms applicable to the exploration and exploitation of outer space are the aforementioned provisions of the Outer Space Treaty of 1967. It therefore follows that the fundamental principles of international space law in force rather advocate a collective exploitation regime of the resources on the Moon and other celestial bodies, at least in principle.

National Approaches to Space Resource Utilization

United States: In the United States, the commercialization of outer space was encouraged by the *U.S. Commercial Space Launch Act* of 1984, according to which "the general welfare of the United States requires that the Administration seek and encourage, to the maximum extent possible, the fullest commercial use of space." At that time, the Act faced only unmanned launch activities, as there was still no question about human space flight. Further, the Act also directed the Federal Aviation Administration (FAA) (through delegations) to encourage, facilitate, and promote commercial space launches and reentries by the private sector, including those involving space flight participants (§ 50903).

On 25 November 2015, the President of the United States signed the *U.S. Commercial Space Launch Competitiveness Act* (H.R. 2262). As far as space mining is concerned, the important part of said Act is its Title IV "Space Resource Exploration and Utilization." According to this part, emphasis is made to the promotion of the right of U.S. citizens "to engage in commercial exploration for and commercial recovery of space resources free from harmful interference, in accordance with such obligations and subject to authorization and continuing supervision by the federal government." Furthermore, "A U.S. citizen engaged in commercial recovery of an asteroid resource or a space resource *shall be entitled to any asteroid resource or space resource obtained, including to possess, own, transport, use, and sell it* according to applicable law, including U.S. international obligations."

It follows that under said Act (in other words, by a unilateral act), U.S. citizens are explicitly entitled, inter alia, to "possess," "own," and "sell" asteroid resources; thus the Act explicitly confers to U.S. citizens property rights to resources in outer space, which is in sharp contrast with the principle of non-appropriation enshrined in article II OST.

In the light of this Act, some experts have advanced arguments in favor of the permissibility of space resource utilization/appropriation. However, the prevailing view on the exact meaning of Article II is that the non-appropriation principle prohibits both the exercise of sovereign rights (by states) and private appropriation (by nongovernmental entities) (Freeland and Jakhu 2009: 50). This conclusion is further strengthened by the clear wording of Article 11 paragraph 3 of the Moon Agreement, according to which "Neither the surface nor the subsurface of the Moon, nor any part thereof or natural resources in place, shall become property of any State, international intergovernmental or non-governmental organization, *national organization or non-governmental entity or of any natural person.*"

In this respect, it is worth mentioning that the question of private property claims in outer space has arisen before U.S. courts in the past, when an individual (Gregory Nemitz) claimed parking fees from NASA regarding use of his own asteroid (!). However, the claim was dismissed on the grounds that the Outer Space Treaty did not permit appropriation of private property. In addition, both NASA and the U.S. State Department rejected the claim, thus adopting a broad interpretation of the non-appropriation principle.

Luxembourg: Luxembourg adopted a law on the exploration of space and the use of space resources in 2017 (*Loi du 20 juillet 2017 sur l'exploration et l'utilisation des ressources de l'espace*). Article 1 of said law states that "space resources are capable of being appropriated," which means that extraction companies will have ownership on the space resources that they extract. Further, pursuant Articles 2–4, Luxembourger corporations or European companies that have their registered office in Luxembourg may extract space resources for commercial use after obtaining approval from the Government of Luxembourg. Again the compatibility of these provisions with the non-appropriation principle is highly questionable.

The Resources' Issue Inside the United Nations Committee on the Peaceful Uses of Outer Space (UNCOPUOS)

Since 2017, the Legal Subcommittee of UNCOPUOS (LSC) has included, as an item in its agenda, a "General exchange of views on potential legal models for activities in the exploration, exploitation and utilization of space resources."

The discussions inside the LSC as well as the practice of states have shown that, at present, there is no consensus among states on the legal regime that should govern the resources issue.

During the 58th session of the LSC (2019), Belgium and Greece submitted a common proposal/working paper (A/AC.105/C.2/L.311/4.3.2019), entitled "Proposal for the establishment of a working group for the development of an international regime for the utilization and exploitation of space resources - Working paper by Belgium and Greece." Said document highlighted the principles of international space law which should govern the future exploitation of the resources of the celestial bodies, in particular:

- That the exploration and use of outer space is a field for all of Humankind
- That such exploration and use is regulated by international law
- That the space treaties require an enhanced international cooperation

As for the impact of the resources' issue on the maintenance of international security on celestial bodies, the working paper mentioned the following (paragraph 16):

> The need for an international legal regime for space resource exploitation also arises from the fact that national approaches to space resource exploitation are bound to result in conflicts between competing players, if left to evolve on their own without international guidance. Hence, even if there is no legal objection to States interpreting at will their international obligations under the Outer Space Treaty when regulating space resources, there is still a clear need for an international institutional framework to regulate competing activities. In order for such a framework to be effective, it would have to be focused on the main purposes described in article 11, paragraph 7, of the Moon Agreement, the value of which is greater and goes beyond any views on the ratification of the Moon Agreement. Those purposes include the following: (a) the orderly and safe development of natural resources from outer space; (b) the rational management of those resources; (c) the expansion of opportunities in the use of those resources; and (d) an equitable sharing by all States in the benefits derived from those resources, whereby the interests and needs of the developing countries, as well as the efforts of those countries which have contributed either directly or indirectly to the exploration of outer space, shall be given special consideration.

Based on these principles, the working paper proposed the creation of an ad hoc working group within the framework of the LSC, having as mandate to investigate the aspects and questions raised with respect to the future use and exploitation of space resources, as well as the preparation of a report. Nevertheless, said proposal did not reach consensus inside the Legal Subcommittee. Instead, the LSC adopted a compromise proposal by the Chair, according to which scheduled informal consultations were to be held at the next session of the Subcommittee, in 2020.

During the 62th session of the COPUOS (2019), a relevant proposal was submitted by the United Arab Emirates (A/AC.105/2019/CRP.17/19.6.2019). The proposal recognized "the importance of establishing a working group that serve as a platform to create a unified strategic conversation among Member States" and considered that said working group "will lead the discussions towards the development of a recommended set of principles governing Space Resources Utilization activities."

The Hague International Space Resources Governance Working Group

Beyond the discussions that take place in the context of the COPUOS, The Hague International Space Resources Governance Working Group is another forum that examined the space resources' issue. The Working Group was established in 2016, in order to assess the need for a governance framework on space resources, and consists of members as well as observers that represent governments as well as industry, academia, science, international organizations, NGOs, and the civil society. Its mandate was to identify and formulate "building blocks for the governance of space resource activities as a basis for negotiations on an international agreement or non-legally binding instrument." The Working Group adopted the final text of the building blocks on 12.11.2019. According to the proposed blocks, the objective of an international framework ("consistent with international law") should be the creation of "an enabling environment for space resource activities that takes into account all interests and benefits all countries and humankind." In this respect, the international framework should propose relevant recommendations and promote "the identification of best practices by States, international organizations and non-governmental entities." It must be stressed that the potential impact of the space resources' utilization issue for the maintenance of space security is recognized in the Building Blocks: According to the proposed Principles, the future international framework "should be designed to: ... prevent disputes arising out of space resources" and "should provide that ...space resources shall be used exclusively for peaceful purposes."

Of course, the aforementioned concerns do not mean that commercial activities in outer space, of a private nature, should be discouraged. Such activities are clearly dictated by a new state of affairs in outer space and everything indicates that they are about to expand in the near future. Nevertheless, in the light of the desire of some states to put the famous "use of outer space" concept in a business perspective, it seems that current international space law might prove insufficient in this respect. This deficiency in the legal field can negatively affect international security with respect to outer space, including the Moon and other celestial bodies.

It must also be kept in mind that other important issues for an effective and secure commercial exploitation of outer space, in general, and the celestial bodies, in particular, – such as the protection of the space environment or the establishment of an effective space traffic management system – should also require, in the near future, the intervention of states through the undertaking of appropriate international action. What is more, new concepts emerge in space affairs: the long-term sustainability of outer space activities and the Space 2030 Agenda constitute additional factors and goals that push toward a rational and equitable use of outer space resources, which cannot be achieved through unilateral initiatives: an interesting model, in this respect, could be found in the International Telecommunication Union (ITU) Constitution, which dictates the "rational, equitable, efficient and economical use of the radio-frequency spectrum by all radiocommunication services" (Articles 12, 44).

Conclusion

Space security is part of a broader security scheme in international affairs, the maintenance of which constitutes the fundamental purpose of the UN Charter. This requirement to maintain international security also applies in the context of space activities. In particular, the security of the celestial bodies presupposes the maintenance of a status quo defined by the current international legal framework. Under this framework, the celestial bodies must be used exclusively for peaceful purposes and constitute global commons, as they are not subject to "national appropriation by claim of sovereignty, by means of use or occupation, or by any other means." What is more, the applicable norms with respect to activities on the celestial bodies are of international character, pursuant the obligation of states to "carry on activities in the exploration and use... [of] the Moon and other celestial bodies in accordance with international law, including the Charter of the United Nations."

Although up to now it has been respected by states, said legal status of the celestial bodies does not seem to be fully adequate. The placement of weapons in empty space as well as on celestial bodies – despite their exclusively peaceful character – constitutes a threat which has preoccupied the international community for a long time. So far, the discussions and deliberations within the framework of the Conference on Disarmament have not been successful in adopting a binding international instrument, as negotiations for the adoption of a treaty on the prevention of the placement of weapons in outer space have failed thus far. This is due to fact that national security concerns are pushing some states to keep the debate on the placement of weapons in outer space open. At present, the adoption of voluntary transparency and confidence-building measures as well as the insistence by the General Assembly on the reference to the issue of the prevention of an arms race in outer space demonstrates that a consensus has already been formed within the international community on the need for multilateral action on this issue.

The growing desire of states and private actors to plan a commercial exploitation of the resources of the celestial bodies, once technology makes this feasible, can also be a source of frustration. Some states have already adopted national legal frameworks that regulate a unilateral exploitation of these resources, despite the international obligation of non-appropriation of the celestial bodies. On the other hand, the only existing collective exploitation regime is provided at present by the Moon Agreement, which has not been universally accepted by states. Consequently, a landscape is being formed in the context of which competition for space resources could trigger tensions and conflicts among states. It is worth mentioning that there are already deliberations in international *fora* (mainly at the COPUOS, but also in the context of The Hague Group) on a potential regulatory model of international character, despite disagreements as to its particular characteristics. Thus, the development of an international legal framework that will govern the commercial exploitation of space resources, ensuring, at the same time, a rational and equitable use of them in accordance with the spirit of Article I of the Outer Space Treaty, is a matter of security.

References

A/48/305 (1993) Prevention of an arms race in outer space. Study on the application of confidence-building measures in outer space. Report by the Secretary-General. 2 October. https://digitallibrary.un.org/record/175346#record-files-collapse-header. Accessed 28 Feb 2020

A/68/189 (2013) Group of governmental experts on transparency and confidence-building measures in outer space activities. Note by the Secretary general, 29 July 2013. https://www.unoosa.org/pdf/gadocs/A_68_189E.pdf. Accessed 28 Feb 2020

A/74/77 (2019) Group of governmental experts on further practical measures for the prevention of an arms race in outer space. Note by the Secretary-General, 9 April 2019. https://undocs.org/A/74/77. Accessed 28 Feb 2020

Agreement Governing the Activities of States on the Moon and Other Celestial Bodies, New York, done 18 December 1979, entered into force 11 July 1984; 1363 UNTS 3; ATS 1986 No. 14; 18 ILM 1434 (1979)

Celestial body. https://www.yourdictionary.com/celestial-body. Accessed 28 Feb 2020

Charter of the United Nations (1945) 1 UNTS XVI. https://www.un.org/en/charter-united-nations/. Accessed 28 Feb 2020

Draft International Code of Conduct for Outer Space Activities, Version 31 (2014) March. https://eeas.europa.eu/sites/eeas/files/space_code_conduct_draft_vers_31-march-2014_en.pdf. Accessed 28 Feb 2020

Draft Treaty on Prevention of the Placement of Weapons in Outer Space and of the Threat or Use of Force against Outer Space Objects (2008) CD/1839. https://undocs.org/en/CD/1839. Accessed 28 Feb 2020

Draft Treaty on the Prevention of the Placement of Weapons in Outer Space, the Threat or Use of Force against Outer Space Objects (2014) CD/1985. https://undocs.org/en/CD/1985. Accessed 28 Feb 2020

Freeland S, Jakhu R (2009) Article II. In: Hobe S, Schmidt-T B, Schrogl K.-U (eds) Cologne Commentary on Space Law, vol 1. Outer Space Treaty, Carl Heymanns Verlag, Köln

Hobe S (2009) Article I. In: Hobe S, Schmidt-T B, Schrogl K.-U (eds) Cologne Commentary on Space Law, vol 1. Outer Space Treaty, Carl Heymanns Verlag, Köln pp. 25–43

Jackson-Preece J (2011) Security in international relations. University of London. https://london.ac.uk/sites/default/files/uploads/ir3140-security-international-relations-study-guide.pdf. Accessed 28 Feb 2020

Jakhu RS, Pelton JN (eds) (2017) Global space governance: an international study. Chapter 12 global governance of space security. Springer, Cham

Kyriakopoulos G (2015) The current legal regulation on the use of ASAT weapons under International Law and Space Law. In: Jakhu RS, Chen K-W, Nyampong Y (eds) Monograph series III: global space governance. McGill Centre for Research in Air and Space Law, Montreal, pp 599–618

Luxembourg, Loi du 20 juillet 2017 sur l'exploration et l'utilisation des ressources de l'espace. Journal officiel, N° 674 du 28 juillet 2017. http://data.legilux.public.lu/file/eli-etat-leg-loi-2017-07-20-a674-jo-fr-pdf.pdf. Accessed 28 Feb 2020

Lyall F, Larsen PB (2018) Space law. A treatise, 2nd edn. Routledge, London/New York

Mosteshar S (2019). Space law and weapons in space. In: Oxford research encyclopedia, May 2019. https://oxfordre.com/planetaryscience/view/10.1093/acrefore/9780190647926.001.0001/acrefore-9780190647926-e-74. Accessed 28 Feb 2020

Petras M (2002) The use of force in response to cyber-attack on commercial space systems – reexamining 'self-defense' in outer space in light of the convergence of U.S. military and commercial space activities. J Air Law Commer 67:1213–1268

Pop V (2001) Celestial body is celestial body is celestial body... In: A proceedings on the law of outer space, vol 44, pp 100–110

Proposal by the United Arab Emirates on the work related to space resources utilization of the committee on the peaceful uses of outer space. A/AC.105/2019/CRP.17/19.6.2019. https://www.unoosa.org/oosa/oosadoc/data/documents/2019/aac.1052019crp/aac.1052019crp.17_0.html. Accessed 28 Feb 2020

Proposal for the establishment of a working group for the development of an international regime for the utilization and exploitation of space resources. Working paper by Belgium and Greece. A/AC.105/C.2/L.311/4.3.2019. https://www.unoosa.org/oosa/oosadoc/data/documents/2019/aac.105c.2l/aac.105c.2l.311_0.html. Accessed 28 Feb 2020

Ranganathan S (2016) Global commons. Eur J Int Law 27:693–717

Sackstender KR, Sanders GB (2007). In-situ resource utilization for lunar and mars exploration. In: 45th AIAA aerospace sciences meeting and exhibit, Reno, 08 January 2007–11 January 2007. https://arc.aiaa.org/doi/pdf/10.2514/6.2007-345. Accessed 28 Feb 2020

Schrogl K-U, Neumann J (2009) Article IV. In: Hobe S, Schmidt-T B, Schrogl K.-U. (eds) Cologne Commentary on Space Law, Vol 1, Outer Space Treaty, Carl Heymanns Verlag, Köln

Space Security Index (2019) Space security index 2019, 16th edn. Waterloo Printing, Waterloo

Takaya-Umehara Y (2010) TCBMs over the military use of outer space. Acta Astronaut 67(9–10): 1299–1305

The Hague International Space Resources Governance Working Group (2019) Building blocks for the development of an international framework on space resource activities, 12 November. https://www.universiteitleiden.nl/binaries/content/assets/rechtsgeleerdheid/instituut-voor-publiekrecht/lucht%2D%2Den-ruimterecht/space-resources/bb-thissrwg%2D%2Dcover.pdf. Accessed 28 Feb 2020

Treaty on Principles Governing the Activities of States in the Exploration and Use of Outer Space, including the Moon and Other Celestial Bodies, adopted on 19 December 1966, opened for signature on 27 January 1967, entered into force on 10 October 1967, 610/U.N.T.S./205

Tronchetti F (2015) Legal aspects of the military uses of outer space. In: Von der Dunk F, Tronchetti F (eds) Handbook of Space Law. Edward Elgar, Cheltenham/Northampton, pp 331–381

UN System task team on the post-2015 UN Development Agenda (2013) Thematic think piece: global governance and governance of the global commons in the global partnership for development beyond 2015. https://www.un.org/en/development/desa/policy/untaskteam_undf/thinkpieces/24_thinkpiece_global_governance.pdf. Accessed 28 Feb 2020

UNGA A/RES/68/50. Transparency and confidence-building measures in outer space activities. 5 December 2013. https://www.unoosa.org/pdf/gares/A_RES_68_050E.pdf. Accessed 28 Feb 2020

Part II

Space Security Policies and Strategies of States

Space Security Policies and Strategies of States: An Introduction

19

Jana Robinson

Contents

Introduction	359
Space Security Policies and Strategies of States	360
Conclusion	364
References	365

Abstract

This chapter provides an introduction to Part 2 of the second edition of the Handbook of Space Security entitled "Space Security Policies and Strategies of States." It covers expert views on space security policies of established spacefaring nations, including the United States, China, Russia, European countries, Japan, and India. It also reviews space security policies of emerging space powers – Brazil, Israel, the UAE, Poland, and Azerbaijan – to showcase a wide range of space policy approaches to this strategic security portfolio. The chapter likewise includes overviews of European and Asia-Pacific approaches to space security issues. These approaches range from strict emphasis on the peaceful uses of outer space (e.g., Brazil and the UAE), to space as a key element of national security and defense (e.g., U.S., Russia, China, and Israel), or a combination of both (e.g., India).

Introduction

In contrast to the Cold War period, the space environment today involves some 88 countries and government consortia with different strategic objectives and levels of economic and technological development (Euroconsult 2019). There are also an ever-growing number of commercial satellite operators. Earth observation,

---

J. Robinson (✉)
Space Security Program, Prague Security Studies Institute (PSSI), Prague, Czech Republic
e-mail: jrobinson@pssi.cz

communications, and satellite navigation, originally supporting mainly military activities, are now part of day-to-day civilian and commercial life. As a result, there is a growing concern regarding how to best preserve safe, stable, and sustainable space operations over the long term.

Established spacefaring nations like the United States, Russia, and some European countries have developed national policies and strategies that address space security issues. This portfolio is also rich with the agenda of other space powers, such as China. Several nations, including Japan, are integrating space security into their broader national security and foreign policy agendas. At a multilateral level, a body of principles and rules governing space activities, including a unique international status of outer space and celestial bodies, was established during the second half of the twentieth century. Implementation of these principles and rules has, however, lagged for a number of reasons, including the "asymmetric" interests in space, the growing connectivity between terrestrial tensions and space security, the lack of a track record on the multilateral management of an incident in space, and other considerations.

This section of the Handbook introduces different approaches to managing space security on the part of select spacefaring nations (i.e., the United States, Russia, China, Japan, India, Israel, Brazil, the UAE, Poland, and Azerbaijan). It also includes overviews of European and Asia-Pacific approaches to space security issues. As it will be evident from the content of various chapters, there is, as yet, no consensus on what issues constitute the space security portfolio. There are also significant differences in the main focus of individual national space programs. There is, however, an evident requirement for an improved dialogue on major space security issues and enhanced international collaboration on the more contested aspects of space operations.

Space Security Policies and Strategies of States

The emphasis on the peaceful uses of outer space coexists with the underlying role of space as an instrument of national security and international prestige. Threats to space assets and activities, both natural and man-made, are multiplying, as evidenced by events such as China's repeated antisatellite (ASAT) tests, Russia and, most recently, India (in March 2019), or close approaches by Chinese and Russian satellites to foreign spacecraft.

Man-made threats include intentional disruption of satellite services and even attacks on space assets. Intentional jamming could have damaging military, political, and commercial knock-on effects. A prominent example of such deliberate interference has been the repeated jamming of Eutelsat satellites by a source located on Iranian territory. Other threats include directed or kinetic energy ASAT attack or cyber assaults. Cyber-attacks against satellites and ground stations are a rapidly growing concern – and vulnerability – and should be added to the list of political and budgetary challenges to enhance security in both the space and the cyber domains. Although there have not, as yet, been any serious consequences (at least confirmed reports) resulting from such incidents, these events have highlighted the need for establishing not only national procedures but also diplomatic processes to facilitate the smooth and efficient management of these types of actions internationally. Although space threats emanating from natural hazards or technical issues warrant genuine concern, the intentional disruption of, or damage to, space assets and systems would generally involve larger – sometimes far larger – geopolitical stakes.

In the *United States*, space policy has remained relatively consistent over the past 60 years with a focus on international cooperation, peaceful intention, and development of outer space for the common good of all humankind. ▶ Chap. 20, "War, Policy, and Spacepower: US Space Security Priorities" by Everett C. Dolman explains how throughout this time, every policy has reserved the right of self-defense in space while interpreting military activity rather broadly. Yet, the 2017 National Security Strategy made a notable shift, in particular with regard to the security aspects and reviving, the National Space Council. The US Space Policy Directives 1 and 2 aim at fostering reinvigorating America's human space exploration, and commercial activities through appropriate regulatory framework, while Space Policy Directives 3 and 4 address the creation of space traffic management and the establishment of a Space Force, respectively. The Space Force has become the sixth branch of the US military, housed within the Department of the Air Force. In addition, the US Space Command was reestablished to organize military space operations. The US leadership now believes that it is irresponsible to rely on the better intentions of its adversaries, international agreements, and deterrence to guarantee its national space interests. The new policy makes clear that space is now considered a warfighting domain, and the USA will prepare itself to fight and win military conflicts that might occur in this domain.

Outer space has also become an important area through which *Russia* aims to rebuild its global status and prestige as a space power, including its efforts to shape global norms. Russia considers outer space predominantly as a strategic region to enhance its military capabilities on Earth, provide intelligence and communication functions, and achieve international esteem. Nicole Jackson describes how Russia reacts to US strategy and continues to develop counterspace technologies (e.g., electronic weapons that can jam satellites) to provide the country with an asymmetrical edge to offset US military advantages. Accordingly, military efforts are but one part of a complex set of tools, employed to navigate what Russia perceives as an increasingly hostile world. Back in 2011, Russia brought about certain institutional modernizations creating the Russian Aerospace Defence Forces to conduct its space security-related activities. Recently, in March 2018, the Russian Defense Minister Sergey Shoigu stated that Russia must deploy a modern fleet of military satellites to support its army and navy. He stated: "only with support from space will it be possible for the Armed Forces to reach maximum effectiveness." However, its economic, military, and technological weaknesses, compared to those of the US and NATO, have led Russia to pursue asymmetrical tactics and focusing on bilateral relations as well as pushing its agenda through the United Nations.

Formerly part of the USSR, *Azerbaijan* has been involved in space activities for some decades. The country has an agency, the Azerbaijan National Aerospace

Agency (ANASA) created in 1992. Its recent activities are part of the space strategy "Azerbaijan 2020" which aims to develop the Azerbaijani space industry and own satellites.

Space security and defense discussions in *Europe* have been supported by an existing space security culture of the Western European Union, as explained by Alexandros Kolovos. Over the past several years, the European Union (EU) has formulated a space security strategy, including in the 2018 proposal for a Regulation for a Space Program for the EU which is based on the 2016 Space Strategy for Europe. One of the main goals of the Space Strategy is to "Reinforce Europe's autonomy in accessing and using space in a secure and safe environment." In order to realize the security objective, the EU Regulation proposes the development of Governmental Satellite Communications (GOVSATCOM) and Space Situational Awareness (SSA) programs to accompany Galileo and Copernicus. At the same time, the European Space Agency (ESA) has more explicitly formulated its space security policy, as reflected in the "Elements of ESA's Policy on Space and Security" and the safety and security program adopted at the ESA Ministerial Council in 2019. In addition to ESA and the EU policy and programmatic developments, NATO Defence Ministers approved in 2019 its first ever space policy. The NATO Secretary General Mr. Stoltenberg stated that the premise for the space policy is that: "Space is part of our daily life here on Earth – it can be used for peaceful purposes, but it can also be used aggressively." The authors of these chapters describe the institutional space and security policy developments in Europe that have taken place in parallel with the policies of the individual European countries. These include the 2019 French Space and Defence Strategy and other trends in the main European space powers, as well as in emerging ones, such as Poland. Overall, European space and security governance is multifaceted, thereby posing a major challenge to effective mutual cooperation among the EU, ESA, and the European States.

Asia is the world's second-largest defense spender while it is becoming increasingly active in space. Japan, China, and India have been the most prominent space actors in this region. Dr. Rajagopalan explains how the geopolitical and military competition in Asia has an impact on the space efforts of these countries. The growing space competition manifests itself by the rapidly growing development of counterspace capabilities, such as kinetic ASAT missiles, electronic and cyber warfare capabilities, as well as new efforts at creating specialized military agencies devoted to space utilization. There are three key drivers to space conflict in Asia, according to her – increasing use of space for military purposes; civilian use that could also lead to conflict because of congestion and competition; and investments in military technologies such as those for ASAT weapons and missile defense.

Concerning *China*, its space ambitions are long-term and strategic, externally emphasizing stability and publishing its 5- and 10-year plans for space. China has been steadily and systematically rising as a major space power and the world has witnessed, among other achievements, the landing of an unmanned mission on the near side of the Moon using its Chang'e 3 in December 2013, and the Chang'e 4 landing on the far side of the lunar surface in January 2019. Over the past several decades, the relationship of space and national security has been evolving as part of a broader ongoing assessment of the role of information in future warfare, as explained by Dean Cheng. Due to the People's Liberation Army's (PLA) attention to information and communications technologies, the centrality of space dominance has grown as well. In 2015, China established the PLA Strategic Support Force (PLASSF) which saw the integration of the PLA space, cyber, and electronic warfare capabilities, and considered a significant domestic achievement concerning China's preparedness for multidomain future warfare. For the PLA and China's security decision-makers more broadly, the Information Age and the Space Age are inextricably linked. Both have been heavily influenced by the growth in computing power and the role of telecommunications. This has gained impetus as Information and Communication technologies (ICT) have become a key element in overall national power and, especially, military capabilities.

Kazuto Suzuki explains how *Japan*'s space policy has been influenced significantly by the overall foreign and security policy. Since the inception of its space activities, Japan has been reluctant to engage in security-related uses of space, largely due to its Constitution. This attitude has evolved over the past few years. In 2013, the country updated the Basic Space Plan which was released in 2009. The revised Plan emphasized the need for new opportunities for the involvement of Japan in international efforts to address the most pressing space security-related challenges of the twenty-first century. At the end, the 2013 update, and its subsequent revision in 2015, marked the reorientation of Japan's space program towards tackling the changes in the regional security environment. The latest version of the Basic Space Plan of 2017 aims at responding to the growing threat of ASAT weapons and the increasing quantity of space debris by putting emphasis on space security. Space security, through the strengthening of security capabilities and the Japan-US alliance, ensures the stable utilization of outer space.

India has acquired multifaceted space capabilities with dual-use applications (i.e., both civilian and military). These include satellite communications, Earth observation, and navigation. Over the years, Indian Space Research Organisation (ISRO)'s program has matured significantly and, at present, India's space program is regarded as one of the important space programs in the world. From launching small satellites to undertaking a successful mission to the Moon and Mars, India has excelled in almost all areas of space experimentations. India is also proposing to undertake its first human space mission by 2022. Ajey Lele describes how India has made significant investments towards establishing its military architecture owing to its strategic needs. Space technologies are finding increasing relevance in strengthening this architecture, essentially as a force multiplier. The March 2019 ASAT test clearly communicated India's intention and capability to use space for military purposes. Soon after the test, India has announced plans to establish a Defence Space Agency marking a shift in the evolution of the Indian space strategy.

Israel, like many other countries, wants to exploit the dual-use aspects of space technologies to advance economic, commercial, security, civil, and foreign policies. Deganit Paikowsky and co-authors describe how, in the past 30 years, Israel developed an indigenous space capability to develop, launch, operate, and maintain satellites in two main areas: Earth observation and communications, including the ground segment of communications satellites. As security in the region has been the country's key concern, Israel's military space program has, throughout its history, been the main driver of the country's space activities. It has, however, also resulted in the growth of the commercial space sector. Israel has expanded, in recent years, its cooperation with international partners as well as established a civilian space policy backed by modest government funding. What began as a purpose-focused defense necessity has blossomed into a diversified amalgam of enterprises, academia, and government dealing with a broad spectrum of space endeavors. Within the context of protecting and encouraging this nationally important ecosystem, Israel considers international space security, safety, and sustainability to be of key importance.

The *United Arab Emirates* (UAE) has realized the increasing importance of the space sector, and supported its continuous growth at a regional and global level. Due to the increasing number of activities in the space industry, and the growing rate of national investments, combined with the technical, economic, and political developments in the sector, the UAE government has pointed towards establishing a national policy and regulatory framework for space activities, aligned with national and international policies and best practices. Naser Al Rashedi and co-authors explain how the UAE actively contributes in utilizing space to achieve long-term sustainability. In line with this mandate, the UAE National Space Strategy 2030 has been reviewed to ensure its alignment with the 17 sustainable development goals (SDGs-2030), the UNCOPUOS Long Term Sustainability Guidelines, and the four pillars defined in the revised zero draft of a "Space2030."

Concerning South America, and despite the challenges to the economy in the region, *Brazil* has managed to sustain growth since the end of 2017 with an industrial production growing slowly. Olavo Bittencourt Neto and Daniel Freire e Almeida describe the evolution of the Brazilian space law, policy, and initiatives related to space security. As an emerging space faring nation, Brazil acknowledges the crucial importance of space activities to the exercise of its national sovereignty. For a territory of 8,514,215.3 km, home of over 200 million people, appropriate satellite coverage is always a challenge. Through the years, successive federal governments have invested time, capital, and people on strategic initiatives focused on local needs, ranging from remote sensing to telecommunications capabilities. Nevertheless, space activities remain an expensive endeavor, and access to crucial technology is often limited by political and legal constraints. Thus, Brazil has recognized that international cooperation is frequently required to secure specific services and expertise, leading to the negotiation of agreements with foreign partners.

Conclusion

The growing ambitions of countries in space have increasingly been accompanied by concerns over the need to protect space systems that enable vital global information flows. All space actors, no matter their level of capability, strive to have a stake in the global evolution of space activities. That said, it remains a reality that there is growing divergence in propositions of how to treat risks and threats to the operational space environment by major space powers and their allies.

The shared interests of spacefaring nations, as well as those of the multitude of beneficiaries of space activities worldwide, can only be accomplished by pursuing a sustainable model of global space governance that is based on transparency and the rule of law. Although different countries emphasize varying space capabilities, strengthening space security should be a priority for all of them.

References

Euroconsult (2019) Government Space Programs. Benchmarks, Profiles & Forecasts to 2028

War, Policy, and Spacepower: US Space Security Priorities

20

Everett C. Dolman

Contents

Principle and Practice in US Space Policy 368
Current US Space Policy 380
Conclusion 383
References 384

Abstract

Official US space policy has remained relatively consistent over the last 60 years. Beginning with Eisenhower, every administration has stressed international cooperation, peaceful intention, and development of outer space for the common good of all humankind. Significantly, every policy has also reserved the right of self-defense in space. In practice, the United States has interpreted these policy constants, particularly regarding military activity, quite broadly. The 2017 National Security Strategy (NSS) made a notable and historically unique public change in emphasis, however. In his preface, President Donald Trump declared that "we are charting a new and very different course" consistent with a larger international policy of "America First." This chapter describes continuity and change in space policy over the last 60 years, summarizes the current administrations space directives, and discusses current proposals for a separate military space force within the U.S. Department of Defense.

Official US space policy has remained relatively consistent over the last 60 years. Beginning with Eisenhower, every administration has stressed international

E. C. Dolman (✉)
Air Command and Staff College, Air University, Montgomery, AL, USA
e-mail: everettdolman@gmail.com

cooperation, peaceful intention, and development of outer space for the common good of all humankind. Significantly, every policy has also reserved the right of self-defense in space. In practice, the USA has interpreted these policy constants, particularly regarding military activity, quite broadly. The 2017 National Security Strategy (NSS) made a notable and historically unique public change in emphasis, however. In his preface, President Donald Trump declared that "we are charting a new and very different course" (National Security Strategy of the United States of America, December 2017, p. i). He asserted the USA would continue to pursue cooperation, but "in an extraordinarily dangerous world" only via a "balance of power that favors the United States, its allies, and our partners" (Ibid, pp. i; ii). To ensure no misinterpretation, just prior to his distinctive signature, he declared: "This National Security Strategy puts America First" (Ibid, p. ii).

The rapidly evolving US space policy is currently in flux. President Trump has issued four Space Policy Directives (SPDs), the last ordering establishment of a separate and equal sixth branch of military service within the Department of Defense (DOD). The new Space Force will be responsible for warfighting operations *in* and *from* space. His administration has not yet released the full text of a National Space Security Strategy (NSSS), but in March of 2018 the White House announced a new National Space Policy (NSP) and released several fact sheets summarizing its contents. In the meantime, the US Congress has yet to debate the details of – or approve – a space force structure or budget. With so little of the coming policy settled, the first task in assessing future probabilities is to understand the past. Accordingly, this chapter begins with a brief survey of previous US military space policies. A review of the space portions of the Trump administration's National Security Strategy (NSS), White House descriptions of the pending Trump NSP, and Trump's SPD-1 through 4 follows. An assessment of the current Space Force proposal completes the chapter.

Principle and Practice in US Space Policy

In his 1960 Farewell Address, President Eisenhower warned of an insidious growth in the partnership between business, technology, and the military – what he termed the military-industrial complex. Just 15 years earlier, as World War II was drawing to a close, the consensus was quite the opposite; a closer relationship was needed. On 12 November 1945, Chief of the Army Air Forces H. H. "Hap" Arnold provided a summary of that view in a report to the Secretary of War: "The conclusion is inescapable that we have not yet established a balance necessary to ensure the continuation of teamwork among the military, other government agencies, industry, and the universities" necessary to successfully prosecute future wars (Ware 2008). This led to the establishment of a scientific think tank, initially based in the Douglas Aircraft company, called Project RAND (a contraction of the term research and development). Tucked into that 1945 report, General Arnold warned the Secretary that ballistic missiles for deterrence and satellites for surveillance and early warning would be the best means for preventing a future Pearl Harbor-like surprise attack (Spires 1998). RAND's first Air Force-sponsored report, released in March 1946, was titled *Preliminary Design for a World-Circling Spaceship* (RAND 1946). War was going to extend to the heavens, and America would not be left behind. Thus began an alternating policy history of pursuing and eschewing military war-making in, to, and from space.

In the 1940s and 1950s, America was obsessed with new technology and enamored with the possibilities for society and war. In the US military, this meant adapting the four great inventions of WW II, Germany's medium range ballistic missile, British radar technology, American electronic computers, and nuclear weapons for what would surely become a radical new way of war (Walter McDougall, in his Pulitzer Prize-winning book, asserts that all the world's space programs were born of these wartime inventions. ...*the Heavens and the Earth: A Political History of the Space Age* (New York: Basic Books, 1985), p. 6). In fact, it was the confluence of these technologies that allowed humanity to achieve the means to go into space. A rocket that could lift an operational payload to orbital altitude, a radar-based guidance system that could place the payload into precise orbit, and a method for making necessary planning and real-time operational calculations were needed. But it was the perceived need for a reliable and certain means to carry the tremendously destructive (and expensive) nuclear bomb to its destination that made the cost of researching, developing, and fielding InterContinental Ballistic Missiles (ICBMs) necessary. The rocket was the motive force that could travel at speeds no defensive system could intercept, while the radar and computer allowed precision targeting over vast distances.

Within the US military, initial development exposed a predominantly Army-centric view of rocket power. Much of the German V-2 development team, to include Director Werner Von Braun, along with equipment and scientific records, were captured in Operation Paper Clip and set to work in the USA to build its own ballistic missile program (Neufeld 2007). The Army perceived the V-2 as long-range artillery, hence within its mandate. Space launch vehicles would be a useful spin-off. The Navy and Air Force emphasized the potential for satellites to support terrestrial operations through intelligence, surveillance, and reconnaissance (ISR), navigation, weather, and communications support, and needed rockets to launch them. Progress for both slowed under the New Look fiscal austerity policies of the Eisenhower administration, as military budgets were tightly squeezed and lower priorities were shelved.

The USA had spent a great deal of money developing nuclear weapons and the means to deliver them through the air. Eisenhower felt a return on that investment should be leveraged. The Massive Retaliation doctrine of the 1950s stated that any transgression against the United States, no matter how slight, *could* be met with a nuclear response. With such a terrifying threat, the administration reasoned that conventional military capabilities could be pared to the bone, as no rational actor would risk such punishment. (Of course, they did, and small-scale threats and conflicts were tolerated so long as they didn't escalate to the need for nuclear strikes. Once the USSR became capable of countering the US with nuclear warheads of its own, larger conventional conflicts became safe from nuclear reprisal so as to not start a global nuclear exchange.) Accordingly, the Air Force retained the largest share of military spending in this period because of its strategic nuclear strike mission, to be carried out entirely by the long-range bombers of Strategic Air Command (SAC). ICBMs were not then as accurate as a bomber, and so it eschewed ballistic missile development. But aircraft were vulnerable to countermeasures, primarily from rapidly advancing Russian Surface-to-Air Missiles (SAMs), whereas a ballistic missile reentering the atmosphere from space was not. Should the Army or Navy achieve a ballistic missile capable of striking deep into Soviet territory with accuracy, the Air Force's strategic strike monopoly, and the percentage of the defense budget allocated to it, would erode. Through a series of bureaucratic and political maneuvers, by the mid-1950s the Air Force usurped much of the Army's ballistic missile mandate and, under the guidance of General Bernard Schriever, began an ICBM program in earnest. (An excellent source is Neal Sheehan 2010.)

Despite more than a decade of interest, the first military satellite development program (WS-117L reconnaissance satellite) was not authorized until 1954, and then only for basic research and preliminary tests (Stares, *Militarization*, pp. 30–3). The following year, a civilian satellite project was announced to coincide with the 1957 International Geophysical Year (IGY) competition calling for a satellite launch by the end of the year (Ibid., pp. 33–5). The Soviet Union proclaimed it would enter as well, proof of its accelerating scientific advancement, but most Americans believed the announcement was pure bravado – their technology was too far behind the USA to be taken seriously. That confidence was shattered on 1 October 1957.

The launch of *Sputnik*-1 caused a national hysteria. The capability to put a satellite into a precise orbit was evidence of the capability to reach out and target any city on the globe. The simple beep-beep transmission that could be heard from *Sputnik* every 90 min as it passed overhead affirmed that no place in America was safe from nuclear war.

Nonetheless, significant evidence has been accumulated to allow some authors to suggest the USA secretively prompted the Soviets into going first (see e.g., Mieczkowski 2013; Johnson-Freese 2007). Under international law, state sovereignty extends into the airspace above its territory. Whether that sovereignty extended into outer space was hotly contested. Recognizing that its illegal U-2 flights over the USSR would end as soon as the Soviets could defend against them, and desperately in need of a replacement for the ability to monitor the closed-off state's nuclear capabilities, spy satellites were the obvious next step (as happened to Gary Powers in 1960, ending airborne reconnaissance of the Soviet Union). Since *Sputnik's* orbit passed over the much of the globe, America immediately declared the action established a precedent – the right of free overflight by orbiting spacecraft worldwide. The Soviets assented. State sovereignty would extend to somewhere between the highest altitude of powered flight and the lowest altitude of sustainable orbit (a line still not established in international law, creating an ambiguous zone between the two criteria). Much like international waters, space would be a global commons where innocent passage was open to all. The event proved so advantageous for US security needs that the Eisenhower administration has been credited with purposely holding back space development to ensure that his "Open Skies" vision would be realized (Johnson-Freese 2009). This position makes the US administration seem too clever by half. The government appeared genuinely unprepared for the public reaction, and the space program – such as it was – lagged significantly behind.

Of note, Von Braun had practically begged to be allowed to use an Army Redstone rocket to place a satellite into orbit as soon as possible (McDougall, *Heavens and Earth*, p. 119). In a 1954 report titled "A Minimum Satellite Vehicle," he requested just $100,000 to accomplish the feat using existing equipment and technology. Because the Redstone was an upgraded version of the V-2 and Von Braun had a shady past in the Nazi Party and SS, to allow him to do so would have put a too militaristic face on American space ambitions. His requests were denied. With *Sputnik's* October surprise, however, and the follow-up launch of a much larger *Sputnik* carrying a live dog (Laika, who as planned burned up on reentry), accelerated the crisis. After the original IGY entrant, Project Vanguard, suffered a series of high profile setbacks and an embarrassing launch failure in December, the Army was finally authorized to launch an alternative satellite on a modified Redstone rocket (dubbed Jupiter-C). Explorer-1, weighing just 3.5 pounds – a fraction of the 135-pound first *Sputnik* – achieved orbit on 31 January 1958.

The close association with ballistic missiles and national space programs was clear. No state has yet developed a space launch vehicle that cannot trace its technological lineage to ballistic missile program. To counter international concern, and recognizing that it was much further behind in space exploration – particularly heavy-lift – than its Cold War rival, America publicly and completely bifurcated its space program into purely civilian and military sectors (Stares 1985). While the civilian side may have perceived scientific competition in space as a route to peaceful cooperation and mutual benefit, the US military had an entirely different view. Access to space was a harbinger of future wars.

Speaking to the National Press Club shortly after *Sputnik*, Chief of Staff General Thomas White declared "whoever has the capability to control space will likewise possess the capability to exert control over the surface of the earth ... We airmen who have fought to assure that the United States has the capability to control the air are determined that the United States must win the capability to control space" (Futrell 1985). Shortly thereafter – perceived by the other services as a bid to ensure that space would be its exclusive purview – Air Force leadership inserted the word *aerospace* throughout its doctrine to cement the notion that air and space were a continuous and indivisible operational medium (Ibid., 64, 68). From a warfighting perspective, this meant developing and deploying *piloted* spacecraft that could mimic established aircraft operations. A push was on to fast-track space fighters, bombers, and transports, a view that would have unsustainable ramifications in the next decade.

The Air Force's vision was not shared by the other services. Army General James Gavin insisted the nation must develop an unmanned satellite interceptor system "... otherwise it will be helpless before any aggressor equipped with armed reconnaissance satellites" (quoted in Stares, *Militarization*, p. 49). Army Air

Defense Artillery (ADA) forces saw space defense as a natural extension of its own inherent capabilities. In 1959, Admiral Arleigh Burke, Chief of Naval Operations, proposed to the Joint Chiefs a combined military space agency, a precursor of future unified space commands, based on the shared "indivisibility of space" (Spires, *Beyond Horizons*, p. 76). The Navy and Army had legitimate needs for space support and capability, and proposed sharing equally in the new mission as a cost-saving measure, but parochial service rivalries were not yet ready for sharing domain responsibilities.

The 1958 National Security Council (NSC) Directive 5824, "U.S. Policy on Outer Space," concerned the development of military reconnaissance satellites and highlighted the need to ensure any political response would be as favorable to the USA as possible. Although the decision had already been made to separate the civilian side of government space duties, to allay potential domestic and international objections, follow-on NSC Directive 5814/1, "emphasizing denying Soviet space superiority," was uncomfortably provocative (Ibid., p. 64).

Early focus was on the development of reconnaissance satellites, with the caveat that any program must be politically reassuring. This left much room for interpretation. In accordance with the logic of the ancient Roman adage *si vis pacem para bellum*, military space activities must be conducted with peaceful *intent*: to deter war. Thus several prospective programs aimed at preparation to fight a future war in space were clustered under the broader program titled Dyna-Soar (a compression of Dynamic Soaring that, in the context of aggressive cutting-edge science, was an unfortunate name). (Eventually eXperimental plane 20 (X-20), a rocket-launched piloted space bomber that could glide back to earth held exclusive title to the name Dyna-Soar. For a full history, see Houchin 2006) Among these was an Air Force satellite interceptor program called SAINT (another compression). Although President Eisenhower would not authorize any actions beyond concept development, his successor would have fewer qualms.

Running against Eisenhower's fiscal policies leading to what he perceived as insufficient funding of non-nuclear military capabilities, Senator John F. Kennedy campaigned on reinvigorating all of America's warfighting capabilities. Speaking on the campaign trail, Kennedy made it clear that war in space was included: "We are in a strategic space race with the Russians and we are losing ... Control of space will be decided in the next decade. If the Soviets control space they can control the earth ... we cannot run second in this vital race" (from an interview in *Missiles and Rockets* magazine, cited at https://history.nasa.gov/SP-4225/documentation/competition/competition.htm). Citing the Soviet's clear lead in space launch capability, Kennedy was especially persuasive on the charge that Eisenhower's administration stood by while the Soviets gained a distinct advantage in ballistic missiles. Famously stoking fears over a growing "missile gap," Kennedy eked out a victory over Eisenhower's political heir, Vice President Richard Nixon (well documented in Christopher Preble 2004).

Despite later admission, there had been no missile gap – or if there had, it was distinctly in favor of the USA – uncertainty over Soviet intentions enhanced military arguments. The Soviet Union was testing warfighting systems *to* space – including the exotic MiG 105 spaceplane – and armed satellites *in* space. A particular concern was indications of a Fractional Orbital Bombardment System (FOBS), an orbiting platform that could release nuclear warheads with trajectories that would bypass American defenses arrayed along the northern Arctic front (Mowthorpe 2004; Stares, *Militarization*, pp. 99–100). Satellites were starting to provide invaluable intelligence, and early-warning was increasingly reliant upon them. In 1962, the Soviets began an intensive political campaign in the UN to prohibit space reconnaissance over sovereign territory (Stares, *Militarization*, pp. 62–71). With the Dyna-Soar Program now focused exclusively on a piloted military spaceplane that could conduct combat operations in space and return safely to the earth and a Manned Orbiting Laboratory (MOL) program the Russians insisted was a military reconnaissance platform, the Kennedy administration pressured the Soviets into dropping their insistence on a satellite reconnaissance ban and to agree with the USA that nuclear weapons should not be based in outer space.

Vice President Lyndon Johnson inherited the Kennedy/McNamara military space program after the tragic assassination of President Kennedy. His administration's official policy continued the general principles of freedom of space and active pursuit of arms control arrangements and international cooperation while developing appropriate and necessary military means to defend American interests in space (Stares, *Militarization*, p. 93). This standard has generally been held throughout all successive administrations, the differences hinging on perceptions of appropriate and necessary military actions.

Johnson also inherited a civilian race to the Moon, a massive conventional arms build-up coupled with a rapidly escalating commitment to a war in Vietnam, and a nuclear arms/ICBM race that had spiraled out of control after the Russian humiliation in the Khrushchev/Kennedy Cuban missile crisis – all putting a colossal strain on the federal budget that jeopardized his signature Great Society socioeconomic transformation. For their part, the Soviets were saddled with similar problems. Despite a foundering economy, they had the same space and arms race commitments as the USA as well as a Vietnam parallel with the continued expense of suppressing anti-Russian sentiments in Eastern Europe. Something had to go, and so by 1964 both sides willingly engaged in serious negotiations to curtail a potentially ruinous race for space weapons.

The 1967 Outer Space Treaty (OST) established outer space as a global commons, free from national appropriation, and formalized the language that would continue to dominate international agreements regarding space policy. Space was the common heritage of all mankind, and the benefits of space exploration should be shared by all. The OST also prohibited the stationing in earth orbit or on celestial bodies of weapons of mass destruction, defined as "nuclear, biological, chemical, and radiological," but did not ban the transit of these warheads through space on the way to their intended targets (www.unoosa.org/pdf/publications/STSPACE11E.pdf).

The USA had by this time cancelled Dyna-Soar and the MOL, and transferred its astronauts to support NASA. For its part, the Soviets declared they had no interest in a Moon race with the Americans, nor any intent to wage war in space.

For public consumption, the two Cold War superpowers had entered into a period of non-military cooperation in space. They even agreed to a few joint exploration events that it was hoped could spill over to cooperation in other, terrestrial areas.

In practice, however, not so much. The nuclear arms race accelerated. From a few hundred medium and intercontinental missiles in 1960, by the end of the decade each side possessed more than 10,000 nuclear warheads, a number that would double again before the end of the 1970s. The USA would continue its war in Vietnam, expanding it throughout Southeast Asia before finally coming home in 1973. The Russians would brutally put down the Czech uprising in 1968. Publically, at least, the race to militarize space was curtailed.

The Richard Nixon/Gerald Ford and James Carter administrations continued the public policies articulated formally under the OST and entered into the first significant Strategic Arms Limitation (SALT I and II) talks with the Soviets. The Nixon administration also concluded a treaty banning anti-ballistic missiles (ABMs) in 1972. With serious economic problems emanating from the previous decade, the Moon programs were cancelled. While policy focused again on fiscal constraint and defusing terrestrial conflicts, in military space, this was a significant era of ASAT experimentation.

The USA had already tested high altitude nuclear detonations in space under the STARFISH PRIME program over the South Pacific, but found the effects to be too indiscriminate and persistent to be of practical warfighting value (Plait 2019). The Air Force co-orbital SAINT interceptor program had been discontinued, but the Navy continued testing its submarine-based Polaris missiles as possible direct ascent ASATs into the 1970s (Stares, *Militarization*, pp. 106–129). The Soviet Union, recognizing the enormous enhancement of US military capability provided by its military satellites, conducted intensive testing of a large co-orbital ASAT from 1968 into the 1980s.

The Soviet ASAT, a massive (2,500-plus kg) conventional explosive launched aboard an SL-11 rocket into the same orbit as its target. Once in sufficient proximity, it detonated, creating a debris field that obliterated the satellite. The Soviets conducted at least 20 on-orbit tests through 1982, successfully destroying the target satellite in more than half of its attempts (Baker 2005). Although it declared a moratorium on further tests in 1983, the Soviet Union then Russia continued to use the ASAT booster to place various military satellites into orbit and maintained the warhead in storage for decades. In 1975, the third Salyut-series space station was equipped with a 23 mm cannon, adapted from a Tupelov bomber tail gun (Zak 2015). The station successfully test-fired the gun remotely that year, and may have equipped follow-on stations with modified air-to-air missiles.

President Nixon could afford little attention to the military ramifications of Russia's offensive space capability. He had run successfully a second time, in both cases on the promise that that he would extract the USA from Vietnam, which he finally did in 1973. But in that second campaign he authorized a break-in of the opposition party's headquarters in the Watergate Hotel. The resulting scandal ultimately forced him to resign. He was succeeded by Vice President Gerald Ford, who ran out Nixon's term. His main priority was the domestic economy, specifically suppression of rampant inflation. Faced with the clear Soviet ASAT asymmetry, in the last month of his term he directed the Defense Department to develop an operational anti-satellite capability (Stares, *Militarization*, p. 175).

President James Carter's administration was keenly interested in negotiations to slow the continuing rise in nuclear arsenals but also stem the growing ASAT threat. The first diplomatic line focused on the nuclear arms race, and resulted in a successfully negotiated arms limitations treaty, SALT II. The second, related line, was an attempt to reduce the operational ASAT imbalance. Not only were the Soviets accelerating their co-orbital ASAT tests, the US intelligence committee identified what it believed was a breakthrough in directed energy technology. A very large laser located in Sary Shagan appeared to overhead reconnaissance to have a primary anti-satellite mission (identified in the U.S. Department of Defense 1990). Since President Ford's directive to build an ASAT capability was already in place, Carter's team decided to use the DOD program as incentive to bring the Russians to the negotiating table and provide leverage once talks were under way (Stares, *Militarization*, p. 183). Both sides accused the other of developing exotic technologies and adapting treaty-allowed anti-ballistic missile systems to ASAT operations, and the talks foundered.

A fundamental change in military space policy followed with the next national election. In an effort to fulfill his campaign promise to restore American military power following its post-Vietnam malaise, defense officials in President Ronald Reagan's administration began circulating support for a fresh look at space organizational structure that would ultimately lead to a new combatant command – US Space Command. The soaring costs and inefficiencies of space systems acquisition, the mess of some 50 uncoordinated military organizations working with pieces of the space enterprise, and the threat of a Soviet program that appeared to be racing ahead in military space warfighting capabilities fueled frustrations that led to a call for action. In September 1981, the Air Force added a fifth subunit to its planning staff, the Directorate of Space Operations, to provide options. Still, as is often the case with large bureaucracies, meaningful change required a push from the outside. In late 1981, House Resolution 5130 required the US Air Force to report to Congress on the feasibility of establishing a service space command (Karas 1983; Spires, *Beyond Horizons*, pp. 188–92).

In arguments that would resurface 35 years later, the DOD strongly opposed the move on the grounds it was not needed, would duplicate bureaucracies, and cost too much. In January 1982, A GAO report undercut those arguments suggesting a separate space command that would coordinate all military space activities could instead result in overall cost savings; specialization was in fact the foundation of organizational efficiency. Thus, in June 1982, the USAF revealed it would establish a subordinate Space Command in Colorado Springs no later than September (Spires, *Beyond Horizons*, pp. 193–205). Two years after, the co-located US Space Command was inaugurated. By the end of the decade, the commander of USSPACECOM (dual-hatted as commander of the North American Aerospace Defense Command) gained authority over all military space operations, took command of land-based ballistic missile responsibilities from Strategic Air Command, secured nascent computer operations (the precursor to today's Cyber Command) authority, inaugurated a warfighting space operations center, and created a Joint Space Intelligence Center (Ibid., pp. 217–8. Authors Note: The commander usually was commander of both US *and* Air Force Space Commands). Such rapid evolution perhaps went too far toward an independent space service and in the process threatened entrenched bureaucratic constituencies.

In 1983, President Reagan announced his Space Defense Initiative (SDI), derisively dubbed Star Wars by an incredulous press. The intent was to reverse what he believed was an immoral lack of protection for Americans. The 1972, ABM Treaty was grounded in the 1960s nuclear theory that Mutual Assured Destruction (MAD) provided by a guaranteed second strike capability was the only way to deter a nuclear exchange. The key to Reagan's approach was that missile defense would be based in space, the only deployment that had the potential to create an effective global shield against ICBMs. To comply with the national policy of peaceful intent in space, Reagan offered to share the protection of a space-based nuclear shield to all the nations of the earth, ridding the world of the only mechanism for global destruction yet devised. Whether SDI was feasible in the manner President Reagan intended, it did cause the Soviet Union to increase its missile defense and ASAT research and development, putting pressure on an economy that had essentially flat-lined, and in so doing may have accelerated its downfall. Regardless, after labeling the Soviet Union the Evil Empire, his strong approach ultimately led to a rapprochement with Soviet Premier Mikhail Gorbachev, in turn leading to the first Strategic Arms Reduction Talks (START) and a lifelong friendship between the two leaders.

The ASAT initiative begun by President Ford in 1977 finally became operational under Reagan in 1985. After two successful proof of concept tests, a modified air-to-ground HARM missile was launched from an F-15 at maximum altitude for the aircraft. The missile intercepted a defunct American scientific satellite, obliterating the target and leaving a debris cloud that remained a navigation hazard for many years. Congress quickly declared that it would no longer fund dedicated ASAT programs and the very limited US capability was set aside.

President George Bush's administration continued the general policies of his predecessors but changed the focus of the SDI program. A completely protective shield was a chimera. Moreover, the MAD doctrine appeared to be working. A guaranteed retaliation on a massive scale could deter a massive scale attack. But Bush was now more concerned with limited attack scenarios that simply could not be deterred; they could only be defended against. There were essentially four: (1) An accidental launch, (2) an irrational national leader from a rogue state that might acquire a nuclear device and use it out of malice or desperation, (3) a terrorist organization gaining control of a nuclear weapon, especially worrisome with the with the collapse of the Soviet Union, and (4) a third-party launch. If India and Pakistan were to engage in a limited nuclear exchange, or Israel and Iraq, Iran or North Korea, the result would not only be tragic, it might well draw the larger nuclear powers in. With these growing possibilities in mind, President Bush pushed ahead with a limited space defense concept developed by the SDI team called Brilliant Pebbles and Brilliant Eyes (see Dolman 2001). The system comprised a network of some 320 satellites in low-earth orbit, each with 24 interceptors capable of engaging a satellite in the boost or intermediate phase, and a dozen reconnaissance satellites in higher orbit that could identify a missile launch from its heat signature and coordinate the best response. If it could be made operational, it was projected to be capable of engaging and destroying up to 100 simultaneous launches anywhere on earth. President William Clinton entered office in 1993 having campaigned against SDI, but was persuaded that the limited defense under Bush had merit. Instead of quickly killing the program, he gradually decreased funding so promising technologies might reach fruition. Although the system had some promising successes in early research and development, it was eventually discontinued.

Significantly, with the demise of the Soviet Union and the possibility of limited nuclear war – much less an earth-killing MAD scenario – off the minds of most Americans, Clinton's administration reduced many of the restrictions that had kept commercial space development under wraps. Because under the OST states were liable for damaged caused to or from space activities, a commercial space launch enterprise had to be licensed by the appropriate national government. More critical in the Cold War, a private space launch could look identical to an out-of-the-blue ballistic missile attack, and the USA was loathed to give any but largest aerospace corporations that accepted continuous oversight and abided by severe restrictions on use any leeway. With nuclear war seemingly a relic of the past, restrictions were loosened under Clinton. The renaissance of private space exploration today, from corporations including Space-X, Blue Origin, and Virgin Galactic (and many smaller start-ups) would not have been possible under Cold War licensing rules.

The spectacularly successful space support debuted in Operations Desert Shield and Desert Storm prompted the Air Force to seek even greater control of the space mission. Not unnoticed, was the fact that by the mid-1990s, the space-specific portion of the DOD budget had approached $10 billion, with upwards of 85% earmarked for the Air Force. A similar amount was distributed to government and intelligence agencies, for an average of $18 billion annually (https://www.af.mil/News/Article-Display/Article/113903/air-force-officials-take-space-budget-acquisition-strategy-to-capitol-hill/). Post-Desert Storm, the US military had grown so reliant on space support that if it were denied for any reason, deployments around the world could become untenable. Former Secretary of Defense Donald Rumsfeld was tasked to lead a study of the problem. The 1998 Rumsfeld Commission Report warned of a "space Pearl Harbor" due to a lack of emphasis on space and recommended a gradual evolution toward a separate Space Corps within the Air Force as an intermediary step on the path to a separate Space Department. Unheeded, USSPACECOM authorities were steadily transferred elsewhere and any notion of a separate service was effectively dispatched. NORAD moved under the newly established North American Command and all duties not already purged were subsumed by Strategic Command. In 2002, USSPACECOM was disbanded and Air Force Space Command became the de facto US Space Force (http://purview.dodlive.mil/2018/10/01/reestablishing-u-s-space-command/).

None of that seemed predestined. Rumsfeld was made Secretary of Defense under President George W. Bush, and in 2001 it appeared that the most pro-military space administration to date was going to dictate the commission's recommendations. The civilian side would be reinvigorated as well, and the slogan "Back to the Moon, and on to Mars!" was touted. The attacks of September 11, 1991 changed all that. Military funds not going to the Global War on Terror (GWOT) were pulled or substantially cut. The succeeding two wars in Afghanistan and Iraq kept focus away from space reform.

The rise of China as a major space faring nation in the twenty-first century did not go unnoticed in the USA, it was simply discounted. The first Taikonauts to go into space on Chinese rockets were a source of immense pride and were quickly followed by a series of spectacular space achievements. To Americans, the Chinese were simply doing the same things the USA had accomplished decades earlier. There was little cause for concern. In China, however, the burgeoning space program was evidence that it was finally taking its rightful place as a modern, first-tier power. In 2007, it could no longer be dismissed. China destroyed one of its aging weather satellites with a direct ascent hit-to-kill ASAT launched on a modified MRBM. The resulting debris cloud caused international outrage, an unexpected reaction as both the Soviets and Americans had done similar damage with their first ASAT tests. But now space was more congested, and much more of the global economy was enabled by space support.

While denying any connection, the USA appeared to respond to the first Chinese test the following year, destroying a deorbiting satellite with an anti-air missile launched from an Aegis cruiser, obviating any debris issues as the target was too low for fragments to remain in orbit. The USA's official position was that the incoming satellite might not burn up completely on re-entry, and its toxic hydrazine fuel could pose a danger to the environment. Regardless, it did appear to counter the Chinese demonstration with an enhanced capability of its own. The Aegis is a mobile platform that can deploy to three-fourths of the globe – a so far unmatched capability.

The official space policy of the Bush administration, while consistent with his predecessors, contained more aggressive rhetoric with regard to America's right to defend its interests in space, though action was postponed. With President Barack Obama's administration a comprehensive national space policy was issued that toned down some of the language but clearly maintained continuity with its predecessors. A more detailed exposition of the Obama NSSS follows below. American involvement in Afghanistan and Iraq, while reduced, still took up a significant portion of the nation's military budget. A globally widening threat from Violent Extremist Organizations (VEOs), most notably the Islamic State of Iraq and Syria (ISIS), and a slow recovery from the profound recession caused by the 2007 collapse of the US housing crisis meant military space priorities remained on the back burner.

China's space capabilities accelerated, and Russia's resurfaced. In 2017, an exasperated Mike Rogers (R-AL), Chair of the House Armed Services Strategic Forces Subcommittee, concerned that America now faced multiple near-peer competitors in outer space had enough. Implicitly accusing the USAF of diverting space funds to priority air projects, and mismanagement of the rest, Rogers and representative Jim Cooper (D-TN) inserted language into the 2018 National Defense

Authorization Act (NDAA) directing an independent Space Corps in the Air Force along the lines of the Marine Corps in the Navy (https://spacenews.com/congress man-rogers-a-space-corps-is-inevitable/). The measure passed the House but was tabled in the Senate pending further study. The Air Force pushed back strongly and marshalled precisely the same arguments the US Army used in its attempts to retain the US Army Air Forces after WW II; it wasn't needed as the purpose of the air arm was to support the fight on the land (and therefore was best overseen and coordinated by the land commander), it would create an unnecessary parallel bureaucracy, and therefore be too expensive. Demonstrating an astonishing lack of historical acumen, the Air Force argued that an independent separate space corps would take away from its primary mission in support of terrestrial forces, it would create an unnecessary parallel bureaucracy, and therefore be too expensive.

Although the president's administration staunchly opposed Rogers' and Cooper's plan, after the mid-term elections President Trump one-upped the Space Corps blueprint and surprisingly announced his intention to create a separate and equal Department of the Space Force. In June of 2018, Vice President Pence detailed the administration's vision (https://www.nbcnews.com/politics/white-house/ trump-says-he-directinh-pentagon-create-new-military-branch-called-n884361). The USAF would immediately begin comprehensive preparations to split off a co-equal Department of the Space Force including a US Marine-style independent organizational structure within the Department of the Air Force. It would draft a plan for congressional budget support, and coordinate with other services and national intelligence space cadres for efficiencies.

Publicly supporting the initiative, USAF leadership continued to privately argue the folly of the move. It would be bureaucratically redundant and wastefully expensive. A memo from the Office of the Secretary of the Air Force was released (or "leaked"; https://www.defenseone.com/politics/2018/09/creating-space-force-will-cost-13b-ove-5-years-air-force-secretary/151312/) stating a 5-year conservative estimate of the additional cost of separating an independent space service would approach $13 billion over 5 years, "likely to be revised upward" (https://spacenews. com/wilson-13-billion-space-force-cost-estimate-is-conservative).

By October, however, Air Force leadership appeared to have dropped even veiled opposition, and submitted a viable, comprehensive transition plan for congressional approval at the end of February 2019. Surprisingly, the anticipated cost of the transition would be quite low, and bureaucratic overlap remarkably lean. The Pentagon requested just $72 million for fiscal 2020 and just $2 billion over the next 5 years to stand up a functioning Space Force within the Department of the Air Force (https://insidedefense.com/daily-news/pentagon-estimates-new-space-force-will-cost-2-billion-over-five-years). Beginning with less than 200 assigned personnel in the first year, the Space Force should grow to approximately 15,000 military and civilian billets by 2025. After that, when the organization is expected to be fully operational, the Pentagon's plan would stabilize the Space Force budget at about $500 million per year, or "about 0.07 percent of the Defense Department's annual budget" (https://abcnews.go.com/Politics/join-space-force-academy/story?id= 61411343).

Current US Space Policy

The National Security Space Strategy (NSSS) is subordinate to the National Space Policy (NSP), which outlines and defines the overall direction and emphases of America's space programs. The most recent published NSP was issued under the Obama administration in 2010. It describes an increasingly diverse space environment but is more representative of a continuation of the rhetoric and substance of the preceding Bush and Clinton administration policies than a "bold new course" (National Space Policy of the United States of America 2010) The future impetus for space development shall emphasize a more international focus, as humanities' reliance on space for an increasingly interconnected global network of finance, trade, production, and security has changed the way we live, "and life on Earth is far better as a result" (Ibid.).

Until the details of the Trump NSSS and NSP are released, guidance for space planning and organizational described in the Obama NSP technically remains in effect. It is not expected that the logic of that document will be completely discarded, and so potentially relevant portions are summarized here as still-current policy. Where the present and previous administrations are not reconcilable is in their underlying premises. President Obama declared the old narrative of national space development, born of conflict and propelled by Cold War challenges, must give way to a new era of cooperation. It called for a renewed commitment to international harmony, the rights of all nations to pursue the peaceful exploration and use of space, and the continuing leadership role of the US in these efforts. But "in this spirit of cooperation," within the five guiding principles put forth for all states to adopt and follow, is the recognition that among the peaceful purposes advocated therein the right of the United States to use space "for national and homeland security activities [and] consistent with the inherent right of self-defense, [to] deter others from interference and attack, defend our space systems and contribute to the defense of allied space systems, and, if deterrence fails, defeat efforts to attack them" is maintained (Ibid., 3). Accordingly, the Secretary of Defense is charged with the development, acquisition, operation, maintenance, and modernization of Space Situational Awareness (SSA) capabilities; Developing capabilities, plans, and options to deter, defend against, and, if necessary, defeat efforts to interfere with or attack USA or allied space systems; Maintaining the capabilities to execute the space support, force enhancement, space control, and force application missions; and Provide, as launch agent for both the defense and intelligence sectors, reliable, affordable, and timely space access for national security purposes (Ibid., 14).

Consistent with 2010 NSP, the Department of Defense published the 2011 National Space Security Strategy. This document "charts a path for the next decade to respond to the current and projected space strategic environment" (Ibid., i). That environment, state the authors, is driven by three trends; "space is becoming increasingly *congested, contested*, and *competitive*" (Ibid., 1. Original emphasis). *Congestion* here refers to the increasing clutter in space, primarily in low-Earth orbit, that has come as a natural result of space launches, satellite deployments, and, not so naturally, from anti-satellite weapons testing. The DOD tracks more than 22,000 objects in orbit that are large enough to be detected by ground sensors, including more than 1,200 active satellites, to assist in payload identification and collision avoidance. Potentially hundreds of thousands of smaller objects also exist in orbit, where the kinetic impact of a pin-head size bit of metal or ceramic could destroy a satellite or puncture a space-suit. In recent years, two events have significantly increased the size of the debris field; a 2007 Chinese ASAT test that obliterated one of its own derelict weather satellites added approximately 3,000 trackable chunks of debris and the 2009 collision of a Russian Cosmos and an American Iridium satellite, resulting in 1,500 additional pieces of observable debris (Ibid., 2). Not only is this effective physical pollution of LEO expanding exponentially, the useable radiofrequency spectrum is increasingly stressed, causing frequent unintentional interference between satellites and reducing bandwidth carrying capability.

Space is also increasingly *contested*. "Today space systems and their supporting infrastructure face a range of man-made threats that may deny, degrade, deceive, disrupt, or destroy assets. Potential adversaries are seeking to exploit perceived space vulnerabilities" (Ibid., 3). The emphasis here is on direct military intervention against US space assets that disrupt the stability and security of the space environment, though it includes unintentional interference through so-called irresponsible behavior.

Competition refers to the declining relative edge in space capabilities held by the USA. The NSSS maintains that America's competitive advantage in space access and market share is dissipating, and its lead in space technology is eroding as more states enter into the strategic environment. Limited access to space is challenging America's "abilities to maintain assured access to critical technologies, avoid critical dependencies, inspire innovation, and maintain leadership advantages. All of these issues are compounded by challenges in recruiting, developing, and retaining a technical workforce" (Ibid.).

The December 2017 National Security Strategy formalized an abrupt shift in emphasis. In his introduction, President Trump opens with the statement "[T]he American people elected me to make America great again" (https://www.whitehouse.gov/wp-content/uploads/2017/12/NSS-Final-12-18-2017-0905.pdf, p. i). He concludes, "This National Security Strategy puts America First" (Ibid., p. ii). The NSP highlights "the growing political, economic, and military competitions we face around the world" and specifically charges China's and Russia's efforts "to erode American security and prosperity" (Ibid., p. 2). Asserting: "An America that successfully competes is the best way to prevent conflict. Just as American weakness invites challenge, American strength and confidence deters war and promotes peace" (Ibid., p. 3). The NSS details four "vital national interests" that undergird the America First strategy (Ibid., pp. 3–4):

- First, our fundamental responsibility is to protect the American people, the homeland, and the American way of life.
- Second, we will promote American prosperity.

- Third, we will preserve peace through strength by rebuilding our military so that it remains preeminent, deters our adversaries, and if necessary, is able to fight and win.
- Fourth, we will advance American influence because a world that supports American interests and reflects our values makes America more secure and prosperous.

The entire section on space, upon which the National Space Policy is presumably based, is less than one page. Following a review of increasing competition in space, and insistence that the US "must maintain our leadership and freedom of action in space," the section concludes that "any harmful interference with or an attack upon critical components of our space architecture that directly affects this vital U.S. interest will be met with a deliberate response at a time, place, manner, and domain of our choosing" (Ibid., p. 31).

Four months later, on 23 March 2018, President Trump's White House announced a new National Space Policy to replace the 2011 Obama administration NSP, opening with the statement: "AMERICA FIRST AMONG THE STARS: President Trump's National Space Strategy works within his broader national security policy by putting America's interests first" (https://www.whitehouse.gov/briefings-statements/president-donald-j-trump-unveiling-america-first-national-space-strategy/). While the full text of the strategy has not been released, the announcement assures that the new strategy "prioritizes American interests first and foremost, ensuring a strategy that will make America strong, competitive, and great" (Ibid.). Accordingly, the NSP promises to maximize cooperation between the military, civil, and commercial space sectors, to include prioritizing "regulatory reforms that will unshackle American industry" (Ibid.). In further challenging almost six decades of cooperative rhetoric that recognizes space as the common heritage of humankind and that its benefits should be shared equally among all nations and people, "[t]he new strategy ensures that international agreements put the interests of American people, workers, and businesses first" (Ibid.).

The announced NSP was followed by four Space Policy Directives (SPDs):

- SPD-1 (11 December 2017) directs NASA to "lead the return of humans to the Moon for long-term exploration and utilization, followed by human missions to Mars and other destinations," instead of an asteroid as the Obama Administration planned (https://spacepolicyonline.com/topics/militarynational-security-space-activities/; https://www.whitehouse.gov/presidential-actions/presidential-memorandum-reinvigorating-americas-human-space-exploration-program/).
- SPD-2 (23 March 2018) "Streamlining Regulations on Commercial Use of Space," directs the Secretary of Transportation to investigate and minimize requirements levied on commercial space activities and licensing of commercial launch and re-entry activities (https://www.whitehouse.gov/presidential-actions/space-policy-directive-2-streamlining-regulations-commercial-use-space/).
- SPD-3 (24 May 2018) "National Space Traffic Management Policy," uses the language of the Obama NSP (space is "congested and contested") to direct steps

to improve SSA and mitigate the effects of orbital debris (https://www.whitehouse.gov/presidential-actions/space-policy-directive-3-national-space-traffic-management-policy/).
- SPD-4 (14 February 2019) specifies the transition plan for creating a military space force (https://www.whitehouse.gov/presidential-actions/text-space-policy-directive-4-establishment-united-states-space-force/).

SPD-4 was released with a full DOD-produced brochure titled "United States Space Force" (https://media.defense.gov/2019/Mar/01/2002095012/-1/-1/1/UNITED-STATES-SPACE-FORCE-STRATEGIC-OVERVIEW.PDF). It is the most far-reaching initiative in military space organization to date, though it is substantially less radical than President Trump's stated intention to establish a *separate Department* of the Space Force, co-equal to the Army, Navy, and Air Force. The brochure was accompanied by a detailed formal reorganization plan for congressional approval establishing Title XII authority for a US Space Force (https://media.defense.gov/2019/Mar/01/2002095010/-1/-1/1/UNITED-STATES-SPACE-FORCE-LEGISLATIVE-PROPOSAL.PDF). The plan will likely go through several revisions before passage, but the key points from all three documents should remain essentially intact.

The Space Force will become the sixth formal branch of the US military, housed within the Department of the Air Force, responsible for Organizing, Equipping, and Training (OTE) all space forces, by fiscal year 2020. A unified combatant command, the United States Space Command, shall also be established, and will be responsible for Joint Force space operations. These organizations will "provide for freedom of operation in, from, and to the space domain; ... provide independent military options for national leadership ... both combat and combat support ... to enhance the lethality and the effectiveness of the space domain" (https://www.whitehouse.gov/presidential-actions/text-space-policy-directive-4-establishment-united-states-space-force/).

Conclusion

The world economy is so intrinsically linked to support from space that should a major outage of satellite capacity occur, financial and trade markets could collapse. A recession spanning the globe would ensue, and security tensions would exacerbate. The increasingly chaotic international environment would be further destabilized by the disastrous incapacitation of US military power. Without the assuredness of space-based surveillance, communications, and navigation support, American and allied military forces would be ordered to hunker down in defensive crouch while preparing to withdraw from dozens of then-untenable foreign deployments.

Such a scenario is not only possible – given the growing investment and reliance on space as a national power enabler – it is increasingly plausible. An attack against low-Earth orbit from a medium range ballistic missile adapted for detonation in space could cause inestimable harm to the national interests of developed and developing states alike. Deterrence may forestall such an attack, but without a space-based defense any decision by an adversary to disrupt space capabilities on-orbit is likely to succeed. The US leadership now believes that it is irresponsible to rely on the better intentions of its adversaries, international agreements, and deterrent threats to guarantee its national space interests. The shift in American space policy makes clear that space is now considered a warfighting domain, and the USA will prepare itself to fight and win military conflicts that might occur there. How the world reacts will ultimately determine the wisdom of these decisions.

References

Baker D (ed) (2005) Jane's spaceflight directory 2004–2005. Jane's Information Group, Alexandria, pp 611–612
Christopher Preble (2004) John F. Kennedy and the missile gap. Northern Illinois University Press, DeKalb
Dolman E (2001) Astropolitik: classical geopolitics in the space age. Frank Cass, London, p 47
Futrell R (1985) Ideas, concepts, doctrine: basic thinking in the United States Air Force 1907–1960, vol I. Air University Press, Maxwell AFB, p 550. https://www.airuniversity.af.edu/Portals/10/AUPress/Books/B_0032_FUTRELL_IDEAS_CONCEPTS_DOCTRINE.pdf
Houchin R (2006) US hypersonic Research and Development: the rise and fall of dyna-soar, 1944–1963. Routledge, New York
Johnson-Freese J (2007) Space as a strategic asset. Columbia University Press, New York, p 44
Johnson-Freese J (2009) Heavenly ambitions: America's quest to dominate space. University of Pennsylvania Press, Philadelphia, p 35
Karas T (1983) The new high ground. Simon & Schuster, New York, pp 17–20
Mieczkowski Y (2013) Eisenhower's sputnik moment: the race for space and world prestige. Cornell University Press, Ithaca, pp 78–79
Mowthorpe M (2004) The militarization and Weaponization of space. Lexington, Lanham, p 58
National Security Space Strategy, Unclassified Summary, January (2011). http://www.defense.gov/home/features/2011/0111_nsss/docs/NationalSecuritySpaceStrategyUnclassifiedSummary_Jan2011.pdf
National Space Policy of the United States of America, 28 June 2010, p. 1. http://www.whitehouse.gov/sites/default/files/national_space_policy_6-28-10.pdf
Neufeld M (2007) Von Braun: Dreamer of Space, Engineer of War. Vintage, New York, ch. 9
Plait P (2019) The 50th Anniversary of Starfish Prime: The Nuke that Shook the World. Online at http://blogs.discovermagazine.com/badastronomy/2012/07/09/the-50th-anniversary-of-starfish-prime-the-nuke-that-shook-the-world/
RAND (1946) Preliminary Design of an Experimental World-Circling Spaceship. RAND, Santa Monica. https://www.rand.org/pubs/special_memoranda/SM11827.html
Sheehan N (2010) A fiery peace in a cold war: Bernard Schriever and the ultimate weapon. Vintage, New York
Spires D (1998) Beyond horizons: a half century of air force space leadership, revised edn. Air University Press, Maxwell AFB, p 9
Stares P (1985) The militarization of space: U.S. policy 1945–84. Cornell University Press, Ithaca, pp 41–43
U.S. Department of Defense annual publication. Soviet Military Power. September 1990, pp 60–61
Ware W (2008) RAND and the information evolution: a history in essays and vignettes. RAND, Santa Monica, p 6
Zak A (November 16, 2015) Here is the Soviet Union's Secret Space Cannon. Popular Mechanics https://www.popularmechanics.com/military/weapons/a18187/here-is-the-soviet-unions-secret-space-cannon/

Russia's Space Security Policy

Nicole J. Jackson

Contents

Introduction	385
Russia's Space Security Policy: Consistency and Change	387
Context	387
Russia's Perceptions of Space: Threats and Opportunities	388
Russia's Space Diplomacy	395
Conclusion	397
References	398

Abstract

Russia's outer space policies fit within its domestic and foreign policy efforts which focus on asserting Russia's authority, status, and prestige and reviving its economy. Outer space has also become an important area through which Russia aims to respond to Western strategy and capabilities and also influence global norms. Military efforts are but one part of a complex set of tools, employed to navigate what Russia perceives as an increasingly hostile world. Its economic, military, and technological weaknesses compared to the USA and NATO have led Russia to pursue asymmetric tactics which include working through bilateral bodies and those affiliated through the UN on space policy. These give it publicity and some legitimacy but little ability to make significant progress on the substantive issues.

N. J. Jackson (✉)
School for International Studies, Simon Fraser University, Vancouver, BC, Canada
e-mail: nicole_jackson@sfu.ca

Introduction

Outer space security may be viewed through a national security (or "militarization of space") approach which focuses on the use of space-based technology to protect against outside threats; from a "peace approach" centered, for example, on using technology for arms control verification, humanitarian relief, and climate change; or from a "security in outer space" approach concerned with how to prevent overcrowding, debris in space, and conflicts over space resources.

Today, the Russian Federation is a major actor in space, and Russia's outer space policy encompasses all three approaches. In this chapter, I focus primarily on Russia's national strategic security and outer space governance. Russia-US international space cooperation, for example, on the International Space Station and their scientific cooperation in the exploration of space, is beyond the scope of the chapter.

Russia considers outer space predominantly as a strategic region to enhance its military capabilities on Earth, provide intelligence and communication functions, and achieve international status and prestige as a space power. It is reactive to US strategy and actions and has developed counterspace technologies (e.g., electronic weapons that can jam satellites) to provide Russia with an asymmetric edge to offset US military advantages. However, Russia's outer space rhetoric and policy are also driven by domestic and identity issues. Outer space strategy is an instrument through which Russia pursues its goal to be a "great power" and to mold the international system more closely to a new multipolar world as it sees it. It could also bring Russia economic benefits while masking internal challenges.

President Vladimir Putin has taken both symmetric and asymmetric actions in outer space. He has increased Russia's investment in new technologies (satellites, electronic warfare, strategic offensive weapons, etc.) and simultaneously pursued diplomatic initiatives to control weapons in space. During the Cold War, despite military tensions and serious concern about a possible arms race in outer space, Russia and the USA negotiated internationally binding agreements related to the governance of space activities. Today, both powers are again a warning of a new arms race in outer space while continuing to strengthen their military capacities in the field.

Since 2000 Russia has actively pursued both binding laws and non-binding norms to ban and control weapons in outer space and has advocated for non-binding, voluntary transparency and confidence-building measures (TCBMs). Sometimes it has done this in cooperation with other states, sometimes in opposition to them. This diplomatic endeavor may seem somewhat at odds with Russia's growing militarization; however, their dual role on outer space fits well within Russia's overall foreign and security strategy which is reactive to US policy and simultaneously pro the United Nations (UN) and consensus-based multilateral negotiations. Russia is strengthening its comprehensive power, including military, diplomatic, and normative global influence, in order to make its voice heard on the international stage. Russia's diplomatic activism is that of an aspirational great power, but it also reflects the limits of its current economic and military capabilities. International negotiations enable Russia to be recognized as a key player in global affairs while also benefiting from an opportunity to highlight the US/West's declining influence and the rise of a multipolar world.

This chapter examines why outer space is so important for Russia. Then, it shows how and why the Russian government's outer space policy, strategy, and capabilities have evolved since the Soviet Union collapsed. The paper concludes with an appraisal of Russia's recent diplomatic initiatives in the field of outer space governance. No longer economically competitive in the race for control of outer space, Russia has attempted several strategies to enable it at least to keep in the running. It has placed its space strategy in the context of defense requirements and state military control. It is using diplomacy – working with international organizations affiliated with the UN – to discuss, cooperate on, and influence the race for the militarization of space. It works with disarmament organizations to influence and promote a collective approach to the problem, rather than one dominated by the richer and more powerful states.

Russia's Space Security Policy: Consistency and Change

Context

Russia's outer space policy should be viewed in the context of the many significant and complex challenges of keeping outer space "secure." For example, there is a growing number of active spacecraft in orbit and well over a thousand satellites operated by more than 70 states and commercial and civil entities. Militaries increasingly rely on space operations and on long-range satellite communications and data transfers to train, navigate, and operate. Satellites and space-based servers are also increasingly important to everyday life. Banking, communications, transportation, and the Internet are all dependent on access to space, as are weather prediction, natural disaster mitigation, and sustainable farming. In sum, space-based infrastructure weaves together the world economy.

At the same time, with globalization and the changing global balance of power, the numbers, diversity, and interdependencies of state and non-state actors involved in outer space are growing. Rapid technological changes provide ever greater access to space. Space has also moved to the forefront of the global security debates because of renewed attention to kinetic anti-satellite (ASAT) capabilities and other non-kinetic denial capabilities, e.g., jammers and lasers. The situation is becoming ever more complex with the evolving role of cyber and artificial intelligence (AI) and growing mistrust between actors. This adds urgency and new challenges to the development of Russian (and other actors') outer space security policy and to multilateral approaches to space governance.

Russia today is second only to the USA as a major space actor. However, many state and non-state actors are increasingly interested in the benefits and threats of outer space and have increased their space capabilities. Each has its own evolving perceptions and priorities. The USA has the largest space budget, a large network of military and commercial satellites, and a growing commercial space sector. China has developed a multifaceted space program. The EU is increasingly investing in its space program and has sought to negotiate an international code of conduct.

There has also been an explosion of commercial space actors who have very specific priorities such as would-be asteroid mining ventures or small satellite operations. And there is a growing competition for access to the most useful parts of the radio-frequency spectrum for satellite operations. Commercial actors are challenging and pushing the boundaries of traditional governance regimes and old Cold War frameworks. They are thus forcing states to rethink regulatory and legal regimes. At the same time, individuals and knowledge institutes now have access to space thanks to decreasing costs of increasingly small satellites (micro, mini, nano).

So far, the dominant powers have shown a significant degree of restraint in their activities. And most states continue to maintain that an arms race in outer space should be prevented, including in annual resolutions adopted by the UN General Assembly. However, this state of affairs is being jeopardized by the ongoing competition between major state actors. Recent assertions that outer space is becoming another "war-fighting domain" have been made by China, Russia, the USA, France, the UK, NATO, and India. Outer space has already been militarized but not weaponized, with passive military use being accepted under a broad understanding of "peaceful purposes" of the Outer Space Treaty of 1967. Today, many militaries, not just Russia's, have indicated a desire for defensive and even offensive space-based capabilities. Many militaries have also initiated efforts to create dedicated military units for space including Russia, the UK, the USA, China, France, India, and Japan.

Russia's Perceptions of Space: Threats and Opportunities

The Russian state defines threats largely in traditional terms of territorial protection from military challenges and views space assets as vital for military communication and defense. Russia's geography highlights the need to protect its extensive borders and military and economic assets and infrastructure scattered over its vast territory. The state has traditionally assessed that it is surrounded by hostile powers and thus needs "buffers" or a "sphere of influence" to protect itself. Today, Russia has expanded this rhetoric of vulnerability to include attacks from outer space. Russians use the term "aerospace" rather than outer space because of the interrelatedness of air space and outer space in the context of contemporary threats and conflicts and because there is no distinct boundary between the two concepts. Russia's rhetoric on outer space broadly mirrors that of the USA, stressing urgency to prepare for a possible future war there.

Rapid technological advancements in the space industry have influenced perceptions that there are economic benefits from being a space power. At the same time, they have given rise to concerns about threats stemming from the militarization of space. For example, the development of cheap miniature satellites promises speedy replacement of disabled satellites in the event of attack. Theoretically, this could allow the US military (or other actors) to use such space constellations to support operations during a conflict. Through technology outer space has become integrated with other domains – land, sea, air, and cyber. Most recently, the first generation of hypersonic weapons has "set the conditions for the merger of air and missiles defense and the air and outer space domains" (Charron and Fergusson 2018). Of course, a healthy space industry also provides strategic resources for a state's military and economy. In Russia's case, the announcement of new technological developments also masks unaddressed structural and systemic weaknesses and confers domestic and international legitimacy on Russia's aspiration to be a "great power."

Russia's official perceptions today are not very different from those of the Soviet period. Outer space has long been significant to Russia, and now it again has the resources to be a major contender. Under Putin, as in Soviet times, Russia seeks global strategic parity with the USA and securitizes the US threat to its nuclear deterrence. Russia perceives a US first strike against its nuclear forces from space-based weapons as the key security threat from space. Its 2010 and 2014 military doctrines classify both the deployment of strategic missile defenses (the intention to place weapons in space) and the deployment of strategic conventional precision weapons as key military dangers to Russia. Other threats listed include impeding state command and control and disruption of strategic nuclear forces, missile early warning systems, and systems for monitoring outer space. Both these doctrines and the 2016 Foreign Policy Concept highlight the USA and NATO as potential enemies at a time of "increased global competition" and conclude that Russia needs to focus on the credibility of its nuclear deterrent but also on conventional and non-conventional elements in a complex toolkit of responses.

Russia has also adamantly opposed US plans for ballistic missile defense (BMD), which it perceives as opening a door toward space-based weapons integrated into BMD architecture and in turn could threaten Russia's strategic missiles forces. The 2002 US withdrawal from the 1972 Anti-Ballistic Missile (ABM) Treaty paved the way for deployment of intercept missiles, and Russians interpreted this move as undermining the consensus on the strictly peaceful use of space. In this context, in 2015 Russia has threatened that "any action undermining strategic stability will inevitably result in counter measures" (Russian Government 2015). Russia's key security preoccupation has been the prospect of space-based interceptors and the US refusal to accept constraints on BMD. It continues to denounce the US withdrawal from the ABM Treaty and argues that the development of US ground- and sea-based missile defense has increased tensions and led to increased missile proliferation which Russia directly links to space-based threats.

The Russian (and Chinese) governments also believe that their missiles and satellites are targeted by US antimissiles, and there is a similar assessment from the USA about Russia. Russia perceives anti-satellite weapon tests (ASATs) by China (2007), the USA (2008), and India (2019) to be precursors to the weaponization of space. ASAT capabilities are those that target an adversary's satellites with the intention of disabling their function – communications; intelligence, surveillance, and reconnaissance (ISR); and positioning, navigation, and timing (PNT) – through interference or damaging/destroying the satellite entirely. The latter creates a second-order effect of creating space debris that threaten other space assets and activities in that spatial region. These trends blur the traditional divide between peaceful purposes and warfare, thus making them challenging for governance and arms control.

The Russian government argues that these multiple developments are leading to a new arms race that disrupts broader arms control and disarmament processes and requires Russia's huge expenses for its space program. In March 2018 Putin announced the development of some 300 new "strategic weapons" which he said was a response to US missile defense capabilities and then unveiled several at the annual Victory Day military parade. (The West is particularly concerned about new anti-access/area-denial (A2/AD) capabilities (air and missile defenses; surface-to-surface ballistic missiles; land-, air-, and sea-launched cruise missile batteries; layered anti-submarine capabilities).) Such showmanship was not new, but Putin's hyperbole and critique of the West have intensified, and Western concern about Russia's intentions and growing, if overblown, capabilities is likely to continue.

1990s: A Period of Retrenchment, Decline, and Dependency

The Soviet Union was a pioneer and military superpower in outer space. Russia inherited the Soviet Union's strategic and tactical nuclear forces. What is more, Russia continues to view its nuclear forces as an equalizer to US power and a way to preserve its great power standing. However, following the Soviet collapse, there was a dramatic decrease in Russia's space budgets, and Russia failed to maintain most of its space assets. Its space industrial complex suffered greatly, and subsequent attempts to rebuild it encountered many structural and budgetary weaknesses. Nevertheless, Russia also inherited unique capabilities (such as early warning satellites) which are vital as the main guarantee for its state security that continued (and continues) to be based on nuclear deterrence.

The 1990s in Russia were characterized overall by economic stagnation, military disintegration, and dependency on the West. It lost not only most of the military capabilities it had in space but also its ability to conduct long-term research and development. In this context, official rhetoric at the time prioritized defensive operations, as it had under Gorbachev.

Russia did continue to exercise full jurisdiction and control over space objects launched earlier by the USSR. However, it lost some infrastructure for satellite control and space surveillance (a crucial part of its early warning system). Losses included stations used to control and receive data from civilian and military spacecraft which were scattered across Russia and also in Ukraine, Kazakhstan, and Uzbekistan. Many of its radar stations also were located outside Russia's territory, and in the 1990s, Russia had to rely on older radars and negotiate the use of alternative stations outside Russia (in Azerbaijan, Belarus, Kazakhstan, Ukraine). Russia also had to rely on Western joint ventures to commercialize its launchers, its combat aircraft used US satellite navigation system, and its Northern Fleet relied on data from Canada's Radarsat-1 satellite.

During the Cold War, the Soviet Union had developed and tested ASATs in the 1960s and 1970s, before it announced a moratorium on testing in 1983. In 1993, Yeltsin continued this policy and warned that any measure to weaponize space would be reciprocated. Russia kept dormant its technological capabilities to develop and operate terrestrial ASAT weapons (Venet 2015). The Soviet satellite navigation system GLONASS (Global Navigation Satellite System) had launched in 1982 for military purposes as counterpart to US GPS (Global Positioning System). This preparation for war included a huge number of low-quality satellites and anti-satellite systems. In the 1990s, however, Russia was unable to keep enough satellites in orbit. Its early warning and "Earth observation" systems were significantly degraded, and it suffered many challenges including lack of funding, short life of satellites, and launch problems.

Russia's 1993 law "On Space Activity" aimed to create a normative legal base to regulate space. It states that Russia shall promote the development of international cooperation in the field of space activity, as well as the solution of international legal problems that may arise in the exploration and use of outer space. It lists the key bodies in charge of space activity – the Supreme Soviet (ratifies international treaties on space activity, adopts legislation about space activities, etc.); the president (issues edicts and executive orders); government (supervises state activities and the Federal Space Program proposed by the Ministry of Defense); the Russian Space Agency (Roscosmos), and the Russian Academy of Sciences.

2000–2008: Russia's Securitization and Militarization of Space

The Russian economic recovery in the 2000s coincided with a political emphasis on space as a strategic sector and subsequent increases in its state budget. Space became a symbol of Russia's revived international standing, and attempts were made to restore its former space glory and prestige. For Putin, space policies became a central tool in Russia's rebirth as a great power and its drive for independence from the West, and he prioritized rebuilding and modernizing Russia's military space capabilities. Clearly, he aimed to reduce Russia's dependencies on West (especially on technology and military data) and ensure strategic autonomy and independent access to space.

Russia began restructuring its space industry in the early 2000s, increasing the role of the state (as it did in other strategic sectors). Space and defense industries reoriented production away from export markets toward national armed forces. Both sectors were placed under the new Military-Industrial Commission (Voenno-promychlennaia komissiia (VPK)). Venet writes that this led to mixed results, including some "spectacular failures," e.g., the loss of military and dual-use satellites, foreshadowing President Medvedev's call for more extensive military reforms (Venet 2015, p. 360). Russia's policies on space militarization (use of space assets to support military actions on Earth) continued to focus on the territory of Russia and the former Soviet Union. According to Venet, Russia could not return to the Soviet global approach (i.e., maintain the "high number of military launches and... extensive constellations of military spacecraft needed for all for military communications, navigation, surveillance, early warning, signals intelligence etc.") (Venet 2015, p. 363).

During Putin's first two terms, many state programs and presidential decrees brought Russia into a leading position in the space industry, developed new public-private partnerships, and expanded international cooperation. (These included three major space policy documents: Federal Space Program (FSP) (2005–2015); Federal Program on Global Navigation Systems (GLONASS) for 2002–2011; Federal Special Program for the Development of Russia's Cosmodromes (DRC) for 2006–2015; and Federal Target Program for GLONASS development 2013–2020.) However, even with steady economic growth during these years, space spending remained precarious, prompting Roscosmos (the Federal Space Agency) to petition President Putin for more funding. In the early 2000s, GLONASS was revived and new satellites were launched. This provided a source of prestige and a symbol of independence from the USA in positioning, timing, information, and navigation.

Concurrently, Russia began to modernize its ground infrastructure (important for satellite control, space surveillance networks, and cosmodromes). It brought ground-based assets back to the Russian territory and militarized space assets already there. For example, Roscosmos took over the Baikonur Cosmodrome in Kazakhstan, and the Plesetsk Cosmodrome in Northern Russia was set to become Russia's major military spaceport. Russia also revived its ASAT system program (not co-orbital ASAT system) and made substantial advances in ballistic missiles, radars, and missile defense interceptors (details below).

2008–2013: Halting Modernization and Growing Ambitions Following the Russia-Georgia Conflict

After Russia's war in Georgia in 2008, and with President Medvedev in power, the push for space modernization resumed, although uncertainty over funding and overambitious plans continued. Russia's space industry did survive the 2008 world economic crisis, declining oil revenues, and foreign capital flight thanks to government subsidies. However, the war in Georgia highlighted the limits of Russia's military capabilities and the failure of its command and control system. Space-based intelligence was deficient, and satellite communications facilities were not useable. There was no situational awareness, and satellite targeting was not operational (for artillery or precision-guided munitions).

Russia's early warning system (crucial for nuclear deterrence) improved in subsequent years but still lacked global detection. Then, it lost its last major satellites in 2014. GLONASS regained full operational capacity only in December 2011 when the second-generation GLONASS-MA entered service, but difficulties with the technology and political disputes over its deployment lingered. However, there is evidence that the Russian government during these years increasingly perceived space-based systems as highly important and essential for integration of command, control, communications, information, surveillance, and reconnaissance (C3ISR) and also for what Russians call the "information-strike operations" – which consist of "information-strike battles, information-weapons engagements and strikes with the goal of disrupting enemy troop command and control of weapons systems and the destruction of its information resource" (Johnson-Freese 2017, p. 44).

The 2011 Presidential Decree on science and technology, which provides the current legal basis for technological development of the Russian economy, included plans for space, information, and communications systems (President of the Russian Federation 2011). However, many of the projected missions and launches were postponed, and a gap between stated goals (e.g., creation of a unified, information, command, and control system) and realities remains. Many experts question the quality of Russia's space-based communication system as well as the required infrastructure on the ground. As a result, it has often been suggested that the Russian armed forces did not evolve relative to new combat realities and that Russia should focus more on reconnaissance, electronic warfare capabilities, command and control, data processing, and information distribution systems (Gareyev 2009; McDermott 2012; Roffey 2013).

2014 and Beyond: Moving to the Offensive?

Since Russia's annexation of Crimea in 2014 and its military involvement in Syria, tensions between Russia and the West have increased dramatically. Russia's 2014 military doctrine included in its list of key external threats: "global strike," the intention to station weapons in space, and strategic non-nuclear precision weapons (Russia's Security Council 2014). (Prompt Global Strike is the Pentagon's strategy of being able to strike anywhere in the world with a conventional warhead in less than an hour.) Russia continues to argue that to preserve the strategic balance of power it must respond to US actions. It therefore seeks to limit the technical superiority of the USA by focusing on counterspace activities such as cyber and electronic warfare while fostering uncertainty about its own intentions. The potential threat to Russia's strategic nuclear deterrent and the US pursuit of "global strike" conventional precision missile systems are frequently cited as the main reasons that "Russia can't consider further reductions of offensive forces at this time." Overall, while Russian rhetoric has become increasingly bellicose, Russia continues to militarize and centralize its policies on outer space with an emphasis on the importance of information. The stated strategic priorities of its current Security Strategy include "Strengthening the country's defense, ensuring the inviolability of the Russian Federation's constitutional order, sovereignty, independence and national and territory integrity' and 'consolidating the Russian Federation's status as a leading world power, whose actions are aimed at maintaining strategic stability and mutually beneficial partnerships in a polycentric world" (The National Security Strategy of the Russian Federation 2015).

Some argue that a close reading of Russia's current space documents reveals confusion about different goals and budgets (Zak 2018). However, the documents also reveal Russia's key principles on space policy, including the protection of state interests such as the right to self-defense; the promotion of economic development, including the development of space assets, launch vehicles, and ground infrastructure; the development and use of space technology and goods and services in the interests of Russia's socioeconomic sphere and the space and rocket industry; and maintenance of Russia's primary position in piloted flights. Public documents include plans to create a new generation of space complexes and systems to be competitive in the world market and, once again, the completion of the GLONASS system.

The Federal Space Program, a long-term planning document (Space Activity of Russia in 2013–2020, 2014) listed three goals: contributing to the development of the economy; enhancing national security and strengthening Russia's position in the world; and increasing the welfare of Russia citizens [my bold]. The subsequent Federal Space Program 2016–2025 continues to prioritize the competitiveness and large-scale use of the GLONASS system as well as ground infrastructure for space activities.

Since 2014 the centralization of Russia's space industry has advanced. It reverted to state ownership in the 2000s, but the heads of companies retained their autonomy and were involved in bureaucratic fights with Roscosmos. Then, Roscosmos merged with state-owned United Rocket and Space Corporation to create the Roscosmos state corporation. This new state corporation has been criticized as being like the old Soviet model with no incentive other than to follow instructions from political leaders. Other recent developments include Russian plans for a new space system including systems for intelligence and warning of air and space attacks and destruction and suppression of forces and means of air and space attacks. In 2015, the Russian Space Forces (established in 1992) was merged as a new branch of Russia's Aerospace Defense Forces responsible for monitoring space objects, identifying potential threats to the nation from space, and preventing "attacks as needed" (Jotham 2018). This branch combines elements of space forces, air forces, as well as air and missile command. Meanwhile, US President Trump has been considering whether and how to separate space activities from the Air Force evidenced in his executive order in June 2018 to create a new Space Force.

Russia has continued to work on ground facilities to control orbit and wage electronic warfare by targeting space communication and navigation systems. It allegedly jammed GPS signals during the Crimean conflict in 2014 (Harrison et al. 2018). Luzin wrote in 2016 that outer space communications and reconnaissance remain the Achilles' heel of the Russian Army (Luzin 2016). However, during the Syrian conflict, Russia used reconnaissance aircraft in addition to Soviet-era Vishnya-class intelligence-gathering vessels (AGIs) and ground-based SIGINT (Signals Intelligence) facilities on Syrian territory (Hendrickx 2017). Nevertheless, for now Russia remains dependent on airborne, sea-based, and ground-based reconnaissance assets to complement satellite data. The US Director of National Intelligence, Daniel Coats, concluded in 2018 that "Russia aims to improve intelligence collection, missile warning, and military communications systems to better support situational awareness and tactical weapons targeting... Russia plans to expand its imagery constellation and double or possibly triple the number of satellites by 2025" (Coats 2018).

Russia also allegedly has, or is developing, new ASAT capabilities including direct energy lasers, interceptor missiles, maneuverable satellites, robotics, and electronic warfare (Weeden 2015; Mizokami 2018). (These are designed to blind US intelligence and ballistic missile defense satellites.) Although many of these remain unverified or denied by the Russian government, US experts believe that the biggest threats from outer space are "non-kinetic threats such as jamming satellite-based capabilities such as GPS and communications" and that Russia has sent microsatellites into space which could be used to ram another satellite or snoop on it to collect data or interfere with its capabilities (Daniels 2017). Russia is also developing ballistic missile defense (BMD) capabilities, which are centered around Moscow with plans for a national missile defense dome. Missile defense capabilities may have dual function as ASATs, particularly systems deployed around the Kremlin. Russia has reportedly carried out the world's longest test of a surface-to-air missile system. Its efforts to develop hypersonic glide vehicles are argued to be explicitly aimed at evading US missile defense systems.

Addressing the Russian Military Academy of the General Staff in March 2018, Army General Valery Gerasimov announced that the next phase of Russia's new high-tech approach will focus on robotics, artificial intelligence, and the information and space spheres as well as on economic and nonmilitary targets (Tucker 2018). This approach is likely to continue. It is reactive to increases in the US military budget, as well as to rapid technological developments, and is part of a larger effort to move possible conflicts into areas of "nontraditional warfare." However, despite all its modernization programs, upgrades, and plans, Russia continues to be weak in many space systems, and many argue that long-term structural weaknesses affecting the broader economy, such as an aging work force, inefficiency, and brain drain, have still not been resolved. (Details about Russia's development of a long-term strategy since 2012 can be found on Anatoly Zak's website RussianSpaceWeb.com.)

Russia's Space Diplomacy

Russia's strategy on outer space security issues since 2000 has also included diplomacy, just as it did in the Soviet period. Previously, the Soviet Union co-sponsored the Partial Test Ban of 1963 and the landmark Outer Space Treaty (OST) of 1967. Soviet delegates were active at the Conference on Disarmament (CoD) to promote discussion on the "Prevention of an Arms Race in Outer Space" (PAROS). (The CoD, a 65-member body in Geneva, is supposed to serve as a United Nations' forum for discussing multilateral agreements on arms control and disarmament. Since 1994, the CoD has been deadlocked due to competing priorities. CoD needs unanimous agreement to move forward on issues and set agenda. PAROS gained near-universal support in annual UN General Assembly resolutions, but the USA has consistently objected arguing that space weapons cannot be defined or effectively verified.) Under Putin, Russia has used the United Nations and its affiliated bodies to attempt to develop binding laws, non-binding norms, and transparency and confidence-building measures (TCBMs) to prevent and to control the use of weapons in space. Russia's official rhetoric is that it cooperates with states that share its goals and preference for inclusive negotiations at the UN. Certainly, Russia's diplomatic activism reflects a desire to participate in international organizations and fora to shape international rules, but it may also be a strategy to hedge its comparative economic and military weaknesses. Russia participates in the UN as part of a broader attempt to develop relations with other states of the emerging "multipolar international system" and as a platform to denounce what Russia perceives as the US role in undermining the international rules of the game (e.g., the US unilateral pulling out of the ABM treaty but also Western military operations that lack the UN Security Council's mandate, Iraq and Kosovo).

Russia introduced a working paper in 2002 and then two more in 2004 at the CoD. These became the basis for the 2008 Russia-Chinese Draft Treaty on "Prevention of the Placement of Weapons in Outer Space and the Threat or Use of Force Against Outer Space Objects" (PPWT). This draft treaty extended the OST prohibitions on placement of weapons of mass destruction (WMD) to all forms of weapons. It sought to ban "any device placed in Outer Space, based on any physical principle, specially produced or converted to eliminate, damage or disrupt normal function of objects" (CD 2008). As Paul Meyer explains, "the termination of the ABM Treaty meant the elimination of the only prohibition on space-based weapons agreed upon beyond the ban on WMD in the 1967 Outer Space Treaty" (Meyer 2016). As mentioned above, Russia equates space weapons with weapons of mass destruction (WMD) and has consistently argued that their deployment would have a destabilizing effect on the global strategic balance. The US and other critics of the PPWT argued that the Russia-China treaty did not include a verification mechanism; it only limited deployment, not building of weapons, did not include terrestrial-based ASAT weapons, and did not resolve the problem of how to define a "weapon." (Although given the inherent ASAT capability of ballistic missile interceptors, any effort to include ground-based systems runs up against US commitment to deploy ballistic missile defense.) In response, in June 2014, Russia and China presented a revised version of the PPWT, which included a new article acknowledging the need for verification measures and suggested that these could be elaborated in a subsequent protocol to the treaty.

However, further consideration of the PPWT was prevented by the general blockage of the CoD, and Russia and China have not taken the draft to another forum. They prefer the CoD which protects their interests and gives them a voice and legitimacy. Unsurprisingly, Russia also opposed the 2008 EU Draft Code of Conduct for Outer Space (and its latest draft of March 2014) arguing that the EU code is undermining the work of the UN on space security. Russia and the BRICS argue that the proper format for such deliberations must be inclusive and consensus-based multilateral negotiations within the framework of the UN, in oder to take into consideration all states' interests.

Since 2005 Russia has also solicited and proposed ideas for nonlegally binding and voluntary TCBMs (transparency and confidence-building measures) at the UN General Assembly. A UN Group of Governmental Experts (GGE), chaired by Victor Vassiliev, head of the Russian delegations to the UN GGE, produced a report in 2013 that enumerates several potential transparency and confidence-building measures, including information exchange, risk reduction measures, visits to space-related facilities, and consultative mechanisms (UN General Assembly 2013). This report led to subsequent UNGA resolutions encouraging states to review it. To quote Vassiliev (2015): "...we tried to put forward proposals that were practical, implementable, did not undermine sovereign rights or security of States." In 2018–2019, a new GGE examined possible legal instruments to prevent an arms race in outer space, including the prevention of the placement of weapons in outer space. Since the CD remains in a state of paralysis, this means that Russia and China showed some creativity and found another platform for official work on their PPWT and other possible legal instruments. However, the GGE was unable to release its report because of last-minute opposition by the US representative.

As a final example of Russia's active diplomacy, Russia has also been pushing a "no first placement of weapons in outer space" resolution which was adopted by the UN General Assembly in 2015. This resolution, critiqued by the USA for not being truly transparent, encourages states to adopt a political commitment not to be the first to place weapons in outer space. In 2016, Russia and Venezuela (after years of Russian loans and weapon deals to Venezuela) released a joint statement to the CoD declaring that they will not be the first to deploy any type of weapon in outer space.

Conclusion

For Russia, outer space has become an important area through which to respond to or negate Western strategy and capabilities as well as influence global norms. Its economic, military, and technological weaknesses relative to the USA and NATO have led it to pursue asymmetric tactics including working through bodies affiliated through the UN which give it publicity and some legitimacy but little ability to make real progress. Asymmetrical tactics adopted to advance their goals include traditional, new, and hybrid military capabilities, use of denial and uncertainty about Russian intentions, information, cyber diplomatic negotiation and cooperation, and legal means (including attempting to develop or reinforce norms) (Jackson 2019). The separation between global security and governance is not distinct. Russia's outer space strategy fits within its security and foreign policy efforts which focus on asserting Russia's authority, status, and prestige. Military efforts are but one part of a complex set of tools, which include not only outer space and new technology such as electronic warfare but also nonmilitary means (negotiation, finance, propaganda, etc.) employed to navigate what Russia perceives as an increasingly hostile world.

This situation is likely to become even more complicated. Outer space is increasingly interrelated with land, sea, and air and cyber domains. It also is increasingly congested with other state and non-state actors, including private companies. This proliferation of actors is taking place just as military strategies are increasingly forced to consider the "battlefield" as a seamless whole.

Today, tensions between Russia and the USA are high, and each recognizes what is sometimes called "an integrated multi-domain threat" coming from the other. This is reinforced by recent technological advances and significant distrust about each other's intentions. The result is the growing militarization of space. A pressing challenge for the future will be how to reconcile different security perceptions of states as well as non-state actors and their understandings about how the laws of armed conflict apply to military (and even civilian) space activities. For example, how does one define proportionality of response to an attack on a satellite? Is radio jamming a use of force or an armed attack? Which activities are legitimate, and which are not? There is a plethora of ambiguity. The goal of creating the sustainable use of space for peaceful purposes and for the benefit of all humankind seems ever elusive.

References

Charron A, Fergusson J (2018) From NORAD to NOR [A] D: the future evolution of North American Defence Cooperation. Policy Paper, May, Canadian Global Affairs Institute, Toronto

Coats D (2018) Worldwide threat assessment of the US intelligence community, congressional testimonies, 13 February. https://www.dni.gov/index.php/newsroom/congressional-testimonies/item/1845-statement-for-the-record-worldwide-threat-assessment-of-the-us-intelligence-community. Accessed 02 Apr 2018

Conference on Disarmament (2008) Draft 'treaty on the prevention of the placement of weapons in outer space, the threat or use of force against outer space objects, February 29, 2008, CoD/1839

Daniels J (2017) Space arms race as Russia, China emerge as 'rapidly growing threats' to US. CNBC, 29 March. https://www.cnbc.com/2017/03/29/space-arms-race-as-russia-china-emerge-as-rapidly-growing-threats-to-us.html. Accessed 2 Apr 2018

Gareyev MA (2009) Issues of strategic deterrence in current conditions. Military Thought 18(2):1–5

Harrison T, Johnson K, Roberts TG (2018) Space threat assessment. CSIS, Washington

Hendrickx, Bart (2017) Russia encounters hurdles in satellite development and expansion. Jane's IHS Markit, 21 June

Jackson N (2019) Deterrence, resilience and hybrid wars: the case of Canada and NATO. J Mil Strateg Stud 19(4):104–125

Johnson-Freese J (2017) space warfare in the 21st century: arming the heavens. Routledge, Abingdon

Jotham I (2018) Russia's 'Space Forces' just completed the construction of two new radar facilities, 1 February, International Business Times. https://www.ibtimes.co.uk/russias-space-forces-just-completed-construction-two-new-radar-facilities-1657957. Accessed on 2 Mar 2018

Luzin P (2016) 'Russia's position in space. Foreign Affairs, September. https://www.foreignaffairs.com/articles/2016-09-21/russias-position-space. Accessed 2 Mar 2018

McDermott, Roger (2012) Russia's 'New Look' military reaches out to space. Eurasia Daily Monitor 9(140)

Meyer P (2016) Dark forces awaken: the prospects for cooperative space security. Nonproliferation Rev 23(2–3):495–503

Mizokami K (2018) Russia, China will have anti-satellite weapons 'within a few years'. Popular Mechanics, 15 February. https://www.popularmechanics.com/military/weapons/a18197465/russia-china-anti-satellite-weapons/. Accessed 2 May 2018

President of the Russian Federation (2011) The presidential decree 'On the approval of the priority directions of science and technologies of the Russian Federation' approved by the President of the Russian Federation on July 7, 2011, no. 899

President of the Russian Federation (2015) The Russian Federation's National Security Strategy, 31 Dec 2015

Roffey R (2013) Russian science and technology is still having problems -implications for defence research. J Slav Mil Stud 26(2):162–188

Russian Government (2015) Russian assessment of the US global ABM defence programme – International security and disarmament, 5 Mar 2015

Russian Security Council (2014) The military doctrine of the Russian Federation, 25 Dec 2014

Tucker P (2018) Russian military chief lays out the Kremlin's high tech war plans, Defense One, 28 March. https://www.defenseone.com/technology/2018/03/russian-military-chief-lays-out-kremlins-high-tech-war-plans/147051/. Accessed 2 Apr 2018

UN General Assembly (2013) Group of Governmental Experts on Transparency and Confidence building Measures in Outer Space Activities, A/68/189, Sixty-eighth session, July 29, 2013

Vasiliev V (2015) Statement at the Joint Ad hoc meeting of the First and Fourth Committees of the 70th session of the UN General Assembly, 22 October 2015. www.reachingcriticalwill.org/images/documents/Disarmament-fora/1com/1com15/statements/22October_SpaceGGE.pdf. Accessed 2 May 2018

Venet C (2015) Space security in Russia. In: Shrogl KU et al (eds) Handbook of space security. Springer, New York, pp 355–370

Weeden B (2015) Dancing in the dark redux: recent Russian rendezvous and proximity operations in space. The Space Review, 5 October

Zak A (2018) Russian space program in the 2010s: decadal review, 6 April. http://www.russianspaceweb.com/russia_2010s.html. Accessed 8 Apr

Development of a Space Security Culture: Case of Western European Union

Alexandros Kolovos

Contents

Introduction	401
The Western European Union Era (1955–2001)	403
The WEU and Space (1979–1999)	405
The Development of a WEU Space Security Program	406
Toward a WEU Space Policy	411
Conclusion	415
References	417

Abstract

This paper deals with the origins of the European space security and defense effort. The main thrust of the argument in this paper runs as follows. All institutional space security and defense discussions were facilitated by an existing space security culture in Europe. This culture goes back several decades. The main institutional actor which paved the way for its creation was the Western European Union. The WEU now may has ceased to exist, but its space security culture it created remained. As such it facilitated the EU to recognize the importance of space activities in its Common Security and Defence Policy.

Introduction

On 15 November 2002, the European Commission (EC) held an informal workshop in Brussels to examine the security aspects of space (European Commission 2002). The meeting was in the framework of a forthcoming Green Paper on the future of

---

A. Kolovos (✉)
Automatic Control, AirSpace Technology, Defence Systems and Operations Section, Hellenic Air Force Academy, Athens, Greece
e-mail: alexandros.kolovos@hafa.haf.gr

© Springer Nature Switzerland AG 2020
K.-U. Schrogl (ed.), *Handbook of Space Security*,
https://doi.org/10.1007/978-3-030-23210-8_80

Europe in space, which would be the base to create a "White Paper" on the subject. This task had been given to the Commission by the Parliament, and among issues to be debated were political sensitive issues including space security and the needed institutional arrangements (The term *space security* is used according to the definition given in the introduction of the first volume of *The Handbook of Space Security (HbSS)* which states: "For the purpose of the HbSS, we understand space security as having two dimensions, security in space and security from space" (Schrogl et al. 2015). This chapter deals only with the second dimension that is security from space.). The Commission was concerned of the security aspects of space, and the reason of the meeting was to give contribution on the development of a Green Paper, which should highlight all the questions that should be answered in a White Paper.

Some of the main questions to be addressed at this workshop were "to define if Space is relevant for Security, how to merge civilian with security (dual-use), if Europe wants to be an actor in the space sector or prefers to buy the technology, or if EU wants to set up a cooperation with the US." Nowadays the first question might raise an eyebrow in surprise. Although the Commission would be the owner of both the Green and White Papers, the reality was at the time that the EC had not been institutionally involved with the security issues before, due to the institutional limitations imposed by the Treaty on European Union.

Maastricht Treaty created an overarching structure to be known as the European Union (EU) comprising three components or *pillars*. The European Commission constituted the first pillar, while the Common Foreign and Security Policy (CFSP) was established as the second pillar. The third pillar dealt with aspects of the area of freedom, security, and justice. Until that time, the Commission has focused more on scientific programs and civilian space applications. Connections between space and security have not been central to most EU documents and were only partially addressed in its official space reports.

At the workshop it has been stressed that the Commission would have been glad if this task could be coordinated internally within the Council General Secretariat (CGS) of the second pillar so that the final outcome could be coherent. CGS could look in the space aspects within CFSP (Common Foreign and Security Policy) and ESDP (European Security and Defence Policy) concerning security.

This approach was chosen due to the fact that while these two policies had been established some years before (CFSP in 1992 and ESDP in 1998), their reflections had not appeared in any of the relevant space documents the Commission had produced. While initially this might considered as an absence of mechanism for policy coordination, actually it was more an institutional duty, due to the limitations of the first (EC) and second pillars (security and defense). Lisbon Treaty of 2009 would make the pillar structure obsolete.

The talk was of "security" but without defining the scope. At a working level, it was clear that the scope was environmental security, as it was evident by the development of the Global Monitoring for Environment and Security (GMES) program. This created a risk that it was perceived at higher levels as also covering those areas that fall within the second pillar. It was clear that the Commission was not on the same footing as the Council General Secretariat within the ESDP (European

Security and Defence Policy) and CFSP (Common Foreign and Security Policy) policies concerning space and security. EC thus acknowledged that no matter how much schooling, it was still difficult to think like the people of a particular culture if one is not working in that environment. The Commission as a responsible actor needed to take a well-prepared collective view.

It was due to this reason that the EC itself wanted some directly involved people to help them to find issues to bring in the security aspects of the Green Paper. Apart from its bodies, the Commission invited the EU Military Staff and representatives from three defense ministries (the UK, Belgium, Greece) that were active in space security (both nationally and in the Western European Union framework).

This specific task was finally implemented successfully, and the Green Paper addressed the sensitive topic of security dimension (European Commission 2003), because the Commission seemed to know exactly who were to ask. And this became possible due to an existing but almost unrecognizable cause, a preexisting space security culture. Cultures are about shared values. Space has always had its big players along the not so advanced players, but the glue in any community is that all of players have something in common even if their institutional circumstances vary.

This space security culture was created at the Western European Union framework, an organization that came into being with the ratification of the modified Brussels Treaty on 6 May 1955. Many people may not be familiar with this, since the relevant literature is rather limited. Key officers enabled in the Green Paper process, from the Council General Secretariat or as national experts, were previously enabled in WEU's space activities.

This chapter pays a long overdue tribute to the previously unknown by many contribution of the WEU in the development of this space security culture. Whereas all political and operational functions of WEU ceased and were transferred to the EU's European Security and Defence Policy in December 1999, its space security culture has been consolidated and preserved. Its influence at the EU was significant in the making of space security policy, as it helped creating, among others, two milestones: shaping the development of the 19 May 2003 General Affairs and External Relations Council (GAERC) decision which for the first time "recognised the importance of space applications and functions needed in order to enhance EU capabilities to carry out crisis management operations" (Council of the European Union 2003), along the Council's decision on "ESDP and Space Policy" (Council of the European Union 2004).

The Western European Union Era (1955–2001)

Before we examine the space security activities of the WEU, it is considered useful in the analysis that will follow to give some basic facts regarding WEU. The Western European Union, as the successor to the Brussels Treaty Organization (founded in 1948), was a defensive alliance composed by seven founding members (Britain, France, West Germany, Italy, Belgium, the Netherlands, and Luxembourg). In its final synthesis (after 1995), WEU consisted of ten member states: the original seven plus Portugal, Spain, and Greece. It was responsible to provide the framework for the creation of a European defense policy until 2001 (European External Action Service 2016). Located in Paris, WEU was created in postwar Europe as the responsible body for European defense and security issues, as these two domains were left outside the European unification attempt that had begun in the 1950s with the creation of the European community (Hatjiadoniu 2000).

In the mid-1980s WEU has been established as the forum to discuss European reactions in several challenges, like the response to a US invitation to Europe to participate in the US "Strategic Defense Initiative" or to the Euromissile crisis which created tensions in the US-European relations. As it was stated in 1986, "WEU has thus become established as the one European body in which space policy is discussed in all its aspects, military and civilian, taking into consideration the diverging viewpoints held by the various European countries" (Voûte 1986). It was a time when both the USA and USSR put heavy emphasis on military space programs, while Europe had no significant military space involvement. On the other hand, in January 1985, the Council at ministerial level of the European Space Agency (ESA) adopted an ambitious long-term European space program, with a purely civilian character.

The Maastricht Treaty, signed in December 1991 (and entered into force in June 1993) which had established the Common Foreign and Security Policy (CFSP), stipulated that it was the Western European Union which would elaborate and implement the development of a European security identity and more specifically the development of a "Common European Security and Defence Policy (CESDP) and a defence role within the European Union." At the same time WEU acted as the European pillar of NATO with a task to enhance the role and responsibility of the European members within the alliance, namely, the "European Security and Defence Identity (ESDI) within the NATO."

The Article J.4 of the Treaty on European Union (Maastricht, 7 February 1992) that followed stated:

> 1. The common foreign and security policy shall include all questions related to the security of the Union, including the eventual framing of a common defence policy, which might in time lead to a common defence. 2. The Union requests the Western European Union (WEU), which is an integral part of the development of the Union, to elaborate and implement decisions and actions of the Union which have defence implications.

The WEU declaration, adopted on that occasion, constituted the initial step in the development of the CESDP. In the declaration WEU member states agreed to strengthen the role of WEU as the defense component of the EU and as the means to strengthen the European pillar of NATO. That led to the initialization of a reflection at the WEU on the question of Europe's future in the security and defense domain with a focus to examine and reinforce common means for action.

Based on the Maastricht provisions, WEU started to formulate and to develop its own crisis management procedures. In this period of time, the WEU entered its "operational" phase, adapting to the new post-Cold War conditions. Elements of relatively successful adaptation include enlargement, missions, and especially its Petersberg tasks. At the Petersberg ministerial meeting in 1992, WEU member states defined the crisis prevention and management missions that could be conducted through WEU by their forces as follows: "military units of WEU member states acting under the authority of WEU could be employed for: humanitarian and rescue tasks; peacekeeping tasks; tasks of combat forces in crisis management, including peacemaking."

In 1997, the Treaty of Amsterdam amended the Treaty on European Union (Maastricht 1992). The relationship between the EU and WEU is contained in the second paragraph of Article 17(1). This states that "The Union shall foster closer institutional arrangements with the WEU with a view to the possibility of the integration of the WEU into the Union, should the European Council so decide. It shall in that case recommend to the member states the adoption of such a decision in accordance with their respective constitutional requirements."

Finally the European Council decided to give the EU the capacity for independent action, backed up by credible military forces. In December 1999, WEU surrendered political and operational work to the EU (and its European Security and Defence Policy – now known as Common Security and Defence Policy (CSDP)) (Bailes and Messervy-Whiting 2011). In 2001 all functions of the WEU have effectively been incorporated into the EU. With the introduction of the Lisbon Treaty (2009), the EU's role was greatly increased, so the European Parliament to avoid duplication called for the abolition of the European Security and Defence Assembly (ESDA)/Assembly of WEU (The Assembly of WEU was founded in 1954 by the modified Brussels Treaty. Composed of 364 national parliamentarians from 28 countries including all EU member states, the Assembly scrutinized European intergovernmental activities in all areas of security and defense. ESDA was debating space- and security-related topics. Following the transfer of WEU's operational activities to the EU, the Assembly also acted as the interim European Security and Defence Assembly.). It was agreed subsequently that WEU would be wound down completely by May 2011 (Remuss 2015).

The WEU and Space (1979–1999)

WEU's interest in space activities goes back to its first parliamentary Assembly report on "Space War or Space Cooperation?" in 1961 (Assembly of Western European Union 1961). A new report on "European Space Organisation" came in May 1962. Then in the mid-1960s, WEU studied the possibilities of setting up a European satellite agency for the verification of arms control agreements, for monitoring crisis situations, and, last but not the least, for intelligence gathering (Assembly of Western European Union 1988a; Heintze et al. 1993).

The WEU Assembly and especially its Technological and Aerospace Committee continued to advise on the evolution of European military and civilian space activities in the framework of European security all through the 1980s (Assembly of Western European Union 1984). A relevant role to a smaller degree to promote public awareness had the WEU's Institute for Security Studies. In 1985 the Technological and Aerospace Committee convened a colloquy regarding the space challenge for Europe (Assembly of Western European Union 1985). The time was right as the ministers of foreign affairs of the then seven countries tried to coordinate their answers to the US President Reagan's proposal to the European members of the Atlantic Alliance to participate in the Strategic Defense Initiative (SDI) which he just had announced.

The seeds of development of the European cooperation in space activities for security purposes have been on WEU's agenda since 1979 (Assembly of Western European Union 1990). The cooperation effort focusing on an earth observation satellite system program started officially on 13 November 1989, when the WEU Ministerial Council Decision in Brussels set up an ad hoc subgroup on space (SGS), the so-called Space Group. Its mandate was to study the possible uses of space technology in three specific areas: the verification of arms control, the monitoring of crises affecting European security, and the monitoring of environmental hazards (Assembly of Western European Union 1988b, McLean 1995). The WEU's Military Staff participated in the Space Group since its inception (The Military Staff was responsible for the implementation of policies and decisions as directed by the Council and Military Committee. It prepared generic and contingency plans, carried out studies, and recommended policy on matters of an operational nature under WEU politico-military control.).

The Development of a WEU Space Security Program

While some major European space-faring nations which were members of the WEU had access to military space systems from the 1970s, the turning point of the institutional European approach in the security and defense domain is found, in our judgment, in two significant incidents: the Chernobyl nuclear reactor accident and the first Persian Gulf War.

On 26 April 1986, a nuclear accident happened at the Chernobyl plant in near Pripyat, Ukraine, in the then USSR. On 29 April 1986, the US Landsat 5 was the first civilian satellite to reveal the burning reactor n°4 caused by the explosion. But its imagery resolution was comparatively poor, being on the order of 30 m. Then 2 days later, on 1 May 1986, just 2 months after its launch, France's SPOT 1 acquired the first 10-m high-resolution imagery, which showed a plume of hot air trailing from the reactor building. Initially the Soviets decided to downplay the catastrophic nature of the event, and only on 14 May 1986, General Secretary Gorbachev spoke about the disaster. At that time the scientific community had realized that the pollution was expanding throughout Europe despite the political messages stating that the pollution will stop at the Ukrainian border. Satellite imagery of Chernobyl marked also the first time the mass media used space surveillance which showed that large-scale activities cannot be hidden or protected by borders. These conclusions were included in two Assembly reports and were sent to the WEU Council. The latter, politely thanked the Assembly for the job done. No action was taken.

But the decisive point for Europeans was the first Persian Gulf war after Iraq occupied Kuwait on 2 August 1990. Coalition Forces, led by the USA, with the cooperation of France and UK forces, began a military deployment which led to the 17 January 1991 active air campaign against Iraq. The Persian Gulf War (1990–1991) international conflict revealed significant lessons to be learnt not only from the British and French personnel involved in the coalition against Iraq but for the Europeans as a whole.

Called also as "the first space war," it highlighted publicly that information from space was a valuable political and military tool for the execution of operations (including Petersberg). It was also an alarm of the weakness shown. On behalf of the Europeans, it was WEU who begun to respond practically by further exploiting the military aspects of space. WEU was first to recognize the relationship between military space activities and their common security. The common political will, shown on the decision for the creation of the WEU Satellite Center 1992, proves that it was also the first time all WEU countries have worked together identifying space as a common mean of action in the security and defense domain.

One outcome of the Gulf War was the wide acceptance of the vital role that satellites played (Kiernan 1991; Gupta 1991). It was in this framework that the European forces felt that their assets and capabilities would not enable them to participate fully and on an equal basis as desired with the Americans. As French Minister of Defense Pierre Joxe put it: "In the Persian Gulf War, without the American intelligence, the european Allies were almost blind. It was the United States who provided, whenever they want it, the necessary intelligence to carry out the war" (De Selding 1991a). Furthermore, the contributing European forces realised the overall US dependence on space.

The Gulf War was the main relevant event where the European Defence Ministers understood how much the EU was dependent to the USA for the global evaluation of a situation and how much space was the solution. After the Gulf War, WEU was convinced that space-based observation represented a strategic capability which was needed to acquire, in order to meet its security and defense responsibilities (Assembly of Western European Union 1991). As such a multipronged approach has been followed:

At the Ministerial Level (Foreign and Defense Ministers)

1. The establishment of the Satellite Center (WEUSC). In April 1991, in Vianden (Luxembourg), WEU Ministerial Council decided the creation of a Satellite Center (SatCen) to be later based at Torrejon Air Base (near Madrid, Spain). The creation of the Satellite Center was for a trial period of 3 years, from 1993 to 1995. The cost of this venture was high (38 million ECU for the first 3-year experimental phase) (Molard 1998). According to its Concept Paper (WEU Council of Ministers 1997), the main function of the Satellite Center (which was officially inaugurated in 1993) was to analyze for security purposes imagery

from satellite and airborne sources relating to areas of interest to WEU and to train European experts (WEU General Secretariat 2000). SatCen depended largely on commercial imagery and on national provision of imagery. SatCen's establishment was a rare political move which gained the consensus of all member states at the time. It was also a unique model in itself in relation with the sharing of satellite-generated information at the EU level. It was the first occasion in WEU which a part of a space program (i.e., the creation of a photointerpretation operational center) preceded the policy meant to implement. At the Ministerial Council in Lisbon (1995), the WEU Satellite Center was declared a WEU permanent subsidiary body and was placed under the authority of the WEU Council. Citizens of the ten full member nations were its staff, while image analysts from member states and associate members could be second personnel for a limited time. Furthermore, SatCen educated national experts via its training activities which also comprised participation in seminars, conferences, and "field" visits.

2. *Toward an independent European space-based observation system for defense.* The Ministerial Council held in Bonn on 18 November 1991 decided to study the feasibility of an independent European space-based observation system for defense purposes in 1992–1993 period. That decision reflected WEU's willingness and determination to move ahead. The President of the Assembly, which had only consultative powers, welcomed these decisions which responded to recommendations the Assembly had been making for several years (Space Policy 1991). Two types of studies were conducted:
   - System concept-related studies consisting of the Main System Feasibility Study at a cost of 4.5 MECU, led by DASA/Dornier company (Assembly of Western European Union 1992), along with seven special studies addressing additional critical aspects of the satellite system (simulation, fusion, sensors) at a cost of 0.7 MECU.
   - Benefit assessment studies aimed at assessing the operational suitability of the most promising satellite system solutions at a cost of 1 MECU. These preliminary studies provided a general overview of what could be offered by a WEU Earth Observation System.

In Kirchberg Declaration on 9 May 1994, "Ministers confirmed the aim of further developing WEU's capability to use satellite imagery for security purposes [...] reaffirmed their will to set up an independent European satellite system. A decision would be taken subject to evaluation of the costs and merits of the proposed system and of other WEU alternatives and affordability."

The final outcome from these studies was that a satellite system fully compliant with the Main System Feasibility Study has been shown to be complex and costly. WEU member states envisaged an interim solution to be proposed after discussions with the user community aimed at refining the requirements. As a trade-off, a new (phase A) study has been proposed, and its estimated cost was in the order of 30 MECU.

At the Space Group Level

1. *Participation in a developing multinational program.* Also WEU's Space Group decided to explore further engagement in activities toward a possible WEU participation in a developing multilateral program (the then Helios-1 program). Helios was the only successful example of a multilateral collaborative military space effort in Europe. French Defense Minister Joxe had repeatedly stated that all WEU member states could participate in the Helios-1 program that was underdevelopment. Of the seven WEU members, Italy and Spain choose to participate.
2. *Procurement of imagery from commercial systems.* The Space Group tasked the SatCen to explore further on other solutions which might overcome limitations on the timeliness of the WEUSC response to tasking. Finally, of several possible options, that chosen was the transmission of images via high data-rate links.

In Lisbon on 5 May 1995, Ministers of Foreign Affairs and Defence of the WEU nations tasked the Space Group "to continue its activities, concentrating on the last three proposed approaches to develop a WEU capability to use satellite imagery for security purposes, namely: The establishment of a WEU earth observation (EO) satellite system, the participation in a developing multinational programme and the procurement of imagery from commercial systems."

Ambassador Horst Holthoff, WEU's Deputy Secretary-General, addressed the main reasons for all these decisions in 1995. An autonomous observation capability to provide appropriate intelligence is a prerequisite for a functioning European policy on security and defense. The EO satellite system is a way for Europe to become an equal partner with the USA in exchanging satellite imagery which might lead to technological cooperation. Also this EO system would support the European aerospace industry, which had experience in participating in European programs and keep it ready for international competition. And finally it can become a tool for accelerating European integration (Holthoff 1995).

WEU's Ambition: An Independent EO System for Defense

For the Earth Observation domain, as a first target, WEU sets the goal of an independent European space-based observation system for defense, enabling the WEU to mount an autonomous operation without recourse to NATO command assets and capabilities placed anywhere on the spectrum of Petersberg missions.

That was very obvious from the following event. In March 1995 WEU organized a colloquy on Gran Canaria, regarding a European space-based observation system (Western European Union 1995). In tshis framework Gil Klinger, then US Acting Deputy under Secretary of Defense (Space), made a point of stressing that the US Department of Defense was keen to increase the level of cooperation between the USA and WEU with regard to space and space systems (McLean 1995). There was not a European consensus to follow this transatlantic cooperation road.

But the road to strategic independence faced obstacles since it could not gather the consensus of all member states. Some governments were hesitant by the order of magnitude of the investment required for the development of an autonomous European Earth Observation System consisting of both optical and RADAR satellites. Since it was evident that there was not a consensus to move ahead, the WEU decided to follow more conservative solutions.

So in Madrid Declaration, the ministers, having in mind the high cost of the establishment of an independent program and knowing of the plans for the Helios-2 program, tasked the Space Group "to continue its activities by defining the basic conditions for possible WEU participation in a developing multilateral European programme, while in the meantime continuing with the procurement of commercial imagery and studying the possible WEU ground segment." Thus, the following actions followed:

1. Regarding the participation in a developing multilateral European program, according to an Assembly report: "WEU's participation in the Helios-2 programme is therefore not only desirable and necessary – it must also be feasible. [...]WEU should not expect to receive an offer from the Helios-2 participating countries but should express its desire to take part in the programme" (Lenzer 1996).
2. The WEU consultations to reach an agreement with the Helios nations (France, Italy, and Spain) on an improved access by the WEU to Helios-1 have been examined since 1992, after France reversed its position that Helios-1 imagery would be available only to its owners (De Selding 1991b). In 1993 the three Helios nations signed a Memorandum of Understanding for the WEU to get access to Helios-1 governmental imagery. It is important to notice that the MoU has been signed the day before the official inauguration of the WEU SatCen and also that this agreement was signed 3 years before the launch of the first Helios satellite. Access was given through a special Helios-1 cell which had been created at SatCen with limited access, due to the classification of Helios imagery at *SECRET* level, which was very complex and difficult to handle. This access to Helios-1 trilateral military photoreconnaissance program gave SatCen a truly European dimension.
3. Also, in 1996 the Space Group tasked SatCen to perform a feasibility study on medium- to long-term opportunities for operational satellite data reception at the WEU, including commercial imagery such as the ones provided by SPOT Image, RADARSAT, etc. (WEU Satellite Centre 1997).
4. Then at the 1999 WEU Luxemburg ministerial meeting, the WEU ministers gave SatCen a mandate to establish, before the end of year 2000, a midterm concept for improving access to satellite imagery (WEU Satellite Centre 2000). A study was launched on the options for setting up a European satellite observation system in the medium and long term.

Ministers in Bremen in May 1999 tasked the Space Group to continue its work on evaluating the possibilities for WEU participation in developing a multilateral European program and on studying questions related to a possible WEU ground segment, taking into account existing ground segments in WEU nations. In so doing, close contacts were maintained between WEU SatCen and the relevant EU authorities, competent in the field of earth observation research and policies.

In November 2000, the relevant study was finalized, and SatCen was expecting a mandate from the WEU Council to initiate in 2001 the procurement process for a suitable receiving, processing, and archiving facility. But since in 2001, all functions of the WEU would have effectively been incorporated into the EU; no immediate decisions could be taken on the issue. Due to the demise of the WEU, these steps in space security program development, originally envisaged for 2001, would not take place under its auspices. The WEU's Satellite Center, along the Institute for Security Studies, had been transferred to the community framework. Also, since the Helios MoU was signed with WEU member states, this MoU was stopped as soon as the WEU SatCen has been transferred to the EU, which had a different list of member states.

Toward a WEU Space Policy

While WEU's space activities were under development, at the institutional level, something was missing. All these actions constituted specific steps toward a WEU space program in the Earth Observation domain. Some steps had been already implemented, while others were at the study level. But, the WEU had not developed firstly an overarching space policy, under which this WEU space program was conducted.

This seemed odd since the Assembly of the WEU started to debate upon the space policy back in 1966, with its "Juridical problems and space policy report" (Assembly of Western European Union 1966). Ten years later, the issue resurfaced in a more general way. In 1976 WEU's Assembly Committee on Scientific, Technological and Aerospace Questions held the first colloquy in Toulouse to study what a European aeronautical policy might be.

Proposal for a WEU's Earth Observation Space Policy

It was in this framework that on January 1998, the WEU Hellenic Presidency presented its program in which there was a conceptual idea for the formulation of a WEU's Earth Observation Space Policy (The Presidency of the WEU was held by member states in rotation for periods of 6 months. The Presidency influences the priorities for work in the WEU and sets the pace of meetings and their agendas.). The third point of Presidency's Programmestated the following:

> In an effort to further develop and enrich the exchange of views among nations on space related questions, Presidency wishes to launch a general discussion on elements that could constitute the framework of a WEU space observation policy. By initiating this exercise, we hope that this group could reach a common understanding on the fundamental guidelines, as well as the medium and long-term goals of a coherent WEU space observation policy.

Before presenting the idea, the Presidency made informal consultations with Germany (the former Presidency) and France, as the dominating power in space issues in Europe (in the framework of the discussions investigating the possibility of Greece's accession to the forthcoming Helios-2 program). Both countries expressed their support. France argued that for a better presentation of the issue, it was necessary to draw up a relevant text to be put on the basis of discussion. For the drafting of the relevant text, the French side proposed further cooperation with the relevant WEU Technological and Aerospace Committee, which resulted in very positive comments.

Similarly, it agreed with the Greek option to include telecommunications and navigation issues in it, including examples of existing European cooperation in these two areas (the tripartite military Trimilsatcom satellite communications between the UK, France, and Germany (later to be failed), as well as the European EGNOS (European Geostationary Navigation Overlay System, under development at the time). But the opinion of the Hellenic authorities was that initiative should focus only in the Earth Observation domain, which was after all the nucleus of the WEU's space activities so far and the mandate of the WEUSC, in order to overcome any objections from other countries if the term "WEU space policy" was raised.

Formulating this document, apart from the relevant body of work from the WEU, and the advice from the Satellite Center and the Secretary of Aerospace and Technological Committee of the WEU General Assembly, the texts of space policies of other countries (France, the UK, Germany, the USA), along with informal consultations from the UK and Belgian delegations, have been taken into account. It was in this framework that on 27 February 1998, the WEU Presidency presented a first working paper entitled "WEU's Earth Observation Space Policy," which proposed a balanced conceptual framework to establish the formulation of such a policy (Hellenic Presidency 1998). This document is presented below for historical and research purposes.

"WEU's Earth Observation Space Policy" [SGS(98), Bruxelles, 27/2/1998]

Introduction For several years (since the Vianden decisions in June 1991), WEU has been engaged in continuing efforts toward a European earth observation space program. The first stage of this program was the establishment of a satellite image interpretation center (which constitutes the ground segment of WEU's space program). A second stage was the study on the possibilities for medium- and long-term cooperation on a European space-based observation system (the space segment). However WEU hasn't yet defined an Earth Observation Space Policy, under which its earth observation space program can be conducted. To this effect, the Greek WEU Presidency undertook the initiative to present a first working paper to the Space Group.

Scope This document provides the first draft for a framework to establish the formulation of WEU's Earth Observation Space Policy (EOSP). The principles and guidelines, with respect to the conduct of WEU's space program and related activities, could provide the basis for further discussions among delegations aiming at the development of EOSP.

Main Objectives The proposed goals of WEU's earth observation space activities are:

- To strengthen its operational role. The possession of up-to-date data is a key element for taking political decisions. and space observation systems can play a key role in this.
- To enhance the scientific, technological, and industrial potential of the WEU participating nations.
- To promote international cooperative activities, taking into account WEU's priorities and interests.

Basic Principles WEU earth observation space activities will be conducted in accordance with the following principles:

- WEU will make use of earth observation space assets for activities in pursuit of its security concerns and its operational needs.
- WEU shall support such activities as surveillance, reconnaissance, and environmental monitoring (including research and development of programs supporting these functions).
- WEU may use both commercial and noncommercial earth observation space systems to meet specific mission requirements.
- WEU should envisage the most effective use of existing earth observation European satellite and/or its participation in the development of an independent system. Such a decision should be based on an in-depth evaluation of the implied costs and benefits.
- WEU should also be ready to use and exploit other existing or emerging commercial space technologies and systems.
- WEU will ensure that its earth observation space program shall be conducted in accordance with its broader mission and policy and the deriving operational requirements.

Policy Guidelines The following ideas provide a framework through which the proposed policy in this document will be carried out.

- To fulfill its missions effectively, the Satellite Center will continue to strengthen its functional links with the Planning Cell and Situation Center.
- The Satellite Center should spare no effort to enhance its operational and technical capabilities, in order to further improve its effectiveness in undertaking the required tasks.
- WEU should continue the further training of its Satellite Center's staff in the development of digital image processing, photointerpretation and gathering, and exploiting accessible data.
- The Satellite Center will pursue the identification and development of appropriate applications deriving from its activities. Such applications will create new capabilities or improve the quality and efficiency of ongoing activities.

- WEU should also follow the developments in the field of European space transportation.
- WEU will encourage multilateral cooperation within its framework for the development, construction, and operational use of space observation systems.
- WEU shall encourage the further development of international cooperation, particularly when this has beneficial effects for European security interests and for European space industry. In this context, WEU should establish relations with the European Space Agency in order to determine possibilities for cooperation in space-based observation. It is hoped that this will develop a synergy between scientific and applications orientated programs.

Drafting a WEU's Space Policy

Of a very general nature in order to gain all member states acceptance, this document has been discussed at various Space Group meetings and revised accordingly to the comments received by three WEU's permanent member states (France, the UK, and Germany). The WEU Military Staff and the WEU SatCen worked closely together in drafting the final paper.

Additionally it has been expanded too. To the original Presidency's initial chapters, three more were added:

- The first (Chap. *V Operational Requirements*) concerned the operational requirements that the WEU had to be fulfilled by satellite systems. They were submitted by the WEU Military Staff, following a remark during the reflections by the Greek Presidency that the operational requirements asked in 1996 by the Ostend Declaration were still pending. The latter in paragraph 30 stated that "the Ministers took note that the further in-depth study could only be continued after having defined WEU's operational requirements." Given this mandate and the time that has passed from 1995 which show the emergence of new technologies, the Presidency believed that the old requirements served its purpose at that time, but it was felt that further work to refine and expand its scope would be necessary, in order to evaluate the various options. WEU Military Staff addressed the issue.
- The next two chapters (*VI Role and Function of the WEU Satellite Centre* and *VII Improvements of The Satellite Centre's Operational Capabilities*) were submitted by Germany in the framework of balancing the relevant space policy text with the Concept Paper, which guided the WEUSC's work internally. Although the approval of the latter has been preceded in time, thematically, it was rather a subset of the text of space policy.

Finally, the Space Group, realizing the good spirit of the reflections between the member states, seized the opportunity to further reflect the importance of space and decided to expand the scope of the initiative, thus covering all aspects of space assets. The new domains of interest that have been added were the developments in navigation and positioning and communications.

The next Italian Presidency finalized the work, the Permanent Council approved the document on the "WEU Space Policy" and finally the ministers on November 1998 in Rome Declaration took note of this policy: "Ministers appreciated the finalization of work on a conceptual framework paper concerning WEU's Space Policy, initiated by the Greek Presidency, and they took note of the relevant document, defining this policy" (WEU Council of Ministers 1998).

This final document has been released at the WEU members with a classification "RESTRICTED." Due to the demise of the WEU, the further implementation of the *WEU Space Policy* could not take place.

Conclusion

In this chapter we argue that the space security culture, which has been emerged in WEU, was consolidated, preserved, and continued serving well the EU, when the latter found itself institutionally responsible for the first time in its history to exercise fully its ESDP policy at the dawn of the new millennium. "The Wise Men" Report stated this: "European Security and Defence Policy is incomplete without a space component. ESDP relies on a mixture of civil and military instruments for crisis and conflict management" (Bildt et al. 2000).

The purely civilian EU space program, developing so far (such as GMES program), was just not enough to cover ESDP needs. The Commission admitted this in its 2000 "Europe and Space: Turning to a new chapter" report, which stated that "Dual-use (civilian-military) aspects of satellite systems have not so far figured highly on Europe's agenda. Through the Satellite Center of the Western European Union (WEU), Europe has gained some experience in dual-use. Integration of the WEU Satellite Center in the EU may open new avenues for shared utilisation" (European Commission 2000).

WEU was there to provide its powerful vision regarding the use of space in the security and defense domain. The realization of the significant role that space plays in European common security resulted in an ongoing strong presence of WEU in the space in the ESDP domain. Even if one can characterize its approach as a little "unorthodox," as it put the space program before its overarching space policy which followed, it was effective.

WEU around 2000 had conquered all the basic steps of the policy cycle in the space domain. These steps in public policy theory usually include agenda setting, policy formulation, policy implementation, and policy evaluation and change. WEU has set the agenda and has formulated its relevant policy, it has taken the decisions, and it has launched a space program which implemented this policy.

Furthermore it created a strong governance for space issues. With the help of an impressive body of work coming from the Assembly of Western European Union, a specialized governing body (the Space Group), along with the excellent leadership and staff gathered at the SatCen and with the cooperation of the European defense industrial technology base, WEU has paved the way for the first robust institutional space program in Europe in the Earth Observation domain for its ESDP (For example, the EUROSPACE Security & Defence working group published quite a few reports demonstrating the capacity and willingness of the European industry to build the space architecture that the EU would need after having worked on its space policy. An evaluation of the cost of such a space system has even been announced: only 5€ per European citizen per year to be invested in a space system for European security. (Personal communication with General Bernard Molard, former Director of WEU SatCen and ASD EUROSPACE Security & Defence WG Chairman (March 3, 2019).).

Evaluating its pioneering studies that has focused on producing specialized and/or expensive satellite systems, WEU exercised a balanced approach in choosing cost-effective solutions when there was not a common political will. WEU and its bodies had the experience to undertake all of the tasks listed above, but not the resources to do them all. Furthermore the economics at that time limited the willingness of MS to contribute more resources. WEU showed a specific restrained culture to not pay any price to achieve the stated goals but chose to expand its capabilities in a very cost-effective approach.

WEU also created a "learning-centered work culture," in which it was safe for members of its bodies to raise doubts. The case of prioritizing first the development of its space program and then its space policy and the issue of moving on with studies without having first defined specific operational requirements are examples of this learning culture of accepting and amending mistakes.

As the WEU has been absorbed by the EU, structures like the Military Staff still remained in the new scheme, along with some of the former WEU staff who participated in them or in other newly created bodies. The Commission wisely sought to benefit from them and from the lessons learned at WEU and considered how the most useful parts of this space security culture, which has been formatted in WEU, might be preserved. It is a task that in this new environment both the EU SatCen and the EU ISS aimed to further contribute (In its website, the EUISS states its aim to foster a common security culture. https://europa.eu/european-union/about-eu/agencies/iss_en.). And the EU was eager to build upon it, at least at this early stage.

The following paradigm is self-explanatory. A 2002 report from WEU's Technological and Aerospace Committee stated that "space applications can provide political decision-makers and in the EU with part of the information they need to carry out an effective European Security and Defence Policy (ESDP). Until now, the absence of political will has prevented the EU from adopting a proper space policy" (Assembly of Western European Union 2002).

Space is a "trans-pillar issue"; one has to deal with it from an overarching architecture. Only 2 months after the EU Hellenic Presidency of 2003 presented its initiative "ESDP and Space" (Kolovos 2009), the Military Staff, tasked by the WEU Military Committee to elaborate upon this issue, prepared an information paper entitled "Space Systems Needs for Military Operations" describing the areas in which space systems could improve EU military capabilities (European Union Military Staff 2003).

This was the first official document produced by a Council body on the ESDP requirements for space assets. The Military Staff was a new structure at the EU, but its space security culture was inherited from the old Military Staff. Then, on 22 November 2004, the Council approved the document "European Space Policy: ESDP and Space," which provides the identified and agreed upon ESDP requirements to be reflected in the global EU space policy and its corresponding European space program.

In 2015, Frank Asbeck, former Director of the WEU SatCen and Principal Adviser for Space and Security Policy at the European External Action Service, stated that "Space has now been clearly recognized as a key element in the EU's security and defence-related activities" (Asbeck 2015).

Verifying this statement in June 2016, the Global Strategy for the European Union's Foreign and Security Policy addresses the issue of space as such, "First, European security hinges on better and shared assessments of internal and external threats and challenges. [...]. This requires investing in Intelligence, Surveillance and Reconnaissance, including [...], satellite communications, and autonomous access to space and permanent earth observation" (EU Global Strategy 2016). Almost 20 years ago, it was the WEU which had decided first to develop an independent European space-based observation system since it understood well this dimension.

Much of this recognition should be credited to the work done by WEU due to its space security culture. The WEU developed its space activities under the assumption that the civilian and military needs for all actions for ESDP purposes are compatible, with potential for synergies. WEU's SatCen is still the only relevant CFSP/ESDP operational tool in the geospatial intelligence. SatCen exploited long before any other structure a "dual-use" approach, which was unique institutionally and paved the way for others to follow. WEU may be gone, but its space security culture established capabilities, and personalities were there to lead the EU in its early steps of recognizing space's role in CSDP, and as such it is not forgotten.

References

Asbeck F (2015) Foreword. In: Schrogl K-U et al (eds) Handbook of space security: policies, applications and programs. Springer, New York

Assembly of Western European Union. Proceedings Vol. III, seventh session, second part, Document 215, December 1961

Assembly of Western European Union. Proceedings Vol. III, twelfth session, second part, December 1966, Document 388, Item no: WEU-22.011, 21/11/1966

Assembly of Western European Union. The Military Use of Space, Report, submitted on behalf of the Committee on Scientific, Technological and Aerospace, Thirtieth Ordinary Session (First Part), Document 976, 15 May 1984

Assembly of Western European Union. The Space Challenge for Europe: colloquy, Munich, Official record. Office of the Clerk of the Assembly of WEU, Paris 18th–20th September 1985

Assembly of Western European Union. Verification: a future European satellite agency: report by Mr Fourré rapporteur, Proceedings Vol. III, Thirty-fourth session, second part, December 1988a, Assembly document 1159

Assembly of Western European Union. Scientific and technical aspects of arms control verification by satellite, report by Mr Malfatti, rapporteur Proceedings, Vol. III, Thirty-fourth session, second part, December 1988b, Assembly document 1160

Assembly of Western European Union. Observation Satellites, a European Means of Verifying Disarmament, Symposium, Technological and Aerospace Committee. Rome,Official Record. Office of the Clerk of the Assembly of WEU, Paris, 27th–28th March 1990

Assembly of Western European Union. Weaponry after the Gulf War-New equipment requirements for restructured armed forces, 37 Ordinary Session, Document 1272, 14 May 1991

Assembly of Western European Union. The development of a European space-based observation system, 38 Ordinary Session, Document 1304, 30th April 1992

Assembly of Western European Union. A European Space-based Observation System,Colloquy, Official Record. Office of the Clerk of the Assembly of WEU, Paris, San Agustin, Gran Canaria, 24th–25th March 1995

Assembly of Western European Union. European space observation vital for EU security, WEU Press Release, 5 June 2002

Bailes A, Messervy-Whiting G. The demise of the Western European Union: lessons for European Defence, International Security Programme Rapporteur Report, 10 May 2011. https://www.chathamhouse.org/sites/default/files/100511_bailes_0.pdf

Bildt C, Peyrelevade J, Späth L (2000) Towards a space agency for the European Union. Report for the Director-General of the European Space Agency. Available from http://esamultimedia.esa.int/docs/annex2_wisemen.pdf

Council of the European Union. Draft GAERC conclusions on ESDP (public), 9174/03, Brussels, 13 May 2003. http://register.consilium.europa.eu/doc/srv?l=EN&f=ST%209174%202003%20INIT

Council of the European Union. European space policy: "ESDP and space", 11616/04, 16 Nov. 2004. http://register.consilium.europa.eu/pdf/en/04/st11/st11616-re03.en04.pdf, par.1

De Selding P. Calls for French Lead in European Spy satellite drive, Space News, May 13–19, 1991a, p 20

De Selding P. Joxe: France to Share Helios Images with WEU, Space News, June 10–16, 1991b, p 22

EU Global Strategy (2016) Shared vision, common action: a stronger Europe. A global strategy for the European Union's foreign and security policy. High Representative of the Union for Foreign Affairs and Security Policy, Vice-President of the European Commission. https://europa.eu/globalstrategy/en/global-strategy-foreign-and-security-policy-european-union

European Commission. Europe and space: turning to a new chapter. Communication from the Commission to the Council and the European Parliament – COM/2000/0597 final

European Commission. Green Paper on the future of Europe in Space, Workshop on Security Aspects Brussels 15th November 2002, Research Directorate-General, Space Policy and Coordination of Research, Brussels, 26/11/2002, D(2002) CB

European Commission. Green paper on European space policy: report on the consultation process – information document to the Joint Task Force 4 September 2003, Sec (2003) 1249 Brussels, 11.11.2003, http://register.consilium.europa.eu/doc/srv?l=EN&f=ST%2014886%202003%20ADD%201. European Commission, European security discussed at Athens, Green Paper consultation, 15 May 2003, https://www.web.archive.org/web/20080205161946; http://www.ec.europa.eu:80/enterprise/space/news/article_598_en.html

European External Action Service. Shaping of a common security and defence policy, 08/07/2016. https://eeas.europa.eu/headquarters/headquarters-homepage/5388/shaping-common-security-and-defence-policy_en

European Union Military Staff. Space systems needs for military operations, 9793/03, 27 May 2003

Gupta V. METEOSAT lifted fog of war to expose reality in Gulf, DEFENCE NEWS, March 18, 1991

Hatjiadoniu A. (2000) The Daedalus European security: the interactions of NATO, EU, WEU. NATO research fellowships programme 1998–2000. https://www.nato.int/acad/fellow/98-00/hatjiadoniu.pdf

Heintze H-J, Wallner J, Hounam D, Nowak M (1993) Remote sensing and strengthening European security. Space Policy 9:65

Hellenic Presidency's Initiative. "WEU Earth Observation Space Policy", SGS(98), Bruxelles, 27/2/1998, presented by Alexandros Kolovos, at the Space Group of WEU

Holthoff H. A European satellite system-a cooperation programme for an autonomous European security policy, Second EUCOSAT Symposium on Satellite System for security: a European multi-user system, Bonn, 20 September 1995

Kiernan V. Satellites play key role in swift Gulf Victory, Space News, March 4–10, 1991

Kolovos A. The European space policy – its impact and challenges for the European security and defence policy. ESPI perspectives 27. European Space Policy Institute, September 2009, p 18. https://espi.or.at/publications/voices-from-the-space-community/publications-of-the-former-espi-perspective-series/send/10-publications-of-the-former-espi-perspective-series/229-the-european-space-policy-its-impact-and-challenges-for-the-european-security-and-defence-policy

Lenzer C. WEU and Helios 2, Document 1525, Assembly of Western European Union, 14 May 1996. http://aei.pitt.edu/53681/1/B0944.pdf

Lisbon Treaty (2009) Amending the Treaty on European Union and the Treaty establishing the European Community, signed at Lisbon, 13 December 2007

Maastricht (1992) Treaty on the European Union signed at Maastricht on 7 February 1992. Official Journal of the European Communities, C 191

McLean A (1995) Integrating European security through space. Space Policy 11(4):239–247

Molard B (1998) How the WEU Satellite centre could help in the development of a European intelligence policy. In: Becher K, Molard B, Oberson F, Politi A (eds) Towards a European intelligence policy. Chaillot paper, 34. Institute for Security Studies of WEU, Paris

Remuss L (2015) Responsive space. In: Schrogl K-U et al (eds) Handbook of space security: policies, applications and programs. Springer, New York

Schrogl K-U et al (eds) (2015) Handbook of Space Security: Policies, Applications and Programs. Springer, New York https://doi.org/10.1007/978-1-4614-2029-3_66

Space Policy. European cooperation in space-based defence, November 1991, p 337

Voûte C, (1986) A European military space community. Space Policy 2(3):206–222

WEU Council Of Ministers. European security: a common concept of the 27 WEU countries, Madrid, 14 November 1995, WEU. http://www.weu.int/documents/951114en.pdf

WEU Council Of Ministers. Concept paper for the satellite centre. Paris, 13 May 1997

WEU Council Of Ministers. Rome declaration, page. 10, para. 5, 16/11/1998

WEU Satellite Centre. Feasibility of Satellite data reception at WEU, SC (97) 6, 1997

WEU Satellite Centre. Mid-Term concept for improving access to satellite imagery, MTC/V.1.0/31.10.2000

WEU Secretariat-General. WEU today. Brussels, January 2000. http://www.weu.int/WEU_Today2.pdf

Strategic Overview of European Space and Security Governance

23

Ntorina Antoni, Maarten Adriaensen, and Christina Giannopapa

Contents

Introduction	422
National Level	423
France	425
Germany	427
Spain	428
Italy	429
United Kingdom	430
Multilateral Level	431
Archetype Model for National Space and Security Governance	442
Concluding Remarks	445
References	446

Abstract

The importance of security is increasing in strategic policies, activities, and programs of European countries, the European Union (EU), and the European Space Agency (ESA). Each European country has a unique national space governance structure that facilitates linkages with the aforementioned and other space-related organizations. National space governance in this chapter means "the way that national authorities and structures are managed within a

---

N. Antoni (✉)
Eindhoven University of Technology, Eindhoven, The Netherlands
e-mail: ntorina.antoni@gmail.com

M. Adriaensen · C. Giannopapa
European Space Agency (ESA), Paris, France
e-mail: maarten.cm.adriaensen@gmail.com; christina.giannopapa@esa.int

© Springer Nature Switzerland AG 2020
K.-U. Schrogl (ed.), *Handbook of Space Security*,
https://doi.org/10.1007/978-3-030-23210-8_146

state in order to supervise space related organizations, space budget, space strategy and policy." These structures determine decision-making power, representative mandates, implementation of programs, and interaction among all organizational actors. The chapter, thus, focuses on the governance structures that enable decision-making processes when it comes to space and security, within which there is ample attention for coordination among and between the national, multilateral, regional, and intergovernmental levels. The chapter concludes with an archetype model of national governance for space and security that facilitates interconnections with relevant organizations.

Introduction

In Europe, the space sector is a particularly interesting and dynamic field, mainly because it includes several actors with varying priorities. Due to the inherent dual-use nature of space activities, responsibility for space has traditionally resorted under a state's sovereign competences. Traditionally, security- or defense-related space programs have been kept at the national level or dealt with bilaterally or multilaterally in *ad hoc* cooperative programs. Only civilian space activities, including Earth observation, telecommunications, human spaceflight, space transportation, and technology development, have been the subject of cooperation at the regional and intergovernmental levels.

However, in the past years the security dimension of space activities has increasingly been coming to the attention of European countries, as well as the European Union (EU) and the European Space Agency (ESA). "Space and security," both in its security *from* space and security *in* space form, is progressively contributing to the further integration of space activities in sectorial policies (Giannopapa et al. 2018). As such, countries conduct space and security activities at the national, multilateral, and intergovernmental levels, but also within EU and ESA contexts. The increasing security challenges the European countries and institutions are facing, together with the political momentum that is favorable for advancing Europe's role in the field of security and defense, may lead to further integration covering the entire dual-use spectrum of space activities.

Europe has a long history of space integration for civil space activities, with tangible successes in the frame of ESA and the EU (Antoni et al. 2018). The rising challenges for Europe's security have caused the EU and ESA to take a more active stance when it comes to security and defense, as elaborated in ▶ Chap. 61, "Institutional Space Security Programs in Europe" in Part 4 of the Handbook. It is noteworthy that in 2019, NATO, although with no operational capabilities of its own, adopted a space policy recognizing that space is essential to deterrence and defense. Simultaneously, European countries are increasingly recognizing the need for joint action to address societal challenges, including in the field of security and defense as described in ▶ Chap. 25, "Space and Security Policy in Selected European Countries" in Part 2 of this Handbook. The different nature and strategic objectives of all organizational actors involved in space and security in Europe lead to a complex space and security governance. This raises challenges with regard to coordination among all actors that affect the governance of space security policy, activities, and programs.

Each European country has a unique national space governance structure responsible for overseeing space related organizations, space budget, space strategy and policy and implementation of space activities. This structure determines decision-making power, representative mandates, implementation of programs, and interaction between actors and stakeholders. The chapter focuses on selected number of countries with the largest space budgets, while considering interaction among the national, multilateral, and intergovernmental levels. In order to advance the understanding of the space and security governance, this chapter investigates the linkages whereof and based on the analysis of these governance structures, it concludes with an archetype model for national space and security governance.

National Level

All European countries have a specific governance structure to organize space- and security-related competences. In all selected ESA Member States, the Ministries of Foreign (and/or EU) Affairs, the Ministry or Ministries responsible for Science, Education and Research, and the Prime Minister's Office, Chancellery, or Council of Ministers are involved in space affairs. In almost all Member States, the Ministries of Defence are included in the national space governance (except for Luxembourg, where the competence for defense resides within the Ministry of Foreign and European Affairs).

A high number of Ministries of Home Affairs (Interior), Ministries of Justice and/or Security, and Ministries charged with environment, sustainable development and/or climate change (sometimes also energy, food, and/or agriculture) are involved in space affairs. This is due to the increasing role of space-based services in the areas of civil protection, crisis management, national security, sustainable development, climate change, and environmental affairs. Because of the importance of research and development (R&D), technology and innovation as well as economy and competitiveness, one or multiple associated ministries per Member State take up an important role of the national space governance.

The Ministry(-ries) competent for transport, infrastructure, telecom, and/or digital affairs are also increasingly active in space activities in the sub-domains of Global Navigation Satellite System (GNSS), Positioning, Navigation and Timing (PNT), Satellite Communication (SATCOM), frequency management, and integrated applications. Depending on the specific geographic location, policy-making approach and strategic priorities in space and other policy areas (for instance maritime or arctic), other Ministries are also involved in space affairs. On average between nine and ten Ministries per country (and not less than seven in the selected European countries) are involved in national space and security governance. Figure 1 below provides an overview of the types of Ministries involved in the national space and security governance.

Regardless of the number of Ministries involved in the national space- and security-related governance, one Ministry typically holds the main responsibility at

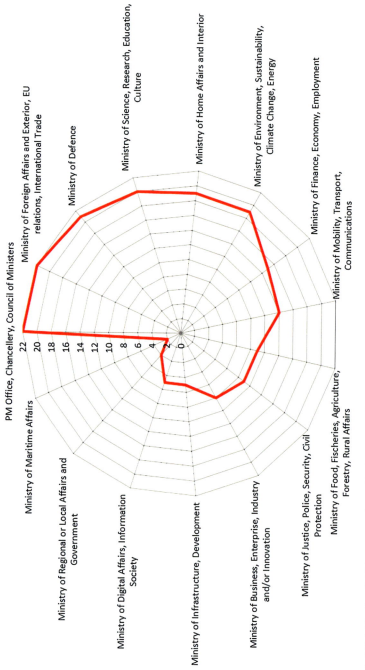

Fig. 1 Ministries involved in space and security for selected European countries

the national level to define, coordinate, and/or implement a national space strategy, space policy, and space programs. Table 1 below provides an overview of the lead Ministries for the 22 ESA Member States.

Additionally, and as a result of the diverse competences that Ministries hold in space and security related areas, a wide variety of governmental bodies and agencies are involved therein. Each governance scheme of the 22 ESA Member States is inherently unique and based on historic remnants, strategic priorities, political compromise, national interests, and geopolitics. The following paragraphs provide some examples of space and security governance. In the following paragraphs, a representative sample of specific European countries is presented in order to show the unique approach followed across Europe in governing space security activities and programs.

France

In France, CNES is the national space agency – a public administration organization with industrial and commercial purposes. CNES is charged with elaborating, proposing, and executing the French space program. CNES is under the joint supervision of the Ministry of Higher Education, Research and Innovation and the Ministry of the Armed Forces (CNES Website).

The Ministry of the Armed Forces is assisted by the *Chef d'état-major des armées* (French Armed Forces High Command), the *Direction générale pour l'Armement* (DGA) and the *Secrétaire Général pour l'Administration* (SGA) (Ministry of the Armed Forces 2017). DGA is the French Defence Procurement and Technology Agency. Within the DGA, the Security and Defence and Information Systems Service within the Security Operations and Information Systems Department and the Service for Preparation of Future Systems and Architecture in the Strategy Directorate (DGA/DS/SPSA/CISR) are responsible for space. DGA, on behalf of the Ministry of Armed Forces, exercises the guardianship over CNES (Ministry of the Armed Forces 2018).

CNES proposes the French space policy under the responsibility of the Minister for Higher Education, Research and Innovation and has a special partnership with the Ministry of the Armed Forces for what concerns military space activities (CNES Website). CNES conducts research and technology for future programs that respond to military needs. DGA delegates the management for military space programs to CNES, and DGA may contribute to the financing of dual-use satellites procured by CNES. Given the dual-use of space assets and the importance of space, and the general acceptance of the use of military capabilities for civilian objectives in France, the French DGA and the General Staff work closely with CNES. The CNES Centre d'Orbitographie Opérationnelle (COO) at Toulouse is the CNES operational orbit tracking center. Coordination of the CNES Defence Team is done by a military advisor (CNES 2017).

The Space Committee and the Interparliamentary Group for Space have further contributed to the French space and security governance. The Space Committee,

Table 1 Coordinating entity and ministries competent for space affairs. (Source: ESA (2017) with updates)

ESA Member State	Coordinating entity	Ministries competent for space affairs
Austria	Austrian Research Promotion Agency (FFG)	Ministry for Transport, Innovation and Technology
Belgium	Belgian Science Policy Office (BELSPO)	State Secretary charged with Science Policy
Czech Republic	Department of Space Activities and R&D	Ministry for Transport
Denmark	Danish Agency for Higher Education and Science	Ministry of Higher Education and Science
Estonia	Estonia Space Office, Enterprise Estonia	Ministry for Economic Affairs and Communications
Finland	Finish Funding Agency for Innovation	Ministry of Economic Affairs and Employment
France	National Centre for Space Research (CNES)	Ministry of Higher Education, Research and Innovation
Germany	German Aerospace Centre (DLR)	Ministry for Economic Affairs and Energy
Greece	5G, Wireless and Space Department	General Secretariat for Telecommunications and Post
Hungary	Department for Space Research and Space Activity	Ministry of Foreign Affairs and Trade
Ireland	Enterprise Ireland	Department of Business, Enterprise and Innovation
Italy	Italian Space Agency (ASI)	Prime Minister/Council of Ministers
Luxembourg	Luxembourg Space Agency	Ministry of Economy
The Netherlands	Netherlands Space Office (NSO)	Ministry of Economic Affairs and Climate
Norway	Norwegian Space Centre (NSC)	Ministry of Industry, Trade and Fisheries
Poland	Space Policy Unit, Innovation Department	Ministry of Economic Development
Portugal	Space Office, Foundation for Science and Technology (FCT)	Ministry of Education and Science
Romania	Romanian Space Agency (ROSA)	Ministry of Research and Innovation
Spain	Centre for the Development of Industrial Technology (CDTI)	Ministry of Economy and Enterprise
Sweden	Swedish National Space Agency (SNSA)	Ministry of Education and Research
Switzerland	Swiss Space Office	State Secretariat for Education, Research and Innovation
United Kingdom	UK Space Agency (UKSA)	Department for Business, Energy and Industrial Strategy

which was established in 1989, is a Committee with representatives from various institutions and bodies charged with the objective of converging to assure the coherence of the national space policy by: analyzing multiannual civil and military space plans and preparing government decision on space policy; assessing the influence of space programs on French and European industry; preparing French position on international cooperation on space activities; and proposing to the prime minister any necessary action. The Committee is composed of representatives from various Ministries and government bodies (Legifrance 1989). Established in 1994, the Interparliamentary Group for Space was renamed the Group of French Elected for Space in 2016, composed of national, regional, and local level representatives that are involved in space activities (GPE Website).

Furthermore, the Ministry of Foreign Affairs plays a key role in space and security governance. Within the Ministry, both the Directorate-General for Political Affairs and Security, the Directorate-General for Globalisation, Culture, Education and International Development, and the Directorate for European Affairs are involved in contributing to space and security activities. In addition to the links established between the Ministries and Implementing Entities, on the one hand, and international/supranational bodies on the other hand, the Ministry of Foreign Affairs is directly involved in the representation of France in international and supranational bodies, for instance through the Permanent Representation to the EU in Brussels. As such, it complements the representation of specific to the Ministry stakeholders.

Germany

In Germany, the Ministry for Economic Affairs and Energy (BMWi) is mainly responsible for Germany's civilian space activities, whereas the Ministry of Transport and Digital Infrastructure (BMVI) oversees selected European Union flagship programs Galileo and Copernicus. The Foreign Office is also active in the space and security realm, as it contributes to international space law discussions. The Ministry of Defence (BMVg) operates Germany's reconnaissance satellites, while the Chancellery oversees the intelligence service (BND) and the Ministry of the Interior (BMI) coordinates the protection of critical infrastructure and national crisis management (Governance Post 2017; BMWI Aerospace Policy). The Federal Government Coordinator of German Aerospace Policy coordinates the Federal Government's measures to strengthen the international competitiveness of Germany's aerospace sector in the fields of research and development (BMWI Aerospace Policy). In addition to the links established between the Ministries, the Government Coordinator and the Implementing Entities, on the one hand, and international and supranational bodies, on the other hand, the Ministry of Foreign Affairs is directly involved in space matters as it often represents Germany in international and supranational bodies.

Within the federal government, the Ministry of Transport and Digital Infrastructure (BMVI) has the lead responsibility for the Copernicus program. The Ministry is responsible for the German participation in the program design vis-à-vis the European Commission and other European players. BMVI is also responsible for the provision of public funds for the German contribution to the Copernicus programs of ESA and EUMETSAT. At the national level, BMVI coordinates the required national accompanying measures related to the programs. Other Ministries, including the Ministry of Defence, also contribute to the Federal Government's positioning. In cooperation with the other Federal Ministries involved, BMVI regularly develops a cross-departmental program where the activities of the Federal Government are defined. The DLR Space Administration monitors the Copernicus Programme on behalf of BMVI. The DLR Space Administration also supports information and public relations activities in Germany for its citizens. As a project management agency to which statutory powers have been transferred, it carries out support measures relating to Copernicus on behalf of the Federal Government. According to its responsibility for implementing the National Programme for Space and Innovation, DLR establishes appropriate linkages between Copernicus and national missions and activities on behalf of the BMWI (BMVI 2017).

The Deutsches Zentrum für Luft- und Raumfahrt (DLR) is the national space agency. It reports directly to the BMWi. DLR was founded in 1969 through the merger of several institutions. Acting on behalf of the Federal Government, DLR Space Administration designs and implements Germany's Space Program, which integrates all German space activities on the national and European level. These activities include Germany's national Space Program and Germany's contributions to ESA and EUMETSAT. In addition, the DLR shapes and monitors the space topic within the European Framework Programme for Research and Innovation. The DLR's principal client is the Federal Ministry for Economic Affairs and Energy. However, the DLR also works for other ministries, mainly in application-related fields such as earth observation, navigation, and satellite communication (DLR Website, BMVI 2017).

Spain

In Spain, the Ministry of Economy and Enterprise (Ministerio de Economía y Empresa) is charged with drafting and implementing government policy on economic matters and reforms to improve competitiveness, industrial development, telecommunications and information society, development of the Digital Agenda as well as the policy of support for the economy and enterprises (Spanish Government 2018). The Ministry has the main responsibility for space affairs in Spain.

Resorting under the Ministry of Economy and Enterprise, the Centre for the Development of Industrial Technology (CDTI) created in 1977 is a public entity which promotes innovation and technological development in Spain. The CDTI manages the space budget for R&D projects and contributes to the improvement of the technological level of the industry at national and international level. Within CDTI there is a dedicated Directorate of European Programmes, Space and Technological Returns that is in charge of liaising with the Copernicus program (CDTI

Website). The CDTI is involved with the Ministry of Defence through agreements in place with the DGAM (Directorate General of Armaments and Material) and Isdefe (Engineering Systems for the Defence of Spain) (CDTI 2011). The CDTI has been supported by the Ministry of Defence to coordinate various aspects of the Spanish participation in the EU SST Consortium and direct interaction with ESA to coordinate the transfer of those aspects related to security and confidential information to its SSA program (Spanish Ministry of Defence 2015, 2017). The Spanish Space Surveillance and Tracking (S3T) system is managed by CDTI (CDTI 2018).

The Interministerial Commission of Industrial and Space Policy constituted in November 2014 has as objective to enhance the existing coordination among the ministerial departments that participate well as promoters or users of space systems or that manage Spanish participation in international organizations such as ESA, EUMESAT, or the EU. This Interministerial Commission was charged in 2015 with preparing the Agenda 2020 of the Spanish space sector, a strategic national planning for space, which was anticipated, which could lead to the establishment of a Spanish Space Agency (Infoespacial 2015).

Italy

Responsibility for space activities has recently been put directly under the supervision of the Prime Minister. The Italian Parliament approved on December 22, 2017, the new space governance law, mandating the establishment of an Inter-ministerial Committee for Space and Aerospace Research Policies, and the President of ASI reporting directly to the Prime Minister's Office. The Law entered into force on February 25, 2018. The Committee is chaired by the President of the Council of Ministers, or by the Undersecretary of State to the Presidency of the Council of Ministers with responsibility for Space and Aerospace Policies, and is composed of the Ministers of Defence, the Interior, Cultural Heritage and Activities and Tourism, Agricultural, Food and Forestry Policies, Education, University and Research, Economic Development, Infrastructure and Transport, Environment and Protection of Land and Sea, Foreign Affairs and International Cooperation, and Economy and Finances, as well as the President of the Conference of Presidents of the Regions and Autonomous Provinces and the President of the ASI (Altalex 2018).

The Italian Space Agency (ASI) is the national public body tasked with promoting, developing, and disseminating scientific and technological research applied to the space and aerospace field, and coordinating and managing national projects and Italian participation in European and international projects in accordance with the government's guidelines as promoted by the Inter-ministerial Committee for related policies space and aerospace research, in the framework of the coordination of international relations assured by the Ministry of Foreign Affairs, taking care to maintain the competitiveness of the Italian industrial sector. ASI reports to the Ministry of Education, University and Research, without prejudice to the powers expressly assigned to the Inter-ministerial Committee (Altalex 2018).

The main players concerned in military exploitation of space are the Stato Maggiore Difesa (SMD, the Joint Defence Staff organisation), the Direzione Nazionale Armamenti (SGD/NAD, National Armaments Directorate), and the Air Force service (IAI 2003).

United Kingdom

The Department for Business, Energy and Industrial Strategy (BEIS) is the parent department of the UK Space Agency. It has the lead for space affairs in the UK. The Minister of State for Universities, Science, Research and Innovation is the responsible minister for space activities in the UK. He is reporting to both BEIS and the Department of Education. BEIS uses satellite data for more accurate mapping through the Ordinance Survey; it also monitors land use through the work of the Land Registry; it delivers space weather and weather forecasting through the Met Office; it coordinates resilience to space weather; and it is in charge of export control (UK Government 2015). Resorting under BEIS, the UK Space Agency helps to bring together UK civil and commercial space programs and interests. It is an executive agency of the Department for Business, Energy and Industrial Strategy and is responsible for all strategic decisions on UK civil space programs. It supports UK space industry initiatives and licenses the launch and operation of all UK satellites, both military and nonmilitary, and has regulatory responsibilities under the 1986 Outer Space Act. It promotes cooperation through participation and contribution to the European Space Programme, and other European initiatives. UKSA represents the UK at ESA. Given the dual-use of space assets and the importance of space, the UK Ministry of Defence works very closely with the UK Space Agency (UK Ministry of Defence 2017).

UK military space capabilities are primarily coordinated and delivered by the Royal Air Force and Joint Forces Command (UK Ministry of Defence 2017). The Ministry of Defence uses satellite services space-enabled services in support of global military operations, from disaster relief response to the employment of precision weapons. The Home Office uses space-enabled capabilities for critical emergency services and law enforcement activities. The National Space Policy clearly articulates cross-government reliance on space-enabled capabilities. From a military perspective, the Royal Air Force (RAF) is recognized as the most significant contributor to space operations, since it retains most of the UK's military space expertise and manages key UK space capabilities (UK Ministry of Defence 2017).

The Foreign and Commonwealth Office is directly involved in the representation of the UK in international and supranational bodies, for instance through the Permanent Representation to the EU in Brussels. The Foreign and Commonwealth Office assists the space sector in capturing new export opportunities. It also facilitates the representation of UK policy in international civil and security partnerships, such as through the United Nations (UK Government 2015).

The Cabinet Office is a key stakeholder of the National Space Security Policy. Its main task is the coordination and management of satellite data for emergency response (UK Government 2018).

Multilateral Level

Moving from the national level to the multilateral level allows to investigate the precise linkages in place between the national authorities representing European countries and the international organizations, initiatives, and *fora* in space, security, and defense. This is an established approach in Europe to develop and acquire high-quality space systems, based on collective needs, pooling of resources, and cost sharing. ESA Member States largely adhere to international legal and regulatory regimes in the field of space, security, and defense. Table 2 below provides an overview of relevant international treaties and initiatives in space and security.

Because of the nature and strategic importance of space, security, and defense activities, European countries are involved in a large variety of international organizations. Every state is represented in a variety of international organizations, international initiatives, and bilateral/multilateral cooperation fora, as presented in Table 3. The latter provides an overview of all relevant space, security, and defense organizations and the level of participation of the 22 ESA Member States. This extensive, yet non-exhaustive list demonstrates the broad range of European and international cooperation in the field of space, security, and defense. Based on the data collected for each European country, this study was able to identify trends on the role of national authorities that are involved in space and security or defense activities.

Each state is represented by the most relevant Ministry or Ministries in the various space and security related organizations. For instance, a Ministry charged with Environment, Sustainable Development and/or Climate Change often represents their respective state in the European Organisation for the Exploitation of Meteorological Satellites (EUMETSAT) and in the European Centre for Medium-Range Weather Forecasts (ECMWF). With regard to ESA and Copernicus/Galileo programs, the Ministries that generally participate are usually those of Education and Science, Transport and Innovation and Ministries of Economy and Trade. For the International Telecommunications Union (ITU) the responsibility typically lies with the Ministry of Transport, Telecom and/or Digital Affairs. The Ministries of Defence have the lead in defense-related organizations, such as the European Defence Agency (EDA), the North Atlantic Treaty Organization (NATO), and the European Union Satellite Centre (SatCen). The Ministry of Foreign (and/or EU) Affairs has the lead role when it comes to space-, security-, and defense-related cooperation with UN-related bodies such as the UN Committee on the Peaceful Uses of Outer Space (COPUOS). For the International Civil Aviation Organization (ICAO), the European

Table 2 Participation of ESA Member States in relevant international treaties and initiatives in space, security, and defense

	Austria	Belgium	Czech Republic	Denmark	Estonia	Finland	France	Germany	Greece	Hungary	Ireland	Italy	Luxembourg	The Netherlands	Norway	Poland	Portugal	Romania	Spain	Sweden	Switzerland	United Kingdom
1967 United Nations Treaty on Principles Governing the Activities of states in the Exploration and Use of Outer Space, including the Moon and other Celestial Bodies (Outer Space Treaty)	✓	✓	✓	✓	✓	✓	✓	✓	✓	✓	✓	✓	✓	✓	✓	✓	✓	✓	✓	✓	✓	✓
1972 United Nations convention on the international liability for damage caused by space objects (Liability Convention)	✓	✓	✓	✓		✓	✓	✓	✓	✓	✓	✓	✓	✓	✓	✓		✓	✓	✓	✓	✓
1974 Convention on Registration of Objects Launched into Outer Space (Registration Convention)	✓	✓	✓	✓			✓	✓	✓	✓		✓	✓	✓	✓	✓	✓		✓	✓	✓	✓
1967 Agreement on the Rescue of Astronauts, the Return of Astronauts and the Return of Objects Launched into Outer Space (Rescue Agreement)	✓	✓	✓	✓		✓	✓	✓	✓	✓	✓	✓	✓	✓	✓	✓		✓	✓	✓	✓	✓

23 Strategic Overview of European Space and Security Governance

Treaty/Regime																				
1979 United Nations Agreement governing the Activities of States on the Moon and Other Celestial Bodies (Moon Agreement)	✓	✓									✓			✓	✓	✓	✓		✓	✓
1963 treaty banning nuclear weapon tests in the atmosphere, in outer space and under water (Limited Test Ban Treaty)	✓		✓	✓		✓	✓	✓	✓	✓	✓	✓	✓	✓	✓	✓	✓		✓	✓
1998 United Nations Comprehensive Nuclear Test Ban Treaty	✓	✓	✓	✓	✓	✓	✓	✓	✓	✓	✓	✓	✓	✓	✓	✓	✓		✓	✓
2002 Inter-Agency Hague Code of Conduct against Ballistic Missile Proliferation	✓	✓	✓	✓	✓	✓	✓	✓	✓	✓	✓	✓	✓	✓	✓	✓			✓	✓
1996 Wassenaar Arrangement on Export Controls for Conventional Arms and Dual-Use Goods and Technologies	✓	✓	✓	✓	✓	✓	✓	✓	✓	✓	✓	✓	✓	✓	✓	✓			✓	✓
Missile Technology Control Regime	✓	✓	✓	✓		✓	✓	✓	✓	✓	✓	✓		✓	✓	✓			✓	✓
UN Conference on Disarmament	✓	✓			✓	✓	✓	✓	✓			✓		✓		✓			✓	✓

Table 3 Participation of ESA Member States to relevant international organizations, initiatives, and fora in space, security, and defense

	Austria	Belgium	Czech Republic	Denmark	Estonia	Finland	France	Germany	Greece	Hungary	Ireland	Italy	Luxembourg	The Netherlands	Norway	Poland	Portugal	Romania	Spain	Sweden	Switzerland	United Kingdom
Arctic council (https://arctic-council.org/en/about/)				✓		✓									✓					✓		
Barents Euro-Arctic Council - BEAC (https://www.barentscooperation.org/en)				✓		✓									✓					✓		
Black Sea economic cooperation – BSEC (http://www.bsec-organization.org/)									✓									✓				
Committee on Earth Observation Satellites – CEOS (http://ceos.org/about-ceos/agencies/)		✓					✓	✓				✓	✓	✓	✓				✓	✓	✓	✓
Coordination Group for Meteorological Satellites – CGMS (https://www.cgms-info.org/index_.php/cgms/index.html)							✓															
COSPAS-SARSAT (https://cospas-sarsat.int/en/about-us/participants)				✓		✓	✓	✓	✓			✓		✓	✓	✓			✓	✓	✓	✓

Organization																
European Organisation for the Exploitation of Meteorological Satellites – EUMETSAT (https://www.eumetsat.int/website/home/index.html)	✓	✓	✓	✓	✓	✓	✓	✓	✓	✓	✓	✓			✓	✓
Eurocontrol (https://www.eurocontrol.int/)	✓	✓	✓	✓	✓	✓	✓	✓	✓	✓	✓	✓		✓	✓	✓
European Agency for the Management of Operational Cooperation at the External Borders – Frontex (https://ec.europa.eu/home-affairs/e-library/glossary/european-agency-management_en)	✓	✓	✓	✓	✓	✓	✓	✓	✓	✓	✓	✓		✓	✓	✓
European Aviation Safety Agency – EASA (https://www.easa.europa.eu/)	✓	✓	✓	✓	✓	✓	✓	✓	✓	✓	✓	✓			✓	✓
European Centre for Medium-Range Weather Forecasts – ECMWF (https://www.ecmwf.int/)	✓	✓		✓	✓	✓	✓	✓	✓		✓	✓			✓	✓
European Defence Agency – EDA (https://www.eda.europa.eu/)	✓	✓	✓	✓	✓	✓	✓	✓	✓	✓	✓	✓	✓	✓		✓
European Economic Area – EEA (https://www.efta.int/eea/eea-agreement)	✓	✓	✓	✓	✓	✓	✓	✓	✓	✓	✓	✓	✓	✓	✓	✓
European Environment Agency – EEA (https://www.eea.europa.eu/countries-and-regions)	✓	✓	✓	✓	✓	✓	✓	✓	✓	✓	✓	✓				✓

(continued)

Table 3 (continued)

Organization	Austria	Belgium	Czech Republic	Denmark	Estonia	Finland	France	Germany	Greece	Hungary	Ireland	Italy	Luxembourg	The Netherlands	Norway	Poland	Portugal	Romania	Spain	Sweden	Switzerland	United Kingdom
European Free Trade Association – EFTA (https://www.efta.int/eea)															✓						✓	
European Gendarmerie Force (https://eurogendfor.org/)							✓												✓			
European Maritime Safety Agency – EMSA (http://www.emsa.europa.eu/about.html)		✓	✓	✓	✓	✓	✓	✓	✓	✓	✓	✓	✓	✓	✓	✓	✓	✓	✓	✓		✓
European Southern Observatory – ESO (https://www.eso.org/public/)	✓	✓	✓	✓		✓	✓	✓				✓		✓		✓	✓		✓	✓	✓	✓
European Space Agency – ESA	✓	✓	✓	✓	✓	✓	✓	✓	✓	✓	✓	✓	✓	✓	✓	✓	✓	✓	✓	✓	✓	✓
European Union – EU Member States	✓	✓	✓	✓	✓	✓	✓	✓	✓	✓	✓	✓	✓	✓		✓	✓	✓	✓	✓		✓
European Union Satellite Centre – Satcen (https://www.satcen.europa.eu/)	✓	✓		✓			✓	✓							✓	✓			✓	✓		✓
European Telecommunications Satellite Organization – EUTELSAT (https://www.eutelsat.int/en/about/member-states/)	✓	✓		✓		✓	✓	✓	✓	✓	✓	✓	✓	✓	✓	✓	✓	✓	✓	✓	✓	✓

23 Strategic Overview of European Space and Security Governance

Organization	C1	C2	C3	C4	C5	C6	C7	C8	C9	C10	C11	C12	C13	C14	C15
Five Eyes Community													✓	✓	✓
Fourteen Eyes Community – SSEUR		✓											✓	✓	
Franco-British Summits and Lancaster House Treaties (https://www.chathamhouse.org/sites/default/files/public/Research/International%20Security/0311pp_gomis.pdf)				✓											✓
Franco-German Summits and Security and Defence Council (CFADS) and Elysée Treaty (https://www.diplomatie.gouv.fr/en/country-files/germany/events/article/franco-german-defence-and-security-council-agreed-conclusions-16-oct-19)							✓		✓						
Franco-Italian Summits							✓	✓							
Group on Earth Observation – GEO (https://www.earthobservations.org/members.phpns)	✓	✓	✓	✓	✓	✓	✓	✓	✓	✓	✓		✓	✓	✓
Inter-Agency Space Debris Coordination Committee – IADC (https://www.iadc-home.org)				✓			✓								✓
International Civil Aviation Organization – ICAO (https://www.icao.int/about-icao/Pages/member-states.aspx)	✓		✓	✓	✓		✓	✓	✓	✓	✓	✓	✓		✓

(continued)

Table 3 (continued)

	Austria	Belgium	Czech Republic	Denmark	Estonia	Finland	France	Germany	Greece	Hungary	Ireland	Italy	Luxembourg	The Netherlands	Norway	Poland	Portugal	Romania	Spain	Sweden	Switzerland	United Kingdom
International Mobile Satellite Organization – IMSO (https://imso.org/member-states/)	✓	✓	✓	✓		✓	✓	✓	✓	✓		✓	✓	✓	✓	✓	✓	✓	✓	✓	✓	✓
International Space Environment Service – ISES (http://www.spaceweather.org)	✓	✓	✓																	✓		
International Telecommunications Satellite Organization – ITSO (https://itso.int)	✓	✓	✓	✓	✓	✓	✓	✓	✓	✓	✓	✓	✓	✓	✓	✓	✓	✓	✓		✓	✓
Nine Eyes Community				✓			✓							✓	✓							✓
Nordic Council (https://www.norden.org/en/nordic-council)				✓		✓									✓					✓		
Nordic Defence Cooperation – NORDEFCO (https://www.nordefco.org/default.aspx)				✓	✓	✓									✓					✓		
North Atlantic Treaty Organisation – NATO (https://www.nato.int/cps/en/natohq/topics_52044.htm?)		✓	✓	✓	✓		✓	✓	✓	✓		✓	✓	✓	✓	✓	✓	✓	✓			✓

Organisation															
Organisation for Joint Armament Cooperation – OCCAR (http://www.occar.int/about-us)	✓												✓	✓	
Organisation for Security and Cooperation in Europe – OSCE (https://www.osce.org/participating-states)	✓	✓		✓	✓		✓	✓	✓	✓	✓	✓	✓	✓	✓
Permanent and Structured Cooperation – PESCO (https://www.eda.europa.eu/what-we-do/our-current-priorities/permanent-structured-cooperation-(pesco))	✓	✓		✓	✓		✓	✓	✓	✓	✓	✓			
United Nations Committee on the Peaceful Uses of Outer Space – UN COPUOS (https://www.unoosa.org/oosa/en/ourwork/copuos/members/evolution.html)	✓	✓			✓		✓	✓	✓	✓	✓	✓	✓	✓	
United Nations International Maritime Organisation – IMO (http://www.imo.org/en/About/Membership/Pages/MemberStates.aspx)	✓	✓		✓	✓		✓	✓	✓	✓	✓	✓	✓	✓	

(continued)

Table 3 (continued)

	Austria	Belgium	Czech Republic	Denmark	Estonia	Finland	France	Germany	Greece	Hungary	Ireland	Italy	Luxembourg	The Netherlands	Norway	Poland	Portugal	Romania	Spain	Sweden	Switzerland	United Kingdom
United Nations International Telecommunications Union – ITU (https://www.itu.int/en/ITU-R/terrestrial/fmd/Pages/administrations_members.aspx)	✓	✓	✓	✓	✓	✓	✓	✓	✓	✓	✓	✓	✓	✓	✓	✓	✓	✓	✓	✓	✓	✓
United Nations World Meteorological Organisation – WMO (https://public.wmo.int/en/about-us/who-we-are)	✓	✓	✓	✓	✓	✓	✓	✓	✓	✓	✓	✓	✓	✓	✓	✓	✓	✓	✓	✓	✓	✓
Visegrad Group			✓							✓						✓						

Union Aviation Safety Agency (EASA), and Eurocontrol, the national agency for civil aviation is the most relevant.

In addition to international cooperation in space and security effected through representation in international organizations, there is a considerable number of initiatives and cooperation taking place at the bilateral and regional levels as seen in Table 3. In this context, and in order to exemplify what types of cooperation take place between European countries, this chapter will take a closer look at a few of these international initiatives and cooperation *fora*. First the Franco-German and Franco-Italian cooperation are addressed as examples of an integrated approach to space, security, and defense at the bilateral level.

The Treaty on Franco-German Friendship (known as the Élysée Treaty), signed on January 22, 1963, by German Federal Chancellor Adenauer and France's General de Gaulle, is the symbol of the relationship forged between France and Germany (France Diplomatie Website). More recently, in the margins of the 19th Franco-German Ministerial Council that took place in July 2017, the French and German Heads of State agreed in the Franco-German Security and Defence Council (CFADS). Through this, they render available the satellite imagery from the French high resolution *Composante Spatiale Optique* (CSO) and the German SARah reconnaissance satellites to the EU Satellite Centre (Satcen) and open the door to partnership with other EU member states. Moreover, France and Germany coordinate their efforts in military Earth Observation. Both countries have called for high security requirements for Galileo, with the objective of ensuring European strategic autonomy for military applications and in order to enhance the international credibility of Galileo (Defense aerospace 2017). During the Meseberg Summit on June 19, 2018 (20th Franco-German Ministerial Council), France and Germany agreed to launch a working group with the task of making reports for the EU that address new challenges in space politics and economics (NewSpace in particular). In the field of launchers, the countries reaffirmed their full support for the ESA Ariane 6 program (Meseberg 2018).

France and Italy enjoy active space collaboration, as reflected by annual summits organized since 1982. In the September 2017 Franco-Italian summit, both countries committed to reinforce their cooperation in space, for example through the development of dual-use technologies and through the identification of opportunities for the next generation of Earth observation and satellite communication satellites. It was agreed to put CSO and COSMO Skymed imagery at the disposal of the European External Action Service (EEAS). Cooperation on Multinational Space-based Imaging System (MUSIS) also continues. France and Italy will update the intergovernmental agreement with which they joined activities since 2007. The governments have mandated CNES and ASI, respectively (Elysee 2017). In addition, the third phase of the SICRAL (*Sistema italiano per communicazioni riservate e allarmi*) is conducted in cooperation between Italy and France. SICRAL became operational after April 2015 with the launch of SICRAL 2, with an estimated operational life span of 15 years. SICRAL 2 is a geostationary satellite able to enhance the capability of military satellite communications already offered by SICRAL 1 and SICRAL 1B and by France's Syracuse System. SICRAL 2 supports satellite communications for the Italian and French Armed Forces, anticipating the needs of growth and development in the next few years. The satellite has an additional backup function to the French Syracuse 3 system and that of SICRAL 1B allocated to NATO communications (Telespazio Website).

Furthermore, there are a number of regional cooperation initiatives that deal with space and defense in Europe. For example, the Five Eyes community (FVEY) brings the United Kingdom, the United States, Canada, Australia, and New Zealand into the world's most complete and comprehensive intelligence alliance. The alliance is also involved in the "aerospace domain" which covers ballistic missile tests, foreign satellite deployments, and the military activities of relevant air forces (UK Defence Journal 2017). In addition, during World War II, the USA developed foreign SIGINT relationships with the United Kingdom and the Dominions of Canada, Australia, and New Zealand. FVEY was originally founded through the signature of the Britain–United States of America (BRUSA) agreement for co-operation in Signals Intelligence (SIGINT), now known as the United Kingdom–United States of America (UKUSA) Agreement, on March 5, 1946. SIGINT capability is space-based, including satellites used for military missions such as earth observation and reconnaissance satellites, electronic signals intelligence satellites, and civil and military communications satellites (Pfluke 2019). The Agreement consolidated the Special Relationship between Britain and the United States. FVEY expanded to also include Canada in 1948 and Australia and New Zealand in 1956. The SIGINT surveillance program was originally known as ECHELON (UK Defence Journal 2017; Lawfare 2019). The FVEY co-operation with Denmark, France, Norway, and the Netherlands is known under the name "Nine Eyes." Additionally, there is the SIGINT Seniors Europe (SSEUR) also known as "Fourteen Eyes" which consists of "Nine Eyes" plus Belgium, Germany, Italy, Spain, and Sweden. SSEUR's primary objective is to coordinate the exchange of military signals amongst its members. NATO Members have additional intelligence cooperation links (UK Defence Journal 2017). Furthermore, the Berne Club is a cooperation framework among Western European internal security services. It is based on periodic meetings attended by the heads of the European Intelligence Services. The Berne Club operates in an informal way (Nomikos 2014).

Archetype Model for National Space and Security Governance

Based on the research on the selected ESA Member States, the organization of space- and security-related activities in countries is multifaceted. The governance of space and security mainly depends on three factors: (1) the applications of space-based assets and services in the fields of Earth observation (EO), Positioning, Navigation and Timing (PNT), Satellite Communications (SATCOM), Space Situational Awareness (SSA), (2) the involvement of a wide variety of involved Ministries based on national competences related to space and security, and (3) the level of programs at the national, multilateral, and intergovernmental levels. National space strategies contribute to this multifaceted nature of space- and security-related activities. Security and Defence applications of space activities (security *in* space and security *from* space) are becoming increasingly important. Both the incorporation of security and defense as a priority in national space strategies and the development of dedicated space security or space defense strategies reflect the increase of this importance. As a result of the increasing importance or security and defense in space, the role of national ministries and agencies, tasked with security and defense activities in national space governance, is gradually increasing. Besides the ministries that traditionally held responsibility for space activities (most often the Ministry of Science, Education and Research or the Ministry of Economy and/or Industry and Innovation), the Ministries of Defence, Foreign Affairs, Home Affairs, Justice and/or Security, and the Prime Minister's Office or Chancellery have an increasingly important role in space governance at all levels.

Figure 2 provides an archetype of space and security governance. It was generated based on recurrent patterns, correlation between ministries, implementing entities, and respective space-related competences of all governance schemes for each ESA Member State and the linkages to space and security relevant organizations. The model reflects the relevance of security and defense actors in national space governance and the observed role of specific Ministries and Implementing Agencies (governmental bodies) in the definition of national space strategy and policy and in the programmatic implementation of space activities. Although space and security governance is inherently unique for each member state, the model in Fig. 2 shows which entities are typically involved in the space and security governance landscape. Because of the increasing role of security and defense applications in space, the Ministry of Defence, the Ministry of Home Affairs, and the Ministry of Foreign Affairs play a crucial role. Their respective roles also imply increasing participation of the national defense procurement agency, the military general staff, defense research agencies, and national intelligence services, both internal, foreign, and military. National civil protection and crisis management services can increasingly benefit from the use of space-based assets and services. The Ministry of Foreign Affairs plays an indispensable role in the context of bilateral, multilateral and supranational space programs (including ESA, EU, and NATO). Although it is primarily the national space agencies that represent their member states at/in ESA, the countries' Ministries of Foreign Affairs are becoming increasingly active (directly) at the ESA Council level and delegate body level. This is because of the increasing strategic importance of space- and security-related activities. As a matter of fact, in an increasing number of countries it is the highest level of government that is involved in space activities, as reflected by the role of the Prime Minister's Office, the national Chancellery, the Council of Ministers, or the Ministry of General Affairs.

As space-based assets and services can be of great help in combating climate change, in furthering sustainable development and in protecting the environment, most member states directly involve the ministries responsible for those competences in the space decision making and implementing landscape. Very often (but no exclusively) the same ministries are also charged with supervision of the national meteorological institute. The national meteorological institutes depend on space-based assets and services and as such often represent their country in the frame of EUMETSAT and ECMWF. The ministry or ministries competent for transport,

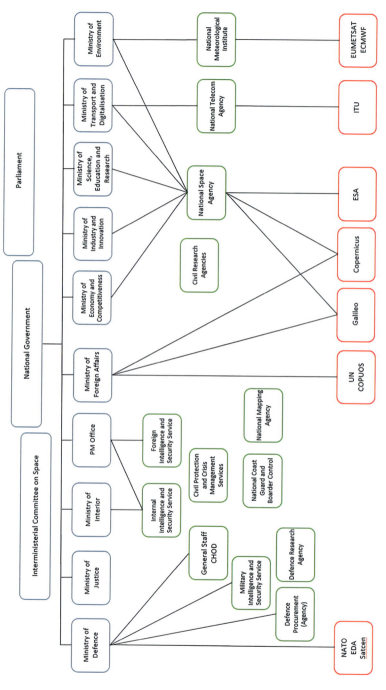

Fig. 2 Archetype space and security governance model (generated based on recurrent patterns of all governance schemes of ESA Member States)

infrastructure, telecom, and/or digital affairs are also increasingly active in space activities. The national telecom agencies, acting under the respective responsible ministry, often represent their country in the frame of the International Telecommunications Union. Space will continue to be an important economic sector, not in the least because of its highly skilled workforce and the innovation that takes place. Depending on the member state, one or more Ministries that are tasked with economic and trade-related subjects, such as competitiveness, enterprise, industry, technology, and/or innovation are therefore directly involved in developing and executing space activities. The Ministries of Science, Education and Research have been involved in space affairs from the very onset of space activities in Europe. For that reason, national space agencies have often been under the direct supervision of the ministry or ministries charged with science, education, and research.

Responsible space entities increasingly respond to the needs of multiple ministries. Multiple funding sources come from different Ministries, increasing the need for coordination of space affairs at the level of the Prime Minister's Office, Chancellery, or Council of Ministers, or the joint supervision by multiple Ministries. This reflects the diversity of Ministerial interests in space activities and applications. Various national (civil or dual-use) research entities are involved in space affairs, as such complementing the role of national space entities. The national mapping entities (or national geo-data or geo-information agencies or cadaster agencies or geographical institutes) increasingly depend on space-based data and provide crucial services for a multitude of Ministries and governmental bodies. In the frame of national security, coast guards and national border protection agencies also increasingly rely on space-based services in the execution of their mandate.

Concluding Remarks

Europe has been taking a more active stance in the field of foreign affairs and security and defense. Member states are increasingly recognizing the need for joint action to address societal challenges together, including in the field of security and defense. Space integration in Europe for civilian uses has a long track record and has demonstrated tangible successes in the frame of ESA and the EU. Contrary to civil space activities, security-related space activities traditionally took place at the national level, or via bi- or multilateral basis. The increasing security challenges the European countries are facing, together with the political momentum favorable for advancing the EU role in the field of security and defense, may lead to further integration covering the entire dual-use spectrum of space activities.

This chapter provides evidence of the increasing importance for security and defense in national space strategy and policy, with an emphasis on the role of international cooperation. This trend runs parallel with an increasing interest in space affairs of Ministries with security- and/or defense-related competences. The participation of the various ministries in space- and security-related international organizations shows that Ministries of Defence are becoming increasingly important stakeholders for national space strategy and resulting programmatic activities. The increasing use of space-based services for civil protection, crisis management, and border control further strengthen the need for participation of the Ministry or Ministries responsible for Home Affairs, Justice and/or Security in the frame of national space policy, governance, and programs. Ministries charged with Environment, Sustainable Development and Climate Change as well as Ministries charged with Transport, Infrastructure, Development, Telecom and Digital Affairs are important user Ministries, fully benefiting from space-based services and applications. Because of the strategic importance of space, security, and defense, the highest national political level is directly involved in space affairs through the PM Office, Chancellery, and/or Council of Ministers.

Disclaimer The contents of this chapter and any contributions to the Handbook reflect personal opinions and do not necessarily reflect the opinion of the European Space Agency (ESA).

References

Altalex (2018) Misure per il coordinamento della politica spaziale e aerospaziale e disposizioni concernenti l'organizzazione e il funzionamento dell'Agenzia spaziale italiana, Legge, 11/01/2018 n° 7, G.U. 10/02/2018, 14 February 2018. Available at: http://www.altalex.com/documents/leggi/2018/02/14/misure-per-il-coordinamento-dellapolitica-spaziale

Antoni N, Adriaensen M, Papadimitriou A, Giannopapa C, Schrogl KU (2018) Re-affirming Europe's ambitions in space: past, present and future perspectives. Acta Astronaut 151:772–778

BMVI (2017) The Copernicus Strategy of the German Federal Government Copernicus for Germany and Europe –the Federal Government's strategy and fields of action for successful implementation of the European Earth Observation Programme, 13 September 2017

CDTI (2011) Spanish strategy for ESA and EU Space programmes "HISNORSAT: encuentro de la industria especial Hispano-Noruega" Jorge Lomba, Jefe del Departamento de industria de la ciencia y el espacio, 3 May 2011. Available at: https://www.cdti.es/recursos/doc/Programas/Aeronautica_espacio_retornos_industriales/Espacio/29437_101110112011135137.pdf

CDTI (2018) Perspectiva CDTI, ESPACIO|Los datos europeos recibidos en el #S3TOC de @CDTIoficial predicen que la reentrada de la estación orbital china #TIANGONG-1 se producirá el 1 de abril de 2018, 28 March 2018. Available at: http://perspectivacdti.es/espacio-los-datos-europeos-recibidos-en-el-s3toc-de-cdtioficial-predicen-que-lareentrada-de-la-estacion-orbital-china-tiangong-1-se-producira-el-1-de-abril-de-2018/

CDTI Website. About CDTI, organizational structure, Comité de Dirección Available at: https://www.cdti.es/index.asp?MP=6&MS=801&MN=3#Juan%20Carlos%20Cort%C3%A9s

CNES (2017) CNES organisation chart, 30 November 2017. Available at: https://cnes.fr/en/cnes-organization-chart

CNES Website. Organigramme du CNES. Available at: https://cnes.fr/fr/organigramme-du-cnes

Defense aerospace (2017) Meeting of the French-German Defense and Security Council: conclusions (excerpt), 14 July 2017. Available at: http://www.defense-aerospace.com/articles-view/verbatim/4/185306/conclusions-of-franco_germandefense-council.html

DLR Website. DLR space administration. Available at: http://www.dlr.de/rd/en/desktopdefault.aspx/tabid-2099/3053_read-4706/

Elysee (2017) Sommet Franco-Italien, 27 September 2017, Lyon. Available at: http://www.elysee.fr/assets/SommetFrancoItalien-FR.pdf

ESA (2017) European Space Agency Council, ESA Member States Strategies and Plans – Countries Overview Information Document, Paris 12 October 2017

France Diplomatie Website. France and Germany, Elysée Treaty. Available at: https://www.diplomatie.gouv.fr/en/country-files/germany/france-and-germany/elysee-treaty/

French Ministry of the Armed Forces (2017) Ministère des Armées, Organigramme simplifié du ministère des Armées, 7 June 2017. Available at: http://www.defense.gouv.fr/portail/ministere/organisation-du-ministere-des-armees/organisation-duministere-des-armees/organigramme-simplifie-du-ministere-des-armees/organigramme-simplifie-duministere-des-armees

French Ministry of the Armed Forces (2018) Ministère des Armées, L'organigramme de la DGA, 19 January 2018. Available at: https://www.defense.gouv.fr/dga/ladga2/organisation/l-organigramme-de-la-dga

Giannopapa C, Adriaensen M, Antoni N, Schrogl KU (2018) Elements of ESA's policy on space and security. Acta Astronaut 147:346–349

Governance Post (2017) After the Election, Germany's Space Policy Must Take Shape, 13 December 2017. Available at: https://www.hertie-school.org/the-governance-post/2017/12/election-germanys-space-policy-must-takeshape/

GPE Website. Le GPE. Available at: http://www.gpespace.fr/spip.php?article1

IAI (2003) Space and Security Policy in Europe, November 2003, ESA Study. Available at: http://www.iai.it/sites/default/files/2003_space-and-security-in-europe.pdf

Infoespacial (2015) Hacia la creación de una Agencia Espacial Española, 15 December 2015. Available at: http://www.infoespacial.com/ie/2015/12/15/opinion-agenda-sector-espacial-espana.php

Lawfare (2019) Newly Disclosed NSA Documents Shed Further Light on Five Eyes Alliance, 25 March 2019. Available at: https://www.lawfareblog.com/newly-disclosed-nsa-documents-shed-further-light-five-eyes-alliance

Legifrance (1989) Décret n°89-508 du 19 juillet 1989 portant création du comité de l'espace, https://www.legifrance.gouv.fr/affichTexte.do;jsessionid=83D1F8EC5C8C56E062B8C8C80DB63E0A.tpdjo12v_1?cidTexte=JORFTEXT000000699675&idArticle=&dateTexte=20101206

Meseberg (2018) Meseberg Declaration, Renewing Europe's promises of security and prosperity, 19 June 2018. Available at: https://archiv.bundesregierung.de/archiv-de/meta/startseite/meseberg-declaration-1140806

Nomikos M (2014) European Union intelligence analysis Centre (INTCEN): next stop to an agency. J Mediterr Balkan Intell 4(2)., December 2014

Pfluke C (2019) A history of the five eyes Alliance: possibility for reform and additions. Comp Strateg 38(4):302–315

Spanish Government (2018) Official State Journal, 7 June 2018. Available at: http://www.boe.es/boe/dias/2018/06/07/pdfs/BOE-A-2018-7575.pdf

Spanish Ministry of Defence (2015) Ministerio de Defensa, D.G.A.M. Plan Director de Sistemas Espaciales, p.5, July 2015. Available at: http://www.defensa.gob.es/Galerias/dgamdocs/plan-director-sistemas-espaciales.pdf

Spanish Ministry of Defence (2017) Programa SST/S3T, October 2017. Available at: http://www.defensa.gob.es/Galerias/dgamdocs/programa-SST_S3T.pdf

Telespazio Website. SICRAL. Available at: http://www.telespazio.com/programmes-programmi/sicral

UK Defence Journal (2017) UKDJ, The Five Eyes – The Intelligence Alliance of the Anglosphere, 14 November 2017. Available at: https://ukdefencejournal.org.uk/the-five-eyes-the-intelligence-alliance-of-the-anglosphere/

UK Government (2015) National Space Policy, December 2015. Available at: https://assets.publishing.service.gov.uk/government/uploads/system/uploads/attachment_data/file/484864/NSP_-_Final.pdf

UK Government (2018) Legislation – Space Industry Act 2018. Available at: http://www.legislation.gov.uk/ukpga/2018/5/contents/enacted/data.htm

UK Ministry of Defence, Joint Doctrine Publication 0–30 (JDP 0–30) – UK Air and Space Power, 2nd edn, December 2017, 113. Available at: https://assets.publishing.service.gov.uk/government/uploads/system/uploads/attachment_data/file/668710/doctrine_uk_air_space_power_jdp_0_30.pdf

European Space Security Policy: A Cooperation Challenge for Europe

24

Jean-Jacques Tortora and Sebastien Moranta

Contents

Introduction	450
Prospects for a European Space Security Policy	452
Stakes Are High for Europe and Will Continue to Increase	452
Socioeconomic Rationale and Service-Oriented Policy	452
European Autonomy and Weight on the International Scene	453
Europe Is Mobilizing But Follows an Approach That Is Called to Evolve	456
Accommodating Different Concerns and Interests: Achievements and Limits	459
Space Traffic Management: A Coordination and Leadership Challenge	461
Conclusions	463
References	464

Abstract

Experts consensually agree that ensuring the security, safety, sustainability, and stability of space activities is a growing challenge and therefore a dominant concern for public and private stakeholders worldwide. At the crossroad of multiple strategic, political, and operational issues, space security also holds a special place in the European context. This chapter reviews major policy stakes for Europe in this field and discusses the achievements and limits of the current approach to space security in Europe. In particular, it examines some recent developments in the space security field such as the major steps announced by the US government in the field of space traffic management and its potential consequences on the ongoing progress of the European agenda in this matter.

J.-J. Tortora · S. Moranta (✉)
European Space Policy Institute (ESPI), Vienna, Austria
e-mail: Jean-Jacques.Tortora@espi.or.at; sebastien.moranta@espi.or.at

© Springer Nature Switzerland AG 2020
K.-U. Schrogl (ed.), *Handbook of Space Security*,
https://doi.org/10.1007/978-3-030-23210-8_127

Introduction

Although sometimes perceived as slow paced and resistant to change, space is actually a fast-changing sector, both at the industry and policy level. Shaped by 60 years of constant governmental efforts in a tense geopolitical context, space has expanded toward privately owned initiatives to become not only a decisive item on governments' agenda but also an area of interest for private companies, entrepreneurs, and investors. In this sense, "New Space," which is usually presented as a sudden change of paradigm, shaking the foundations of a long-established and rigid industry, can rather be seen as a natural step in the evolution (maturation) of a budding sector. Notwithstanding, the rise of new actors and innovative ways of conducting space activities that is currently witnessed is undeniably opening new, sometimes unforeseen, prospects for the sector. These opportunities trigger, in turn, the emergence of new, and somewhat unexpected, challenges.

A top transverse issue in this new space era is, and will increasingly be, to continue ensuring the security, safety, sustainability, and stability (4S) of space activities and to safeguard our very capacity to deploy and operate systems in space. Indeed, space systems are exposed to an increasing level of "man-made" threats in addition to a naturally hazardous space environment (e.g., geomagnetic storms, solar radiations, etc.). This includes both unintentional hazards stemming from human activity (e.g., debris, interferences, etc.) and capacities to deliberately disrupt space systems or services (e.g., anti-satellite technologies, signal jamming, cyberattacks, etc.). Overall, an assessment of space security threats shows that they are (ESPI 2018):

- *Multiple and diverse* in nature and origin and, as a consequence, require a set of different mitigation and protection measures
- *Interrelated and interdependent* and therefore require a coherent and holistic approach adhered to by all space stakeholders
- *Ubiquitous and inclusive,* although some systems are less exposed or vulnerable to specific threats
- *Intensifying,* driven by endogenous and exogenous trends including a growing space activity, an increasing number of governmental and commercial actors, the emergence of new concepts, technologies and capabilities (e.g., mega-constellations, CubeSats, on-orbit services, etc.), an ever more connected and critical space infrastructure (i.e., making it a strategic target), or the rehabilitation of a "space warfare" doctrine encompassing activities to develop "space control" capabilities

As summarized in the US National Security Space Strategy in 2011, "space, a domain that no nation owns but on which all rely, is becoming increasingly congested, contested, and competitive" (U.S. Department of Defence 2011). This situation, expected to further deteriorate in the future, should be appraised in the context of a growing strategic and socioeconomic significance of space. As the use of space applications becomes more pervasive, brings more benefits, and becomes part of the business-as-usual routine in many sectors, our dependence on space intensifies, which creates new vulnerabilities for the economy and society at large.

This deteriorating situation has already been widely acknowledged by European stakeholders including national governments, European institutions, and the private industry. The recognition of strategic interests in operating space systems in a secure, safe, and sustainable environment is now well integrated in key political documents. The Space Strategy for Europe clearly puts forward, "Europe's autonomy in accessing and using space in a secure and safe environment" as a top priority European Commission 2016.

To date, the European approach to space security builds upon a multilayered structure of diverse frameworks and activities. Organized at national, intergovernmental, and European level, institutional efforts span across the spectrum of capacity-building programs, legal and regulatory measures, and diplomatic initiatives. Public undertakings are complemented by industry-led endeavors such as the Space Data Association which brings together satellite operators worldwide "to enhance safety of flight via sharing of operational data and promotion of best practices across the industry" (Space Data Association).

Despite multiple initiatives and noticeable achievements, European ambitions remain somewhat limited in comparison with other space-faring nations, especially the United States. In the field of Space Surveillance and Tracking (SST), for example, European resources for capacity building, from system development to service delivery, are 10–20 times lower than those available across the Atlantic for comparable activities. Here, differences in space budgets do not explain, alone, such a massive gap. Across the Atlantic, securing space assets has long been established as a strategic priority. The launch of a $1.6 billion US Space Fence program aiming to further increase domestic Space Situational Awareness (SSA) capabilities and the recent publication of a National Space Traffic Management (STM) policy reminded the strong will of the US government to lead (or even rule) in this domain.

Along such recent developments, in the United States as well as on international scene, European policy-makers are realizing the criticality and urgency of setting up a more ambitious framework supported by adequate means commensurate to the level of challenges at stake. This already translated into an increase of the budget allocated by the EU to a European SSA program for the period 2021–2027 and into additional investments by France, among others, to support the development of national SST capabilities as part of a more ambitious *Stratégie Spatiale de Défense* adopted in July 2019 (Ministère des Armées 2019). Along these lines, more developments can be expected over the coming years including revised upward objectives and new support funds (i.e., European Defence Fund, Horizon Europe, ESA and national budgets, etc.).

While European resources for space security-related activities are likely to substantially increase over the next period, some structural issues may persist, in particular for what concerns coordination and leadership. Further progress in this direction is complex since it implies finding a new, suitable, and acceptable compromise on an appropriate balance between the need to preserve national sovereignty over sensitive matters and ways to limit duplication of efforts under a TBD European leadership. Difficulties in this field are closely related to the overarching challenges of the European construction but also intimately linked to the multifaceted nature (individual/common, military/civil, institutional/commercial, etc.) which brings together various communities having different objectives and requirements to converge on joint solutions.

Nonetheless, stakes are high for Europe.

This chapter examines how the current European model successfully accommodated the interests of multiple stakeholders but may progressively reach some limits that could affect the place of Europe on the international scene. With a forward-looking approach, the chapter also discusses how some expected developments could impact the relevance and suitability of this model and shape its further evolution.

Prospects for a European Space Security Policy

Stakes Are High for Europe and Will Continue to Increase

The strategic significance of space security for Europe arises from its overarching ambition to "promote its position as a leader in space, increase its share on the world space markets, and seize the benefits and opportunities offered by space" (European Commission 2016). Achieving these objectives has a number of security-related implications.

In a recent study, the European Space Policy Institute suggested that public action in this domain is justified by four key rationales (ESPI 2018):

- *Secure the results of the continuous and substantial investment* made by public and private actors.
- *Protect the European economy and society* against risks related to its pervasive and sizeable dependence on the space infrastructure.
- *Contribute to a service-oriented policy* by assuring the ability of the infrastructure to deliver a service that can justifiably be trusted, in particular for users in defense and security.
- *Guarantee European autonomy and freedom of action* in the field of security in outer space and in the space domain at large.

Socioeconomic Rationale and Service-Oriented Policy

Multiple studies and case studies demonstrated that space applications bring substantial socioeconomic benefits across numerous sectors (European Commission 2017).There is, however, a negative corollary to this: as the use of space-based solutions becomes more pervasive and part of business-as-usual, the dependence of governments, businesses, and consumers on space deepens, creating new risks. A study of the European

Commission estimated that an incapacitation of space systems – intentional or not – would lead to a significant economic loss of up to EUR 50 billion per year of gross added value and put up to 1 million jobs at risk in Europe (Ibid.). This is comparable, in size, to the entire European air transport sector. Furthermore, with the expected use of space capabilities (e.g., PNT signals, bandwidth, EO data) for new promising terrestrial technologies (e.g., autonomous vehicles, smart cities, precision agriculture), these risks will most likely increase in the future.

The criticality of mitigating the risks associated with a potential outage of space systems took a new meaning in Europe over the 2014–2020 period. With EU space infrastructure (i.e., Galileo, EGNOS, and Copernicus) fully operational and planning to complement these programs with a new initiative in the field of secure governmental communication (i.e., GOVSATCOM), the next Multiannual Financial Framework (MFF) (2021–2027) will give a high priority to exploitation, maintenance, and upgrade of space systems but also to the promotion of the services uptake in order to maximize the benefits of public investment in these programs. This entails the adoption of a service-oriented approach to build up user confidence through (1) a proven or certified level of performance, (2) the long-term availability of services, and (3) a service that can justifiably be trusted. This last condition translates into the need to take appropriate measures to protect the infrastructure against faults and threats. This requirement is a prerequisite for governmental and defense users that the EU seeks to support to reinforce and leverage synergies between civil and defense-oriented applications.

In short, the more Europe invests in and benefits from space, the more critical its responsibility to safeguard the space environment and infrastructure gets.

The socioeconomic rationale and service-oriented policy are already reasonable arguments for Europe to position security high in the space policy agenda. It can be legitimately expected that both of these drivers will further gain in prominence in the European space policy debate over the next Multiannual Financial Framework (MFF) period. As a means of illustration, the recent partial outage of the Galileo system brought to the front the sensitivity of space infrastructure security. Consequences of the outage were limited since – it must be stressed – the Galileo system is still under deployment. However, the event gave an early notion of the potential crisis that might have emerged in case of an unintentional, indelicate, or deliberate service disruption and underlined the importance of preventive measures to mitigate such risks.

European Autonomy and Weight on the International Scene

Eventually, ensuring space security (i.e., mitigating risks associated with the exploitation of space systems) requires a variety of measures targeted to (1) mitigate threats to space systems on one hand and (2) reduce vulnerability of space systems on the other (see Fig. 1) (Note: Complementary measures targeting directly the source of the threats such as actions against cybercrime, space disarmament policies or radio spectrum management are not included here.):

Fig. 1 Space security concepts and the relationship between them

As a consequence, it is clear that space security is a broad domain that encompasses many complementary areas and in particular:

- *Space Situational Awareness (SSA)*: Current and predictive knowledge and understanding of the outer space environment including space weather and location of natural and man-made objects in orbit around the Earth
- *Space Environment Protection and Preservation (SEPP)*: Preventive and curative mitigation of the negative effects of human activity in outer space on the safety and sustainability of the outer space environment
- *Space Infrastructure Security (SIS)*: Assurance of infrastructure ability to deliver a service that can justifiably be trusted despite a hazardous environment (i.e., security-by-design, security in operation)
- *Space Traffic Management (STM)*: Planning, coordination, and on-orbit synchronization of activities to enhance the safety, stability, and sustainability of operations in the space environment. (Note: As defined by the U.S. Space Policy Directive 3 on a National Space Traffic Management. Alternative, broader definition of STM exist such as the one from the IAA which describes STM as a wide encompassing concept (i.e. "the set of technical and regulatory provisions for promoting safe access into outer space, operations in outer space and return from outer space to Earth free from physical or radio-frequency damage").)

Despite various past, current, and planned initiatives and measures across the board, Europe has not yet established a leadership in this field to express its concerns and represent its interests on the international scene and instead finds itself increasingly pressured by the proactivity of some other major players.

In the SSA domain, for example, a considerable share of European actors relies on data sharing agreements signed with the United States. This includes national ministries or armies (i.e., Belgium, Denmark, France, Germany, Italy, the Netherlands, Norway, Spain, the United Kingdom), European intergovernmental organizations (i.e., ESA, EUMETSAT), and commercial satellite operators and launch service providers. From a US perspective, SSA sharing agreements are powerful leadership instruments that aim to support transparency on operations in outer space, promote cooperation for security and safety, enhance the availability of information among the partners, and improve the quality of US SSA information (Helms 2010; USSTRATCOM Public Affairs 2017). From a European perspective, they are now a critical input for SSA capabilities at large and, by extension, for the safe operation of space systems.

No need to remind here how much Europe greatly benefits from the open policy of the US government. The importance of transatlantic cooperation, in particular in defense and security domains, cannot be challenged. However, the gap between European and American capabilities, expected to increase with the deployment of the US Space Fence, creates a situation of reliance/dependence for European stakeholders and a major imbalance in cooperative arrangements. This state of affairs is in many respects beneficial for Europe but has also some strategic implications:

- *SSA data and service restrictions:* Although SSA data sharing agreements enhance the availability of information among the partners, restrictions exist. Because of its intrinsic military nature, a lack of transparency or a delay in information provision can occur for a variety of motives related to US national security. (Note: Specifically, the DoD resistance to open SSA data sets, algorithms, and processes to external review and scrutiny results in the uncertainty of the data and in possible false positive. See: https://www.ida.org/idamedia/Corporate/Files/Publications/STPIPubs/2016/P-8038.ashx.)
- *Reliability and accountability:* Although the most advanced worldwide, the US system is not flawless and may provide wrong information due to measurements or processing errors, in particular for smaller objects that Europe cannot track (Froeliger 2017). If European operators are blind, not being able to verify the data, they find themselves exposed to single points of failure.
- *Uncertainty on future access:* The US Government holds the right to terminate the agreement at any time for any reason, to limit both access duration and data amount, to deny access to SSA data and information, and to change or modify the terms and conditions at any time, and without prior notification (Space-Track).

From this perspective, although access to foreign SSA data and services is a relevant and effective way to augment domestic SSA capabilities, such access, even deemed unrestricted, free, and guaranteed, shall not sustainably remain a critical input on which they depend. For this reason and comparably to other fields such as access to space or critical technologies, Europe seeks to enhance its strategic autonomy in the SST/SSA domain, which is a pillar to establish any leadership in the space security domain. The objective to reach an "appropriate level of European autonomy" is addressed in most, if not all, policy documents setting the route for Europe in this matter. There is, however, no clear definition of a minimum, required, level of capabilities that would be strategically acceptable. Member States may also have different views on this question.

Overall, European stakeholders converge on the assessment that a fully effective approach to space security can only be envisioned as the outcome of a coherent and inclusive global effort and that cooperation with third countries, in particular the United States, is essential for many reasons. Autonomy is therefore not sought at the expense of cooperation with key partners, but Europe shall ensure its capacity to control its level of reliance on third parties and to maintain it within acceptable boundaries.

Since space security will constitute an increasing and central share of the so-called space diplomacy, it can be reasonably argued that guaranteeing European autonomy and freedom of action in this field is a prerequisite to establish Europe's credibility and legitimacy in the global space arena. Europe's ambition to *promote its position as a leader in space* necessarily entails to play a central role (even as initiator) in international dialogues and negotiations as a promoter of a clear, united, and consistent "European way." From a more practical standpoint, to be positioned as a key player on the international space scene, Europe must contribute its share to this endeavor and ensure a balanced cooperation with other key players including, prominently, the United States.

Europe Is Mobilizing But Follows an Approach That Is Called to Evolve

Space Security in Europe: A Multilayered Framework

Space security is a complex issue because of the diversity of potential threats and mitigation or remediation measures that, in one way or another, impact all actors in space. As far as Europe is concerned, the multiplicity of stakeholders brings an additional layer of complexity.

In Europe, space security activities are primarily led and organized at national level, comparably to other security and defense domains:

- *National governments are the owners and operators of the main European SST systems.* These systems are operated in different ways according to the nature of ownership (civil/military) and to their purpose (scientific/operational). The diverse data they produce are also handled according to different protocols and data policy.
- *National space laws govern activities in space* by setting the requirements, conditions, and restrictions for licenses authorizing organizations to conduct launch and space operations. Some of these legal regimes contain provisions that actively contribute to space environment protection and preservation through obligations in the domain of space objects registration, space debris mitigation, or space systems reentry (Froehlich and Seffinga 2018).
- *National delegations hold voting rights in international diplomatic frameworks* such as the United Nations Committee on the Peaceful Uses of Outer Space (UN COPUOS) and the Conference on Disarmament (CD).

Despite a convergence of interests between different policy domains in the field of space security, the military and national security dimension remains prevalent at national level. (Note: The cross-domain aspect of space security is best illustrated by the United Kingdom's National Space Security Policy (2014), which resulted from cooperation between the Ministry for Universities and Science, the Ministry for Defence Equipment, Support and Technology, the Ministry of State, and the Ministry for Immigration and Security.) Consequently, the most active countries in military space programs (i.e., France, Germany, Italy, the United Kingdom, Spain) are also the most active in space security. These countries may have different objectives, although overall convergent, that are set by national policy documents (space policy, defense policy). Different concerns and approaches gave rise to different governance models involving several ministries and organizations concerned by space security matters among which, of course, national space agencies. Efforts also remain uneven among the most proactive countries, in particular in the field of SST capacity building.

Although national sovereignty remains, so far, a structural component of space security in Europe, national action in this field has been flanked, since long, with the development of bi- and multilateral agreements for data sharing, resources pooling, operational coordination, or diplomatic collaboration among European countries. Dispositions to cooperate stem from a programmatic and geopolitical environment facilitating and promoting cooperation between European countries (i.e., European Union, ESA, etc.) but also from practical considerations concerning the capacity of each country to achieve, on its own, objectives requiring considerable resources. From this standpoint, national strategies underline a growing readiness and willingness to build on European cooperation in the field of space security. The involvement of European institutions in space security, namely, the European Union and the European Space Agency, obviously contributed to give raise to a new dimension in European cooperation. Building on bilateral and pan-European cooperation, the French *Stratégie Spatiale de Défense* declares in this respect that "beyond existing projects, a Europe of space must emerge to contribute directly to the construction of European security and defense on the continent. To do so the Franco-German motor must federate energies, in particular in the frame of a European SSA project" (Ministère des Armées 2019).

In this context, the role of the European Union in space security has grown within a broader and more political framework and at the crossroad between developments of the EU mandate and ambitions in the space domain on one hand and in the security and defense domain on the other. In this regard, the Lisbon Treaty (2009) was a stepping-stone for both domains, establishing shared competences between Member States and the European Union, but it is certainly the significant progress of EU space programs together with Jean-Claude Juncker's Security Union ambitions that contributed more recently to make of the European Union an increasingly relevant actor in the space security field.

Notwithstanding a noticeable progress of its perimeter and involvement, the role of the Union in this field remains, so far, limited to support actions and diplomatic initiatives:

- *Support to R&D projects*, funded under the 7th and 8th Framework Programmes for Research and Technological Development (FP7, H2020) for a total contribution estimated around €68 million between 2007 and 2017. These projects cover a

wide range of different developments from SSA-related technologies to active debris removal solutions, including, among others, autonomous collision avoidance or space weather forecast.
- *Support to European SST cooperation,* with the establishment of a Space Surveillance and Tracking Support Framework in 2014 (European Commission 2014) to support the networking and operations of SST assets owned by EU countries and provide EU SST services with the EU Satellite Centre acting as front-desk and interface with users.
- *Diplomacy and international cooperation,* with the establishment of Space Policy Dialogues with key partners; active participation to international organizations and committees such as the International Telecommunication Union (ITU), the International Committee on Global Navigation Satellite Systems (ICG), or the UN COPUOS; the ill-fated proposal to establish an International Code of Conduct for Outer Space Activities; or the promotion of initiatives such as the Principles of Responsible Behavior for Outer Space (PORBOS) in the Conference on Disarmament.

The European Space Agency also plays an important role in the field of space security and safety, in particular for capacity building, R&D, and regulations/standards. Despite a convention that did not envision (but did not forbid either) activities in the field of space security and safety, ESA developed a recognized competence and expertise in the scientific and research dimension, with somewhat limited connection to national defense and security strategies. Eventually, the Agency "has evolved to conduct security related projects and programmes and to address the threats to its own activities" (Giannopapa et al. 2018). This evolution has been marked by the setting up of a comprehensive regulatory framework including ESA's security agreement and security regulations and implementing procedures and facilities. ESA now has the capacity to receive, store, and produce classified information as well as to exchange classified information with third parties such as the EU Council, marking a step forward in the role that the Agency could play in the field of space security in the future.

Beyond security-related responsibilities in the management of space programs and the safe operation of its own space systems, activities of the Agency in the field of space security include noticeably:

- *An SSA program* funded by 19 ESA Member States through 2020 at approximately €200 million for the period 2009–2020 and dedicated to a variety of capacity-building projects including research and technology, set up and operation of data and coordination centers, and systems development and procurement across three main segments: Space Surveillance & Tracking, Space Weather and Near-Earth Objects.
- *The Clean Space Initiative,* which promotes an eco-friendly and sustainable approach to space activity throughout the entire life cycle of space systems from conceptual design to end of life and up to removal of debris with the

development of industrial materials, processes, technologies, and standards that are both Earth and space environment-friendly.
- *Participation to international fora* that work in different ways on space security including the International Astronautical Congress (IAC), the Committee on Space Research (COSPAR), the Inter-Agency Space Debris Coordination Committee (IADC), or the UN COPUOS of which ESA became an observer in 1972.

In addition to these central institutional actors and activities, the European space security landscape also encompasses other minor public activities as well as some private-led initiatives such as the participation of European operators to the Space Data Association, a non-profit association of satellite operators that supports the controlled, reliable, and efficient sharing of data that is critical to the safety and integrity of satellite operations.

Accommodating Different Concerns and Interests: Achievements and Limits

As a result, the European framework for space security, which involves national and European (bilateral and multilateral) layers, civil and military organizations, public and private interests, and a broad range of concerns and needs, may seem intricate.

The overall European situation in the space security domain, including achievements and limits, is best illustrated by the SST/SSA component, which crystallizes many of the challenges ahead of Europe in this field.

Back in 2012, at the end of the Preparatory Phase of the ESA SSA program, several Member States decided to withdraw or reduce their contribution to the program component related to SST, with the objective of managing it through a different arrangement, more adequate to their needs. The main reason behind this decision was related to the prominence of military and operational functions in the SST domain and concerns about ESA capacity to properly manage security aspects, including compliance with defense requirements and handling of classified information and data. Some Member States, especially France and Germany, also raised some concerns related to national sovereignty. German and French military organizations negotiated and prepared a non-paper submitted to the European Commission in December 2012 which requested the establishment of a new mechanism at EU level with a specific governance scheme where Member States could maintain control over their assets while cooperating in SST activities. The objective was then to promote a framework that would foster pan-European cooperation and improve cost efficiency (e.g., by avoiding unnecessary duplication of efforts) to deliver EU SST services while complying with national concerns resulting from the specific nature of SST systems and data. In compliance with the principle of shared competences between the European Union and Member States in the field of security and space, the framework would also enable to transfer a share of the financial burden to the European Union. Eventually, the Space Working Party in Brussels started to negotiate the contents of what became Decision No 541/2014/EU of the European Parliament and of the Council of 16 April 2014 establishing a Framework for Space Surveillance and Tracking Support (European Commission 2014).

The outcome of this framework are, at this stage, mixed. The framework has been instrumental to reinforce European coordination (i.e., complex systems networking, delivery of European services, common database of unclassified data, enhanced bilateral agreements for classified data, etc.) and to build confidence among partners but does not seem to provide, in its current form, the best configuration to avoid duplication of efforts and support an efficient development of EU SST capabilities, necessary to close the capability gap and achieve an acceptable level of European autonomy while meeting the needs of a variety of users throughout the entire space mission life cycle (European Commission 2018).

To put things into perspective, it is important to remind that the current arrangement allowed, first and foremost, to accommodate the requirements, sometimes divergent, of different parties. In essence, the intergovernmental model is meant to mitigate two conflicting objectives:

- *Leverage cooperation* among the most motivated Member States to share efforts and results between partners to enhance efficiency and performance of European capabilities
- *Preserve national interests* by safeguarding national sovereignty over dual capabilities development and control over SST systems and data

Beyond issues related to the level of resources available to support European objectives in space security, many blocking points arise from the unresolved question of the relative weight of national interests and European cooperation and therefore of the sharing of responsibilities between Member States and with the EU. For example, the role of the European Commission is rather limited in the current configuration, even though it is the principal funding source and holds the responsibility over the security of the EU space program. From a very practical perspective, two immediate risks come to mind:

- *A risk of divergence of interests among stakeholders* (between Member States and with the European Union), hindering the capacity to implement a coordinated policy. This risk is growing as stakeholders' concerns and positions on space security issues tend to progress faster than European integration and leadership.
- *A risk of duplication of efforts and reduced cost-effectiveness*, if motives to develop specific national capabilities surpass the willingness (and readiness) to focus on distribution and complementarity across Europe. This risk is also growing as the need to optimize resource utilization increases with the overall cost of required capabilities to provide necessary coverage and precision.

An important step forward is the preparation of the next Multiannual Financial Framework 2021–2027 of the European Union. Yet, although an increased level of financial resources can be expected to support SST/SSA (from multiple sources), the current version of the regulation (still to be officially endorsed) proposes mostly to maintain the current model with an enlargement of contributing Member States. With the objective to develop the performance and autonomy of capabilities, the SST component of the EU space program should still build on a network of mostly national SST sensors and support operation and delivery of SST services. The principles of complementarity and the necessity to avoid duplication are recalled but not formally enforced since room is left to develop redundant capabilities.

Space Traffic Management: A Coordination and Leadership Challenge

The limits of the current approach to space security, in particular those related to European coordination and leadership, have recently been accentuated by the announcements of the US government regarding space traffic management which invite other space-faring nations, and in particular Europe, to take position in this domain as well.

With the recent adoption of a National Space Traffic Management Policy, the United States made a major step forward in recognizing the severity of issues at stake and the urgency of setting up a framework to address space security challenges, in particular those related to the expected boom/change in space traffic (deployment of mega-constellations, non-maneuverable CubeSats, in-orbit services, etc.). In this context, the US policy aims to "develop a new approach to space traffic management that addresses current and future operational risks" (White House 2018). Such approach necessarily encompasses a wide range of measures for the development of space traffic monitoring capabilities, the requirements related to space traffic rules, and the coordination of various stakeholders (see Fig. 2).

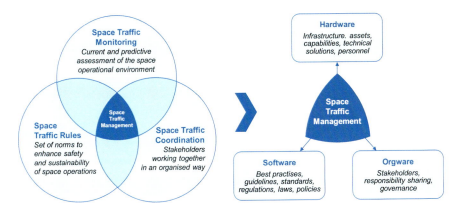

Fig. 2 Representation of space traffic management components

Although national, the policy does not inherently challenge the relevance of multilateral efforts in this field and actually recalls that "it is a shared interest and responsibility of all spacefaring nations to create the conditions for a safe, stable, and operationally sustainable space environment" (Ibid.). However, it opens the door to the development of multiple, possibly divergent, national and regional STM frameworks in parallel to international discussions. It also sets the intention of the US government to take leadership in this field through different means including enabling greater SSA data sharing, developing and promoting space safety standards and best practices across the international community, and encouraging and facilitating US commercial leadership.

The rising need to address new safety and sustainability challenges associated with the development of space activities and the US determination in this domain should create a strong impetus for Europe to take up the issue now and move forward with the definition and implementation of its own policy for space traffic management. It is also a matter of protecting Europe's industrial and commercial interests to set adequate standards and norms for future STM services markets as well as means to foster Europe's credibility on the international scene.

Setting up an effective STM framework is a serious challenge, and even though the US directive sets the principles, goals, roles and responsibilities, as well as guidelines to be followed, the concrete implementation of such policy proves to raise major difficulties. Much of the issues faced across the Atlantic have to do with the transverse challenge of bringing together the views and capabilities of multiple stakeholders (See: Hitchens, See also: Weeden). Greater coordination, harmonization, and convergence between actors are at the heart of the various STM components:

- *To enhance SSA data* and reach the appropriate level of coverage and precision to safely manage operations in orbit, STM involves a better pooling and sharing (or purchasing) of data and services. This implies the necessity to address collective information management (collection, fusion, distribution) to safeguard data quality, availability, integrity, and confidentiality. It also implies more transparency from operators.
- *To specify and promote safety and sustainability norms,* STM involves the definition of best practices, technical guidelines, safety standards, or regulations addressing the full life cycle of space systems and applicable (possibly tailored) to a wide range of actors. This implies to make the interests of different actors converge in a balanced framework. It also implies some kind of arbitration between conflicting views.
- *To distribute roles and responsibilities,* STM involves to clearly delineate domains of activity (military/civil, public/private, national/international) and to establish the appropriate channels between them. This is essential to simultaneously achieve national security, commercial, and diplomatic objectives.

Comparable coordination challenges will arise in Europe as well. Given the state of affairs previously described, the lack of top-down leadership and the national/regional dimension will likely bring an additional layer of complexity.

Conclusions

As Europe is contemplating its next 7-year period, some forward-looking reflection seems in order. Incidentally, the projections are clear: if the space sector continues to mature and develop along anticipated trends, coming years will be marked by a steep increase in space traffic, a diversification of actors, the emergence of new concepts, possibly rising tensions, and, in general, a serious degradation of safety conditions in space putting at risk the sustainability of in-orbit operations in the long term. In front of these challenges, a proportionate response is necessary if Europe wants space to continue to serve as a strong socioeconomic engine but also if it intends to engage in a more assertive approach to security and defense and to hold its place around the table of negotiation of international space affairs.

European stakeholders share, today, the awareness of the growing urgency and criticality of issues at stake. They also certainly share the conviction that a successful response to space security challenges should be articulated around a comprehensive and coherent approach building on an appropriate level of resources and on European cooperation in view of a global solution. This ideal approach will, as usual, need to be confronted to the complexity inherent to the conduct of European affairs. However, the fast pace of evolution of the situation, both in orbit and on the ground, presses for decisions to be made in the short term.

In this respect, the European approach to space security shall be objective driven, rather than dictated by pre-existing processes and institutional constrains. Given the wide-encompassing range of safety and sustainability issues at stake, the definition of a European Space Traffic Management regulation and operational setup, as a first step toward a full-fledged European space security policy, probably requires dedicated and innovative decision-making processes. This a matter of complementing the current bottom-up approach building on operational collaboration with a top-down leadership, empowered to define principles, goals, roles and responsibilities, as well as guidelines to be followed in Europe.

While the current European setup – designed to accommodate the interests of various stakeholders – allowed to progress substantially on many technical and cooperation challenges, questions arise on its capacity, in its current form, to tackle the emerging challenges that will continue to intensify over the next 7 years and beyond. In light of the rising stakes and challenges for Europe in the field of space security, moving from an "accommodation" to a "convergence" of interests – while preserving core national sovereignty concerns – seems to be an increasingly pressing issue. To this end, further progress shall be made in terms of balancing national interests against the merits of a European cooperation and therefore in terms of sharing of responsibilities among Member States and with the EU.

Such progress cannot be isolated from the broader foreign and security political context. While space has long been nested in European cooperation, it is not the case of the foreign and security policy, which has been, for the largest part and so far, left aside of the EU framework. We are observing at the moment the premises of a movement in this direction, which will have profound political and cultural implications at European level, as well as on the global scene.

In this context, space security might actually serve as an adequate test case for the European Union and its Member States (in collaboration with relevant organizations such as ESA) to experiment new forms of arrangements giving more room to European leadership and cooperation to serve shared interests:

- Space security is a comprehensive issue that involves a wide range of key concerns (i.e., programmatic, financial, regulatory, diplomatic, etc.) that make it a good candidate to be considered not only as a transverse matter but also as a possible full-fledged programmatic and policy domain.
- Space security is inherently a dual topic offering promising perspectives of synergy between space and defense. At the crossroad of military and civil domains with already existing interactions, it also involves competitiveness and innovation challenges and opportunities for the European industry.
- Space security already has strong European cooperation foundations, and various structures – meant to evolve along with policy objectives – are in place, including for SST capabilities development or for the security component of programs of direct military significance such as Galileo. For these different structures, there is a strong European added-value, both to deliver operational services and optimize cost-effectiveness.
- Space security will likely be subject to profound changes in the next few years, expected to ease up some of the current difficulties in this matter. For example, several experts anticipate a trend toward more transparency between space players, including military ones. In this respect the development of new capabilities, commercial in particular, may lead to a paradigm shift in the value of data confidentiality versus availability. Another interesting trend is the growing readiness of many public and private stakeholders to consider more stringent regulations.

Last but not the least, space security involves a strong link between internal and external action. That could make it an excellent candidate to support a more strategic, more assertive, and more united Europe in the World, as the incoming President-elect of the European Commission, Ursula von der Leyen, called for in her mission letter to Josep Borrell, High Representative of the Union for Foreign Affairs and Security Policy (von der Leyen 2019).

References

ESPI. Security in outer space: rising stakes for Europe – report 64, August 2018
European Commission (2014) Decision no 541/2014/EU of the European Parliament and of the council of 16 April 2014 establishing a framework for space surveillance and tracking support

European Commission. Communication from the Commission to the European Parliament, The Council, The European Economic and Social Committee and the Committee of The Regions. Space Strategy for Europe. COM (2016) 705 final, October 2016

European Commission, Dependence of the European economy on space infrastructures: potential impacts of space assets loss, March 2017

European Commission (2018) Report from the commission to the European Parliament and the council on the implementation of the space surveillance and tracking (SST) support framework (2014–2017)

Froehlich A, Seffinga V (2018) National space legislation. European Space Policy Institute (ESPI)/Springer, Vienna

Froeliger JL (2017) Greater industry cooperation needed to avoid space collisions. INTELSAT. Retrieved from: http://www.intelsat.com/news/blog/greater-industry-cooperation-needed-to-avoid-space-collisions/

Giannopapa C, Adriaensen M, Antoni N, Schrogl K-U (2018) Elements of ESA's policy on space and security. Acta Astronaut 147(2018):346–349

Helms S. Space situational awareness. Power point presentation for the United Nations Committee on Peaceful Uses of Outer Space (COPUOS), June 3, 2010

Hitchens T. #SpaceWatchGL Op'ed: New space debris rules stalled by year-long interagency spat. Retrieved from https://spacewatch.global/2019/09/spacewatchgl-oped-new-space-debris-rules-stalled-by-year-long-interagency-spat/

Ministère des Armées, Stratégie Spatiale de Défense, July 2019

Space Data Association website. http://www.space-data.org/sda/

Space-Track. User agreement. Retrieved from https://www.space-track.org/documentation#/user_agree

U.S. Department of Defence. National Security Space Strategy, January 2011

USSTRATCOM Public Affairs (2017) U.S. strategic command, Norway sign agreement to share space services, data. Retrieved from U.S. Strategic Command Peace is our Profession Retrieved from http://www.stratcom.mil/Media/News/News-Article-View/Article/1142970/us-strategic-command-norway-sign-agreement-to-share-space-services-data/

von der Leyen U. Mission letter to Josep Borrell, High representative of the Union for Foreign Policy and Security Policy/Vice-President designate of the European Commission, 10 September 2019, Brussels

Weeden B. Time for a compromise on space traffic management. Retrieved from http://www.thespacereview.com/article/3673/1

White House (2018) Space policy directive-3, National Space Traffic Management Policy

Space and Security Policy in Selected European Countries

25

Ntorina Antoni, Maarten Adriaensen, and Christina Giannopapa

Contents

Introduction	467
National Space and Security/Defense Strategies	469
France	473
Germany	475
Italy	476
Spain	478
The United Kingdom	479
Priorities and Trends in National Space and Security/Defense Strategies	481
Conclusion	482
References	482

Abstract

This chapter presents space and security policies of selected European countries and indicates the main priorities and trends thereof. In particular, it addresses relevant strategic documents that show how certain states in Europe use space assets and applications to ensure security policy objectives. Recent developments at the national level are linked with space policymaking in the frame of the European Union and the European Space Agency.

N. Antoni (✉)
Eindhoven University of Technology, Eindhoven, The Netherlands
e-mail: ntorina.antoni@gmail.com

M. Adriaensen · C. Giannopapa
European Space Agency (ESA), Paris, France
e-mail: maarten.cm.adriaensen@gmail.com; christina.giannopapa@esa.int

© Springer Nature Switzerland AG 2020
K.-U. Schrogl (ed.), *Handbook of Space Security*,
https://doi.org/10.1007/978-3-030-23210-8_86

Introduction

Europe is a continent consisting of a multitude of national, intergovernmental, and supranational organizations with diverse space policy priorities, thereby creating a multifaceted and dynamic space sector. By and large, the European space sector has been interwoven with the European integration process, which has been central in addressing rising security challenges including climate change, migration, and cybersecurity. Space policies have contributed to reinforcing European cooperation and integration in order to tackle these challenges. In addition, the surge in the number of private commercial actors as well as the emergence of a civil-military paradigm has resulted in the transformation of the space sector. The advent of a new era in space in combination with particular dynamics in the European space sector has put space security at the forefront of policy and regulatory debates.

The main actors in Europe engaging in space and security activities are the European countries, the European Union (EU), and the European Space Agency (ESA). The EU Global Strategy, adopted by the European Council in June 2016, the European Commission Space Strategy for Europe, launched in October 2016, and the European Defence Action Plan 2016 all stress the importance of space security. More recently, in June 2018, the EU presented the Proposal for a Regulation for a Space Programme for the EU (European Commission 2018). Furthermore, ESA has increasingly contributed to space security "in" and "from" space, as reflected in the Council Document "Elements of ESA's Policy on Space and Security," issued in June 2017 (Giannopapa et al. 2018). In December 2019, the Council at the Ministerial level adopted the "safety and security pillar" along with associated programs. With regard to ESA–EU relations, in October 2016, they signed the Joint Statement on Shared Visions and Goals, while in May 2019 they convened the first joint Space Council in 8 years (Spacewatch.global 2019). It is also worth mentioning that the North Atlantic Treaty Organization (NATO) has increased its interest in the use of space assets for defense, with the most notable milestone being the adoption of space policy in June 2019 (Euractive 2019). For further information about the institutional and various space and security programs in Europe, please see the respective chapters in this Handbook.

The different space security policies of the European countries are to a large extent determined by national needs and priorities as well as their participation in relevant space and security organizations. Figure 1 below visualizes the current status of the countries' membership to ESA, EU, EDA, and NATO (adapted from Papadimitriou et al. 2019). In the present chapter, the European countries with the largest ESA annual budget and their defense expenditure as share of their gross domestic product (GDP) are presented. Namely, these are: France, Germany, Italy, Spain, and the United Kingdom (UK).

The aforementioned countries may be distinguished not only on the basis of their membership to these organizations but also based on their space budget. In the absence of an official grouping of these countries within any of the aforementioned organizations, their ESA annual budget and their defense expenditure as share of their GDP are used in this chapter to classify them into three groups, as seen in Figs. 2 and 3.

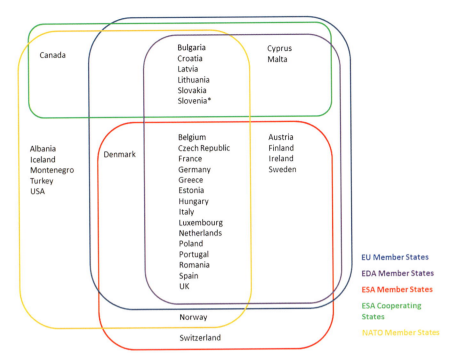

Fig. 1 EU, ESA, EDA, and NATO Member States (Papadimitriou et al. 2019) (*Slovenia is an ESA Associate Member State*)

National Space and Security/Defense Strategies

The current priorities and trends in national space and security strategies of the mentioned European countries derive from space and security elements stipulated in strategic documents. All countries included in this study have established their strategic priorities in the field of space and security mainly in the following types of documents: National Defense Strategy and Doctrine, National Defense Procurement Strategy and Policy, National External or Internal Security Strategy, and National Space Strategy or Policy. Depending on the country, more specific documents complement the strategic landscape, for instance, through a dedicated space security strategy or through the inclusion of space and security aspects in strategy documents covering other policy areas. For example, space and security aspects can be found in maritime strategies and arctic strategies that also stress the importance of space-based assets and applications in these domains. An overview of space and security strategic documents, which have been studied to analyze the priorities and trends in the European countries with the largest space and security budget – France, Germany, Italy, Spain, and the United Kingdom – is presented in Table 1.

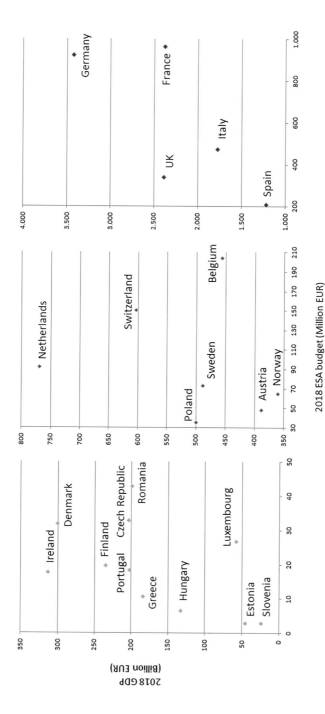

Fig. 2 ESA Member States 2018 GDP versus 2018 ESA space budgets categorization in three groups (ESA and IMF World Economic Outlook)

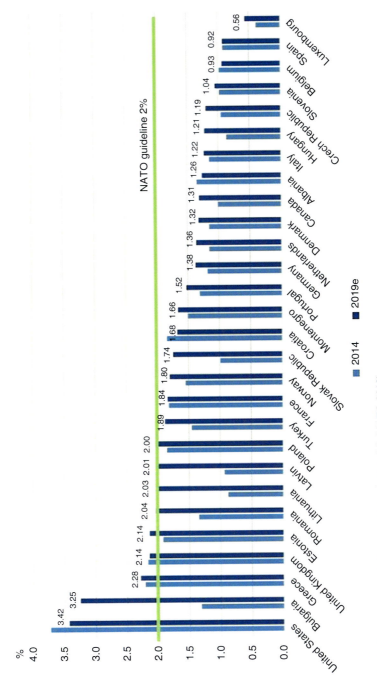

Fig. 3 NATO defense expenditure as share of GDP (%) (NATO 2019)

Table 1 Strategic documents on space and security for selected European countries

	National Space Law	National Space Strategy/Policy	National Security and Defence Strategy/Policy (other relevant domains)
France	2008 Law concerning Space Operations (French Space Operations Act)	2019 Space Defence Strategy 2015 Ambition 2020–2015, Space for the climate 2012 French Space Strategy	2017 Defence and National Security Strategic Review 2015 French National Strategy for the security of maritime areas 2013 French White Paper on Defence And National Security and Bill on Military Planning 2014–2019 (updated 2019–2025)
Germany	2007 Satellite Data Security Act 1990 Law Governing the Transfer of Responsibilities for Space Activities	2010 making Germany's space sector fit for the future – the space strategy of the German Federal Government	2018 High-Tech Strategy 2025 2016 White Paper on German Security Policy and the future of the Bundeswehr
Italy	2018 Space Bill containing measures for the coordination of the space and aerospace policies along with important regulations concerning the organization and functioning of the Italian Space Agency-ASI	2019 Government guidelines on space and aerospace 2016 Strategic Vision Document 2016–2025	2019 National Security Strategy for Space 2018 Plurennial programmatic document 2015 White Paper for International Security and Defence
Spain	1995 Royal Decree 278/1995 establishing in the Kingdom of Spain the Registry foreseen in the Convention adopted by the United Nations General Assembly on 2 November 1974	Spanish Strategy for ESA and EU Space Programs 2007–2011	2017 IDS (infodefensa) R&D strategic approach 2013 National Security Strategy – sharing a common project 2012 National Defence Directive 1/2012
United Kingdom	1986 Outer Space Act	Defence Space Strategy (under development) 2015 National Space Policy 2015 Space Innovation and Growth Strategy 2014–2030 2014 National Space Security Policy 2013 Strategy for Earth Observation from Space (2013–2016) 2012 UK Civil Space Strategy	2015 National Security Strategy and Strategic Defence and Security Review

France

Space and Security Status
France is a permanent member of the United Nations (UN) Security Council. France has been at the forefront of the elaboration of the EU Common Security and Defense Policy (CSDP). France is a member of the G7 and G20. France was one of the founding members of the NATO in 1949. France is a member of EDA and the European Union Satellite Centre (SatCen). France is a member of the Organisation for Joint Armament Cooperation (OCCAR). France has established and maintained extensive bilateral relations with European partners on security and defense, notably with Germany, Italy, and the UK. France was one of the founding member states of ESA in 1975. Also, Paris hosts the ESA Headquarters. Created in 1961, under the joint supervision of the Ministry of Higher Education, Research and Innovation and the Ministry of the Armed Forces, the National Centre for Space Studies (CNES) is the national space agency, a public administration institution with industrial and commercial purpose (*établissement public à caractère industriel et commercial*), charged with developing and executing the French space program. CNES prepares the French space policy under the responsibility of the Minister for Higher Education, Research, and Innovation, while it has a partnership with the Ministry of Defense for military space activities.

Space and Security Elements in Strategic Documents
Space operations in France are regulated by the 2008 Law concerning space operations, the "French Space Operations Act" – *Loi sur les Opérations Spatiales*. The French Space Operations Act sets up a national regime to authorize and control space operations based on international commitments of the French government. The Act sets out an authorization regime with specific conditions, procedures, and technical requirements along with a control regime and mechanism (French Republic 2008). The Act lays down safety and security standards for the Guyana Space Centre and defines the liability regime for space activities. The 2012 French Space Strategy published by the Ministry of Research establishes a directorate of innovation, applications, and science to support new space. It also identifies security and defense as one of the specific applications and orientation areas for the French space policy (French Republic 2012). In particular, the development axes of the 2012 Space Strategy include very high resolution (VHR), secure satellite communications (SATCOM), electronic intelligence (ELINT), and detection and early warning for ballistic missiles. Defense aims to fully benefit from the dual-use nature of space systems (French Republic 2012). On top of that, the CNES ambition 2015 stresses the relevance of VHR optical observation, electronic intelligence, ultra-secure telecommunications, and space situational awareness (SSA) for defense purposes (CNES 2015).

In terms of security and defense, the French 2017 Defense and National Security Strategic Review emphasizes "the need to develop space situational awareness and to ensure the resilience of space capabilities." It also refers to space as "a provider of essential navigation, communication, meteorological and imagery services, while also a domain of confrontation where some states can be tempted to use force to deny access or threaten to damage orbiting systems." The list of operational capabilities, required for a coherent and full-spectrum model, includes imagery intelligence (IMINT), Electronic Intelligence (ELINT), Command and Control (C2), and the protection and security of space assets requiring adequate SSA. It stipulates that "space is of crucial importance to defense capabilities, while becoming an increasing source of vulnerability for C2 and surveillance assets. Monitoring objects in low earth orbits – and watching geostationary orbits, as planned for by the aerospace operations command and control system (AOCCS) – is essential to ensuring the security of our space-based assets and conducting operations [. . .] An early warning capability would enable better characterization of ballistic threats, determination of the source of a launch and prediction of the target area." The strategic review further stresses the relevance of enhancing satellite-based surveillance for maritime security (French Ministry of the Armed Forces 2017).

Moreover, the 2013 French White Paper on Defence and National Security and the Bill on Military Planning 2014–2019 (updated 2019–2025) highlight the strategic function of space-based systems: "Outer space has become crucial to the operation of essential services (French Ministry of the Armed Forces 2013). In the military field, strategic autonomy is dependent on free access to and use of space, which make it possible to preserve and develop the technological capabilities on which the quality of our defense system and, not least, the credibility of our nuclear deterrent, depend." The White Paper refers to increasing threats from space debris promulgation and offensive space weaponry. The Bill foresees the delivery of three *Composante Spatiale Optique* (CSO) satellites no later than 2021 and plans the launch of a third Syracuse satellite by 2030.

Space, security, and defense policies in France are merged under the recently released 2019 Space Defence Strategy that intends to respond to challenges from the emergence of New Space (French Ministry for the Armed Forces 2019). As stated, "this new environment implies a Space Defence Strategy founded on the protection of French capabilities. That involves first and foremost improving Space Situational Awareness (SSA), especially in order to detect and attribute unfriendly or hostile acts in all orbits of interest and defend against them." In order to guarantee France's capacity to act in space, the armed forces aim to "(i) strengthen a space doctrine which establishes the ground rules for and typology of military space operations, (ii) overhaul military space governance, and (iii) ensure that they have appropriate capabilities and human resources." Accordingly, "military space operations consist in operating space capabilities that provide services in support of government authorities and military operations, thus helping to increase the effectiveness of action. They contribute to national security, the robustness of economy and protection of the population. They also include action taken in space to protect assets and discourage any aggression. They are organized around four functions: space service support; space situational awareness; operations support; and active space defense" (French Ministry for the Armed Forces 2019). The 2019 Space Defense Strategy is aligned with the previously published security/defense and space strategy documents.

Last but not the least, in the maritime domain, the 2015 French National Strategy for the Security of Maritime Areas addresses the relevance of space-based resources for the monitoring of French maritime areas, including Satellite – Automatic Identification System (SAT-AIS), long range tracking and identification (LRIT), vessel monitoring systems (VMS), and Earth observation imagery (French Republic 2015). In that regard, France considers requirements for satellite surveillance in maritime areas. The strategy highlights the weaknesses in resilience and security of SAT-AIS. The strategy refers to the role of industry in providing solutions for the governments' requirements for maritime surveillance including radar or optical satellite imaging, spatial AIS, and satellite communications.

Germany

Space and Security Status

Germany is not a permanent member of the UN Security Council. Germany is a member of the G7 and G20. Germany was one of the founding members of the NATO in 1949. Germany is a member of EDA and SatCen. Germany is a member of the OCCAR. Germany and France have a strong history of cooperation in defense. Germany also works closely with the Netherlands in the defense realm. Germany was one of the founding member states of the ESA in 1975. The German Ministry for Economic Affairs and Energy (*Bundesministeriums für Wirtschaft und Energie – BMWi*) is mainly responsible for Germany's civilian space activities.

Space and Security Elements in Strategic Documents

Space activities are governed by the Law of 8 June 1990 Governing the Transfer of Responsibilities for Space Activities (*Raumfahrtaufgaben Übertragungsgesetz*) and the 23 November 2007 Satellite Data Security Act (German Federal Ministry of Justice 1990, 2007). The Law delegates the responsibility for the management of the German space program from to DLR (German Aerospace Center – *Deutsches Zentrum für Luft- und Raumfahrt*).

The 2010 German Space Strategy addresses space security based on the approach of "whole-of government security preparedness," which includes not only space-based early warning for impending crises but also increased sensitivity for the importance of an unhindered use of satellite systems for national security (BMWi 2010). Accordingly, "Satellite data and services make a vital contribution, notably to disaster relief and management, environmental and climate protection, to warning of threats, development aid, border monitoring, and arms control. In this regard, it is stated that:

- "Military operations are now inconceivable without the support of space-based systems;
- Space systems in the domains of communication, navigation and Earth observation make a decisive contribution to the ability to conduct an effective foreign and security policy and to achieve whole-of-government security preparedness;

- Wherever possible, exploit synergies between civil developments and dual-use technologies when further developing system capabilities and strategically important competences in key technologies" (BMWi 2010).

In the context of security and defense, the 2016 White Paper on Security Policy and the Future of the *Bundeswehr* notes that satellite systems are a fundamental component of Germany's critical infrastructure. All aspects of national and international communication and navigation decisively depend on them. The White Paper thus stresses the need for monitoring these critical systems. In line with the EDA Capability Development Plan (CDP), Germany's common security priorities in the White Paper include UAVs, air-to-air refueling, satellite communication, cyber protection, and cyber defense (German Federal Government 2016). The 2015 Joint Concept for Space, commissioned by the Ministry of Defense, acknowledges that space is an operational domain of its own. Space support to operations has become an indispensable military function and as such, it requires the establishment of situational awareness and of the necessary command and control functions for the space domain (Japan Space Forum 2015).

What's more, the White Paper on German Security Policy and the Future of the *Bundeswehr* from 2006 highlights the capability category of intelligence collection and reconnaissance and the procurement of the SAR(Synthetic Aperture Radar) LUPE space-based reconnaissance system. Using a joint approach: efficient command, control, and information systems of the armed forces ensures the capability to exercise command and control worldwide. The White Paper describes joint that networkable radio equipment and "SATCOMBw," satellite-based communications system, are important prerequisites for network-enabled operations (German Federal Government 2016).

Finally, the 2025 high-tech strategy emphasizes using the potential of key technologies for the benefit of the industry including space technologies (German Federal Government 2018). In particular, "to ensure technological sovereignty in the area of satellite infrastructure," the strategy elaborates on the establishment of the Institute for Satellite Geodesy and Inertial Sensors of DLR in Hanover and Bremen, the DLR Institute for Quantum Technologies in Ulm, and the DLR Galileo Competence Center in Oberpfaffenhofen. The most important objective is to support industry in the transfer of research results from the field of quantum technology and artificial intelligence into practical application including the aerospace sector (German Federal Government 2018).

Italy

Space and Security Status

Italy is not a permanent member of the UN Security Council. Italy is a member of the G7 and G20 and was one of the founding members of the NATO in 1949. Italy is a member of EDA and SatCen, as well as OCCAR. At the bilateral level, Italy and France have active collaboration on space, reflected in annual summits organized since 1982. Italy was one of the founding member states of ESA in 1975. In Italy, responsibility for space activities has recently been put directly under the supervision of the Prime Minister.

Space and Security Elements in Strategic Documents

The Italian Parliament approved on 22 December 2017 the new 2018 Space Bill containing measures for the coordination of the space and aerospace policies along with important regulations concerning the organization and functioning of the Italian Space Agency – ASI. ASI reports directly to the Prime Minister's Office. The Bill mainly provides that the management and coordination of space and aerospace policies are assigned to the Presidency of the Council. The Bill also established an inter-ministerial Committee that will be responsible for defining the government's orientations in the sector (Italian Parliament 2018).

In July 2019, the National Security Strategy for Space was presented to promote a "systemic" strategy for national security (Italian Presidency of the Council of Ministers 2019). The strategy was based on the March 2019 Government guidelines on space and aerospace (Italian Prime Minister's Office 2019). The strategic objectives of the National Security Strategy for Space are: (a) "to ensure the security of space infrastructures (according to the two Anglo-Saxon terms, safety and security), regarded as enablers of the national infrastructure as a whole; (b) to safeguard national security, including through space, by ensuring access to and use of national security capabilities in any given situation; (c) to strengthen and protect the institutional, industrial and scientific sectors, also with a view to protecting national classified information; (d) to promote a space governance capable of ensuring sustainable, safe and secure space operations at international level; (e) to ensure that the development of private initiatives in the space sector (upstream and downstream) is consistent with the country's overriding interests" (Italian Presidency of the Council of Ministers 2019).

According to the Strategy, *safety* is defined as "a set of measures put in place to ensure protection against unintentional events," while *security* is defined as "a set of measures to guarantee security against malicious activities or actions carried out by opposing parties."

Additionally, the ASI Strategic Vision Document 2016–2025 highlights the use of space assets also for security and defense purposes (ASI 2016): "Earth observation data for needs of crisis management, security, defense, and disaster monitoring:

- Space Situational Awareness (SSA), Space Traffic Management, Confidence building measures, Code of Conduct;
- Promotion of institutional services: climate, environment, managing the cycle of risks and emergencies, weather-sea and atmosphere monitoring, national security with maritime surveillance, border control and humanitarian aid;
- Development of domestic GNSS activities, Public Regulated Service (PRS) security center, interference monitoring center, PRS terminals, interfaces and networks for domestic users; Ensure Italy's role and participation in the processes of multilateral international coordination in various international contexts of a

global nature, like the United Nations Committee on the Peaceful Uses of Outer Space (UNCOPUOS) in strategic sectors like Earth observation (Group on Earth Observations – GEO; Committee on Earth Observation Satellites – CEOS), exploration of the universe (International Space Exploration Coordination Group – ISECG), Medicine and Microgravity (International Space Life Sciences Working Group – ISLSWG), the analysis of space debris (Inter-Agency Space Debris Coordination Committee – IADC), satellite navigation (International Committee on Global Navigation Satellite Systems – ICG) and security" (ASI 2016).

Concerning security and defense, the Italian White Paper for International Security and Defence from 2015 regards space as "a strategic element for preserving the safety of the 'national system' and increases the solidity of the political, economic and social structures" (Italian Ministry of Defence 2015). In addition, the Plurennial Programmatic document (*Documento Programmatico Pluriennale* – DPP) that provides the multiyear plan of the Italian Defence, includes investments to be made in the space component (Italian Ministry of Defence 2018). Space funding includes €212 million in the Cosmo SkyMed second-generation synthetic aperture radar satellite over 4 years and €18,4 million in the Multinational Space-based Imaging System (MUSIS) – Common Interoperability Layer over the 3 years covered by the DPP.

Spain

Space and Security Status
Spain is not a permanent member of the UN Security Council. Spain is not a member of the G7 but is invited as a permanent guest member to the G20. Spain has been a member of NATO since 1982. Spain is a member of EDA and SatCen. Spain joined OCCAR in 2005. Spain became ESA Member State in 1979. In Spain, the Ministry of Economy and Enterprise (*Ministerio de Economía y Empresa*) is in charge of drafting and implementing government policy on economic matters and reforms, including space affairs.

Space and Security Elements in Strategic Documents
The Royal Decree 278/1995 of 24th February 1995 on space exploration, established the Registry of Objects Launched into Outer Space as provided for in the Convention adopted by the United Nations General Assembly on 12 November 1974 (Spanish Prime Minister's Chancellery 1995).

The 2013 National Security Strategy acknowledges that outer space has become a domain where confrontation is possible (Spanish Government 2013). In addition, the 2015 Ministry of Defence Master Plan of Space Systems highlights the importance of space capabilities for the development of military operations (Spanish Ministry of Defence 2015). This master plan proposes a set of actions to guarantee the maintenance of the existing space capabilities (communications, observation of the earth, navigation and positioning). In addition, it has classified space systems in the following four families: communication – earth observation – navigation and positioning – surveillance and monitoring. From a technological and industrial point of view, it emphasizes the importance of maintaining effective coordination between public institutions and between these institutions with the industry, avoiding duplications to optimize the benefits from the made investments (Spanish Ministry of Defence 2015).

The 2017 IDS (infodefensa) R&D strategic approach for the defense industry stresses that "it is possible that new programs in other areas, such as space, unmanned aircraft, in-flight refueling aircraft and electronic warfare [..] will be implemented in the short term, and that they will undoubtedly represent an opportunity for Spanish companies" (IDS 2017). The Spanish space industry is primarily involved in contracts of high added value in the areas of qualification of flight and ground equipment and the development and operation of satellite systems. There are also several space centers located in Spain, the most important being namely the European Space Astronomy Centre (Madrid), the Madrid Deep Space Communication Complex (NASA), and Boeing's European Centre for Research and Technology (also located in Madrid).

The United Kingdom

Space and Security Status
The UK will hold the presidency of the G7 in 2021. The UK was one of the founding members of the NATO in 1949. Brexit has ended British obligations that include permanent involvement in the institutional structures and decision-making processes of the Common Foreign and Security Policy (CFSP) and The Common Security and Defence Policy (CSDP) in the EU. However, this does not automatically preclude any future involvement in defense and security cooperation. The UK is not involved in the Permanent Structured Cooperation (PESCO) in the field of defense. The UK is a permanent member of the UN Security Council and a member of the G7 and G20. The UK is currently a member of EDA and SatCen. It is also a member of OCCAR. The UK has developed strategic alliances with various international partners through multilateral and bilateral cooperation. Since the 1940s, the UK and the USA have been close military allies relishing the so-called the Special Relationship. The UK was one of the founding member states of the ESA in 1975. The Department for Business, Enterprise, and Industrial Strategy (BEIS) is the parent department of UK Space Agency and has the lead for space affairs in the UK.

Space and Security Elements in Strategic Documents
The 1986 Outer Space Act governs and regulates national space activities, including launch and operations of space objects. The Outer Space Act 1986 is the legal basis for the regulation of activities in outer space carried out by organizations or individuals established in the UK or one of its overseas territories or crown dependencies (UK Government 1986).

The UK Government published the National Security Strategy and the Strategic Defence and Security Review: "A Secure and Prosperous UK," on 23 November 2015 (UK Government 2015a). The two documents comprise the Government's strategic decisions on defense and security. The former focuses on the country's "ends" or objectives, while the latter addresses the "ways" and "means" to achieve them. The Strategic Defence and Security Review 15 key investments include the establishment of a Space Operations Control Centre and *Skynet 5 Beyond Line* of sight satellite communications before 2025, and the upgrade of the Space Operations Control Centre and investment in the next generation of secure strategic communications after 2025. Accordingly, the Royal Air force will improve its capabilities in Space Surveillance and Tracking (SST). It will also develop a high-altitude communication relay capability. The Joint Forces Command will provide new satellite communications and "future proof" the navigation and targeting services based on space assets (UK Government 2015a).

Space is an official part of the UK's critical national infrastructure, as declared by the UK Centre for the Protection of National Infrastructure. The UK Air and Space Doctrine recognizes that the UK relies not only on space for its national security interests but also for its economic prosperity (UK Ministry of Defence 2017). UK space capabilities are inherently dual-use, provided that the same environment, technology and infrastructure, are used to meet both military and civil operations. Space services play a pivotal role in contributing to UK national security, the strength of the economy, and the delivery of public services (UK Ministry of Defence 2017). According to the air and space doctrine, space power makes a pivotal contribution to the potency of UK military power, both as an enabling domain and, increasingly, as an operating domain (in its own right). It is also the domain which makes the most significant contribution to the effectiveness of all the instruments of national power. Space power is defined as "exerting influence in, from, or through, space." Diplomatic, military and economic credibility, together with a coherent strategy, play a large part in the ability to influence (UK Ministry of Defence 2017).

The Ministry of Defense is expected to release its first dedicated defense space strategy, a plan in line with its ambition for developing and improving Britain's military space capabilities. The ministry outlined four objectives which may become the guiding priorities of the DSS later this year: "to enhance the resilience of space systems; to improve operational effectiveness; to enhance space support to frontline troops; and to support wider government activities" (UK Ministry of Defence 2019).

Furthermore, the Strategic Defense and Security Review called for a ministerial committee on security and prosperity and the development of the UK National Space Policy. The mandate included the process to "mitigate space weather impacts, improve forecasting, protect the space environment by using civil and military capability, work with international partners, recognize the criticality of satellite navigation, need for enhanced resilience, and innovation in the field of resilient satellite communications" (UK Government 2015a). The Strategic Review refers to the December 2015 UK National Space Policy which has a dedicated section on safety and security of space (UK Government 2015b).

The overarching 2015 National Space Policy links the UK civil space strategy (July 2012) with the National Space Security Policy (April 2014). It aligns with the UK Government's Science and Innovation Strategy and the National Security Strategy. In addition, the 2015 updated Space Innovation and Growth Strategy sets out a partnership between industry, government and academia in order to develop, grow and exploit new space-related opportunities. The 2015 National Space Policy clearly articulates cross-government reliance on space-enabled capabilities. Space is the environment which makes the most significant contribution to the effectiveness of all the instruments of national power (UK Government 2015b).

Finally, the UK Space Agency published in October 2013 a strategy for earth observation from space (2013–2016) in the context of the National Space Policy, the National Space Security Policy (April 2014), and expands on the themes of the UK Civil Space Strategy (July 2012). The strategy concentrates on civil Earth observation requirements but recognizes that some civilian space systems could be dual-use in nature and be capable of supporting national security requirements (UK Space Agency 2013).

Priorities and Trends in National Space and Security/Defense Strategies

Security and defense-related aspects of space activities cover a wide range of activities, mainly divided into two categories: priorities for security *from* space and priorities for security *in* space.

The "Security from Space" priorities consist of disaster management, resource management, transport and communications, environment, climate change and sustainable development, external security including foreign policy and border surveillance, internal security including support to justice and home affairs, military, and financial. The "Security in Space" priorities consist of defensive space security and control, offensive space security and control, space surveillance and tracking, space weather, near earth objects, orbital debris mitigation, space traffic management, active debris removal, and access to space.

The European countries' military space capability priorities point toward several key areas in line with the identified priorities for space in the frame of the EDA Capability Development Plan 2018. National defense strategies and policies define satellite systems as critical infrastructure (enabling domain) and identify space as a separate operating domain. Military operations are inconceivable without the support of space-based systems. Space-based assets and applications are essential to navigation, communication, meteorological, and imagery services, early warning and ballistic missile interception. Space systems make a decisive contribution to the ability to conduct an effective foreign and security policy and to achieve whole-of-government security preparedness including defense aspects in a holistic approach. Strategies and policies increasingly call for the full exploitation of synergies between civil and military developments and dual-use technologies when further developing system capabilities and strategically important key technology competences.

What is more, space is perceived as a domain of socioeconomic development and is essential for security and defense. The list of operational capabilities, required for a coherent and full-spectrum force model include inter alia: IMINT, ELINT, C2, and the protection and security of space assets. SSA underpins all other space roles, as it provides an understanding of the space environment. It enables the timely assessment of and response to space threats, risks, and events, both natural and man-made. Defense White Papers and Strategies are explicitly referring to the increasing threats from space debris promulgation, space weather, offensive space weaponry, as well as the inherent vulnerability of space-based systems from interference and cyber-attacks. In this context, multiple member states are in the process of elaborating dedicated space defense strategies and revising their organizational structures.

Conclusion

This chapter provides evidence of the trend towards increasing relevance for security and defense in national space policy initiatives of the presented European countries. The identified national space, security, and defense strategies demonstrate an evolution of European states priorities to dual-use utilization. The policy developments at national level, in combination with the recent EU and ESA policy-making initiative, recognize the important role of space for security and defense. They also demonstrate the increasing relevance of security and defense in Europe, in general, which to some extent could be framed as the necessity for Europe to further enhance its own security and defense.

Disclaimer The contents of this chapter and any contributions to the Handbook reflect personal opinions and do not necessarily reflect the opinion of the European Space Agency (ESA).

References

CNES (2015) Ambition 2020–2015, Space for climate. Available at: http://www.cnes-csg.fr/auto mne_modules_files/standard/public/p11636_99f9990e67f7bbf293ff282b37d9bc1eCNES_plk_instit_2015_171214_GB.pdf

Euractive (2019) NATO braces for the new space age, 21 November 2019. Available at: https://www.euractiv.com/section/global-europe/news/nato-braces-for-the-new-space-age/

European Commission (2018) Proposal for a regulation of the European Parliament and of the Council establishing the space programme of the union and the European Union Agency for the space programme and repealing regulations (EU) No 912/2010, (EU) No 1285/2013, (EU) No 377/2014 and Decision 541/2014/EU, COM(2018) 447 final 2018/0236 (COD), 6 June 2018, Brussels. Available at: https://ec.europa.eu/commission/sites/beta-political/files/budget-june2018-space-programme-regulation_en.pdf

French Ministry for the Armed Forces (2013) French White Paper on Defence and National Security and Bill on Military Planning 2014–2019 (updated 2019–2025). Available at: https://www.defense.gouv.fr/english/dgris/defence-policy/white-paper-2013/white-paper-2013

French Ministry for the Armed Forces (2017) Strategic Review of Defence and National Security of 13 October 2017. Available at: https://www.defense.gouv.fr/english/dgris/defence-policy/strategic-review-of-defence-and-national-security-2017/strategic-review

French Ministry for the Armed Forces (2019) Space defence strategy, Report of the "Space" working group, July 2019. Available at: https://www.defense.gouv.fr/english/dgris/defence-policy/defence-strategy/defence-strategy

French Republic (2008) Law concerning space operations – *Loi sur les Opérations Spatiales* (French Space operations act). Available at: https://www.unoosa.org/pdf/pres/lsc2009/pres-04.pdf

French Republic (2012) French space strategy (*Stratégie spatiale française*). Available at: https://www.vie-publique.fr/sites/default/files/rapport/pdf/124000161.pdf

French Republic (2015) National strategy for the security of maritime areas – Adopted by the interministerial sea committee on 22 October 2015. Available at: https://www.gouvernement.fr/sites/default/files/contenu/piece-jointe/2016/01/strategie_nationale_de_surete_des_espaces_maritimes_en_national_strategy_for_the_security_of_maritime_areas.pdf

German Federal Government (2016) White paper on German security policy and the future of the Bundeswehr. Available at: http://www.gmfus.org/publications/white-paper-german-security-policy-and-future-bundeswehr

German Federal Government (2018) High-tech strategy 2025. Available at: https://www.bmbf.de/upload_filestore/pub/The_High_Tech_Strategy_2025.pdf

German Federal Ministry of Economics and Technology-BMWi (2010), Making Germany's space sector fit for the future – the space strategy of the German Federal Government, November 2010. Available at: https://www.dlr.de/rd/en/Portaldata/28/Resources/dokumente/Raumfahrtstrategie_en.pdf

German Federal Ministry of Justice (1990) Law governing the transfer of responsibilities for space activities of 8 June 1990 as amended. Available at: http://www.esa.int/About_Us/ECSL_European_Centre_for_Space_Law/National_Space_Legislations

German Federal Ministry of Justice (2007) Satellite data security act. Available at: https://www.bmwi.de/Redaktion/DE/Downloads/S-T/satdsig-hintergrund-en.pdf?__blob=publicationFile&v=1

Giannopapa C, Adriaensen M, Antoni N, Schrogl K-U (2018) Elements of ESA's policy on space and security. Acta Astronaut 147:346–349

Italian Ministry of Defence (2015) White Paper for International Security and Defence, 2015. Available at: http://www.iai.it/sites/default/files/iaiwp1734.pdf

Italian Ministry of Defence (2018) Plurennial Programmatic document (Documento Programmatico Pluriennale, DPP). Available at: https://www.iai.it/en/pubblicazioni/italys-defence-expenditure-what-impact-eu-defence-cooperation

Italian Parliament (2018) Space bill containing measures for the coordination of the space and aerospace policies along with important regulations concerning the organization and functioning of the Italian Space Agency-ASI. Available at: https://www.altalex.com/documents/leggi/2018/02/14/misure-per-il-coordinamento-della-politica-spaziale

Italian Presidency of the Council of Ministers (2019) National Security Strategy for Space, July 2019. Available at: http://presidenza.governo.it/AmministrazioneTrasparente/Organizzazione/ArticolazioneUffici/UfficiDirettaPresidente/UfficiDiretta_CONTE/COMINT/NationalSecurityStrategySpace.pdf

Italian Prime Minister's Office (2019) Government guidelines on space and aerospace, March 2019. Available at: http://presidenza.governo.it/AmministrazioneTrasparente/Organizzazione/ArticolazioneUffici/UfficiDirettaPresidente/UfficiDiretta_CONTE/COMINT/DEL_20190325_aerospazio-EN.pdf

Italian Space Agency – ASI (2016) Strategic vision document 2016–2025. Available at: https://www.researchitaly.it/en/projects/space-strategic-vision-2016-2025-document-published-by-asi/

Japan Space Forum (2015) The German interagency approach to SSA, presentation, 2015 Japan Space Forum SSA symposium

NATO (2019) Defence expenditure of NATO countries (2013–2019) press release, PR/CP (2019) 123, November 2019. Available at: https://www.nato.int/cps/en/natohq/news_171356.htm

Papadimitriou A, Adriaensen M, Antoni N, Giannopapa C (2019) Perspective on space and security policy, programmes and governance in Europe. Acta Astronaut 161:183–191

Spacewatch.Global (2019) first EU-ESA Space Council in Eight Years Held, discusses Space as enabler, 28 May 2019. Available at: https://spacewatch.global/2019/05/first-eu-esa-space-council-in-eight-years-held-discusses-space-as-enabler/

Spanish communication company specialized in Defence and security – IDS (infodefensa) (2017) Spain Defence & Security Industry, R&D, the strategic approach 2017. Available at: https://www.infodefensa.com/archivo/files/spain2017(ing)OK%20.pdf

Spanish Government (2013) The National Security Strategy – sharing a common project. Available at: https://www.lamoncloa.gob.es/documents/estrategiaseguridad_baja_julio.pdf

Spanish Ministry of Defence (2015) Master plan of space systems. Available at: http://www.defensa.gob.es/Galerias/dgamdocs/plan-director-sistemas-espaciales.pdf

Spanish Prime Minister's Chancellery (1995) Royal Decree No. 278/1995 of 24 February 1995, Establishment in Spain of the Registry of Objects Launched into Outer Space as provided for in the Convention adopted by the United Nations General Assembly on 12 November 1974. Available at: http://www.unoosa.org/oosa/en/ourwork/spacelaw/nationalspacelaw/spain/royal_decree_278_1995E.html

UK Government (1986) Outer space act. Available at: http://www.unoosa.org/res/oosadoc/data/documents/2017/aac_105c_12017crp/aac_105c_12017crp_21_0_html/AC105_C1_2017_CRP21E.pdf

UK Government (2015a) National Security Strategy and Strategic Defence and Security Review 2015-A Secure and Prosperous United Kingdom, November 2015. Available at: https://assets.publishing.service.gov.uk/government/uploads/system/uploads/attachment_data/file/478933/52309_Cm_9161_NSS_SD_Review_web_only.pdf

UK Government (2015b) National Space Policy., 13 December 2015. Available at: https://www.gov.uk/government/uploads/system/uploads/attachment_data/file/484865/NSP_-_Final.pdf

UK Ministry of Defence (2017), Joint Doctrine Publication 0–30 (JDP 0–30) – UK Air and Space Power, December 2017. Available at: https://assets.publishing.service.gov.uk/government/uploads/system/uploads/attachment_data/file/668710/doctrine_uk_air_space_power_jdp_0_30.pdf

UK Ministry of Defence (2019) Towards a defence space strategy. Available at: https://assets.publishing.service.gov.uk/government/uploads/system/uploads/attachment_data/file/712376/MOD_Pocket_Tri-Fold_-_Defence_Space_Strategy_Headlines.pdf

UK Space Agency (2013) Strategy for earth observation from Space 2013–16, October 2013. Available at: https://assets.publishing.service.gov.uk/government/uploads/system/uploads/attachment_data/file/350655/EO_Strategy_-_Finalv2.pdf

Poland and Space Security

26

Małgorzata Polkowska

Contents

Introduction	486
The Polish Space Agency (POLSA)	486
Strategic Priorities in Space Law and National Security and Defense	487
Polish Security and Defense Strategy	487
Polish Space Law and Space Strategy	489
National Programs	490
Multilateral Programs	492
Bilateral Cooperation	492
European Union SST Consortium and Role of Poland	493
ESA and Poland	496
Conclusions	497
Acknowledgement	497
References	497

Abstract

The main objectives of this chapter are to address the importance of space security in Poland, as mainly demonstrated through the activities of the Polish Space Agency (POLSA). This agency has played an important role in the Space Situational Awareness (SSA) program for safety and security purposes. Poland decided to put in place an SSA system as part of the national and regional security law and policy. The proper implementation of the SSA program is necessary for all states. Poland, as a full member of the European Space Surveillance and Tracking (SST) Consortium, is willing to participate in European projects. Thus, the regional and multilateral cooperation in this matter is crucial.

M. Polkowska (✉)
University of War Studies, Warsaw, Poland
e-mail: m.polkowska@akademia.mil.pl

© Springer Nature Switzerland AG 2020
K.-U. Schrogl (ed.), *Handbook of Space Security*,
https://doi.org/10.1007/978-3-030-23210-8_130

Introduction

Space security has always been an important component of national policy. The Space Situational Awareness system is crucial for a state's security, as well as for its international responsibility to implement it correctly into national policy and law. This chapter presents some legal challenges occurring in this process within the European Union, with particular emphasis being put on Poland. The proper governance of space security is important for states, not only at international but also at regional and national level. That is why the role of the national regulator should be significant in incorporating space security in the policy and strategy of the state. Regular cooperation between states at international and bilateral level is necessary. Although Poland is not a major space-faring nation, the security issue has always been treated as a priority.

The Polish Space Agency (POLSA)

Poland has become more and more active in space and also in space security, in particular since the Polish Space Agency (POLSA) was created. This agency is a governmental executive body, subject to the Prime Minister. It consists of civilian and military personnel. POLSA was established by the Act of 26 September 2014 and became fully operational at the end of 2015 (Polish Government 2014). The agency participates in fulfilling the strategic goals of the Republic of Poland by supporting the utilization of satellite systems and the development of space technologies. The main tasks of POLSA cover the following five areas: coordinate the activity of the Polish space sector at national and international level; represent Poland in relation to international space sector organizations; support national science and business projects associated with space technologies; popularize the use of satellite data by public administration; and increase the defensive capabilities of the country. The agency is executive in nature in accordance with the Act of 27 August 2009 on public finance (art. 1.1 – Act of 26 September 2014), and it can create local branches. The headquarters of the agency are located in Gdansk (Art. 1 par. 3).

The activities of the agency are under the auspices of the President of the Council of Ministers (Art. 2). The duties of the agency are written in Art. 3. The POLSA Council consists of representatives of the government – one from each administration and four representatives of scientists and the industry with recognized achievements in research or business – which are chosen based on their knowledge competence in areas concerning POLSA activities (Art. 14) (Polkowska 2016). One of the main areas of POLSA activities is international cooperation. POLSA is committed to multilateral cooperation in the framework of the European Space Agency (ESA) and the European Union (EU). POLSA supports especially Polish actors who apply to the space programs. They have already started efforts to integrate the national industrial sector in projects implemented by the European Organization for the Exploitation of Meteorological Satellites (EUMETSAT). POLSA provides also active bilateral cooperation with ESA Member States, the EU, and other countries, primarily in the field of space exploration.

POLSA supports the Polish space sector and facilitates exchanges, by organizing competitions for advisory services. The entities receive professional support in the form of consultation with experts. Entrepreneurs from SMEs receive support in applying to competitions organized by the European Space Agency. POLSA activities aim at contributing to the growth of innovation and competitiveness of Polish companies in the space sector. POLSA encourages the involvement of high-tech or IT operators in the space industry, and it also promotes solutions supporting the Polish state administration at central and local government. This results in enhanced efficiency of the administrative work by using everyday services based on satellite data and satellite technology, including Earth observation, navigation, and telecommunications. With regard to Research and Development (R&D), POLSA supports the Polish scientific institutions and companies who are active in R&D in the field of fundraising for scientific and industrial research. POLSA assists in conducting work on space applications and space technology development.

In addition, POLSA carries out educational activities in the field of popularization of knowledge of space research at secondary education and high school in Poland. It covers the subject of space engineering and astronomy, as well as it initiates and supports with its expertise the creation of new courses of higher education. Last but not least, one of the main priority tasks of the Polish Space Agency is to provide security for Poland and its citizens by increasing Polish defense capabilities through the use of satellite systems. To this direction, the Polish Space agency aims to ensure the security of the state and its citizens and to contribute to the Polish defense potential through the use of satellite systems. Therefore, an important area of the agency's activity is the coordination of activities aimed at the effective use of space technologies and satellite applications for defense purposes.

Strategic Priorities in Space Law and National Security and Defense

Polish Security and Defense Strategy

The tasks resulting from the abovementioned priorities are carried out by the Vice President's Division for Defense Affairs. The Division consists of the Department of Military Satellite Technologies and the Department of Defense Projects. The Military Satellite Technologies Department consists of the Earth and space satellite recognition team, the satellite navigation team, and the satellite communications team. The Military Satellite Technologies Department is responsible for the following tasks (Polish Space Agency www.polsa.gov.pl.):

- Use of satellite products and services for national security and defense; ensuring the state's capacity in the field of Earth observation, satellite navigation, and satellite communications, in particular for the needs of the Armed Forces of the Republic of Poland
- Identification of military needs in the field of satellite and space technologies and recommendation of their implementation within the framework of the plan for technical modernization of the Polish Armed Forces
- Cooperation with the institutions within the Ministry of National Defense who manage military equipment in the field of Earth observation, communication, and satellite navigation, including the determination of the scope of this cooperation
- Developing concepts for the development and application of satellite technologies for national security and defense, including the development of documentation for the preparation of feasibility studies
- Analyzing and issuing opinions on projects prepared and implemented by Polish entities for national and international programs within the framework of military satellite technologies from the technical point of view
- Monitoring and analyzing progress in the global development of satellite technologies and methods, as well as supporting Polish scientific, research, and industrial entities that develop satellite technologies for potential application in the Armed Forces of the Republic of Poland

The Department of Defense Projects has the following responsibilities:

(a) Supporting the needs of the Polish Armed Forces and other bodies and institutions responsible for national security through:

- Implementation of programs and projects as well as management of scientific research or development works aimed at acquiring new technologies for the purposes of national defense and security
- Monitoring defense and security programs and projects while developing recommendations on how to manage their effects in order to achieve new capabilities resulting from space exploration and development of satellite technologies
- Preparation of analyses, reports, recommendations, and advice in the field of space exploration, as well as the use of space and space technologies for national defense and security purposes
- Representation of national defense and security interests in the area of space exploration and uses of outer space

(b) Increasing Poland's participation in international programs of significance for the national defense and security objectives through:

- Activities aimed at the increasing involvement of scientific centers, research centers, industry, and the Ministry of National Defense in international programs

- Supporting, in consultation with the Minister of National Defense, the representation of national defense interests in the field of space exploration and use in international programs
- Participation in the consultation process concerning Polish activity in the field of space exploration and use for defense purposes, including by appointing advisors and experts to Polish delegations and representations in European Union institutions and bodies of international organizations, in particular ESA, EDA, and NATO
- Monitoring activities in the field of initiatives and participation of the Polish Armed Forces in international programs
- Taking care of the cohesion of the Polish space policy within the framework of implemented international programs and recommending the possibility of implementing the effects of this cooperation
- Supporting works aimed at preparing and updating the National Space Sector Development Program

Polish Space Law and Space Strategy

Poland is very active in developing a policy and legal framework concerning space security and Space Surveillance and Tracking (SST). In February 2017 the Polish Space Strategy, published by the Polish Ministry of Economic Development, entered into force (Polish Space Agency www.polsa.gov.pl.). The objectives stipulated in the strategy include, among others, increasing the competitiveness of the Polish space sector and its share in turnover (increasing participation in the EU space programs through the EU SST Consortium); development of satellite applications; strengthening capacities in the area of security and defense using space (establishment of SSA); and creating favorable conditions for the development of space sector in Poland or building resources for the Polish space sector. The strategic aim is to obtain 3% of the EU market in 2030.

In 2018 the National Space Plan was established for the years 2019 and 2021 (www.polsa.gov.pl). One of the main ambitions is the establishment, development, and operation of national Space Situational Awareness (SSA) system in relation to the EU SST Consortium. The objective of the project is to enhance the security of citizens and infrastructure (Earth and space) in the context of space threats, to build national SSA capabilities, and to prepare for commercial exploitation of services provided in the area of SSA. The first phase of the activity is to launch basic functions of the national SST system, inter alia, through the development of infrastructure and capabilities enabling the implementation of tasks envisaged within the framework of Poland's future membership in the European SST Consortium. On 19 December 2018, Poland joined the European SST Consortium related to the tracking of space debris threatening infrastructure in space and on Earth (Polkowska 2019, Malacz etal. 2018).

The Polish Space law is still waiting for parliamentary approval. Several versions of the draft have been developed. Currently, the Government Legislation Centre website has published a draft law on the regulation of space activities and the National Register of Space Objects (Polish Government, Government Legislative Process https://legislacja.rcl.gov.pl/.). Earlier, however, the amendment of the Act on POLSA will be processed. The changes proposed in the draft act aim to clarify the scope of the Polish Space Agency's tasks, as an executive agency that provides the expertise and technological knowledge to other public administration bodies involved in space activities and is responsible for the preparation and coordination of the National Space Program implementation; adapt the supervision of POLSA to the solutions in place in other European countries, especially in the Member States of the European Space Agency (ESA); as well as introduce improvements in the organization of POLSA.

National Programs

In December 2018 the Polish National Space Program was introduced. At the moment it is still in public consultations. The Polish Space Agency (POLSA) will be responsible for the implementation of the program. POLSA has considered a few areas of public support within the program, such as "Development of satellite systems" – with one of the priority projects being the "Space Situational Awareness" system. The vital goal of the project is to ensure a long-term access to the European and national space infrastructure by providing services crucial for securing its operations. As a consequence, a network of sensors (telescopes, lasers, radars) responsible for space object observation and tracking will function on the territory of Poland, and personnel will be trained in order to perform tasks in the frame of SST (Polkowska 2019). Poland is very motivated to follow the European legislation steps and supports EU activities at the SSA level. Poland joined the European SST Consortium related to the tracking of space debris threatening the infrastructure in space and on Earth. At the end of 2018, Poland became a full member of the European Space Surveillance and Tracking Consortium. The accession agreement was signed on 19 December 2018 at the seat of the Polish Space Agency in Warsaw. Joining the consortium will enable national entities to participate in projects financed by the European Union, whose budget in the current and future financial perspective may amount to more than EUR 350 million.

Membership in the consortium will allow for faster development of the Polish SST system, which will provide Poland with data necessary to protect the planned missions of Polish satellites and will support national security and defense in monitoring threats from artificial space objects. Participation in the European program also brings great scientific and business potential. Ensuring the operability of the observation sensors forming the Polish SST infrastructure, the possibility of their modernization, and the demand for new ones – all this will facilitate a faster growth of competence in the area of SST and optical and radar observations for Polish entities, which already today gain experience by implementing projects under the optional SSA program in ESA.

Starting from 2019 Poland, as a full member of the SST Consortium, shares benefits from services at national level and especially adapts nationally to the possible future contribution. That is why to both stimulate and secure the economic growth and to protect European citizens, there is a strong need for continuous dialogue to improve the boundaries for a more comprehensive future Space Situational Awareness program. Joining the SST Consortium will allow to enhance national capabilities related to observation and awareness building in space, increase the national space sector competences and their role in current and future programs of the European Union and the European Space Agency, as well as strengthen Poland's position in the international arena. Therefore, the establishment, development, and exploitation of the national system of situational awareness in space has been included as one of the five large projects in the National Space Program for the years 2019–2021.

Poland – as a member of the SST Consortium – will ensure the use of a number of telescopes located worldwide (including Poland, Argentina, Australia, Chile, South Africa, and the USA) belonging to:

The Nicolaus Copernicus Astronomical Center of the Polish Academy of Sciences with research groups: High Energy Astrophysics, Stellar Astrophysics, etc.

The Adam Mickiewicz University Astronomical Observatory Institute in Poznań: research on dynamics of artificial satellites and space debris or optical observations and laser ranging of artificial satellites and space debris

The Baltic Institute of Technology Research Foundation in Gdynia: robotic telescopes for the detection and monitoring of space debris and satellites and companies

The Sybilla Technologies: a software development company and integrator of turn-key robotic optical observatories, specializing in the "from sensor to TDM" chain of SST services

The 6ROADS: a focused company that owns and operates a global network of telescopes to provide high-quality data in the SST domain

The infrastructural core of 6ROADS is composed of six optical observatories located across the globe. As a company 6ROADS was established in 2016, although its experience can be traced back to 2003 and the laser station of the Space Research Centre of the Polish Academy of Sciences from the Observatory in Borowiec. The Borowiec satellite laser ranging station belongs to the Space Research Centre of the Polish Academy of Sciences (SRC PAS), and it is the only such device in Poland and one of the few ones in the world working since the mid-1980s. The Borowiec satellite laser ranging station (CSLRB) started to track typical space debris targets, cooperative inactive/defunct satellites, and uncooperative rocket bodies in the middle of 2016. Currently, the station is actively involved in the development of the Space Surveillance and Tracking program, developed by European Space Agency (Space Safety program), as well as programming tools and competences related to the processing and analysis of SST data (Polish Space Sector catalogue of selected entities, POLSA 2018 and Konacki et al. 2019).

Multilateral Programs

Bilateral Cooperation

In April 2019 during the 35th Space Symposium in Colorado Springs, POLSA signed an agreement with US Strategic Command (USSTRATCOM) for sharing SSA data. Due to this agreement, Poland will gain access to Space Situational Awareness services and data, including alerts concerning collisions, fragmentation, or uncontrolled satellite reentries into the Earth's atmosphere. The data will be provided by the US Air Force Space Command's 18th Space Surveillance Squadron on the basis of the agreement signed. Poland joins 18 nations (the Netherlands, Brazil, the UK, the Republic of Korea, France, Canada, Italy, Japan, Israel, Spain, Germany, Australia, Belgium, the United Arab Emirates, Norway, Denmark, Thailand, and New Zealand), 2 intergovernmental organizations (the European Space Agency and the European Organization for the Exploitation of Meteorological Satellites), and more than 77 commercial satellite owners, operators, and launchers already participating in SSA data-sharing agreements with USSTRATCOM (US Embassy and Consulate in Poland https://pl.usembassy.gov/ssa/.). The collaboration falls within areas of SWE, NEO, and SST.

Poland understands the SSA influence on national security in space, which is a priority for Polish politicians and legislative bodies. In October 2019 the leaders of POLSA and the US National Aeronautics and Space Administration (NASA) signed a landmark Joint Statement of Intent for Space Cooperation during the 70th International Astronautical Congress in Washington, D.C. This agreement highlighted the countries' respective interests in cooperative human and robotic space exploration in all spheres.

During just the past 40 years, over 80 instruments designed and constructed by Polish scientists and engineers have been employed in various international space missions, and 7 years of participation in the European Space Agency (ESA) have resulted in the dynamic development of the Polish space sector, in which more than 350 Polish enterprises operate. These companies and institutions cooperate with ESA and other national agencies, including NASA, DLR (German Space Agency), JAXA (Japanese Aerospace Exploration Agency), and CNSA (China National Space Administration). These include several sensors and robotic probes on NASA's Mars Curiosity and InSight landers, among other significant accomplishments.

Poland has established relations with European states as well in space. In 2000 started the relations with EUMETSAT as a cooperating member and gained full membership in 2009. The Act on Ratification of the Convention on the Establishment of the European Organization for the Exploitation of Meteorological Satellites (EUMETSAT) drawn up in Geneva on 24 May 1983 was adopted by the Parliament of the Republic of Poland on 9 January 2009. The Institute of Meteorology and Water Management, the National Research Institute, is responsible for relations with the organization and participates in the work of EUMETSAT bodies. Thanks to its membership in the organization, Poland has full access to satellite data from its satellites. The images provided by EUMETSAT are primarily used by the state hydrological and meteorological services to prepare forecasts, hydrometeorological protection of society and economy, and monitoring of climate change by the Polish Armed Forces, universities, and research institutes. Images from meteorological satellites may also be used in agriculture, forestry, land management, spatial management, investment planning, and monitoring of water areas (including the Baltic Sea) and atmosphere, including, among others, trace gas and aerosol measurements. Furthermore, the Polish Institute of Geodesy and Cartography is collaborating in the project COSMO-SkyMed.

European Union SST Consortium and Role of Poland

The European SST Support Framework related to detection, tracking, and monitoring of orbital movements of active and inactive satellites and space debris is a key operational program of the European Union in the space area, in addition to the Galileo satellite navigation program and the Copernicus Earth observation program. The initiative is based on the 2014 Decision of the European Parliament and the Council of the European Union to establish a Space Observation and Tracking Support Framework to contribute to the long-term availability of European and national space infrastructure and services necessary for the security of Europe's economy, society, and citizens (Decision No 541/2014/EU of the European Parliament and of the Council of 16 April 2014 establishing a Framework for Space Surveillance and Tracking Support, 27.5.2014, L 158/227.).

The main objectives of the SST program concern assessing and reducing the risks to European spacecraft launches and operations resulting from possible collisions; studies, assessments, and warnings against uncontrolled entry into the atmosphere of space objects and debris threatening the safety of citizens and ground-based infrastructure; or exploring ways of preventing the spread of space debris (Polish Space Agency https://polsa.gov.pl/en/.). The European SST program related to the detection, tracking, and monitoring of orbital movement of active and inactive satellites and space debris is a key operational program of the European Union in the space area, alongside the Galileo satellite navigation program and the Copernicus Earth observation program. In 2015 the European Commission set up the SST Consortium to coordinate this initiative, whose task is to pool the capabilities of European countries in order to secure European and national space infrastructure (Faucher 2019).

Member States have contributed to the consortium national optical and radar sensors capable of observing artificial objects moving around the Earth and the ability to analyze the data provided by these sensors. On the basis of processed data, SST services shall be implemented in the form of risk assessment, information, and alerts on actual and predicted space events involving artificial space objects. Such events can be, e.g., collisions and fragmentations of objects in orbit or uncontrolled entry into the Earth's atmosphere of artificial space objects. Information is made available to stakeholders, including EU institutions, Member States, and satellite operators. Initially, the members of the consortium were leading European countries, Germany, France, Spain, Italy, and the UK, represented mainly by national space agencies. At the end of 2018, Romania, Portugal, and Poland joined the group (Faucher 2019). In view of progressing the commercialization of products related to situational awareness in space, domestic entities providing solutions and services in this area will be able to direct their offer also to the global market, which will grow as a result of the New Space trend, the increasing number of small satellites, the planned development of mega-constellations and new areas such as satellite in-orbit servicing, or, in the longer term, the sourcing of raw materials from celestial bodies (Polish Space Agency https://polsa.gov.pl/en/.).

Having in mind the provisions of the Space Strategy for Europe and longstanding dialogue in the space domain in Europe, as being performed among others within the SST Committee, Poland supports the SST Support Framework and its further evolution. Space technologies, operational spacecraft, and relevant services are the key elements to sustain, support, and enhance the EU policies, economy, security, and technology leadership which are nowadays already of great importance and undoubtedly will be much more in the future. Poland joining the SST Consortium is the result of several years of activity from the Polish side, including the Ministry of Enterprise and Technology, the Ministry of National Defense, and the Ministry of Science and Higher Education and the Polish Space Agency, which prepared the accession application submitted to the European Commission and, together with specialists from the Ministry of Science and Technology, conducted negotiations on the most favorable membership conditions. As part of the preparatory works of the Polish Space Agency, a feasibility study of the Architecture of the Space Awareness System in Poland (SSA) project was commissioned, with particular emphasis on the space object observation and tracking subsystem (SST).

Participation in the Consortium brings for Poland a number of other obligations of organizational nature (e.g., creation of a network of declared sensors and a national operational center), human resources (e.g., provision of specialists for the functioning of the national SST system), financial (allocation of national resources for the launch of the SST system and absorption of EU funds), and legal obligations (adjustment of the SST system). The first stage of work resulting from Poland's membership in the SST Consortium and provided for in the National Space Program will be to launch basic functionalities of the system based on declared sensors, inter alia, through the development of national infrastructure and capabilities enabling the implementation of tasks and projects within the European SST Consortium and increasing the participation of Polish entities in the SSA (Space Situational Awareness) program of the European Space Agency.

Poland shares the main point of the Commission recognizing the aim of current Decision from 2014 "to ensure the long term availability of European and national space infrastructure facilities and services which are essential for the safety and security of the economies, societies and citizens in Europe" and have a will to extend it to the issues of "sustainable access to space and its usage" and "protection of EU and its citizens against any natural and artificial hazards coming from space" by the possible intervention in the following domains: Space Surveillance and Tracking (SST), Space Weather (SWE), and Near-Earth Object (NEO). It seems that the future steps of Commission might also consider space traffic management (STM), active debris removal (ADR), and space mining (SM). SST, SWE, and NEO seem to be of high priority as directly influencing safety and security of European citizens and very challenging to tackle at national level. In those areas EU seems to have the necessary prerogatives to take necessary actions on European level. However, SWE and NEO should be considered to be continued and enhanced based on ESA accomplishments and with ESA involvement at least in the Research and Development (R&D) phase. To fulfil the objectives of the current Decision from 2014 and having in mind the magnitude of necessary intervention, Poland is in favor of developing the idea of broadening the current SST Framework Support, eventually like main EU space program as Galileo or Copernicus. Only such an approach may properly correspond to the challenges of contemporary space environment such as (Chimicz 2018):

1. Increase in congested and contested space environment, among others taking into account planned mega-constellations, new space entrants, suborbital flights challenges, maturing, and usage of ASAT technologies
2. Possible incorporation of space into air traffic management
3. Emergence of new industry branches in the domain (e.g., sensors technologies as a commodity, commercialization of SST delivery)
4. Growing interest of space weather for aviation, transport, and other domains
5. Increase of security and defense issues of space
6. Enforcing international regulation and responsible behavior in space (COPUOS and IADC guidelines)

Moreover, in case of financial constraints, it should be considered to use also other tools available to the European Commission to enable a coherent intervention. In practice, the overall program may be accomplished by the main financial line (e.g., SSA) with additional supporting lines for, e.g., SWE/NEO coming from other programs (e.g., Copernicus, Galileo, or H2020). Poland strongly recognizes the two-dimensional aspects of the programs, i.e., the military and civilian. From this perspective the European intervention and existing and future regulations must respect national interests and issues related to sovereignty of countries, in particular the assets ownership, as well as recognize the program as a kind of contribution to European Common Security and Defense Policy. As today's EUSST program should encompass strong cooperation with the USA and parallel allow to include (if applicable) the option to cooperate with other non-EU counties able and willing to contribute to it (e.g., allied nations as well as space faring nations). Additionally, other ways of cooperation between potential EUSST/SSA Consortium member and other EU states (if really profitable) should be considered (EUSST https://www.eusst.eu.).

ESA and Poland

ESA is a basic partner for the majority of Polish companies from the space sector. Importance of the cooperation with ESA lies not only on the opportunities of funding development work but also on a large transfer of know-how from ESA to Polish entities such as SAT-AIS-PL, in the frame of which Polish entities build a microsatellite. This work is supervised by ESA experts (Brona 2018). Regarding the trends of the Space 4.0 era, the niches occupied by the Polish space sector are primarily:

- Optical observations
- SSA program systems and subsystems
- SSA data processing software
- Space robotics
- Photonics
- Optoelectronics
- Small satellite systems and subsystems (power supply systems, on-board computers, satellite receivers for GPS and Galileo systems, integrated microwave modules, satellite propulsion systems, and exploration missions)
- EGSE (Electrical Ground Support Equipment) and MGSE (Mechanical Ground Support Equipment) systems
- Aggregation and data processing systems
- EO and GNSS data processing applications

The Polish Space Agency, together with state institutions, takes part in the European Space Surveillance and Tracking (EUSST) clusters and PERASPERA (Plan European Roadmap and Activities for Space Exploitation of Robotics and Autonomy), as well as in the European network ENTRUSTED (Governmental Satellite Communication (GOVSATCOM) program). Additionally, a primary national space-focused project is the SAT4ENVI platform, concerning the operational collection of sharing and promotion of digital satellite information on the environment. Polish companies and research institutes are already implementing ambitious ESA projects developing satellite applications or creating products present in orbit around the Earth and used in scientific missions in the Solar System. Poland allocated a total of EUR 39 million for optional programs, engaging in seven programs: European Exploration Envelope Program (E3P), Space Safety Program, Earth Observation, program of Advanced Research in Telecommunication Systems (ARTES 4.0), Navigation (NAVISP), General Support Technology Program (GSTP), and PRODEX.

Poland will take part in Earth observation programs, which in recent years have proved to be a very popular and fully exploited area for the Polish space sector. Among the optional programs that Poland has decided to support is also the Space Safety Program, which gives Polish companies the opportunity to achieve a high degree of specialization in space security technologies. The Polish Institute of Geodesy and Cartography has launched several projects within ESA. Current projects are ESA EOStat (2018–2020) – Agriculture Poland: Services for Earth Observation-based statistical information for agriculture (Coordinator: PhD Jedrzej Bojanowski, cooperation: Institute of Geodesy and Cartography, Space Research Centre of Polish Academy of Sciences, Statistics Poland); and ESA IRRSAT (2017–2019) – Irrigation Factor 4 potato growth using Sentinel-1 and Sentinel-2 data (Coordinator: Industrial Institute of Agricultural Engineering, cooperation: IGiK Remote Sensing Centre, WUT Institute of Electronic Systems) (Polish Government, Ministry of Development, Poland on Space 19+, https://www.gov.pl/web/rozwoj/polska-na-space19—perspektywy-rozwoju-sektora-kosmicznego-na-najblizsze-trzy-lata.).

Conclusions

As it was already mentioned in national strategies, societies increasingly depend on space technologies. Poland and its Space Agency are convinced that security in Space is a crucial issue that needs to be tackled. The practical example of the international use of SSA is the creation of the EU Consortium (with Poland on board), which will require from the Member States internal and external cooperation for security purposes. Due to the fact that Poland is more active in space, security has become more and more important. That is the reason why there is a need for increased cooperation among Member States in Europe and worldwide and information exchange about the best achievable ways forward.

Acknowledgement

This publication was financed under a project implemented in the Research Grant Program of the Ministry of National Defense, Republic of Poland.

References

Brona G (2018) Mapa rozwoju rynków i technologii dla sektora kosmicznego w Polsce, 2019, Polska Agencja Przedsiębiorczości 2018

Chimicz A (2018) POLSA, 13th military SSA conference presentation, April 2018, London, enhancing space capabilities for the polish MOD: SSA System Development

EUSST. https://www.eusst.eu/

Faucher P (2019) EU space surveillance and tracking framework a consortium of member states safeguarding European space infrastructure and orbital environment, presentation made by chairman, SMI conference, 2 Apr 2019

Konacki M et al (2019) Optical, laser and processing capabilities of the new polish space situational awareness centre presentation done at AMOS conference, September 2019

Malacz A, Konacki M, Malawski M (2018) Information of POLSA about polish accession to the EUSST consortium. Presentation at the April 2019 SSA conference, London

Polish Government (2014) Act of 26 September 2014, Dz. U., 2014, poz. 1533

Polish Government. Government Legislative Process. https://legislacja.rcl.gov.pl/

Polish Government. Ministry of Development, Poland on Space 19+ https://www.gov.pl/web/rozwoj/polska-na-space19%2D%2D-perspektywy-rozwoju-sektora-kosmicznego-na-najblizsze-trzy-lata

Polish Space Agency-POLSA. https://polsa.gov.pl/en/

Polkowska M (2016) Polish Space Agency pursues task of developing country's space expertise. Room Space J 2(8):68–69

Polkowska M (2019) European challenges in SSA. Poland example, presentation at the Space Situational Awareness Workshop: perspectives on the future, directions for Korea, Seoul 24–25 Jan 2019

POLSA (2018) Polish Space Sector catalogue of selected entities. https://polsa.gov.pl/images/polski_sektor_kosmiczny_katalog_pl_eng/PODGLAD_PAK-KATALOG_EN_small.pdf

US Embassy and Consulate in Poland. https://pl.usembassy.gov/ssa/

Space Security in the Asia-Pacific

27

Rajeswari Pillai Rajagopalan

Contents

Introduction	500
Emerging Space Security Dynamics in Asia	500
Indicators of Conflict	502
Drivers of Space Conflict	507
What Can Be Done?	509
Conclusion	511
References	511

Abstract

Asian countries are becoming increasingly active and more serious players in outer space. On the other hand, political and military competition in Asia are also impacting the Asian space efforts. The growing space competition is indicated by the rapid increase in counter-space capabilities such as anti-satellite (ASAT) missiles, electronic and cyber warfare capabilities as well as new efforts at creating specialized military agencies devoted to space utilization. There are three key drivers to space conflict in Asia – increasing use of space for military purposes; civilian use that could also lead to conflict because of congestion and competition; and investments in military technologies such as ASAT and missile defense. Existing global governance mechanisms are clearly inadequate to manage these challenges. While there are a number of different ideas including the Prevention of Arms Race in Outer Space (PAROS), political disagreements have so far prevented any progress. Unless the dangers inherent in this competition is recognised, the Asian space security efforts will only aggravate, which will further add to the challenges in developing global rules of the road. Key Asia-Pacific

R. P. Rajagopalan (✉)
Nuclear and Space Policy Initiative, Observer Research Foundation, New Delhi, India
e-mail: rajeswarirajagopalan@gmail.com

© Springer Nature Switzerland AG 2020
K.-U. Schrogl (ed.), *Handbook of Space Security*,
https://doi.org/10.1007/978-3-030-23210-8_100

space players should make all efforts at developing legal measures, norms, Group of Governmental Experts (GGE), and codes of conduct.

Introduction

Asia has been going through a strategic flux owing to the changing regional and global military balance. The growing trends of militarization and securitization of political and territorial issues mean there is a greater emphasis on hard power. This is beginning to play out in outer space affairs as well. The Asia-Pacific space landscape itself has been changing rapidly in recent times. Renewed anti-satellite (ASAT) tests starting with China's first successful test in January 2007 highlight a new phase in space security competition among the Asian great powers. Other spacefaring powers are re-examining their options, strategies, and capabilities, assigning a greater security role to their space programs. For instance, after the Chinese ASAT test, the USA undertook its own test in 2008. The Chinese test also sparked a new debate in India as to how it should defend its own satellites and whether it requires New Delhi to demonstrate its own ASAT capability. India's decision to finally have a demonstrated capability resulted in its first successful ASAT test in March 2019, following which India has conducted a tabletop war game called the "IndSpaceEx" with all the stakeholders including its scientific establishment and the military. Along with the capability development, India is also shaping its institutional architecture to develop the long-term strategies for space utilization and its integration into the armed forces.

These developments are merely a sign of the increasing insecurities on Earth and how outer space has become one more domain where the terrestrial politics are playing out (Pekkanen 2015). The growing indicators of space insecurity in the Asia-Pacific mean both use of space for terrestrial conflict and conflicts about use of space (Rajagopalan 2018a). In fact, the first could also lead to the second with greater integration of space in conventional conflicts leading to more insecurities and additional conflicts. But the impact of the changing global balance of power equations cannot be ignored. The changing power dynamics in both the Asia-Pacific and at the global level have had an effect on the region. Given that many of the rising powers are in Asia and some of them are challenging the prevailing geopolitical equilibrium, there is a big focus on acquiring hard power capabilities including in the outer space realm. While there are several internal and external factors pushing countries to respond in a more aggressive manner, the consequences are difficult to predict unless states are willing to adopt measures that will bring about a certain restraining effect on irresponsible behavior in outer space.

Emerging Space Security Dynamics in Asia

The twentieth-century space competition played out between the USA and the Soviet Union, and the technology was restricted to a handful of countries, but the situation has changed dramatically. There are around 80 active space players today, including also non-state, private sector commercial players. While the private sector participation in space is still a Western phenomenon, this is changing in the Asia-Pacific, including in countries such as China and India, which have traditionally not shared space with private sector. Japan has remained somewhat more open to private sector participation as compared to China and India. Entry of the private sector is important because they bring about innovation, thereby reducing the cost to access space. Access to outer space over the last couple of decades has become much more democratized with the spread of technology beyond the handful of countries who cherished the uses of space earlier. This however is not the problem. Given the significant utilities of outer space to a number of civil applications in changing the lives of ordinary people, more and more countries have pursued and will continue to develop space programs. The growing number of countries with space programs in Asia is also phenomenal. A look at the number of countries with independent launch capabilities in Asia – China, India, Iran, Israel, Japan, and North Korea – is a sign of the increasing interest and sophistication of space capabilities in Asia (Lele 2012). In addition, there are a growing number of emerging space players including Australia, Bangladesh, Indonesia, Iran, Israel, Malaysia, North Korea, Pakistan, Singapore, South Korea, the United Arab Emirates, and Vietnam (Moltz 2011).

Space security dynamics have become more complex partly due to the crowded and congested nature of space environment. A direct issue that arises from a crowded outer space realm is space traffic management and orbital debris that have implications for safe, secure, and sustainable use of outer space. But space is a dual-use asset, and many countries are beginning to use space for a variety of military and security applications. Radio-frequency interference, for instance, could be a result of the increasingly crowded nature of outer space, but there are also deliberate attempts to jam or otherwise impede radio signals (Powell et al. 2018; Weeden). Laser dazzling and blinding, or using cyber means to create temporary disruptions as well as denial of services, or to generate interference in command and control systems and logistics network, are worrisome (Attacking Satellites). These are a lot cheaper and more easily accessible, and there is plausible deniability with these technologies, making them more attractive as an option. In general, the increasing willingness to develop and possibly use counter-space capabilities is quite worrying, and the trend is particularly evident in the Asia-Pacific (Rajagopalan 2019a; Weeden and Samson 2019; Harrison et al. 2019). This is fueled by the fact that the Asia-Pacific houses some of the fastest growing economies, which in turn has aided higher military spending including in military space programs. All of this comes against the backdrop of an Asia that has already been bickering for several different reasons. Major power relations in Asia are one of the most contested ones, with these countries having gone into war with each other, and there are also unresolved sovereignty and border and territorial issues among these great powers. This historical background adds to the reality that three of the rising powers today are in Asia, which makes the political and security issues ever more challenging. Nevertheless, it is China's rise that has been the most spectacular and the most consequential in strategic terms.

In this regard, China's systematic rise as a major space power with consequences for regional and global security cannot be lost sight of (Columba Peoples 2013). China landed an unmanned mission on the near side of the Moon using its Chang'e 3 in December 2013, and in January 2019, its Chang'e 4 landed on the far side of the lunar surface, a feat not attempted by any other country until now (Jones 2019a). The next logical step for China in this regard would be to have a human lunar mission – Beijing has already stated its future Moon missions will take it closer to establishment of a possible research base on the Moon. China also plans to develop and operate its own version of the International Space Station (ISS) in low Earth orbit by 2024, by which time the current ISS will be wound up unless additional funding comes along to keep it running (Jones 2019b).

China's growing sophisticated space capabilities, in certain niche areas such as these, have driven the other two Asian space powers also to respond. India's recent attempt to soft land on the lunar surface faced a setback; nevertheless, it was a reflection of the budding Asian space competition. A successful mission would have made India the second Asian power to successfully carry out an unmanned lunar mission (Rajagopalan 2019b). Five years ago, India, in its first attempt, became the first Asian country to send an orbiter around the Mars, called the Mangalyaan (Rajagopalan 2013). India has more ambitions for its future including a mission to study the Sun, called the Aditya mission, a second mission to the Mars, and a Venus mission (For details, see Indian Space Research Organisation, http://www.isro.gov.in). But the emerging competition is not just in terms of attempting their own "firsts," but it is moving to the security domain as well, with more serious implications. India tested an ASAT weapon in March 2019, an effort to catch up with a capability that China had already demonstrated. After the Chinese ASAT test in January 2007, India had to possess its own deterrent capability to ensure its assets are protected (Rajagopalan 2011). Japan too has had impressive achievements such as a lunar orbiter mission in 2007 although it has followed more commercially viable projects such as the Hayabusa mission, the first time a spacecraft landed on an asteroid and brought back samples (Kodama and Hoshi 2019). Therefore, Asia will continue to witness both of these races playing out, and both of these are driven more by terrestrial power politics than any other factor (Rajagopalan 2019c). Asia will continue to have important achievements, but the tense geopolitics of the Asia-Pacific will further intensify the space security dynamics as well.

Indicators of Conflict

Over the last decade, international tensions have risen in the Asia-Pacific region. This is partly the result of China's growth but also the consequence of China's behavior. China has found itself in conflict Space security:China's conflict with many countries in the region including with Japan, South Korea, Singapore, Vietnam, Australia, Malaysia, the Philippines, New Zealand, and India. Partly as a consequence of this, there is also strategic collaboration between Japan and Australia and between India and a number of countries in the region such as with Japan, South

Korea, Singapore, Vietnam, and Australia (Shaw 2019a). Again, as a consequence of this, there is also strengthening of relations between the USA and many of its partners in the region, which China in turn sees as a concern.

While some of the regional competition is giving way to more cooperative ventures, this is also driving states to develop more military space programs. Given that military operations are extremely net-centric ones, using shorter timeframes in a high-tech environment, integration of space has become absolutely critical. Operation Desert Storm and Iraqi Freedom demonstrated the "force-multiplier" nature of space, and since then major militaries have studied options to make their militaries exploit space as well. At the very least, more and more militaries are moving to develop space-based intelligence, surveillance, reconnaissance (ISR), positioning navigation and timing (PNT), satellite communications (SATCOM), information gathering, weather, environment, and terrain observation. Space has come to have a multiplier impact in terms of gaining greater sense of predictability of the operating environment for military missions on Earth, thereby reducing uncertainty and facilitating better command decisions. But these very same benefits become vulnerabilities as well – China, for example, has been investing in counter-space technologies because it sees space as USA's Achilles' heel because of the heavily networked US nature of US military operations.

One of the more direct manifestations of these growing tensions is the establishment of specialized space security institutions. Of course, the establishment of the US Space Force by President Trump received astonishing publicity by way of newspaper headlines. However, the reality is that the US effort is just the latest in a series of such actions and other major space powers have been setting up such forces as well. Even earlier, in 2011, Russia brought about certain institutional modernizations creating the Russian Aerospace Defence Forces which are meant for space security-related activities (Bodner 2018). Similarly, in 2015, China established the PLA Strategic Support Force (PLASSF) which saw the integration of the PLA space, cyber, and electronic warfare capabilities, which is considered a significant achievement considering the future of warfare that would see the interface between all these different capabilities (Kania 2018a; Davis 2019). India has its own plans to establish a Defence Space Agency. The establishment of these kinds of specialized space units has important implications for security in the Asia-Pacific. As space becomes more integrated and assumes more direct roles in conventional military operations, it can be expected that states would seek improved coordination of the military and security functions, which call for developing such specialized institutions.

A useful step will be to acknowledge that more and more countries are in the path to establishing such units. Without acknowledging this, it will be difficult to logically understand the rationale for such moves and the security conditions that are pushing countries to take such steps. As space gets further integrated into conventional military operations, the need to bring better coordination through institutions such as space force is real. While space force may sound like a fighting force, in reality it mostly seeks to achieve greater integration of functions and improved coordination levels among the several agencies involved. Nevertheless, it is important to devote attention to these institutional changes and bureaucratic innovations because of their implications in the national security realm. Also, bringing to the open as much details as possible is important because in the absence of information, public debates become unnecessary alarmist and shrill, which is not helpful. Absence of information can also heighten the security dilemma which could further prompt more countries to go down this path. A few counties doing this and lack of information on why they are doing it could produce detrimental negative effects for all. The establishment of space-specific organizations such as "space commands" clearly does not mean we are about to enter a shooting war in space. Therefore, the usefulness of full appraisal of such institutions and their purposes are important in removing the alarmism that prevails and create more transparent and sensible debates on their realistic roles. Open debates of sorts on these issues can be an important transparency and confidence building measure (TCBM). With the prevailing great power politics, dialogue and transparency measures are important to stem the current trends where countries feel that the establishment of specialized units is a way to secure themselves. Especially since these new forces and commands are coming up primarily to aid greater coordination, openness and transparency, to a great extent, should be doable. These can go a long way in removing misperceptions and miscommunication among the major spacefaring nations. It is obvious that it is in the interests of every spacefaring nation to push forward certain norms and broad rules and regulations that guide the activities of the space forces and commands. But we are at the beginning of the process, and only a handful of states have so far established such specialized institutions. There is still time to regulate this space and not make it dangerous. Even until a decade ago, space was not implicated by terrestrial geopolitics, but that cannot unfortunately be said today. The growing insecurities among the Asian powers are visible including in outer space affairs.

As for the US Space Force, it is being established through the Space Policy Directive-4 of February 2019 that would facilitate the functioning of a joint-service combat command, envisaging greater integration with the air force, army, navy, marine corps, and other national security-related institutions (For more details, read Weeden 2019; Erwin 2019; Wang 2019). Specific satellite-based services that would need to see fuller coordination and collaboration among these agencies include GPS, satellite communications, missile warning, reconnaissance, and weather (Shaw 2019b). Meanwhile, a Space Command has been established, again, as a division within the Department of Defense on August 29, 2019, similar to the other 11 unified combatant commands that the USA has. The idea of a Space Command is not new – in fact, it existed in the past, established by President Reagan in 1985, but it was wound up in the early 2000s following the terrorist attacks on September 11, 2001. President Trump has, in essence, re-established the Command as a precursor to establishing a full-fledged Space Force as the sixth branch of the US military (Reichert 2019). The logic of the Space Command or a Space Force is very clear – to deal with competition from Russia and China. In a speech to the Pentagon, Vice President Mike Pence said, "Both China and Russia have been conducting highly sophisticated on-orbit activities that could enable them to manoeuvre their satellites into close proximity of ours, posing unprecedented new dangers to our space systems. Both nations are also investing heavily in what are known as hypersonic missiles designed to fly up to five miles per second at such low altitudes that they could potentially evade detection by our missile-defense radars" (Remarks by Vice President Pence on the Future of the U.S. Military in Space 2018). He added that "China and Russia are also aggressively working to incorporate anti-satellite attacks into their warfighting doctrines. In 2015, China created a separate military enterprise to oversee and prioritize its warfighting capabilities in space. As their actions make clear, our adversaries have transformed space into a warfighting domain already. And the United States will not shrink from this challenge." On the rationale for the Space Force, Pence said, "America will always seek peace in space as on the Earth, but history proves that peace only comes through strength and in the realm of outer space the United States Space Force will be that strength in the years ahead" (Remarks by Vice President Pence on the Future of the U.S. Military in Space 2018). While these statements are absolutely true, it is also a function of the worsening general security situation and also a result of the great power politics.

China has also consolidated its efforts at bringing greater integration of space functions. China's establishment of the PLA Strategic Support Force (PLASSF) in 2015 became a significant institutional innovation as the PLASSF brought an effective amalgamation of space, cyber, electronic, and psychological warfare capabilities that were spread across other branches of the PLA and its former general departments (Costello and McReynolds 2018); for a detailed appreciation of the PLASSF and the role of outer space in PLA's military operations see (Pollpeter et al. 2017). The Chinese efforts appear to be a Gold Nicholson moment for China aimed at bringing true integration of the PLA ground, naval, air forces, and the rocket wing, as per a *People's Daily* report (Li 2016). In fact, a few months after the establishment of the PLASSF, Senior Colonel Yang Yujun, spokesperson of China's Ministry of National Defense, said that the reorganization is "mainly to integrate the various types of support forces with strong strategic, basic and supportive functions. . . . The establishment of a strategic support force is conducive to optimizing the structure of military forces and improving comprehensive support capabilities. We will adhere to system integration, military-civilian integration, strengthen the construction of new combat forces, and strive to build a strong modern strategic support force. . . . It is conducive to the adjustment of functions and streamlining of institutional personnel by the military commission" (Ministry of Defense Spokesperson 2016). The Senior Colonel went on to add that "The Strategic Support Force is a new-type combat force for safeguarding national security. It is an important growth point of the military's new combat capability. It is mainly formed for the functional integration of various types of support forces with strong strategic, foundational and supportive functions. The establishment of the Strategic Support Force is conducive for optimizing the military's force structure and improving integrated support capabilities. [The PLA] will persist with system integration, military-civilian integration, the construction of new combat forces, and will strive to build a strong and modern strategic support force" (Ministry of Defense Spokesperson 2016). According to some of the sinologists tracking this institutional innovation, it will be years before China can consolidate its reorganization process. They argue that "SSF is still in the process of consolidating, reorganizing and integrating the assorted capabilities and organizations that have fallen under its banner" (Ni and Gill 2019). Nevertheless, Adam Ni and Bates Gill have been able to put together a comprehensive picture of the SSF including the organizational structure, leadership, operational thinking, capabilities, and facilities. The two authors conclude that the PLASSF is "an important step in the PLA's journey towards realizing integrated information operations and deploying an integrated strategic deterrent," with a big focus in consolidating China's military space and information warfare capabilities. Another analyst has gone on to say that the "future trajectory will be a critical bellwether of the PLA's progress towards fulfilling its ambition of emerging as a 'world-class' military by mid-century" For a detailed Q&A on the PLASSF's future directions including the doctrine, strategy, concept of operations, see (Kania 2018b)

Russia's institutional innovation saw the birth of the Aerospace Defence Forces on December 1, 2011, combining the Air Defence and Space Forces. According to Russian reports, establishment of the Aerospace Force will be "the next logical step" with a goal to "organise military operations of multiservice force groupings in a common system of combat under a single leadership, in the new theatre of military operations" (Vekshin 2015). By way of providing the context, former Russian Air Force Commander, General Pyotr Deinekin, said, "Over the last decade the armed struggle all over the world has been actively shifting from near-Earth space to outer space," and this has been the imperative to bring the air and space under a single command. In August 2015, Russia took the next step in further consolidating its conventional military forces by creating a new branch of service – the Aerospace Forces. According to Russian reports, these have been part of the efforts to modernize and integrate existing commands with a big focus on possible enemy air and space attacks against Moscow (McDermott 2015). In August 3, a few days after the establishment of the Aerospace Forces, Defense Minister Sergei Shoigu stated that the new organizational structure was "prompted by a shift in the center of gravity of the armed struggle toward the aerospace sphere" (See Ministry of Defense of the Russian Federation 2015). However, Russian scholars argue that this is not a mere combining of air and space assets under one single command, but President Putin's August 1 decree makes the Aerospace Forces a full-fledged branch of the military, giving it equal importance like the army (McDermott 2015).

India over the last decade has begun to respond to changes in the neighborhood and beyond (Rajagopalan 2019d). The changing nature of warfare along with greater integration of space assets into conventional military operations has driven India to adapt with its own capacity building and institutional architecture. This has not come easy for India, a country that has traditionally maintained that space must be for peaceful purposes alone. At the same time, India cannot ignore the developments in its neighborhood and beyond – India's concerns around space militarization and the early trends toward weaponization of space are real. But the new developments have made New Delhi approach space from a more nuanced perspective. For instance, India's utilization of space assets for passive military applications such as surveillance and intelligence gathering is a case in point. India has increasingly come to rely on space for its Command, Control, Communications, Computers, Intelligence, Surveillance, and Reconnaissance (C4ISR) requirements. India is increasingly mindful of the fact that as militaries get dependent on space, there are also growing vulnerabilities with states developing capabilities to interfere with, disrupt, or damage these assets. The worsening trends in the overall security environment have further pushed India to get proactive in the last couple of years. India's ASAT in March 2019 is a consequence of this nuanced approach to space security. Further, India has gone on to establish the Defence Space Agency, a front runner to a full-fledged aerospace command (Lele 2019). These institutional structures are much required to bring greater synergy between the Department of Space, Ministry of Defence, and the military. India took the first step in this direction in 2008 with the establishment of the Integrated Space Cell within the Integrated Defence Headquarters, but the usefulness of the DSA will be to bring about better integration. While the DSA will be dealing with the strategy and policy questions affecting space utilization, India is now in the process of establishing a Defence Space Research Organisation (DSRO), along the lines of the Defence Research and Development Organisation (DRDO). But India must declare a space policy, which will be important as a tool for messaging to one's friends and foes alike. It can also bring about greater clarity in the domestic context which could also strengthen resource allocation.

Drivers of Space Conflict

There are three broad drivers to space conflict – increasing use of space for military purposes; civilian use that could also lead to conflict because of congestion and competition; and investments in military technologies such as ASAT and missile defense. Each of the three drivers is playing out in the Asia-Pacific region given that the region also hosts new and rising powers.

In the Asia-Pacific region, there are not only a growing number of space players, but the region also hosts some of the most advanced military space programs. The geopolitical competition, for instance, between India and China and China and Japan, is becoming new imperatives for these countries to develop outer space capabilities beyond pure civil space programs. Given the growing military dependency on outer space, countries are also developing counter-space capabilities. China has grown by leaps and bounds in this area, since it has been looking at the USA as the primary competition. The rapidly advancing counter-space capabilities including electronic and cyber warfare capabilities are concerning to India, Japan, Australia, the USA, and other Asian powers. Even though no country has placed weapons in outer space, India and other spacefaring powers are concerned about China's anti-satellite capabilities that could function as an effective A2/AD especially during conflicts. It is quite evident that no country will want to place weapons in outer space, but using ground-based assets such as ASATs to destroy a satellite or deny satellite-based services is quite a real threat in Asia. But space security threats could also be a function of civil space activities. For instance, the crowded and congested nature of outer space poses serious problems for most Asian nations, as they are still expanding their programs. Increasing space debris with the possibility of risking space assets is a serious issue. Along with the amount of space debris, a new problem is the proliferation of small satellites – mini, micro, and nano satellites – which makes the challenge of monitoring and detection of these satellites a bigger problem. In fact, there is a global trend toward breaking bigger satellite constellations into smaller ones, given that they are easier to launch and deploy, if interfered with or destroyed, for instance. At the same time, satellites have become a requirement for well-coordinated and synchronized tactical capability, integrating weapons systems, radars and sensors, and missiles, aerial capabilities, and logistical capabilities in vast geographical spaces like the Asia-Pacific. But complicating these is the dual-use nature of space and the entry of private sector in the military space arena. Launches carrying mixed payloads make it an even bigger a challenge.

The growth and dispersal of military technologies has meant that many more countries are capable of investing in areas of technology that were previously limited to the superpowers. In addition, the spread of both missiles as well as increasing dependence on space for civilian and military purposes has meant that development of missile defenses and ASAT weapons has become more prevalent. A large number of countries now possess ballistic missiles – the spread of ballistic missile technology has meant a simultaneous pursuit of defenses against these missiles or at least the pursuit of such technologies. This search for a defense against missiles has accelerated in recent years. The increasing sophistication of surface-to-air missiles has led to adaption of such missiles also in antiballistic missile roles. Thus, the two prominent antiballistic missile systems, the American-built variants of the Patriot systems and the Soviet S-400 systems, have become increasingly popular. But countries like India (and Israel) have also pursued indigenous antiballistic missile systems because of the significant ballistic missile threats that both countries face. So, India has pursued Ballistic Missile Defence (BMD) at least since the mid-1990s, alternating between seeking to develop indigenous BMDs and acquiring it from abroad. India has shown interest in the Israeli Arrow and American BMD systems while also developing a domestic BMD program, built around the Prithvi missile system. Its capabilities remain unclear which is one reason why India has decided to acquire the S-400 system from Russia. China also has pursued BMD systems indigenously, but it is also buying the S-400s from Russia.

The increasing dependence on space-based systems as well as the growing capabilities of rocket engineering technology and more widely available sensor technology has meant increasing interest in ASAT systems. China demonstrated an ASAT capability in January 2007 which was possibly driven by China's pursuit of asymmetric capabilities against the far more capable American military forces. China's ASAT capabilities allow it to attack a vulnerable link in American military capability, but China's pursuit of ASATs, even if directed at the USA, clearly has implications for other powers such as Japan and India, countries that are also dependent on space for both civilian and military purposes. China's demonstration of its ASAT capability drove India immediately to begin its own ASAT program because it could not allow such a vulnerability given the competition between India and China. Though neither country might actually deploy an ASAT system or use it in combat, demonstrating the capability was necessary for the purpose of deterrence. Following China's test in 2007, India also demonstrated its capability in early 2019.

What Can Be Done?

The renewed emphasis on space security has brought the attention of all the key spacefaring powers to work toward some semblance of stability with regard to outer space activities. It is an entirely different issue that these countries are so far apart in identifying and agreeing upon major threats and the ways to address them. The existing treaties and other global instruments have regulated outer space activities to a great extent, but the re-emergence of counter-space as well as other new threats cannot be addressed effectively by the existing global mechanisms. There are gaps and ambiguities that should be addressed if one has to ensure safe, secure, and continued access to outer space (Rajagopalan 2018b). Given the worsening geopolitical trends, technologies with peaceful applications such as satellite inspection, refueling, and repair (on-orbit satellite servicing) or technologies to clean up space junk can be used for nefarious purposes. China has ground-based direct-ascent missiles that can physically destroy a satellite, jammers that can interfere, and lasers that can be used to dazzle or perhaps even blind imaging satellites and has also done a series of tests of on-orbit proximity and rendezvous operations, even though this is not indicative of explicit offensive capabilities. Similarly, Russia has a ground-based direct-ascent system known as Nudol, an airborne laser dazzler system known as the A-60 as well as GPS jammers. It has also engaged in a series of on-orbit proximity and rendezvous operations demonstrations, both in low Earth and geosynchronous orbits, and has shown high priority in integrating electronic warfare into military operations. The USA too has done multiple tests of technologies for close rendezvous operations; however, it does not have a declared direct-ascent ASAT program. Nevertheless, it possesses the capability to develop co-orbital ASATs should there be a decision to do so. It likely also has the capability to jam global navigation satellite service receivers such as GLONASS, BeiDou, and other regional navigation systems. Based on the growing inventory of counter-space capabilities among the key spacefaring nations, the Office of the US Director of National Intelligence, *2018 Worldwide Threat Assessment of the US Intelligence Community*, said, "We assess that, if a future conflict were to occur involving Russia or China, either country would justify attacks against US and allied satellites as necessary to offset any perceived US military advantage derived from military, civil or commercial space system." Adding to the complexities is the dual-use nature of space assets.

In the absence of successful multilateral efforts, states will be forced to rely on deterrence as a way of defending themselves. Deterrence could produce cascading effects because if one state relies on deterrence, others will be forced to as well and the net consequence will be negative for all. This could lead to increasing suspicions that will make cooperation difficult. There is still time left because the deterrence model has not yet become policy for any state regarding space, and therefore it is possible to prevent it. One might add that it is necessary too, before states proceed down this path. Therefore, there is a small window of opportunity to halt the process of states pursuing deterrence as a state policy, but the bigger responsibility rests on the shoulders of all the major spacefaring powers to develop certain new rules of the road to address the more contemporary threats in outer space.

Yet the debates on the global governance aspects have not progressed much, and there are broadly two schools of thought. One school suggests that developing legal instruments are the way to address the gaps in the existing global mechanisms, whereas the second school argues that under the current international political climate, legal measures are unlikely and therefore they want to pursue political instruments such as a transparency and confidence building measures (TCBMs). But the EU Code of Conduct for Outer Space, a TCBM proposed a few years back, could not be adopted, even though the Code ran into problems mostly on process issues rather than on substantive issues on which there appeared to be significant agreement. Therefore, a point to be emphasized is that the process is as important as the substantive measures that are being developed. An inclusive process in developing a code or any TCBMs gives a large number of states a sense of ownership and responsibility to see that it becomes a success (Rajagopalan 2012). Accordingly, one might argue that even if the document is less than the ideal, it is more important to have a large number of state stakeholders to ensure its durability. However, in order to develop a large support base among the Asia-Pacific countries, it is important not only to have a critical number of states being party to it, but also the critical states need to be bought in. The lack of consensus among great powers has hampered the process of developing effective global measures. Within the Asia-Pacific again, there is no unity of approach to global rules of the road on space. For instance, Russia and China proposed a draft Treaty on the Prevention of the Placement of Weapons in Outer Space, the Threat or Use of Force against Outer Space Objects (PPWT), first in 2008 and brought out a renewed text in 2014. But that has not gathered much momentum despite the fact that countries like India had generally preferred a legally binding mechanism (Listner and Rajagopalan 2014). In an effort to make some progress, there are those who have argued for a middle path: legally binding TCBMs. However, this has not generated much support either. Nevertheless, given the more tense space security environment today, there is a need to work on all possible tracks including legal measures and codes. The Conference on Disarmament (CD), where space security and arms control issues are debated, has been in a state of stalemate, and therefore the key spacefaring powers need to find innovative ways of approaching the global governance debates including the venue of such debates.

Irrespective of the outcome of the global governance front, there is a need for developing an effective Space Situational Awareness (SSA). Traditionally, the USA has maintained the largest SSA network with its radars and sensors, followed by Russia, which has a better coverage of the southern hemisphere. Europe has also developed certain capabilities in this regard. In the Asia-Pacific region, China, India, Australia, and Japan are developing SSA capabilities, and these need to be interlinked with other global networks. This is a relatively noncontroversial area of cooperation that can instill greater confidence among major players in Asia who otherwise have difficult relations. But any player that wants to be a relevant stakeholder in this domain should develop these capabilities in order to have better awareness on how it impacts on their activities in outer space. The three essential activities – tracking of objects in space, monitoring space weather, and characterization of space objects – can go a long way in avoiding collisions (Secure World Foundation). SSA is useful also in addressing space weather – solar storms and explosions of charged particles that can damage satellites or even power grids on Earth.

Conclusion

The challenges of space security are most pronounced in the Asia-Pacific. There is clearly no defense against ASATs. Moreover, there are no arms control measures or even TCBMs to address ASATs and other counter-space capabilities. Key Asia-Pacific space powers need to explore all ways to control these trends including legal measures, norms, Group of Governmental Experts (GGE), and codes of conduct. Given the continuing security threats and lack of progress on global governance, states need to invest in building better redundancy, hardening of space capabilities, and enhancing the security of backups.

References

Attacking satellites is increasingly attractive – and sangerous. The Economist, 18 July 2019. https://www.economist.com/briefing/2019/07/18/attacking-satellites-is-increasingly-attractive-and-dangerous

Bodner M (2018) As trump pushes for separate space force, Russia moves fast the other way. Defense News, 21 June 2018. https://www.defensenews.com/global/europe/2018/06/21/as-trump-pushes-for-separate-space-force-russia-moves-fast-the-other-way/

Columba Peoples (2013) Introduction: reading East Asia's space security dilemmas. Space Policy 29:95–98

Costello J, McReynolds J (2018) China's strategic support force: a force for a new era. National Defense University, Washington, DC

Davis M (2019) China's plans to dominate space. The National Interest, 15 April 2019. https://nationalinterest.org/blog/buzz/chinas-plans-dominate-space-52562

Erwin S (2019) Air force nominee barrett calls for assertive US Posture on Space, says Space Force is a 'Key imperative'. Space News, 12 September 2019. https://spacenews.com/air-force-nominee-barrett-calls-for-assertive-u-s-posture-on-space/

Harrison T, Johnson K, Roberts TG (2019) Space Threat Assessment 2019, Center for Security and International Studies, April 2019. https://csis-prod.s3.amazonaws.com/s3fs-public/publication/190404_SpaceThreatAssessment_interior.pdf

Jones A (2019a) Chang'e-4 returns first images from lunar farside following historic landing. Space News, 3 January 2019. https://spacenews.com/change-4-makes-historic-first-landing-on-the-far-side-of-the-moon/

Jones A (2019b) Chinese space station core module passes review but faces delays. Space News, 11 September 2019. https://spacenews.com/chinese-space-station-core-module-passes-review-but-faces-delays/

Kania EB (2018a) China Has a 'Space Force.' What are its lessons for the pentagon? Defense One, 29 September 2018. https://www.defenseone.com/ideas/2018/09/china-has-space-force-what-are-its-lessons-pentagon/151665/

Kania EB (2018b) China's strategic support force at 3. The Diplomat, 29 December 2018. https://thediplomat.com/2018/12/chinas-strategic-support-force-at-3/

Kodama S, Hoshi M (2019) Asteroid landing a triumph for Japan's Space Industry. Nikkei Asian Review, 28 February 2019, https://asia.nikkei.com/Business/Science/Asteroid-landing-a-triumph-for-Japan-s-space-industry

Lele A (2012) An Asian space race? Space News, 21 November 2012. https://spacenews.com/an-asian-space-race/

Lele A (2019) India needs its own space force. Space News, 28 May 2019. https://spacenews.com/op-ed-india-needs-its-own-space-force/

Li J (2016) PLA's new support force to be hub of China's joint military operations. South China Morning Post, 24 January 2016. https://www.scmp.com/news/china/diplomacy-defence/article/1904756/plas-new-support-force-be-hub-chinas-joint-military

Listner M, Rajagopalan RP (2014) The 2014 PPWT: a new draft but with the same and different problems. Space Review, 11 August 2014. http://www.thespacereview.com/article/2575/1

McDermott R (2015) Russia reforms aerospace defense structures – again. Eurasia Daily Monit 12 (151), August 11, 2015. https://jamestown.org/program/russia-reforms-aerospace-defense-structures-again/

Ministry of Defense of the Russian Federation (2015) Russian defense ministry army general sergey Shoygu holds regular teleconference. August 3, 2015 cited in Boston S, Massicot D The Russian way of warfare: a primer. RAND Corporation, 2017. https://www.rand.org/content/dam/rand/pubs/perspectives/PE200/PE231/RAND_PE231.pdf

Ministry of Defense spokesperson accepts media interview on issues related to Deepening national defense and military reform. 1 January 2016. http://www.mod.gov.cn/info/2016-01/01/content_4637926.htm

Moltz JC (2011) Asia's space race: national motivations, regional rivalries, and international risks. Columbia University Press, New York

Ni A, Gill B (2019) The people's liberation army strategic support force: update 2019. China Brief 19(10), 29 May 2019. https://jamestown.org/program/the-peoples-liberation-army-strategic-support-force-update-2019/

Pekkanen SM (2015) Asia's simmering rivalries are shifting to outer space, where anything goes. Forbes, 27 March 2015. https://www.forbes.com/sites/saadiampekkanen/2015/03/27/asias-simmering-rivalries-are-shifting-to-outer-space-where-anything-goes/#405f1feb2090

Pollpeter KL, Chase ML, Heginbotham E (2017) The creation of the PLA strategic support force and its implications for Chinese Military Space Operations. Rand Corporation, Santa Monica

Powell TD, Lubar DG, Jones KL (2018) Bracing for impact: terrestrial radio interference to satellite-based services. Center for Space Policy and Strategy, The Aerospace Corporation, January 2018. https://aerospace.org/sites/default/files/2018-05/BracingForImpact_0.pdf

Rajagopalan RP (2011) India's changing policy on space militarization: the impact of China's ASAT test. India Review 10(4):354–378

Rajagopalan RP (2012) Writing the rules on space: why inclusion matters. Space News, 23 January 2012, https://spacenews.com/writing-rules-space-why-inclusion-matters/

Rajagopalan RP (2013) India's race to mars goes way beyond science. The Wall Street Journal, 5 November 2013. https://blogs.wsj.com/indiarealtime/2013/11/05/indias-race-to-mars-goes-way-beyond-science/

Rajagopalan RP (2018a) Asia in space: cooperation or conflict? Policy Forum, Asia & the Pacific Policy Society, Australian National University, 10 October 2018. https://www.policyforum.net/asia-space-cooperation-conflict/

Rajagopalan RP (2018b) Space governance. Oxford Research Encyclopedias, Planetary Science Policy and Planning, Oxford University Press, August 2018. https://oxfordre.com/planetaryscience/view/10.1093/acrefore/9780190647926.001.0001/acrefore-9780190647926-e-107

Rajagopalan RP (2019a) Electronic and cyber warfare in outer space. Space Dossier 3, United Nations Institute for Disarmament Research, May 2019. http://www.unidir.org/files/publications/pdfs/electronic-and-cyber-warfare-in-outer-space-en-784.pdf

Rajagopalan RP (2019b) Chandrayaan 2 moon landing: the sign of a mature space mission. Hindustan Times, 6 September 2019. https://www.hindustantimes.com/india-news/chandrayaan-2-moon-landing-the-sign-of-a-mature-space-mission/story-aTed8LJKpDVe3FenVeYiJI.html

Rajagopalan RP (2019c) China extends terrestrial rivalries into orbit with new space race. Nikkei Asian Review, 23 August 2019. https://asia.nikkei.com/Opinion/China-extends-terrestrial-rivalries-into-orbit-with-new-space-race

Rajagopalan RP (2019d) India's strategy in space is changing. Here's why. World Economic Forum, 14 August 2019, https://www.weforum.org/agenda/2019/08/indias-strategy-in-space-is-changing-heres-why/

Reichert C (2019) Trump formally establishes US space command. CNET, 29 August 2019. https://www.cnet.com/news/trump-establishes-the-us-space-command/

Remarks by Vice President Pence on the Future of the U.S. Military in Space (2018) The Pentagon, 9 August 2018. https://www.whitehouse.gov/briefings-statements/remarks-vice-president-pence-future-u-s-military-space/

Secure World Foundation. Space situational awareness. https://swfound.org/media/205304/ssa-brochure-2015.pdf

Shaw N (2019a) India and Japan awaken to risks of superpower space race. Nikkei Asian Review, 8 January 2019. https://asia.nikkei.com/Spotlight/Asia-Insight/India-and-Japan-awaken-to-risks-of-superpower-space-race

Shaw J (2019b) The US space force must be independent but not insular. Space News, 13 September 2019. https://spacenews.com/op-ed-the-u-s-space-force-must-be-independent-but-not-insular/

Vekshin R (2015) Air force becomes an integrated new aerospace force. Russia Beyond, 1 September 2015. https://www.rbth.com/economics/defence/2015/09/01/air-force-becomes-an-integrated-new-aerospace-force_393913

Wang K (2019) US needs a space force to counter China. Asia Times, 24 April 2019. https://www.asiatimes.com/2019/04/opinion/us-needs-space-force-to-counter-china/

Weeden B (2019) U.S. Space Command and Trump's Space Force can seem funny. But space threats are deadly serious. NBC News, 4 September 2019. https://www.nbcnews.com/think/opinion/u-s-space-command-trump-s-space-force-can-seem-ncna1049251

Weeden B. Radio frequency spectrum, Interference and satellites fact sheet. Secure World Foundation. https://swfound.org/media/108538/swf_rfi_fact_sheet_2013.pdf

Weeden B, Samson V (eds) (2019) Global counterspace capabilities: an open source assessment. Secure World Foundation, April 2019. https://swfound.org/media/206408/swf_global_counterspace_april2019_web.pdf

Chinese Space and Security Policy: An Overview

28

Zhuoyan Lu

Contents

Introduction	516
Security in Space: Practices and Movements in China	517
Peaceful Uses and Exploration of Outer Space	517
Civil-Military Integration Policy in Space	518
Sustainability in Space: Endeavors from China in Promoting Space Stability and Preventing Conflicts	520
Space Debris Mitigation Measures in China	520
Radio Frequency Management Regime	521
Governance of Space: Space Policy-making in China	523
Conclusions	524
References	525

Abstract

China's space policies have expanded in numerous areas and have generated enormous documents during the past decades. It is impossible to cover all topics within one chapter; however, during the past decade, particularly the last 5 years, some fundamental space policies have played a pivotal role in China's space security and sustainability development. Consequently, this chapter concentrates on these prominent movements and provides an overview of these policies.

Z. Lu (✉)
International Space University (ISU), Strasbourg, France
e-mail: zhuoyan.lu@isunet.edu

© Springer Nature Switzerland AG 2020
K.-U. Schrogl (ed.), *Handbook of Space Security*,
https://doi.org/10.1007/978-3-030-23210-8_133

Introduction

China's journey towards space capability building began in 1956 with the establishment of the Aviation Industry Commission – to supervise and manage the Chinese aviation and space industry – in pursuit of national space development. China established its first launch site in 1958 and launched the first satellite in 1970. China tested the first unmanned spacecraft, Shenzhou-1, in 1999 and, later, in 2003 sent the first astronaut onboard spacecraft Shenzhou-5. China started the Moon exploration in 2007 by launching the Chang'e-1 satellite and deployed the first spacelab Tiangong-1 in 2011. At the same year, China completed the docking of Shenzhou-8 and Tiangong-1 and in 2013 Yutu rover landed on the Moon. In 2018, China released the first issue of Blue Book of China Aerospace Science and Technology Activities envisioning 40 launches in 2020, including the first Mars exploration mission and initiating the deployment of China's space station. Most recently, China successfully launched CZ-5 Y3 rocket, the heaviest launch vehicle ever, playing a significant role for the so-called super 2020 space plan.

China's aerospace industry has its origins on national defense. At the very beginning, aerospace technology was utilized for scientific exploration, technology development, military reconnaissance, and for other purposes that are mostly strategic to the State. However, as a space-faring nation, China is gradually taking more responsibilities for the international space governance and the maintenance of space for peaceful uses. This is reflected in the recently enacted 2015 National Security Law with the first ever reference to the peaceful uses and exploration of space.

Over the past 5 years, the Chinese space industry has been witnessing an increasing number of private space actors. This is in stark contrast with the situation prior to 2015, where China's aerospace industry was mainly dominated by two companies, the China Aerospace Science and Technology Corporation (CASC) and the China Aerospace Science and Industry Corporation (CASIC). This change in the space industry is closely related with the civil-military integration policy and its profound influence on China.

For international and national space development, not only does security play an essential role, but so does the sustainability aspect. The issue of space debris has raised international concerns, since it increases the chances for collisions in space and poses hazards to space objects in space as well as to the safety of property and life on the Earth. China has launched and deployed a considerable number of space objects and, therefore, space debris mitigation is one of the long-term policies and regulatory requirements. Nowadays, with the private space actors entering the space industry, some new measures and provisions are provided for space debris mitigation.

Radio frequency is another important element for consideration. To prevent harmful interference and to ensure access to and equal use of space, China has implemented a bunch of instruments for the management of radio frequency spectrum. This is critical for the sustainable development of space, as in the future the demand for frequency will increase while the availability of the resource will be a challenge for the government.

Space sector developments are supported and stimulated by national space policies, strategies, and regulatory frameworks. During the past decade, the space sector in China has experienced some significant moments, with some of them steering the space industry of the country towards a new page, some strengthening the responsibility of the country for peaceful exploration of space, and others reinforcing the duty of the country for space sustainability. This chapter will further illustrate these policies and regulatory making processes that cover space security and sustainability.

Security in Space: Practices and Movements in China

Peaceful Uses and Exploration of Outer Space

International space treaties comprise of five major instruments created in the 1960s and 1970s. The principle of exploration and use of outer space for "peaceful purposes" is ingrained in these treaties and followed by the international community. China is the State Party to four of the five treaties, namely, the Outer Space Treaty (1966), the Astronaut Agreement (1967), the Liability Convention (1971), and the Registration Convention (1974), yet China is not a State Party to the Moon Agreement (1979). The international space treaties have a binding effect upon State Parties and China as one of the State Parties to these treaties has to adhere to the rules and regulations set out in them.

The "Space White Paper" is the most important document for China's space policy. Starting with the first issue of the Paper in 2000, the fundamental principles of the international space treaties are codified in China's national space policy. In the Paper, it is stated that China promotes peaceful exploration and use of outer space, since outer space shall be the province of all mankind. The development objectives of China's space industry are to:

- Explore outer space and expand the understanding of the universe and Earth
- Promote human civilization and social development for the benefit of all mankind by peaceful uses of outer space
- Meet the growing needs of economic construction, national security, scientific and technological development, and social progress
- Safeguard national interests and enhance comprehensive national strength

It also points out that the fundamental mission for China is to develop its economy and constantly promote the modernization process of the country. Therefore, the important role of space in safeguarding national interests and implementing national development strategies determine the purposes and principles of China's space policy.

Although the principle of peaceful uses and exploration of outer space is constantly affirmed in China's space policy, it is not enshrined in its national legislation.

In 2015, the National Security Law of the People's Republic of China was enacted and it clearly stipulates – for the first time ever in law – that China shall:

- Maintain peaceful exploration and use of outer space
- Enhance international cooperation
- Advance capabilities of scientific investigation and exploitation
- Protect the security of space activities, assets, and other interests

In fact, outer space does not fall under the jurisdiction of States' sovereignty. However, the particular legal status of the area has an influential impact on security, the latter shall be linked to the peaceful uses and exploration of outer space. This shall also be reflected in national legislation in order to further benefit international space activities.

Civil-Military Integration Policy in Space

The space industry has long been dominated by States. In China, the turning point came with a decisive document in 2014 when the State Council released the "Guidance on Innovation of Investment and Financing Mechanisms in Key Fields to Encourage Private Investment." As such, private actors are stimulated to participate in the space industry and private capitals are encouraged to engage in this endeavor as well. Subsequently, 2015 marked the first year of China's commercial space with a "green light" to private entities, meaning less policy barriers in the upstream and downstream industry.

Further on, in 2015, the document on "Suggestions for the Thirteenth Five-year Plan for National Economic and Social Development" proposes the development of civil-military integration and requires the establishment of innovation in several fields including space. This accelerates the development of private space engagement in China. This policy is further detailed in the "Opinions on the Integrated Development of Economic and National Defense" whereby the civil-military integration is promoted to national strategic level. The latter allows for satellite data sharing, remote sensing resources integration, and dual-use navigation services. Later on, the "White Paper: China's Space Activities in 2016" reaffirms the policy and provides for management measures fostering the commercial space industry. A rationale behind the change is that the commercial space sector has been playing a considerably important role in the global space industry.

In order to further promote the civil-military integration, the "Central Civil-Military Integration Development Committee" was established in 2017, with two meetings held until now, one in June 2017 and another in March 2018. These meetings further set forth a central-local government system for civil-military integration: the central government provides top-level overall planning and the local government provides in-depth implementation measures and policies applicable to each administrative jurisdiction. Accordingly, more than 28 local governments have released their local policies in promoting civil-military integration (Security RSS 2019). Most of the local policies include support of supplementary facilities, financial support for civil-military projects and programs investment, manufacturing, establishment, and the like. However, these policies vary from each other and apply only locally.

The "Civil-Military Integration" policy has not only been able to attract private capital in space sector but also has been successful in stimulating private space enterprises in China. For instance, in 2018 only, more than 70 private investors invested in over 30 space start-ups (EO Intelligence 2019). Statistics from Future Aerospace show that till the end of 2018 the number of commercial space companies registered are 141. Among them, 123 are private space enterprises, and 61 out of the 123 are founded in the recent 3 years (National Business Daily 2019). In addition, one private space company, founded in 2016, has been capable of three successful launches including two suborbital launches and one orbital launch within three years. The space business of private space enterprises covers a wide range of industry clusters, from launch to various satellite applications. Geographically they are grouped around Beijing, Xi'an, Shanghai, and other administrative jurisdictions, while among them Beijing accounts for over one half of the total.

In view of the successful outcome of the policy in promoting commercial space industry, the China National Space Administration (CNSA) continuously encourages and progressively promotes the commercial space industry in China by:

(a) Fostering a friendly environment for commercial space enterprises
(b) Improving the administrative mechanism and better serving the industry
(c) Framing a policy and regulatory regime for the maintenance of a good order in space sector
(d) Pushing the governing authorities to get actively engaged (CNSA 2019)

Apart from the impact on commercial space development, the civil-military integration policy also influences profoundly the scoping of regulation-making framework. Previously, both military authorities, the State Council and Ministries under the Council, were empowered to establish regulations and this mechanism has caused several conflicts during the past years. The drafting process of the Aviation Law and National Defense Mobilization Law are some examples. The ambiguity of regulatory scope, if not improved, will further deepen the gap. This situation led to the amendment of the Law that passed by the National People's Congress (NPC) in 2015 (NPC 2015a, b). The amendment adds that national defense matters may be issued jointly by the State Council and the Central Military Commission, where previously the article stipulated that administrative regulations shall be issued solely by the State Council. Moreover, the civil-military integration policy further pushes relevant authorities to revisit the current regulatory framework and to abolish or update regulations that are out of date, like these two regulations: "Measures for Military Products Pricing" and "Measures for National Defense Research Project Pricing," born at the 1990s, yet not applicable to the current commercial market.

Sustainability in Space: Endeavors from China in Promoting Space Stability and Preventing Conflicts

Space Debris Mitigation Measures in China

Space sustainability is a fundamental theme of China's national space policy provided that space environmental problems such as orbital congestion and the proliferation of space debris have become ever prominent. The Chinese government places great importance on it and steadily advances the work related to space debris mitigation, including the implementation of long march vehicles passivation, and undertaking post-mission disposal operations for many geostationary orbit satellites such as Fengyun meteorological satellite. The government of China is also actively engaged in international cooperation on space debris matters and holds a close relationship with the Inter-Agency Space Debris Coordination Committee (IADC).

In order to promote sustainable development of space activities, the Chinese government initiated a Space Debris Operation Plan for 2006–2020 with the following objectives:

- Generate earth- and space-based space debris monitoring network
- Design space debris protection program
- Prepare for space environmental protection

Space debris matters stated in the Plan are repeated in the 2016 White Paper "China's Space Activities," where one of the visions is related to space debris and it confirms that China will continuously improve the technology of space debris monitoring, early warning, and mitigation to safeguard spacecraft in orbit. It also aims to achieve technology breakthrough in spacecraft protection design.

As a follow-up to the policy, the Chinese government issued the "Aerospace Industry Standard of the People's Republic of China – Requirements for space debris Mitigation" in 2005. The latter aims to reduce the generation of space debris and bring down the risk of potential collisions and damages caused by space debris. In 2009, the government published the "Interim Measures for Space Debris Mitigation and Spacecraft Protection" and formulated more than 10 supporting regulations to push industries to follow space debris mitigation rules. The Interim Measures at the time of formulation referred to the Space Debris Mitigation Guidelines published by the United Nations Committee on the Peaceful Uses of Outer Space (UNCOPUOS) and IADC.

In addition to the abovementioned efforts, space debris mitigation requirements are provided in regulations as well. As a State Party to the Registration Convention, China released the "Measures for Space Object Registration" in 2001. The document is the first regulatory instrument in China that governs space matters. Accordingly, all space objects, either launched in the territory of People's Republic of China or launched in a foreign country, yet China is a joint launching state, shall be registered with the State Administration of Science, Technology and Industry for National Defense (SASTIND). It is required that SASTIND shall maintain a space object registry containing necessary features of space objects. The registration regime has a great advantage in space object identification and collision risk analysis, and it will further facilitate mitigation during the post-mission disposal in a long run.

Later on, China's "Interim Measures on the Administration of Licenses for Civil Space Launch Projects" came out in 2002. It requires civil space launch applicants to provide, alongside their applications, supplementary documents on measures to avoid the generation of space debris and prevent contamination of space environment. The same requirement is reiterated in the "Notice on Promoting the Progressive Development of Commercial Rocket Transportation" that was issued in 2019, and it also states that launch activities shall not impair national security and public interests.

The requirements of space debris mitigation measures are not only required for launch activities but also for satellite activities. This is based on the "Interim Measures for the Administration of Civil Satellite Projects" which requests entities involved in satellite industry to undertake their activities in accordance with the "Interim Measures for Space Debris Mitigation and Spacecraft Protection." The SASTIND supervises and manages the work of space debris mitigation and safety of civil spacecraft and launch vehicles, and also organizes the formulation of relevant measures and standards. However, the application scope of this instrument is limited to civil scientific and commercial satellites, and other engineering projects approved by the State Council or relevant Departments of the State that fully or partly use central financial funds.

In addition to regulations and requirements of space debris mitigation measures, in practice, the government of China also addresses scientific research for space debris. In 2016 the Ministry of Finance (MOF) together with the SASTIND issued the "Interim Measures for the Administration of Post-Subsidy for National Defense Science, Technology and Industry Research Projects" to further encourage innovation and developments in national defense science, technology, and industry. The implementation of the document will enable legal entities, research institutions, and universities that are registered in China to obtain funds from a central budget. It further forms 10 scientific fields applicable for this funding, with space debris scientific research being one of them.

In this manner, the government of China completes the national governance of space debris issues and with an emphasis on international cooperation for space debris mitigations, the country is trying to contribute to the capacity building of space sustainability and to comply with its international responsibilities. However, current circumstances of space debris and the condition of conjunction in space will remain and even get worse. The number of large constellations of small satellite keeps growing and thus space debris mitigation shall be maintained as a long-term strategy for the international community.

Radio Frequency Management Regime

Radio spectrum is not only a limited resource but also a nonexhaustive one. On one hand, the continuous development of satellite technologies and applications brings an increasing demand on the use of radio spectrum since the resource is limited and it is therefore required to effectively manage and efficiently assign and allocate the resource. On the other hand, the occupancy of a certain frequency and spectrum is not permanent, which makes it available for reuse after a period of time and consequently the supervision of efficient use and authorization of proper occupancy are critical for the resource management. In addition, supervision and authorization could enable the avoidance of harmful interference and maintain space sustainability for all.

From a policy and strategy perspective, the Ministry of Industry and Information Technology (MIIT) released the "National Radio Management Plan" (2016–2020). The Plan states the following working targets for the end of 2020:

(a) Allocate radio spectrum in a more scientific manner
(b) Manage radio activities more efficiently
(c) Increasingly serve economic and social development, and national defense

In the Plan, it is reinforced that the government shall also work to improve the legal environment that governs radio spectrum activities. These include amending the Radio Regulation, cooperating with the judicial authority for the interpretation of article 288 of the Criminal Law of People's Republic of China, and pushing the legislation of radio activities when appropriate.

China has developed a comprehensive regime governing the use of the radio-frequency spectrum, including both legislative and regulatory frameworks. In general, the legislations cover critical issues and activities with disruptive influences, while the regulations provide detailed rules accordingly. At legislative level, the "Property Law of the People's Republic of China" was adopted by the National People's Congress (NPC) in 2007 and it stipulates that the ownership of radio frequency spectrum resources shall be regulated by the State. For illegal use of the resource, the Criminal Law of the People's Republic of China in 1997 covers the crime of disturbing the radio communication. In 2015, it was amended to include activities of setting up or using a radio station or a radio frequency without authorization, and in violation of State regulations, to interfere with the order of radio communication; such an activity causes severe consequences and is subject to penalties. Moreover, the Administrative License Law of the People's Republic of China (2019 amendment) regulates the general establishment and implementation of the administrative license related to radio management and the Public Security Administrative Punishments Law of the People's Republic of China (2012 amendment) stipulates that "the violation of relevant regulations and deliberately interfering with the operation of normal radio communication, or producing harmful interference to any normally operating radio stations, without eliminating the interference after receiving a warning from the authority, and causing severe consequences, shall be detained for punishments."

The major regulations for radio spectrum management in China comprise of the Radio Regulation, the Radio Control Provisions, the Radio Spectrum Allocation Provisions, and other regulations, provisions, measures, and notices from the Ministries of the State Council. Currently there are more than 50 regulations and normative documents issued by the State Council, which are important supplements to the regulations on radio administration. For instance, despite the provision provided in the Radio Regulation amendment of 2016 that the feasibility verification for the intended radio frequency shall be made during the planning phase of the satellite manufacturing. However, the regulation does not provide any further guidance in detail, such as how to carry out the verification and what shall be the feasibility verification elements in the process. In consequence, the "Measures for Feasibility Verification of Satellite Radio Frequency Occupancy" is introduced by MIIT in 2020 to bridge the regulatory gap. The "Measures" document explicitly specifies that the operator of the satellite is responsible for the feasibility verification process, the MIIT is in charge of supervision and guidance in the process, and the National Radio Spectrum Management Center provides technical support to MIIT.

Furthermore, remote sensing and space science satellites are one of the key areas of national civil space infrastructure. At the time of writing, China's remote sensing and space science satellites have entered a new stage of development with a dramatic increase in the demand from the industry for radio frequency spectrum and satellite orbit resources. Reasonable and efficient utilization has therefore become essential for the government. The "Radio Frequency Utilization Plan for Remote Sensing and Space Science Satellites" (2019–2025) was created in 2019 with the aim to formulate plans for the use of satellite radio frequency and orbit resources, so as to guide relevant bodies to reasonably declare the use of satellite radio frequency and orbit resources. Such an instrument can be more responsive and effective for the needs and development of radio activities and has a profound influence in national radio spectrum management.

In China, the radio frequency matters are managed through a top down approach, central-local, separating civil from military uses. For the use of non-military radio system, the central management is under the following two authorities. One is the Ministry of Industry and Information Technology (MIIT) National Radio Administration Bureau (NRSA) and the Bureau is responsible for the preparation of radio spectrum plan, radio frequency allocation and assignment, supervision of radio station and radio interference, coordination of satellite orbits and the local-military radio management. The other one is the MIIT State Radio Monitoring Center (SRMC) who takes the responsibility of testing and certifying radio equipment, monitoring radio signals, positioning the interference radio and providing technical support for national radio management authorities and local radio management departments. For the management at the local government level, each local Radio Administration Office is responsible for the implementation of national radio policy, regulation, and provision, and is entitled to provide local radio management rules and licenses for radio stations, and to coordinate radio management matters within local jurisdiction.

Governance of Space: Space Policy-making in China

The space policy-making in China involves various governmental departments from central government to local governments. The military system maintains a separate policy-making regime apart from the central-local governments. Depending on the level of the policy-making authority each policy may vary from each other. A space policy from the State Council has more persuasive authority than the policy from the

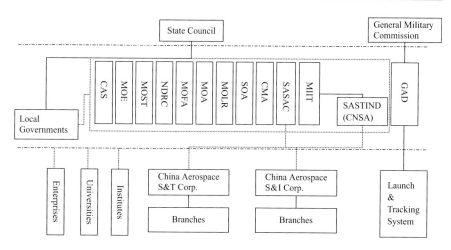

Fig. 1 source from Handbook of Space Security 1st edition (2015) Chapter 24 Fig. 24.4

Ministry of the Council, while policies of different State Council Ministries are equivalent. For local government, a local regulation is prior to a local rule but is equal to a Ministry rule. In the previous edition of the Handbook (Schrogl et al. 2015), a detailed structure of space governance is provided and it shows how the policy-making mechanism is working in China.

In general, the State Council is the top policy-making body within the system, to which all the ministries and agencies are affiliated. Major ministries and agencies involving in space policy making include the State Administration of Science, Technology and Industry for National Defense (SASTIND), the former Commission for Science and Technology and Industry for National Defense, (COSTIND), the National Development and Reform Commission (NDRC), and the Ministry of Science and Technology (MOST). In particular, SASTIND is responsible for making the space industry policy, development plans, regulations, and standards, as well as for organization and coordination of major space programs. The NDRC – in charge of macroeconomic planning, operation, and adjusting – also gets involved in space industry policy making and commercialization of space technologies. The MOST is in charge of Science and Technology policy making and program management. Other governmental ministries and agencies involved in space policy making according to their functions and responsibilities are: the Ministry of Education (MOE), the Ministry of Foreign Affairs (MOFA), and the State-owned Assets Supervision and Administration Commission of the State Council (SASAC) (Schrogl et al. 2015) (Fig. 1).

Conclusions

In the past decade, the space sector has been through several major changes. Private space industries have expanded almost across all the industry chain, from rocket manufacturing to satellite operations, and from ground infrastructures to service providers. Hence, commercial space enterprises come in great numbers and attempt to position themselves in the market. These changes require national policies to regulate the activities of private actors. National policies attempt to create a dynamic and competitive environment for commercial space industries. The rapid development of the private space in the near future will lead to more favorable policies and governing measures. Accordingly, the role of government in space activities will gradually shift from the main rule-making and supervision authority to a more cooperative and open body.

Space is the frontier of science and technology and is an important area for national security. The international space order was formed in the 1960s and 1970s and the peaceful exploration and use of outer space remains fundamental for all nations and for all mankind. The reaffirmation of China's position in favor of peaceful uses and exploration of outer space represents the principles set forth in international treaties and indicate its responsibility as a space faring nation. In the new space era, the development of China's space industry will, as always, follow the established purposes mentioned in its policy and regulatory framework.

References

Agreement Governing the Activities of States on the Moon and Other Celestial Bodies (1979) RES34/68. https://www.unoosa.org/pdf/gares/ARES_34_68E.pdf. Accessed 30 Jan 2020

Agreement on the Rescue of Astronauts, the Return of Astronauts and the Return of Objects Launched into Outer Space (1967) RES 2345 (XXII). https://www.unoosa.org/pdf/gares/ARES_22_2345E.pdf. Accessed 30 Jan 2020

CASC (2020) Blue book of China Aerospace Science and Technology Activities 2019 (in Chinese). http://www.spacechina.com/n25/n2014789/n2014804/c2827360/content.html. Accessed 30 Jan 2020

CNSA (2019) China National Space Administration will support commercial space sector and promote the development of Commercial Space Industry (in Chinese). http://www.cnsa.gov.cn/n6759533/c6806141/content.html. Accessed 30 Jan 2020

Convention on International Liability for Damage Caused by Space Objects (1971) RES 2777 (XXVI). https://www.unoosa.org/pdf/gares/ARES_26_2777E.pdf. Accessed 30 Jan 2020

Convention on Registration of Objects Launched into Outer Space (1974) RES 3235 (XXIX). https://www.unoosa.org/pdf/gares/ARES_29_3235E.pdf. Accessed 30 Jan 2020

CPC Central Committee, State Council, Central Military Commission (2016) Opinions on the Integrated Development of Economic and National Defense (in Chinese). http://www.gov.cn/xinwen/2016-07/21/content_5093488.htm. Accessed 30 Jan 2020

EO Intelligence (2019) Research on space industry in China 2019 (in Chinese). https://www.iyiou.com/intelligence/reportPreview?id=0&&did=663. Accessed 30 Jan 2020

MIIT (2016) National radio management plan (2016-2020) (in Chinese). http://www.gov.cn/xinwen/2016-08/29/5103177/files/1d69478c03ae488fb1d97dffd62d892c.pdf. Accessed 30 Jan 2020

MIIT (2018) Radio spectrum allocation provisions of the people's republic of China (in Chinese). http://www.miit.gov.cn/n1146285/n1146352/n3054355/n3057735/n4699781/c7025032/content.html. Accessed 30 Jan 2020

MIIT (2020) Measures for feasibility verification of satellite radio frequency occupancy (in Chinese). http://www.srrc.org.cn/article24153.aspx. Accessed 30 Jan 2020

MIIT, SASTIND (2019) Radio frequency utilization plan for remote sensing and space science satellites (2019–2025) (in Chinese). http://www.miit.gov.cn/newweb/n1146300/n7121908/n7121930/c7309243/content.html. Accessed 30 Jan 2020

MOF, SASTIND (2016) Interim measures for the administration of post-subsidy for National Defense Science, Technology and Industry Research Projects (in Chinese). http://www.mof.gov.cn/zhengwuxinxi/caizhengwengao/wg2016/wg201610/201703/t20170317_2559983.html. Accessed 30 Jan 2020

National Business Daily (2019) The development of commercial aerospace is in the ascendant with the help of new policies (in Chinese). http://www.nbd.com.cn/articles/2019-06-11/1342005.html. Accessed 30 Jan 2020

NPC (2007) Property law of the people's republic of China (in Chinese). http://www.gov.cn/flfg/2007-03/19/content_554452.htm. Accessed 30 Jan 2020

NPC (2015a) Legislation law of the people's republic of China (2015 Amendment) (in Chinese). http://www.npc.gov.cn/zgrdw/npc/dbdhhy/12_3/2015-03/18/content_1930713.htm. Accessed 30 Jan 2020

NPC (2015b) National security law of the people's republic of China (in Chinese). http://www.npc.gov.cn/wxzl/wxzl/2000-12/05/content_4581.htm. Accessed 30 Jan 2020

SASTIND (2002) Interim measures on the administration of licenses for civil space launch projects (in Chinese). http://www.miit.gov.cn/n1146295/n1652858/n7280902/c3554585/content.html. Accessed 30 Jan 2020

SASTIND (2016) Interim measures for the administration of civil satellite projects (in Chinese). http://www.scio.gov.cn/xwfbh/xwbfbh/wqfbh/35861/36552/xgzc36558/Document/1549898/1549898.htm. Accessed 30 Jan 2020

SASTIND, Central Military Commission Equipment Department (2019) Notice on promoting the progressive development of commercial rocket transportation (in Chinese). http://www.gov.cn/xinwen/2019-06/17/content_5400951.htm. Accessed 30 Jan 2020

SASTIND, Measures for Space Object Registration (2001) (in Chinese). http://www.scio.gov.cn/xwfbh/xwbfbh/wqfbh/2013/20131216/xgzc29845/Document/1354170/1354170.htm. Accessed 30 Jan 2020

Schrogl K-U et al (eds) (2015) Handbook of space security: policies, applications and programs. Springer, New York

Security RSS (2019) A collection of civil-military policies from local governments (in Chinese). https://www.secrss.com/articles/9381. Accessed 30 Jan 2020

State Council (2014) Guidance on innovation of investment and financing mechanisms in key fields to encourage private investment (in Chinese). http://www.gov.cn/zhengce/content/2014-11/26/content_9260.htm. Accessed 30 Jan 2020

State Council, Central Military Commission (2010) Radio control provisions of the people's republic of China (in Chinese). http://www.gov.cn/zwgk/2010-09/06/content_1696833.htm. Accessed 30 Jan 2020

State Council, Central Military Commission (2016) Radio regulation of the people's republic of China (in Chinese). http://www.gov.cn/zhengce/content/2016-11/25/content_5137687.htm. Accessed 30 Jan 2020

State Council Information Office (2006) White Paper: China's space activities in 2006 (in Chinese). http://www.scio.gov.cn/zfbps/ndhf/2006/Document/307876/307876.htm. Accessed 30 Jan 2020

State Council Information Office (2016) White Paper: China's space activities in 2016 (in Chinese). http://www.scio.gov.cn/wz/Document/1537090/1537090.htm. Accessed 30 Jan 2020

Treaty on Principles Governing the Activities of States in the Exploration and Use of Outer Space, including the Moon and Other Celestial Bodies (1966) RES 2222 (XXI). https://www.unoosa.org/pdf/gares/ARES_21_2222E.pdf. Accessed 30 Jan 2020

Chinese Concepts of Space Security: Under the New Circumstances

29

Dean Cheng

Contents

Introduction	528
Evolution of Chinese Thinking About Military Space	528
Space and Local Wars Under Modern, High-Technology Conditions	528
Space and Informationized Local Wars	530
Chinese Space Capabilities: A Brief Review	531
Chinese Concepts of Military Space Operations	533
Space Dominance and Information Dominance	536
Mission Areas Associated with Space Operations	537
Space Deterrence (kongjian weishe; 空间威慑)	538
Space Blockade (kongjian fengsuo zuozhan; 空间封锁作战)	540
Space Strike Operations (kongjian tuji zuozhan; 空间突击作战)	540
Defensive Space Operations (kongjian fangyu zuozhan; 空间防御作战)	542
Space Information Support Operations (kongjian xinxi zhiyuan zuozhan; 空间信息支援作战)	543
Space and Information Dominance "Under the New Circumstances"	544
Creation of the PLA Strategic Support Force (PLASSF)	546
Civil-Military Integration of Space Industrial Capabilities	547
Conclusions	549
References	551

Abstract

Chinese thinking about the relationship of space to national security has been evolving over the past several decades, as part of a broader ongoing assessment of the role of information in future warfare. As the People's Liberation Army has accorded growing importance to information and communications technologies, the centrality of space dominance has grown as well.

D. Cheng (✉)
The Heritage Foundation, Washington, DC, USA
e-mail: dean.cheng@heritage.org; deanbcheng@gmail.com

© Springer Nature Switzerland AG 2020
K.-U. Schrogl (ed.), *Handbook of Space Security*,
https://doi.org/10.1007/978-3-030-23210-8_132

Introduction

Since Xi Jinping came to power in 2012, he has been propounding the "China Dream" involving the "great revival of the Chinese people." A central part of this "China dream" is a strong military. This effort to modernize and strengthen the Chinese People's Liberation Army (PLA) occurs within a broader context of an evolving Chinese view of future warfare. Assessing other peoples' wars, the Chinese concluded that future wars would include space warfare as an integral part of operations. This has not been so much because of the importance of space systems as such, as their growing role in providing the information support necessary for the successful conduct of future local wars, which are characterized as occurring under "informationized conditions."

Indeed, PLA assessments of American and Russian military operations concluded that space-based information played an outsize role in the conflicts of the 1990s, and therefore, in the event of a conflict with the PRC, it must strive to deny an adversary the ability to use space freely. Thus, space warfare and information warfare have long been intimately linked in the Chinese mind.

Evolution of Chinese Thinking About Military Space

While China's space program dates from the 1956 founding of the Fifth Academy of the Ministry of Defense, it is not clear how the PLA thought about space, if at all, in the early years. This is likely due, in part, to the skepticism Mao exhibited toward military professionalism. For many years, it was more important for the PLA to be "red," i.e., ideologically reliable, than "expert," proficient in the military arts.

This was compounded by the limited space capabilities available to the PLA. China only launched its first satellite in 1970. In the ensuing decades, it only orbited a handful of communications and reconnaissance satellites. Chinese military planners therefore did not necessarily have a full understanding of the potential capabilities space provided. Moreover, space was seen as more a political gesture than a vital part of the economic or military arena and therefore was a lower priority, especially for Deng Xiaoping. After he succeeded Mao, Deng made it clear that the Chinese space program needed to focus less on gaining prestige and headlines and instead "concentrate on urgently needed and practical applied satellites" (Li 1999).

Space and Local Wars Under Modern, High-Technology Conditions

The coalition performance against Iraq in 1990–1991 during Operation Desert Shield/Desert Storm served as a wake-up call for the PLA. This included highlighting the importance of space as one of the key high-technology areas that would influence the course of future wars, since it is a key enabler of joint operations.

As envisioned by the PLA, joint operations would involve multiple services operating together across significant distances. The Gulf War, for example, sprawled across some 140 million square kilometers and included forces ranging from armored units to aircraft carriers and long-range bombers (Wang and Zhang 2000, p. 400). The ability to coordinate such diverse forces spread across a variety of domains would therefore require not only extensive communications but also precise navigation and positioning information, both for units and for the growing plethora of precision munitions. Space-based platforms would play an essential role in the command and control of joint operations.

In this light, space capabilities were recognized as playing an essential role in any effort to wage a "local war under modern, high-tech conditions." According to PLA estimates, the 70 satellites that were ultimately brought to bear against Iraq provided the United States with 90% of its strategic intelligence and carried 70% of all transmitted data for coalition forces (Gao 2001, p. 54). Indeed, these assets were the first to be employed, since they were essential for the success of all subsequent campaign activities. As one Chinese analysis observed, "Before the troops and horses move, the satellites are already moving" (Gao 2005).

The growing importance of space was not immediately recognized, however. In the 1997 *PLA Military Encyclopedia*, the discussion for "space warfare (*tianzhan*; 天战)" explicitly states that space is *not* a decisive battlefield – the key to wartime victory would remain in the traditional land, sea, and air realms. "It is impossible for it [space warfare] to be of decisive effect. The key determinant of victory and defeat in war remains the nature of the conflict and the human factor" (PLA Encyclopedia Committee 1997, p. 602). Space was seen as a supporting, not a leading, player.

By 2002, this view had evolved. In that year's supplement to the *PLA Encyclopedia*, a very different assessment is made of the importance of space. In a discussion of the "space battlefield (*taikong zhanchang*, 太空战场)," the chapter concludes with the observation that the impact of the space battlefield on land, sea, and air battlefields will become ever greater and the space battlefield "will be a major component of future conflict" (PLA Encyclopedia Committee 2002, p. 455). It is clear that, in the intervening 5 years, the perception of space had changed and was now seen as a substantially more important arena for military operations.

This progression may have been partly due to the intervening NATO conflict in the Balkans. The ability to defeat Belgrade through airpower, seemingly on its own, clearly caught Beijing's attention. In their analyses of that conflict, the Chinese accorded great prominence to the role of space power. NATO forces are assessed to have employed some 86 satellites (Zhang et al. 2005). Another Chinese analysis concluded that NATO space systems provided 70% of battlefield communications, 80% of battlefield surveillance and reconnaissance, and 100% of meteorological data and did so through all weather conditions, 24 h a day (Jiang 2013, p. 65). These provided a dense, continuous flow of real-time data, allowing the NATO forces to establish precise locations for Serbia's main military targets for sustained, coordinated strikes, with 98% of the precision-guided munitions employing space-based information (Liu and Wang 2008, p. 44; Jiang 2013, p. 65). Thus, airpower could achieve its goals only because it was supported by substantial space power.

Given the importance of such support from space systems, victory in future "local wars under modern, high-technology conditions" was already recognized as requiring not only one's own unfettered access to space but also the denial of the same ability to the adversary. The focus, however, was not so much on space systems as the information gathered by transmitted via those systems.

By preventing the enemy from obtaining the amount of information they required, it would be far more difficult for them to coordinate their forces and operations. As important, by preventing them from operating in the manner to which they were accustomed (and had trained), they would be far less efficient and flexible and therefore more vulnerable to Chinese actions. In effect, by degrading adversary space capabilities, the enemy would suffer from a slower OODA (observe-orient-decide-act) loop. Space information support was therefore not only a vital part of joint operations but would increasingly be complemented by offensive space operations (which somewhat aligns with Western concepts of counter-space operations).

Space and Informationized Local Wars

This shift may also have been a reflection of the ongoing development of Chinese concepts of future warfare. As part of the PLA's "new historic missions," Hu Jintao in 2004 made clear that the PLA must secure China's interests in outer space, as well as the electromagnetic spectrum (Hu 2004). (For further discussion of the "new historic missions," see (Hartnett 2008).) The incorporation of the space domain into the specific range of PLA responsibilities reflected the steadily growing emphasis placed upon establishing space dominance as part of the larger effort to secure information dominance.

Indeed, as the PLA shifted from preparing to fight "local wars under modern, high-technology conditions" to fighting "local wars under informationized conditions" and then to "informationized local wars," space has been increasingly seen as part of those "informationized conditions." As PLA writings noted, "informationized conditions" do not simply refer to computers and cyberwarfare. Instead, it involves the acquisition, transmission, and exploitation of all forms of information. Space plays a central role in all these tasks. In the 2006 edition of *The Science of Campaigns*, it is specifically stated that "the space domain daily is becoming a vital battle-space.... Space has already become the new strategic high ground" (Zhang 2006, p. 87).

This is exemplified in that volume's revised version of "campaign basic guiding concept." This concept of "integrated operations, precision strikes to control the enemy (*zhengti zuozhan, jingda zhidi*; 整体作战, 精打制敌)" has even more need for information support from space-based assets than the previous version of "integrated operations, key point strikes." For example, precision strikes involve the use of precision munitions to attack vital targets. The goal is not only destroying key targets but also precisely controlling the course and intensity of a conflict (Zhang 2006, p. 81). It also entails disrupting the ability of the enemy's systems (and systems of systems) to function normally. The focus is on disruption leading to paralysis, not just destruction of the adversary's weapons or forces (Wang and Zhang 2009, pp. 202–203). Information from space systems facilitates the conduct of such precision operations. "Establishing space dominance, establishing information dominance, and establishing air dominance in a conflict will have influential effects" (Zhang 2006, p. 83).

Similarly, in the 2013 edition of *The Science of Military Strategy*, space is deemed the "high ground in wars under informationized conditions," tied to the struggles in network space and the electromagnetic spectrum as key future battlegrounds (Academy of Military Science Military Strategy Research Office 2013, pp. 146–147). In the Chinese conception, space is important for the advantage it confers with regard to the ability to collect, transmit, and exploit information, rather than for its own sake. As other Chinese analysts conclude, "space operations will be a core means of establishing information advantage" (Yuan 2008, p. 324).

Chinese Space Capabilities: A Brief Review

China's overall space capabilities have expanded significantly during the past two decades. Indeed, its growth during this period is in sharp contrast to its first 20 years. From 1956 to 1976, China enjoyed only very limited advances in its space capabilities, due to a lack of financial, technological, and trained human resources, as well as repeated political upheavals that disrupted research efforts. Even after orbiting its first satellite, in 1970, space development remained limited, with only a handful of satellites orbited before Mao died in 1976.

Deng Xiaoping initially did little to promote space development for either the military or civilian sectors. For his first several years in power, rather than committing further resources toward space, Deng focused on developing the civilian economy, forcing the space industrial sector to fend for itself through conversion to products with civilian demand.

Support for China's overall space program did not improve until 1986, when Deng, at the urging of a number of top Chinese scientists, authorized Plan 863, formally termed the National High-Technology Research and Development Plan (*guojia gao jishu yanjiu fazhan jihua*; 国家高技术研究发展计划). (Material drawn from *Guojia Gao Jishu Yanjiu Fazhan Jihua 863*, in FBIS-CHI (July 21, 2000). For further discussion of the creation of Plan 863, see (Feigenbaum 2003, pp. 141–143.) Plan 863, which remains an ongoing effort, was seen as providing the scientific and technological research foundations essential for a modernizing economy. Aerospace, along with information technology, and later telecommunications were seen as key areas of high technology, justifying substantial, sustained resource investment.

In the 1990s, with renewed support from senior leaders, China's space program benefited from expanded investment and intensified high-level support. Under Jiang Zemin (1992–2002), China deployed both low-Earth orbit and geosynchronous weather satellites (the *Fengyun* series) and improved geosynchronous communications satellites (the *Dongfanghong*-3 series), as well as recoverable satellites with varying payloads (the *Fanhui Shi Weixing* series).

Chinese Earth observation capabilities also improved during this period. In cooperation with Brazil, China in 1999 deployed the China–Brazil Earth Resources Satellite (CBERS), its first electro-optical imaging satellite capable of beaming its pictures directly down to Earth. China subsequently launched several similar satellites with no Brazilian involvement; these are known as the *Ziyuan* series, to distinguish them from the CBERS satellites.

In 2000, China became only the third country to deploy a navigational satellite system, launching two *Beidou* regional navigation satellites into geosynchronous orbit. This system also has a communications function, which was employed during the 2008 Sichuan earthquake (Lu 2008).

Jiang Zemin's successor, Hu Jintao, maintained support for China's space program. During his two terms, China deployed a variety of new satellite systems, including remote sensing satellites (the *Yaogan* series), microsatellites such as the *Shijian* series, as well as improved versions of the *Fengyun* and *Ziyuan* satellites. Under Hu, China also orbited several manned spacecraft (the *Shenzhou* program), as well as initiated a lunar exploration program, launching the *Chang'e*-1 and -2 lunar probes.

Chinese leader Xi Jinping has placed even greater emphasis upon the development of China's aerospace capabilities. He has repeatedly stated that China must become a "major aerospace power (*hangtian qiangguo*; 航天强国)." Indeed, the Chinese official news agency Xinhua at one point released a series of quotations from Xi from 2013 through 2019 on this very issue, under the title "Xi Jinping Repeatedly Supports Building a Major Space Power" (Xi Jinping 2019).

Chinese space capabilities have steadily improved under Xi. China inaugurated its *Gaofen* series of high-resolution Earth observation satellites in 2013. The second Chinese space lab, *Tiangong*-2, was launched in 2016, reflecting China's continued interest in manned space missions. China's Chang'e-3 mission deployed the lunar rover *Yutu* (Jade Rabbit) in 2013, the first mission to land on the Moon since Apollo 17. In 2018, China landed Chang'e-4, the first probe to ever explore the lunar far side and polar regions.

In keeping with Deng Xiaoping's admonition that China's space program must serve the broader goal of national economic development, many of China's satellites are dual-purpose, supporting urban planners and agricultural programs as well as the military. For many of its satellite programs, including Earth observation satellites, position and navigation systems, and weather satellites, the focus has been more on providing suitable information to support Chinese economic development objectives than necessarily producing cutting-edge capability.

This has also been a two-way street. The Chinese space industrial complex has benefited from the steady investment of resources to develop space capabilities. Much of China's space technology has been indigenously developed; Chinese satellites, launch vehicles, and ground support equipment are largely domestically produced. Two major aerospace conglomerates, the China Aerospace Science and Technology Corporation (CASC) and the China Aerospace Science and Industry

Corporation (CASIC), manufacture the full range of space systems, including launch vehicles, satellites, and ground equipment, and the associated sub-systems and support items.

Similarly, there is recognition that a strong national space infrastructure will benefit the military as well. Military space systems, by their exquisite nature and extreme capabilities, are expensive and therefore necessarily limited in number. Civilian space assets are likely to be more robust in numbers and are also in many cases developing faster than their military counterparts. Consequently, Chinese analysts conclude that a comprehensive set of civilian space systems can usefully augment military space forces, at least in terms of space information support and space monitoring (Ma 2013, p. 220).

Chinese Concepts of Military Space Operations

China's space program is not solely devoted to civilian use, however. It also provides the PLA with key pieces of information, deemed essential for "informationized local wars." Moreover, the military plays an outsize role in Chinese space activities, as the PLA runs China's space facilities. (Up until December 31, 2015, the PLA was managed by several General Departments which oversee all the armed forces, including all the services. These were the General Staff Department (GSD), the General Political Department (GPD), the General Logistics Department (GLD), and since 1998 the General Armaments Department (GAD). These Departments comprised the membership of the Central Military Commission (CMC) until 2004, when the PLA Navy, PLA Air Force, and Second Artillery were added to the CMC.)

Under Hu Jintao, the PLA began to demonstrate overt space combat capabilities. The PLA tested its direct ascent, kinetic kill anti-satellite (ASAT) system in January 2007. Launched from Xichang Satellite Launch Center, the Chinese ASAT destroyed a defunct Fengyun-1C weather satellite in low orbit. In the process, China also generated a massive amount of space debris (David 2007). Almost precisely 3 years later, in January 2010, China engaged in what was termed an anti-missile test, involving "two geographically separated missile launch events with an exo-atmospheric collision also being observed by space-based sensors," according to the US Department of Defense (China: Missile 2010). This test also helped Chinese scientists improve their ASAT system. And in August 2010, two Chinese microsatellites were deliberately maneuvered into close proximity and apparently "bumped" each other (Matthews 2010).

This effort at developing anti-satellite systems has been sustained under Xi Jinping. In May 2013, the Chinese conducted another anti-satellite test. This weapon, however, is assessed as demonstrating an ability to threaten targets as far as the geosynchronous belt, over 26,000 miles away (Weeden 2014). This is the first time that any nation has tested a weapon explicitly intended to hold satellites in that orbit at risk. Described by one senior US military officer as the "most valuable orbit," the geosynchronous region is populated by not only large numbers of communications satellites but also strategic early warning satellites as well as weather satellites

(Gruss 2015). The ability to destroy such satellites would be a major step toward establishing information dominance.

As with other Chinese military activities, the PLA's approach to space operates within the context of "guiding thoughts." The "guiding thoughts" for space are "active defense, all-aspects unified, key point is dominating space (*jiji fangyu, quanwei yiti, zhongdian zhitian*; 积极防御，全维一体，重点制天)" (Jiang 2013, p. 40). Each of these phrases embodies a number of essential concepts.

Active defense is integral to all Chinese military strategies and, as noted earlier, is not limited to space-related operations. While assuming the strategic defensive, the PLA concept of "active defense" emphasizes the importance of seizing the initiative at the tactical and operational level. In the context of space operations, "active defense" again assumes a more strategically defensive stance, although one which nonetheless seeks to deter aggression and maintain national security and interests. Chinese military writings assume that in space, as terrestrially, the Chinese would not be the party responsible for precipitating a war. At the same time, however, the "active defense" expects the PLA to undertake space combat preparations so as to be able to seize the initiative in space-related operations. In particular, it presumes "offensive actions at the campaign and tactical level to secure strategically defensive goalsx" (Jiang 2013, p. 40).

"*All aspects unified*" refers to the need to unify thinking about a number of different aspects of space operations. In the first place, it entails viewing space as a holistic environment, encompassing not just the satellites that are in orbit but also the terrestrial mission control; launch; and tracking, telemetry, and control (TT&C) facilities that allow those satellites to operate, as well as the data links that bind the entire structure together. In striving to achieve space dominance, the Chinese envision attacking and defending all three components. The destruction of mission control facilities, the jamming of TT&C links, or the entry of instructions that turn off the satellite at key moments can be every bit as effective as launching an ASAT against a given satellite.

"All aspects unified" also requires viewing the various domains of military activity, including not only outer space but land, sea, air, and the electromagnetic spectrum (e.g., cyber and electronic warfare operations), in a joint fashion, with operations in each domain contributing to, and requiring support from, the other domains. Since space operations are an integral part of joint operations and especially information operations, it is essential to adopt a joint perspective, not only to forge an organic, integrated whole but also so that each component force supplements the others (Li 2012, p. 98). Space operations support terrestrial operations, while land, sea, air, and computer network operations can help achieve space superiority. All of these operations, in turn, are ultimately aimed at achieving predetermined political ends.

Similarly, "all aspects unified" requires seeing all the various wartime activities, including offensive and defensive operations, provision of information support and fire support, and hard- and soft-kill methods, in an integrated or unified fashion, rather than as discrete phases, tasks, or methods. Thus, proper conduct of space operations should involve the application of "soft-kill" methods such as dazzling or jamming, in coordination with "hard-kill" methods such as direct ascent kinetic kill vehicles. Space operations should be coordinated with terrestrial operations, not only for the provision of meteorological, positioning and navigation, and communications information from space systems but also for air, land, and sea attacks on an enemy's space launch and mission support facilities. As with cross-domain operations, the various methods and activities should be seen holistically, all contributing to the goal of establishing space dominance, in support of establishing information dominance, while also serving the larger, strategic ends of the overall campaign.

To this end, command and control of space operations plays a central role. Not only must the various space activities, including offensive and defensive operations, be closely controlled, but competing demands for reconnaissance and early warning, communications, navigation, and various other space information support assets must also be managed. This encompasses not only military space assets but civilian and commercial systems as well. Space operations must therefore be integrated into the larger joint campaign plans to help achieve terrestrial objectives. Command and control of space operations must reconcile space-related requirements, timing, and structure with those of the overarching joint campaign (Jiang 2013, p. 43). This integrated command and control network, capable of drawing upon military, civilian, and commercial resources, is a vital means of achieving "all aspects unified."

"*Key point is establishing space dominance*" in part builds upon the PLA's emphasis on striking the enemy's "key points (*zhongda yao hai*; 重打要害)," especially those nodes within the enemy's "combat system-of-systems (*zuozhan tixi*; 作战体系)." As a "key point" is to establish space dominance, the PLA commander is expected to concentrate his best forces and capabilities to precisely strike key targets with a combination of hard- and soft-kill weapons, with the goal of paralyzing the adversary.

The massive advances in information technology have meant that forces and weapons are now much more networked across the land, sea, air, space, and electromagnetic/cyber domains. While this has significantly advanced combat effectiveness by creating synergies across various forces, it has also introduced a new set of vulnerabilities. Chinese writings note that, given the importance of space systems for navigation, positioning, and timing, disruption of associated networks will result in the disruption of the OODA loop. Therefore, key point strikes in joint operations should seek to disrupt the enemy's information collection and transmission nodes and command and control networks, through complementary hard and soft means. In the Chinese view, such attacks will cause the adversary's integrated systems of systems to decohere.

The concept of "key point is space dominance" consequently emphasizes the importance of securing space dominance, through the comprehensive application of various types of tactics and forces, in a variety of ways, including interference, obstruction, disruption, and destruction of enemy space-related systems (including terrestrial facilities and data links). The objective is both to prevent the enemy from operating their space systems for as much of the course of the conflict as possible and also ensure that one's own space systems can operate effectively. To this latter end, establishing space dominance also encompasses the exploitation of space, whether in the provision of information support to terrestrial operations, undertaking space deterrence or engaging in operations against remaining enemy space assets (Jiang 2013, p. 44).

"Key point is space dominance" therefore has two meanings. On the one hand, it is reminding PLA officers and staff that an important priority must be securing space dominance over an opponent. Therefore, resources must be applied against an enemy's space systems (terrestrial facilities, orbiting platforms, data links) to disrupt and deny an opponent the ability to exploit space. Moreover, these attacks must be sustained throughout the course of the conflict, but special attention should be paid to the first battle, where it is important to secure maximum effect as it is likely to influence the entire course of the conflict.

As important, one must also be prepared to defend one's own space infrastructure, since the enemy is likely to be striving to secure space dominance as well. This will entail incorporating both hard and soft defenses, including deceptive measures and maintaining secrecy to mislead an opponent's allocation of offensive measures, as well as hard defenses aimed at countering their attacks directly.

However, even with the full range of national space assets at one's command, there remains only a limited resource base. Chinese analysts recognize that space systems are fragile; as important they are extremely expensive, so even wealthy nations are unlikely to have a substantial reserve of platforms. Nor do many nations have a multiple redundant terrestrial space launch and mission control network. (In this regard, it is worth noting that, with the inauguration of the Hainan Island spaceport, China will have four space launch facilities.) Therefore, the other aspect of "key point is space dominance" is that space operations need to be focused, with a specific focus, a key point, and not scattershot. Attacks against adversary space infrastructure need to be carefully coordinated and undertaken at essential moments in the overall campaign, to maximize effect.

Space Dominance and Information Dominance

As information dominance has assumed a higher importance in Chinese thinking and as China has developed its own array of satellites, the relationship between establishing space dominance and information dominance has evolved. Chinese analysts have long recognized, since at least the first Gulf War 25 years ago, that space is a key means of providing information support to terrestrial forces. Consequently, the emphasis upon establishing space dominance, as part of the struggle for information dominance, has become more explicit.

Several PLA analyses, for example, have observed that space is the "strategic high ground (*zhanlue zhigao dian*; 战略制高点)" in informationized warfare. They conclude that the ability to dominate space will have greater impact on informationized warfare than any other domain, because it will provide:

- Real-time, global monitoring and early warning, such that no major military activity can occur without being spotted
- Secure, long-range, intercontinental communications

- Positional and navigational information that will support long-range, precision strike, including against targets that are over the horizon

All of these will occur without restriction from political borders, physical geography, or weather conditions and time of day (Ye 2007, p. 154; Chi and Xiao 2005, pp. 38–39).

Space dominance entails not only the ability to provide information support to the PLA but also to deny an adversary the ability to exploit space to gain information. The American reliance on space systems, in particular, has been remarked upon. One Chinese assessment notes high levels of American investment in military communications satellites, navigation satellite, reconnaissance and surveillance satellites, ballistic missile early warning satellites, and environment monitoring satellites (Xu 2013, p. 50). These satellite constellations, moreover, will be complemented by an array of terrestrial and aerial systems, to provide a complete, overlapping array of surveillance capabilities. The expectation is that the United States is preparing to disrupt, degrade, deny, and destroy adversary space systems in the effort to establish information dominance and conversely that the Americans are also preparing to face such attacks against their own systems.

Nor is American dependence upon space unique, in the Chinese view. PLA writings indicate that they are also closely observing other nations' space developments. Russian space developments in particular seem to garner heavy Chinese attention. The Chinese military textbook *Military Astronautics* discusses Russian as well as American aerospace forces (Chang 2005, pp. 219–220). The 2013 edition of *The Science of Military Strategy* observes that Russia has made space a major focus of its military refurbishment effort and that Moscow has increased its investments in the space sector as the Russian economy has improved (Academy of Military Science Military Strategy Research Office 2013, p. 180). In particular, Russian dependence on space systems has been noted. One Chinese volume related the Russian observation that "If Russia did not have an advantage in space, then it would not have reliable communications and reconnaissance, in which case, it would lack modernized information systems," leaving Russia blind and deaf (Wu 2004, p. 102).

This will make the struggle for space dominance that much more pointed. Chinese authors believe that without space dominance, one cannot obtain information dominance and aerial dominance, and therefore one cannot achieve land or maritime dominance. Space will therefore inevitably be a battleground, if only in order to deny an adversary the ability to use it freely (Ye 2007, p. 154). Consequently, the space arena will be one of the very first scenes of conflict, as the two sides struggle for control of space. Neither side can afford to neglect this theater, as it will be a central determinant of who will secure information dominance (Chi and Xiao 2005, pp. 38, 39).

Mission Areas Associated with Space Operations

PLA analysts believe that military space operations are likely to entail five broad "styles (*yangshi*; 样式)" or mission areas: space deterrence, space blockades, space strike operations, space defense operations, and provision of space information support. (This section draws upon Jiang (2013, pp. 126–154).) It is important to recognize that such operations will most likely not be undertaken by themselves but in the context of a larger, joint campaign. Nonetheless, the purpose of all such operations is ultimately to affect information dominance by securing space dominance.

Space Deterrence (kongjian weishe; 空间威慑)

Space deterrence is the use of space forces and capabilities to deter or coerce an opponent, preventing the outbreak of conflict or limiting its extent should conflict occur. Space deterrence is possible because of the growing importance of space-derived information in not only military but economic and social realms. By displaying one's own space capabilities and demonstrating determination and will, the PLA would hope to induce doubt and fear in an opponent over the prospect of loss of access to information gained from and through space and the resulting repercussions. This, in turn, would lead the adversary to either abandon their goals or else limit the scale, intensity, and types of operations (Zhou and Wen 2004; Academy of Military Science Military Strategy Research Office 2013, p. 181).

It is important to note here that the Chinese concept of space deterrence is not focused on deterring an adversary from conducting attacks against China's space infrastructure, per se. Instead, it is focused on employing space systems as a means of influencing the adversary's overall perceptions, in order to dissuade or compel them into acceding to Chinese goals. Thus, it is not so much deterrence *in* space, as deterrence ***through space means***.

Space capabilities are seen as contributing to overall deterrent effects in a number of ways. One is by enhancing other forces' capabilities. Thus, conventional and nuclear forces are more effective when they are supported by information from space-based platforms, such as navigational, reconnaissance, and communications information. This makes nuclear and conventional deterrence more effective and therefore more credible.

In addition, though, space systems may coerce or dissuade an opponent on their own. Space systems are very expensive and hard to replace. By holding an opponent's space systems at risk, one essentially compels them to undertake a cost-benefit analysis. Is the focus of Chinese deterrence or coercive efforts worth the likely cost to an adversary of repairing or replacing a badly damaged or even destroyed space infrastructure? Moreover, because space systems affect not only military but economic, political, and diplomatic spheres, damage to space systems will have wide-ranging repercussions (Li and Dan 2002). Is the target of Chinese deterrent or coercive actions worth the impact of the loss of information from space-based systems on other military operations or on financial and other activities? The Chinese clearly hope that the adversary's calculations would conclude that it was better not to challenge Chinese aims. Even the threat of interference and disruption of space systems "will impose a certain level of psychological terror, and will generate an impact upon a nation's policy-makers and associated strategic decision-making" (Academy of Military Science Military Strategy Research Office 2013, p. 181).

PLA teaching materials suggest that there is a perceived hierarchy of space deterrence actions, perhaps akin to an "escalation ladder" involving displays of space forces and weapons; military space exercises; deployment or augmentation of space forces; and employment of space weapons.

Displays of space forces and weapons (*kongjian liliang xianshi*; 空间力量显示) occur in peacetime or at the onset of a crisis. The goal is to warn an opponent, in the hopes of dissuading them from escalating a crisis or pursuing courses of action that will lead to conflict. Such displays involve the use of various forms of media to highlight one's space forces and are ideally complemented by political and diplomatic gestures and actions, such as inviting foreign military attaches to attend weapon tests and demonstrations.

Military space exercises (*kongjian junshi yanxi*; 空间军事演习) are undertaken as a crisis escalates, if displays of space forces and weapons are insufficient to compel an opponent to alter course. They can involve actual forces or computer simulations and are intended to demonstrate one's capabilities but also military preparations and readiness. At the same time, such exercises will also improve one's military space force readiness. Examples include ballistic missile defense tests, anti-satellite unit tests, exercises demonstrating "space strike (*kongjian tuji*; 空间突击)" capabilities, and displays of real-time and near-real-time information support from space systems.

Space force deployments (*kongjian liliang bushu*; 空间力量部署) are seen as a significant escalation of space deterrent efforts. It occurs when one concludes that an opponent is engaged in preparations for war and involves the rapid adjustment of space force deployments. As with military space exercises, this measure is not only intended to deter an opponent but, should deterrence fail, is seen as improving one's own preparations for combat. Such deployments, which may involve moving assets that are already in orbit and/or reinforcing current assets with additional platforms and systems, are intended to create local superiority of forces so that an opponent will clearly be in an inferior position. It may also involve the recall of certain space assets (e.g., space shuttles), either to preserve them from enemy action or to allow them to prepare for new missions. This may be akin to the evacuation of dependents from a region in crisis, as a signal of imminent conflict.

The Chinese term for the final step of space deterrence is "space shock and awe strikes (*kongjian zhenshe daji*; 空间震慑打击)." If the three previous, non-violent deterrent measures are insufficient, then the PLA suggests engaging in punitive strikes, so as to warn an opponent that one is prepared for full-blown, comprehensive conflict in defense of the nation. Such strikes are seen as "the highest, and final technique (*zuigao xingshi he zui hou shouduan*; 最高形式和最后手段)" in seeking to deter and dissuade an opponent. Employing hard-kill methods, soft-kill methods, or a combination, one would attack an opponent's physical space infrastructure or data links, respectively. If this succeeds, opposing decision-makers will be psychologically shaken and cease their activities. If it fails, an opponent's forces will nonetheless have suffered some damage and losses.

Space Blockade (kongjian fengsuo zuozhan; 空间封锁作战)

Space blockades involve the use of space and terrestrial forces to prevent an opponent from entering space and from gathering or transmitting information through space. There are several different varieties of space blockade activities. One is to blockade terrestrial space facilities, including launch sites; tracking, telemetry, and control (TT&C) sites; and mission control centers. They can be disrupted through the use of kinetic means (e.g., special forces, missiles) or through computer and information network interference.

Another means is to obstruct orbits. This can include actually destroying satellites that are in orbit or else obstructing orbits, such as by creating clouds of space debris or deploying space mines. By threatening the destruction of adversary satellites (without necessarily doing so), one might limit the function of those satellites (e.g., by limiting their maneuvers). The risk, however, is that either such step might damage third-party space systems, which in turn could lead to strategic consequences. Therefore, this approach to imposing a space blockade imposes very high requirements for precise control, extremely detailed space situational awareness, and highly focused, limited deployment.

Another method is the obstruction of launch windows. If one can delay a launch, whether through interfering with its onboard systems or otherwise disrupting the schedule, then a satellite may not be able to reach its proper orbit. In the past, some American space launches have been delayed because fishing and pleasure boats were present down-range (Atlas 3 2000; Orwig 2014). This alternative also includes the possibility of a boost-phase intercept of a space launch vehicle.

Finally, one can impose an information blockade. By interfering with and disrupting an opponent's data links between terrestrial control stations and the satellite, one can effectively neutralize an orbiting satellite by hijacking the satellite's control systems or preventing ground control from issuing instructions. Alternatively, one can interfere with the data that the satellite is transmitting. That is, rather than tampering with the satellite's controls, one can contaminate or block the data that is passing through the satellite. A third form of information blockade involves "dazzling" a satellite, using low-powered directed energy weapons against sensors or other systems. In each case, the intent is to affect a "mission kill," whereby the satellite cannot perform its functions, but is not necessarily destroyed.

Space Strike Operations (kongjian tuji zuozhan; 空间突击作战)

Essential to the credibility of space deterrence and the undertaking of space blockades is the ability to conduct space strike operations. Space strike operations involve space and other forces pursuing offensive operations against an enemy's space-related targets, whether in space, or on land, at sea, or in the air. They are therefore not strictly limited to attacks against the space infrastructure, and certainly not only against orbital platforms. In general, space strike operations are expected to be against vital strategic and operational space-related targets, i.e., "key points" (Jiang 2013, p. 137).

Space strike operations, in the Chinese view, are marked by "integrated operations; stealth and surprise; key point strikes; and rapid, decisive action." Integrated operations reflect the need to coordinate space strike operations with land, sea, and air operations, to forge "integrated combat power (*zhengti weili*; 整体威力)." They should also be undertaken at key moments when the enemy least expects it, exploiting stealth. They should also incorporate unexpected methods and tactics, so as to not only maximize material damage but also to undermine the enemy's morale. By employing a mixture of hard- and soft-kill methods, one can maximize stealth and generate additional surprise by confusing an opponent, making it harder to defend against.

Key point strikes are part of what might be the guiding thought for space operations in general. Here, the emphasis is on tightly focusing space operations and concentrating space forces along the main direction, at key times, against key targets in the enemy's combat systems of systems. The goals should be to disrupt, attrit, and paralyze the enemy's combat systems of systems, to prevent them from generating integrated capabilities (Jiang 2013, p. 142). This also requires carefully assessing an adversary's space system and identifying those key systems and vulnerabilities, since neither side is likely to field large numbers of space systems.

Rapid, decisive action denotes the need to use space strikes to seize the overall initiative in a campaign. By overwhelming an opponent and then sustaining strikes afterward, one can not only retain the initiative but ideally achieve operational goals and conclude the conflict. At the same time, due to the limited numbers of space platforms and weapons likely to be available, their fragility, and their expense (which limits numbers acquired), space strike operations are likely to be of relatively limited duration.

In the Chinese conception, space strike operations involve attacks against the full range of enemy space-related systems. One central element is the enemy's various satellite constellations. These can be targeted by a variety of hard-kill methods, such as directed energy weapons, kinetic kill vehicles (such as the one used in the 2007 anti-satellite test), and space mines and co-orbital anti-satellite systems (Chi and Xiao 2005, p. 39).

Equally important are such information warfare methods as "space electronic warfare and space network warfare (*taikong dianzi zhan he taikong wangluo zhan*; 太空电子战和太空网络战)." The application of integrated network electronic warfare (INEW) methods in space can interfere and disrupt the enemy spacecraft's various systems, including its onboard computers and other electronic components. Such methods can achieve a "mission kill," effectively neutralizing the platform, without generating the physical debris associated with collisions by kinetic kill vehicles and other physical attacks. Such soft-kill methods are seen as a vital means of conducting space information combat.

In addition, Chinese analyses suggest that striking at both space and terrestrial targets is necessary to establish local space superiority (Li et al. 2003). Such integrated attacks are comparable to traditional attacks against enemy command nodes or military bases (Hong and Liang 2002). Such attacks carry the additional advantage of retarding an opponent's ability to reinforce or replace damaged or destroyed orbiting systems. Therefore, Chinese analyses indicate that vital targets for securing space dominance include the ground components of the adversary's space systems, such as space launch vehicles and their launch sites and the attendant data and communications systems that link them together. Air, naval, ground, and special operations forces are therefore part of the arsenal of offensive space weapons, alongside ASATs and laser dazzling systems (Ma 2013, p. 220).

Chinese authors, however, also recognize that attacks against terrestrial targets, especially those based in the enemy's home territory, are likely to have significant strategic implications and potential repercussions. Therefore, attacks against strategic space targets require the direction of the highest-level political authorities.

Chinese analysts also believe that space strike operations will eventually include space-to-ground offensive operations, that is, the use of space-based weapons to bombard terrestrial targets. Some Chinese commentators, for example, have posited that the X-37B unmanned space vehicle might serve as a basis for a prompt global strike capability (Zhi and Li 2014). In this regard, they clearly see parallels between the development of space power and air power, i.e., the steady move from providing information support (aerial artillery observation, space-based reconnaissance and surveillance) to attacks against the adversary's information support systems, to the provision of fire support.

Defensive Space Operations (kongjian fangyu zuozhan; 空间防御作战)

At the same time as conducting space information operations and space strike operations, the PLA also expects to undertake defensive space operations. These defend one's own space systems (including orbiting satellites, terrestrial facilities, and the associated datalinks) from attacks by enemy space or terrestrial weapons and also protect national strategic targets from attacks from space systems or ballistic missiles (Zhang and Li 2005).

Defensive space operations involve a combination of passive and active defensive measures. Passive measures involve making Chinese satellites harder to track or determine their function. Chinese writings suggest that space systems should, as much as possible, incorporate camouflage and stealthing measures, so as to hide the nature and functions of the spacecraft from opposing observation and probes (Chang 2005, p. 316). Other passive measures include deploying satellites into orbits designed to avoid enemy detection; employing political, diplomatic, and other channels to mislead opponents of real operational intentions or otherwise influence enemy decision-making; and deploying false targets and decoys, to overload opponents' tracking capacities.

Because it is difficult to hide objects in space for very long, the Chinese have also shown an interest in resilience, i.e., extending survivability of space systems even after they are discovered. Some Chinese writings have discussed the deployment of small- and microsatellites in networks and constellations, rather than single large systems. According to one Chinese analysis, employing larger numbers of much smaller satellites may yield the same or greater capability than deploying a smaller number of larger, individually more capable systems, with less vulnerability (Bei et al. 2002). Larger satellites should be capable of functioning autonomously, so that even if their ground links are severed, they would nonetheless be able to continue operations (Chang 2005, p. 320). In addition, ground controllers should be prepared to move satellites, if there are indications that they might be attacked. Chinese planners may also be considering the incorporation of sufficient autonomy into satellites that they might be capable of altering their orbits on their own in order to evade perceived attacks.

Another set of survivability measures is the incorporation of hardening. This can only go so far, however, since spacecraft are very fragile. This is an inherent function of concentrating a number of sub-systems into a small volume and the extremely hostile environment of outer space, so that any damage to the spacecraft is likely to lead to substantial cascading effects (Xie and Zhao 2009). Similarly, while some ground facilities, including mission control facilities, might be physically hardened, the requirement for large antennae to handle telemetry imposes limits on how much physical hardening is possible.

In the Chinese view, defensive space operations cannot be solely reactive measures. As one PLA article notes, one can, and should, also employ offensive means and seek the initiative in the course of space defensive operations. More active defenses might include targeting enemy anti-satellite weapons, such as adversary co-orbital ASATs. Both offensive and defensive means, moreover, should be undertaken by not only space forces but also by land, sea, and air forces (Hong and Liang 2002). In the PLA's view, a combination of electronic and physical measures, including firepower strikes, may disrupt and suppress enemy space systems. By attacking terrestrial support components such as the TT&C facilities, the enemy's ability to conduct any kind of space operations, whether attacks against satellites or even provision of information support from space-based platforms, will be disrupted, thereby allowing one's own side to achieve space dominance.

It should be noted that the Chinese concept of "defensive space operations" does not necessarily parallel "defensive space control," as laid out in US Joint Publication 3–14 *Space Operations*. Indeed, some aspects would seem to overlap with that of "offensive space control" in the American sense (Joint Chiefs of Staff 2009, p. II-6).

Space Information Support Operations (kongjian xinxi zhiyuan zuozhan; 空间信息支援作战)

In the 2005 edition of *Military Aerospace*, a PLA textbook on military space activities, provision of information support by space systems was listed as the second task, after space deterrence (Chang 2005, pp. 304–309). In PLA teaching materials published in 2013, provision of information support by space systems was now the fifth of five tasks. This would suggest that space information support operations, while still important, are being eclipsed by more active space offensive and defensive measures. Indeed, as one Chinese assessment observes, as space resources become ever more important, and as military aerospace technology, especially those related to offensive space operations, steadily develops, space force development will shift from providing information support toward securing space dominance (Tan 2012, p. 170).

Nonetheless, in the context of informationized warfare, provision of space information support will be one of the greatest benefits of achieving space dominance. As the 2013 edition of *The Science of Military Strategy* notes, "space information support is now and for a long time into the future the main form (*zhuyao fangshi*; 主要方式) by which various nations apply space strength" (Academy of Military Science Military Strategy Research Office 2013, p. 181). As the PLA continues to emphasize joint operations, it will increasingly depend upon space-based systems to provide information support, especially as Chinese forces move farther and farther away from Chinese territory (and therefore, land-based information support infrastructure).

Key tasks within this mission area of "space information support (*kongjian xinxi zhiyuan*; 空间信息支援)" to the ground, air, and naval forces include:

- Space reconnaissance and surveillance
- Early warning of missile launches
- Communications and data relay
- Navigation and positioning
- Earth observation, including geodesy, hydrographics, and meteorology

While the priority accorded space information support may have fallen somewhat relative to other tasks such as space deterrence, providing this information to the other parts of the Chinese joint force remains essential. These capabilities are essential enablers for the PLA's ability to coordinate forces, engage in precision strikes, and assess damage.

So long as the PLA's main contingencies are on its immediate periphery (e.g., Taiwan, the Sino-Indian border), senior commanders will have a plethora of additional resources to gain information, ranging from unmanned aerial vehicles (UAVs), aircraft, and fishing boats to radio direction finding to terrestrial communications and sensor networks. But as the PLA's activities extend farther from China's shores (e.g., the Indian Ocean and Central Pacific), space will provide an increasing proportion of the information needed by Chinese military planners.

Space and Information Dominance "Under the New Circumstances"

Chinese military analysts and planners have long emphasized "informationized conditions" will dominate future warfare, but even they did not foresee the extent to which information would evolve and permeate all aspects of conflict. The sustained growth in information and communications technology (ICT) has led to what is termed "the new circumstances (*xin xingshi*; 新形势)." These "new circumstances" pose a range of challenges for defense planners, as weapons have become more precise, more intelligent, and more stealthy and are often autonomous. Moreover, the future battlefield will involve an ever-closer linkage between physical systems and electronic and virtual ones. "There will be basic changes in traditional concepts of time and space on the battlefield, as combat forms increasingly shift from mechanized towards informationized" (The Growing Importance 2016).

Especially important is the impact of these in ICT on the conduct of joint operations. The ability to establish dominance across the various domains, including land, sea, air, and the electromagnetic domain as well as outer space, will be much more difficult. Chinese writings have noted the importance of "multi-domain dominance" as part of the approach to future conflicts. But if commanders are able to achieve this, then their situational awareness will be far greater, given the various sensors and information systems at their disposal. At the same time, the ability to share information among the various participating forces, creating synergies among them, will make the side with multi-domain dominance virtually unassailable.

Consequently, the need to establish multi-domain dominance, centering around information dominance, is a central part of "the new circumstances." Key ICT technologies, including artificial intelligence, big data, and cloud computing, are shaping these "new circumstances." At the same time, however, space is a fundamental enabler of these ICT technologies, because of its central role in the acquisition and movement of information.

The development of massive constellations of Earth observation, communications, and data relay satellites, each with hundreds or even thousands of small-sats, will fundamentally alter the role of space in military operations. Such massive constellations may provide mobile access to broadband, allowing even more constant and rapid data flow. Proliferation of Earth observation satellites will make it much harder to conceal forces or hide development of new capabilities. These capabilities will further enhance terrestrial joint operations.

As important, denying space to an adversary will also be a growing priority, as it prevents an adversary from undertaking effective joint operations. Indeed, the growing significance apparently accorded space deterrence and space blockade, relative to space information support operations, in Chinese writings, suggests that PLA analysts consider space forces increasingly capable of exerting a *strategic* impact in their own right. In particular, the PLA appears to believe that by threatening adversary space capabilities, one can employ compellence strategies against those adversaries, in ways comparable to conventional and nuclear deterrence. This would appear to be yet another consequence of the "new circumstances."

At the end of 2015, in order to address these "new circumstances," the PLA undertook some of the most extensive and fundamental reorganizations since its founding in 1927. These embodied the ideas of "The Central Military Commission would manage the overall, the war zones would focus on warfighting, and the services would manage construction (*junwei guan zong, zhanqu zhuzhan, junzhong zhujian*; 军委管总, 战区主战, 军种主建)" (Xi Jinping 2015; Zhang 2016):

- The Central Military Commission, which manages the overall Chinese military, was reorganized from four General Departments to 15 departments, offices, and commissions. In the course of this reorganization, the General Staff Department has been renamed the Joint Staff Department, underscoring the heightened importance of joint operations for the PLA.
- The previous seven military regions (*junqu*; 军区) were consolidated into five theater commands or war zones (*zhanqu*; 战区). Where the previous military regions had been primarily peacetime entities with no wartime role, the new war zones are the operational-level command structures that are expected to undertake joint operations within their respective geographic areas of responsibility.
- The creation of three new services. The Second Artillery, previously a "super-branch," was elevated to the level of a full-blown service. The ground forces, which had previously been the default, now became a separate service (the PLA ground forces or PLA Army), in a sense losing political clout as a result. And the PLA Strategic Support Force (PLASSF) was created.

The PLASSF, in particular, will affect how the PLA implements space operations.

Creation of the PLA Strategic Support Force (PLASSF)

The PLASSF combined the PLA's electronic warfare, network warfare, and space warfare capabilities. This included what had previously been specific departments under the General Staff Department (GSD), such as the GSD Third Department (responsible for signals intelligence) and the GSD Fourth Department (responsible for electronic intelligence and electronic warfare). It also involved the transfer of key space facilities that had been part of the General Armaments Department (GAD), including China's launch sites; satellite control centers; tracking, telemetry, and control facilities; and China's fleet of space surveillance ships. Because of the Chinese emphasis on influencing adversary commanders and staffs, some political warfare elements from the General Political Department have also been incorporated into the PLASSF.

One reason for the establishment of the PLASSF appears to be the shift from a task or mission-oriented approach to warfare (e.g., reconnaissance, strike) to one more focused on specific domains (Costello and McReynolds 2018, p. 12). The PLASSF, as a service, will be responsible for planning, force construction, and operations within the information domain, including space operations. While the GAD had space responsibilities, it was neither a military service nor a warfighting entity; the GAD's main tasks were supporting military research and development, including new weapons, as well as managing China's nuclear and space facilities. Creating the PLASSF effectively created a service that was more focused on space warfighting doctrine and forces, rather than space systems and capabilities.

Moreover, this new service is "intended to create synergies between disparate information warfare capabilities, in order to execute specific types of strategic missions" (Costello and McReynolds 2018, p. 5). By reorganizing the PLA's basic structures and combining various information-related departments, offices, and bureaus across the PLA, many of the organizational stovepipes that impeded programmatic and doctrinal coordination were effectively defanged. In the case of space operations, the PLASSF's Space Systems Department now oversees GAD space facilities but also units responsible for space-based C4ISR (such as space-based remote sensing) that had resided in the General Staff Department (Costello and McReynolds 2018, p. 20).

At the same time, by embedding the Space Systems Department alongside the Network Systems Department (the other main subordinate entity within the PLASSF), there is greater ability to integrate space operations with network warfare (including cyber warfare) and electronic warfare operations. Chinese writings emphasize the importance of electronic and network warfare as key means of establishing space dominance, as soft-kill (e.g., laser dazzlers, cyberattack methods against TT&C facilities and onboard systems) approaches are an essential complement to hard-kill (e.g., direct ascent anti-satellite missiles, co-orbital anti-satellite systems) ones. By placing all of these capabilities in the same service (albeit in separate subordinate departments), PLA space dominance efforts will benefit from enhanced coordination and integration.

Civil-Military Integration of Space Industrial Capabilities

Another aspect of the "new circumstances" is the recognition that there must be greater civil-military integration, at not only the industrial level but also in terms of the broader organization of the nation's resources. This is reflected in the shift in Chinese terminology from "civil-military linkage" or "civil-military integration" (*junmin jiehe*; 军民结合) to "civil-military fusion" or "civil-military melding" (*junmin ronghe*; 军民融合). The idea of "fusion" or "melding" underscores the need to move beyond just linkages between the civilian and military sides. Instead, there must be a broader reorganization of the overall national economy and industry, so that the two sides are mutually served by a common economic base (Innovation Department 2019).

This has particularly significant importance in the space context, as the PRC has begun to support the development of additional commercial space capacity. Until recently, all Chinese aerospace industrial activity was undertaken by two state-owned enterprises: China Aerospace Science and Technology Corporation (CASC) and China Aerospace Science and Industry Corporation (CASIC). Beginning in 2015, however, the Chinese began to promote the development of non-state-owned space companies, with the promulgation of the "National Civilian-Use Space Basic Facilities Mid- and Long-Term Development Plan (2015–2025) (*guojia minyong kongjian jichu sheshi zhong chang qi fazhan guihua*; 国家民用空间基础设施中长期发展规划)." (It is important to note that "civilian use (*minyong*; 民用)" does not only mean **nonmilitary** use but can also mean **nongovernmental** use.)

This Plan was formulated by the National Development and Reform Commission, the Ministry of Finance, and the State Administration of Science, Technology, and Industry for National Defense (SASTIND), the entity responsible for the military industrial complex (Xie et al. 2017).

The Plan laid out several specific goals for the coming decade:

- Supporting construction of civilian and commercially financed and supported satellite manufacturing and associated research and development programs.
- Linking the public interest and commercialization. Commercial ventures should rely on societal (i.e., nongovernmental) investment.
- Encouraging and supporting investment by wealthy enterprises in planned satellite systems (Mao 2017).

The Plan expects that the main areas of commercial development will be in satellite remote sensing, communications satellites, navigation satellites, and associated applications. Chinese writings suggest that they foresee especially robust growth in the communications satellite sector over the next 5–10 years (Guo 2017). These elements will be integrated into a unified set of space-ground networks that will support global provision of space-derived services and information.

Another area of interest will be commercial launch services. The development of the Kuaizhou solid rocket booster for launching small satellites is seen as facilitating a broader Chinese role in commercial space launch. The January 2017 Kuaizhou-1A launch of three satellites is described as the first time the Chinese have commercially launched more than one satellite at a time.

As important, some Chinese writers see a "Kuaizhou model" for promoting broader use of Chinese launch systems. The model emphasizes:

- Low cost.
- Customized or tailored launch services. Clients can either choose to purchase an entire rocket for themselves (the Kuaizhou-1) or be part of a group of launched satellites (the Kuaizhou-1A).
- Different commercial approaches. One can first sign a purchase order and then cooperate, or one can engage in cooperative development and then sign a purchase order (Rao 2017).

Some of these themes were further explored at a conference held on September 15, 2015. Described as China's first conference on commercial space flight development, it reportedly brought together key departments and ministries, scientific research organizations, academics, as well as representatives from civilian and commercial enterprises. One of the conclusions from the conference was that commercial space travel and exploitation would be difficult, if not impossible, with current Chinese approaches to research and development (China Commercial 2015).

The PRC seems to be pursuing four broad paths for expanding its aerospace industrial capacity along increasingly commercialized lines: traditional state-owned enterprises (SOEs), traditional research and development facilities and institutions, private commercial enterprises, and Internet-associated enterprises. These efforts at developing an expanded commercial capacity will be overseen by the State

Administration for Science, Technology, and Industry for National Defense (SASTIND) Since SASTIND is the entity also responsible for overseeing the overall Chinese military industrial complex, it is likely to ensure that Chinese military requirements and capabilities are not slighted by any of these new commercial ventures.

For example, Chinese e-commerce giant Alibaba has begun to probe possible investments in space. In August 2015, Alibaba reportedly joined with NORINCO Corporation to establish a joint venture to explore possible linkages between the Internet and the Beidou position, navigation, and timing system (Feng 2015). Through cloud computing and advanced data management techniques, this would help improve Alibaba's services. NORINCO is known as China North Industries Group Corporation, Limited, and is also known as China Ordnance Industries Group Limited. It is a major manufacturer and exporter of military as well as civilian goods.

Conclusions

For the PLA and Chinese security decision-makers, the Information Age and the Space Age are inextricably linked. Both have been heavily influenced by the growth in computing power and the role of telecommunications. This has gained impetus as ICT has a growing role in overall national power and especially military capability.

Chinese analyses of recent wars underscore the intimidate relationship between these two realms when it comes to warfighting. Modern wars have demonstrated the close relationship between information and space, where space systems play a central role in the collection, transmission, and exploitation of information. Consequently, as one PLA analysis observes, "seizing the space information advantage as a high ground is the first decisive condition for seizing information dominance, space dominance, air dominance, naval dominance, land dominance, and therefore the initiative in wartime" (Lanzhou Military Region Headquarters 2003).

By dominating space, one gains an enormous advantage in terms of access to information and managing information flow. The side that dominates space thereby gains several key advantages.

- *The battlefield is much more transparent.* Combat forces can therefore be much more effective, since enemy and friendly dispositions will be known.
- *Command and control is much more precise and capable.* Because the battlefield is more transparent, commanders can respond in real-time or near-real time enemy actions, and widely separated units drawn from a variety of services can act in a highly integrated manner.
- *It makes noncontact, nonlinear warfare possible.* By dominating space, one has secured the most important portion of the battlefield that of information space. The more transparent battlefield, the facilitated command and control, enables long-range, precision strikes, against which the adversary is gravely disadvantaged in trying to counter. Friendly casualties are reduced, while one's actions are much more effective.

For Chinese military planners, these advantages are further enhanced by certain geographic and strategic realities. China, even now, is not oriented toward mounting extensive military operations far from its shores, but remains focused on such flash points as Taiwan, the Korean peninsula, the South China Sea, and the Sino-Indian border. For the PRC, the consistent concern since the 1980s has been on "local wars." Such wars are not only limited in means but also are expected to occur mainly on China's periphery.

Consequently, the PLA can bring to bear substantial resources, drawn from across the entire country if necessary, in order to establish information dominance. Mobilized civilian assets, ranging from fishing boats for maritime surveillance to the militia for camouflage and deception operations, can supplement regular PLA forces. Shorter-range assets from fast attack craft to older fighter aircraft can similarly be employed to deny and counter adversary forces, including information collection platforms. Communications can be sustained through fiber-optic cable (which is difficult to monitor), cell phones, line-of-sight radios, as well as satellite communications, enhancing communications security and providing redundancy. In many ways, China does not need space for the PLA to operate in accordance with its doctrine.

By contrast, the United States is an expeditionary military, operating far from American shores. In time of conflict, it is therefore much more reliant upon space-based systems even for operational communications, especially to coordinate disparate, separated forces, as well as for intelligence collection against targets typically halfway around the globe. As important, American military planners have chosen to rely on space-based assets for positioning, navigation, and timing, whether it is aircraft routing, shipborne navigation, or weapons guidance. The combination of geostrategic conditions and weapons acquisition policies makes American forces much more dependent upon space.

In short, in the struggle for information dominance, because of the asymmetric strategies and starting conditions, there is a resulting asymmetric dependence on space.

At the same time, it is important to recognize that the various space architectures are microcosms of the larger information battlefields. Space networks, encompassing the satellites, terrestrial support structures such as mission control, and the data links connecting them are themselves systems of systems bound together via information networks. Satellites require communications and data links, not only to carry information about various targets or to provide navigation and other updates but also to allow mission control to monitor the satellites' status, adjust their orbits, update their software, and otherwise manage and control their operations. It is this tracking, telemetry, and control (TT&C) network, and the information flow over it governing the satellite constellations and linking them back to Earth, that allows space systems to operate. Damaging or affecting that flow can effectively neutralize the constellation or even allow an adversary to take control of one or more satellites. Its preservation is as important for securing information dominance as having the constellation provide information support to the various terrestrial struggles. The ability to deny the adversary not only information gathering and transmission but satellite monitoring and control, that is, denying them information about oneself or information about their space assets, can be equally damaging to their larger ability to establish information dominance.

It is for this reason that the Chinese emphasize that space dominance entails not only targeting satellites but ground facilities such as mission control sites and the data links connecting them. The struggle for space dominance is, in fact, a part of the larger struggle for information dominance. It is the facet that occurs within the confines of the two sides' space architectures.

References

Academy of Military Science Military Strategy Research Office (2013) The science of military strategy. Military Science Publishing House, Beijing
Atlas 3 Scrubbed to Tuesday. Space Daily, May 21, 2000. http://www.spacedaily.com/news/eutelsat-00g.html
Bei C, Yang J, Zhang W (2002) Nanosatellite distributor design proposal. Zhongguo Hangtian Bao, August 23, 2002, p 4, in FBIS-CHI
Chang X (2005) Military astronautics, 2nd edn. National Defense Industries Press, Beijing
Chi Y, Xiao Y (2005) Essentials of informationized warfare and information operations theory. Military Science Publishing House, Beijing
China Commercial Spaceflight Development Conference Opens in Beijing (2015). Aerospace China, no 10
China: Missile Defense System Test Successful. USA Today, January 11, 2010. http://www.usatoday.com/news/world/2010-01-11-china-missile-defense_N.htm
Costello J, McReynolds J (2018) China's strategic support force: a force for a new era. INSS China strategic perspectives paper no 13. National Defense University, Washington, DC
David L (2007) China's antisatellite test: worrisome debris cloud encircles earth. Space.com, February 2, 2007. http://www.space.com/3415-china-anti-satellite-test-worrisome-debris-cloud-circles-earth.html
Feigenbaum E (2003) China's techno-warriors. Stanford University Press, Stanford
Feng C (2015) China's Alibaba teams up with state-owned arms maker to develop positioning services as Beijing shuns GPS. South China Morning Post, August 19, 2015. https://www.scmp.com/tech/science-research/article/1850807/online-shopping-king-alibaba-working-chinas-state-owned
Gao Y (chief editor) (2001) Joint campaign course materials. Academy of Military Science Publishing House, Beijing
Gao Q (2005) Aerospace reconnaissance characteristics and limits in high-tech local wars. J Acad Command Equip Technol XVI(1)
Gruss M (2015) Space surveillance satellites pressed into early service. Space News, September 18, 2015. http://spacenews.com/space-surveillance-sats-pressed-into-early-service/
Guo J (2017) Our nation's commercial aerospace shifts towards industrialization. Economics Daily, September 5, 2017. http://www.gov.cn/xinwen/2017-09/05/content_5222713.htm
Guojia Gao Jishu Yanjiu Fazhan Jihua 863. FBIS-CHI, July 21, 2000
Hartnett D (2008) Towards a globally focused Chinese military: the historic missions of the Chinese armed forces. CNA Corporation, Alexandria
Hong B, Liang X (2002) The basics of space strategic theory. China Military Science, no 1
Hu J (2004) See clearly our military's historic missions in the new period of the new century, December 24, 2004. http://gfjy.jxnews.com.cn/system/2010/04/16/011353408.shtml
Innovation Department, Beijing University Science and Technology Park (2019) An outline of the development of our nation's civil-military fused enterprises. China New High-Technology

Enterprise Guiding Paper, April 15, 2019. http://www.chinahightech.com/html/paper/2019/0415/521151.html

Jiang L (2013) Space operations teaching materials. Military Science Publishing House, Beijing

Joint Chiefs of Staff (2009) Space operations, JP 3-14. Department of Defense, Washington, DC

Lanzhou Military Region Headquarters Communications Department (2003) Space information support and its influence on future terrestrial operations. Military Art, no 10

Li D (1999) A survey of the development of space technology in China. China Aerospace, June 1999, pp 16–19, in FBIS-CHI, September 21, 1999

Li Y (2012) Joint campaign teaching materials. Military Science Publishing House, Beijing

Li J, Dan Y (2002) The strategy of space deterrence. China Military Science, no 1

Li D, Zhao X, Huang C (2003) Research on concepts of space operations and its command. J Acad Equip Command Technol XIV(5)

Liu K, Wang X (2008) The first conflict won through airpower: the Kosovo war. Academy of Military Science Publishing House, Beijing

Lu J (2008) Lu Jin: satellite communications – the information bridge during earthquake relief operations. Speech before the Chinese Communications Studies Association, September 26, 2008. http://www.ezcom.cn/Article/8591

Ma P (2013) Joint operations research. National Defense University Publishing House, Beijing

Mao L (2017) Development state of the main domestic commercial aerospace enterprises. Satellite Applications, no 10

Matthews W (2010) Chinese puzzle. Defense News, September 6, 2010. http://www.defensenews.com/story.php?i=4767907

Orwig J (2014) A rocket launch Monday was delayed because of a boat. Business Insider, October 28, 2014. http://www.businessinsider.com/why-rocket-launch-delayed-by-a-boat-2014-10

PLA Encyclopedia Committee (1997) Chinese military encyclopedia. Military art, vol III. Academy of Military Science Publishing House, Beijing

PLA Encyclopedia Committee (2002) Chinese military encyclopedia, supplemental volume. Military Science Publishing House, Beijing

Rao H (2017) Commercial space: a necessary choice for major aerospace nations. National Financial Report, no 12

Tan R (2012) Operational strength construction teaching materials. Military Science Publishing House, Beijing

The Growing Importance of the Basic Factors in Strengthening National Defense and Army Building Under the New Circumstances. PLA Daily, August 2, 2016. http://www.xinhuanet.com/politics/2016-08/02/c_129198048.htm

Wang H, Zhang X (chief editors) (2000) The science of campaigns. National Defense University Publishing House, Beijing

Wang W, Zhang Q (2009) Discussing military theory innovation with Chinese characteristics. National Defense University Publishing House, Beijing

Weeden B (2014) Through a glass darkly: Chinese, Russian, and American anti-satellite testing in space. Secure World Foundation, Washington, DC

Wu R (2004) Theory of Informationized conflict. Military Science Publishing House, Beijing

Xi Jinping (2015) Xi Jinping: form a structure where the central military commission manages the overall, the war zones focus on warfighting, the services manage construction. PLA Daily, November 26, 2015. http://news.sohu.com/20151126/n428330471.shtml

Xi Jinping Repeatedly Supports Building a Major Aerospace Power. Xinhuanet, April 12, 2019. http://www.xinhuanet.com/politics/xxjxs/2019-04/12/c_1124357478.htm

Xie Z, Zhao D (2009) On the fundamental features of the military space force. China Military Science, no 1

Xie P, Zhou X, Wang Y, et al (2017) A review of aerospace enterprise development during the '12th Five Year Plan'. National Development and Reform Commission, Department of High Technology Industry. http://gjss.ndrc.gov.cn/zttp/xyqzlxxhg/201708/t20170802_856978.html

Xu G (2013) Research on our military's information operations strength construction. Military Science Publishing House, Beijing

Ye Z (2007) Concepts of informationized operations. Military Science Publishing House, Beijing

Yuan W (2008) Science of military information. National Defense University Publishing House, Beijing

Zhang Y (chief editor) (2006) The science of campaigns. National Defense University Publishing House, Beijing

Zhang X (2016) There is no natural boundary between war zones focus on warfighting, services manage construction. PLA Daily, March 21, 2016. http://www.xinhuanet.com/mil/2016-03/21/c_128816639.htm

Zhang Q, Li X (2005) Space warfare: from vision to reality. China Military Science, no 1

Zhang Y, Dong Z, et al (2005) Informationalized warfare will make seizing the aerospace technology 'high ground' a vital factor. People's Liberation Army Daily, March 30, 2005

Zhi T, Li W (2014) X-37B, the mysterious space fighter. China Youth Daily, November 3, 2014. http://jz.chinamil.com.cn/n2014/tp/content_6209028.htm

Zhou P, Wen E (2004) Developing the theory of strategic deterrence with Chinese characteristics. China Military Science, no 3

Historical Evolution of Japanese Space Security Policy

30

Kazuto Suzuki

Contents

Introduction	556
The Diet's 1969 Resolution on "Exclusively Peaceful Purposes"	556
The End of the Cold War Paradigm	557
The Information-Gathering Satellite Program: Treading a Narrow Path Through a Legal Jungle	558
Kawamura's Initiative to Modify the 1969 Resolution	559
Legalizing the Strategic Objectives of Space Policy	560
Regional and Global Security	562
Changes of the Role of the Ministry of Defense	563
JAXA's View on Space Security	565
Japanese Reaction to the Code of Conduct	566
Conclusion	568
References	569

Abstract

Japanese perspective on space security has begun with a very unique setting. The 1969 Diet Resolution has put heavy constraints on its space activities, and interpretation of "non-military" approach has refrained Japan from anything related to security. However, the 1998 Taepodong launch and subsequent reform of space policy eventually created Basic Space Law in 2008. Although the organizational culture and history still influence on the decision-making process, changing security environment and the role of Japan in the Asia-Pacific region made Japan to be more active and committed to both security by and of space.

K. Suzuki (✉)
Hokkaido University, Hokkaido, Japan
e-mail: kazutos@juris.hokudai.ac.jp

© Springer Nature Switzerland AG 2020
K.-U. Schrogl (ed.), *Handbook of Space Security*,
https://doi.org/10.1007/978-3-030-23210-8_15

Introduction

The concept of space and security has been inconsistent for a long time in the history of Japan. Any space activities should be considered as peaceful one, which meant to be a place where military shall not play any part of it. This extreme interpretation of "exclusively peaceful purpose" has been challenged by many incidents concerning the security environment around Japan, and eventually reinterpreted through the discussion on the Basic Space Law in 2008. However, the older interpretation of "non-military use of space" is still persistent in Japan (Aoki 2008).

This chapter discusses the reasons why space and security were not compatible in Japan and analyzes the recent developments on the issues of space security. This chapter defines that the concept of "space security" does not limit to the issue of "security of space environment" but also includes "security on the earth through space systems."

The Diet's 1969 Resolution on "Exclusively Peaceful Purposes"

Japan has restricted itself from using space for its security needs. As one of the most advanced industrialized country, Japan possesses technological and industrial capability to use space for its national security. Many non-Japanese space experts may wonder why it has not done so, if only for non-aggressive purposes.

The main reason for Japan's reticence is its pacifist constitution, which is interpreted to prohibit using space for security purposes. In 1969, the Japanese parliament, the Diet, passed a resolution "Concerning the Principle of the Development and Utilization of Space," popularly known as "the exclusively peaceful purposes resolution." It stipulates that Japan's space programs may be conducted by the civilian sector, not the defense sector, and only for the research and development of new technology for exclusively peaceful purposes (Suzuki 2005).

The principle of "exclusively peaceful purposes" is not new, as it appears in the Outer Space Treaty and the ESA Convention. The Japanese application of this principle, however, was unique. While debating the resolution in the Diet in 1969, the Diet members argued that it should be applied to the development and use of space in the same way that nuclear technology had been. As dual-use technology, they can be developed simultaneously for both civilian and military purposes. In addition, because Japan's Science and Technology Agency (STA, currently MEXT) was in charge of both nuclear and space technology, the Diet felt that the development of space should be restricted as tightly as that of nuclear technology was. Ever since the horror of the nuclear holocausts in Hiroshima and Nagasaki, the Japanese people have been skeptical of using nuclear technology even for peaceful purposes, and therefore the Diet stipulated that it be used only for civilian purposes and that the military not be involved administratively, financially, and politically in its development and operation. (Response by Masao Yamagata, member of the Space Activities Committee, in the Special Committee of the Promotion of Science and Technology,

Lower House, minutes, April 16, 1969.) Accordingly, this notion of "exclusively peaceful purposes" was applied to space as well.

Based on the interpretation of the Diet resolution, all of Japan's operations in space have been conducted for scientific and technological purposes. The strategic goal of Japan's space policy thus has been to "catch up" with the technology of other advanced countries such as the United States and European nations. Thus, the goal of most of Japan's space programs, even that of those for communication, broadcasting, and meteorology, has been technological excellence. For many politicians, space was the "necktie of advanced countries" (Matsuura 2004), suggesting that Japanese space policy should aim at gaining national prestige.

The principle of "non-military" use of space has changed, however, in accordance with the Basic Space Law which passed the Diet in June 2008. The aim of the law is to redefine the purpose and rationale for Japan to invest in space, and for the first time, the term *security* appears in an official document pertaining to space. Why has the notion of security suddenly appeared in the draft law, and how is it likely to change Japan's space policy?

The End of the Cold War Paradigm

For many years, particularly during the Cold War, Japan's strictly "non-military" use of space was not challenged. The reason was that the alliance between Japan and the United States already provided the necessary infrastructure for telecommunication and intelligence gathering from space. In addition, Japan's pacifist Constitution was interpreted as prohibiting its Self-Defense Forces (SDF) from being deployed beyond its national border. When the Cold War ended, however, Japan was forced to begin thinking about changing its space strategy (Suzuki 2007a).

With the end of the Cold War, the threat of Communism had waned, and so the reason for stationing US troops in Japan also had become less compelling as well. Although the United States still maintains a need for forward deployment bases in Japan, it is no longer a condition of the alliance. The unilateral collective defense – according to which the United States is obliged to defend Japanese territory but Japanese forces are not obliged to protect American territory or even US Forces – has now become too great a burden, so the US government wants the Japanese government to share more of responsibility for global security. Accordingly, Japan has enlarged its participation in the UN peacekeeping operations and the war on terror, particularly by deploying naval forces to support the multinational anti-terrorist operations in Afghanistan and its own ground troops in Iraq.

It is through these operations that the Self-Defense Forces have come to realize their technological shortcomings. Because the SDF are restricted from developing and operating their own space capabilities, they have had to rely on commercial satellite communication and commercial imagery services. But until now, because the SDF have not been permitted to be deployed outside Japan's borders, they have not needed long distance communication or imagery of foreign countries other than its neighbors. However, the SDF have recognized the gap in Japan's military technology, particularly in regard to the United States' military transformation in recent years. Given the increasing possibility of Japan's sharing its security burden and joint operations with US forces, the SDF and the Japan Defense Agency (JDA, now the Ministry of Defense) now acknowledge the importance of developing their own space capability.

In addition, the Japanese people's perception of their security has been dramatically changed by North Korea's launch of a Taepodong missile over Japanese territory in 1998. The alarmed Japanese public thereupon demanded that the government take measures to protect them, and the government immediately decided to launch what is known as the Information-Gathering Satellite (IGS) program.

The Information-Gathering Satellite Program: Treading a Narrow Path Through a Legal Jungle

At the start, the IGS program faced the serious legal constraints of Japan's space policy. Although it was clear that the purpose of the IGS program was to monitor the military activities of its neighbors, including North Korea, this was concealed under the guise of a "multipurpose" satellite program. This way the civilian nature of the program was implied in order to comply with the 1969 Diet resolution.

But this arrangement ran into problems. In the 1980s when Japan and the United States were sparring over trade, the US government pressured Japan to open its public procurement market in order to reduce the US trade deficit. The industry targeted was the space satellite industry, in which Japanese companies enjoyed exclusive contracts with the National Space Development Agency (NASDA, now the Japan Aerospace Exploration Agency, or JAXA). The US government maintained that it was unfair to exclude its satellite industry from competitive bidding and if Japan does not open up satellite procurement process to international bidders, the US government threatened to use its "Super 301" measure (imposing retaliatory tariff on Japanese products according to the Omnibus Foreign Trade and Competitiveness Act of 1988), according to which the US government could impose punitive tariffs on Japanese imports. In response, the Japanese government enacted the 1990 Accord on Non-R&D Satellite Procurement. The accord obliged the Japanese government to open the procurement market for civilian satellites to international competitive bidding, and as a result, 18 of a total 19 civilian non-R&D satellites in orbit were contracted to American companies, and only 4, the MTSAT-2 (meteorological and navigation satellite), Himawari 8 and 9 (Meteorological satellites) and Superbird-7 (commercial communication satellite), was contracted to a Japanese company. However, the launch of Quasi-Zenith Satellite System (QZSS) of which the first satellite was launched in 2009 has changed the dynamics. They are non-R&D satellites but procured by the government. The US government encouraged Japanese government to develop QZSS as supplementary system for its GPS.

As a result of the 1990 accord, the civilian multipurpose non-R&D satellites under the IGS program were part of the open procurement procedure as well, which put the Japanese government in a difficult position. If it wanted to avoid applying the 1990 accord, it would have to admit that the "multipurpose" satellites under the IGS program actually had a military mission, a position that would violate the 1969 Diet resolution.

This problem was resolved by careful legal interpretation. The government placed control of the satellites not under the Japan Defense Agency (JDA) but under the Cabinet Secretariat, a small office with a national intelligence-gathering mission and crisis management functions. The IGSs then were formally designated as "crisis management satellites" with both civilian and military purposes (Sunohara 2005).

This incident made politicians realize that the legal constraints of the "exclusively peaceful purposes" resolution allowed no room for maneuver and that in the changing security environment of the post-Cold War period it seemed counterproductive to maintain such a rigid pacifist position.

The Koizumi government's decision in 2003 to participate in the missile defense program raised another problem for Japanese space and security community. If JDA depended completely on US intelligence for initiating the deployment of counterattack missiles, it might mistakenly shoot down hostile missiles flying toward US territory. And because shooting down hostile missiles aimed at US territory would be an exercise of the right of collective self-defense by Japan, which is considered unconstitutional, it needed to have its own early warning satellite to verify the US satellite intelligence. Thus, many people in the Liberal Democratic Party (LDP), particularly those interested in defense issues, demanded a reconsideration of the "exclusively peaceful purposes" clause of the 1969 Diet resolution.

Kawamura's Initiative to Modify the 1969 Resolution

Despite the increasing demand to modify the resolution and the mounting pressure to reduce the space budget, neither the government nor the politicians took any action until the end of 2004. In early 2005, Takeo Kawamura, LDP politician and a former Minister for Education, Culture, Sports, Science, and Technology (MEXT), took the initiative for change. While he was Minister, Kawamura witnessed the failure of the H-IIA no. 6 launch carrying two IGSs. Although he was responsible for the actual launch on the H-IIA and not the IGS program, both the public and the government accused him for not properly supervising a strategically important satellite project like the IGS. From Kawamura's standpoint, JAXA was responsible only for research and development, and it is acceptable, for R&D agency, to fail launch attempts because it is R&D rather than operational program. However, the Cabinet Secretariat, the operator of the IGS, was extremely furious about the failure of launching two IGSs which are vitally important for the national security. Yasuo Fukuda, Chief Cabinet Secretariat and No. 2 in the government, blamed Kawamura and JAXA for being irresponsible, and he claimed that it is not acceptable for JAXA to have excuse that launcher was only for R&D purposes. To Kawamura, this was a critical failure of Japan's national strategy, and he was convinced that something had to be done (Suzuki 2007b).

As soon as Kawamura left his post as Minister of MEXT in September 2004, he convened an informal study group, the Consultation Group for National Strategy for Space, popularly known as the Kawamura Consultation Group. It was made up of members of LDP, including vice ministers in MEXT, the Ministry of Economy, Trade and Industry (METI), the JDA, and the Ministry of Foreign Affairs (MoFA). The Kawamura Consultation Group considered the problems of Japanese space policy, including the modification of the 1969 resolution and several public-private partnership programs such as the Quasi-Zenith Satellite System (QZSS), as well as the privatization of H-IIA (Suzuki 2006).

After ten meetings, the Kawamura Consultation Group issued an over 100-page report in October 2005, which argued that Japan's space policy lacked a coherent strategy and a clear institutional arrangement (Consultation Group for National Strategy for Space 2005). That is, because Japan's space policy was dominated by the Science and Technology Agency and MEXT, it did not have a plan for using space to pursue its national strategic objectives. The result was a lack of competitiveness in the Japanese space industry and the difficulty which Japan was facing in assuming a larger role on the international stage.

The report therefore suggested drafting a new space policy and an institutional framework within which a more coherent space policy could be formulated. First, it proposed that the government create a new minister for space in the office of the Cabinet, who would serve as the center for Japanese space strategy. Furthermore, the report pointed out that Japanese space policy had been to develop new technology but that along the way it had neglected the users' needs and demands. The report thus recommended that the new minister for space should bring relevant ministries into the policy-making process and refer their needs to the R&D program. The minister of space also should be authorized, the report continued, to use Japanese space assets to advance foreign and security policy purposes under the current constitutional framework.

The Kawamura Consultation Group's report also proposed the modification of the 1969 resolution. It suggested that any modification of the resolution would have to come from the Diet because the resolution was unanimously taken by the Diet, and the only way to legitimatize the modification should be done through new resolution or legislation. Consequently, members of the LDP and the government were pleased with the report, as it paved the way for Japan to change its space policy.

Legalizing the Strategic Objectives of Space Policy

Kawamura also submitted this report to the LDP's Policy Research Council. With support from Hidenao Nakagawa, then the Chairman of the Policy Research Council and the third-ranking member of the LDP, Kawamura established the Special Committee on Space Development (SCSD) with himself as its leader. In July 2006, when North Korea conducted a second missile test, the SCSD got another boost of support as public opinion quickly shifted from guarding its pacifist principles to demanding a more flexible interpretation of the 1969 resolution. In this atmosphere, the SCSD decided to submit to the Diet its draft Basic Space Law in June 2007. The bill was sponsored not only by the LDP but also Komeito, its coalition partner, and Democratic Party of Japan (DPJ), the largest opposition. The reasons of DPJ sponsoring the bill was largely due to the belief of Kawamura that space is a national strategic issue and should not become a subject of partisan politics. He took initiative to invite Yoshihiko Noda, then the leader of Science and Technology policy group of DPJ and later became Prime Minister, to join the LDP sponsorship and Noda agreed. The participation of DPJ, largely composed by liberal and pacifist politicians, was important because it would guarantee that this bill will remain the pacifist nature of the space policy. Also, bipartisanship gave stability and continuity of space policy after the change of government in 2009. The bill passed the Diet in June 2008.

The first and the most important feature of the Basic Space Law is its institutional renovation, in accordance with the Kawamura Consultation Group's report. It proposes establishing a new Minister for Space and a Strategic Headquarters of Space Policy (SHSP), which will serve as Cabinet-level inter-ministerial coordination and decision-making body. The idea of establishing high-level political body is to make sure that the R&D activities and utilization of space systems will be seamlessly coordinated. For many years, Japanese space programs were decided based on the technological interests and "catch-up" to other space-faring nations. However, under the heavy budgetary constraints, the government could not afford to spend large sum of money only for the sake of technological development. In order to secure and sustain space budget, the government has to prove that the investment in space is effectively contributing to the policies and services to the people. The Minister for Space will be coordinating space-related policies of various ministries. One such ministry is the Ministry of Foreign Affairs (MoFA), which formulates policies to make Japan as leading state for setting new international rules by taking advantages of Japan's space technology as part of its foreign policy.

The second feature of the Law pertains to security. Article 2 of the Law states that "Our space development shall observe the Outer Space Treaty and other international agreements and shall be conducted in accordance with the principle of pacifism upheld in the Constitution." In other words, the traditional interpretation of "exclusively peaceful purposes" as "non-military" should no longer apply. Instead, the policy should be to adopt the international standard interpretation of the "peaceful use" of space as the "non-aggressive" or "non-offensive" use of space. The new Law would accordingly enable the Japanese defense authority to become involved in the development, procurement, and operation of space systems.

In addition, Article 3 states that "the government shall take necessary measures to promote space development that will contribute to international peace and security and also to our nation's security." Because this statement is so general, Article 3 could be interpreted as allowing the government to use space systems for aggressive purposes. But because Article 2 stipulates that the use of space systems for national and international security comply with both the framework of international agreements and Japan's constitution, it implies that Japan may use its space assets for crisis management and disaster monitoring in Asia and for peacekeeping missions outside its territory. Article 2 also suggests that Japan can use early warning satellites for its missile defense, as this falls into the category of self-defense (Aoki 2009).

The Law therefore is designed to strengthen Japan's capability in settling disputes and managing crises by peaceful means and is intended to change only the interpretation of the Diet resolution, preventing any use of space by Japan's military authority.

Regional and Global Security

With the Basic Space Law, the government will at last have a legal base for using space to strengthen its national security and expand diplomatic activities. This combination of security and diplomacy is important for two reasons. First, one of Japan's primary objectives of using space for security is to acquire the capability to defend its own country, particularly by means of a missile defense system. Given the small size of Japan's territory, space is not a very useful tool. It may not require a constant surveillance and communication capability. However, in the context of Japan's expanding role in international security and the Japan-US alliance, the SDF operations far from home would require long distance telecommunications and satellite intelligence. Such needs were confirmed by the Maritime SDF ships sent to the Indian Ocean to support the United States or to protect commercial vessels from piracy in the Gulf of Aden, and allied operations in Afghanistan as well as the Ground and Air SDF troops sent to Iraq.

The first SDF forces deployed outside Japan were sent to Cambodia in 1992 for UN peacekeeping missions. Since then, Japanese troops have been sent to places such as the Golan Heights, Mozambique, Zaire, East Timor, and South Sudan. Now the majority of Japanese no longer doubt their country's intention to contribute to international security and peace through UN operations, and consequently, the 1969 resolution has become both awkward and irrelevant. Although it states that space should be used for "exclusively peaceful purposes," during their UN operations, the SDF have not been allowed to use Japan's space assets to maintain "peace." The new law would not only enhance the scope of operation and capability of Japanese contribution to global security but it would also increase the efficiency and effectiveness of its participation in multinational operations.

Second, the combination of security and diplomacy in the new Law is important because Japan would be able to change Asia's security environment in Asia. A number of issues are causing instability and threatening the security of the region, particularly North Korea's nuclear and missile tests, tensions between Taiwan and China, China's opaque security strategy and defense budget, China's ASAT test in January 2007 (Hagt 2007), and various territorial and resource disputes. Although these conflicts have been contained by larger international organizations like the Association of Southeast Asian Nations (ASEAN) or ASEAN Regional Forum (ARF), they still need to be closely monitored to develop confidence building measures to ensure stability in the region and to seek peaceful solutions. Japan is, of course, the concerned party in some of these conflicts and needs to participate in such regional forums as ARF, the Asia-Pacific Economic Cooperation (APEC), and the East Asian Summit. Although it is committed to providing ideas and resources for their development, Japan needs first to prioritize the promotion of regional interests as well as its own domestic interests. Japan also uses its technological advantages to assume a leadership role in the region. To date, it has been providing its technological expertise through the Asia-Pacific Regional Space Agency Forum (APRSAF). Japan has been playing a central role in APRSAF, but this was mostly done by the MEXT and JAXA without coordinating with the Ministry of Foreign Affairs. Thus, Japanese initiative focused purely on technical cooperation and has not fully incorporated with Japanese diplomatic strategy (Suzuki 2012a). However, the establishment of the Basic Space Law changed this lack of coordination. The Ministry of Foreign Affairs has created the "Office of Space" under the Foreign Policy Bureau and assigned several diplomats to dedicate their efforts to utilize space systems and activities for the diplomatic affairs.

Changes of the Role of the Ministry of Defense

Due to the establishment of the Basic Space Law, the role of the Ministry of Defense (MoD) has changed dramatically. It has been excluded any space activities until 2008, except the use of satellite communications, broadcasting, and weather imageries. Although the 1969 Resolution has prohibited the JDA and later MoD to develop, own, operate, and use space systems, the government expressed its interpretation of the resolution in 1985 as follows:

> The clause "exclusively peaceful purpose" in the Diet resolution means, of course, the SDF would not use satellites as lethal or distractive means, but also the use of satellite which is not generally available. In this context, the use of satellite, which is generally available and utilized, or possessing equivalent function to commercially available satellites, can be used by the Self-Defense Force. (Kato 1985)

Under this interpretation, the SDF was able to "use" satellite systems but still not allowed to "develop, own, and operate" them. However, the Basic Space Law urges that the government shall use space system "to ensure international peace and security and also to contribute to our nation's security" (Article 3). This suggests that the MoD may develop, own, operate, and use space system for the purpose of international peace and security such as UN peacekeeping operations or national security. Nevertheless, the MoD was not given full-fledged space capability. The Article 2 defines that "Our space development shall observe the Outer Space Treaty and other international agreements and shall be conducted in accordance with the principle of pacifism upheld in the Constitution." In other words, space activities of MoD shall remain within the framework of Japanese Constitutional constraints, which limits its military capability solely for self-defense. It means that the new interpretation of the use of space system for security purpose remains for passive use of space such as telecommunications, surveillance, and navigation, and even the case of surveillance and reconnaissance, it would not be acceptable to use satellite intelligence for aggressive purposes.

The executives of the MoD were not very enthusiastic to invest in space for several reasons. First, under heavy budgetary constraints, MoD had to invest in a lot of new defense equipment to counter the modernized and upgraded Chinese forces. Also, the missile defense programs needed to be upgraded for the increasing threats from North Korea. In order to meet these challenges, there was no luxury to increase spending for unfamiliar domain of space. Second, due to the long period of refraining from the investment in space, MoD has almost no staff or technical expertise in space technology that eventually make MoD to depend on JAXA. Given the secretive nature of MoD, it would be difficult to depend on civilian agency to develop military sensitive technology. Although there is gradual rapprochement between MoD and JAXA, the level of cooperation is not satisfactory. For these reasons, the role of MoD was limited even after the establishment of the Basic Space Law.

Nevertheless, there were some pressing needs for MoD to take action in space domain. The first priority was to develop telecommunication satellites. Since 1985 government decision to allow SDF to use generally available satellite services, MoD has been using commercial satellite telecom services. The Sky Perfect JSAT, commercial operator, had satellite with X-band transponder dedicated to the use of SDF, but this satellite will reach the end of life by 2015. Thus, MoD needed to replace this satellite capacity with some other new services. Given the budgetary constraints and the lack of expertise, MoD decided to follow the British Skynet military communication satellite procurement model. The Skynet system was procured by British MoD as Private Finance Initiative (PFI) that means the commercial operator develop, manufacture, launch, and operate the satellite system and the MoD spend money only on services (Suzuki 2006). In this way, the MoD does not have to invest in developing human resources and technical expertise for building and operating satellites. The contract was awarded to the consortium led by Sky Perfect JSAT in 2012 and the satellite was launched in 2017 (now called "Kirameki").

The second priority is the Space Situational Awareness (SSA). It seems strange for MoD to pay attention of SSA because it does not own or operate space assets. Even for civilian programs, most of JAXA's assets are developed as R&D satellites, so that they are not operational satellites (even the satellites such as ALOS series are classified as technology developing satellites). There are commercial and operational satellites for telecommunications, broadcasting, and meteorology, but these are located at Geostationary Orbit and the MoD does not take them as the assets to be protected. However, in the context of US-Japan alliance, SSA became important issue for MoD to deal with. Since the Basic Space Law passed the Diet in 2008 – 1 year after the Chinese ASAT test – the US government welcomed the new approach that Japan would invest in security-related programs in space and asked to participate in the construction of international network of SSA. Japan is strategically located to monitor airspace of West Pacific/East Asia, and the SSA data from Japan would complement the US's own SSA stations (Suzuki 2012b).

In this circumstance, MoD approached the SSA issue reluctantly. On the one hand, it recognizes that the SSA capability is important for strengthening the US-Japan alliance, while on the other hand, it would spend certain portion of its budget for not substantially important to protect Japanese assets. The final decision was made at the top of the political level. The Prime Minister has ordered that SSA should be included in the Basic Space Plan which defines mid-term space policy guidelines. MoD has begun working on developing deep space radar and telescope in Yamaguchi Prefecture for building SSA capabilities in cooperation with JAXA. SSA was also included in the National Defense Program Guidelines, which was adopted in 2018 (Ministry of Defense 2018).

The third priority is the early warning satellite system. Japanese government made a decision to construct missile defense system in cooperation with the United States when the Taepodong flew over Japanese territory in 1998. Furthermore, two Hwasong-12 missiles flew over Hokkaido and splashed into Pacific Ocean in 2017. These incidents made Japanese security policy community as well as the nation as a whole recognized that Japan is under serious threat, and something has to be done within the limit of the Constitution. Thus, together with the development of IGS, it became imperative to build missile defense capabilities. Japanese Theater Missile Defense System has constructed with four Aegis frigates and PAC-3 surface-to-air missiles, and there will be addition of two batteries of Aegis Ashore in early 2020s. However, Japan lacked early warning satellite system to detect missile launch. Currently, Japanese Missile Defense System entirely depends on the United States for early warning signal intelligence, but it has some problems.

In 2009, North Korean regime prepared to test its launcher, and Japanese SDF prepared to intercept if it falls on its territory. The launch itself was not successful and there was no harm in Japan, but the SDF ground-based radar has misinterpreted signals from Korean Peninsula and sent false alarm. Contrary in April 2012, when North Korea failed again to launch, Japanese government did not issue warning for the people living in the flight path because it wanted to double check if the early warning signal was not false alarm. The government was heavily criticized for both incidents and it was considered that the failures occurred due to the lack of early warning intelligence. As a result, there was a strong argument for having early warning satellite of its own would improve the detection capability and verification of the United States' early warning signal. Although MoD understood the importance of having early warning satellites, it is still reluctant to move forward because of the cost of developing these satellites. The MoD decided to develop Infrared sensors which will be mounted on Quasi-Zenith Satellite in Geostationary Orbit, but it is only for R&D purpose. There is no decision to develop operational early warning satellite yet.

JAXA's View on Space Security

For many years, JAXA (formerly NASDA and ISAS) were prohibited to develop satellite or launcher technology explicitly aiming to improve military capability. However, space technology in general is dual-use technology, which does not discriminate military and civilian use. There has been always suspicion that JAXA would have a hidden agenda to develop space technology for military purpose, despite the 1969 Diet Resolution. In order to eradicate the suspicion, JAXA had to emphasize its programs were designed, developed, operated, and used exclusively peaceful purpose. As a result, JAXA has averted anything that relates to security issues.

Thus, even after the Chinese ASAT test became public, JAXA was reluctant to call this issue as "space security" issue; but instead, it preferred to use the term "long-term sustainability of space environment." In other words, JAXA wanted to frame the issue of space security within the issue of space debris regardless of whether the debris was created intentionally or not.

However, given the rapid increase of debris population especially in Low Earth Orbit, JAXA became concerned about the risk of collision with debris. The JAXA's safety analysis procedure now contains the risk assessment of debris collision probability. Nevertheless, JAXA, as an R&D agency, demonstrated its interest not in the development of international regulation or rule of the road but in the development of new technology for debris removal. JAXA is increasing its budget allocation for the study of Active Debris Removal (ADR) technology. In 2020, JAXA concluded a commercial partnership program with Astroscale, commercial ADR venture, for experimental operation for debris removal.

The problem here is that any technology of ADR has possibility to be perceived as space weapon development, and also there are many political, economic, and legal problems on ADR technology even if JAXA would successfully develop one. JAXA's legal department has been working on the possible interpretation of Outer Space Treaty and other space-related international agreements but has not taken any initiative to provide legal framework for debris removal activities.

Japanese Reaction to the Code of Conduct

After the ASAT test by China, the global space community has moved toward establishing a general rule to prevent similar action which created large number of debris. In April 2007, the United Nations Committee on the Peaceful Uses of Outer Space (UNCOPUOS) has adopted Space Debris Mitigation Guidelines, based on the Inter-Agency Space Debris Coordination Committee (IADC) Space Debris Mitigation Guidelines, which was published in 2002. Although these guidelines are not legally binding documents, they set up an ethical and social benchmark on what should be done to secure the safety of space environment. Furthermore, the delegates of European Union (EU) in the Conference on Disarmament (CD) in Geneva submitted Code of Conduct for Outer Space Activities. This document took a step further to set up norms of behavior of spacefaring states to protect space environment and prevent intentional as well as unintentional creation of space debris.

Since most of Japanese spacecrafts were R&D-oriented (not operational) ones, Japanese space community paid little attention to the risks of orbital environment. The risks of colliding with space debris or being targeted from hostile parties were considered minimum because Japanese spacecrafts were not in military or public operations. In other words, the damages of losing these satellites were not important since the objectives of developing these spacecrafts were to test and demonstrate Japanese engineering capability.

Japan did not make an explicit action when the European Union proposed the Code of Conduct. Not only because Japan has not been ready yet to engage in the negotiation with the EU but also because the United States, Japanese major ally, did not express its position against the Code. Japanese government was well aware of the US national space policy which was published in 2006 during the Bush Jr. administration which explicitly rejected any international agreement that binds the freedom of American space activities. Although the Code assumed voluntary participation, the commitment to the Code would make the US space policy more constrained.

For Japan, the Code seems to be very suitable for the world after the 2007 ASAT test. Its voluntary nature would be essential for inviting as many nations as possible. Initially, the EU intended to make the Code a voluntary one in order to include spacefaring countries (the United States under Bush Administration in particular). But since the United States under Obama Administration preferred to establish international platform for negotiation based on the Code of Conduct, the dividing line emerged between Western countries (EU, the United States, Japan, and Australia) and China and Russia. By avoiding the division among spacefaring countries and making the Code as international norm, the voluntary nature of the Code is utterly important. Of course, it would be much effective if the Code is legally binding, but there is a trade-off. In order to include the countries such as China and Russia, it should be based on voluntary participation, at least in the beginning (Suzuki 2012c).

The most important clause in the Code of Conduct from Japanese point of view is the Clause 4.3, which states that "when executing maneuvers of space objects in outer space, for example to supply space stations, repair space objects, mitigate debris, or reposition space objects, the Subscribing States confirm their intention to take all reasonable measures to minimize the risks of collision." This allows Japan to invest in the development of the technologies to remove space debris. JAXA in particular is very interested in developing debris mitigation and removal technology, but it is well-known that the removal technology can be used as space weapons. Thus, it is extremely important that the Code explicitly allows debris removal activities with good intention.

These interpretation underlines that Japanese perspective on the Code is based on its diplomatic and technological concerns, not on its military and security needs. Since MoD is not engaged in the process of decision-making for the Code, it would be difficult to assume that Japan would commit to the Code for security purposes. Although Japan expressed its interest to participate in the process of negotiation for drafting the International Code of Conduct, its objective is not purely driven by the needs for securing healthy environment for space activities. Rather, its objective is driven by the interest of JAXA and MoFA.

Currently, the negotiation on the Code of Conduct has completely stalled and it is impossible to foresee that there will be consensus to adopt the Code as international standard. Nevertheless, it is important for Japan and any spacefaring nations to have international rules to govern increasingly crowded orbital environment. There are new states starting to launch their own satellites, and numerous constellation businesses are deploying thousands of satellites in already crowded orbits. The security and safety of orbital traffic is the primary importance of space security. Japan is willing to participate in any discussion for setting up rules of the road, but the political climate, especially that of US-China relationship, might not provide opportunity to even put the Code or any proposal for international rule-making on the negotiating table.

Conclusion

Japanese perspective on space security has begun with a very unique setting. The 1969 Diet Resolution has put heavy constraints on its space activities, and long and enduring interpretation of "non-military" approach has refrained Japan from anything related to security. However, the 1998 Taepodong launch and subsequent reform of space policy eventually created Basic Space Law in 2008.

Although the new legal framework provides opportunity for Japan to play much bigger role in the field of space security, it seems that Japan is still taking steps cautiously. It is largely because the history and culture of Japanese space activities. The understanding of "exclusively peaceful purpose" as "non-military" nature of space has been long and persistent not only among the people in space community but also the people working in the security field. The Ministry of Defense and Self-Defense Forces do not consider themselves as the actor in space policy making, and they refrained from investing in space because they have constructed Japanese defense system without relying on space assets. Similarly, JAXA is reluctant to take security programs as its central mission. Under severe budgetary constraints, JAXA had to give up some of its pet projects for the sake of security program. Thus, it has resisted changing its status as civilian R&D agency.

Nevertheless, the security environment as well as the role of Japan is changing. On the one hand, the emergence of China as military superpower and the territorial dispute over Senkaku/Diaoyu Islands urge Japan to improve its surveillance capability and space-based infrastructure for SDF operations in the West Pacific and East China Sea. Furthermore, the successful launch of North Korean missiles over Japan increased awareness for building robust Missile Defense System. Japan can no longer have a luxury to stay away from the discussion on the space for security purposes.

On the other hand, increasing number of countries in Asia-Pacific region began using space systems as their national infrastructure, which provides indispensable services to their socioeconomic activities. Japan as one of the leading spacefaring nations in this region is taking a leadership to formulate regional space cooperation framework which includes a forum to discuss space security. The Ministry of Foreign Affairs took initiatives with Australia to hold a workshop within the ASEAN Regional Forum and APRSAF. These initiatives cannot be capitalized if Japan would not play active role for securing space environment.

Japanese space policy towards space security – both national and international security through space and security of space environment – is gradually changing thanks to the Basic Space Law and changing security environment. There has been strong political intervention from Prime Minister to focus on SSA and missile defense which enabled Japan to participate in global SSA networks. The National Defense Program Guideline, for the first time, took up space as important domain to maintain superiority against other countries. These changes are now shifting the mindsets of MoD and JAXA, two most important but reluctant agencies for space security. Since the negotiation on the Code of Conduct is stalled and there will be no international consensus on the governance rules in space for a foreseeable future, Japan needs to develop its own plan and strategy to maintain space capabilities for itself and its allies. It is unthinkable in these days that military operations can be conducted without space infrastructure, and Japanese investment in space security would improve not only the survivability of space system for Japan but also for the allied countries.

References

Aoki S (2008) Japanese perspective on space security. In: Logsdon J, Moltz JC (eds) Collective security in space: Asian perspective. The Elliot School of International Affairs, The George Washington University, Washington, DC, pp 47–66

Aoki S (2009) Current status and recent developments in Japan's National Space Law and its relevance to Pacific Rim space law and activities. J Space Law 36(2):335–364

Consultation Group for National Strategy for Space (2005) Toward a construction of new institutions for space development and utilization. Liberal Democratic Party, Tokyo

Hagt E (2007) China's ASAT test: strategic response. China Secur 3(1):31–51

Kato K (1985) Government official statement. Budget Committee, House of Representatives, Feb 6, 1985

Matsuura S (2004) Kokusan rokketo wa naze ochirunoka (Why do Japanese launchers fail?). Nikkei BP Publishers, Tokyo

Ministry of Defense (2018) Heisei 31 Nendo ikou ni kakaru boueikeikaku no taikou ni Tsuite (National Defense Program Guidelines after 2019). Ministry of Defense. Retrieved from http://www.mod.go.jp/j/approach/agenda/guideline/2019/pdf/20181218.pdf

Sunohara T (2005) Tanjo kokusan supai eisei (The birth of national spy satellites). Nikkei BP Publishers, Tokyo

Suzuki K (2005) Administrative reforms and policy logics of Japanese space policy. Space Policy 22(1):11–19

Suzuki K (2006) Adopting the European model: Japanese experience in implementing a public–private partnership in the space program. Paper presented to the Council for European studies, Fifteenth biennial international conference, Chicago, Mar 31, 2006

Suzuki K (2007a) Transforming Japan's space policy-making. Space Policy 23(2):73–80

Suzuki K (2007b) Space: Japan's new security agenda. RIPS policy perspectives no. 5. Research Institute for Peace and Security, Tokyo

Suzuki K (2012a) The leadership competition between Japan and China in the East Asian context. In: Morris L, Cox KJ (eds) International cooperation for the development of space. Aerospace Technology Working Group, Charleston, pp 243–259

Suzuki K (2012b) Japan, space security and code of conduct. In: Lele A (ed) Decoding the international code of conduct for outer space activities. Institute for Defence Studies and Analyses, New Delhi, pp 94–96

Suzuki K (2012c) Space code of conduct: a Japanese perspective. Analysis, Observer Research Foundation, July 25, 2012

India in Space: A Strategic Overview

31

Ajey Lele

Contents

Introduction	572
India's Space Architecture	572
Space and National Power	577
Space for National Security	579
Military Specific Space Systems	581
India's ASAT Test (Lele 2019)	583
Conclusion	586
References	587

Abstract

India could be said to have begun its space program during early 1960s by undertaking launching of sounding rockets. A structured approach towards evolving the space agenda for the nation could be said to have begun with the establishment of the Indian Space Research Organisation (ISRO) on August 15, 1969. The prime objective of ISRO is to develop space technologies to cater for various societal needs. Subsequently, the Department of Space (DOS) and the Space Commission were set up in 1972 which oversee planning and implementation of India space agenda. India launched its first satellite during 1975 and earned the space-faring nation status during 1980. Over the years ISRO's program has matured significantly and at present Indian space program is regarded as one of the important space programs in the world. From launching small satellites to undertaking a successful mission to Moon and Mars, India has excelled in almost all areas of space experimentations. India is also proposing to undertake its first human space mission by 2022. India is a nuclear weapon state and has made significant investments towards establishing its military

A. Lele (✉)
Institute for Defence Studies and Analyses, New Delhi, India
e-mail: ajey.lele@gmail.com

© Springer Nature Switzerland AG 2020
K.-U. Schrogl (ed.), *Handbook of Space Security*,
https://doi.org/10.1007/978-3-030-23210-8_85

architecture owing to its strategic needs. Space technologies are finding increasing relevance towards strengthening this architecture, essentially as a force multiplier. This chapter analyzes various significant aspects of India's space program.

Introduction

The word "strategic," broadly gets identified as long-term and is mostly associated with the broad purposes and interests of an organization on enduring basis. The progression of strategic planning, generally involves establishing the definite purpose behind planning for a particular mission, the strategy for achieving the desired objective, and various processes involved towards that. The processes involve setting up of the priorities and allocation of resources accordingly. This chapter undertakes a strategic overview of India's space program. For long, the word strategic has established semblance to the military domain and gets often used linking to the realization of overall or long-term military policies and strategies. This chapter also specifically critically focuses on these aspects of India's space program too.

India's Space Architecture

India launched its first sounding rocket on November 21, 1963. This launch took place from a location called Thumba in the southern parts of India. A church building at a place called Thumba village (in the Kerala province which is at the southern tip of Indian peninsula) is the place of birth of India's space program. This particular location was selected since the geomagnetic equator passes through Thumba and the church was selected because it was the only properly constructed building in that village (Das 2007). The sounding rocket launch was assisted by the National Aeronautics and Space Administration (NASA), which provided Nike-Apache rocket along with other required equipment. These were early days and India had limitations in regard to both financial and technological resources. During last five to six decades India has made a significant progress in the space domain and has also successfully conducted missions in deep space, to Moon and Mars.

During 1962, Indian government established Indian National Committee for Space Research (INCOSPAR). Subsequently, during 1969 INCOSPAR was superseded and Indian Space Research Organisation (ISRO) was established. Presently, this organization is the torchbearer of India space program and has brought many laurels to the country by undertaking various successful programs. In the changing world ISRO did realize the need for opening up their expertise for business. This led to the institution of Antrix Corporation Limited (ANTRIX) during 1992 a government of India Company. Both ISRO and ANTRIX are under the administrative control of Department of Space (DOS) which got established during 1972. At the same time a space commission was also setup. The department of space and the space commission directly reports to the Prime Minister of India. Few public sector organizations like the Hindustan Aeronautics Limited (HAL) also play an important role towards assisting ISRO's various programs. The private industry also plays its role towards assisting ISRO in various projects. However, their involvement is more at subsystem level. India is under the process of establishing a full-grown eco system for private space industry. Defence Research and Development Organisation (DRDO) also plays some role essentially assisting India's defense establishment in the area of space.

Two scientists deserve a credit for conceptualizing and executing India's space vision in the early years. They are Dr Vikram Sarabhai (12 August 1919–30 December 1971) and Prof Satish Dhawan (25 September 1920–3 January 2002). It was Dr Sarabhai who gave the initial vision which was implemented and expended upon by Prof Dhawan. In simple terms, the articulated vision then was "to use space for socioeconomic development" which continues to remain relevant in the twenty-first century too.

India launched its first satellite in 1975 with assistance from the erstwhile USSR. Within 5 years after the launch of first satellite with the outside assistance, India became a spacefaring state during the 1980 by launching made-in-India satellite with the Indian rocket. Since then, India has made significant progress in the space domain and today India has earned a reputation of a committed and serious space player. Till date India has sent a range of satellites to different orbits. However, particularly to launch satellites into the geostationary Earth orbit (GEO) India is required to take the assistance from other agencies. This dependence is seen reducing by 2018 with India successfully developing a launch vehicle to send heavy (say 4–6-t category) satellites to GEO.

ISRO has come a long way from the Sounding Rockets (1963) to having capability launch heavy satellites with geostationary satellite launch vehicle (GSLV) Mark III (2018). Initially, ISRO began with Satellite Launch Vehicle-3 (SLV-3) program as India's first experimental satellite launch vehicle. India became the sixth spacefaring state in the world on July 18, 1980 when Rohini, a 40-kg satellite was placed in the Low Earth Orbit (LEO) by SLV-3. The successful culmination of the SLV-3 project showed the way for various advanced future launch vehicle projects. The next vehicle for ISRO was the Augmented Satellite Launch Vehicle (ASLV), which was followed by the Polar Satellite Launch Vehicle (PSLV) and the Geosynchronous Satellite Launch Vehicle (GSLV).

The Augmented Satellite Launch Vehicle (ASLV) Program was designed to augment the payload capacity to 150 kg, thrice that of SLV-3, for Low Earth Orbits (LEO). While building upon the experience gained from the SLV-3 missions, ASLV proved to be a low-cost intermediate vehicle to demonstrate and validate critical technologies for the future launch vehicle development projects like strap-on technology, inertial navigation, heat shield, vertical integration, and closed-loop guidance.

Polar Satellite Launch Vehicle (PSLV) is the third generation launch vehicle of India. It is the first Indian launch vehicle to be equipped with liquid stages. The PSLV-C45 mission on April 01, 2019 was its 47th mission and the PSLV rocket has witnessed only two failures till date since its first successful launch in October 1994. The rocket is fondly known as the most reliable workhorse of ISRO. Till April 01, 2019, PSLV has launched 46 Indian satellites, 10 satellites built by students from Indian Universities and 297 international customer satellites. Even India has used this rocket for its first mission to moon, the Chandrayaan-1 in 2008 and for the Mars Orbiter Mission (MOM) in 2013.

The PSLV mission configuration constitutes four stages using solid and liquid propulsion systems alternately. The first stage carries solid propellant, and the second stage uses the indigenously developed Vikas Engine which carries liquid propellant. The third stage is a solid stage and the fourth stage is a liquid stage with a twin-engine configuration. PSLV can take up to 1,850 kg of payload to Sun-Synchronous Polar Orbits of 600 km altitude. It has also carried payload of 1,425 kg to Sub GTO. Due to its unmatched reliability, PSLV has also been used to launch various satellites into Geosynchronous and Geostationary orbits, like satellites from the IRNSS (NAVIC) constellation. There are different variants of PSLV depending on the nature of the mission and the weight of the payload to be carried. Four to six ground-lit strap-on boosters are added (attached) to the rocket based on the requirement.

The flexibility of the PSLV system is amazing. This one rocket which ISRO is using for multiple orbit launches in a single mission. Also, ISRO has successfully started using the fourth stage of the rocket for carrying out scientific experiments. It is important to take note of two new features of the PSLV system which ISRO has started experimenting with since 2015. One, single mission undertaking two/three orbit launches and two, using the fourth stage of the rocket as a platform for experimentation. Developing capabilities to launch satellites into different orbits in a single mission gives ISRO more flexibility to manage their commercial interests. ISRO has developed and aptly demonstrated its capability to launch multiple satellites in a single mission. The PSLV-C37 mission on February 15, 2017 created a world record by successfully launching 104 satellites in single mission (during May 1999, with PSLV-C2 mission ISRO had launched more than one satellite in a single mission). Now, this added capability of launching satellites in differ orbits in one mission could attract more international customers.

Another less debated but a significant feature is about ISRO demonstrating its capability to convert the fourth stage of the rocket into an experimentation platform. Space is a medium where scientific and strategic communities are keen to undertake various experimentations for many decades for various purposes. Mainly such experiments are important since they are carried out in (almost) zero-gravity atmosphere. Such experiments are useful also from the point of view of learning for the futuristic space (both manned and unmanned) missions.

The International Space Station (ISS), a station with 16 member states, is known to serve as a microgravity and space environment research laboratory. This station has been inhabited continuously since the year 2000. Here the astronauts from the member states are conducting experiments in biology, physics, astronomy, meteorology, and few other fields. Now, China is also going to establish its own space station in near future. Against the backdrop of this, India's efforts to conduct experiments in space could look infinitesimal. ISRO is expending the fourth state of the PSLV, as a laboratory for conduct of experiments. Here there is no human element involved towards the conduct of such experiments, but still that in no respect reduces the relevance of such unique experimentation. The maturing of the UAV technology has demonstrated that to a certain extent and for specific purposes an alternative on a manned flying platform is feasible and available. Today, India may not have its own space station, but the innovation of ISRO indicates that developing such techniques (extremely cost effective) of converting the fourth stage of the rocket into a laboratory could definitely offer India a moderate alternative to the space station.

For many years, the major success achieved by ISRO towards developing a reliable launch vehicle for LEO to carry around 2-t verity satellites could not be repeated for geostationary orbit. ISRO struggled for many years to fully establish its Geosynchronous Satellite Launch Vehicle (GSLV) program. There has been significant amount of dependence of ISRO on outside agencies for launching 4–6-t verity of satellites in the geostationary orbit. The first launch of GSLV mission was launched during April 2001 with a payload mass of 1,540 kg. Till date, the maximum lift-off mass of 3,423 kg has been put in Geosynchronous Transfer Orbit (GTO, GSLV Mk III-D2/GSAT-29 Mission, November 14, 2018) by ISRO. GSLV is a three-stage system with solid, liquid, and cryogenic stage.

The major challenge which India had faced was about mastering the art of cryogenic technology. In fact, a Cryogenic Study Team was set up at ISRO as early as 1982; however, this idea got neglected. Finally, around 1990s when the need arose, India got in the technology transfer agreement with Soviet Union/Russia. However, this arrangement could not work owing to the international (read the US) pressure. The USA and its partners in the then 18-member Missile Technology Control Regime (MTCR) contend that the Indo-Russian deal is inconsistent with the MTCR, which was set up in 1987 to prevent the transfer of missile technology to non-member countries. The Russian republic was not a signatory to the accord but has agreed in principle to abide by its guidelines since it had applied for what is called "adherent" status (Russia joined MTCR during 1995). Also, the National Defense Authorization Act of the USA (1991) had provisions for penalties for such transfers. Thus, both Glavkosmos of Russia and the ISRO would have got blacklisted if the deal had gone through. At the hindsight it could be said that the Russian leadership then (Boris Yeltsin) capitulated to the US pressure and the transfer of cryogenic engine technology deal was cancelled. Finally, Russia was allowed to supply only seven cryogenic engines to India.

Broadly, it could be argued that Cryogenic engines and associated technological knowledge has almost no relevance for missile launches in war. This is because the procedure of fueling this system requires few months' time. Even missile system with liquid fuel stages are not preferable since the crucial time gets wasted and particularly the process of fueling requires to be undertaken at the launch site (almost close to the battlefield). However, the major powers appear to have taken the advantage of MTRC provisions by undertaking selective (mis)interpretation. Finally, India decided to develop these engines on its own, but it took considerable amount of time for the development. However, the development of the indigenous cryogenic engine took much time (almost three decades) and only on June 05, 2017 GSLV Mk III-D1 launched GSAT-19, a 3,136 kg bird (ISRO 2017a; Chengappa 2013; Krishnan Simha 2013). For all these years the basic limitation of India's space program has been its inability to develop a cryogenic engine, but now that challenge has been overcome.

Keeping an eye on the future commercial market for launch of small satellites, ISRO is also developing a Small Satellite Launch Vehicle (or SSLV) for an approximate payload capacity of 500 kg and expected to undertake the first flight by mid of 2019.

Satellite technology development is a complicated task. Also, satellites once launched are required to survive in space for number of years. Hence, their manufacturing requires components and systems with very high reliability. India understood the need of outside assistance in this field, if they have to make progress. During its formative years India took significant foreign inputs to build threshold capabilities in complex systems. Initially, ISRO took assistance from developed space powers to build different kinds of payloads for sounding rocket experiments (Baskaran 2001). Over the years India has acquired good capability to build very complex and world-class satellites for remote sensing, weather, and communications. India has also developed good capability in other sectors like navigation and various categories of small satellites.

The first important structure established by India towards working on satellite technology was the Experimental Satellite Communication Earth Station (ESCES) at the city of Ahmedabad in 1967. During 1970s, Indian engineers got trained in France. ISRO started developing sensors for airborne remote-sensing surveys and processing of imageries provided by NASA. Subsequently, few satellites were built and launched. This was basically an experimental phase for the scientists. This was followed by the conceptualization of two major projects which actually would be viewed as the beginning of a very systematic and organized space agenda. The socioeconomic focus articulated earlier was evident. These two satellite programs included: the Indian Remote Sensing Satellite system (IRS) and the Indian National Satellite system (INSAT), for commercial operations. Probably, ISRO was confident about their capabilities in the remote sensing arena but not that much about the telecommunication field which incidentally was also a high priority area. Hence, after a realistic appraisal of its capabilities, it was decided that the IRS–1 series would be indigenously built and the INSAT–1 series would be brought from abroad. It was also decided that the INSAT–2 series would be built indigenously. There were reasons behind these decisions. As INSATs were communications satellites, they were more complex and ISRO needed longer development time. INSAT is a unique experiment which has not been replicated elsewhere. During 1977, ISRO defined INSAT as a multipurpose system consisting of telecommunication, meteorological, and TV broadcasting elements. It was the world's first geo-stationary satellite system to combine these three elements (Baskaran 2001). That was the period when India was not in a position to afford specific satellites for specific purposes, hence decided to have one satellite which could be multipurpose.

Today, India has a technologically mature remote sensing satellite program. There is an array of Indian Earth observation (EO) satellites with imaging capabilities in visible, infrared, thermal, and microwave regions of the electromagnetic spectrum, including hyper-spectral sensors. The imaging sensors have been providing spatial resolution ranging from 1 km to less than 1 m. India has launched various other satellites for communication, education, meteorology, astronomy, and navigation purposes. India's own navigational system is a regional system with seven satellites. All satellites have already been placed in orbit and the system is expected to become fully operational in the near future. A ground network of several nodes for data gathering has been set up under various collaborative mechanisms to establish a global chain of command and control for its space assets.

India has also put in place a well-articulated Deep Space agenda. India's first Moon mission, Chandrayaan-1 (2008–2009), was successful. This mission was instrumental towards the discovery of water on the Moon. The second lunar mission encountered delays, given uncertainties in Russia, which was assisting India with a rover and lander system for this mission. Presently, India is undertaking this mission with an indigenously designed and developed rover and lander system and is expected to get launched during the second half of 2019. India's first mission to Mars, Mars Orbiter Mission (MOM), was also successful. MOM which was designed for 6 months of stay, has been orbiting Mars since September 2014. India is the only country in the world till date which had successfully entered into the Martian orbit in the first attempt. India proposes to undertake its first human space mission by 2022.

There has been some amount of criticism with respect to India, as a developing state for having made investments in the programs like Moon and Mars. However, one of the chief architects of India's space program Prof S Dhawan (1996) had argued to the effect that, "it is moral to planets and stars in spite of having hunger, poverty and misery on earth because various programs which explore the planets enhance human's capacity to face the unknown and severe to survive in any environment. Humans reach to space for solving problems on earth" (In a lecture delivered at Astronomical society of India at the city of Bangalore on September 6, 1996).

Space and National Power

Normally, it has been observed that space programs of various states are born out of their ballistic missile programs. However, that is not the case with India. Since initiation of its space program it has been witnessed that India is having a focused attention on developing and using space for societal purposes. Satellite systems used for the purposes like meteorology, remote sensing, and communications were seen exclusively used for the purposes like weather forecasting, TV, education, resources mapping, and medicine.

It is also important to factor in the strategic milieu at global level during the formative years of India's space program. It was a period of Cold War and mostly the world was divided into two power blocks. India was one of the few countries which were not under the influence of any superpower. Since independence (1947), the then political leadership had taken significant amount of interest towards investing in science and technology. Political leadership, policy makers, and scientific community were found working in unison. It was a considered view that if India has to progress, then investments in science and technology are a must. There is no evidence to show that Indian state had any interest to use their investments in technology as a tool to display power. At the same time, it needs to be acknowledged that immediately after independence both from economic and strategic perspective India was not in a position to project power.

Satellite technology is inherently dual-use in nature and in a limited way could even be used for military requirement by any power which has control over it. During Cold War period, space technology could be seen as a currency of that of force (technology) projection. Subsequently, the 1991 Gulf War ended up showcasing the importance of space technologies in the warfare. Actually, this war was an eye-catcher for many states in respect of utility of space systems in the warfare and India was no exception. At the same time, it is important to put in context the level of proficiency achieved by India in the space domain then.

ISRO was established on August 15, 1969, while a month before that on July 20, 1969, Apollo 11 had already landed on the moon. On August 29, 1991, India has launched IRS 1-B, a 975 kg satellite from the Baikanur Cosmodrome Kazakhstan using Launch vehicle Vostok. This remote sensing satellite had three solid state Push Broom Cameras with resolutions ranging from 72.5 to 36.25 m. During the 1991 Gulf War, the USA and allied forces had range of remote sensing satellites like SPOT (France) with 10 m resolution and other satellites KH 11 (Block I and Block II) and few Landsat series satellites (Marcia and Smith 1991). During the year 1991, NASA had undertaken six human missions to space by using the Space Shuttles (models used Atlantis, Discovery, and Columbia) while India yet to undertake its first human mission to space (could happen by 2022). The USA had fully operational GPS satellite navigational system during the 1991 Gulf War and India is yet to (by April 2019) operationalize its regional satellite navigational system. All this indicated that the use of space technologies during 1991 Gulf War could have been much learn for countries like India; however, technologically India was nowhere close to the capabilities of the allied forces in the space domain. More importantly Indian policy makers were fully aware that any military investments in space from their side should be threat specific and there is no requirement to raise the ante just because the demonstration of technologies have happened some other theatre of war.

It is well understood that national power and aspects of national security is not only about military influence but is also about political and economic influence. The core of national power is also about national performance and management of natural resources. The realization of national performance involves ensuring socioeconomic development. Under such framework, there has been major and continuing focus by India towards using space for communication, meteorology, education, tele-medicine, disaster management, linking of cities and villages, scientific research, and navigation. One of the major utilities of the satellites for India has been in the field of meteorology essentially because India is an agricultural economy and timely weather information is critical. Also, remote sensing satellites have much of utility for resources planning and management. Such satellites also provide useful information on range of issues from forest cover to inputs for river and ground water management. Education and telemedicine are two important areas where India is known to be using space technologies very effectively. In addition, various communication satellites are catering for the needs of television broadcasting and other sectors. India is also looking at space as a tool for foreign policy. On May 05, 2017, India has launched a communication satellite (GSAT-9) which is providing assistance to the South Asian states, namely, Bangladesh, Afghanistan, Nepal, Bhutan, Sri Lanka, and Maldives. Space is an important part of India's various multilateral and bilateral arrangements for many years.

Realizing the importance of commercial facets of space sector, way bank during 1982 India had established the commercial arm of ISRO called ANTRIX for promoting products, services, and technologies developed by ISRO. However, India has not been able to make a significant dent to global commercial space sector, yet. One of the key commercial focuses for ISRO has been to provide satellite launching services to the foreign customers. ISRO could be said to have a specialization in the field of small satellites (mini, micro, and nano) launching. However, for all these years ISRO has launched just about 300 satellites for the international customer (till Apr 2019, it had launched 297 satellites). The market size of launch services currently is about $5.5–6 billion globally and India has some 7% of this market share. However, ISRO is keen to increase its overall market share significantly. Currently, the revenue of ANTRIX is Rs 20 billion (20 billion INR is equal to 288,400,000 USD) and the company expects to double it within next 5 years (*Sun, September 092018*). ISRO is also working towards privatizing its launch services. As the name suggests, it is but obvious that the main business of ISRO is not business but to undertake research and development. Hence, presently ISRO is working towards commercializing the PSLV launching services. It is expected that as and when the small satellite launch vehicle becomes operational, it would also be fully commercialized. Apart from launch services, ANTRIX also provides other services like data, imagery, and ground infrastructure construction.

Space for National Security

It is important to highlight broad security challenges which India is facing before identifying the military-related investments made by India in the military domain. India has two advisories in the region, namely, China and Pakistan with whom conventional wars have been fought in past. Basically, the differences are owing to the unresolved boundary disputes with these states. They exist since India got independence from the British power during 1947. China and India share a border of over 4,000 km, with nearly all of it founded on colonial-era settlements and surveys and much of it still under dispute. China has certain claims over Indian Territory and India has some on the territory under the Chinese control. India shares more than 3,300 km land border with Pakistan and have some unresolved border issues. Kashmir dispute is the key security challenge which India is facing presently. Owing to this problem, continuous bloodshed is happening in the region for last few decades. One of the recent confrontations witnessed between India and China was the crisis of Doklam from June to August 2017. However, luckily in the last 40 years, not a single bullet has been fired because of the border issue. There are few other differences amongst these three states, like water issues and China's new economic/security project called Belt and Road (BRI) initiative. India has a major objection to the part of the BRI project called the CPEC, the China–Pakistan Economic Corridor. Here, China is trying to advance this project without concerns for India's unresolved border problems with Pakistan.

Essentially, India could be said to have saddled in a typical security scenario with unresolved border issues for more than seven decades. At the same time, India also faces few internal security challenges owing to both inter and intra state security problems. India's paramilitary forces play an important role to address such challenges and it is important to note that they also require (and want) assistance from the satellite systems.

India has fought four wars with neighbors since independence. British government during 1947 divided the then united India which they were ruling into two differ states: one remained as India and a portion of India was declared as Pakistan. Immediately, after that the first war was fought between India and Pakistan over the Kashmir issue. Owing to the differences with China on boundary issue, a war was fought during 1962. Subsequently, India and Pakistan have fought two more wars during 1965 and 1971. The last conflict fought between these two powers in the Kargil conflict (May and July 1999), few consider this conflict as a half-war. Since Pakistan was not able to win any of the wars/conflicts against India for last few decades, they are found using Terrorism as a tool for possible conflict resolution (!).

China conducted its first nuclear test during 1964, and by 1967 they have also conducted their hydrogen bomb test. India and Pakistan became nuclear weapon states by 1998. All these three states have advanced missile programs and various types of missiles in their inventory. These states also have made investments into different category of missiles which could be launched from land, air, and submarine-based platforms. China has Nuclear Triad in place while India is almost there. Pakistan could take some more time to establish a Nuclear Triad. China and India have Intercontinental Ballistic Missiles (ICBMs) in their inventory. China has tested multiple independently targetable re-entry vehicle (MIRV) missiles while Pakistan is known to be developing this technology. China has developed hypersonic weapons and India is also known to have interest in MIRV and hypersonic technologies. Both China and India have Russian made S-400 Ballistic Missile Defence (BMD) systems. India is developing its own missile defense architecture.

India's strategic area of interest is not restricted to Pakistan-China. Geographically, India's location at the base of continental Asia along the Indian Ocean places the state at a vintage point in relation to maritime trade. India has a strong stake in the security and stability of these waters since a large percentage of Asian oil and gas supplies is shipped through the Indian Ocean. (From the speech delivered by Mr. Pallaim Raju in the PC Lal Memorial Lecture on March 19, 2007, at New Delhi. Mr. Raju was then the minister of state for Defence in the government of India) As per the United Nations Department of Economic and Social Affairs (UN DESA 2017) report (UN DESA 2017), India has one of the largest diaspora populations in the world with over 15.6 million. A major part of Indian diaspora mainly constituting of unskilled or semiskilled workers is largely found employed in Middle East and African region. India feels responsible for security and safety of this population. India has conducted more than 30 evacuation operations across Africa, Asia, and Europe, including its largest-ever civilian airlift of 110,000 people from the Persian Gulf in 1990 (Constantino 2017). During 2015 in one of the major evacuation operations (Op Raahat) by sea, India evacuated 5,600 people, including 4,640 Indian nationals and 960 nationals from 41 countries, from Yemen (Ians 2015). Indian Air Force and Indian Navy have played a crucial role in these operations. For major natural disasters help of Indian armed forces is always sought. During 2004 Indian Ocean tsunami, particularly Indian Navy have played a major role towards search and rescue operations and had helped many affected countries. Even naval amphibious warfare vessels and landing craft were put in use. All these indicate that the role of the Indian armed forces is not restricted only towards guarding the state. Also, on various occasions Indian troops are deployed for contributing towards the United Nations peacekeeping missions. In conduct of all such operations space technologies have an important role.

Space technologies have an inherent duel use character. Obviously, various Indian satellite systems could be viewed to have some utility for the Indian armed forces. It is also important to note that modern-day warfare is about remaining ready to address various types of contests. Such tasks could involve remaining ready for conventional warfare, nuclear warfare, asymmetric warfare, and hybrid warfare. Modernizing the armed forces in terms of technology, weapon platforms, and weapon system is a dynamic process. Present day conceptualization of Revolution in Military Affairs (RMA) demands changes in doctrines to suit new technologies, equipment, and tactics. For a state like India, to bring radical changes in equipment and arms inventory is not possible essentially owing to financial issues. India is expected to follow a hybrid RMA approach were at any given point in time both old and new weapon systems and fighting platforms would be available. Investments in space would be from a point of view of a "force multiplier" to the existing military architecture.

Military Specific Space Systems

So far India has launched few military specific satellites. Essentially, it is found that officially such satellite systems are identified more as systems with strategic utility than actually calling them as military satellites. India has been launching remote sensing satellites for many years. On October 22, 2001 ISRO had launched a Technology Experiment Satellite (TES), weighing 1,108 kg in 572 km Sun Synchronous orbit. This satellite gets recognized as one of the first satellites launched for experimental as well as strategic purposes.

The first satellite exclusively claimed by the Indian establishment as a satellite for military purposes is the communications satellite (GSAT 7) launched for Indian Navy during 2013. This satellite is also known as Rukmini and has nearly 2,000 nm "footprint" over the Indian Ocean region and provides significant assistance to Indian Navy. Another communications satellite was launched during 2018 called GSAT-7A for Indian Air Force. This satellite is used to interlink ground radars, unmanned aerial vehicles (UAV) airbases, and Airborne Warning and Control System (AWACS).

During 2015, a communication satellite GSAT-6 got launched which has been also described as a system for strategic use. Owing to topographical challenges (India has various features from oceans, deserts, snow to thickly vegetated jungles), soldiers on many occasions encounter breaks in commutations. This geostationary satellite with S-Band antenna is used for gathering information over Indian mainland and very small handheld devices are used for data, video, or voice transfer.

It is expected that few more satellites would be launched to cater for the requirements of India army and other agencies. GSAT 7B is expected to be launched for Indian Army in near future. There are some proposals like development of GSAT 7D and 7E. But this is no official confirmation in this regard. Also, it is expected that some E/O satellites would be launched in the future.

India's expertise in the remote-sensing arena is coming handy to establish a network of reconnaissance satellites. This activity could be said to have started with the TES launch during 2001. Now India has satellites with sub-meter resolution which essentially are dual-purpose satellites. India has also launched (with Israeli assistance) two Synthetic Aperture Radar (SAR) satellites called RISAT II (2009) and RISAT I (2011) essentially to address terrorism-related threats. Also, a Hyper-Spectral Imaging Satellite (HysIS) was launched during 2018. Following table provides details about Cartographic satellites:

Name of satellite	Launch date	Resolution	Remarks
Cartosat-1	5 May 2005	2.5 m	
Cartosat-2	10 Jan 2007	Less than 1 m	
Cartosat-2A	28 Apr 2008	80 cm	Perceived to be for the Indian armed forces
Cartosat-2B	12 Jul 2010	Less than 80 cm	
Cartosat-2C	22 Jun 2016	"	Used for weather mapping too
Cartosat-2D	15 Feb 2017	"	

(continued)

Name of satellite	Launch date	Resolution	Remarks
Cartosat-2E	23 Jun 2017	"	
Cartosat-2F	12 Jan 2018	Could be less than 50 cm	For mapping, enhance disaster monitoring & damage assessment

These satellites are proving very useful for the Indian Armed Forces. However, there is very less real-time data availability. Particularly, owing to cross-border terrorism, India is forced to have a continuous vigilance of its border. Existing space-based resources are not adequate to cater for various security challenges. India requires a major space-based surveillance network. As per the assessment carried out by experts, India would require a constellation of 24 small satellites in LEO for meeting ISR needs during times of crises. Also, this report identifies the need for a constellation of 40 satellites in LEO that provide Internet services for the military (Chandrashekar 2015).

In the field of navigation, India has developed an Indian Regional Navigation System (IRNSS/NavIC) (ISRO 2017a). The space segment of this system consists of the constellation of eight satellites: three satellites in suitable orbital slots in the geostationary orbit and the remaining four in geosynchronous orbits with the required inclination and equatorial crossings in two different planes. Currently, all these satellites have been positioned in their respective locations and the system is expected to become operational shortly.

NavIC is designed to provide accurate position information service (primary service area) to users in India and the region extending up to 1,500 km from its boundary. There is also an Extended Service Area covering more area. NavIC would provide two types of services: Standard Positioning Service (SPS) and Restricted Service (RS), which is an encrypted service provided only to the authorized users. This IRNSS System is expected to provide a position accuracy of better than 20 m in the primary service area. Possibly for military users the accuracy could be 10 m or less.

India's ASAT Test (Lele 2019)

On March 27, India conducted Mission Shakti, an anti-satellite missile test. This was a technological mission carried out by the Defence Research and Development (DRDO). During this test, India targeted one of its own satellites with a ground-based missile. With this successful demonstration, India becomes the fourth country to test an ASAT after China, Russia, and the United States.

The satellite used in the mission was one of India's existing satellites, Microsat-R, operating in a low orbit about 300 km high. Such tests require an extremely high degree of precision and technical capability and, with the success of the test, India has demonstrated such capabilities. This test also demonstrates the maturation of India's missile defense program.

Looking at the types of missile interceptors being used for these tests broadly, it could be argued that such tests are the offshoot of ballistic missile defense programs of the respective nations.

ASAT testing is not a new phenomenon. During the Cold War period, the United States and former Soviet Union conducted a number of such tests. More recently there have been two ASAT tests. In January 2007 China conducted an ASAT test, the first such test conducted in the post-Cold War era. This test was followed by the US test in February 2008 when they destroyed an out-of-control intelligence satellite at an approximate altitude of 250 km.

These two tests and the test conducted by India were essentially hit-to-kill or direct ascent systems or a KKV (Kinetic Kill Vehicle) missions. Here the warhead of a missile is not an explosive but rather a piece of metal. This metal warhead hits the satellite and, owing to the impact velocity and the kinetic energy thus generated, the satellite is broken up. The Indian test used DRDO's Ballistic Missile Defence interceptor, which is a part of India's ongoing ballistic missile defense program. Reports indicate the test has generated at least 250–300 pieces of trackable debris. Such debris is expected to re-enter the atmosphere within next 1–2 months because of the low altitude of the satellite struck by the ASAT.

The 2007 Chinese test involved the destruction of an old weather satellite. This 750-kg satellite was orbiting at altitude of about 850 km. China used ground-based midcourse missile interception technology in that test. The problem with the Chinese test was that since it was conducted at higher altitudes, much of the debris created remains in orbit today. Moreover, it is even increasing in numbers as debris strikes each other or other objects in orbit. The US test a year later was conducted by using a modified Standard Missile-3 interceptor, essentially designed to counter short to intermediate-range ballistic missiles. This missile was launched from a ship-based platform. The debris created by this test re-entered, mostly within weeks of the test.

Looking at the types of missile interceptors being used for these tests broadly, it could be argued that such tests are the offshoot of ballistic missile defense programs of the respective nations. ASAT weapons of the KKV variety are useful only for hitting targets in low Earth orbit, up to about 2,000 km. DRDO is confident that they can hit a satellite by a ground-based interceptor up to a distance of 1,000 km. There is no authentic information available with regard to capabilities of countries like Russia, the USA, and China about the orbits they could reach with their missile interceptors. Some reports indicate that China is testing kinetic interceptors that can reach satellites in the geostationary orbit, 36,000 km high.

Over last two decades India has steadily and thoughtfully increased its investments in the space domain. At present, India has about 50 operational satellites in different orbits. Most of satellites are communications (19) and Earth observation (17). Obviously, India needs to ensure that their satellites are safe. India's space research organization has been working on satellite hardening technologies, while scientists and policymakers are trying to ensure that redundancy would be built-in in various systems as such. Possibly, owing to geostrategic compulsions, India's government felt the need to display the technological capabilities related to anti-satellite weapons.

One important geopolitical aspect of this test was that there was an official announcement of the test by Indian prime minister. This implies that India wants to be transparent in all activities it wants to undertake in space. Space security is an important issue for India and the highest level of decision-making structure is handling the issues concerning space.

There are differing opinions globally regarding the rationale behind this test. There is a need to situate Indian test in the overall security matrix of the region. India shares borders with China and Pakistan and there are some unresolved import border issues. Terrorism is a major challenge for the region. Unfortunately, no immediate solution to this problem appears to be in sight. All three nations are nuclear power states and have various missile systems in their inventory. The last classical war that this region witnessed was the 1971 war between India and Pakistan. At that time both the states were non-nuclear weapon states. India and Pakistan became nuclear weapon states by 1998. There have been various major security-related disagreements in the region over the years, but luckily no major war has broken out. Hence, there is a case to argue that nuclear deterrence has delivered.

In the missile domain, in spite of all these nations conducting various tests, no untoward incident has happened so far. India and Pakistan have a treaty that requires both nations to give advance warning to each other in respect of their proposed ballistic missile tests. This arrangement is working well.

In the space arena, China has put in place a major space program. India also has reasonable capabilities in the space arena. In a relative sense, Pakistan's investments are limited, but they have some sort of "space umbrella" from China. China demonstrated its ASAT capabilities more than a decade ago. The general notion that testing of military systems is destabilizing is found bit misplaced in this region. In fact, the acts of terrorism in the region have been more destabilizing that testing of any military systems. It is obvious that India's ASAT would be criticized by both of its adversaries, but there is a space for such noise in international politics. However, in the longer run India's ASAT testing is expected to emerge more as a stabilizing action for the region. Such a demonstration of technological capabilities is expected to deter potential adversaries.

More importantly, India's test is unlikely to increase the space debris problem. Experts had mentioned that there is no threat to the International Space Station since the Indian test took place well below the station's 400-km altitude. However, as per NASA's assessment some part of the debris had reached to the higher orbit, but many calculations show that there is very less probability of ISS coming in the impact zone. Also, since the debris created by the 2008 US test disappeared within days after the test, the same is expected to happen in this case.

Interestingly, the US Strategic Command chief General John E. Hyten has defended India before members of the Senate Armed Services Committee, saying that the country had tested the anti-satellite missile because it needed the capability to defend itself in space. The general called for international norms of behavior in space to curtail the dangerous debris issue (Sputnik News 2019).

Also, there is a need to take a note that India is the only state that has officially announced its ASAT testing. This announcement was made by none other than the

Indian Prime Minister Narendra Modi himself. Presently, India is in the grip of election fever. However, there is a need to look beyond domestic politics and assess the importance of the prime minister owing the test. With his announcement of this mission, it becomes clear that India wants to be transparent in all activities it wants to undertake in space. Space security is an important issue for India.

The possible weaponization of space is an issue of major concern for many nations, including India. Unfortunately, the space arena has very limited globally accepted multilateral treaty mechanisms, and such available mechanisms are mostly issue-centric and could not be viewed as all pervasive. For example, the 1967 Outer Space Treaty (OST) is basically about the banning of testing of weapons of mass destructions in outer space. For last decade or so, some efforts have been made to address this issue, such as the European Union and its International Code of Conduct.

India fully supports the formulation of universal and nondiscriminatory transparency and confidence-building measure, although such measures have limited relevance since they typically are not legally binding. Nonetheless, India believes such mechanisms have a useful complementary role and could become an "appetizer" for formulation of any future treaty. India has participated actively in the consultations called by the EU since 2012 to discuss a draft Code of Conduct for Outer Space Activities.

The possible weaponization of space is an issue of major concern for many nations, including India. Unfortunately, the space arena has very limited globally accepted multilateral treaty mechanisms.

Resolution 69/32, titled "No First Placement of Weapons on Outer Space" and adopted in the United Nations General Assembly on December 2, 2014, has the full support by India. However, India feels that there is a need to grow beyond such ideas and decide on a released and legally binding treaty. In this context, India is ready to give consideration to the revised PPWT (Treaty on the Prevention of the Placement of Weapons in Outer Space, the Threat or Use of Force against Outer Space Objects) presented by Russia and China in the Conference on Disarmament. There has been a total rejection of this proposal by some major powers. However, India is of the opinion that such ideas need to be discussed under the UN umbrella.

Conclusion

India has made significant progress in the outer space area and has earned a global reputation for its professionalism. Since the inception of its space program, India has followed the policy of the use of space for socioeconomic development, and this agenda remains valid today. India is also keen to develop its space industry. ISRO has earned good reputation in the area of satellite launch market particularly in the small satellite sector and is keen to expand further. India is also effectively using space as tool for diplomacy.

India does not have a well-articulated Military Space Programbut has launched some space assets for the Military. It is known that satellites provide various benefits, but substantial vulnerabilities too, owing to both natural and man-made threats. Owing geopolitical realities, India needs to ensure that its assets in space are secure. Hence, to demonstrate that its defenses are ready, India has undertaken an ASAT test. It is expected that the way nuclear and missile deterrence has worked in the region, the ASAT deterrence would also deliver and weaponization of space would not happen.

India fully understands that space is an extremely important area for human survival and should not be tinkered with unnecessarily. Modern-day life is totally dependent on assets in space. India's growth story, scientific and economic, also involves the contributions made by its space agency and space industry. Today, space offers a major soft-power potential for India, and India believes that it is in nobody's interest to weaponize space. The need of the hour is to evolve a rule-based and transparent mechanism for protecting space.

References

Baskaran A (2001) Competence building in complex systems in the developing countries: the case of satellite building in India. Technovation 21:109–121

Chandrashekar S (2015) Space, war & security– a strategy for India. National Institute of Advanced Studies, Indian Institute of Science Campus, Bengaluru, pp 6–8

Chengappa R (2013) US blocks critical cryogenic deal, forces India to indigenise. India Today Press. https://www.indiatoday.in/magazine/science-and-technology/story/19930815-us-blocks-critical-cryogenic-deal-forces-india-to-indigenise-811389-1993-08-15. Accessed 10 Apr 2019

Constantino X (2017) India's expatriate evacuation operations: bringing the diaspora home. Carnegie India Press. https://carnegieindia.org/2017/01/04/india-s-expatriate-evacuation-operations-bringing-diaspora-home-pub-66573. Accessed 03 June 2019

Das SK (2007) Touching lives: the little known triumph of the Indian space programme. Penguin Books, New Delhi, p 1

Ians (2015) India evacuates 4,640 nationals, 960 others from Yemen. One India Press. https://www.oneindia.com/india/india-evacuates-4640-nationals-960-others-from-yemen-1711703.html. Accessed 03 June 2019

Indian Space Research Organisation (2017a) IRNSS programme. https://www.isro.gov.in/irnss-programme. Accessed 03 June 2019

Indian Space Research Organisation (2017b) Launchers. https://www.isro.gov.in/launchers/. Accessed 10 Apr 2019

Krishnan Simha R (2013) How India's cryogenic programme was wrecked. Russia Beyond Press. https://www.rbth.com/blogs/2013/12/04/how_indias_cryogenic_programme_was_wrecked_31365. Accessed 10 Apr 2019

Lele A (2019) The implications of India's ASAT test. The space review. http://www.thespacereview.com/article/3686/1. Accessed 03 June 2019

Marcia S, Simth (1991) Congressional Rsearch Service, The Library of Congress

Smith SM (1991) Military and civilian satellites in support of allied forces in the Persian Gulf War. Congressional Research Service (CRS) report for the US Congress

Sputnik News (2019) India's ASAT justified, but space debris still a concern says Pentagon – report. Press. https://sputniknews.com/asia/201904121074075266-pentagon-justifies-indian-asat/. Accessed 03 June 2019

UN DESA (2017) International migrants stock. https://www.un.org/en/development/desa/population/migration/data/estimates2/estimatesgraphs.asp. Accessed 03 June 2019

Israel's Approach Towards Space Security and Sustainability

32

Deganit Paikowsky, Tal Azoulay, and Isaac Ben Israel

Contents

Introduction	590
An Overview of Israel's Space Activities	591
Israel's Perspectives on Space Security and Sustainability	596
Conclusion	598
References	598

Abstract

In the last 30 years, Israel developed an indigenous space capability to launch, develop, operate, and maintain satellites in two main niche areas: Earth observation and communications, including the ground segment of communications satellites. Israel's space program was born out of national security needs. However, it has led to the growth of a commercial space sector. Recent years have seen Israel expand its cooperation with international partners as well as establish a civilian space policy backed with modest government funding. Space-related academia has begun partnering more with government and businesses, and space start-ups have sprouted, a notable though noncommercial example being SpaceIL. What began as a purpose-focused defense necessity has blossomed into a diversified amalgam of enterprises, academia, and government dealing in a broad spectrum of space endeavors. Space activities are seen as contributing significant and cross-cutting benefits to Israeli society from early

---

The views expressed are solely the authors' and do not reflect an official policy or opinion of any Israeli entity.

D. Paikowsky (✉) · T. Azoulay · I. B. Israel
Yuval Neeman Workshop for Science, Technology and Security, Tel Aviv University, Tel Aviv, Israel
e-mail: deganit.paik@gmail.com; tal.azoulay.ssp11@gmail.com; itzik@post.tau.ac.il

© Springer Nature Switzerland AG 2020
K.-U. Schrogl (ed.), *Handbook of Space Security*,
https://doi.org/10.1007/978-3-030-23210-8_17

education to an advanced economy and all points in between. Within the context of protecting and encouraging this nationally important ecosystem, Israel considers international space security, safety, and sustainability to be of importance. As such Israel actively promotes this position through its own space activities and increasingly in international forums.

Introduction

Since its establishment, Israel has suffered from acute security threats beyond its immediate borders. Israel's needs to relate to a broad circle of states which surround it demand an orientation towards space. As a result, Israel's space program was developed to fulfill acute national security needs. This mainly involves early warning, intelligence, deterrence, and self-reliance in advanced technologies. As a small country, which suffers from limited resources, the country adopted a pragmatic approach to space.

Israel's pragmatic approach contends that Israel's space program includes the capability to build, operate, and launch remote sensing satellites into space, as well as develop and operate communication satellites. Israel does not undertake to build all systems entirely on its own. It has, for example, no navigation or weather satellites, and has no indigenous human spaceflight missions. However, Israel increasingly cooperates with international partners on projects of this nature, as well as scientific projects.

It is important to note that the overall space activity of Israel is much broader than national security activity. Almost a decade ago, the Israeli government adopted an official civil space policy and began modestly funding civilian space activities. Israel also has a strong scientific sector as well as commercial space activity. In this regard, Israel has long-established space industries but also start-ups and innovative space initiatives, including educational initiatives. For example, numerous nano-satellites designed and built by high school students. Some have already been launched and successfully operated, while many others are in development.

Israel's long legacy as a spacefaring nation, and the development of its space ecosystem, demands an orientation towards space security and sustainability. Therefore, Israel attributes great importance to securing the space environment for peaceful uses for all nations. This interest extends beyond security needs; should outer space become inaccessible and unsafe, this will negatively impact Israel's overall space activities.

This chapter analyzes Israel's overall approach to space security and posits that Israel's approach to space security may be described as threefold: (1) promoting a robust and diversified space sector that provides for Israel's national security needs and protecting and safeguarding Israel's space assets, systems, and capabilities; (2) competing in the global space market and encouraging new space capabilities and activities; and (3) maintaining a safe and sustainable space environment for all users.

This chapter contains two primary sections. The first section provides a detailed overview of Israel's space activity. The second section provides an analysis of the Israeli perspective related to space security.

An Overview of Israel's Space Activities

Israel first built its presence and strength in space in accordance with priorities that correspond to its national and security needs. Therefore, understanding Israel's perception of space and of space security demands an analysis of Israel's strategic conception in relation to space (Ben Israel and Paikowsky 2017).

Israel's security conception is based on the profound understanding that it suffers from a significant quantitative inferiority against its rivals. To overcome this numeric disadvantage, Israel's leadership has chosen to focus its efforts on the development of a qualitative edge. In this perspective, Israel's space program plays a significant role in the country's overall answer to its strategic challenges. First, Israel's space program provides significant tangible capabilities to deal with the threats imposed by Israel's enemies. Second, and equally important, a national space program, which includes the ability to develop and to launch satellites into space, indicates very advanced national capabilities. Israel's achievements in space, whether civilian or military, project a clear message of national might. They emphasize the qualitative gap between Israel and its neighbors; they contribute to the country's accumulated achievements, aimed at deterrence; and philosophically, they reinforce the image of the "Iron Wall" in the eyes of its enemies, i.e., a power which cannot be overcome easily. All of that is accomplished without articulating an explicit military threat, which could provoke an unwanted and dangerous chain reaction in the region (Ben Israel and Paikowsky 2017).

More specific and tangible is the role the space program fulfills in mitigating the challenge of Israel's lack of strategic depth and acute need of early warning caused by Israel's narrow borders. Under these geopolitical circumstances, it was necessary to avoid the elements of strategic surprise and sudden attack. For these reasons, Israel's security doctrine demands advanced intelligence capabilities for early warning, as well as combat capabilities for a rapid transfer of battle away from Israel's population centers. The orientation towards space assists Israel in coping with the challenges presented by the lack of strategic depth and need to provide early warning.

With this in mind, the major impetus leading to the decision to embark on an independent Israeli space program was the 1979 Egypt-Israel Peace Treaty, and the perceived need to protect Israel, including through the need to verify Egypt's compliance with the treaty. The treaty did not neutralize Israel's concerns of hostile Egyptian aspirations. Moreover, Israel was to withdraw from the Sinai Peninsula. The greater distance from Egyptian territory meant that the Israeli military lost much of its early-warning intelligence-collection capabilities, including the ability to carry out manned reconnaissance flights over the Sinai Peninsula, now part of the

Egyptian sovereign state. (Such flights were considered a violation of Egyptian sovereignty and were a very sensitive issue in the embryonic relations between the two countries.) Therefore, there was a clear need for intelligence on what was happening in Egypt without violating its sovereignty. One of the potential solutions to the early-warning problems was using reconnaissance satellites. (For a review of the history of Israel's space effort, see: Paikowsky, D., "From the Shavit-2 to Ofeq-1- A History of the Israeli Space Effort", *Quest*, Vol. 18, No. 4, Fall 2011, pp. 4–12.) In 1981, the Israeli space program was established out of a pragmatic approach aimed to satisfy national security needs of early warning, deterrence, and self-reliance in advanced technologies. In 1988, Israel successfully launched its first satellite – Ofeq-1. "Ofeq" would become a very successful line of increasingly advanced earth observation capabilities.

The opportunity to observe Earth from space is a technological solution which enables Israel to cope with threats from hostile countries directly bordering the country, as well as those that threaten Israel but are located farther away geographically. At the 2013 Ilan Ramon Space Conference, the then Israeli Air Force Commander, Major General Amir Eshel, clearly stated the value of Israel's space capabilities as relates to strategic depth. "The threats we must deal with come from the border fence and from far away. Today, space is our strategic depth and it is what allows us to maintain our qualitative advantage. Thanks to our indigenous satellites, our ability to operate at any distance increases tremendously." (Chanel and Michael 2013).

The Israeli space program is recognized as a critical component of its independent intelligence capability. However, the issue of Israel's self-sufficiency is a complex one, and as a small country, Israel cannot be completely self-reliant. Nevertheless, in the field of intelligence, Israel has a great deal of autonomy.

Possession of independent intelligence capabilities has many implications for Israel beyond the field of intelligence. It enhances the power of the state and the image of Israel in the eyes of its opponents as well as its allies. It provides flexibility, both in its ability to collect information and the resulting autonomy in decision-making. Independent capabilities also permit the country to conceal its operational plans and areas of interest and to collect information unhindered. The space program is an important building block of this capability. To achieve this independence, Israel has continued to build its space program, especially the capability to develop and launch reconnaissance satellites.

Israel's space program also contributes to its deterrence. The following statement by Major General (Ret.) David Ivri, former Air Force Commander (1977–1982) and later Director General of the Ministry of Defense, provides valuable insight into the role of the Israeli space program in Israel's deterrence strategy:

> The perception of one's capabilities and one's willingness to use those capabilities are important components of deterrence. The perception of space capabilities is one of the primary components in Israel's future deterrence. Therefore, Ofeq 1, 2, and 3 contributed far more than anyone estimated. Imaging resolution is not the strategic measurement. Rather, the strategic measurement is the perception of capabilities that the State of Israel displays.

Not what we possess, but rather what the enemy estimates that we possess. The gaps in capabilities and information, in the tactical field, miniaturization field, and others are an immeasurably important component in the dimension of our strategic deterrence. (Ivry 2006)

Despite the obvious attention that national defense reconnaissance satellites receive, the fact that Israel was able to attain independent launch capacity is justifiably considered a significant achievement. Israel remains one of very few countries in the world with this capability. This is despite the fact that Israel is the only country which launches its' rockets to the west, against the rotation of the earth, to avoid launching eastward over neighbors with which it has strained relations. Launching to the west incurs a "cost" of approximately one-third of boost efficiency which leads to significant constraints on payload weight. Israel overcame this disadvantage by developing expertise in miniaturization of components. This expertise is one of many which eventually were able to serve the development of commercial space presence.

In the 1990s, Israel's space industry followed in the footsteps of many other technological sectors that were originally related to defense and began commercial spin-offs. As such, Israel developed commercial platforms such the Amos communication satellite series, EROS remote-sensing electro optical series, sub-systems, and other equipment.

In the last decade, the national space activity of Israel underwent a comprehensive process of re-evaluating its goals, objectives, and policies. In November 2009, a national task force was appointed to reexamine the Israeli space program and recommend a new framework. (The task force was headed by Mr. Menachem Greenblum, Director General Ministry of Science and Technology and Prof. Isaac Ben Israel, Chairman of the Israeli Space Agency.) The main objective of the task force was to focus on civilian applications and scientific activity that would allow Israel greater industrial scale and competitiveness in the growing world space market. The task force submitted its report and recommendations in June 2010 (Paikowsky and Levi 2010). The report outlines Israel's strengths, weaknesses, opportunities, and challenges for achieving its goals in space. Scrutinizing all of these parameters, the task force argued that Israel has great potential to lead in space technology in specific areas, but because of insufficient investments, Israel is in danger of gradually losing its competitive edge. In order to upgrade the scale of the local space industry, it was suggested that the Israeli government prioritize a national civilian space program focused on developing and renewing infrastructures, supporting academic research, and promoting international collaborations with other spacefaring nations, as well as with developing nations. In 2011, the Israeli Parliament established a subcommittee dedicated to space matters. In December 2012, after careful review by the treasury officials, the Finance Ministry approved an investment of $50 million for Israel's new civil space program (The Marker 2012).

On the basis of this new funding, the space agency began implementing a new space program, geared towards R&D and modernization of its civilian space activities. Among the program's objectives are advancing the local space industry, strengthening academic research, and raising the Israeli public's awareness of space activities and research, as well as reinforcing and expanding international cooperation. For example, in 2017, Israel and France launched Venus, a jointly developed and run environmental satellite. Following the success of Venus, the Israeli and French space agencies signed a statement of intent in 2018 to develop a new environmental project called C3IEL, which will involve a constellation of three nano-satellites focusing on climate research (CNES 2018). Among its efforts in these directions is also hosting of prestigious international conferences. In 2015, Israel hosted the annual International Aeronautical Congress and in 2016 hosted the Space Studies Program of the International Space University.

In September 2016, the communication satellite Amos-6 was lost on SpaceX launch-pad. Consequently, the Ministry of Science and Technology initiated a national task force to review the state of Israel's communication satellites. The report was submitted in December 2016 pointing to the need for a comprehensive and long-term program to upgrade capabilities to develop the new generation of Israeli made communication satellites. (For a detailed overview of Israel's Communication satellites, as well as the report of the 2016 Task-force, please see National Report by the state comptroller number 69A published in October 2018. Available at: https://www.mevaker.gov.il/he/Reports/Report_642/5caecf12-5145-4007-a355-83731735b0c1/2018-69a-304-Lavyanim.pdf. Accessed on April 14, 2019.) Two years later, in September 2018, Israel's government approved the development of Amos-8. The development of future satellites is yet to be decided. (For additional information about the decision, see: https://www.space.gov.il/news-space/131329. https://www.space.gov.il/news-space/131329. Accessed April 14, 2019.)

Israel is often referred to as the Start-Up Nation, and this is true as well for the space field. Following the impressive defense-related space advances and subsequent commercial space enterprises, a vibrant community of space start-ups has sprung up in Israel. The activities of a number of start-up companies, like SpacePharma and Effective Space Solutions, are noteworthy. (For additional information about these companies, see: http://www.effective-space.com and http://www.space4p.com/) Effective Space Solutions, in particular, is proposing solutions that are directly linked to increasing the safe extension of lifespan for satellites.

Aside from these initiatives, various educational projects have evolved as well, such as the Herzliya Science Center Space Lab, which is located in a high school. In 2014, this project launched its first nano-satellite, Duchifat-1, which was built by high school students (Winer 2017). Duchifat-2 was launched 3 years later in 2017. A year later in 2018, the Israel Space Agency initiated a new project dedicated for high schools in which seven high schools around the country will be able to plan and build their own innovative nano-satellites. The project will enable students to participate in the design and construction of these satellites. All seven satellites will be launched at the end of the process as one constellation that will serve as a communications network for transmitting information from space. (More information is available on the Israel Space Agency website: https://www.space.gov.il/news-space/131327. Accessed on April 15, 2019.) Recently, this initiative evolved into an even larger project and to an establishment of a dedicated research center for small satellites at Tel Aviv University.

Another particularly profound educational initiative is SpaceIL. This project began as a competitor in the GoogleX Prize to land a spacecraft on the moon. Despite the later cancellation of the GoogleX Prize, SpaceIL continued and in February 2019 launched its spacecraft aboard a SpaceX Falcon-9 launcher. Several weeks later, it succeeded putting its spacecraft "Beresheet" into lunar orbit, making Israel only the seventh country to achieve this feat. Unfortunately, the landing was not successful and the craft crash-landed on the lunar surface on April 11, 2019. Not to be deterred, SpaceIL has already begun working on a follow-up mission. In addition to the mission goal of landing on the moon, the SpaceIL team sees their mission also as educational and cultural. SpaceIL team and volunteers consistently lecture at schools and other public events to encourage youth involvement in STEM subjects in order to advance "a Beresheet Effect" in the Israeli society. It is important to note that SpaceIL is a private-commercial initiative and not governmental. (According to GoogleX Prize, governmental funding had to be very limited.)

There are also a number of more "formal" educational initiatives aiming to provide special space-oriented programs to schools around Israel from elementary school up through high school. The Ministry of Science and the Israeli Space Agency collaborates with the Ministry of Education to encourage interest in STEM subjects based on the understanding that space serves to inspire youth (and young professionals).

The increasing place of commercial and civil space activities is also noticeable in recent conferences and forums taken place in Israel. The Ilan Ramon International Space Conference has developed into an important date on the international space community's calendar, consistently attracting the attendance of many heads of space agencies, industry leaders, and decision-makers. Its most recent annual meeting (14th meeting), which took place in January 2019, focused on commercial space (Ramon14.forms-wizard.net 2019). In addition, for the first time, the conference was preceded by a special workshop dedicated to space entrepreneurship.

On top of that, the Ramon Foundation, which is a nonprofit organization set up in memory of Israel's first astronaut Ilan Ramon and his son Asaf Ramon, has set as one of its main objectives "to help engineers and entrepreneurs in their first steps in the space industry." To this end, the Foundation assists in networking, knowledge sharing, hackathons, and conferences (Ramon Foundation 2019).

To conclude this part, the rational for Israel's engagement in space activities described above reveals that Israel, which operates a successful space program on a modest budget, views space as a significant opportunity, especially as a force multiplier projecting the quality of force over its quantity in the most broadly manner. Retired Brig. General Amnon Harari, Head of Space Programs in the Israeli Ministry of Defense, stated clearly "Beyond the defense needs for which our satellites provide solutions, the Israeli space industry represents an important component [for Israel] in terms of the economy, education, advancing technology, science, small businesses, start-ups and more." (Shoval 2018).

This opportunity is accompanied by significant challenges, especially in maintaining its' qualitative advantage and preserving Israel's position at the forefront of technology. The significance of space in Israel's strategic conception and its long term and growing space capabilities in the defense, civilian, and commercial fields leads Israel to look for ways to protect its satellites, as well as shape its perspective on space security. The next section provides an overview and an analysis of Israel's perspectives and activities regarding space security and sustainability.

Israel's Perspectives on Space Security and Sustainability

Israel attributes great importance to space security and sustainability. This interest extends beyond security needs and derives also from the interest to compete in the global space market and encouraging new space capabilities and activities, as well as maintaining a safe and sustainable space environment for all users. Should outer space become inaccessible and unsafe, this will negatively impact Israel's overall space activities.

At the heart of Israel's approach to advancing space security and sustainability is promoting a robust and diversified space sector that provides for Israel's national security needs and protecting and safeguarding Israel's space assets, systems, and capabilities. For example, the following statement was made by Commander of the Israeli Air Force, Eliezer Shkedy, at the 2007 Ilan Ramon Annual Space Conference: "the operational importance of space is increasing constantly. Why is this field critical? There exists a concern that others who recognize its importance will try to attack space assets. We must consider defense measures, against physical harm, jamming, blinding, or any other technique. One of the greatest surprises that can happen in the modern world, in advanced countries with space assets, is a situation in which a country is surprised to find its space assets damaged." (Shkedy 2007). Shkedy's statement is an example of the growing recognition in Israel of the importance of the need to protect space systems. In this regard, one direction recognized by many in Israel is the likelihood of soft interference in space systems, especially the ground segment, through cyberattacks and therefore the need to protect space systems against this threat. (For more information on Israel's cyber policies, see: Tabansky and Ben Israel (2015).)

Another effort on this regard is in the field of Space Situational Awareness (SSA). SSA is one of the fields in which Israel looks for international cooperation. In 2015, Israel signed a data-sharing agreement with the US STRATCOMM for Space Situational Awareness. This agreement enables Israel to benefit from the collective data of dozens of countries and many commercial entities to avoid collisions with satellites and space debris (Pomerleau 2019).

Due to this kind of collaborations, Israel's perspective on space security and sustainability is broad and includes concern for the continued smooth operation of those other parties' capacities as well as cooperation and collaboration with some of those partners. Therefore, maintaining a safe and sustainable space environment for all users is also an important objective of Israel's perspective regarding space security. In order to achieve this goal of greater space security and sustainability for all users, Israel is actively looking to contribute to a sustainable space environment. In this capacity, Israel shares the idea that achieving the goal of space security and sustainability requires international collaboration and development of best practices for responsible behavior. In this regard, Israel recognizes the significance of contributing to multinational efforts. Joining the US efforts to improve Space Situational Awareness for debris tracking serves as an example of Israel's general perspective that international efforts by responsible players will have positive effect on the sustainability of outer space (Azoulay 2019). By taking this step, Israel not only improves the protection of its own satellites but demonstrates that it sees itself as a party to the international effort of sustaining the space environment for global stability and security.

Another example to Israel's efforts to promote responsible, peaceful, and safe use of space is its greater involvement in UN-COPUOS. In 2015, Israel was voted in as a regular member of the UN Committee on the Peaceful Uses of Outer Space [COPUOS] (United Nations 2015). By 2017, Israel was voted to serve on the six-member Steering Bureau of COPUOS (Israeli Ministry of Foreign Affairs 2017). These developments speak not only to Israel's decision to increase its international cooperation but also the positive feedback and support Israel is receiving from the international space community. In this capacity, Israel is contributing to diverse activities of COPUOS. One such activity is Planetary Defense. "Israel Space Agency is making efforts to find unique ways to contribute to planetary defense efforts. Specifically, Israeli researchers are conducting studies in order to contribute the world's effort." (Statement to UN-COPUOS by Israel's Space Agency delegate at the STSC 2019.)

Last but not least, due to the understanding in Israel of the need for international cooperation to ensure that space remains accessible and sustainable for the future, Israel favorably views legally nonbinding efforts towards space sustainability. For example, The European Union proposed an Outer Space Code of Conduct. (For the updated version of the European initiative of the Outer Space Code of Conduct dated June 2012, please see: http://www.consilium.europa.eu/media/1696642/12_06_05_coc_space_eu_revised_draft_working__document.pdf.) In 2012, an interministerial team of space experts was put together to discuss the code and its implications for Israel. An Israeli delegation actively participated in the open-ended consultation, which took place in Luxembourg in May 2014, as well as at the negotiations which took place in New York in July 2015.

Israel was also favorable towards more recent efforts to advance space as a secure and sustainable environment taken by UN-COPUOS. The following Israeli statement reflects on its overall approach: "Israel perceives space as a global commons and therefore aspires to contribute to a secure and sustainable space environment. Our small country acknowledges the worldwide use of space for supporting sustainable activity for development, as well as to promote contribution to UN system and its Sustainable Development Goals. Israel, through the Israel Space Agency, seeks greater international collaboration and cooperation, especially among democratic spacefaring nations, in maintaining space as a peaceful environment for the benefit of all." (UN-COPUOS 2018 General exchange of views, Israel delegation.)

Conclusion

In conclusion, Israel's space program was launched in response to national security needs. Over the years, with Israel's development and evolution as a country, its needs and capabilities have also evolved. Today, Israel has commercial, scientific, and civilian space assets and is expanding its involvement in international space cooperation. These developments, combined with increasing reliance on space for day-to-day activities and the nation's continuing security issues, make space security a concern for Israel and demand orientation to space sustainability on a diverse range of perspectives and activities. For example, as the local civil and commercial space activity of Israel grows and flourishes, there will be a greater need to update national regulations regarding space security and sustainability. On the international level, as these global trends of advancing space technologies and reliance on space systems and capabilities continue to grow, the need of maintaining the space environment safe and secured will only rise. Hence, Israel will continue to support global efforts to promote responsible behavior and pursue partnerships of this kind.

References

Azoulay Y (2019) Yisrael v'Arhab Yishatfu Peula Kdai L'mnoa Hitnagshuyot Bein Lavyanim c'Halal]Israel and USA will cooperate to prevent crashes between satelites]. [online] Globes. http://www.globes.co.il/news/article.aspx?did=1001061882. Accessed 4 Feb 2019

Ben Israel I, Paikowsky D (2017) The iron wall logic of Israel's space programme. Survival 59(4):151–166

Chanel L, Michael T (2013) Ha Halal Hu Ha'Omek Ha'astrategi [Space as strategic depth]. [online] Israeli Air Force. http://www.iaf.org.il/4391-40401-he/IAF.aspx. Accessed 12 Jan 2019

CNES (2018) France-Israel Space Cooperation. https://presse.cnes.fr/en/france-israel-space-cooperation-cnes-14th-ilan-ramon-conference-newspace-venms-and-c3iel-focus. Accessed 15 Feb 2019

Israeli Ministry of Foreign Affairs (2017) Israel elected to UN Space Committee. [online] https://mfa.gov.il/MFA/PressRoom/Pages/Israel-to-Lead-.aspx. Accessed 18 Apr 2019

Ivry D (2006) Hahalal Kezira Estrategit Ba'avar, Bahove, Veba'atid [Space as a strategic arena in the past, present and future]. Fisher Institute for Air and Space Strategic Studies, Hertzliya, Israel, p 50

Paikowsky D, Levi R (2010) Space as a national project – an Israeli space program for a sustainable Israeli space industry, presidential task-force for space activity final report. Israel Ministry of Science and Technology, Jerusalem

Pomerleau M (2019) Stratcom expands space surveillance with Israel agreement – Defense Systems. [online] Defense Systems. https://defensesystems.com/articles/2015/08/13/us-stratcom-israel-space-surveillance-agreement.aspx. Accessed 9 Jan 2019

Ramon Foundation (2019) Ramon Foundation. [online] http://ramonfoundation.org.il/. Accessed 9 Jan 2019

Ramon14.forms-wizard.net (2019) 2019 conference agenda. [online] https://ramon14.forms-wizard.net/website/index. Accessed 20 Feb 2019

Shkedy E (2007) Address at the 2007 Ilan Ramon annual space conference, Fisher-Institute for Strategic Air and Space Studies, Herzliya

Shoval L (2018) Images revealed of Israel's first satellite launch. [online] Israelhayom.co.il. https://www.israelhayom.co.il/article/587527. Accessed 24 Jan 2019

Tabansky L, Ben Israel I (2015) Cybersecurity in Israel. Springer,

The Marker (2012) State will invest 90 million shekels in international space cooperation. https://www.themarker.com/news/macro/1.1826164. Accessed 2 Mar 2019

United Nations (2015) Fourth Committee approves four draft texts, concludes general debate on questions relating to information | Meetings Coverage and Press Releases. [online] http://www.un.org/press/en/2015/gaspd593.doc.htm. Accessed 2 Feb 2019

Winer S (2017) Nanosatellite built by Israeli high-schoolers blasts into space. [online] Timesofisrael.com. https://www.timesofisrael.com/nanosatellite-built-by-israeli-high-schoolers-blasts-into-space/. Accessed 1 Apr 2019

Policies and Programs of Iran's Space Activities

33

Hamid Kazemi and Mahshid TalebianKiakalayeh

Contents

Introduction	602
International Space Policy	604
National Space Policy	604
Vision Plan	604
The Comprehensive Scientific Map of Iran	606
The Comprehensive Document of Aerospace Development	607
The Five-Year Iranian Economic, Social and Cultural Development Plans	610
Supervisory and Regulatory Structure for Implementing Space High-Level Policy Documents	612
Challenges on Implementing High-Level Iranian Space Policy Plans and Documents	614
A Glance at the Space Security of Iran	615
Conclusion	617
References	618

Abstract

Iran pursues its space activities, through international and national space policies, in accordance with international space treaties and ensuring the peacefulness of its space activities. At international level, Iran has prioritized international cooperation and compliance with international space regulations. It has always taken steps toward the peaceful use of space by playing an effective role in the United Nations Committee on the Peaceful Uses of Outer Space (UNCOPUOS), signing international space law treaties, membership in international and regional space organizations, and cooperation in the implementation of joint international space

H. Kazemi (✉)
Research Institute, Department of Air and Space Law, Tehran, Iran
e-mail: h.kazemi@ari.ac.ir

M. TalebianKiakalayeh
Graduate of Islamic Azad University, Tehran North Branch, Tehran, Iran
e-mail: mahshid.talebian@yahoo.com

projects. Its national space policy which was triggered by the commencement of its national space activities has been defined accordingly. In addition, principles for the regulation of space activities have been codified and implemented based on high-level national documents including the Vision Plan, the Comprehensive Scientific Map of the Country, the Five-Year Development Plan Rules (Fourth to Sixth), and the Comprehensive Document of Aerospace Development. These documents outline the scope of space activities and technologies while securing space security and ensuring the peaceful uses of outer space. The responsible authorities for space are proceeding to develop space technologies and activities based on the provisions in these documents. According to the policy, Iran ultimately aims for technology development and satellite operation as well as sounding rockets that provide civil services and also ensure the presence of Iranian astronauts in outer space. These goals are based on the principle of international cooperation and peaceful uses and exploration of outer space.

Introduction

Since the development of the space industry, space security has been a core matter of attention for the international community, maintaining throughout the years a prominent role in space activities and the economic growth of countries (Bockel 2018). In recent years, the development of space science, space technologies, and space applications for civilian purposes have had significant effects on people's daily lives (Concini and Toth 2019). In this context, the Islamic Republic of Iran (Iran) has put a particular focus on the space industry and on the use and exploration of outer space for national scientific, economic, social, and cultural development. Consequently, Iran has focused its national policies on this issue.

Iran has taken the potential of space into consideration as reflected in the Iranian national and international policy making. The competent space authorities have actively specified the space strategy and goals that contribute to the development of space technology and space activities. At international level, the space authorities promote the peaceful exploration and use of outer space in pursuit of ratifying and applying international space provisions. In parallel they attend international fora and in particular the UN Committee on the Peaceful Uses of Outer Space (UNCOPUOS). At national level, the space strategy concerning Iran's space activities has been codified in its high-level national documents. Iran's international strategic space policy, developed in discussions taking place at international fora such as the United Nations, has placed particular emphasis on the peacefulness of space activities. The involvement of Iran since the dawn of the space age in the international community – by becoming the 24th member of the UNCOPUOS in 1961 – can be attributed to its early motivation and aspiration to participate in the space sector (UNCOPUOS membership).

Iran has brought two main aspects of space activities into focus of its national strategic space policies. First, it takes the development of space activities into consideration so as to ensure national security and sovereignty of Iran over its territory. Second, by developing such activities, it intends to provide better services to the citizens through the use of space applications. Namely, surveillance and communication satellites are able to provide appropriate services in sectors such as agriculture, telecommunications, and transportation (Blount 2010). The allocation of dedicated budgets to authorized space organizations, as set out in Iran's high-level national documents, indicates the importance of space for the strategic position of Iran. Long-term plans for the space sector have been outlined in high-level legal documents that outline the implementation of development programs. Relevant documents include the 20-Year Perspective Document, the Comprehensive Scientific Map of the Country, the Five-Year Development Plan Rules (Fourth to Sixth), and the Comprehensive Document of Aerospace Development, which will be discussed in the next section (Islamic Republic of Iran 2013).

Since the late 1990s, with the aim to develop peaceful space activities, Iran has actively pursued the development of satellite communications and sounding rockets (Global Security 2010). To achieve international cooperation in space, Iran began cooperating with Russia in the 1990s regarding the development of civilian satellites for communication and imagery, followed by cooperation with China and Thailand in 2008 concerning the launch of a satellite for surveillance and response to natural disasters. In addition, Iran has cooperated with other countries in the region, particularly allied countries, as well as European countries. In 2003, Iran, as the first Islamic country in the region aiming to develop space activities, decided to launch a satellite into orbit. Iran operated its first satellite, Sina-1, which launched with a Russian satellite launch vehicle from Russia. Since then, Iran has declared itself as the 43rd country with its own satellite capacity, which was a landmark in the history of Iranian space activities.

Iran has stated that the satellite is used for imagery of Iran, in particular for controlling and monitoring natural disasters. During this period, another satellite, Omid, was presented as the first made Iranian satellite. On February 2, 2009, this satellite was launched by Safir Satellite Launch Vehicle (SLV), a domestic launch vehicle (Tarikhi 2009). This was a big step forward for Iran that it enabled it to enter the league of space-faring nations. Omid was equipped with experimental satellite control devices and power supply systems. It was also designed for gathering information and testing equipment. Iran's space program up to that point had been based on ground stations that relayed Intelsat communications and received Landsat data (Tarikhi 2009).

With regard to Sounding Rocket Vehicles (SRV), in February 2007, Iran tested an SRV for research purposes which was followed by SRV 1. SRV 2, which was successfully launched into space, provided the opportunity for Safir SLV to launch the first national satellite, Omid. Iran's space program up to that point had been based on ground stations that relayed Intelsat communications and received Landsat data (Tarikhi 2015). The success of this launch placed Iran alongside the eight countries with the ability to put their own space-built objects in orbit. It has been explicitly stated that this satellite is used for peaceful purposes and in particular for agricultural and economic purposes. The launch of SRV 3 took place in 2010, demonstrating a new capacity in space technology development for Iran (Harvey et al. 2010). As a result, Iran has made its space activities debut as one of new space farers with access to space through of the development and launch of satellites and SRVs directly and indirectly. The access of Iran to space has been recognized as a necessity for the protection of national interests. This research seeks to investigate the goals, strategies, and long-term space policies that have been outlined in Iran's high-level national documents.

International Space Policy

Taking into consideration the central role of UNCOPUOS in space, Iran has been involved in deliberations for the formulation of international space treaties in the 1960s and the resolutions that followed. Iran has signed the Outer Space Treaty 1967 and the Registration Convention 1975 and has also ratified the Liability Convention 1972 and the Rescue Agreement 1968. Iran has not signed the Moon Treaty 1979 (UN Treaties). As a regional influential country, Iran has always placed importance on peaceful space activities. The 1970s was the dawn of Iran's space activities. The most important initial steps marked the establishment of the Remote Sensing Organization, the registration of orbital slots in 1974, the membership to the International Telecommunication Union in 1967, and the membership to UNCOPUOS in 1969 (UN Treaties). The current status of Iran's participation in international space treaties and its membership in international and regional organizations related to space activities are shown in Tables 1 and 2.

National Space Policy

The body of general policies of the regime is a new foundation which was contained in the Constitutional Law (Islamic Republic of Iran 1989) after the Islamic revolution of Iran in 1989. Determining the national policies falls under the responsibility of the Supreme Leader. After consultation with the Expediency Discernment Council, their implementation is announced. The general policies are determined based on national principles and goals that lay down the guiding framework for the orientation of the country in all governmental areas, including space activities. This system allows for a reasonable approach toward the accomplishment of its goals. Long-term space policies and programs are developed in three high-level national documents including the Vision Plan, the Map, and the Document of Aerospace Development, as well as the Five-Year Development Plans that are approved and implemented by the Cabinet.

Vision Plan

The 20-Year Vision Plan of Iran (Islamic Republic of Iran 2003) is a document determining Iran's development in numerous aspects consisting of cultural, scientific, economic, political, and social fields which are codified by Expediency

Table 1 Status of Iran's participation in international treaties of outer space

United Nations International Space Treaties	Action	Name of treaties/conventions	Date of notification/deposit
	Signature	Outer Space Treaty	1967
	Signature and ratification	Rescue Agreement	1968
	Signature and ratification	Liability Convention	1972
	Signature	Registration Convention	1975
	–	Moon Treaty	1979
Related agreements	Signature and ratification	Treaty banning nuclear weapon tests in the atmosphere, in outer space, and under water (NTB)	1963
	–	Convention relating to the distribution of program–carrying signals transmitted by satellite (BRS)	1974
Cooperation organizations' conventions	Signature and ratification	International telecommunications satellite organization (ITSO)	1971
	–	International system and Organization of Space Communications (Intersputnik)	1971
	–	European Space Agency (ESA)	1975
	–	Agreement of the Arab Corporation for Space Communications (ARB)	1971
	–	Agreement on cooperation in the exploration and use of outer space for peaceful purposes (INTC)	1971
	Signature and ratification	International Mobile satellite organization (IMSO)	1971
	–	European telecommunications satellite organization (EUTELSAT)	1982
	–	European Organization for the Exploitation of meteorological satellites (EUMETSAT)	1983
	Signature and ratification	International telecommunication constitution and convention (ITU)	1992

Discernment Council of the country (Islamic Republic of Iran 2013, Article 9). In implementing the Plan from 2005 to 2025, Iran aims to obtain the first economic, scientific, and technological position in the South West Asian region and to be recognized as a developed country equipped with advanced knowledge, security, independence, power, and autonomy. The Plan is the basis of Five-Year Development Plans and regulations which outline the overall direction of activities in different aspects. It falls under the Constitutional Law and regulates the framework within which the general polices are regulated. It should be considered as a source of

Table 2 Status of Iran's membership in other international governmental and intergovernmental organizations

Membership of Iran	Headquarter location	Authorized organization	Number of states parties	Name of organization	Established
	Vienna-Austria	Ministry of Foreign Affairs and Iranian Space Agency	86	COPUOS	1959
	Geneva-Switzerland	Ministry of Information and Communications Technology of Iran	143/193	ITU	1865
	Beijing-China	Iranian Space Agency	9	APSCO	2005
	France-Paris	Aerospace Research Institute	66 out of 300	IAF	1951

the state's comprehensive policy through which the approach of Iran's space activities should be organized in developing a strategic industry.

According to the Plan, Iran shall take steps to increase its role in the world's scientific outputs aiming at organizing and mobilizing the country's capabilities and facilities. The aim is to strengthen the software movement and promote research capacity as well as acquisition of technology, especially new technologies including sub-technologies, biotechnologies, information and communication, environment, and aerospace. As it can be witnessed, the Plan, which is the most important document for the future policy making of the country, has addressed Iran's contribution to producing science in the world and acquiring new technologies, particularly aerospace technology.

There is no reservation that active presence in outer space is one of the great achievements that has been predicted in the Vision Plan. The objectives of the Plan are met with sufficient investment, the efforts of national experts, and cooperation with other countries. Furthermore, the Plan outlines the prospect of reinforcing national authority and security with an emphasis on scientific and technological growth. In other words, this Plan recommends taking steps to enhance security and promote the global position in parallel to the development of science and technology, such as the aerospace sector. Regarding the implementation of the plan in the aerospace sector, a number of additional fundamental high-level documents have been developed, including the Comprehensive Scientific Map of the Country, the comprehensive aerospace development document and development plans.

The Comprehensive Scientific Map of Iran

Living in the age of knowledge, countries need a plan to promote science exploitation; otherwise they will not be able to enhance or maintain their position. The Plan has revealed the general frontiers of Iran movement toward the future. In order to achieve the goals, policy makers decided to chart Iran's scientific roadmap: a Comprehensive Scientific Map of the Country which clarifies the paths to access these goals. In other words, the required paths to achieve the outlined goals in the plan are determined by this Map, which was approved for science and technology areas. The Map was ratified by the Supreme Cultural Council of the Islamic Republic of Iran in 2011 (Islamic Republic of Iran 2011a). It is described as "a coordinated and dynamic set of goals, policies, and structures, planning requirements for strategic transformation of science, technology and innovation based on Iranian Islamic values to achieve the goals of the Plan." All executive organizations including state entities and private entities shall comply with its provisions. The main orientations or outputs of the Map should be implemented over the next two decades. The Map priorities focus on cognitive sciences, aerospace, and oil and gas, for which each of these specific documents has been prepared and approved (Islamic Republic of Iran 2011a).

The Map addresses two main issues including sending humans into space (human spaceflight) and designing, manufacturing, and operating satellites. According to that (Islamic Republic of Iran 2011a, No 9, Para 2–2), achieving these goals is mirrored in international cooperation and the Islamic world cooperation with other countries as the subset of national long-term technological and scientific targets. In addition to that, aerospace technology is prioritized as an A level among three levels of A, B, C (Islamic Republic of Iran 2011a, Paragraph 2–3: technological and scientific priorities of a country). It is noteworthy that the amount of allocated resources and consideration of competent authorities are the center of this prioritization. Implementation of main priorities usually requires the attention and support of superior competent authorities and executives. Aerospace technology and space activities are one of the top and main priorities under support of superior competent authorities. The supervision of the implementation of the provisions and principles of the Map is carried out by a Strategic Committee. Its mission is to take care of executing the Map in a timely manner. The Committee is responsible for monitoring the proper implementation of air and space policies, which are provided in the Comprehensive Document of Aerospace Development, which will be reviewed as follows:

The Comprehensive Document of Aerospace Development

The Comprehensive Document of Aerospace Development was adopted in 2012 in accordance with the Map. It seeks to implement the fifth chapter of the Map. Since Iran has a special status, the document emphasizes the characteristic capacities and capabilities of Iran for space activities such as capabilities in manufacturing, operating, and deploying satellites and spacecraft in outer space and exploitation of outer space and training aerospace experts in universities and research institutes based on a global large scale. The details of the Document, consisting of the development of aerospace-based science, technology, and knowledge-based industries of aerospace sector in Iran, familiarize us with the approach adopted by the legislators and policy makers. The Document not only governs all institutions and sectors, including military and nonmilitary (governmental and nongovernmental), but also has the ability to define the exploration of air and space applications. The latter include media, telecommunications activities, imaging, remote sensing, and cargo transportation in air and space. The Document also concludes activities in the areas of education, research, acquisition and development of technology, industrial activities, and service provision (Islamic Republic of Iran 2013).

Due to the high speed of technological developments and the limited time to put these developments into practice, one of the most important features of the space industry would be the knowledge base feature. Since a large part of space activities is knowledge-based, the economic importance of the space industry in the region for the coming years has been amplified due to its estimated million-dollar labor market. The Document covers all ground-based and space-based services and products related to the use or exploration of outer space, including telecommunications, earth observation, remote sensing, navigation, space security, bio space, and space science. The scope of aerospace and defense sector has been defined to include all required operations and services in the aerospace sector that protect and defend the security of Iran promoting its military authority. In principle, it deals with strategies, fundamental values, and long-term aerospace policies that are common to the both the air and space sectors. The goals and strategies of each of them will be stated in order to examine common principles particularly in the space sector (Islamic Republic of Iran 2013).

Common Principles in Air and Space Activities

The Document in this part deals with strategy part, fundamental values, and long-term aerospace policies.

Air and Space Strategy

Promoting national security and authority, surpassing in technological and scientific advancements in other fields, and benefits resulted from the spreading developed or localized technologies flowing to other sectors are the factors that will globally convert an Iran's aerospace sector to the strategic one. In this regard, recent achievements, particularly in designing, manufacturing, and operating satellites and other aircrafts, have increased national dignity and self-esteem which could ultimately pick the rank of Iran in front of international observers (Islamic Republic of Iran 2013).

The Document declares that space policy making should be pursued in a way through which Iran can achieve the following capabilities:

- Has an effective authority in national security.
- Be efficient and reliable in fulfilling the strategic and current needs of world and Iranian society.
- Compatibility with culture and Islamic values of Iranian.
- Ability to creating budgets for designing, developing, manufacturing products, and providing aerospace services.

- Being inspired in expanding the frontiers of knowledge and developing aerospace technology.
- Playing a pioneer role in other areas of science, technology, industry, and services.
- Knowledge based and capable to apply the recent scientific, research, and technological achievements (Islamic Republic of Iran 2013).

Fundamental Values

The Document, like the map, emphasizes Iran's space activities when it expresses the of the national aerospace sector. The most important of these values can be centralized justice in the development of aerospace sector and the benefits of its achievements; fostering talent, creativity, innovation, risk-taking, and courage, paying attention to the principle of rationality in the long-term management of space domain; coordinating science and technology with the environment; promoting international cooperation, raising the spirit of cooperation, partnership, the responsibility of academics and related institutions; and recognizing all space resources including celestial bodies, orbits, and their exploitation under the justified legal system as a common heritage of mankind (Islamic Republic of Iran 2013).

Long-Term Aerospace Policies

The Document outlines long-term Iranian aerospace policies. The most crucial of these policies are as follows:

Emphasis on production of indigenous science and development of related sciences, utilization of maximized internal capacity, enabling capacities; taking responsibility for maintaining and improving the environment; creating the maximum cooperation of private sector and addressing them with an emphasize on a role of government in policy making and supervision, observing diplomacy frameworks of national technology in international aerospace partnerships and activities; exploiting the surplus age of aerospace technology in other industrial and manufacturing sectors; observing passive defense and protecting of this technology to develop aerospace sector, absorbing elites as a major factor in technological developments; and last but not least creating the basis for economic and social activities in this sector maintaining sustainable security of Iran (Islamic Republic of Iran 2013).

Particular Principles in Space Activities

The Document has specifically targeted goals to achieve the abovementioned purposes. These orientations can be categorized as follows:

- Reaching first place in the region in conquering space and dominance in space through relevant science and technology based on capabilities of national universities and research centers.
- Research and development of manned space missions.

- Designing, manufacturing, and operating national satellites in orbit with various applications such as communications and remote sensing with a priority on international cooperation.
- Access to space-based communications infrastructure to meet national regional, universal, public, and commercial requirements compatible with ground-based telecommunications platform.
- Achieving required technologies for remote sensing and ground observation with accuracy below 10 m.
- Cooperation in positioning, navigation, and timing on national and regional level in accordance with global and competitive quality.
- Designing, manufacturing, and operating satellites and carriage systems of defense and security satellites which are needed to insure national space security.
- Maximum usage of all state institutions and private sector to develop and promote space activities for long-term space programs of the country.
- Supporting privatization and providing necessary framework for space knowledge-based industries and companies.
- Supporting space educational and research activities.
- Development of international collaborations and interactions to carry out space programs.
- Applying space achievements to understanding the universe and developing astronomy (Islamic Republic of Iran 2013).

It further refers to the development of basic space sciences and the promotion of space knowledge, space technology, and space achievements among young scholars. It also specifically addresses a number of policy issues for national space security, the most important of which are:

- Consideration of regional and international defense treaties and, if necessary, accession to them.
- Effective membership in regional and international defense organizations within the framework of foreign policy and based on national security priorities.
- Recognizing space threats ahead and methods of dealing with them.
- Improving collaboration with universities or industries and integration of aerospace defense networks.
- Improvement and development of indigenous and national standards of space strategic systems in products and services and increasing their reliability and intelligence (Islamic Republic of Iran 2013).

The Five-Year Iranian Economic, Social and Cultural Development Plans

The Five-Year Iranian Economic, Social and Cultural Development Plans (Five-Year Development Plans) are known as a set of interim plans, which are prescribed by the state for 5 years and approved by the Parliament. Six development plans have been approved after the Islamic Revolution of Iran in 1978. Since 2004, by regulating the Fourth Development Plan Act, aerospace and space security activities have also been considered. The focus on space industry and activity in development plans is actually the explanation of high-level national documents such as the Vision Plan, the Comprehensive Scientific Map of Iran, and the Comprehensive Document of Aerospace Development for 5 years (Islamic Republic of Iran 2013).

The Fourth Development Plan Act

The Fourth Development Plan Act was approved by the Parliament in 2004 (Islamic Republic of Iran 2004).The state is obliged to make necessary schemes for the extreme exploitation of national and regional capacities in the aerospace sector with regard to the importance of knowledge role, technology, and ability as the main factors of value added in recent economy. Moreover, according to the Act (Islamic Republic of Iran 2004), the state should promote new or emerging technologies and data applications in utilizing defense systems. They should be made to strengthen defense and military capabilities of the state in order to protect security, territorial integrity, national or vital interests, and resources of the country in preparing against any threats. As it can be seen, the Act does not specifically address space high-level policies and strategies. But it has directly obligated the state to apply and develop national and regional aerospace capabilities (Islamic Republic of Iran 2004).

The Fifth Development Plan Act

The Fifth Development Plan Act was approved by the Parliament in 2011 and has significant arrangements in the space domain. The Act declares (Islamic Republic of Iran 2011b, January 5) that the state may take necessary steps to create and develop infrastructure related to designing, operating, and testing space satellite systems, ground station, and satellite launch vehicle due to the ever-increasing importance of the space sector and achieving new or emerging technologies to ensure required space applications and services in Iran. In addition to that, it explains that the state should maintain orbital slots for Iran and predict approaches so as to establish the infrastructure and implement national satellite projects to protect such slots.

In terms of defense and security, it refers to the establishment of command, control, data, and advanced supervision systems in different aspects including space, air, sea, and land in order to promoting awareness and being prepared to recognize threats and taking measures to encounter them effectively (Islamic Republic of Iran 2011b, January 5). It seems that the Act comparing to other national space-related law and regulations is unique, because as a high-level national document, it not only sets general requirements in the field of space activity development but also has posed specific tasks on the state, such as preserving national orbit slots.

The Sixth Development Plan Act

The Sixth Development Plan Act has been approved in 2017 (Islamic Republic of Iran 2017). It has prioritized the space sector, and it considers the development of space activities. In the General Policies section of the Act, first it refers to prioritizing industrial strategic areas such as aerospace and increasing the technology influence, and then it addresses the development of space technologies (Islamic Republic of Iran 2017, paragraphs 28 and 38 of the General Policies section). All these policies are detailed in the Act (Islamic Republic of Iran 2017, Article 40), according to which, the state is obliged to take the following steps in order to develop air and space industry and acquire new science and technology in that field:

1. The necessary support for the development of infrastructure and industries related to the designing, manufacturing, testing, and operation of space systems, satellite launch vehicles, satellite, and ground stations.
2. Maintaining orbital slots which belong to Iran and also predicting necessary arrangements to establish and implement national satellite designs so as to maintain the aforementioned slots (Islamic Republic of Iran 2017, Article 41).

Obviously, the strategy, policy, and fundamental measures for space activities are dealt within the Act. Regarding the space strategy, it provides for a balanced development of activities the highest utilization of space infrastructure and also the maintenance of orbital slots. In the field of policy making, the Act states:

- Utilizing the full capacity of domestic space actors and transfer of advanced technologies along with maintaining systematic integration of capabilities.
- Cooperating with foreign space operators and manufacturers.
- Cooperating with the nongovernmental sector (national and foreign) in order to purchase national satellites.

Concerning fundamental measures, the state is expected to manufacture telecommunication satellites, remote sensing satellites, and to purchase national satellites. In the Act, a separate section is devoted to the science and technology in the space industry. This section focuses on designing, manufacturing, testing, operating, and deploying space systems and maximizing the maintenance and utilization of orbital slots in national space policy so as to provide development opportunities for Iran's space technology (Islamic Republic of Iran 2017, Article 42). The quantitative objectives for the development of space activities in the Act can be seen in Table 3:

Supervisory and Regulatory Structure for Implementing Space High-Level Policy Documents

There are two important regulatory bodies regarding the implementation of space policies set forth in the high-level space policy documents in Iran: the Supreme Space Council and the Iranian Space Agency. Until 2003, there was no dedicated institution for space activities in Iran. In that year, the Parliament established the Supreme Space Council, chaired by the President and the relevant space entities who are members. In the same year, the decree of the Iranian Space Agency was approved. According to the Comprehensive Document of Aerospace Development

33 Policies and Programs of Iran's Space Activities

Table 3 Quantitative objectives of Sixth Plan in the space domain

General objective	Quantitative objective			Status at the end of year 2015	Years of Sixth Plan				
	Subject		Unit		2016	2017	2018	2019	2020
Increase the level of technology, infrastructure, space applications, and utilization or maintenance of orbital slots for a country	Increase the capability of national technology, manufacturing remote sensing satellite, and telecommunication satellite to 50%		Percent	1	5	15	25	35	50
	Purchase and operate two telecommunication and remote sensing satellites		Number	–	–	1	2	–	–
	Offer 15 new space services considering value added		Number	–	2	4	7	11	15

approved by the Supreme Council of the Cultural Revolution, the Supreme Space Council is the highest policy-making authority in the space domain of the country and has responsibilities to execute, lead, supervise, and approve major space-based plans and strategies. The Statute of Council was approved by the Cabinet in 2004 (Ataafar 2009). The Supreme Space Council is responsible for the policy making and for determining the general outline within which the space activities regime in Iran is developed. The Iranian Space Agency is responsible to implement the policy developed by the Supreme Space Council. Since the decision of the latter is inclusive and has both domestic and international policy dimensions, the obligations of the Iranian Space Agency are also comprehensive. Although the Iranian Space Agency manages the governance of the development of space technology and space activities for peaceful purposes, it should be noted that universities and aerospace research institutes practically develop space science and technologies in Iran (Ataafar 2009).

Challenges on Implementing High-Level Iranian Space Policy Plans and Documents

The aforementioned high-level documents on space activities have been regulated and approved by the responsible authorities. According to the Vision Plan, space defense and security capabilities and technology policies should be defined and regulated. Therefore, various documents on air and space fields have been approved according to this key document. Alongside the Vision Plan conveyed by the Supreme Leader, other high-level documents such as the Comprehensive Scientific Map and the Comprehensive Document of Aerospace Development are approved by the Supreme Council of the Cultural Revolution. These high-level documents contain elements for strategy and policy making adopted by the competent authorities to achieve specific goals or obtain appropriate means for those purposes. The documents are implemented by the various executive entities and can be enforced by the legislator if necessary.

Most of the governmental and nongovernmental agencies are facing challenges concerning the specific legal status and the implementation of the mentioned documents. Generally, these documents contain program elements that are in the form of guidelines, meaning that they do not have the content and legal status of the documents or programs approved by the Parliament or the Cabinet. In fact, these documents lack the necessary binding status and are mainly used to guide the formulation of government development plans approved by the Parliament. Iran has attempted to regulate its long-term policies and programs in these documents in various domains, including space activities. In the space sector, however, non-allocation of funds or deviation of priorities in scientific, economic and technological aspects can affect the goals set out and their implementation. The implementation of such documents generally depends on various political and economic conditions. Besides, the Parliament or the Cabinet shall approve the relevant regulations for the government agents to find a legal way to accomplish the purposes thereof.

Two important factors are added to the emerging challenges of implementing these documents in the space domain:

1. Iran's space activities face a lack of integration. An integration of space activities and organizational integrity should be considered in these documents. There are various space institutions in Iran, including executive, scientific, and technology related that are involved in implementing the high-level space policy documents. However, it is not legally clear which entity is responsible to apply what part of the policy. The responsibilities of any military, governmental, and non-governmental entity in the conduct of space activities should be clearly stated. There is a need that space policy documents outline that the responsibilities of institutions and authorized entities based on research, scientific, and technological requirements.
2. Besides the necessity of codifying regulations to specify the tasks of each space entity, it is also required to create a coherent structure to apply and control high-level space policies. Although according to law, the Iranian Space Agency and the Supreme Space Council are responsible to implement the document of Aerospace Development, to determine main authorities, and to consider priorities in the allocated budget, they have not succeeded in the process. Firstly, the Supreme Space Council has rarely and disorderly convened since its establishment. Secondly, the Iranian Space Agency has administrative activities that have prevented it to fulfill its sovereign responsibilities. Thirdly, Iran has not yet approved a national space law. Codifying such law could facilitate the way for governmental and nongovernmental institutions to engage in space activities. According to the provisions of national space law, organizations, private companies and individuals could access their position in implementing high-level national space policy documents.

A Glance at the Space Security of Iran

Outer space plays a pivotal and determining role in ensuring the security of countries and international community. Space security is recognized as one of the most fundamental components of relations among countries. Therefore, countries tend to set long-term goals in order to expand their space activities. The balance of power constitutes the main reason which may result in security challenges in space. The arms race in space might create benefits or hindrances to the balance of power. The balance of power is a major phenomenon brought about by the structure of an international system relating to the external behavior of countries. By the end of bipolar order between the two countries, the USA and the former Soviet Union, space age entered into a new phase providing an opportunity for new space farers to participate in space activities. Along with the continued traditional use, namely, research, military, and quasi-military, the use of space expanded to commercial and civilian activities, with both governmental and nongovernmental actors particularly in the development of communication, commercial, and telecommunication satellites (Moltz 2019).

Given the mentioned circumstances, Iran has decided to introduce itself as a new space-faring nation. Iran's approach to regulating space is transparent. International peace and security is very important to Iran. Iran desires to gain the benefits that space technology has brought to other space-faring nations. It has gradually developed its presence in space activities over the past two decades and continues to do so by taking steps forward. The expansion of space activities with the acquisition of advanced space technology in recent years has positioned Iran among countries that actively participate in outer space activities. Iran is one of the new space actors which could develop the capability to launch its satellites and sounding rockets and eventually join the few space-faring nations that have accessed space (UNCTAD 2016).

There was no organized structure in the space policy of Iran till 2004. The establishment of the Iranian Space Agency in 2004 enabled Iran to systematically develop space activities in the different fields of communication, telecommunication, remote sensing, meteorology, geology, exploitation, and so on. The launch of the Omid satellite could mark Iran's engagement with the development of a space industry. Aiming to expand its space activities and its presence in outer space, Iran space policy includes the commercialization of space applications, Earth observation and environmental changes, mapping and weather forecasting, as well as the promotion of international cooperation and the involvement of private entities. Therefore, it can be said that space policy in Iran was based on crucial steps especially in terms of designing and manufacturing of satellite (Malmiran 2009).

Although Iran's space activities are just beginning, they exert positive influences on Iran's space security policy at national and international level. The space security policy of Iran revolves around four themes:

1. At international and national level, Iran has always emphasized the peacefulness of space activities carried out by Iran or other countries. Iran has signed four of the United Nations treaties on outer space. The space policy of Iran is actively being developed at all international and regional meetings related to space subjects, including space security (Zargar 2010).
2. Iran, since the establishment of UNCOPUOS in 1959, views international cooperation as central to space security. In the 1970s, the pursuit of space activities and the development of satellite communications were considered important for Iran. Iran is one of the 65 members in the Conference on Disarmament (CD) which supported the "Prevention of an Arms Race in Outer Space." This approach of Iran has shown its attention to outer space and space technologies, despite the fact that the country didn't have space activity back then. Iran plays an important role in the strategic cooperation for outer space activities in Asia and is also a member of the Asia-Pacific Space Cooperation Organization (APSCO) for multilateral cooperation in technology and space applications along with China, South Korea, Thailand, Bangladesh, Indonesia, Peru, and Mongolia. Iran is also a member of the Islamic Space Science and Technology Network (ISNET) (Khosravi et al. 2013).

3. Iran seeks to develop its space activities in accordance with official aerospace strategic documents that have been developed within the last two decades. As stated above, it aims to codify a strategy that assures space security within its national policy. According to the provisions of high-level national space policy documents, the access to space through the active presence of Iran has been recognized as an essential matter in line with Iran's national interests. Iran intends to achieve the first place in economic, scientific, and technological fields such as space in the Middle East. In the development plans of the country, producing space science and access to new space technologies is one of the top priorities (Gorwitz 2011).
4. Iran has affirmed the exploitation of space to ensure Iran's security by reducing threats and using space to defend the country's territorial integrity and enhance its positioning. Iran also seeks to develop space activities for peaceful purposes among which operating meteorological satellites are the examples. In 2009, the Fourth Development Plan Act was codified in which the highest importance was placed on space technology applications as an effective means for the sustainable development of the country. Based on the Act, there is a need of acquiring space science and technology as well as cooperation with the international community (Kamran and Nami 2008).

Conclusion

The development of space technologies has made available various space applications in remote sensing, communications, telecommunications, earth observation, use and exploitation of celestial bodies, and so forth. Moreover, due to the expansion of space technology, the number of space-faring nations has been increasing far beyond the initial few countries with the involvement of many new countries entering the space industry. There has also been a surge in the number of private entities that engage in space activities. Iran has increased its space activities over the last couple of decades. At the same time, policy makers and legislators have approved Iran's space policies and strategies aiming to regulate space activities. Pursuant to the Constitutional Law, the long-term space policies of Iran have been codified in the "Vision Plan," followed by the "Comprehensive Scientific Map" and the "Comprehensive Document of Aerospace Development." The aerospace policies of the country have been enacted and implemented under the Five-Year Development Plans as well.

Iran's space policy is also reflected in its international presence. At international level, Iran's approach is not only to ratify and enforce international space treaties but also to maintain an active presence in UNCOPUOS and an effective role in determining resolutions and guidelines issued in the form of soft law. Iran also desires to expand its international cooperation with other countries through multilateral, regional, and bilateral agreements on space activities, especially in space science and research. The same approach to international cooperation applies also to space security and space defense as explained in Iran's space security policy. The peaceful use of outer space has always been highlighted in Iran's policy on space security as reflected on the development of space science and research. Firstly, the emphasis is on designing, manufacturing, producing, and operating national satellites for various peaceful uses. Secondly, human presence in space is declared as a feature of national autonomy. The development of space technologies and activities aimed at defense affairs for countering probable cyber and space attacks, are emphasized by the national security policy.

References

Ataafar A (2009) Vision of development and fundamental revolutions required for universities. Iranian Higher Education Quarterly, no 4, 1st year, at 27

Blount PJ (2010) The development of international norms to enhance space security law in an asymmetric world. Proceedings of the 52nd Colloquium on the Law of Outer Space, p 2

Bockel JM (2018) The future of the space industry, NATO Parliamentary Assembly, General Report 173 ESC 18 E fin | Original: French | Available at: https://www.nato-pa.int/

Concini A, Toth J (2019) The future of the European space sector How to leverage Europe's technological leadership and boost investments for space ventures, European Investment Bank. Available at: https://www.eib.org/attachments/thematic/future_of_european_space_sector_en.pdf

Global Security (2010) In 1997, Iran made plans to develop satellites called Mesbah (Lantern). Available at: GlobalSecurity.org. http://www.globalsecurity.org/space/world/iran/multi.htm

Gorwitz M (2011) Iranian dual-use space related research, Iranian dual-use science and technology bibliography, vol VII, Space Related Research. Available at: https://fas.org/nuke/guide/iran/biblio5.pdf

Harvey B, Smid HHF, Pirard T (2010) Emerging space powers. The new space programs of Asia, the Middle East and South-America. Springer

Islamic Republic of Iran (1989) Iranian constitutional law 1989. Article 110

Islamic Republic of Iran (2003) The Vision Plan of Iran by 1404 (2024), approved in 2003 November 4. Available at: https://rc.majlis.ir/fa

Islamic Republic of Iran (2004) The Fourth Economic, social and Cultural Development Plan Act, approved by Parliament in 2004 September 1. Available at: https://rc.majlis.ir/fa

Islamic Republic of Iran (2011a) The Comprehensive scientific map of Iran approved by Supreme Cultural Council of the Islamic Republic of Iran in 2011 January 4. Available at: http://www.nlai.ir/documents/

Islamic Republic of Iran (2011b) The Fifth Economic, Social and Cultural Development Plan Act by 2012–2016, approved by Parliament in 2011 January 5, no 73285/419. Available at: https://rc.majlis.ir/fa

Islamic Republic of Iran (2013) The Comprehensive document of aerospace development approved by Supreme Cultural Council of the Islamic Republic of Iran in 2013 January 8, no 728. Available at: http://www.nlai.ir/documents/10184/198216/

Islamic Republic of Iran (2017) The sixth Economic, Social and Cultural Development Plan Act by 2018–2022, approved by Parliament in 2017 March 4, no 20995 RooznamehRasmi. Available at: www.dastour.ir

Kamran H, Nami MH (2008) Space, power, security. J Iranian Geogr Soc 16, 17, 6th year. Available at: http://ensani.ir/

Khosravi I, Joday J, Jafari SA (2013) Impacts of armed arena on military security of Islamic Republic of Iran. Defense Policy Magazine 22(85). Available at: https://journals.ihu.ac.ir/article_203188.html

Malmiran H (2009) Space, technology and its role in National Security of Islamic Republic of Iran. Strateg Stud Q 26:149–174. 7th year

Moltz JC (2019) The politics of space security; strategic restraint and the pursuit of National Interests, 3rd edn. Stanford University Press

Tarikhi P (2009) Iran's space program: riding high for peace and pride. Space Policy, at 3

Tarikhi P (2015) The Iranian space endeavor, ambitions and reality. Springer, Cham, pp 140–141

UN Treaties. United Nations Treaty Collection. Outer Space. Note Table 1 and also Table 2. Available at: https://treaties.un.org/Pages/ViewDetails.aspx?src=TREATY&mtdsg_no=XXIV-2&chapter=24&clang=_en

UNCTAD (2016) United Nations conference on trade and development. Science, technology and innovation policy review – Islamic Republic of Iran, United Nations publication UNCTAD/DTL/STICT/2016/3. United Nations. Switzerland. Available at: https://unctad.org/en/PublicationsLibrary/dtlstict20163_en.pdf

Zargar A (2010) Space and National Security; Iran's space developments enhancing National Security. Foreign Policy Q 3. Available at: http://fp.ipisjournals.ir/article_9512.html

UAE Approach to Space and Security

34

Naser Al Rashedi, Fatima Al Shamsi, and Hamda Al Hosani

Contents

Introduction	622
The Importance of Space to the UAE	623
National Regulatory Framework for the UAE Space Sector	625
The UAE's Contribution to International Space Cooperation	627
The UAE Space Economy	628
The UAE Contribution to Sustainability	629
The UAE Space Strategy's Contribution to Space2030 Agenda	629
The UAE Space Sector's Contribution to the 17 SDGs	634
The UAE Space Policy Contribution to the LTS Guidelines	642
LTS Guideline A: Policy and Regulatory Framework for Space Activities	643
LTS Guideline B: Safety of Space Operations	645
LTS Guideline C: International Cooperation, Capacity-Building, and Awareness	646
LTS Guideline D: Scientific and Technical Research and Development	649
Conclusions	650
References	651

Abstract

Space exploration and utilization have been able to enrich the human knowledge of the universe and contribute to facilitate and improve the daily lives of humans in different aspects, such as communication, earth observation, weather, and navigation. The leadership of the UAE has realized the increasing importance of the space sector and supported its continuous growth at a regional and global level.

Due to the increasing number of activities in the space industry, and depedency on the space applications, as well as the growing rate of national investments, combined with the technical, economic, and political developments in the sector,

---

N. Al Rashedi (✉) · F. Al Shamsi · H. Al Hosani
Space Policies and Legislations Department, UAE Space Agency, Masdar City, Abu Dhabi, United Arab Emirates
e-mail: N.alrashedi@space.gov.ae; F.alshamsi@space.gov.ae; H.Alhosani@space.gov.ae

© Springer Nature Switzerland AG 2020
K.-U. Schrogl (ed.), *Handbook of Space Security*,
https://doi.org/10.1007/978-3-030-23210-8_115

the UAE government decided to further consolidate, and grew these into a sustainable national space program; and establishing a federal authority to oversee, support, and promote the space sector in the UAE.

The UAE Space Agency is focused on establishing a national space policy and regulatory framework, building national human capacitiy in various space professional feilds, supporting R&D, expanding national space activities, and strengthening national and global partnerships to support the space program and sector. Moreover, the Agency actively works towards increasing and spreading the use of space technology and applications in the UAE, with aim of maximizing the benefits from them towards other sectors needs and national interests. As such, the agency helps the nation achieve its diversification plans and supports the creation of a knowledge-based economy. The UAE actively contributes in utilizing space to achieve long-term sustainability. In line with this mandate, the UAE National Space Strategy 2030 has been reviewed to ensure its alignment with the 17 Sustainable Development Goals (SDGs), the Long-Term Sustainability Guidelines, and the four pillars defined in the revised zero draft of the "2030" Agenda.

In this chapter, the UAE is proud to show the strong and diverse space sector, which over the past three decades managed to develop capacities and expertise qualifying it to compete globally and to move on to a new era in its national space program.

Introduction

Outer space refers to the expanse starting around 60 miles from the Earth's surface and reaching the other celestial bodies. It is made of a hard vacuum and contains low-density particles, mainly the plasma of hydrogen and helium. Besides that, outer space is a source of electromagnetic radiation, magnetic fields, cosmic rays, and a number of other elements that can pose threats. Nonetheless, space activity refers to any kind of human-carried activity in outer space.

It has been more than six decades since the Soviet Union launched the first satellite Sputnik 1, which marked the start of space activities for various purposes, including communication, scientific experiments, military uses, navigation and global positioning, and intelligence gathering, just to name the few. As the time passes, a vast majority of countries are becoming progressively more dependent on global satellite capabilities for national and international infrastructures. Without satellites, it would be impossible to achieve routine tasks such as airspace and ships' navigation, financial transactions, as well as to reap the benefits of modern telecommunications and the Internet. Hence, satellites have become essential tools, enabling social, economic, and scientific activities and bolstering private and public sectors' infrastructure. Outer space recognizes no concept of national borders, allowing all countries not only to utilize satellites for different purposes but to also work on boosting satellites' capabilities (Outer Space and Security (2014)).

Nevertheless, outer space is considered to be the most fragile system that exists due to its limited ability of self-recovery. Near-Earth space, which is the atmosphere, is the only known part of the outer space being able to take out satellites from orbit, while the objects stuck in the proximity of more than 800 km from the atmosphere will remain in the position for hundreds of years. Due to the ever-increasing volume of space activity, and resulting space debris, "near-Earth space" requires proper attention and protection for the benefit of all humankind (Moltz 2002).

Events such as deliberate destruction of obsolete satellites or other space objects without the consent of the other countries could further increase the amount of debris and pose a significant threat to the integrity of other satellites. One such event took place in 2007 when China conducted ASAT (anti-satellite test) to destroy one of its satellites. That event corroborated even further the need for stable use of the outer space and highlighted the critical challenges that can emerge for space activities (Issues in the International Community 2014). Similar activity was also conducted by India in March 2019, when launched the Ballistic Missile Defence, Interceptor missile, in ASAT missile test (Mission Shakti) engaging an Indian orbiting target satellite in Low Earth Orbit (LEO), and consequently turning the object into debris.

It is noted, however, that the first steps for outer space protection have been defined in the Outer Space Treaty of 1967. The treaty states that the exploration and use of outer space, including the moon and other celestial bodies, should be only conducted for the benefit of all peoples regardless of their economic or scientific development. On top of that, it encourages free access to all areas of space for all countries, as well as international cooperation, while it forbids the use of outer space for installation of any kind of weapons of mass destruction. Furthermore, it proclaims all States members bear responsibility for national space-related activities carried out by either governmental agencies or nongovernment entities. Also, each State Party to the Treaty that launches or procures the launching of an object into the outer space is liable for damage to another State Party (Outer Space Treaty of 1967 n.d.).

Ever since the launch of the first satellite, there has been an ever-increasing competition in space activities between developed nations. Moreover, influence and participation in space activities reflect an element of governmental power for these nations. In turn, this leads to the increasing interest for participation and investment in space activities among emerging and developing nations. Nowadays, there are more countries willing to invest in space activities than ever before.

The Importance of Space to the UAE

Space utilization and discoveries have been able to enrich the human knowledge of the universe and contribute to improving the daily lives of human beings. As such, communication, broadcasting, earth observation, and navigation applications using satellite data contribute remarkably to the conduct of daily human activities. In addition, space technology supports vital societal services and sectorial policies.

Moreover, the use of space-based assets plays a vital role in monitoring the weather, climate and environmental change, management of natural resources, crises and disaster management, as well as rescue and humanitarian aid programs. Also, many capacities and advanced technologies that were developed for space have been widely used in other domains such as medicine, energy, and manufacturing. Hence, the space sector has become a source of innovation and inspiration for human beings and, particularly, for the generations to come.

The good leadership of the United Arab Emirates (UAE) has realized the increasing importance of the space sector and its continuous growth at a regional and global level. This is why it has been investing in space since the early 1990s. Today the UAE is proud to have a strong and diverse space sector. The UAE has managed over the past three decades to develop space capacities and expertise qualifying it to compete globally and to move on to a new phase in its national space program. Thus, in 2014, the UAE Space Agency was created and the UAE Mars exploration program was announced.

Provided the increased national investments and growing activities in the field of space, combined with the technical, economic, and political developments in the space industry, the UAE government aims at establishing a national policy and regulatory framework for space activities. Such intention follows suit the international developments and is aligned with the UAE government's ambitions and higher interests that have manifested in the national agenda of the UAE Vision 2021 and Centennial 2071. The latter are aimed at enhancing the role of the space sector and, thereby, making the UAE among the best countries in the world with a stable and diverse economy providing for current and upcoming Emirati generations. This entails the transition toward a knowledge-based economy centered on innovation, high-level education, and increasing national expertise and qualifications.

Since its establishment, an annual budget is assigned to the UAE Space Agency by the UAE Federal Cabinet with the aim to improve the lives of its citizens, national security, crisis management, discovery of national resources, climate monitoring, diversification of the UAE economy, and further strengthen cooperation with the other States. In 2017, the UAE Space Agency made an agreement with Luxembourg on cooperation and exchange of information and expertise in the areas of space science, research, and technology (The Economic Times 2017). In addition, the UAE has taken a part in international cooperation initiatives for space activities led by China. As a matter of fact, the UAE has become part of "the Belt and Road" initiative and has consequentially become integrated in China's space-based infrastructure services, such as BeiDou Satellite Navigation System, satellite communications, and remote sensing (Hui 2018).

The UAE has also become a part of the "International Charter Space and Major Disasters." Moreover, the Mohammed Bin Rashid Space Centre has joined Sentinel Asia with the goal of supporting disaster management in the Asia-Pacific region by means of providing high-resolution satellite images. In November 2016 and 2017, in partnership with UNOOSA, the UAE held the High Level Forum on using space as a driver for sustainable development and signed an agreement on the peaceful uses of outer space (Space Security Index 2019).

The UAE has also become the first Arab country to fully manufacture its own satellite, KhalifaSat, which is one of the most advanced remote sensing satellites, allowing the UAE to establish its position on the global map of space activities (MBRSC 2019).

National Regulatory Framework for the UAE Space Sector

As per the UAE Federal Decree No. 1 of 2014, the UAE Space Agency has the mandate of overseeing and promoting the country's space sector and activities. Under such mandate the agency is responsible for the development of the so-called "National Space Framework" consisting of four main components, namely, Space Policy, Space Strategy, Space Law, and Space Regulations.

In 2016, the agency launched the National Space Policy, which clearly states the UAE principles and ambitions for its space program. The policy addresses the reasons that the UAE is investing in space and what the ultimate goals are. It consists of six main sections that summarizes the importance of outer space and explains the main purpose of the policy. It also explains how the space sector will contribute to the achievement of the national vision, priorities, and goals and defines the capabilities, guidelines, and governance system needed to support the achievement of the policy goals.

In March 2019, the National Space Strategy 2030 was approved, setting the general framework for the UAE's space industry and activities carried out by public and private entities for the years leading up to 2030. Following the Cabinet approval, the UAE Space Agency launched the National Space Strategy 2030, which translates the principles and ambitions in the policy into a set of national programs and initiatives to be implemented by the United Arab Emirates space sector by the year 2030. The UAE throughout the 18 programs of the National Space Strategy will work on achieving 71 initiatives. The strategy is structured around six strategic objectives, namely, competitive and leading space applications and services, advanced R&D and manufacturing capabilities, inspiring space scientific and exploration missions, high level of space awareness and expertise, effective local and international partnerships and investments, and adopting and enabling frameworks and infrastructure.

In December 2019, the National Space Law was approved and came into effect, hence setting the regulatory basis for space activities by covering the organization and objectives of space projects undertaken by the country, including peaceful space exploration and the safe use of space technologies. It also addresses new and complex concepts, such as the right to own resources found in space and organizing manned space travel and other commercial activities, such as asteroid mining.

Also, the Space Investment Promotion Plan aims to achieve these Policy directions by defining in a high level the UAE approach to facilitate more investments in space industry in the UAE, and contribute towards the attainment of UAE goals stated in the UAE Vision 2021, the UAE 2071, and the UAE Plan on the 4th Industrial Revolution. In particular to diversify and ensure sustainability of the UAE Economy; to promote highly productive knowledge-based economy; to provide incentives for R&D and innovation, and to encourage entrepreneurship and

SMEs role and opportunities in private sector. The Space Investment Promotion Plan is structured around four objectives:

1. Sustain the growth of the UAE Space Industry.
2. Increase the UAE space sector contribution to the diversification of national economy, and to knowledge-based economy.
3. Support other national strategic interests.
4. Promote partnerships at national and international levels.

The plan is further supported by two complementary pillars allowing UAE to propose a desirable and unique offering for Early- and Later-Stage space companies looking for VC funds compared to other locations worldwide. The first pillar is on creating an attractive environment for space industry and ecosystem, whereas the second pillar is on creating an investment vehicle and the need supporting entities through setting up a catalyst, then a platform to facilitate investment, creating UAE Space Angels to invest in seeds, and in creating Space Accelerator to help start-ups and entrepreneurships. Additionally, to support and enable the space sector, in 2020, the UAE Space Agency developed and approved the Remote Sensing Space Data Policy Guidelines for Institutional UAE Missions, where it:

1. Recalls principles, goals, and ambitions stemming from the UAE National Space Policy and elaborates these in the context of UAE Institutional RS space missions and data provision.
2. Contains voluntary guidelines that aim at providing a reference to UAE Institutional RS space mission owners/operators to develop their own data policies for their own missions.

These data policy guidelines leverage international best practices and include forward looking considerations. They have been developed for RS Institutional (civil or dual-use) satellite missions generating remote sensing data. They provide guidance to UAE RS institutional satellite owners/operators to develop their own data policies. Whenever meaningful, each main data policy topic is addressed in these guidelines by providing a best practice characterization from both an Open and a Restricted data policy approach. An UAE RS institutional satellite owner/operator reading these guidelines should be able to clearly see which data policy approach is most suitable to its mission(s) and find useful references that can be of assistance, or inspiration, in shaping its own data policy. A data policy always needs tailoring to the specificities of a given mission and its owner/operator objectives. Furthermore, the UAE Space Agency is working towards establishing a comprehensive legal framework and principles to regulate the national space sector, which aims to clearly define the requirements and the legislation for conducting space activities and the activities supporting the space activities in the UAE. Hence, the Agency is currently working in drafting a set of regulations and guidelines including the following: Regulation on Authorization of the Space Objects, Regulation on Registration of the Space Objects, Regulation on Human Spaceflight Activities, Space Debris Mitigation Guidelines, and Third Party Liability Insurance Guidelines.

The UAE's Contribution to International Space Cooperation

The United Arab Emirates acknowledges the significance of supporting international cooperation in the field of exploration and peaceful uses of outer space. This comes with placing importance on strategic discussions at international level that ensure aligned directions toward achieving the common goals.

With the aim to accelerate the dialogue on the role of space science and technology in achieving global development, the first High Level Forum (HLF) was held in Dubai, United Arab Emirates, during the period of 20 to 24 November 2016. The event was co-organized by the United Nations Office of Outer Space Affairs and the government of the United Arab Emirates. The forum provided a platform for the space community to exchange recommendations for the UNISPACE+50 blueprint and to share insights in regard to the global governance and the role of space science and technology in supporting the achievement of the global sustainable development in each of the four thematic pillars identified by the office, namely, space economy, space society, space accessibility, and space diplomacy.

The United Nations/United Arab Emirates High Level Forum resulted with the Dubai Declaration, which recommended the forum to continue to serve as a platform for the space community to exchange views on connecting the four pillars of the UNISPACE+50 and Space2030 and to encourage collaboration with the United Nations Office of Outer Space Affairs. There were a number of discussion sessions addressing the following (United Nations/United Arab Emirates High Level Forum: Space as a driver for socio-economic sustainable development 2017):

- The importance of coming up with solutions addressing the need to guarantee access to space and the availability of a regulatory instrument mechanism for cooperation to ensure benefits for global societies
- The achievement of goals set by the 2030 Agenda for sustainable development with the support of space and its role as a driver for social and economic sustainable development
- The importance of joining global efforts in the development of space, taking various aspects into account, such as space technology data and facilities
- The role of regulatory frameworks and international mechanisms at the national and international level in achieving cooperation in the peaceful exploration and use of outer space

In June 2018, the international community gathered in Vienna for UNISPACE +50, a special segment celebrating 50 years after the first United Nations Conference on the Exploration and Peaceful Uses of Outer Space during the 61st session of the Committee on Peaceful Uses of Outer Space (COPUOS). It offered an opportunity for the international community to agree on the upcoming steps toward better global space cooperation for the benefit of mankind.

The UAE followed the United Nations Committee on the Peaceful Uses of Outer Space (UNCOPUOS) endorsement for the UNISPACE+50 resolution "international cooperation in the peaceful uses of outer space," which was adopted by the General Assembly on 7 December 2018. The United Nations resolution A/RES/73/91 called for "strengthened international cooperation in the peaceful use of outer space and the global governance of outer space activities, and encouraged coordination to ensure that space science, technology and applications serve the Sustainable Development Goals." The resolution represents affirmation by the global community of the critical role that space can play in attaining the UN Sustainable Development Goals (SDGs) and moreover, of the needed for collaboration and partnerships at all levels in order to expand the opportunities and maximize and reap the benefits space could provide in this regard.

Therefore, the UAE government played a significant role in the preparation of the UNISPACE+50 and the development of the above remarkable UN resolution. The continued efforts of the UAE comes from its deep value and recognition of the essence of space and its significant impact on improving everyday lives, as well as its role in general as a driver for the socioeconomic sustainable development of the country, the Arab region, and the world as a whole.

The UAE Space Economy

It's worth noting that the UAE space sector has accomplished a number of significant developments at national, regional, and international levels. Most of those achievements were enabled by the continuous process of regulation and organization of the sector by the UAE Space Agency toward achieving its strategic goals. Besides, the UAE Space Agency is focused on raising national capabilities and the use of space technology in the UAE. As such, the agency helps the nation achieve its diversification plans and supports the creation of a knowledge-based economy.

What is more, the UAE has developed the most diverse space sector in the Middle East and North Africa (MENA) region, thus, representing a regional hub for space activities, services, technology, education, and events. Its sector consists out of over 52 entities contributing to the space economy, Including six main players, namely, the UAE Space Agency, the private entity YahSat/Thuraya, the National Space Science and Technology Center (NSSTC), Mohammed Bin Rashid Space Centre (MBRSC) Sharjah Academy of Astronomy, Space sciences & Technology (SAASST), and Khalifa University. One of the UAE's most ambitious missions, as well as the first of the kind in the region, is the Mars exploration mission, named "The Hope" (Building UAE Space Capabilities 2017).

When it comes to measuring the socioeconomic impact of the space industry, the space economy survey has been developed and conducted in partnership between the UAE Space Agency and the Federal Competitiveness and Statistics Authority since 2018. The purpose of the survey was to collect data meant to measure the National Agenda indicators in accordance with the UAE Vision 2021, and the UAE Centennial 2071, as well as a number of sectorial and strategic Key Performance Indicator (KPIs) concerning the space sector. The survey methodology followed best practices adopted by OECD. The following parameters have been analyzed: the contribution of the space sector to the UAE economy, the total number of staff employed in the

UAE space sector, expenditures on the UAE space sector, and expenditures on research and development (R&D) in the UAE space sector.

The survey provides important data for decision-makers in the country to take the appropriate action regarding monitoring and enhancing the competitiveness of the UAE space sector. The study indicates positive trends in expenditures on R&D and space exploration, where 38% and 63% increases have been recorded on the expenditures of the R&D and space exploration assets, respectively. Moreover, the study shows 41% increase in expenditures on commercial projects in the space sector, as well as significant increase in the percentage of women working in the 12 national space operators, which totaled 47% at the time when the study was conducted. Ultimately, the average community awareness of national space institutions is estimated at 65%, whereas awareness among students has been reported to be significantly above the average, equaling to 81%.

The UAE Contribution to Sustainability

The UAE Space Strategy's Contribution to Space2030 Agenda

In 2015, the world leaders adopted the 17 Sustainable Development Goals and their objectives for the sustainable development plan for 2030 (SDGs). It is worth mentioning the important role of the UAE in supporting these objectives and the role that space applications and services can play in achieving a number of these goals. Accordingly, in January 2017, the National Committee on SDGs was formed by the UAE Cabinet Decree No. 14 of 2017. The committee was chaired by Her Excellency Reem bint Ebrahim Al Hashimy, Minister of State for International Cooperation, and the Federal Competitiveness and Statistics Authority serving as vice chair and secretariat. As of 2018, the committee includes 17 members, which is led by FCSA and includes federal government entities, working toward the successful implementation of the SDGs. The mandate includes:

- Aligning the SDGs with the UAE's national development priorities and serve as a coordination body to implement the SDGs
- Undertaking regular follow-up and review of progress on implementation
- Managing domestic and international stakeholder engagement
- Coordinating the collection of official statistics, identifying new data sources, and assisting the National Statistics System to build capacity to monitor and report on SDG indicators
- Managing ad hoc SDG-related projects assigned by the Cabinet

In line with this mandate, the UAE National Space Strategy 2030 has been reviewed to ensure its alignment with the 17 Sustainable Development Goals (SDGs), the Long-Term Sustainability Guidelines, and the four pillars defined in the revised zero draft of the "Space2030" agenda, namely, space economy, space society, space accessibility, and space diplomacy. For each of the abovementioned pillars, the following table contains implementation areas and actions in the UAE space sector.

Space2030 Agenda pillar	Implementation area	Actions in the UAE
Space economy *(Development of space-derived economic benefits)*	Increased involvement of international financial institutions	National Space Investment Promotion Plan and UAE NewSpace Innovation Program. Government investment institutions for various projects (Mubadala-Dubai Future Foundation, ADHC, etc.)
	Global partnership in space exploration and innovation	The Mohammad Bin Rashid Space Centre is currently working in cooperation with the UAE Space Agency and in the University of Colorado on manufacturing the Hope probe for Mars exploration
	Governmental support for the stimulation of start-ups to trigger increased private investments	There are incubators and accelerators for start-ups and entrepreneurs Special support funds for different projects (angel investors, venture capital funds, private equity funds) Public and private funding Several special economic zones (such as free economic zones)
Space society *(Evolution of society and societal benefits stemming from space-related activities)*	Better links to be created within the space society and in support of existing efforts (user needs)	The contribution of the space sector to other sectors such as transportation and agriculture sectors The contribution of the space sector to entities working in other sectors

	Capacity-building and awareness-raising efforts	Four research centers specific to space sciences and astronomy:
		   Mohammed Bin Rashid Space Centre
		   Sharjah Academy of Astronomy, Space sciences and Technology
		   National Space Science and Technology Center
		   Center for Space Science, NYU Abu Dhabi

	Gender balance in the space sector	National gender balance policy (under progress) New initiatives in the National Space Strategy: to stimulate and retain space sector personnel

(continued)

Space2030 Agenda pillar	Implementation area	Actions in the UAE
		while taking gender balance into consideration
	Inclusion of youth	The ratio of youth working in space sector is high Summer space camps for students
Space accessibility *(All communities using and benefiting from space technologies)*	Establishment of a capacity-building network	The UAE has developed national capacities and expertise in satellite manufacturing, assembly, integration, and testing Partnerships with international prestigious manufacturing companies to support the educational purposes Establish partnerships to build capacities (the UAE Ministry of Education, ICT fund) Host summer camp for students in collaboration with international institutes (e.g., Mission to Mars summer camp in Space School Hamilton College, Australia)
	Conducting of space exploration and innovation in an inclusive manner	The UAE Astronaut Programme
	Support and allow access to space for developing countries	Supporting other nations to establish space agencies and programs
Space diplomacy *(Building partnerships and strengthening international cooperation in space activities)*	Enhancing the safety, security, and sustainability of outer space activities	The UAE hosts space security-related events (Space Security Forum)
	Strengthening the global and regional presence	The UAE hosts several world conferences that include the participation of different groups of the space society, to share the developments and enhance cooperation regarding space activities United Nations High Level Forum on the role of space as an engine for sustainable development Space Congress 2019 Seminar of the United Nations Office for Outer Space Affairs and ICAO on aviation-space regulatory issues Space Risk Forum

(continued)

Space2030 Agenda pillar	Implementation area	Actions in the UAE
		Flight and Space Summit The membership of the UAE in the Committee of the Peaceful Uses of Outer Space, International Telecommunication Union, Space Exploration Coordination Group, Space Research Committee, International Astronautical Federation (IAF), and Space Navigation Coordination Group
	Participation in the process of establishing new elements of international outer space policy and governance	Aligning with: 17 Sustainable Development Goals and their 169 purposes for the SDGs Long-Term Sustainability Guidelines International space exploration roadmap for 2040 (ISECG Roadmap) Principles of space exploration
	Efforts to maintain space governance	The UAE is part of 10 out of the 15 major international conventions governing the use and exploration of outer space including: The 1967 treaty on principles governing the activities of states in the exploration and use of outer space, including the moon and other celestial bodies (Outer Space Treaty) The 1968 agreement on the rescue of astronauts, the return of astronauts, and the return of objects launched into outer space (Rescue Agreement) The 1972 convention on international liability for damage caused by space objects (Liability Convention) The 1975 convention on registration of objects launched into the outer space (Registration Convention) ITU CS/CV ITU RR and ITRs 1971 ITSO 1976 IMSO 1976 ARABSAT

(continued)

Space2030 Agenda pillar	Implementation area	Actions in the UAE

		The UAE has developed the following national instruments:
		  National Space Policy 2016
		  National Space Strategy 2019
		  National Space Law 2019
		  Regulations and Procedures on Space Objects Registration 2018
		  Regulation on Human Space Flight 2019
		  Space Funding Policy 2017
		  Space Investment Promotion Plan 2019
		  Federal Decree for Application of Rescue Treaty 2017
		  Regulation on Space Activity Authorization [2020]
		  Regulation on Incident and Accident Investigation [2020]
		  Regulation on Technical Audit of operators [2020]
		  Guidelines for Space Data Policy [2020]
		  Space Debris Mitigation Guidelines [2020]

	Update existing instruments/legal documents to reflect the new space era	Revised Regulations and Procedures on Space Objects Registration Revised Regulation on Human Space Flight
	Build constructive, knowledge-based partnerships	The UAE Space Agency cooperates with several international and regional space agencies: the USA, France, China, India, Japan, South Korea, Italy, Germany, Kazakhstan, Saudi Arabia, Egypt, Bahrain, and others
	International efforts to address space debris threats	Development of guidelines to mitigate space debris Propose incentives to comply with space debris mitigation guidelines and practices.
	Increased information exchange for effective space debris mitigation	New initiatives in the National Space Strategy Effective participation in the development of international space policies, decisions, laws, and regulations with focus on space debris mitigation

The UAE's National Committee on SDGs conducted a mapping exercise between the UAE National Agenda, which includes a set of national indicators in different sectors, and the 2030 Agenda. At a thematic level, the 17 goals can be mapped to the pillars of the national agenda as shown in the following table:

#	National agenda	SDGs
1	Competitive knowledge economy	
2	Sustainable environment and infrastructure	
3	First-rate education system	
4	Cohesive society and preserved identity	
5	World-class healthcare	
6	Safe public and fair judiciary	

The UAE Space Sector's Contribution to the 17 SDGs

The UAE space sector is an active contributor toward achieving the sustainable development goals, where for each SDG, an indicator has been identified along with contribution description.

SDG 1: No Poverty

Indicator 1: Improved Communications as Driver for Growth
The United Arab Emirates owns and operates a number of communication satellites, namely, Thuraya Sats 1–3 for mobile satellite communications from 2000 to 2008; YahSAT Y1A for fixed communications and broadcasting, in addition to YahSAT Y1B for broadband services from 2011 to 2012; and YahSAT 3 in 2018 for broadband and broadcast with coverage of South America.

Indicator 2: Better Monitoring of Climate and Environment to Predict Crisis
An important breakthrough has been made in the area of modeling weather patterns. With the aid of supercomputer and data obtained by Himawari-8 satellite, Japanese scientists are able to measure the height of the top of clouds, which is of the essence for predicting wind and temperature. The program could have a strong impact in improving weather warnings, thus providing more time for evacuations (Japan forecasting breakthrough could improve weather warnings (2018)). Moreover,

Japan provided support in the creation and launching of Central America's first indigenously manufactured satellite used for monitoring carbon emissions in forests, as well as GhanaSat-1 cubesat, used for conducting research on monitoring illegal mining, water use, and deforestation (Successful deployment of five "BIRDS project" CubeSats from the "Kibo" 2017).

The UAE Space Agency and Mohammed Bin Rashid Space Centre are active members in the International Charter Space and Major Disasters. The International Charter is composed of space agencies and space system operators from around the world who work together to provide satellite imagery for disaster monitoring purposes. Types of disasters include cyclones, earthquakes, fires, floods, snow and ice, ocean waves, oil spills, volcanoes, and landslides.

In addition, the United Arab Emirates owns and operates a number of remote sensing satellites which supports the monitoring of environment, namely, the UAE remote sensing satellite DubaiSat-1, launched in 2009 with 4m resolution; the UAE remote sensing satellite DubaiSat-2, launched in 2013 with 1m resolution; and the UAE remote sensing satellite KhalifaSat, launched in 2018 with 0.6m resolution.

Furthermore, the UAE Space Agency is currently working along with academic institutes and research centers on a number of satellites and cubesats to support the monitoring of climate and environment such as MeznSat.

Indicator 3: Better Logistics Management
Earth observation and remote sensing applications provide important information and data for policy-making and decision-making, particularly in the areas related to the protection of the environment and the management of disasters and crisis. Today the UAE owns and operates a number of Earth observation satellites where the latest is KhalifaSat which was launched in 2018 with 0.6m resolution. Nonetheless, there are ground station facilities for transmitting space-related data and providing value-added service. There are also other supporting services for rapid communication and modern information technologies and applications; transportation of all types, especially by air; availability of cargo and logistics services; customs services such as import and export; and travel and tourism.

Indicator 4: Offering Businesses and Jobs Opportunities
Reference to the economic survey carried out by the UAE Space Agency and the Federal Competitiveness and Statistics Authority, the following results indicate that 22 entities working in the space sector have positively contributed to increasing business offerings and job opportunities (results for 2017) with a total workforce of 1513 employees, 37.4% of them UAE Nationals, 15% females, and 82.5% in specialized posts.

SDG 2: Zero Hunger

Indicator 1: Optimized Agriculture
The new 813 hyperspectral satellite is funded by the UAE Space Agency and will be developed by Arab engineers at the National Space Science and Technology Center at the United Arab Emirates University in Al Ain. The development of the satellite

will take 3 years and will have a lifespan of about 5 years. The planned launching year is 2023–2024. It will also have a polar orbit of 600km.

The data will be sent to a ground station in the UAE and receiving stations in some Arab countries for the benefit of a number of environmental authorities, municipalities, and institutions concerned with the agricultural sector and urban planning industry.

Indicator 2: Better Emergency and Aid Plans and Responses

Space utilization plays a vital role in monitoring the weather, climate, and the environment; management of natural resource; crisis and disaster management; and rescue and humanitarian aid programs. Also, effective utilization of available space capacities in the UAE is in constant improvement; this is through enhancing coordination among local institutions that offer space services and applications, and the governmental entities concerned with natural disasters and national crisis management.

Indicator 3: Better Resources Management

The UAE National Space Strategy 2030 includes an initiative that aims at enhancing national utilization of space services and capacities through integration between different space applications such as communications, Earth observation, remote sensing, and navigation. It seeks to enhance their integration with ground applications of communication, navigation, remote sensing, and others. It also aims to reach new applications and innovative solutions that support governmental, commercial, and research interests in different fields, such as transportation of all kinds, natural resources management, surveillance, energy, and the environment. Some of these initiatives are already reflected in several national projects such DubaiSat, KhalifaSat, and DM SAT to support better management of resources including urban planning.

SDG 3: Good Health and Well-Being

Indicator 1: E-Health, Including Telemedicine and Medical Teletraining and Learning

There are many health portals that enable patients in the United Arab Emirates to view their health profiles, medical results, and book appointments online. An example is the smart patient portal system provided by the Ministry of Health and Prevention. A second example is e-Malaffi portal by Seha in Abu Dhabi. In addition to Salama Electronic Medical Record System which is under process by Dubai Health Authority (Healthcare providers 2019).

Indicator 2: Monitoring Public Health via EO Applications

With the adoption of the 17 Sustainable Development Goals, and considering the important role of the UAE in supporting these objectives, as well as the role that space applications and services can play in achieving public health, efforts are being made toward enhancing the coordination between space service providers for Earth observation, communications, and broadcasting, together with the concerned entities in the UAE in the fields of environment, climate change, natural resource management, health, human aid, and sustainable development in general. This is both to identify the needs of these entities and to identify the opportunities to utilize the space applications and services in addressing these needs.

SDG 4: Quality Education

Indicator 1: Telelearning (Distance Learning)

In October 2018, the United Arab Emirates government announced the official launch of Madrasa, a distance e-learning platform that provides 5000 free Arabic-translated videos in basic science, math, biology, chemistry, and physics. It also provides 11 million words on educational content to students from kindergarten to grade 12. Through satellite communications, more than 50 million Arab students from around the world are able to access to the online platform, hence establishing the foundations of self-learning, along with traditional educational institution, in addition to creating a new generation of qualified Arab researchers, scientists, and innovators that are capable of building knowledge-based societies and better future for their countries (eLearning, mLearning and distance learning 2019).

Indicator 2: Driver for STEM/ STEAM Education

The space industry is closely linked to science, mathematics, engineering, technology, and design. Therefore, the development of specialized space competencies requires appropriate and attractive educational programs in these areas or the so-called "STEAM education." The space industry is also an attractive area for students to participate in science, math, engineering, technology, and design activities as the industry is inspired by innovative ideas.

The UAE Space Agency supported and funded the creation of a remote sensing data clearing center at the UAE Space Agency. The goal is to establish a state-of-the-art remote sensing data clearing center at Zayed University to lead the development of Emirati scientists that will be in a position to utilize the data generated by Emirates Mars Mission as well as other sources of Earth observation data (multispectral and hyperspectral imaging) to solve environmental challenge.

SDG 5: Gender Equality

Indicator 1: Women's Active Role in Space Exploration, Science, Industry, Policy, and Diplomacy

The UAE National Space Strategy 2030 includes an ongoing initiative focusing on stimulating and retaining space sector personnel while taking gender balance into account. The space professionals acquire high and unique skills and expertise that are compatible with space projects and various practical experiences. This initiative aims at preserving these competencies in the space sector especially the female competencies in the field as they constitute more than 40% of the workforce.

Reference to the economic survey carried out by the UAE Space Agency and the Federal Competitiveness and Statistics Authority (results for 2017), 47% of workforce in the top 12 UAE local entities in the space sector are females.

SDG 6: Clean Water and Sanitation

Indicator 1: Water Management, Detection, Pollution Monitoring, and Distribution (Network Planning and Monitoring Logistics)

The satellites images obtained through remote sensing have numerous applications about the water, for example, measuring water temperature. As such, the images that will be obtained from the hyperspectral satellite (813), which will be developed by engineers from the Arab region, will be used for that purpose among others.

SDG 8: Decent Work and Economic Growth

Indicator 1: Space Services Enable Other Businesses Opportunities

Recently, there has been an increasing number of emerging space programs in African and Latin American countries focusing on socioeconomic development and environmental monitoring. Examples of those programs include the launch of Algerian Alcomsat-1 satellite used for providing the Internet for neighboring countries and satellites launched by countries including Argentina, Costa Rica, and Ghana used for environmental and disaster monitoring. Moreover, four new countries (Australia, Egypt, Kenya, and New Zealand) have launched their national space agencies, thus accomplishing the first step of "space technology ladder." Other countries making significant investment in their space program include Canada, South Korea, and Saudi Arabia (West and Stocker 2019).

SDG 9: Industry, Innovation, and Infrastructure

Indicator 1: Private Sector Providing Various Space Services and Products

The GeoTech Innovation Program is a collaborative project between the UAE Space Agency, and KryptoLabs is an innovation program that encompasses the selection of high potential entrepreneurs that can develop an application using satellite data in areas such as urban and rural land management, crisis and disaster management, and coastal border security.

Another program is the NewSpace program, which is a global space industry accelerator that aims to provide the support for entrepreneurs and start-ups in the space industry to align with the UAE Space Agency goals and objectives. Therefore, the program covers the value chain of entrepreneurship and innovation from ideation to growth to maximize the value-added for all beneficiaries and ensure high impact and returns.

Indicator 2: Innovative Launch Technologies

The global space sector is witnessing significant developments in the capabilities and techniques in the field of launch vehicles in terms of reuse of vehicles or whether these vehicles are very light or heavy. The small satellite revolution, the increase in interest by governments in space-related activities, the increase in demand for launch services, and the long waiting period in many cases have contributed to the remarkable commercial/private sector involvement in launch activities. Also, during the next phase, an increase in competition between the new and established institutions in the field is expected.

The Nation Space Strategy 2030 includes an initiative that aims to enhance the opportunity for the UAE to benefit from this boom in launch activities and to examine the best opportunities for investment whether in the development of launching vehicles, launching platforms, or operational capabilities in order to achieve economic return as well as guarantee and enhance access to space-related technologies.

Indicator 3: Innovative EO, Telecom, Navigation Applications

The USA has been working on steadily improving the next-generation Global Positioning System, and the newly launched SV01 is expected to offer improved future connectivity worldwide for commercial and civilian use. Similarly, recently launched ESA's Galileo constellation of satellites is expected to deliver improved global coverage and more accurate pinpointing of Earth's locations. Chinese BeiDou satellites, which can offer accuracy comparable with GPS, were launched for providing communication and navigation services. Furthermore, as of 2017, India was in a process of completing its plans of launching its own GPS-type services for mobile users. In cooperation with GPS, the new Japanese positioning system, developed by joint efforts of JAXA and Mitsubishi Heavy Industries, is promising to reduce positioning errors to just a few inches (New orbiters for Europe's Galileo satnav system 2017) (India Plans to Roll Out National GPS Next Year 2017) (Kaneko and Perry 2017).

SDG 11: Sustainable Cities and Communities

Indicator 1: EO Data for Safety, Disaster Management, Pollution and Climate Change Monitoring, Energy Management and Land Use Planning, and Monitoring Cultural Heritage

The Italian-Argentine System of Satellites for Emergency Management is expected to use their SAOCOM 1A to create risk maps of plant diseases and flood recovery plans and detect humidity levels (Pacheco 2018). On top of that, there have been a significant number of the other recently launched satellites having the mutual goal of improving disaster planning and agriculture, as well as for controlling deforestation and climate changes. Some of them include South African EOSat1, Venezuelan VRSS-2, and Algerian Alcomsat-1.

Mohammed Bin Rashid Space Centre was established in 2015, encompassing previously existing Emirates Institution for Advanced Science and Technology (EIAST) as one of its affiliated institutions. So far, it has developed and launched three satellites, namely, DubaiSat-1, DubaiSat-2, and KhalifaSat, as well as Nayif-1, the first ever cubesat designed by Emirati engineers. Besides being the hub for promoting space science and scientific research in the UAE and the whole region, it also provides significant support for other causes, including natural disasters management, rescue missions, and environmental monitoring (The UAE Space Sector 2018).

Indicator 2: Satcom for Telecom Services

Al Yah Satellite Communications Company was established in 2007 and currently owns 5 fully operational satellites covering more than 160 countries. Its mission is to provide exceptional commercial, governmental, and military services, while at the same time focusing on further growth, empowerment of human capital, and quality enhancement. Thuraya Telecommunications Company was established in 1997 with the goal of providing global mobile communication services. It covers two third of the Earth, and its most recent achievement is the production of the world's first Android based satellite and GSM phone (Thuraya 2019).

SDG 12: Responsible Consumption and Production

Indicator 1: Sat/Nav for Logistics Management in Production

The UAE has support services for the space industry and its activities such as communications, transportation, shipping, logistics, customs clearance, import and export, and travel and tourism.

Organizations operating in the UAE space sector, including research and development centers, rely on the availability of services that support their work to carry out their activities and space projects. Such services include high-speed telecommunication services, modern information technology and transportation, aviation, logistics, customs services, import and export, and travel and tourism services. As for start-ups, it is important to provide incubation facilities and accelerators for innovation and entrepreneurship. The quality and low cost of these services are attractive elements for these institutions. For that, the UAE National Space Strategy 2030 includes an initiative focusing on evaluating and providing recommendations on current and future needs for infrastructure, facilities, and other logistical services supporting institutions and space activities.

SDG 13: Climate Action

Indicator 1: Earth Observation Data Key for Climate Change, Pollution Monitoring, and Mitigation Strategies

The research in predicting harmful space weather has been active in the recent time as well. Agencies such as NOAA, USAF, and ESA are actively working on studying space weather events. NOAA and USAF jointly run the center for potential disturbances of people and equipment in space environment, while ESA operates a warning network for monitoring solar storms. The US space weather program currently has a technology for predicting and warning about severe solar storm half an hour before they happen. The World Meteorological Organization (WMO) is in the process of integrating space weather effort into its integral part of its work and facilitation of coordination with external players, which should result in improved space weather service capabilities (Sutherland 2015).

SDG 14: Life Below Water

Indicator 1: Space Science and Technologies for Efficient Use of Water Resources Including Preservation

Programs such as Copernicus, jointly run by ESA and European Commission and Italy's COSMO-SkyMed, utilize the space-based capabilities for observing Earth's surface, ocean levels, and climate (Zosimovych and Chen 2018). The European Global Navigation Satellite Systems Agency holds a responsibility for the operation of Galileo satellites used for global navigation. In cooperation with Japan, Europe is leading BepiColombo mission with the aim of examining Mercury's magnetic fields and polar regions for the presence of water (DEPICOLOMBO n.d.).

SDG 15: Life on Land

Indicator 1: EO Data for Biodiversity Monitoring, Pollution Monitoring, and Land Use Management and for Compliance and Policing

The UAE Space Agency is working on a land cover and land use mapping for the whole UAE in collaboration with a number of national stakeholders. The goal of this project is to develop a land cover and land use map product for the entire UAE using the high resolution of satellite images. This will lead to the development of local capabilities in the field and positive contribution toward updating the developed maps.

SDG 17: Partnerships for the Goals

Indicator 1: PPP at Local Levels (Public + Industry + R&D + Academia)

There are a number of partnership projects between local UAE entities. First project is between the UAE Space Agency and Khalifa University in manufacturing a scientific and educational satellite for studying the Earth's atmosphere (MiznSat) where the launching year is 2020. The second project is between the UAE Space Agency and UAE University in manufacturing a scientific and technology testing satellite focusing on Global Navigation Satellite System (GNSS) to be launched in 2020. Third is a project between the UAE Space Agency and Khalifa University and the National Authority for space science in Bahrain in manufacturing a scientific and educational satellite to be launched in 2020 as well. Fourth is a project between YahSat and Khalifa University in manufacturing MySat, an educational satellite to be launched in 2021. Fifth is a project between Mohammed Bin Rashid Space Centre and Dubai Municipality in manufacturing DMsat, an educational and services satellite to be launched in 2019. Sixth is a project between the UAE Space Agency and the National Space Science and Technology Center in manufacturing an educational and services hyperspectral satellite. Planned launching year is 2023–2024.

Indicator 2: Space and non-Space Actors

The UAE generally enjoys strong cooperation and partnerships at the regional and international levels, especially in political, economic, and commercial terms, hence supporting its space-related cooperation and companies.

Also, reference to the economic survey carried out by the UAESA and FCSA (results for 2017), the number of sectors benefiting from space services and activities is 17. Examples of benefiting sectors include telecommunications, education, energy, tourism and leisure, high-tech industries, and agriculture. It is worth noting that the number of contracted entities benefiting from space services and activities is 709, where 268 are within the UAE and 441 are outside UAE.

Indicator 3: International Collaboration and Partnerships

The UAE Space Agency cooperates actively with several international and regional space agencies such as the USA, France, China, India, Japan, South Korea, Italy, Germany, Kazakhstan, Bahrain, and others. The UAE have also signed more than 25 MoUs/agreements with international space partners for carrying out projects feeding toward the national space program, and its initiatives, and is a member of the Committee of the Peaceful Uses of Outer Space, International Telecommunication Union, Space Exploration Coordination Group, Space Research Committee, International Astronautical Federation (IAF), and the Space Navigation Coordination Group

The UAE had the privilege to host a number of prominent satellite events such as the United Nations High Level Forum on the role of space as an engine for sustainable development back in 2015, the seminar of the United Nations Office for Outer Space Affairs and ICAO on aviation and space regulatory issues, the Space Risk Forum, and the Flight and Space Summit, including others.

The UAE Space Policy Contribution to the LTS Guidelines

The long-term sustainability of outer space activities is defined as the ability to maintain the conduct of space activities indefinitely into the future in a manner that realized the objectives of equitable access to the benefits of the exploration and use of outer space for peaceful purposes, in order to meet the needs of the present generations while preserving the outer space environment for future generations (Guidelines for the Long-term Sustainability of Outer Space Activities 2018). As such, in June 2016, the committee on the peaceful uses of outer space agreed to a first set of guidelines for the long-term sustainability of outer space activities addressing policy, regulatory, operational, safety, scientific, technical, international cooperation, and capacity-building aspects. In June 2019, a set of nine additional guidelines for long-term sustainability of outer space activities of the Committee of the Peaceful Uses of Outer Space were adopted (Long-term Sustainability of Outer Space Activities 2019).

UNCOPUOS encourages states and international intergovernmental organizations to voluntarily take measures to ensure that the guidelines are implemented to the greatest extent. Accordingly, the United Arab Emirates policy efforts in achieving the guidelines are highlighted below.

LTS Guideline A: Policy and Regulatory Framework for Space Activities

Guideline A.1: Adopt, Revise, and Amend, as Necessary, National Regulatory Frameworks for Outer Space Activities

The open nature of space requires a robust international legal framework to ensure the harmonization of domestic space laws among nations with space programs. The UAE Space Agency works to represent and reflect the UAE's interests to the international community through active participation and contribution to key international organizations and forums related to outer space. Simultaneously, the UAE will continue to respect international laws and norms when developing and maintaining domestic legislation and regulations.

The UAE is currently a signatory to four out of the five main international space treaties, namely:

- Outer Space Treaty 1967: Treaty on Principles Governing the Activities of States in the Exploration and Use of Outer Space, including the Moon and Other Celestial Bodies
- Rescue Agreement 1968: Agreement on the Rescue of Astronauts, the Return of Astronauts and Return of Objects Launched into Outer Space
- Liability Convention 1972: Convention on International Liability for Damage Caused by Space Objects
- Registration Convention 1975: Convention on Registration of Objects Launched into Outer Space

The UAE has issued more than eight national space policy and regulatory documents, including the National Space Policy, the National Space Strategy 2030, the Federal Law on Regulating the Space Activities, and several others.

Guideline A.2: Consider a Number of Elements when Developing, Revising, or Amending, as Necessary, National Regulatory Frameworks for Outer Space Activities

The UAE Space Agency is responsible for drafting the legal framework for the UAE's space sector. The National Space Law will be the first of its kind in the region, providing a legislative and legal framework for the space sector that is in line with the UAE federal policies and international laws and regulations.

The National Space Law, which was approved in December 2019, covers the organization and objectives of space projects undertaken by the country, including peaceful space exploration and the safe use of space technologies. It also addresses new and complex concepts, such as the right to own resources found in space, organizing manned space travel, and other commercial activities, such as asteroid mining.

The following considerations were taken into account when drafting the UAE National Space Law:

- The provisions of the General Assembly Resolution 68/74, on recommendations on national legislation relevant to peaceful exploration and use of outer space
- The Space Debris Mitigation Guidelines of UNCOPUOS.
- Potential risks associated to implementing space activities
- Promoting regulations and policies that support sustainability of space and the Earth
- Guidance contained in the Safety Framework for Nuclear Power Source Applications in Outer Space
- Potential benefits of using existing international technical standards, including those published by the International Organization for Standardization (ISO), the Consultative Committee for Space Data Systems, and national standardization bodies
- Legal capacities of imposed regulations
- Advisory input from affected national entities during the development process
- Existing relevant legislations

Guideline A.3: Supervise National Space Activities

The UAE Space Agency is a federal agency that was created under Federal Law by Decree No. 1 of 2014. The agency has the mandate to oversee the space sector activities and develop and roll out national space polices and legislations.

The UAE Space Agency works to represent and reflect the United Arab Emirate's interests to the international community through active participation and contribution to key international organizations and forums related to outer space. Simultaneously, the UAE continues to adhere to and respect international laws and norms when developing and maintaining domestic legislation and regulations.

Guideline A.4: Ensure the Equitable, Rational, and Efficient Use of the Radio Frequency Spectrum and the Various Orbital Regions Used by Satellites

The UAE has been able to develop national competencies specialized in the areas of spectrum management in general and the management of space frequencies and orbits, in particular. That is done through the programs and initiatives of the National Space Strategy 2030, which includes program 6.3 that focuses on effective management and coordination of the interests of the space sector regarding the radio spectrum and orbital positions. The program includes a number of initiatives such as the identification of frequency bands and orbital positions of priority to the UAE's space-related activities and the development of standards, procedures, and capacities to increase the efficiency of spectrum and orbital use as well as situational awareness.

Guideline A.5: Enhance the Practice of Registering Space Objects

In order to maintain an up-to-date record of UAE space objects and to fulfill the obligations under the 1976 Registration Convention, the UAE Space Agency is responsible for the development and maintenance of the National Register of Space Objects and also responsible for notifying, via the Ministry of Foreign Affairs, the United Nation Office of Outer Space Affairs (UNOOSA) for the registration and updating the statuses of the UAE space objects in the International Space Object Register.

LTS Guideline B: Safety of Space Operations

Guideline B.1: Provide Updated Contact Information and Share Information on Space Objects and Orbital Events

In 2018, the UAE issued the Space Objects Registration Regulation, and in 2019, the UAE Space Agency ensured the 100% registration of the UAE Space Objects at the UNOOSA Register of Space Objects.

Guideline B.2: Improve Accuracy of Orbital Data on Space Objects and Enhance the Practice and Utility of Sharing Orbital Information on Space Objects

The UAE, when sharing orbital information on space objects and operators, ensures using commonly used method that is compliant with international standards, and hence facilitating wider distribution and awareness of current and predicted location of space objects.

As such, and in regard to space debris object WT1190F, an airborne observing campaign was organized to practice the rapid response to announced small asteroid impacts by the UAE Space Agency and in collaboration with International Astronomical Center, NASA, and ESA. Also, in 2016, the UAE Space Agency established the Science, Technology and Innovation Roadmap where space debris and meteorites were flagged as an area of interest.

Guideline B.3: Promote the Collection, Sharing, and Dissemination of Space Debris Monitoring Information

Due to the ever-increasing volume of space activity, and resulting space debris, nearby outer space already requires a special treatment and protection from the entire humankind. Consequently, the UAE Space Agency completed the development of guidelines to mitigate space debris. Also, the National Space Strategy 2030 includes an initiative focusing on effective participation in the development of international space policies, decisions, laws, and regulations with focus on space debris mitigation.

Guideline B.4: Perform Conjunction Assessment During All Orbital Phases of Controlled Flight

In compliance with Article VI of the 1967 Outer Space Treaty, the UAE ensures that conjunction assessment is conducted by all space craft operators during all orbital phases of controlled flight for their current and planned spacecraft trajectories.

Guideline B.7: Develop Space Weather Models and Tools and Collect Established Practices on the Mitigation of Space Weather Effects

The National Space Strategy 2030 confirms the United Arab Emirates commitment for the next decade toward enhancing space R&D activities, capacities, and efforts, namely, in launching research projects in high national priority fields with a focus on conducting research on best practices and technologies to manage space risks and space traffic and to track and reduce the effects of space debris and space weather.

Guideline B.9: Take Measures to Address Risks Associated with the Uncontrolled Reentry of Space Objects

The National Space Strategy 2030 confirms the United Arab Emirates commitment in ensuring a supporting legislative framework and infrastructure to match the future developments in the sector, namely, in establishing an investigation mechanism for space-related accidents and incidents during their launch or reentry. The mechanisms investigate these cases efficiently and effectively to identify their causes, the lessons learned from such incidents, and the opportunities for improvement.

Guideline B.10: Observe Measures of Precaution when Using Sources of Laser Beams Passing Through the Outer Space

Through the National Space Strategy 2030, program 6.3 on effective management and coordination of the interests of the space sector regarding the radio spectrum and orbital positions includes an initiative focusing on the development of standards, procedures, and capacities to increase the efficiency of spectrum and orbital use as well as situational awareness. As such, the UAE working team will consider issues related to the management of space frequencies, such as the use of laser and frequency management on the surfaces of other celestial bodies.

LTS Guideline C: International Cooperation, Capacity-Building, and Awareness

Guideline C.1: Promote and Facilitate International Cooperation in Support of the Long-Term Sustainability of Outer Space Activities

In March 2019, the Arab Space Cooperation Group was announced, setting a huge milestone toward sharing of experiences related to long-term sustainability of outer space activities, expertise, and information exchange. The prime objective of the group is to exchange knowledge, boosting the Arab space industry and working on joint projects. The group's first project "813" will be a remote sensing/Earth observation satellite built by Arab space specialists from all countries in the group. It will aim to tackle climate and environmental issues in the Arab world and other parts of the globe.

The National Space Investment Promotion Plan, announced in 2019, provides a high-level approach to facilitate more investments in the United Arab Emirates space industry and contribute toward the alignment of the government goals stated in the UAE Vision 2021, the UAE 2071, and the UAE Plan on the fourth industrial revolution, in particular, to diversify and ensure sustainability of the UAE economy, to promote highly productive knowledge-based economy, to provide incentives for R&D and innovation, and encourage entrepreneurship, and SMEs role and opportunities in the private sector. Additionally, The UAE NewSpace Innovation Program is a joint initiative by the UAE Space Agency and Krypto Labs to accelerate the growth of four tech businesses in the field of NewSpace, which refers to the rise of the private spaceflight industry that aims to make space more accessible, affordable, and commercial, for scientists and the general public. The program falls under the National Space Investment Promotion Plan which aims to heighten the role of the space industry in contributing to the economy of the UAE, while encouraging a culture of interest in the space sector, in efforts to establish a knowledge-based competitive national economy built on innovation and the latest technologies.

Moreover, and in accordance with the 2018 space economy survey conducted by the UAE Space Agency, the total number of entities working in the United Arab Emirates space economy is 52, with more than 3100 people working in the UAE space economy; 40% of them are below 35 years. To date, in more than 17 sectors, 1232 contracted entities are benefiting from the UAE space services and applications, in particular more than 750 from inside the state, whereas 480 and from international entities. Nevertheless, the main and most benefiting sectors are telecommunications, education, data analytics and LBS, science and R&D, tourism, transportation, and urban planning.

Guideline C.2: Share Experience Related to the Long-Term Sustainability of Outer Space Activities and Develop New Procedures, as Appropriate, for Information Exchange

In alignment with the United Arab Emirates efforts to promote international cooperation to enable all countries through the exchange of expertise in developing national space policies and regulations in support of these guidelines, a number of knowledge transfer workshops were conducted during 2019, namely, with space agencies in Egypt, Kingdom of Saudi Arabia, and Republic of Azerbaijan.

The National Space Strategy 2030 confirms the United Arab Emirates commitment for the next decade toward achieving relevant strategic objectives, namely, in creating space culture and expertise with 3 supporting programs and 11 initiatives, as well as in promoting effective local and global partnerships and investments in the space industry by outlining 3 programs and 15 initiatives.

To date, the UAE Space Agency has signed more than 30 agreements and cooperation MOUs to promote and facilitate international cooperation with space agencies and organizations in support of the long-term sustainability of outer space activities. These agreements create the basis for further collaborative activities between countries in mutually agreed projects, and priority areas.

Guideline C.3: Promote and Support Capacity-Building

The Hope Mars Mission is a planned space exploration probe mission to mars which is funded by the UAE Space Agency and is built by the Mohammed Bin Rashid

Space Centre, the University of Colorado, and Arizona State University. The mission is set for launch from Japan in the summer of 2020 with the aim of enriching the capabilities of Emirati engineers and increasing human knowledge about the Martian atmosphere. This is An example of a collaborative scientific project with foreign research institutions, in contribution toward a knowledge-based economy.

The UAE Space Agency undertakes efforts to support capacity-building initiatives and promote new forms of national, regional, and international cooperation to improve space-related expertise and knowledge. The UAE government sponsored more than 100 distinguish students to study space fields in and outside the UAE, conducted in 2019 alone more than 17 space-specialized trainings and education camps and sessions, launched rocket assembly competition with 167 participants,

sponsored qualified students to join the NASA International Internship Program (NASA I^2), which provides an environment for US and non-US university undergraduate-level or graduate-level students to work collaboratively on NASA-relevant research with a NASA mentor, and also conducted a number of awareness and training workshops under the future scientists program in collaboration with Airbus, and provided sponsorship for several students to attend the Space Generation Congress and the science workshops.

Guideline C.4: Raise Awareness of Space Activities

The UAE is committed to increasing the effectiveness and diversity of its awareness-raising activities for the space industry in the state as well as for the space industry in general. It is ensured that all awareness activities target different segments of the public and highlight the achievements in this scientific and inspiring field.

It also seeks to coordinate efforts to achieve efficiency, diversity, and integration of awareness-raising programs. As such, the UAE hosted several world conferences that include the participation of different groups of the space society, to share the developments and enhance cooperation regarding space activities. To name few, in 2019, the UAE was the first Arab country to host the international conference Young Professionals in Space (YPS) which provided a valuable opportunity for Arab youth to connect with experts and scientist and share insights on future contributions to solutions for humanity.

Moreover, strategic global space industry leaders and more than 600 key space agencies, commercial space organizations, academic institutions, and users of space services were brought together on the sidelines of the Global Space Congress in 2019, an event hosted by the UAE Space Agency offering a unique experience on the biggest opportunities in the space sector by providing exposure to worlds most vibrant and energetic space programs.

Also, in 2019, the Humans in Space Symposium was organized by the Mohammed bin Rashid Space Centre where experts from around the world gathered to discuss topics such as the challenges in of future space flights, biology and biotechnology in space, and Mars exploration. The symposium offered a multidisciplinary discussions involving the exchange of research results and ideas in the field of space science, as well as sharing of UAE's recent experience and expertise in the area of human spaceflight.

In October 2020, the UAE will be the first county in the entire region to host one of the most premiere space events, the 71st International Astronautical Congress. The Mohammed bin Rashid Space Centre is committed toward making this event a sounding success with the attendance of world decision-makers in an iconic meeting of minds, working toward the advancement of space science and technology, and ultimately the betterment of all humankind.

The UAE will continue to raise awareness of space activities through initiative implementation of strategic objective 4 in the National Space Strategy 2030, namely, creating a high level of national space culture and expertise. One example is the development of an informative program to spread knowledge about the national achievements in space and about the space sector in general. This is achieved through the development of a diverse and renewable informative program to spread general knowledge about space among the different social groups, as well as the launch of an international campaign to promote awareness about the national space sector.

LTS Guideline D: Scientific and Technical Research and Development

Guideline D.1: Promote and Support Research into and the Development of Ways to Support Sustainable Exploration and Use of Outer Space

The United Arab Emirates is continuously working on the development and sustainability of ambitious activities in space sciences and exploration. It reconfirms its commitment in increasing the state contribution to the space science community.

The Emirates Space Innovation Group (ESIG) is a UAE Space Agency project designed to increase opportunities for national space activities and strengthen engagement throughout the local space sector with agency-funded missions and projects. ESIG aims to advance science, technology, and innovation on a local level and coordinate projects being implemented by the various space sector actors and stakeholders. ESIG is made up of government agencies, companies, universities, and research centers throughout the state. The group meets on a quarterly basis to discuss the latest updates related to a number of space projects and activities taking place nationwide. Meetings also cover ongoing projects and future funding of missions and research.

In 2017, The United Arab Emirates government announced its ambitious inspiring scientific space exploration project that aims at building the first complete inhabitable human settlement on the surface of Mars in 2117, with the objective of planting the seed today in the hope for future generations to reap the benefits, driven by its passion to learn to unveil a new knowledge. In order to achieve this ambitious program with excellence, a practical plan is being developed in partnership with national capacities and expertise, regional and international partnerships, as well as international scientists and experts opinions on the subject.

In 2018 alone, a total of 64 scientific research papers were published, and up to 24 space inventions were registered. To further support the sector, the National

Space Strategy 2030 includes a strategic goal with 3 programs and 11 initiatives, namely, launching inspiring space scientific and exploration programs.

In 25 September 2019, the UAE celebrated the launch of its first Emirati astronaut, as part of the UAE Astronaut Program, to the ISS where he conducted more than 16 scientific experiments. In July 2020, the UAE intends to launch the Hope Probe to Mars, the first deep space exploration mission in the Arab and Islamic world. The Mission aims to study the atmosphere of the red planet.

As part of the Mars 2117 Program, launched during the fifth World Government Summit, the Mars Science City project falls within the United Arab Emirate's objectives to contribute to the global scientific race to take people to Mars. This project will include advanced laboratories that stimulate the red planet's terrain and harsh environment through advanced 3D printing technology and heat and radiation insulation. It seeks to attract the best scientific minds from around the world in a collaborative contribution in the state to human development and the improvement of life. It also seeks to address global challenges such as food, water, and energy security on Earth.

Guideline D.2: Investigate and Consider New Measures to Manage the Space Debris Population in the Long Term

The UAE is committed toward investigating new measures to manage the ever-increasing accumulation of space debris in outer space. As such, the UAE Space Agency has completed two detailed studies on the topic where a comparative analysis of technical and legal considerations was conducted, resulting in reports containing a set of recommendations and worldwide best practices. In addition, assessing the development of new space-related activities technologies and applications, the necessary authorization system for these activities is being developed. Such procedures would also take into account measures necessary to maintain safety and stability in the space environment which will include the development of guidelines to mitigate space debris.

Conclusions

The present chapter has outlined the UAE's efforts and contributions in achieving space security, the sustainable development goals of Agenda 2030, and the long-term sustainability of outer space.

Space utilization and discoveries have been able to enrich the human knowledge of the universe and contribute to improving the daily lives of humans. The leadership of the UAE has realized the increasing importance of the space sector and has supported its continuous growth on a regional and global level, justifying its investment in the sector since the early 1990s. Today the UAE is proud to have a strong and diverse space sector, which over the past three decades has managed to develop capacities and expertise qualifying it to compete globally and to move on to a new phase in its national program for space.

The National Space Strategy 2030 aims at supporting the achievement of this national vision by the space industry with its different sciences, technologies, applications, and services. It also translates the space policy issued in 2016 into a group of programs and initiatives that the space sector in the state will work on executing during the coming decade, with the purpose of reaching the national ambitions drawn by the space policy.

The National Space Law, which was approved in December 2019, covers the organization and objectives of space projects undertaken by the country, including peaceful space exploration and the safe use of space technologies. It also addresses new and complex concepts, such as the right to own resources found in space, organizing manned space travel, and other commercial activities, such as asteroid mining.

Nevertheless, The UAE will continue to work on the development of regulatory environment attractive to the different space activities. Such would be transparent and futuristic, one that balances the safety, security, and environmental needs on one side and the economic, commercial, and innovative needs on the other. It will also enhance the efforts internationally in decision-making and the alignment between local and global regulations. It will continue to work alongside its national and international partners in implementing the necessary measures in achieving space sustainability. As a responsible national organization, the UAE Space Agency is committed to foster and dedicate its local competences and technology for the betterment of humankind.

References

Building UAE Space Capabilities (2017) UAE Space Agency. Retrieved from http://www.unoosa.org/documents/pdf/hlf/1st_hlf_Dubai/Presentations/51.pdf?fbclid=IwAR2DNXEd_kqhrehL1s Gi_q1XzaRfKR36ud44WzVAlUrz4vhpM8I4bhIW9i8

DEPICOLOMBO (n.d.). Retrieved from https://sci.esa.int/web/bepicolombo

eLearning, mLearning and distance learning (2019, September 08) Retrieved from government.ae: https://www.government.ae/en/information-and-services/education/elearning-mlearning-and-distant-learning

Guidelines for the Long-term Sustainability of Outer Space Activities (2018, June 27) Retrieved from United Nations Office for Outer Space Affairs: https://www.unoosa.org/res/oosadoc/data/documents/2018/aac_1052018crp/aac_1052018crp_20_0_html/AC105_2018_CRP20E.pdf

Healthcare providers (2019, December 12) Retrieved from government.ae: https://government.ae/en/information-and-services/health-and-fitness/healthcare-providers

Hui J (2018) The Spatial Information Corridor Contributes to UNISPACE+50. Retrieved from http://www.unoosa.org/documents/pdf/copuos/stsc/2018/tech-08E.pdf

India Plans to Roll Out National GPS Next Year (2017, July 07) Retrieved from GPS Daily: http://www.gpsdaily.com/reports/India_Plans_to_Roll_Out_National_GPS_Next_Year_999.html

Japan forecasting breakthrough could improve weather warnings (2018, January 17) Retrieved from Space Daily: http://www.spacedaily.com/reports/Japan_forecasting_breakthrough_could_improve_weather_warnings_999.html

Kaneko K, Perry M (2017, June 01) Japan launches its version of GPS satellite to improve location positioning. Retrieved from Reuters: https://www.reuters.com/article/us-japan-satellite/japan-launches-its-versionof-gps-satellite-to-improve-location-positioning-idUSKBN18S3WG

Long-term Sustainability of Outer Space Activities (2019) Retrieved from United Nations Office for Outer Space Affairs: https://www.unoosa.org/oosa/en/ourwork/topics/long-term-sustainability-of-outer-space-activities.html

MBRSC (2019) Retrieved from KhalifaSat: https://www.mbrsc.ae/satellite-programme/khalifasat

Moltz JC (2002) Future security in space: commercial, military, and arms control trade-offs. University of Southampton/Center for Nonproliferation Studies, Monterey. Retrieved from http://www.nonproliferation.org/wp-content/uploads/2016/09/op10.pdf

New orbiters for Europe's Galileo satnav system (2017, June 22) Retrieved from GPS Daily: http://www.gpsdaily.com/reports/New_orbiters_for_Europes_Galileo_satnav_system_999.html

Outer Space and Security (2014) In: J. M. Defense, Defense of Japan. pp 105–108. Retrieved from https://www.mod.go.jp/e/publ/w_paper/pdf/2014/DOJ2014_1-2-4_web_1031.pdf?fbclid=IwAR21XzkwrU2ZKjknxlfH6ox9NnaW8VK8qgS3WCfcWz2s0msqBtmgwZDBgig

Outer Space Treaty of 1967 (n.d.). Retrieved from https://history.nasa.gov/1967treaty.html

Pacheco C (2018, January 04) SpaceX is launching an Argentine satellite in 2018, and here's what it'll do. Retrieved from https://www.thebubble.com/spacex-is-launching-an-argentine-satellite-in-2018-and-heres-what-itll-do

Space Security Index (2019) Canada: Spacesecurityindex.org. Retrieved from http://spacesecurityindex.org/wp-content/uploads/2019/10/SSI2019ExecutiveSummaryCompressed.pdf

Successful deployment of five "BIRDS project" CubeSats from the "Kibo" (2017, July 07) Retrieved from Japan Aerospace Exploration Agency: http://iss.jaxa.jp/en/kiboexp/news/170707_cubesat_birds.html

Sutherland S (2015, November 6) New plans emerge to deal with looming risks of space weather. Retrieved from The Weather Network: www.theweathernetwork.com/news/articles/whitehouseesa-plans-promise-better-response-to-space-weather-risks/59482

The Economic Times (2017, October) Retrieved from UAE Space Agency signs MoU with Luxembourg: https://economictimes.indiatimes.com/news/science/uae-space-agency-signs-mou-with-luxembourg/articleshow/61024984.cms

The UAE Space Sector (2018, April). Retrieved from https://www.sasic.sa.gov.au/docs/default-source/5th-sa-space-forum-presentations/1015%2D%2D-naser-al-hammadi-and-abir-khater.pdf?fbclid=IwAR15eszd2AkXEEEDLYyK00u8gGzwfEjFavDjVx2WL_OYAH8FtYfXkSROtZY

Thuraya (2019). Retrieved from https://www.thuraya.com/x5-touch

United Nations/United Arab Emirates High Level Forum: Space as a driver for socio-economic sustainable development (2017). Retrieved from http://www.unoosa.org/documents/pdf/hlf/HLF2017/AnnouncementHLF2017.pdf

West J, Stocker W (2019, October) Space Security Index. Retrieved from http://spacesecurityindex.org/wp-content/uploads/2019/10/SSI2019ExecutiveSummaryCompressed.pdf

Zosimovych N, Chen Z (2018, September) CubeSat Design and Manufacturing Technique Analysis. Retrieved from IOSR Journal of Engineering: http://www.iosrjen.org/Papers/vol8_issue9/Version-5/A0809050106.pdf

Space Security in Brazil

Olavo de O. Bittencourt Neto and Daniel Freire e Almeida

Contents

Introduction	654
Space Security and Emerging Space Faring Nations	656
Brazilian Space Policy: An Overview	656
Domestic Regulatory Instruments	659
Brazilian Space Situational Awareness Initiatives	662
Prevention of Arms Race in Outer Space: Brazilian Perspectives	664
Conclusions	665
References	666

Abstract

The present chapter provides updated information regarding Brazilian space law, policy, and initiatives related to space security, considering applicable perspectives. The secure and sustainable access to, and use of, outer space is obviously important to the international community in general, but of particular relevance to emerging space faring nations, which tend to experience overreliance in relation to a small selection of space assets. Appraising the Brazilian perspective on space security contributes to a better understanding of the challenges faced by emerging space faring nations on that regard, acknowledging their particularities as well as possible synergies. To that end, this chapter, after a brief introduction, clarifies the importance of space security for emerging space faring nations. Then, an overview of related Brazilian space policy is presented, followed by an examination of space safety related regulatory instruments. Brazilian position on the prevention of an arms race in outer space is appraised, considering current international debates. Finally, reference is made to local space situational awareness initiatives, and future prospects are considered.

O. d. O. Bittencourt Neto (✉) · D. Freire e Almeida
Catholic University of Santos, Santos, Brazil
e-mail: olavo.bittencourt@unisantos.br; danielfreire@unisantos.br; da616@georgetown.edu

© Springer Nature Switzerland AG 2020
K.-U. Schrogl (ed.), *Handbook of Space Security*,
https://doi.org/10.1007/978-3-030-23210-8_131

Keywords

Space security · Space situational awareness · Space law and policy · Brazil

Introduction

Nowadays, space security emerges as a global concern, not only to governments but also to the private sector as well, decurrent from increasing worldwide dependability over a wide rage space activities, not only to governments, leading to the support of different but interrelated initiatives toward long-term sustainability of outer space (Casella 2009). Ranging from domestic legislation to industrial guidelines, those frameworks focus on major threats to space-based assets and systems, most specifically relating to condition and knowledge of the space environment, access to and use of outer space by different actors and security of space systems themselves.

One may affirm that space activities in general, whether encompassing space vehicles, artificial satellites, launching centers, collection or remittance of data, and their related applications, would clearly benefit from governance initiatives designed to address relevant vulnerabilities. Emerging space faring nations are no stranger to those complications, enhanced by insufficient redundancy, as far as space objects are concerned; therefore, appropriate measures have increasingly been under consideration, in different regulatory levels, deserving to be properly accessed in view of possible synergies.

Domestic guidelines on space security initiatives, as verified in certain emerging space faring nations, represent relevant initiatives fully supported by International Law. In fact, the fundamental framework provided by applicable multilateral treaties is in a straight line with the prerogative of States in relation to their rights of self-defense and of autonomous regulation of national space activities.

A brief assessment of certain international instruments is justified for our study, starting from the United Nations Charter, of 1945. Accordingly, the first purpose of this important international organization is solemnly stated as being "*to maintain international peace and security, and to that end: to take effective collective measures for the prevention and removal of threats to the peace, and for the suppression of acts of aggression or other breaches of the peace, and to bring about by peaceful means, and in conformity with the principles of justice and international law, adjustment or settlement of international disputes or situations which might lead to a breach of the peace*" (Article 1, para. 2 of the United Nations Charter 1945).

Peaceful resolution of international disputes emerges as a fundamental principle at Article 2, para. 4 of the United Nations Charter, which provides that "*all Members shall refrain in their international relations from the threat or use of force against the territorial integrity or political independence of any state, or in any other manner inconsistent with the Purposes of the United Nations.*" (Monserrat Filho 2007).

Additionally, the United Nations Charter regulates self-defense in Article 51, stipulating:

Nothing in the present Charter shall impair the inherent right of individual or collective self-defence if an armed attack occurs against a Member of the United Nations, until the Security Council has taken measures necessary to maintain international peace and security. Measures taken by Members in the exercise of this right of self-defence shall be immediately reported to the Security Council and shall not in any way affect the authority and responsibility of the Security Council under the present Charter to take at any time such action as it deems necessary in order to maintain or restore international peace and security.

Consequently, as far as defensive measures are concerned, States reserve their right of self-defense against aggressive conducts *vis-à-vis* their space objects and related space-based systems, in accordance with and within the limits provided by the United Nations Charter (Shaw 2003).

Space Law treaties, celebrated under the auspices of the United Nations Committee on the Peaceful Uses of Outer Space (UNCOPUOS), (Cheng 1997) also support domestic space security initiatives, as provided by its undeniable "Magna Carta," i.e., the 1967 Treaty on Principles Governing the Activities of States in the Exploration and Use of Outer Space Including the Moon and Other Celestial Bodies ("OST").

Early in its text, said treaty specifically recognizes *"the common interest of all mankind in the progress of the exploration and use of outer space for peaceful purposes."* Further, in accordance with Article I OST, it is stated that *"the exploration and use of outer space, including the Moon and other celestial bodies, shall be carried out for the benefit and in the interests of all countries, irrespective of their degree of economic or scientific development, and shall be the province of all mankind."*

The importance of this provision to emerging space faring nations is clarified by (Masson-Zwaan and Hofmann 2019):

This clause was included to accommodate the developing countries, united in the 'Group of 77', who wished to be more involved in space activities. But what exactly the term 'province of mankind means is not very clear; the treaty itself provides no further hint or explanation, and its meaning is subject to debate.

Irrespective of conflicting interpretations in relation to the definition of specific terms in the OST such as "peaceful purposes" and "province of mankind," continuous access to, as well as exploration and use of outer space, is granted therein to all nations, provided applicable rules are observed and general interests of humankind are acknowledged.

Finally, since international responsibility for all national space activities, whether governmental or nongovernmental, contemplated by Article VI OST, domestic regulation of space activities is justified, devising proper mechanisms for authorization and supervision.

Appraising Brazilian initiatives towards space security contributes to a more comprehensive understanding of emerging space faring nations perspectives on that regard, representing a case study capable of contributing to the further development of not only domestic initiatives, related to space policy and law, but also to international cooperation mechanisms in general.

Space Security and Emerging Space Faring Nations

The secure and sustainable access to, and use of, outer space, free from hazardous threats, is obviously important to the international community as a whole, but of particular relevance to emerging space faring nations (Space Security Index 2018). Due to insufficient technical redundancy, limited scientific capabilities, and pure economic constraints, those nations tend to experience overreliance in relation to a small selection of space assets. Therefore, whether associated with space environment or related to irresponsible or dangerous conduct from other actors, those risks are to be suitably comprehended, in order to support efficient planning and appropriate courses of action.

Around the world, developing nations look for space capabilities to secure their economic, political, and social development (Space Security Index 2018). Artificial satellites represent strategic tools to address national demands from the vintage point of Earth's orbit, supporting strategic governmental initiatives towards local concerns. Thus, through the use of space objects, it has become possible for many nations of the so-called "global South" to support continuous contact with remote locations, as well as to monitor and survey frontiers, some of which are still to be properly delimited. Urgent measures to protect fragile biomes, in face of natural disasters and ecological challenges, are often only made effective by reliance on space capabilities. As far as crucial economic activities are concerned, the exploitation of natural commodities growingly depends upon weather and remote sensing satellites.

As a de facto emerging space faring nation, Brazil acknowledges the crucial importance of space activities to the exercise of its national sovereignty. For a territory of 8,514,215.3 km, home of over 200 million people, appropriate satellite cover is always a challenge. Through the years, successive federal governments have invested time, capital, and people on strategic initiatives focused on local needs, ranging from remote sensing to telecommunications capabilities. Nevertheless, space activities remain an expensive endeavor, and access to crucial technology is often limited by political and legal constraints. Thus, Brazil has recognized that international cooperation is frequently required to secure specific services and expertise, leading to the negotiation of agreements with foreign partners.

Brazilian space assets are to be considered not only crucial to its national development, but even precious, irrespective of their economic worth or technological capabilities, due to the relatively remarkable efforts required for acquisition and use. Relying on an arguably insufficient satellite network, Brazilian authorities should certainly reserve more attention to space security initiatives. Nevertheless, official assessments and academic researches on the topic remain few and far between, contributing to a somewhat challenging local scenario, similar to the one faced by other emerging space faring nations.

Brazilian Space Policy: An Overview

Brazilian space activities have been developed in accordance with a comprehensive and ambitious space program, first envisaged in the early 1960s. Since then, the South American giant has successfully conducted a wide range of initiatives, contracted launching services for many "home-built" satellites and even for sending an astronaut into Earth's orbit, in an overarching effort towards achieving complete national space autonomy (Bittencourt Neto 2011).

National launching facilities were developed in Brazilian territory, including the Alcantara Launching Center ("Centro de Lançamento de Alcântara," CLA), designed in 1983. Strategically located in a privileged geographical region, near the Equator line (Paubel 2002), more than 200 sounding rockets have been launched from CLA (Costa Filho 2002). Due to its global position, launchings benefit from the Earth's rotation in order to achieve greater speed, allowing fuel economy and increased payload capacity. Furthermore, space objects can be launched from CLA into equatorial and polar orbits, without passing over inhabited regions, due to its proximity to the Atlantic Ocean (Monserrat Filho 2002). The launching center's economic potential has been recognized internationally, leading to related agreements being accorded with foreign partners such as Ukraine and USA.

A relevant number of space objects were designed and assembled in Brazil, in full or in part, usually emerging from the technological laboratories of the National Institute for Space Activities ("Instituto Nacional de Pesquisas Espaciais," INPE), at São José dos Campos. As devised, INPE's main mission is to produce high quality of science and technology in the space and terrestrial environment areas and to offer unique products and services for Brazilian benefit (INPE 2019).

The final segment of the Brazilian space program revolves around the development of a national launching vehicle, thus providing independent access to outer space. Named VLS (for Satellites Launching Vehicle, "Veículo Lançador de Satélites" in Portuguese), the program has faced budgetary and technical burdens since its conception, in the late 1970s. Without fruitful international cooperation, the VLS suffered from delays and complications (Paubel 2002). Despite notable constraints, several sounding rockets have successfully been conceived, most notably from series VS-30, VS-40, and VSB-30, allowing Brazilian scientists and engineers to progressively advance towards a national launching vehicle (Azevedo 2007).

International agreements, accorded with both space-faring and emerging space-faring nations, have allowed Brazil to achieve further capabilities, through joint missions involving satellites and rocketry. Important south-south partnerships have been accorded, most importantly with China, through the CBERS Program (China Brazil Earth Resources Satellites), focusing on remote sensing capabilities, which led to the joint development and launching of satellites CBERS-1 (1999), CBERS-2 (2003), CBERS-2B (2007), and CBERS-4 (2014). The latest satellite of the CBERS program was expected to be launched in December 2019 (INPE 2019).

National space activities have qualified Brazil, by fact and by law, as a Launching State, term defined by Article I, c, of the Convention on the International Liability for Damage Caused by Space Objects (LIAB), of 1972, as including those which launch or procures the launching of a space objects, as well as those from whose territory or facility a space objects is launched, irrespective of the success of its mission (Article I, c, LIAB).

To improve coordination of the Brazilian space program and stress its peaceful purposes, the Brazilian Space Agency ("Agência Espacial Brasileira," AEB) was created by Law 8.854, of February 10, 1994. A civilian entity headquartered in

Brasilia, it enjoys financial and administrative autonomy (Article 2). AEB coordinate the efforts of INPE, especially in relation to satellites, with those of Aerospace Technical Center (Centro Técnico Aeroespacial – CTA), regarding launching vehicles and launching centers. The Brazilian space agency was conceived to improve management of the national space program, while stressing the civil inclination of its many programs and initiatives (Costa Filho 2002).

The peaceful nature of the national space program is in line with the Brazilian Constitution of 1988, which states that the country shall be governed in its international relations, inter alia, by the defense of peace, peaceful settlement of disputes and cooperation among peoples for the progress of humankind (Article 4). Indeed, Brazil has been part of several relevant disarmament instruments, including the Treaty of Tlatelolco (1967), which bans nuclear weapons in Latin America and the Caribbean, as well as Treaty on the Non-Proliferation of Nuclear Weapons (1968), of universal application. More recently, Brazil's membership in the Missile Technology Control Regime (MTCR) was approved, in 1995.

The goals and objectives of the Brazilian space program are determined by the National Program for Development of Space Activities ("Programa Nacional de Desenvolvimento das Atividades Espaciais," PNDAE), established by Decree 1.332, December 8, 1994. Accordingly, said document clarifies that the Brazilian space program aims at fostering national capability in relation to space activities capable of addressing national concerns, while benefiting the whole local people.

In order to plan and promote the PNDAE, the National Program of Space Activities ("Programa Nacional de Atividades Espaciais," PNAE), a periodically reviewed instrument, was contemplated. Currently in its current fourth edition, applicable to the period of 2012–2021, the PNAE stresses the peaceful purposes of the Brazilian space program. Additionally, the importance of acquiring sensitive technologies applicable to launching vehicles and satellites is mentioned, in order to obtain national space autonomy. As stated therein (Brazilian Space Agency 2012):

> *The PNAE aims to enable the country to develop and use space technology in solving national problems and for the benefit of the Brazilian society, contributing to improve the quality of life, through the generation of wealth and job opportunities, improvement of scientific activities, expanding awareness of the national territory and better perception of environmental conditions.*

Emerging space faring nations often face complex barriers to develop nationally integrated space initiatives. In accordance with José (Monserrat Filho 1997), the greatest challenge faced by the Brazilian Space Program is *"to raise the number of specific programs, taking advantage of growing possibilities and needs of international cooperation in the space sector, with participation of many nations."*

To overcame those difficulties, the National System for Development of Space Activities ("Sistema Nacional de Desenvolvimento das Atividades Espaciais," SINDAE) was constituted, through Decree 1.953, of July 10, 1996, aiming at the coordination of space activities, not only between the federal bodies and entities but also with private actors (Article 3, I, II and III).

Domestic Regulatory Instruments

As other Latin American nations, Brazil still awaits the enactment of a specific law regarding space activities, which, as per Article 22, X, of Brazilian Constitution of 1988, would be subject to federal exclusive jurisdiction. Nevertheless, reflecting the intention of opening the country to the international space market and based on CLA's potential to attract foreign partners, AEB enacted administrative edicts regulating launching activities taking place on Brazilian territory.

The referred instruments, namely, Edict 27, of June 20, 2001 (which revoked the Edict n. 8, of February 14, 2001), regarding launch licenses and Edict 5, of February 21, 2002, relating to the authorization of space launch activities on Brazilian territory, are not restricted to CLA only, enjoying vast jurisdiction. However, both edicts stress that their rules do not apply to "*space launching activities which could be carried out by Brazilian governmental organizations or bodies*" (Article 1, Paragraph 2, Edict 27/2001, and Article 1, Paragraph 2, Edict 5/2002).

As far as the Brazilian legislative structure is concerned, edicts enacted by federal autonomous organizations such as AEB are not located in a prime position; even so, they are enforceable and do provide rights and obligations, as long as there are no conflicts with superior norms (Bittencourt Neto 2009; Von der Dunk 2003). Frans G. VON DER DUNK highlighted the importance of such edicts, affirming that, with Edict 27/2001, "*Brazil became the ninth nation world-wide to establish a national space law in the narrow sense of the world – an act focusing exclusively on space activities and prominently including a system for encapsulating private participation in such activities within the State jurisdiction, international responsibilities and international liabilities.*"

On this point, more recently, the AEB Edict 120, of 26 August, 2014, approved an updated regulation on procedures and definition of the necessary requirements for the application, evaluation, dispatch, control, monitoring, and supervision of license to perform space launching activities on Brazilian territory, therefore revoking Edict 27/2001.

In addition to the referred instruments, further regulatory mechanisms were advanced, including the Space Sector Safety Regulations, issued by AEB Superior Council in 2007, applicable to space launchings conducted from Brazilian national territory. Providing technical rules and standards for execution of certain space activities, the Space Safety Regulations cover all actions encompassed from the design stages up to the operational phases. General and specific rules are defined therein dependent on the nature of the system.

According to a comprehensive set of interrelated instruments, provided by the AEB Resolution 71, of December 5, 2007, now in force, the Space Sector Safety Regulations describe the conditions to be complied with for "*the protection of people, property and the environment with respect to operations for the sending of commercial space artifacts.*" Applicable requirements include manufacturing operational processes, rocket assembly and satellite integration, launches, and environmental impact studies.

The Space Sector Safety Regulations are arranged in two volumes of rules, encompassing diverse but interconnected provisions. The first volume covers the "General Regulations for Space Safety," while the second consists of seven different technical regulations, ranging from space safety to areas such as environmental safety, launch and flight safety, payload safety, vehicle launch complex safety, and intersite area safety. The third Volume of the Brazilian Space Safety Regulations, destined to "Accident Investigation and Prevention Regulation," is currently under preparation by the local authorities.

The first set of regulations, provided by volume I, establishes the general safety requirements for commercial space activities in Brazil. For this reason, any entity intending to develop space activities in and from Brazilian launching centers shall meet those regulations, in mandatory bases. Likewise, the conditions agreed upon in the international instruments in force, for a foreign participation in activities from launching centers licensed by AEB, are also applicable.

The General Regulations for Space Safety addresses a total of 22 sections, supporting the topic of security certificates. In fact, it is up to the AEB to constitute or designate a government technical entity, the Space Certification Body ("SCB"), responsible for conducting the space certification processes and the issuance of the "Safety Certificates" established in the regulation. Indeed, Safety Certificates constitute the evidence of compliance with the security requirements necessary to issue applicable licenses and authorizations.

Accordingly, in the topic concerning the building and operation of a launching complex, the technical approval of the SCB project is mandatory. The applicant shall demonstrate that the engineering design meets the requirements of the Brazilian General Regulations and the Technical Regulations, including space safety, environmental safety, launch and flight safety, payload safety, launch complex safety, launching vehicle safety, and intersite area safety. After complying with the mentioned requirements, the "Launch Complex Project Safety Certificate" shall be issued by the SCB.

Moreover, said document establishes the requirement of a "Certificate of Technical and Operations Safety" for space launching activities. The Certificate shall be issued by the SCB only after the organization responsible for the launching complex demonstrates compliance with the requirements of the Technical Safety Regulations for the Launch Complex, which is in part 5 of volume 2 of the Brazilian Space Sector Safety Regulations (Superior Council of the Brazilian Space Agency 2007).

Brazil has also focused efforts on developing the "Launch Vehicle Safety Certificate of Conformity." Again, the document must be issued by the SCB, in accordance with specific conditions. In this case, the entity responsible for the launching vehicle design shall prove that it meets the requirements of the Launching Vehicle Safety Technical Regulation. This regulation is the part 6 of volume 2 of the Brazilian Space Sector Safety Regulations (Superior Council of the Brazilian Space Agency 2007).

Any change in the design of the safety-related systems of the launching vehicle shall require the application for a new certificate, including all technical documentation, such as test and analysis memorials, material specification, descriptive drawings, production processes, the tests, transport procedures, assembly, operation, and procedure of integrating the vehicle with the payload.

The significance granted to SCB certification procedure is illustrated by the applicant's obligation of obtaining a "Certificate of Production Compliance of the Launch Vehicle." Essentially, assembly, transportation, and integration activities of each launching vehicle can only be performed after SCB issuing of a Certificate, in order to assure the necessary safety of operations.

Regarding payload security, another certificate of conformity must be issued by the SCB on that regard. Thus, the entity responsible for each space launching, after proving that its payload meets the "Technical Regulations for Payload Safety" requirements, shall be awarded such specific certification (Superior Council of the Brazilian Space Agency 2007).

In consideration of the launching center's technical and operational safety, as found in volume 2 of the Brazilian Space Sector Safety Regulations, the launching center entity is required to hold a valid "Certificate of Technical and Operational Safety of the Launching Center."

Likewise, as reflected in other instances, the certificate will be issued by the SCB after proof of compliance with the general character of the Technical Regulations for Launch and Flight Safety (part 3, volume 2 of the Brazilian Space Sector Safety Regulations). It is worth mentioning that, according to the standards, "general" requirements of the Technical Regulation are those not directly related to a specific mission or launching.

It must be stressed that safety for space launching and flight is major concern for Brazilian authorities. In fact, the SCB shall issue a "Certificate of Safety for Space Launch and Flight" whenever it verifies that it meets the requirements of the Regulations for Launch and Flight Safety, directly related to a specific mission or launch (Superior Council of the Brazilian Space Agency 2007). Note should be granted to the fact that each launching shall require a new, additional Certificate.

Accordingly, the authorization for space launching can only be granted after the fulfillment of important provisions imposed by the Brazilian Law. The entity responsible for the space launching shall demonstrate that all applicable requirements and certificates are valid, under supervision of SCB. Indeed, it is possible to affirm that current Brazilian regulations do meet the essential safety standards for launchings from the national territory, in line with the established international instruments and guidelines. The Space Certification Body of Brazil, as provided, may follow any certification application by entities responsible for a space launching, including proof of compliance with regulations and those applications for technical approval of a component, software, or device.

One should notice that, during the certification process, SCB is entitled by law to require access to technical data, necessary to approve the entity responsible. Some important questions may arise during such process, since the public authority may demand further contact to the entity and all their suppliers, including manufacturing facilities. To properly address rights and duties, the treatment of those data should not, by any means, characterize or lead to public disclosure of trade secrets, nor impact in the related intellectual property rights.

Technical certificates issued by foreign States may be accepted by the SCB, with adoption of some procedures, such as a mutual recognition agreement, identification of rules applied to the certification process, and any applicable additional requirements, among other instances eventually appropriate to that end, as long as compatible with the Brazilian Space Sector Safety Regulations (Superior Council of the Brazilian Space Agency 2007). SCB may cancel any certificate if the risk associated with some requirement is above of acceptable levels established by Brazilian Space Safety Regulations, including situations in emergency scenarios. Finally, the holder of a certificate shall storage all data related to any fact or event that results in exposure to risks that exceeds the acceptable levels of the regulations, for possible consultation.

The new dimension brought by Internet, with the digitalization of data, can potentialize breaches and negative disclosure. It is appropriate to address, at this point, that Brazil has a new law (n. 13.709, of August 14, 2018), designed to foster data security and protection (Brazil 2019). In this context, the Space Safety Regulations in Brazil also allows international agreements, which Brazil is part of, to provide access to data. Essentially, we consider that is very important to establish international channels to exchange data between States and companies working with the development, launch, and operation of satellites.

The before mentioned law, known as the General Personal Data Protection Act (LGPD), which will come into force in 2020, is intended to regulate the processing of personal data of customers and users by public and private companies, with important topics on international data transfer and data security, a sensitive issue as far as specific space activities are concerned.

The Brazilian "Data Law" applies to any processing operation carried out by a legal entity of public or private law, irrespective of the country in which its headquarter is located or the country where the data sits. On this regard, one should notice that international transfer of personal Brazilian data shall only be allowed to countries or international organizations capable of providing a level of protection deemed adequate in accordance to the provisions of the Brazilian Law. Additionally, international transfer of personal data is authorized when resulting in a commitment undertaken through international cooperation, therefore eventually encompassing space cooperation agreements.

Brazilian Space Situational Awareness Initiatives

In order to better protect its national space objects against potential threats, Brazil has been developing space situational awareness initiatives, including through international cooperation. Space Situational Awareness (SSA) has been defined as *"the comprehensive knowledge of space objects and the ability to track, understand and predict their future location,"* in accordance with the Space Generation "Space Safety & Sustainability Working Group" (2012).

The purpose of SSA initiatives is to safeguard space-based systems, currently recognized as fundamental assets to the sustainable development of every nation.

Nowadays, those actions are of particular relevance since, as observed by (Bobrinsky and Del Monte 2010), *"the destruction of even part of space infrastructure can have heavy consequences for the safety of citizens and economic activities."* For emerging space faring nations like Brazil, understanding potential threats to their satellites and related space-based systems is highly advised, considering the difficulties involved in supporting strategic orbital capabilities.

The major concerns of SSA initiatives may be summarized as including, first and foremost, tracking and surveillance of space objects, since the increasing population of space debris represents relevant risks of in-orbit collisions and interferences. Also, space weather capabilities are taken into consideration, due to the fact that solar storms and explosions of charged particles can seriously damage space objects and power grids on Earth. Finally, near-earth objects, i.e., small natural bodies attracted by Earth's gravitational field, are of particular interest for representing orbital or terrestrial dangers.

Thus, improved awareness of the space environment is important to secure safe and continuous development of space activities. For an emerging space faring nation as Brazil, those initiatives may be considered even more relevant, providing information necessary to address potential threats to the continuous service of operational space objects.

In 2007, the Brazilian National Institute for Space Research (INPE) assembled a task force named EMBRACE, acronym for Brazilian Studies and Monitoring of Space Weather ("Estudo e Monitoramento Brasileiro de Clima Espacial," in Portuguese), which culminated in the establishment of an official space weather program. EMBRACE's objective is *"to monitor the Solar-Terrestrial environment, the magnetosphere, the upper atmosphere and the ground induced currents to prevent effects on technological and economic activities."*

The relevance of space weather initiatives is clarified by (Pelton 2013), who explains that *"solar flares and coronal mass ejections (CMEs) could be much larger problems than previously thought with modern electronic and electrical systems particularly at risk."*

EMBRACE is financed by the main governmental funding agencies, including the National Council for Scientific and Technological Development (CNPq), the São Paulo Research Foundation (FAPESP), the Brazilian Innovation Agency (FINEP), and the Brazilian Ministry of Science, Technology, Innovation and Communication (MCTI). Such support clarifies how decisive and strategic space weather monitoring has been recognized by local public authorities. Currently, EMBRACE is the official interlocutor of Brazil in space weather at the World Meteorological Organization (WMO).

As far as space debris are concerned, the Brazilian National Laboratory of Astrophysics ("Laboratório Nacional de Astrofísica," LNA), with support from the Brazilian Space Agency, celebrated an agreement with the Russian Space Agency (ROSCOSMOS) to install a specialized telescope in the Pico dos Dias, Brazópolis. Strategically located in the Southern Emisphere, it enhanced Russian space situational awareness capabilities, while also granting access to Brazilian authorities regarding potential threats represented by man-made derelict, or uncontrolled space objects, including their related parts, to national satellites.

Recently, the Brazilian Ministry of Defense signed, in December 12, 2018, a Space Situational Awareness (SSA) agreement with USA Defense Department, as part of a larger effort to increase safety of space operations. Data cross-check among applicable parties provides essential support for space activities, such as satellite launching and decommissioning, satellite maneuver planning, support for in-orbit anomalies, electromagnetic interference investigation, and in-orbit conjunction assessments (Barreto 2019).

As Brazilian space activities naturally evolve, and more national space objects are launched in orbit, further space situational awareness initiatives will be required, whether developed with autochthony or under international cooperation agreements. Commons goals and concerns may in fact lead to the development of regional and even global initiatives, based on further synergies among the involved parties.

Prevention of Arms Race in Outer Space: Brazilian Perspectives

Brazilian delegations have, at least until recently, sustained a consistent positioning at the United Nations Conference on Disarmament (CD), in Geneva, as far as the topic of prevention of an arms race in outer space (PAROS) is concerned. Alternate governments and regimes in the country did not impair Brazilian engagement in debates, denouncing growing concern in relation to the weaponization of outer space.

As explained by (Tronchetti 2013), *"due to the increasing importance of space assets and the consequent need to protect them, there is a widespread concern that states might eventually weaponize outer space. Considering that the space treaties do not impose with any substantial limit to such a weaponization, initiatives aimed at creating legal barriers to such an option have been launched."*

Debates on PAROS at CD dates back to the early 1980s, gaining traction after the establishment of a specific ad hoc committee in 1985, designed to *"to examine as a first step at this stage, through substantive and general consideration, issues relevant to the prevention of an arms race in outer space"* (Conference on Disarmament, Report of the Ad Hoc Committee on Prevention of an Arms Race in Outer Space, document CD/642, December 4, 1985, para. 1).

More recently, in accordance with the United Nations General Assembly Resolution 72/250, of 2017, on further practical measures for the prevention of an arms race in outer space, the Secretary-General appointed a Group of Governmental Experts from 25 Member States, including Brazil. At its first session, the Group elected the Brazilian Ambassador Guilherme de Aguiar PATRIOTA as its chair, leading its sessions in Geneva.

The latest report of the Group of Government Experts on further practical measures for the prevention of an arms race in outer space was presented by the chair in 2019, informing about the meetings, devoted to interactive discussions and share of views on the following topics: the existing legal regime in outer space and elements of general principles; elements of general obligations; elements related to monitoring, verification, and transparency and confidence-building measures; and elements related to international cooperation, institutional arrangements and final provisions (A/74/77, p. 7).

The group was guided by an indicative timetable, prepared to focus discussions on various thematic areas considered of relevance, including the international security situation, existing legal regime applicable to the prevention of an arms race in outer space and the application of the right of self-defense in outer space (A/74/77, pp. 8–9).

Regarding other intergovernmental initiatives related to the prevention of an arms race in outer space, reference should be made to the Brazilian position on the European Union's Draft International Code of Conduct for Outer Space Activities. A draft text was first presented to the international community in 2012 as an independent and ad hoc initiative, not related to debates at CD.

Brazil presented, at the time, serious reservations in relation to the Code of Conduct, most importantly for the fact that the document was not widely and openly discussed at the CD or COPUOS, and that it did lacked legitimacy due to the fact that countries from other continents, such as Brazil, did not have opportunity to discuss or propose amendments. Debates in relation to the Code of Conduct have lose momentum with time, including among Brazilian officials.

Conclusions

Emerging space faring nation are increasingly concerned with the development of space security initiatives. Brazil, a Latin American country with a long-standing space program, which lead to the development of many space objects, two launching centers and aims at the development of a national launching vehicle, cannot be left out of the debates on this subject matter.

Indeed, among various international fora, Brazil has presented its contribution to the discussions, with relevant engagement at the United Nations. Domestically, various normative instruments and guidelines have been enacted regarding important space security aspects, although still in a sparse and somewhat complex formulation. The enactment of a national federal law on space activities, contemplating space security provisions, is certainly advised.

Space situational awareness capabilities are certainly important in our days to support the continuous operation of valuable space-based assets, providing significant space security advantages. Whether via national initiatives or, even more importantly, through international cooperation, initiatives are to be continuously by the Brazilian Space Agency to the furtherance of its space program.

Multilateralism allows States, in international relations, to reduce coordination costs, while achieving greater legitimacy, thus supporting global governance. As a global concern, space security benefits from open international exchanges, reflecting national concerns and regional perspectives.

International cooperation remains a fundamental cornerstone for the peaceful development of space activities and, as such, should never be taken for granted. For an emerging space faring nation as Brazil, multilateral collaboration with foreign partners should be perceived, as well mentioned by (Forman 2002), as *"no longer a choice. It is a matter of necessity, and of fact."*

References

Azevedo JLF (2007) A Capacidade Brasileira de Acesso ao Espaço. Espaço Brasileiro, a. 1, n. 2, Brasília, Brazil, March, April, May, 2007, pp 15–17

Barreto, Andréa (2019) US and Brazil Share Information on Space. Available at: https://dialogo-americas.com/en/articles/us-and-brazil-share-information-space, accessed on 24 October, 2019

Bittencourt Neto Olavo de O (2009) Private launching activities on Brazilian territory: current legal framework. Z Luft- Weltraumrecht 58(3):429–449. Cologne

Bittencourt Neto Olavo de O (2011) Direito Espacial Contemporâneo. Juruá, Curitiba

Bobrinsky N, Del Monte L (2010) The space situational awareness program of the European Space Agency. Kosm Issled 48(5):402–408

Brazil (2019) General Personal Data Protection Act. Law n. 13.709/2018. Available at: http://www.planalto.gov.br/ccivil_03/_ato2015-2018/2018/lei/L13709.htm, accessed on 23 October, 2019

Brazilian Space Agency (2012) Programa Nacional de Atividades Espaciais: PNAE: 2012 – 2021. Agência Espacial Brasileira. Brasília: Ministério da Ciência, Tecnologia e Inovação, Agência Espacial Brasileira, 2012

Brazilian Space Agency (2020) The Embrace Program. Available at: http://www3.inpe.br/50anos/english/presentation.php, accessed on January 30, 2020

Brazilian Space Agency (2020) Acordo entre Brasil e Rússia garante monitoramento de lixo espacial. Available at: http://portal-antigo.aeb.gov.br/acordo-entre-brasil-e-russia-garante-monitoramento-de-lixo-espacial/, accessed on January 30, 2020

Brazilian National Institute for Space Research (INPE) (2019) Half a century conquering the space to care for Earth. Available at: http://www3.inpe.br/50anos/english/presentation.php, accessed on September 22, 2019

Casella PB (2009) Direito Internacional dos Espaços. Atlas, São Paulo

Cheng B (1997) Studies in international space law. Clarendon Press, Oxford, UK

Committee on Prevention of an Arms Race in Outer Space Conference on Disarmament (1985). Report of the Ad Hoc Committee on Prevention of an Arms Race in Outer Space, document CD/642, December 4, 1985

Costa Filho E (2002) Política Espacial Brasileira. Revan, Rio de Janeiro

European Union (2014) Draft International Code of Conduct for Outer Space Activities, Version March 31, 2014

Forman S (2002) Multilateralism and US foreign policy: ambivalent engagement. Lynne Rienner, London

Masson-Zwaan T, Hoffmann M (2019) Introduction to space law, 4th edn. Kluwer, Alphen aan den Rijn

Monserrat Filho J (1997) Introdução ao Direito Espacial. SBDA, Rio de Janeiro

Monserrat Filho J (2002) Brazilian launch licensing and authorizing regimes. Capacity Building in Space Law – UN/IIASL, The Hague

Monserrat Filho J (2007) Direito e Política na Era Espacial. Vieira & Lent, Rio de Janeiro

Paubel EFC (2002) Propulsão e Controle de Veículos Aeroespaciais: uma Introdução. UFSC, Florianópolis

Pelton JN (2013) Space debris and other threats from outer space. Springer, Dordrecht

Rosenau J (1992) Governance without government: order and change in world politics. Cambridge University Press, Cambridge, UK

Shaw M (2003) International law, 5th edn. Cambridge University Press, Cambridge, UK

Space Safety & Sustainaibility Working Group (2012) Space situational awareness. SSS educational series. Available at: https://www.agi.com/resources/educational-alliance-program/curriculum_exercises_labs/SGAC_Space%20Generation%20Advisory%20Council/space_situational_awareness.pdf. Accessed on 25 Oct 2019

Tronchetti F (2013) Fundamentals of space law and policy. Springer, Dordrecht

Von Der Dunk, FG (2003) Lauching Alcântara into the global space economy – the 2001 Brazilian National Space Law. Revista de Direito Aeronáutico e Espacial – SBDA, n. 86, Rio de Janeiro, p 23

Weeden B (2018) Space security index 2018, 15th edn. Waterloo, Waterloo Printing

Space and Security Activities in Azerbaijan

36

Tarlan Mammadzada

Contents

Introduction	668
Azercosmos: the Satellite Operator of Azerbaijan	670
Satellite Imagery Sources and Analysis Methodology	672
Results of Satellite Imagery Analysis	674
Environmental Damage	674
Permanent Infrastructure Changes	675
Exploitation and Pillage of Natural Resources	679
Destruction of Public and Private Property, Including Historical and Cultural Heritage	682
Exploitation of Agricultural and Water Resources	687
Implantation of Settlers and Construction of Permanent Social Infrastructure	694
Conclusion	697
References	698

Abstract

Earth observation satellites are the source of valuable data for military intelligence, strategic planning, and development. Azerbaijan continuously monitors its occupied territories by means of remote sensing techniques. These territories constitute 20% of Azerbaijan, including the Nagorno-Karabakh region, the seven adjacent districts, and some exclaves. Despite the resolutions, adopted by UN Security Council in 1993, which reaffirmed the Nagorno-Karabakh region as a part of Azerbaijan and called for immediate, complete, and unconditional withdrawal of the occupying forces from all territories of Azerbaijan, the occupant still continues its activities, which lead to catastrophic changes in the occupied areas.

T. Mammadzada (✉)
"Azercosmos" OJSCo, Baku, Azerbaijan
e-mail: tarlan.mammadzada@azercosmos.az

© Springer Nature Switzerland AG 2020
K.-U. Schrogl (ed.), *Handbook of Space Security*,
https://doi.org/10.1007/978-3-030-23210-8_139

Azerbaijan owns a high-resolution Earth observation satellite Azersky, which in constellation with SPOT 6 is used to continuously monitor these areas, revealing the unacceptable changes of physical, economic, demographic, and cultural character in the occupied territories, such as environmental damage, permanent infrastructure changes, exploitation and pillage of natural resources, destruction of public and private property, implantation of settlers, and construction of permanent social infrastructure.

This chapter contains information about the remote sensing capabilities and techniques, as well as high-resolution satellite imagery, acquired by "Azercosmos" OJSCo, the satellite operator of Azerbaijan, which provides the overview of activities in the occupied territories of Azerbaijan from space.

Introduction

Satellite imagery, from the very beginning, seems to have found its main domain of application in the military field and until today has been used mostly for military strategic planning, intelligence, and development (Van Persie et al. 2000). It is essential, for military purposes, to have an accurate representation of a specific zone of interest anywhere in the world. According to Euroconsult, the leading global consulting firm specializing in space, military is the largest consumer of remote sensing data (more than 60% of all data) (Euro consult Research 2012).

This is particularly important for Azerbaijan, which has 20% of its territories occupied by Armenia, including the Nagorno-Karabakh region, the seven adjacent districts, and some exclaves (Fig. 1). The fact of the occupation of the territories of Azerbaijan using military forces has been consistently deplored by the international community. In 1993, the UN Security Council adopted resolutions 822 (1993), 853 (1993), 874 (1993), and 884 (1993), condemning the use of force against Azerbaijan and the occupation of its territories and reaffirming the sovereignty and territorial integrity of Azerbaijan and the inviolability of its internationally recognized borders. In these resolutions, the Security Council concluded that the Nagorno-Karabakh region is part of Azerbaijan and called for immediate, complete, and unconditional withdrawal of the occupying forces from all occupied territories of Azerbaijan. (See UN Doc. A/67/875–S/2013/313, 24 May 2013.) However, the occupant disregards the international community to this day, continuing to violate international law.

Azerbaijan owns an Earth observation satellite Azersky (formerly SPOT-7), operated by "Azercosmos" OJSCo (hereafter referred to as Azercosmos) since 2014. This satellite orbits regularly over the occupied region, producing high-resolution images of huge areas. Using this satellite, Azerbaijan continuously monitors its occupied territories, observing unacceptable catastrophic changes.

The activities in the occupied territories of Azerbaijan are leading to permanent changes in the economic, physical, cultural, and demographic character of the occupied territories, which was also confirmed by Organization for Security and Co-operation in Europe (OSCE) missions in the occupied territories in 2005 and 2010 (Since 1992 the Organization for Security and Cooperation in Europe (OSCE)

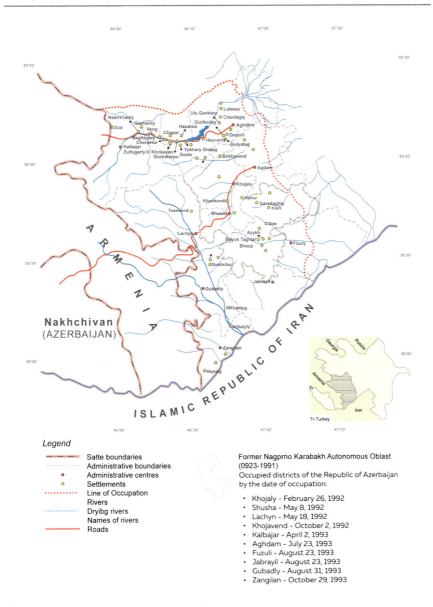

Fig. 1 The occupied territories of the republic of Azerbaijan

has engaged in efforts to achieve a settlement of the conflict under the aegis of its Minsk Group, currently under the co-chairmanship of the French Republic, the Russian Federation and the United States of America; See "Letter of the OSCE Minsk Group Co-Chairs to the OSCE Permanent Council on the OSCE Minsk Group Fact-Finding Mission (FFM) to the occupied territories of Azerbaijan Surrounding Nagorno- Karabakh (NK)", UN Doc. A/59/747–S/2005/187, 21 March

2005, annex I, pp. 4 and 5. For the Executive Summary of the OSCE field assessment mission report and Press release of the Ministry of Foreign Affairs of the Republic of Azerbaijan, UN Doc. A/65/801–S/2011/208, 29 March 2011.). The Ministry of Foreign Affairs of the Republic of Azerbaijan produced a report "Illegal Economic and Other Activities in the Occupied Territories of Azerbaijan" in 2016 (hereafter referred to as the "2016 Report"), which presented well-documented evidence of dramatically expanding illegal activities in the occupied territories. In 2019 Azercosmos jointly with the Ministry of Foreign Affairs of Azerbaijan published the book "Illegal Activities in the Territories of Azerbaijan under Armenia's Occupation: Evidence from Satellite" (hereafter referred to as the "2019 Report"), where the illegal activities in the occupied territories were demonstrated by means of satellite imagery.

The present chapter contains the information about the remote sensing technologies and techniques, as well as high-resolution satellite imagery of the occupied territories, which provides sufficient and convincing evidence of ongoing activities in the occupied territories. The key findings of "2016 Report" and "2019 Report" also presented here to provide overall context in the following areas:

1. Environmental damage
2. Permanent infrastructure changes
3. Exploitation and pillage of natural resources
4. Destruction of public and private property, including historical and cultural heritage
5. Exploitation of agricultural and water resources
6. Implantation of settlers and construction of permanent social infrastructure

Azercosmos: the Satellite Operator of Azerbaijan

Azercosmos – the national satellite operator of Azerbaijan – was established in 2010 to provide satellite-delivered telecommunication and Earth observation services to customers in both the public and the private sectors. It strives to establish Azerbaijan as one of the driving forces of the global space industry and is committed to providing customized solutions based on advanced technologies for peace and prosperity. The activities of Azercosmos are carried out in five main directions: supporting the socioeconomic development of Azerbaijan, supporting national security, expanding commercial activities, supporting space R&D activities, and representing the country in the international space arena. Azercosmos operates three satellites: two telecommunication satellites, Azerspace-1 and Azerspace-2, and the Earth observation satellite, Azersky.

In February 2013 Azercosmos launched its first telecommunication satellite – Azerspace-1, which was manufactured by Orbital Sciences Corporation. Located at 46° East longitude, the satellite has a wide coverage area including Europe, Africa, Central Asia, and the Middle East. It is designed to provide broadcasting and telecommunications services and create platforms to meet government and corporate customers' demands.

In December 2014, within the framework of the strategic partnership with Airbus DS, Azercosmos took over the rights to operate and commercialize SPOT-7 (later rebranded to Azersky), a high-resolution optical Earth observation satellite, and started its commercial activities. Azersky acquires direct and unlimited high-resolution satellite imagery from any part of the world on a daily basis. Azercosmos has been carrying out projects for the stimulation of Earth observation services in Azerbaijan and a number of Commonwealth of Independent States (CIS) countries. Within these projects, both independent researchers and organizations, engaged in research activities, are provided with satellite imagery obtained via Azersky free of charge to implement their innovative, scientific research covering areas such as geography, dendrology, botany, zoology, soil science, and agrochemistry.

In September 2018, Azercosmos launched its third satellite – Azerspace-2. Built for Azercosmos by Space Systems Loral (SSL), Azerspace-2 is a telecommunication satellite with a 35 Ku-band active transponder payload. The satellite enhances the capacity, coverage area, and spectrum of service offerings of Azercosmos to support the demand for government and network services in Europe, Central and South Asia, the Middle East, and Sub-Saharan Africa. Azerspace-2 is ideally designed for smaller antenna and has cross connectivity between East, West, and Central Africa, as well as Europe and Central Asia.

Besides their direct commercial benefits, satellite projects also serve as the basis for the transfer of advanced technologies to the country and acquisition of knowledge, skills, and practices for the independent implementation of space-related projects as a next step. To share accumulated experience and grow the knowledge and skills necessary for the development of satellite components in Azerbaijan in the years to come, Azercosmos, together with the Ministry of Education of the Republic of Azerbaijan, annually holds CanSat Azerbaijan satellite modeling competition. Furthermore, the rocket modeling festival – Rocketry Azerbaijan – that took place within CanSat competition in April 2019 was the first of its kind in the country, where students of top local technical universities were involved in the design and launch of rocket models at the Main Satellite Ground Control Station of Azercosmos. Azercosmos was one of the main partners to hold the international ActInSpace Hackathon, organized by The National Centre for Space Studies (CNES) and European Space Agency (ESA), in Baku for the first time in 2018. Considering students' great interest to innovative projects, Azercosmos organized the NASA Space Apps Challenge in Azerbaijan in October 2019, aiming to engage coders, scientists, designers, technologists, and space enthusiasts in developing innovative solutions to NASA's toughest challenges.

The Research and Development Center of Azercosmos conducts scientific research in the field of space and astronautics. The Center develops various software for the satellite operations; conducts research on automatic recognition of objects and changes on satellite imagery through artificial intelligence techniques; and designs small satellites, small launchers, and unmanned aerial vehicles. Moreover, the team of the Research and Development Center of Azercosmos successfully solved the assigned task at the Global Trajectory Optimization Competition and made the top 20 in June 2019.

Azercosmos officially represents the Republic of Azerbaijan at the United Nations Committee on the Peaceful Uses of Outer Space (UN COPUOS), the International Telecommunications Satellite Organization (ITSO), the European Telecommunications Satellite Organization (EUTELSAT), and International Organization of Space Communications (INTERSPUTNIK). It is also a member of the International Astronautical Federation (IAF), EMEA Satellite Operators Association (ESOA), World Teleport Association (WTA), and Smart Africa Alliance. Azercosmos is actively involved in a wide variety of international space-related projects with government, industry, and academia of Asian, European, African, and North and South American countries aimed at using outer space for peaceful purposes. Furthermore, in order to increase engagement within international organizations and support global activities in the space environment, Azercosmos became a member of the Advisory Committee of the European Telecommunications Satellite Organizan (EUTELSAT IGO) in April 2019 and joined the Space Climate Observatory (SCO) initiative in June 2019. Azercosmos attaches a great importance to the development and strengthening of bilateral relations. For instance, in July 2018 Azercosmos and the National Center for Space Research (CNES) of the French Republic signed a framework agreement on space cooperation, and in February 2020 Azercosmos and the Italian Space Agency signed a memorandum of understanding on the use of outer space for peaceful purposes.

On October 25, 2019, during the elections held at the General Assembly of the International Astronautical Federation in the capital city of the United States, Washington, D.C., Baku has got the right to host the International Astronautical Congress (IAC) in 2022. Singapore, Rio de Janeiro, and New Delhi were among the bidding candidate cities to host this prestigious event.

Satellite Imagery Sources and Analysis Methodology

Development of Earth observation technologies and techniques is a matter of high priority for the government of Azerbaijan. On November 15, 2018, the "State Program for the Development of Earth Observation Satellite Services in the Republic of Azerbaijan for 2019–2022" was approved by the president of Azerbaijan. The main goal of the state program is to support the socioeconomic and technological development of Azerbaijan through the application of remote sensing services. The program creates opportunities for development of the latest technologies in various areas, such as defense and security, agriculture, environmental protection, urban planning, emergency response, exploration of natural resources, maritime surveillance, and cartography.

Azersky is placed on the sun-synchronous orbit in constellation with SPOT 6, providing daily revisit over the occupied region. The high-resolution optical instrument consists of panchromatic band (0.45–0.75 μm) and four multispectral bands, red (0.62–0.69 μm), green (0.53–060 μm), blue (0.45–0.52 μm), and near-infrared

(0.76–0.89 μm), and is able to acquire up to 3 million km² area daily with a resolution of 1.5 m suitable for 1:25,000 scale topographic mapping, wide coverage capacity of 60 km swath width at Nadir, and location accuracy of 10 m (CE90) (Fig. 2).

The cooperation with Airbus DS also includes operations of SPOT 6 high-resolution (1.5 m imagery) optical Earth observation satellite. SPOT 6 and Azersky are twin satellites that compose a constellation operating on the same orbit and phased 180° from each other. This orbit phasing allows the satellites to revisit any point on the globe daily, which is very effective for anticipating risks and covering large areas (Table 1).

In addition, the ground segment, constructed in Azerbaijan, allows to access images of Pleiades 1A and 1B, very high resolution (0.5 m imagery) optical Earth observation satellites of Airbus Defence and Space. This satellite constellation provides the opportunities for many applications and specifically for security, providing the latest images within an unprecedented time frame.

The analysis of the images is performed using various digital change detection methods, which aim to detect differences in the state of objects on the images over time. There are two mainly used techniques, the traditional/classical pixel-based, which employs an image pixel as a fundamental unit of analysis, and object-based, which emphasizes creating image objects and then using them for further analysis. Pixel-based technique includes the following approaches:

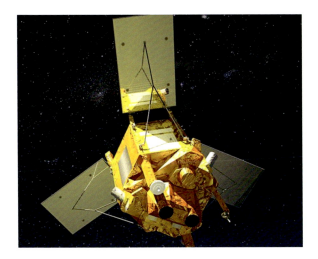

Fig. 2 Azersky, high-resolution Earth observation satellite © airbus.com

Table 1 Satellites viewing angles

Viewing angle	SPOT 6 or Azersky only	SPOT 6 and Azersky
<5 degrees	26 days	13 days
<20 degrees	7 days	4 days
<30 degrees	5 days	2 days
<45 degrees	2 days	1 day

- Direct comparison (image differencing, image rationing, regression analysis)
- Transformation from image (vegetation index differencing, change vector analysis, etc.)
- Classification based (post-classification comparison, multi-date direct comparison)
- Machine learning (AI network, support vector machine, decision tree)
- GIS (GIS integration)
- Advanced methods (spectral mixture analysis, fuzzy change detection, multi-sensor data fusion)

Object-based classification is performed in the following ways:

- Direct object comparison based (objects extracted from one image and then assigned to or searched from image data from second acquisition)
- Object classification comparison based (two segmentation created separately and compared)
- Multi-temporal object change detection (stacked bi-temporal images)

The changes may be detected also due to the differences in illumination, atmospheric conditions, sensor calibration, ground moisture conditions, and other effects. These factors also should be considered and corrections should be implemented (Deer 1995; Hussain et al. 2013).

Results of Satellite Imagery Analysis

Environmental Damage

The forest cutting and the depredatory exploitation of natural resources in the monitored areas are seriously affecting the environment, which leads to the disappearance of various species of trees and millions of tons of tailings in tailing dumps across the occupied territories. The pollution of the territories and damage to the ecosystem is also resulted by the leaks from tailing dumps and ponds at the mining sites. Ferocious exploitation of farmlands in the occupied territories for decades has led to their extreme depletion. (See Section XV (Cutting of rare species of trees for timber and other damage to the environment) of the Report of the Ministry of Foreign Affairs of the Republic of Azerbaijan (2016).)

The following examples of environmental damage are presented:

- Burned area affecting villages of the occupied Jabrayil district ($39°\ 12'\ 32''$ N, $46°\ 59'\ 11''$ E)
- Burned area affecting villages of the occupied Fizuli district ($39°\ 36'\ 07''$ N, $47°\ 08'\ 44''$ E)

- Tailing dump caused by exploitation of Gyzylbulag underground copper-gold mine near Heyvaly village in the occupied Kalbajar district (46° 35′ 43.645″ E, 40° 8′ 34.632″ N)
- Burned area covering 26 km² affecting Jilan and Bunyadli villages of the occupied Khojavand district and Khalafly, Khybyarli, Kurds, and Qarar villages of the occupied Jabrayil district (39° 28′ 29.04″ N, 46° 47′ 18.24″ E)
- Forest cutting for construction of water canal near the Sarsang Water Reservoir in the occupied part of the Tartar district (46° 30′ 30.882″ E, 40° 8′ 45.486″ N)
- Expansion of tailing dump caused by exploitation of Demirli open-pit copper molybdenum mine near Demirli, Gulyatag, and Janyatag villages in the occupied part of the Tartar district (46° 46′ 16.713″ E, 40° 9′ 2.801″ N)
- Deforestation caused by mining activities near Chardagly village in the occupied part of the Tartar district (46° 41′ 33.965″ E, 40° 14′ 9.393″ N)
- Burned area covering 347 km² and affecting 25 villages in the occupied Fuzuli district stretching 22 km from the South to the Northand 17 km from the Eastto the West (39° 36′ 24.92″ N, 47° 06′ 26.71″ E) (Figs. 3, 4, 5, 6, 7, 8, 9, and 10)

Permanent Infrastructure Changes

The changes in social, energy, agriculture, and transport infrastructure, including the construction of irrigation networks, roads, electrical transmission lines, water supply systems and other economic and social facilities, including the construction and reconstruction of network of roads, which are linked to the supporting of settlements in the occupied territories, are continuously monitored. (See Section XII (Extensive

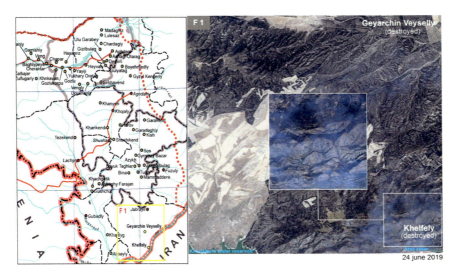

Fig. 3 Burned area affecting villages of the occupied Jabrayil district (39° 12′ 32″ N, 46° 59′ 11″ E)

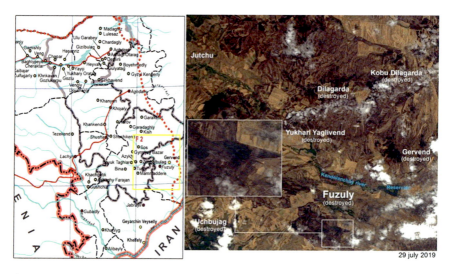

Fig. 4 Burned area affecting villages of the occupied Fizuli district (39° 36′ 07″ N, 47° 08′ 44″ E)

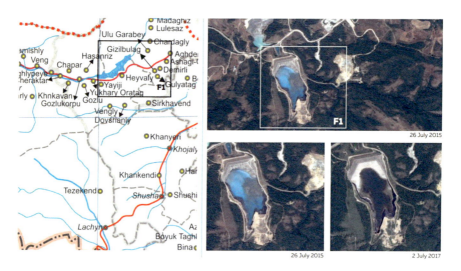

Fig. 5 Tailing dump caused by exploitation of Gyzylbulag underground copper-gold mine near Heyvaly village in the occupied Kalbajar district (46° 35′ 43.645″ E, 40° 8′ 34.632″ N)

exploitation of agricultural and water resources) of the Report of the Ministry of Foreign Affairs of the Republic of Azerbaijan (2016).) Obviously, this infrastructure allows the occupant to pervasively control the entire economic and commercial system in the territories, including maintaining and supporting the economic resources and trade flows.

The following examples of permanent infrastructure changes are presented:

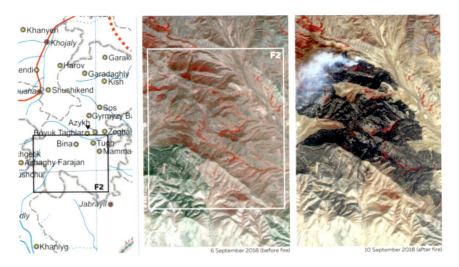

Fig. 6 Burned area covering 26 km^2 affecting Jilan and Bunyadli villages of the occupied Khojavand district and Khalafly, Khybyarli, Kurds, and Qarar villages of the occupied Jabrayil district (39° 28′ 29.04″ N, 46° 47′ 18.24″ E)

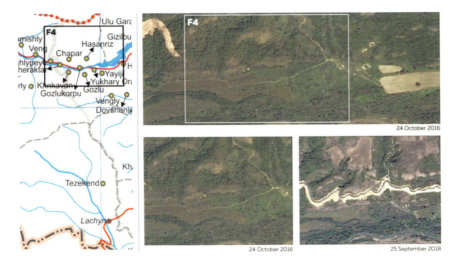

Fig. 7 Forest cutting for construction of water canal near the Sarsang Water Reservoir in the occupied part of the Tartar district (46° 30′ 30.882″ E, 40° 8′ 45.486″ N)

- Reconstruction of Goris (Armenia)–Khankandi road passing through the occupied Lachyn district (39° 36′ 17″ N, 46° 32′ 43″ E)
- Construction of Vardenis (Armenia)–Aghdara highway passing through the occupied Kalbajar district (40° 13′ 18″ N, 45° 58′ 59″ E)

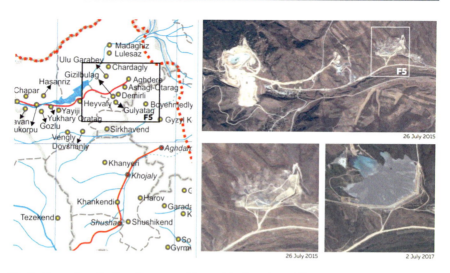

Fig. 8 Expansion of tailing dump caused by exploitation of Demirli open-pit copper molybdenum mine near Demirli, Gulyatag, and Janyatag villages in the occupied part of the Tartar district (46° 46′ 16.713″ E, 40° 9′ 2.801″ N)

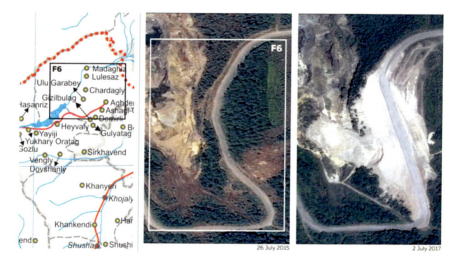

Fig. 9 Deforestation caused by mining activities near Chardagly village in the occupied part of the Tartar district (46° 41′ 33.965″ E, 40° 14′ 9.393″ N)

- Hydroelectric power plant in the west of the Sarsang Water Reservoir in the occupied part of the Tartar district (46° 32′ 0.862″ E, 40° 8′ 38.201″ N)
- Control point on the Goris (Armenia)–Khankandi road near Zabukh village in the occupied Lachyn district (39° 35′ 36″ N, 46° 32′ 10″ E)

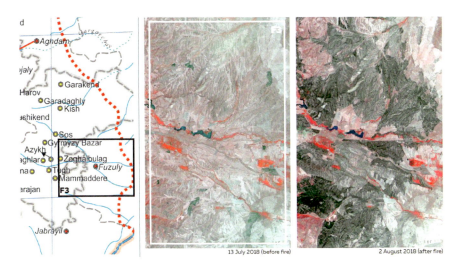

Fig. 10 Burned area covering 347 km^2 and affecting 25 villages in the occupied Fuzuli district stretching 22 km from the South to the North and 17 km from the East to the West (39° 36′ 24.92″ N, 47° 06′ 26.71″ E)

- Infrastructure in support of exploitation of copper-gold and molybdenum mine near Demirli village in the occupied part of the Tartar district (40° 08′ 59″ N, 46° 47′ 12″ E)
- Infrastructure in support of exploitation of Gyzylbulag underground gold and copper mine near Heyvaly village in the occupied Kalbajar district (40° 08′ 43″ N, 46° 35′ 50″ E) (Figs. 11, 12, 13, 14, 15, and 16)

Exploitation and Pillage of Natural Resources

The environmental and economic damage, resulted by the high scale of exploitation and pillage of natural resources in the occupied territories of Azerbaijan, such as mining of minerals and metals utilizing heavy engineering machinery and equipment, is monitored and assessed by means of remote sensing techniques. (See Section XIII (Systematic pillaging, exploitation of and illicit trade in assets, natural resources and other forms of wealth in the occupied territories) of the Report of the Ministry of Foreign Affairs of the Republic of Azerbaijan (2016).)

The following examples of exploitation and pillage of natural resources are presented:

- Mining activities on 859.7 ha near Janyatag village in the occupied part of the Tartar district (40° 09′ 08″ N, 46° 46′ 14″ E)
- Mining activities on 44.78 ha near Chardagly village in the occupied part of the Tartar district (40° 14′ 04″ N, 46° 41′ 49″ E)

Fig. 11 Reconstruction of Goris (Armenia)–Khankandi road passing through the occupied Lachyn district (39° 36′ 17″ N, 46° 32′ 43″ E)

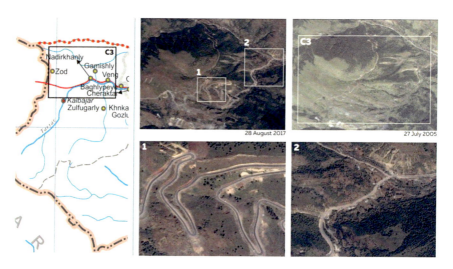

Fig. 12 Construction of Vardenis (Armenia)–Aghdara highway passing through the occupied Kalbajar district (40° 13′ 18″ N, 45° 58′ 59″ E)

- Mining and ore processing facilities on 63.78 ha of the Gyzylbulag underground mine near Heyvaly village in the occupied Kalbajar district (40° 08′ 29″ N, 46° 35′ 38″ E)
- Mining activities in the Tutkhum gold-molybdenum deposit in the occupied Kalbajar district (40° 01′ 14″ N, 46° 09′ 36″ E)

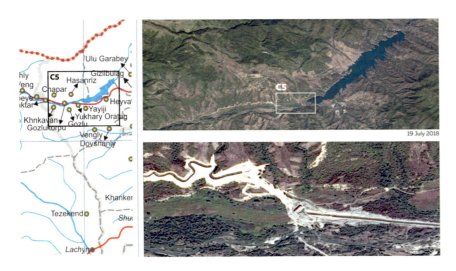

Fig. 13 Hydroelectric power plant in the west of the Sarsang Water Reservoir in the occupied part of the Tartar district (46° 32′ 0.862″ E, 40° 8′ 38.201″ N)

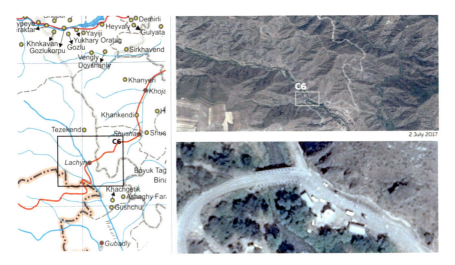

Fig. 14 Control point on the Goris (Armenia)–Khankandi road near Zabukh village in the occupied Lachyn district (39° 35′ 36″ N, 46° 32′ 10″ E)

- Mining activities on 442.17 ha of Soyudlu gold mine in the occupied Kalbajar district (40° 14′ 01″ N, 45° 58′ 17″ E)
- Mining and ore processing facilities on 70.10 ha of gold mine near Vejnaly village of the occupied Zangilan district (38° 55′ 52″ N, 46° 31′ 40″ E) (Figs. 17, 18, 19, 20, 21, and 22)

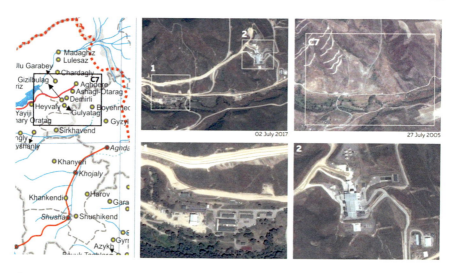

Fig. 15 Infrastructure in support of exploitation of copper-gold and molybdenum mine near Demirli village in the occupied part of the Tartar district (40° 08′ 59″ N, 46° 47′ 12″ E)

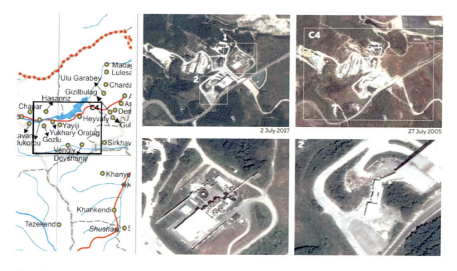

Fig. 16 Infrastructure in support of exploitation of Gyzylbulag underground gold and copper mine near Heyvaly village in the occupied Kalbajar district (40° 08′ 43″ N, 46° 35′ 50″ E)

Destruction of Public and Private Property, Including Historical and Cultural Heritage

Extensive destruction and appropriation of public and private property, deconstruction of houses and buildings for use as constructive materials, construction of new buildings on lands of Azerbaijani displaced people, and altering and destruction of

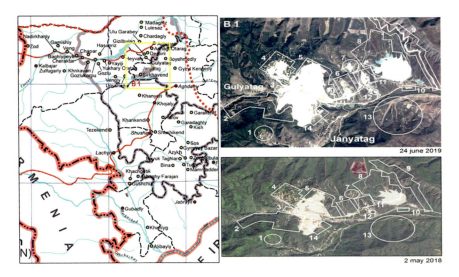

Fig. 17 Mining activities on 859.7 ha near Janyatag village in the occupied part of the Tartar district (40° 09′ 08″ N, 46° 46′ 14″ E)

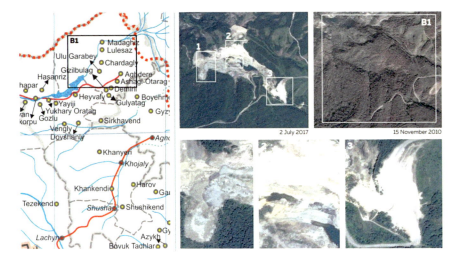

Fig. 18 Mining activities on 44.78 ha near Chardagly village in the occupied part of the Tartar district (40° 14′ 04″ N, 46° 41′ 49″ E)

historical and cultural features, including the archeological, cultural, and religious monuments, are also the catastrophic results of the occupation of the territories. (See Section XIII (Systematic pillaging, exploitation of and illicit trade in assets, natural resources and other forms of wealth in the occupied territories) and Section XVI (Archaeological excavations, embezzlement of artefacts, altering of cultural

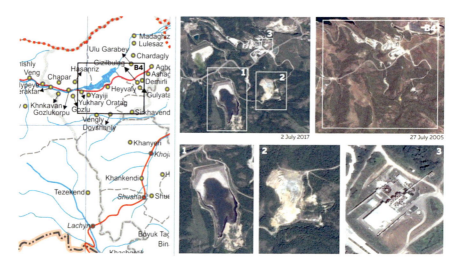

Fig. 19 Mining and ore processing facilities on 63.78 ha of the Gyzylbulag underground mine near Heyvaly village in the occupied Kalbajar district (40° 08′ 29″ N, 46° 35′ 38″ E)

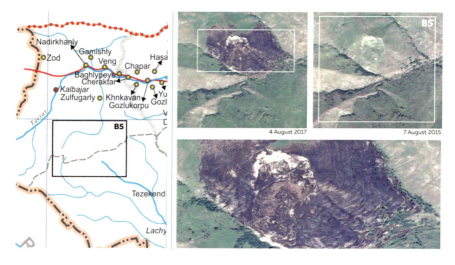

Fig. 20 Mining activities in the Tutkhum gold-molybdenum deposit in the occupied Kalbajar district (40° 01′ 14″ N, 46° 09′ 36″ E)

character of the occupied territories) of the Report of the Ministry of Foreign Affairs of the Republic of Azerbaijan (2016).)

The following examples of destruction of public and private property, including historical and cultural heritage, are presented:

- Destroyed town of Fuzuli of the occupied Fuzuli district (39° 35′ 56″ N, 47° 08′ 49″ E)

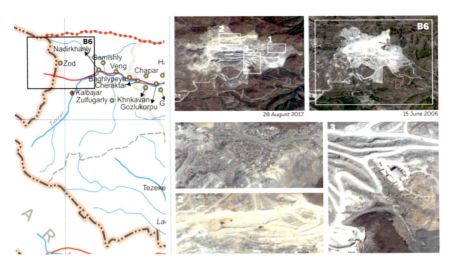

Fig. 21 Mining activities on 442.17 hectares of Soyudlu gold mine in the occupied Kalbajar district (40° 14′ 01″ N, 45° 58′ 17″ E)

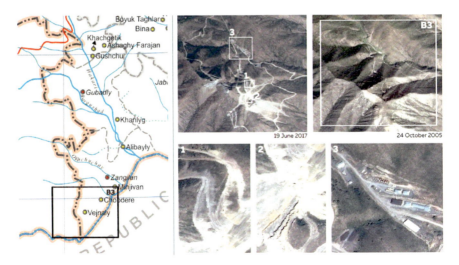

Fig. 22 Mining and ore processing facilities on 70.10 ha of gold mine near Vejnaly village of the occupied Zangilan district (38° 55′ 52″ N, 46° 31′ 40″ E)

- Destroyed town of Jabrayil of the occupied Jabrayil district (39° 23′ 56″ N, 47° 01′ 39″ E)
- Severely damaged Saatly Mosque in the town of Shusha in the occupied Shusha district (39° 45′ 45.26″ N, 46° 45′ 3.49″ E)
- Destroyed Ashaghy Govhar Agha Mosque in the town of Shusha in the occupied Shusha district (39° 45′ 42.31″ N, 46° 45′ 14.22″ E)

- Destroyed Juma Mosque in the town of Aghdam in the occupied Aghdam district (39° 59′ 35.91″ N, 46° 55′ 53.88″ E)
- Ruins of Ismailbayli village of the occupied Aghdam district (39° 56′ 02″ N, 47° 00′ 19″ E) (Figs. 23, 24, 25, 26, 27, and 28)

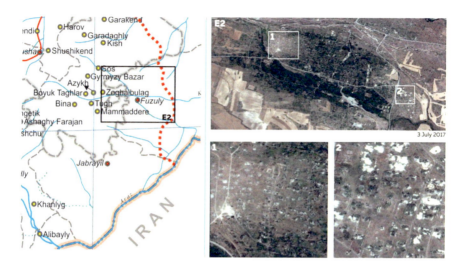

Fig. 23 Destroyed town of Fuzuli of the occupied Fuzuli district (39° 35′ 56″ N, 47° 08′ 49″ E)

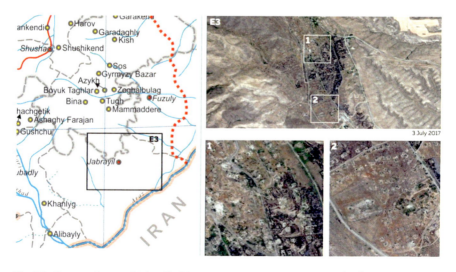

Fig. 24 Destroyed town of Jabrayil of the occupied Jabrayil district (39° 23′ 56″ N, 47° 01′ 39″ E)

36 Space and Security Activities in Azerbaijan

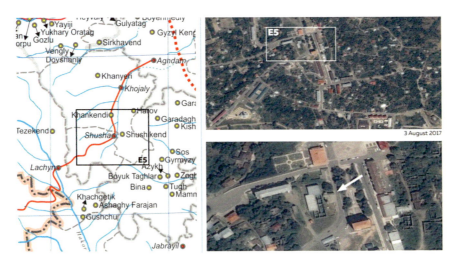

Fig. 25 Severely damaged Saatly Mosque in the town of Shusha in the occupied Shusha district (39° 45′ 45.26″ N, 46° 45′ 3.49″ E)

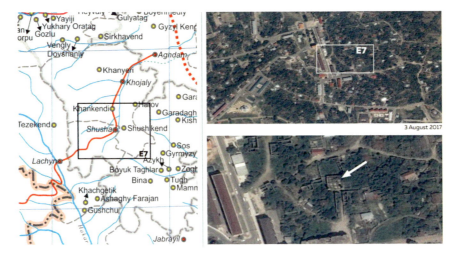

Fig. 26 Destroyed Ashaghy Govhar Agha Mosque in the town of Shusha in the occupied Shusha district (39° 45′ 42.31″ N, 46° 45′ 14.22″ E)

Exploitation of Agricultural and Water Resources

The settlements in the occupied territories rely primarily on agriculture development and water resources, which is pursued for economic and demographic reasons. Farmlands in Zangilan, Gubadly, Jabrayil, and other occupied districts, abandoned

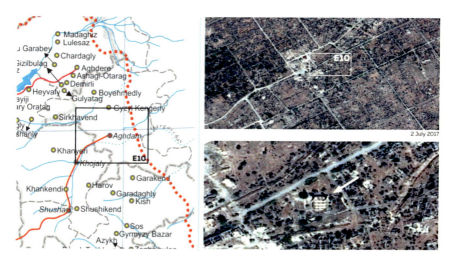

Fig. 27 Destroyed Juma Mosque in the town of Aghdam in the occupied Aghdam district (39° 59′ 35.91″ N, 46° 55′ 53.88″ E)

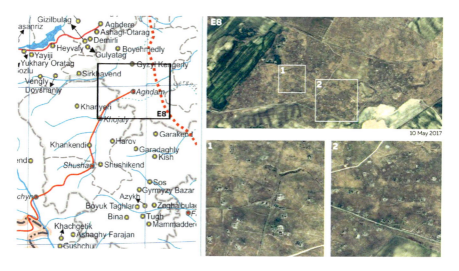

Fig. 28 Ruins of Ismailbayli village of the occupied Aghdam district (39° 56′ 02″ N, 47° 00′ 19″ E)

by Azerbaijani people, have been extremely exploited. The waters of Araz River and other rivers are diverted, and the existing and new artesian wells, pump stations, and irrigation canals are serving the settlements and farming areas. (See Section XII (Extensive exploitation of agricultural and water resources) of the Report of the Ministry of Foreign Affairs of the Republic of Azerbaijan (2016).)

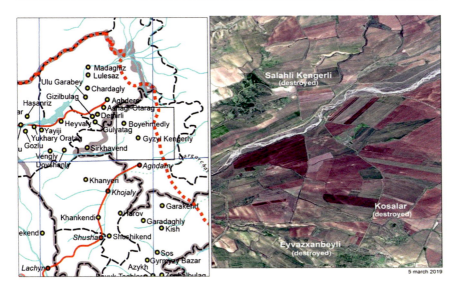

Fig. 29 Agricultural activities near the Salahli Kengerli village in the occupied Aghdam district (46° 56′ 34.56″ E, 40° 06′ 49.37″ N)

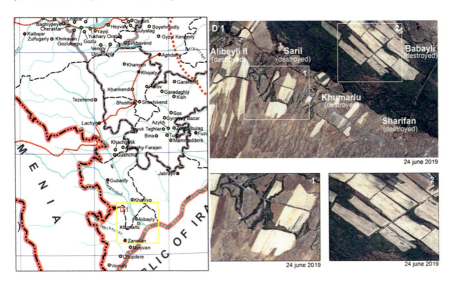

Fig. 30 Agricultural activities near the Khumarlu village in the occupied Zangilan district (39° 08′ 38″ N, 46° 46′ 45″ E)

The following examples of exploitation of agricultural and water resources are presented:

- Agricultural activities near the Salahli Kengerli village in the occupied Aghdam district (46° 56′ 34.56″ E, 40° 06′ 49.37″ N)

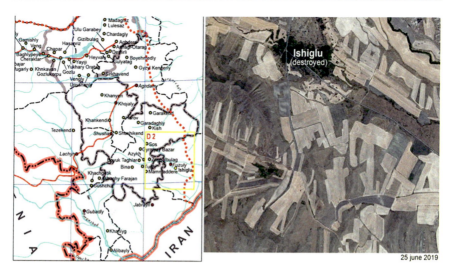

Fig. 31 Agricultural activities near the Ishiglu village in the occupied Fizuli district (39° 08′ 38″ N, 46° 46′ 45″ E)

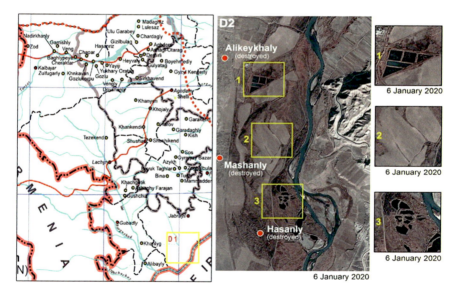

Fig. 32 Agricultural activities near the Mashanly village in the occupied Jabrayil district (39° 13′ 32″ N, 47° 03′ 04″ E)

- Agricultural activities near the Khumarlu village in the occupied Zangilan district (39° 08′ 38″ N, 46° 46′ 45″ E)
- Agricultural activities near the Ishiglu village in the occupied Fizuli district (39° 08′ 38″ N, 46° 46′ 45″ E)

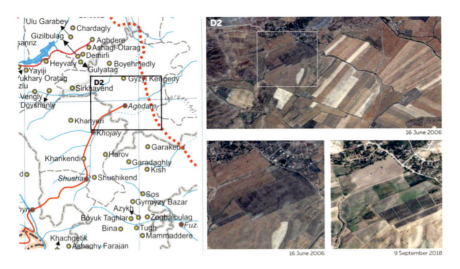

Fig. 33 Newly sown agricultural lands in the southwest of Shelly village in the occupied part of the Aghdam district (46° 53′ 35.324″ E, 39° 56′ 58.337″ N)

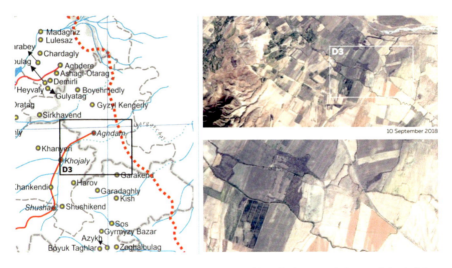

Fig. 34 Newly sown agricultural areas near Shelly village in the occupied part of the Aghdam district (46° 54′ 46.658″ E, 39° 57′ 18.185″ N)

- Agricultural activities near the Mashanly village in the occupied Jabrayil district (39° 13′ 32″ N, 47° 03′ 04″ E)
- Newly sown agricultural lands in the southwest of Shelly village in the occupied part of the Aghdam district (46° 53′ 35.324″ E, 39° 56′ 58.337″ N)
- Newly sown agricultural areas near Shelly village in the occupied part of the Aghdam district (46° 54′ 46.658″ E, 39° 57′ 18.185″ N)

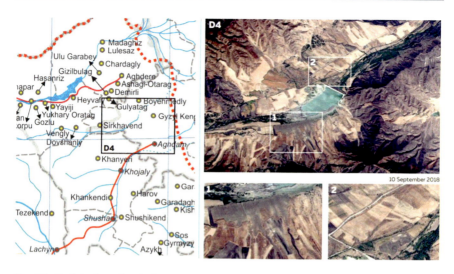

Fig. 35 Exploitation of agricultural lands along the Khachinchay River near the Khachinchay Water Reservoir in the occupied part of the Aghdam district (46° 49′ 44.735″ E, 40° 2′ 45.579″ N)

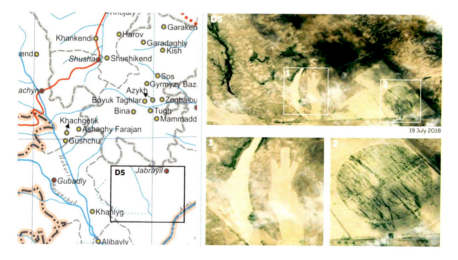

Fig. 36 Exploitation of agricultural lands along the Araz River in the occupied Jabrayil district (47° 2′ 8.634″ E, 39° 15′ 52.574″ N)

- Exploitation of agricultural lands along the Khachinchay River near the Khachinchay Water Reservoir in the occupied part of the Aghdam district (46° 49′ 44.735″ E, 40° 2′ 45.579″ N)
- Exploitation of agricultural lands along the Araz River in the occupied Jabrayil district (47° 2′ 8.634″ E, 39° 15′ 52.574″ N)

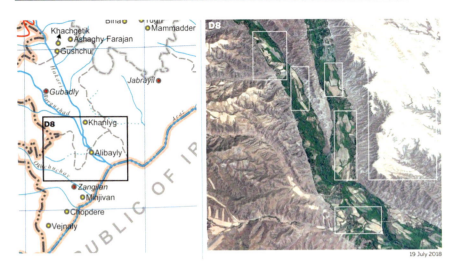

Fig. 37 Exploitation of agricultural lands along the Hakari River in the occupied Zangilan and Gubadly districts (39° 36′ 17″ N, 46° 32′ 43″ E)

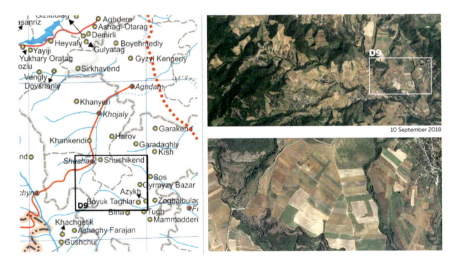

Fig. 38 Exploitation of agricultural lands near Chanagchi, Khanyeri, Khanabad, and Shushikend villages in the occupied Khojaly district (46°50′27.382″E, 39°43′54.474″N)

- Exploitation of agricultural lands along the Hakari River in the occupied Zangilan and Gubadly districts (39° 36′ 17″ N, 46° 32′ 43″ E)
- Exploitation of agricultural lands near Chanagchi, Khanyeri, Khanabad, and Shushikend villages in the occupied Khojaly district (46°50′27.382″ E, 39°43′54.474″ N) (Figs. 29, 30, 31, 32, 33, 34, 35, 36, 37, and 38)

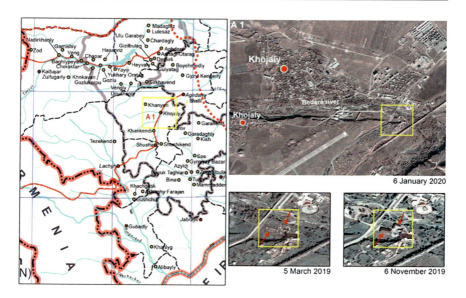

Fig. 39 Construction of new buildings in the occupied Khojaly city (39° 54′ 29″ N, 46° 48′ 20″ E)

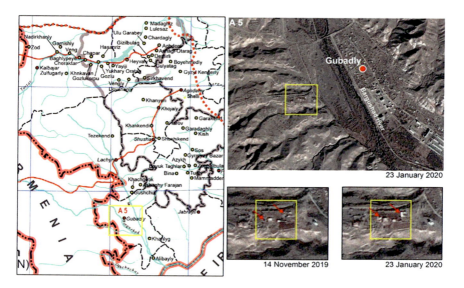

Fig. 40 Construction of new buildings in the occupied Gubadly city (39° 20′ 30″ N, 46° 34′ 23″ E)

Implantation of Settlers and Construction of Permanent Social Infrastructure

The scale of construction and renovation of residential buildings and other associated social infrastructure has been critically increased, which is carried out for the

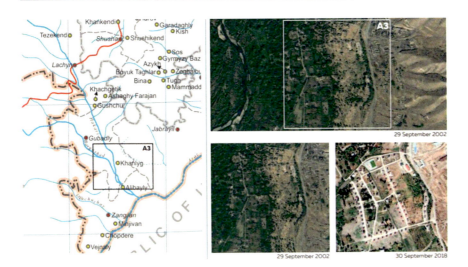

Fig. 41 New settlement in Khanlyg village in the occupied Gubadly district (39° 16′ 3.59″ N, 46° 43′ 13.12″ E)

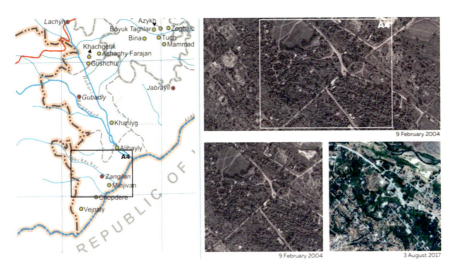

Fig. 42 Newly built houses in the town of Zangilan in the occupied Zangilan district (39° 4′ 55.67″ N, 46° 39′ 24.07″ E)

support of continuous settlement activities. (For more details see Section VIII (Implantation of settlers from Armenia and abroad in the occupied territories) of the Report of the Ministry of Foreign Affairs of the Republic of Azerbaijan (2016).)

The following examples of implantation of settlers and construction of permanent social infrastructure in support of settlement activities are presented:

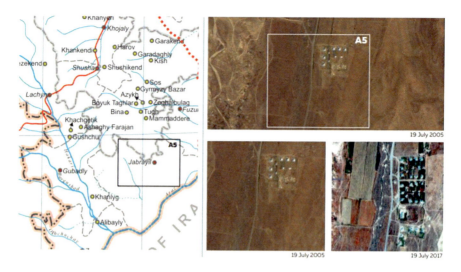

Fig. 43 Newly established "Arajamugh" settlement in the occupied Jabrayil district (39° 21′ 34.45″ N, 47° 2′ 3.54″ E)

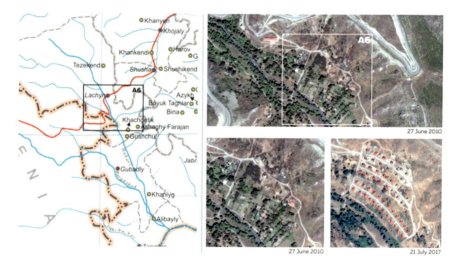

Fig. 44 Newly established "Ariavan" settlement near Zabukh village in the occupied Lachyn district (39° 35′ 27″ N, 46° 32′ 36.33″ E)

- Construction of new buildings in the occupied Khojaly city (39° 54′ 29″ N, 46° 48′ 20″ E)
- Construction of new buildings in the occupied Gubadly city (39° 20′ 30″ N, 46° 34′ 23″ E)
- New settlement in Khanlyg village in the occupied Gubadly district (39° 16′ 3.59″ N, 46° 43′ 13.12″ E)

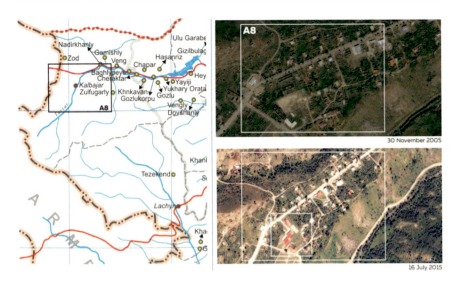

Fig. 45 Newly constructed buildings in the town of Kalbajar in the occupied Kalbajar district (40° 6′ 35.50″ N, 46° 2′ 42.07″ E)

- Newly built houses in the town of Zangilan in the occupied Zangilan district (39° 4′ 55.67″ N, 46° 39′ 24.07″ E)
- Newly established "Arajamugh" settlement in the occupied Jabrayil district (39° 21′34.45″ N, 47° 2′ 3.54″ E)
- Newly established "Ariavan" settlement near Zabukh village in the occupied Lachyn district (39° 35′ 27″ N, 46° 32′ 36.33″ E)
- Newly constructed buildings in the town of Kalbajar in the occupied Kalbajar district (40° 6′ 35.50″ N, 46° 2′ 42.07″ E)
- Newly constructed buildings in the town of Kalbajar in the occupied Kalbajar district (40° 6′ 35.50″ N, 46° 2′ 42.07″ E) (Figs. 39, 40, 41, 42, 43, 44, 45)

Conclusion

Remote sensing by means of satellites is one of the main sources of data for military strategic planning and intelligence. Azerbaijan owns a high-resolution Earth observation satellite Azersky, which is used to continuously monitor the occupied territories, which constitute 20% of its territories, including the Nagorno-Karabakh region, the seven adjacent districts, and some exclaves.

The Azersky satellite in constellation with SPOT 6 is orbiting over the occupied territories of Azerbaijan, providing high-resolution images of specific zones of interest. The satellite imagery, acquired by Azercosmos, has provided the evidence of illegal activities on the occupied territories from Space, such as environmental damage, permanent infrastructure changes, exploitation and pillage of natural resources, destruction of public and private property, implantation of settlers, and construction of permanent social infrastructure.

Despite the resolutions, adopted by UN Security Council, which reaffirmed that the Nagorno-Karabakh region is part of Azerbaijan and called for immediate, complete, and unconditional withdrawal of the occupying forces from all occupied territories of Azerbaijan, the occupant continues its illegal activities on the occupied territory. Azerbaijan continuously monitors its territories, revealing and providing the evidence of violation of international law from Space. (Illegal Activities in the Territories of Azerbaijan under Armenia's Occupation: Evidence from Satellite, "Azercosmos" OJSCo & Ministry of Foreign Affairs of the Republic of Azerbaijan, 2019.)

References

Deer, P., 1995. Digital Change Detection Techniques in Remote Sensing. Technical Report, DSTO-TR-0169. Department of Defence, Australia, p. 52

Euro consult Research Report: Satellite-Based Earth Observation. Market Prospects to 2021, October 2012.

Hussain M, Chen D, Cheng A, Wei H, Stanley D (2013) Change detection from remotely sensed images: From pixel-based to object-based approaches. ISPRS J. Photogramm. Remote Sens. 80:91–106

Van Persie, M., Noorbergen, H.H.S., Van Den Broek, A.C., Dekker, R.J., (2000). Use of Remote Sensing Imagery for Fast Generation of Military Maps and Simulator Databases, National Aerospace Laboratory NLR, NLR-TP-2000-397.

Further Reading

http://www.osce.org/mg/76209?download=true
http://mfa.gov.az/files/file/MFA_Report_on_the_occupied_territories_March_2016_1.pdf
https://president.az/articles/30731
https://spacenews.com/42840airbus-sells-in-orbit-spot-7-imaging-satellite-to-azerbaija2n/